Principles and Practice of X-Ray Spectrometric Analysis

W. C. Röntgen
(1845-1923)
Discoverer of x-rays

H. G. J. Moseley
(1887-1915)
Founder of x-ray spectrometry

Principles and Practice of X-Ray Spectrometric Analysis

Eugene P. Bertin

RCA Laboratories
David Sarnoff Research Center
Princeton, New Jersey

Plenum Press · New York–London

Photograph of W. C. Röntgen from Otto Glasser, *Dr. W. C. Röntgen* (2d ed.; Springfield, Illinois: Charles G. Thomas, 1958); courtesy of Mrs. Otto Glasser and the publisher.

Photograph of H. G. J. Mosely from Mary Elvira Weeks, *Discovery of the Elements* (6th ed.; Easton, Pennsylvania: Journal of Chemical Education, 1960); courtesy of the publisher.

Library of Congress Catalog Card Number 74-107536
SBN 306-30435-X

A Division of Plenum Publishing Corporation
227 West 17th Street, New York, N.Y. 10011

Distributed in Europe by Heyden & Son Ltd.
Spectrum House, Alderton Crescent,
London N.W. 4, England

Printed in the United States of America

Preface

The first commercial x-ray secondary-emission (fluorescence) spectrometer became available about 20 years ago. Within a decade, x-ray spectrometry had become a firmly established method of instrumental analysis for elements having atomic number down to 12 (magnesium) in concentrations down to a few tenths of a per cent. During the last decade, several thousand commercial instruments have been placed in service in laboratories and factories, and the applicability of the method has been extended to trace and micro analysis and to substantially the entire periodic table. However, although many colleges and universities offer full courses in optical and electrical methods of instrumental analysis and in x-ray diffraction, very few offer courses in x-ray spectrometry. Proficiency in this method must be acquired by self-instruction, on-the-job training and experience, the workshops held by the instrument manufacturers, and the one- or two-week courses offered by a few universities.

This book is intended primarily as a textbook for three groups of readers: (1) students; (2) technicians newly assigned to the x-ray spectrometric laboratory; and (3) technicians having practical experience, but little or no formal instruction, and therefore lacking a basic understanding of the method. The book is intended particularly for those who are not content simply to operate the instrument and perform the required analyses, but who want to understand the method and instrument. It is hoped that the book will also be useful to expert x-ray spectrochemists and to scientists and technicians in other disciplines who want to evaluate the applicability of x-ray spectrometry to their fields.

Pursuant to these objectives, the book emphasizes the practical aspects of x-ray secondary-emission spectrometry, but also gives the principles underlying the method, instrumentation, and analytical techniques. Explanations are given primarily in descriptive rather than mathematical terms, but all equations of practical or tutorial value are included.

The applications of x-ray spectrometry are so numerous and varied that it would be hopeless to attempt even to summarize them in a book of reasonable length. Moreover, the writer has always believed that an understanding of the principles and instrumentation of the method leads to proficiency in application of the method to specific problems. Consequently, rather than dealing with specific applications and materials, the book presents in general terms the widest range of analytical methods, techniques, concepts, and ideas possible in the available space. Specific examples are used only for illustration.

The book is concerned primarily with conventional x-ray secondary-emission spectrometry as applied on commercially available spectrometers and accessories. However, extensive treatment is also given to unconventional modes of operation of the instrument and to other x-ray analytical methods that can be conducted on commercial instruments. Related x-ray methods requiring their own instrumentation are also discussed.

An understanding of the instrumentation is necessary to full realization of its potentialities. Accordingly, the x-ray tube and generator, spectrogoniometer, and detection, measurement, and readout components are described schematically in considerable detail. However, specific commercial instruments are not described, except for special features.

The 21 chapters of the book fall into seven groups as follows.

Chapters 1 and 2 deal with *x-ray physics*—the origin and nature of continuous and characteristic x-ray spectra, primary and secondary excitation, fluorescent yield, absorption, scatter, and diffraction.

Chapters 3–9 constitute a detailed description of *the x-ray spectrometer, its components, and their operation*. A general description of the instrument and of the method and its scope, advantages, and limitations is given in Chapter 3. Chapter 4 (Excitation) gives the principles of secondary excitation and detailed descriptions of the x-ray tube and generator. Chapter 5 (Dispersion) is concerned primarily with conventional flat-crystal dispersion. However, most of the other flat- and curved-crystal dispersion arrangements and instruments based on them are also described. Chapter 6 (Detection) considers the construction, function, performance, properties, and principles of operation of proportional and scintillation counters. However, the relatively new solid-state detectors and other less common detectors are also described. Chapter 7 (Measurement) describes the instrumentation and methods for measurement of line and background intensities. Chapter 8 is devoted to pulse-height analysis and "nondispersive" methods. Chapter 9 describes complete conventional manual and automatic spectrometers as well as special instruments having primary or radioisotope excitation and instruments for the ultralong-wavelength region.

Chapter 10 deals with *qualitative and semiquantitative analysis.*

Chapters 11–13 consider *performance criteria* and other features of the method: error, precision, and counting statistics (Chapter 11); specific, nonspecific, special, and unusual absorption-enhancement effects (Chapter 12); and sensitivity, resolution, and spectral-line interference (Chapter 13).

Chapters 14 and 15 present the *basic methods of quantitative x-ray spectrometric analysis*: standard addition and dilution; calibration standardization; internal standardization; standardization with scattered x-rays; matrix dilution; thin-film methods; special experimental methods; and mathematical correction of absorption-enhancement effects. Chapter 15 is devoted entirely to this last method.

Chapters 16 and 17 are devoted to *specimen preparation and presentation* of solids, small parts, powders, briquets, fusion products, liquids, and supported specimens.

The remaining chapters describe *unconventional modes of operation* of the spectrometer and *related x-ray methods* of analysis. Chapter 18 deals with measurement of composition and thickness of films and platings. Chapter 19 treats the analysis of small specimens and of small selected areas on large specimens by "x-ray probe" techniques. Chapter 20 describes polychromatic and monochromatic absorptiometry, absorption-edge spectrometry, absorption-edge fine structure, microradiography, and analytical methods based on scattered x-rays. All these methods are treated with emphasis on their applicability to commercial spectrometers. This chapter also describes Auger-electron and x-ray-excited photoelectron spectrometry and x-ray-excited optical fluorescence. Chapter 21 is a brief description of the instrumentation and method of electron-probe microanalysis.

Each chapter is substantially self-contained, with many cross references to other pertinent sections and with some replication of material covered in other chapters.

Some readers may be disappointed by the absence of a chapter on micro and trace analysis. However, these subjects are dealt with in Chapters 16–19.

The Appendixes contain tables of wavelengths of spectral lines and absorption edges, excitation potentials, and mass-absorption coefficients. The Bibliography provides a list of books, reviews, bibliographies, and tables that should be available in the x-ray spectrometric laboratory, and the 513 papers and reports referred to throughout the book. In general, papers published after 1968 are not included.

The writer gratefully acknowledges that in his task he benefited substantially from access to the excellent books on x-ray spectrometry already in print, especially those by Adler (*1*), Birks (*6*), Liebhafsky *et al.* (*16*), and, particularly, Jenkins and de Vries (*14*). Finally, the writer is most grateful to Thomas J. Cullen, who read the entire manuscript and made many valuable suggestions.

Eugene P. Bertin

Harrison, New Jersey
November 1969

Contents

Chapter 1

Excitation and Nature of X-Rays; X-Ray Spectra

1.1. HISTORICAL

By the late 1800's, many eminent physicists had come to believe that the nature of the physical world was substantially understood and that there remained only to acquire detail and refine data. This complacency was decisively shattered by an unprecedented series of truly epochal discoveries, all within a period of about two decades: radio (Hertz, 1887); an entire, wholly unsuspected group of inert chemical elements (Ramsay, Rayleigh, and Travers, 1895–98); x-rays (Roentgen, 1895); radioactivity (Becquerel, 1896; the Curies, 1898); the electron (Thomson, 1897); quantum theory (Planck, 1900; Einstein, 1901); relativity theory (Einstein, 1906); and cosmic rays (Hess, 1910).

Roentgen's discovery was as significant as any of these, and the invaluable potentialities of x-rays in the theory and practice of science, technology, and medicine were realized immediately. X-rays were widely applied in medical (including dental) and industrial radiography and fluoroscopy within a year of their discovery. X-ray diffraction and spectrometry were applied in many university laboratories within two decades; however, the present widespread application of these disciplines had to await the develop-

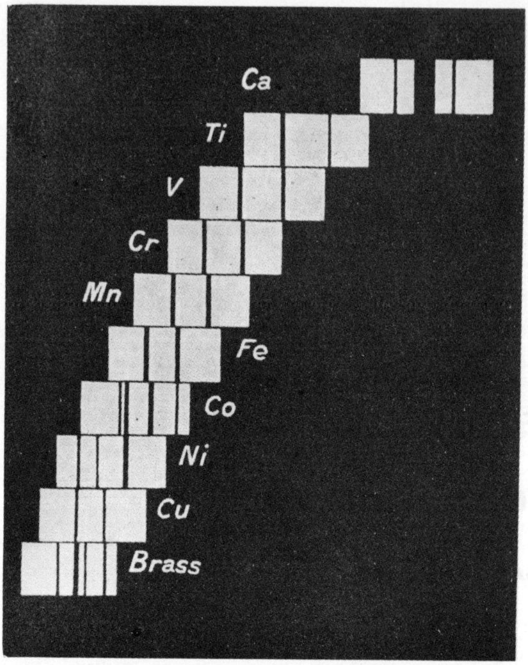

Fig. 1.1. Moseley's first published photograph of x-ray spectra, showing lines of calcium through zinc. [H. G. J. Moseley, *Philosophical Magazine* **26**, 1024 (1912); Taylor and Francis, Ltd., London; courtesy of the publisher.]

ment and commercial availability of reliable x-ray tubes and generators and electronic detection and readout equipment.

Moseley established the basis for qualitative and quantitative x-ray spectrochemical analysis in 1913 (*M32*). Figure 1.1, his first published photograph of x-ray spectra, shows (1) the relationship between wavelength of x-ray spectral lines and atomic number; (2) contamination of cobalt by iron and nickel; and (3) greater intensity of copper lines than zinc lines in the x-ray spectrum of brass (70Cu–30Zn).

Other outstanding contributions in the development of x-ray spectrometry are listed chronologically in Table 1.1, which shows that the Friedman–Birks instrument (*F17*)—the prototype of the first commercial instrument—did not appear until 1948. However, since then, the development of x-ray spectrometric instrumentation and technique has proceeded at a phenomenal rate and continues unabated today.

TABLE 1.1

Chronological Development of X-Ray Spectrometry

1895	W. C. Roentgen[a] discovered x-rays (*R8*)
1896	J. Perrin measured x-ray intensity using an air ionization chamber
1909	C. G. Barkla noted evidence of absorption edges (*B5*)
1911	C. G. Barkla noted evidence of emission line series, which he designated *K*, *L*, *M*, *N*, etc. (*B4*)
1912	M. von Laue, W. Friedrich, and E. P. Knipping demonstrated diffraction of x-rays by crystals (*F19*)
1913	W. L. and W. H. Bragg built the Bragg x-ray spectrometer (*B52*)
1913	H. G. J. Moseley[a] showed the relationship between wavelength of x-ray spectral lines and atomic number, noted that copper lines are stronger than zinc lines in the x-ray spectrum of brass; thereby established the basis of qualitative and quantitative x-ray spectrochemical analysis (*M32*)
1913	W. D. Coolidge introduced the hot-filament, high-vacuum x-ray tube (*C41*)
1913–23	M. Siegbahn did his classic work of measuring wavelengths of x-ray spectra of the chemical elements
1922	A. Hadding first applied x-ray spectra specifically to chemical analysis (of minerals) (*H1*)
1923	D. Coster and G. von Hevesy discovered hafnium, the first element to be identified by its x-ray spectrum (*C45*, *C46*)
1923	G. von Hevesy proposed quantitative analysis by secondary excitation of x-ray spectra
1924	W. Soller constructed an x-ray spectrometer using parallel-foil collimators (*S26*)
1928	R. Glocker and H. Schreiber applied x-ray secondary-emission (fluorescence) spectrometry (*G6*)
1928	H. Geiger and W. Muller developed the gas-filled detector tube to a high degree of reliability (*G2*)
1948	H. Friedman and L. S. Birks built the prototype of the first commercial x-ray secondary-emission spectrometer (*F17*)
1949	R. Castaing and A. Guinier built the first electron-probe x-ray primary-emission spectrometer

[a] See Frontispiece.

1.2. DEFINITION OF X-RAYS

All types of radiation have a dual nature in that they exhibit some properties best explained in terms of particles, and other properties best explained in terms of waves. Nevertheless, most radiation has either *predominantly* corpuscular or wave properties, and it is convenient to classify all types of radiation in these two categories.

The corpuscular radiation includes alpha (α) rays, or helium nuclei

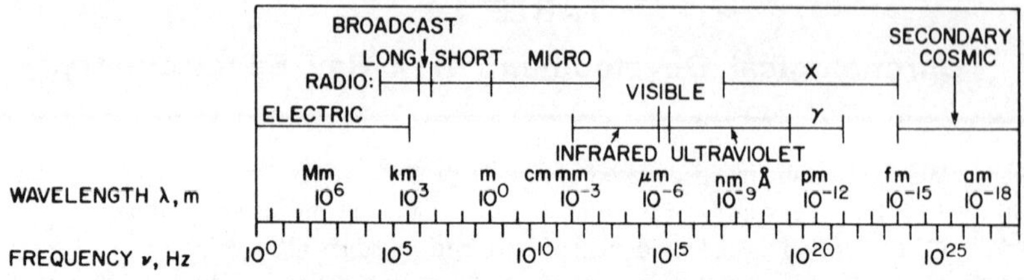

Fig. 1.2. Electromagnetic spectrum.

(He^{2+}); beta ($\beta-$) rays, or electrons ($e-$); positrons ($\beta+$), or positively charged electrons ($e+$); neutrons (n); and the primary cosmic rays, which consist mostly of high-energy protons ($p+$), which are hydrogen nuclei (H^+).

Wave radiation comprises the electromagnetic spectrum, which is divided into overlapping regions, as shown in Figure 1.2. The visible region (4000–7500 Å) is defined by human visual response. The gamma (γ) region comprises high-energy electromagnetic radiation originating in the nuclei of atoms undergoing radioactive decay. The secondary cosmic radiation results from interaction of the primary cosmic radiation (corpuscular) with terrestrial matter. The other spectral regions are defined somewhat arbitrarily on the basis of wavelength or, more realistically, the technology used to generate, transmit, detect, and apply the radiation in each region.

X-rays may be defined as electromagnetic radiation of wavelength $\sim 10^{-5}$ to ~ 100 Å produced by deceleration of high-energy electrons and/or by electron transitions in the inner orbits of atoms. Both these processes are discussed in detail below. The 10^{-5}-Å radiation is produced in betatrons operating at ~ 1 GV; 100 Å represents the *K*-band spectra of the lightest elements. In conventional x-ray spectrometry, the spectral region of interest is ~ 0.1 Å (U $K\alpha$) to ~ 20 Å (F $K\alpha$). In ultrasoft x-ray spectrometry, the region of interest is ~ 10 Å to ~ 100 Å (Be $K\alpha$).

1.3. PROPERTIES OF X-RAYS

Table 1.2 summarizes the properties of x-rays, the phenomena occurring on interaction of x-rays with matter, and the types of radiation resulting from this interaction. X-rays exhibit the corpuscle–wave duality mentioned above: corpuscular properties include photoelectric absorption, incoherent scatter, gas ionization, and scintillation production; wave properties include velocity, reflection, refraction, diffraction, polarization, and coherent scatter. When electromagnetic radiation exhibits the corpuscular aspect of its dual nature, the individual "particles" are termed *photons* or *quanta*.

TABLE 1.2

Properties of X-Rays and Interaction Phenomena of X-Rays with Matter

General properties

- Occur as continuous spectra
- Occur as characteristic line spectra
- Occur as characteristic band spectra
- Produce characteristic absorption spectra
- Propagate at velocity of light
- Propagate in straight lines
- Propagate without transfer of matter
- Are undeflected and otherwise unaffected by electric and magnetic fields
- Are invisible and otherwise undetected by human senses

On encountering matter, x-rays may undergo any of the following phenomena:

- Transmission unaffected
- Reflection
- Refraction
- Polarization
- Diffraction (by pinhole, slit, or crystal)
- Coherent scatter
- Incoherent scatter
- Photoelectric absorption
- Pair production (at energy >1.02 MeV)

On absorption of x-rays, matter may undergo any of the following phenomena:

- Increase in temperature
- Alteration of dielectric properties
- Alteration of electric properties
- Ionization (especially gases and liquids)
- Radiolysis (photolysis) with consequent decomposition, chemical reaction, precipitation, gas evolution, free-radical production, etc.
- Photographic effect
- Production of color centers and other lattice defects
- Radiation damage
- Fluorescence and/or phosphorescence (infrared, visible, ultraviolet)
- Excitation of secondary characteristic x-ray line and band spectra
- Excitation of photo, Auger, and recoil electrons
- Stimulation, damage, genetic change, or death (biological tissue)

Interaction of x-rays and matter may result in emergence of any of the following types of radiation:

Corpuscular

- Ions
- Photoelectrons excited by x-rays (primary and secondary)
- Photoelectrons excited by visible and ultraviolet fluorescence
- Auger electrons
- Recoil electrons
- Electron–positron pairs (from x-rays having energy >1.02 MeV)

Electromagnetic

Incident x-radiation (continuous or line)

- Transmitted (unaffected or imprinted with the absorption spectrum of the specimen)
- Reflected
- Refracted
- Polarized
- Diffracted
- Coherently scattered
- Incoherently scattered

Emitted x-rays

- Diagram spectral lines or bands
- Nondiagram (satellite) spectral lines
- Continuum (*Bremsstrahlung*) excited by photo, Auger, and recoil electrons

Other

- Infrared (thermal)
- Infrared, visible, and ultraviolet fluorescence and phosphorescence
- Annihilation radiation (0.51-MeV photons) produced by recombination of electron-positron pairs
- Cerenkov radiation from very-high-speed electrons produced by very-high-energy x-rays

1.4. UNITS OF X-RAY MEASUREMENT

The x-ray spectrochemist is concerned with four physical quantities in measurement of x-rays: frequency ν (nu), wavelength λ (lambda), energy E, and intensity I.

1.4.1. Frequency

Frequency is expressed in vibrations per second (s^{-1}) or, preferably, hertzes (Hz); 1 Hz $= 1\ s^{-1}$. Other units, used principally in optical spectroscopy, are wave number ν', which is the number of vibrations per centimeter (cm^{-1}), and the fresnel, which is 10^{12} Hz.

1.4.2. Wavelength

X-ray wavelength is expressed in centimeters only in physics. In x-ray spectrometry, wavelength was formerly given in X-units (X or X.U.) defined as 1/2814.00 of the (200) interplanar spacing of rock salt crystal (NaCl) at 18°C. Actually, the kilo-X-unit (kX or kX.U.) was more commonly used; 1 kX = 1000 X. Today, x-ray spectrochemists use the angstrom (Å), defined as exactly 10^{-10} m. Wavelengths in kilo-X-units λ_{kX} are converted to angstroms $\lambda_{Å}$ by

$$\lambda_{Å} = 1.00202\lambda_{kX} \tag{1.1}$$

The angstrom is a nonsystematic unit, and wavelengths should be expressed in nanometers (nm); 1 nm $= 10^{-9}$ m $= 10$ Å. However, the angstrom is so well established that its abandonment in favor of the nanometer is most unlikely. Another unit of wavelength, commonly used in optical spectroscopy, is the millimicron (mμ). This term is to be avoided in favor of the systematic nanometer, to which it is equal.

The relationships among units of wavelength and frequency are:

$$\lambda_{cm} = c/\nu \tag{1.2}$$

$$\lambda_{Å} = (c/\nu) \times 10^8 \tag{1.3}$$

$$\nu = c/\lambda_{cm} \tag{1.4}$$

$$\nu' = 1/\lambda_{cm} \tag{1.5}$$

$$\nu_{fresnel} = \nu \times 10^{-12} \tag{1.6}$$

All these quantities are defined in the foregoing text except c, which is the velocity of light (3×10^{10} cm/s). X-ray wavelength regions are sometimes designated as ultrahard (<0.1 Å), hard (0.1–1 Å), soft (1–10 Å), and ultrasoft (>10 Å).

Throughout this book, the symbol λ indicates wavelength in angstroms, except where another unit is indicated by subscript.

1.4.3. Energy

Energy of an x-ray photon in ergs is given by

$$E_{\text{erg}} = h\nu = hc/\lambda_{\text{cm}} \tag{1.7}$$

where c, ν, and λ_{cm} have been defined already; E_{erg} is photon energy in ergs, and h is Planck's constant. Of greater utility is x-ray photon energy in electron volts (eV), given by

$$E_{\text{eV}} = hc/\lambda_{\text{cm}}e \tag{1.8}$$

where h, c, and λ_{cm} have been defined already; E_{eV} is photon energy in electron volts, and e is electron charge in electrostatic units. Substituting values for h (6.6×10^{-27} erg · s), c (3×10^{10} cm/s), and e (4.8×10^{-10} esu), and inserting conversion factors 10^8 Å/cm and 1/300 esu/V, one gets

$$E_{\text{eV}} = 12{,}400/\lambda \tag{1.9}$$

where λ is in angstroms.

Throughout this book, the symbol E indicates photon energy in electron volts, except where another unit is indicated by subscript.

1.4.4. Intensity

Intensity of an x-ray beam is defined in physics in terms of energy per unit area per unit time, usually ergs per square centimeter per second. However, in x-ray spectrometric analysis, intensities are almost always given in "counts" per unit time—that is, x-ray photons per unit area per unit time. The unit area is usually the useful area of the detector and, being constant for all measurements in a group, is not included in the unit. The unit time is usually the second, the minute, or the preset counting time. Many workers use the symbol R (for count *rate*) for intensity in counts per second or minute. Throughout this book, the symbol I indicates intensity in counts per second.

α-rays, and ions, also generates continuum, but these sources are of little practical importance in x-ray spectrochemical analysis.

The continuum cannot be generated by secondary excitation, that is, by irradiation of matter with high-energy x-rays. However, continuum may appear in secondary x-ray beams due to scatter of incident continuum from the specimen and to generation by photo, recoil, and Auger electrons resulting from interaction of the incident x-rays with the specimen.

1.5.3. Short-Wavelength Limit

Consider an electron moving from filament to target in an x-ray tube operating at (peak) potential V. The shortest wavelength $\lambda_{\min}$ that this electron can possibly generate is emitted if the electron decelerates to zero velocity in a single step, giving up all its energy as one x-ray photon. The energy E_e gained by the electron between filament and target is

$$E_e = eV \tag{1.10}$$

where e is the electronic charge in electrostatic units, and V is potential in volts. The energy of the x-ray photon E_x is [see Eq. (1.7)]

$$E_x = hc/\lambda_{\text{cm}}$$

If the electron gives all its energy to the photon,

$$E_x = E_e \tag{1.11}$$

and

$$hc/\lambda_{\text{cm}} = eV \tag{1.12}$$

giving

$$\lambda_{\text{cm}} = hc/eV \tag{1.13}$$

Substituting the same values and conversion factors as were used in Equation (1.8), one gets

$$\lambda = 12{,}400/V \tag{1.14}$$

where λ is in angstroms, V in (practical) volts.

Equation (1.14) is the Duane–Hunt law (*D16*). It permits the calculation of the short-wavelength limit $\lambda_{\min}$ (angstroms) for an x-ray tube operating at any specified potential V (volts), and of the minimum potential at which an x-ray tube must be operated to produce a specified wavelength. However, if this wavelength is to be generated at high intensity, the x-ray tube must be operated at a potential several times that indicated by the equation (Section 1.5.5). When Equation (1.14) is applied to an x-ray tube operating from a full-wave supply, V is peak potential. The analogy of Equations (1.14) and (1.9) is evident. Many workers use the symbol λ_0 or λ_{SWL} (for short-wavelength limit) instead of λ_{min}.

1.5.4. Origin of the Continuum

Stepwise deceleration accounts for the generation of a continuum, rather than a single wavelength λ_{min} corresponding to the x-ray tube potential. Most electrons give up their energy not in a single step, but rather in numerous unequal increments ΔV, each resulting in an x-ray photon of wavelength $12{,}400/\Delta V$ [Equation (1.14)]. The net effect of large numbers of electrons undergoing such stepwise deceleration is a continuous band of wavelengths $\geq \lambda_{min}$.

Many other phenomena contribute to the formation of the continuum. Consider a strictly monoenergetic electron beam bombarding a monatomic-layer target. Many of the electrons would pass through unaffected, and the others would undergo only one deceleration. But these decelerations need not necessarily be to zero velocity, so a continuum would appear, like that shown for layer 1 in Figure 1.3. An infinitely thick target may be regarded as a stack of such thin targets. The electrons lose energy as they penetrate deeper into the target, so that each successive layer receives electrons having average energy slightly less than those incident upon the layer above. Thus, λ_{min} and the entire spectral distribution from each successively deeper layer are displaced to longer wavelength, as shown in Figure 1.3. Moreover, because the electrons undergo decelerations of various magnitudes, each layer receives electrons having a greater energy spread than those incident upon the layer above. Further complexity arises from lateral scatter of electrons in the target.

In practice, the electrons arriving at the target are far from monoenergetic. With full-wave rectification, V varies from zero to the peak value each half cycle. With constant potential, V fluctuates slightly with residual ripple, but even with ripple-free constant potential, the electrons have some energy spread because they leave the filament with a range of thermal energies and in random directions. Finally, x-rays generated in successively deeper layers have greater difficulty in emerging from the target surface because (1) being of slightly longer wavelength, they are more readily absorbed, and (2) they have a greater thickness of overlying target to penetrate.

α-rays, and ions, also generates continuum, but these sources are of little practical importance in x-ray spectrochemical analysis.

The continuum cannot be generated by secondary excitation, that is, by irradiation of matter with high-energy x-rays. However, continuum may appear in secondary x-ray beams due to scatter of incident continuum from the specimen and to generation by photo, recoil, and Auger electrons resulting from interaction of the incident x-rays with the specimen.

1.5.3. Short-Wavelength Limit

Consider an electron moving from filament to target in an x-ray tube operating at (peak) potential V. The shortest wavelength $\lambda_{\min}$ that this electron can possibly generate is emitted if the electron decelerates to zero velocity in a single step, giving up all its energy as one x-ray photon. The energy E_e gained by the electron between filament and target is

$$E_e = eV \tag{1.10}$$

where e is the electronic charge in electrostatic units, and V is potential in volts. The energy of the x-ray photon E_x is [see Eq. (1.7)]

$$E_x = hc/\lambda_{\text{cm}}$$

If the electron gives all its energy to the photon,

$$E_x = E_e \tag{1.11}$$

and

$$hc/\lambda_{\text{cm}} = eV \tag{1.12}$$

giving

$$\lambda_{\text{cm}} = hc/eV \tag{1.13}$$

Substituting the same values and conversion factors as were used in Equation (1.8), one gets

$$\lambda = 12{,}400/V \tag{1.14}$$

where λ is in angstroms, V in (practical) volts.

Equation (1.14) is the Duane–Hunt law (*D16*). It permits the calculation of the short-wavelength limit $\lambda_{\min}$ (angstroms) for an x-ray tube operating at any specified potential V (volts), and of the minimum potentia at which an x-ray tube must be operated to produce a specified wavelength However, if this wavelength is to be generated at high intensity, the x-ra

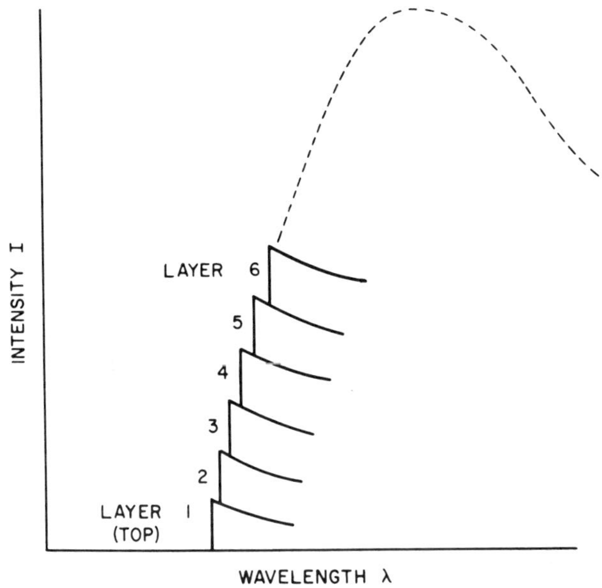

Fig. 1.3. Continuous spectrum from an infinitely thick target regarded as the summation of the individual continua from a stack of very thin targets.

1.5.5. Effect of X-Ray Tube Current, Potential, and Target

Of the several expressions relating the spectral distribution in the continuum to excitation conditions, perhaps the following are the most useful (*K20*, *K21*):

$$I_\lambda = CZ\left(\frac{1}{\lambda^2}\right)\left(\frac{1}{\lambda_{\text{min}}} - \frac{1}{\lambda}\right) + BZ^2\frac{1}{\lambda^2} \tag{1.15}$$

$$I_\lambda \propto iZ\left(\frac{\lambda}{\lambda_{\text{min}}} - 1\right)\frac{1}{\lambda^2} \tag{1.16}$$

$$I_{\text{int}} = (1.4 \times 10^{-9})iZV^2 \tag{1.17}$$

where I_λ is the intensity at any wavelength λ; I_{int} is the integrated intensity of the continuum; Z is the atomic number of the x-ray tube target; V and i are x-ray tube potential and current, respectively; λ_{min} is the short-wavelength limit; and B and C are constants. In Equation (1.15), $C \gg B$, so that the second term on the right is relatively insignificant.

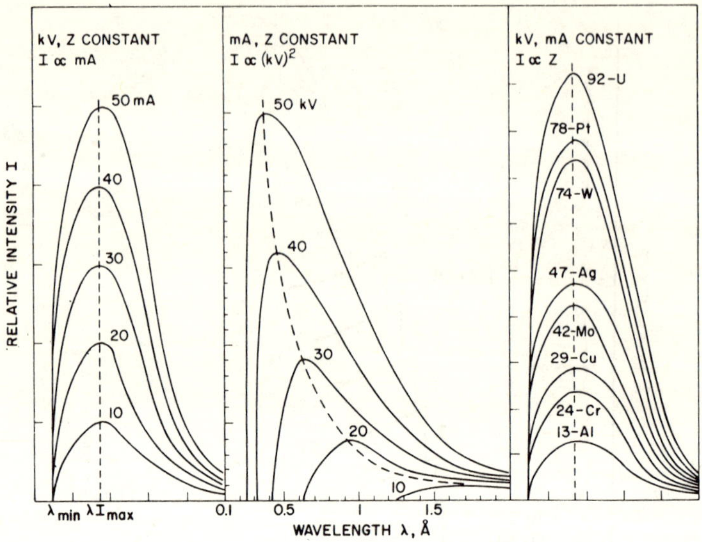

Fig. 1.4. Effect of x-ray tube current, potential, and target atomic number on the continuous spectrum.

Figure 1.4 shows the effects of x-ray tube current, potential, and target on the continuum. It is evident that none of these parameters substantially affects the general profile of the continuum. A change in x-ray tube current causes a proportional change in intensity of the continuum because the number of electrons arriving at the target from the filament is directly proportional to current. The atomic number Z of the x-ray tube target has much the same effect, because the number of orbital electrons in each target atom—and therefore the total number of electrons in the target—is proportional to Z. It is these electrons that cause the stepwise deceleration of the filament electrons. However, it is evident that the wavelengths in the continuum are in no way characteristic of the target element.

X-ray tube potential V has by far the most marked effect. As V is increased: (1) The intensity at any wavelength already present is increased because the filament electrons are accelerated to higher velocity and can undergo more decelerations; (2) λ_{min} and, consequently, $\lambda_{I_{max}}$ are displaced to progressively shorter wavelengths in accordance with Equation (1.14); and (3) the intensity at and near $\lambda_{I_{max}}$ increases rapidly with V, as much as 16-fold by doubling V. It must be remembered that with full-wave rectification the continuum shifts back and forth in wavelength each half cycle and attains its shortest λ_{min} and $\lambda_{I_{max}}$ only at the peak. With constant potential, the continuum remains substantially constant with time.

1.5.6. Significance

In x-ray spectrochemical analysis, the continuum provides the principal source of specimen excitation and of background. The continuum may also provide incident radiation for polychromatic and monochromatic absorption and bracketing wavelengths for absorption-edge spectrometry. A wavelength from the continuum may be used for the background-ratio technique. All these applications are discussed in detail in later sections.

1.6. THE CHARACTERISTIC LINE SPECTRUM

1.6.1. Atomic Structure

An understanding of the nature and origin of characteristic x-ray line spectra requires some knowledge of atomic structure. All atoms consist of a dense central nucleus containing Z protons and $M - Z$ neutrons, with Z electrons revolving around this nucleus like planets around the sun. (Z and M are atomic number and mass number, respectively.) The electrons are grouped in shells designated K, L, M, N, etc. in order of increasing distance from the nucleus. The shells were thus designated by Barkla (*B4*) to allow for possible shells inside the innermost shell known at the time. The electrons in each shell are classified further with respect to angular momentum and direction of spin. Each of these parameters—shell, momentum, spin, etc.—is designated by a quantum number, which may have only certain values, and no two electrons in an atom may have an identical set of quantum numbers (Pauli exclusion principle).

Table 1.3 lists these quantum numbers, their significance, and the allowed values that each may assume for an electron in an atom. The magnetic quantum number m has little significance in this discussion and is included only for completeness. Figure 1.6 lists the n, l, and j quantum numbers for all possible electrons in the first four atom shells (K, L, M, N) and in the first five subgroups of the O shell. It is evident that on the basis of these n, l, and j quantum numbers these electrons fall into subgroups—one K, three L, five M, and seven N. These are the subgroups of principal significance in x-ray spectrometry. The number of subgroups in each shell corresponds to the allowed values of j in that shell: 1/2, 3/2, 5/2, 7/2, etc. (Table 1.3). Figure 1.6 also lists the number of electrons in each subgroup for gold atoms, and these numbers happen also to be the maximum numbers of electrons in each subgroup for any atom. Gold atoms also have one electron in the PI subgroup. Of course, if *all* the quantum numbers of each

TABLE 1.3
Quantum Numbers

Symbol	Name	Significance	Allowed values	Selection rules
n	Principal	Principal binding energy; indicates shell	$1, 2, \ldots, n$ $K, L, M, N, \ldots$	$\Delta n \neq 0$
l	Azimuthal	Orbital angular momentum; determines shape of orbital	$0, 1, \ldots, (n-1)$ $s, p, d, f, \ldots$	$\Delta l = \pm 1$
m	Magnetic	Projection of angular momentum (l) on magnetic field; indicates orientation of orbital in a magnetic field	$\pm l, 0$	—
s	Spin	Direction of spin—clockwise or counterclockwise	$\pm \frac{1}{2}$	—
j	Inner precession	Vector sum of l and s	$l \pm \frac{1}{2}$, except $j \neq 0 - \frac{1}{2}$	$\Delta j = \pm 1$ or 0

electron were considered, it would be seen that no two electrons in an atom have the same complete set of quantum numbers.

1.6.2. Nature and Origin

1.6.2.1. *General*

The characteristic line spectrum consists of a series of discrete wavelengths—spectral lines—characteristic of the emitting element and having various relative intensities. The line spectrum of an element originates when electrons are expelled from the inner levels of its atoms, and electrons from levels farther out fall into the vacancies. Each such transition constitutes an energy loss which appears as an x-ray photon. For example, on creation of a *K*-shell vacancy, a succession of spontaneous electron transitions

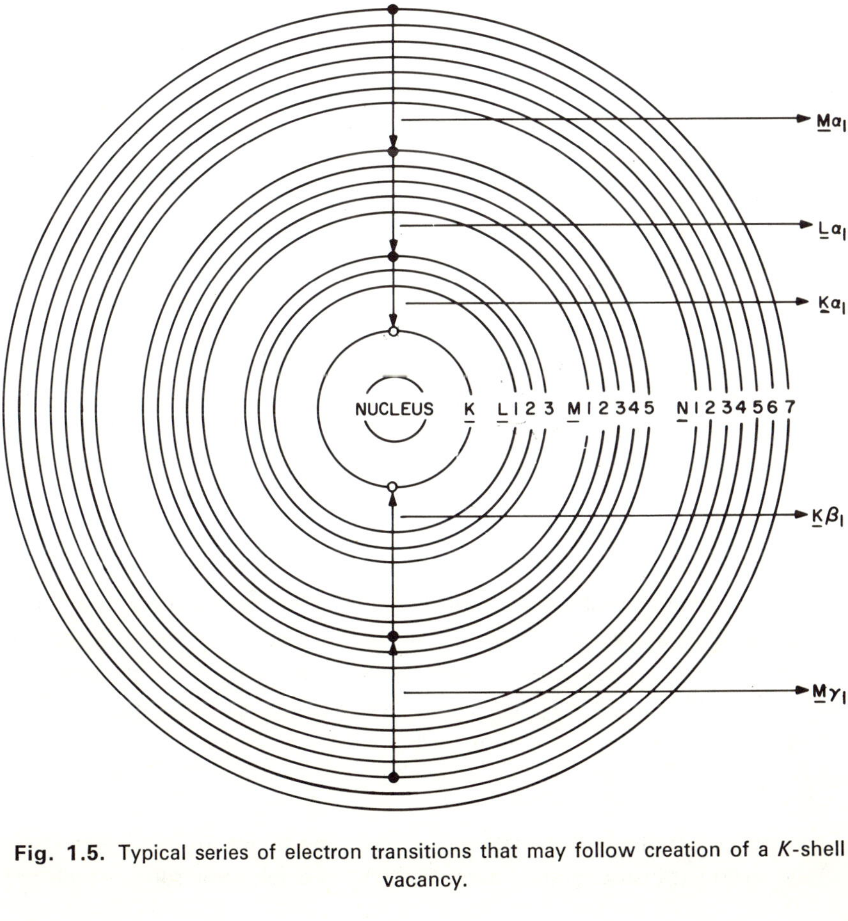

Fig. 1.5. Typical series of electron transitions that may follow creation of a *K*-shell vacancy.

follows; each fills a vacancy in a lower level, with resultant emission of an x-ray photon, but also creates a vacancy in a level farther out. Figure 1.5 shows two of the many possible series of transitions that may follow creation of a *K*-shell vacancy. The result of such processes in large numbers of atoms is the simultaneous generation of the *K*, *L*, *M*, etc. series of x-ray spectra of that element.

Figure 1.6 shows schematically all *K*, *L*, *M*, and *N* levels and the first five *O* levels, giving for each the *n*, *l*, and *j* quantum numbers and the optical spectroscopy notation. The arrows show the electron transitions giving rise to vacancies in each level, and to all *K*, *L*, and *M* lines of relative intensity ~1 or more. Each transition is labeled with the symbol of the resultant line. The weakest of these lines are rarely encountered in x-ray secondary-emission spectrometry. However, in the electron-probe x-ray microanalyzer, where primary excitation and curved-crystal optics combine to give high efficiency, some of these lines may appear. At the top of the figure are the approximate relative intensities of the lines and their wavelengths and energies for gold. Physicists would prefer Figure 1.6 with the arrows pointing in the opposite direction, indicating transitions in atom state, rather than electron transitions. For example, the figure shows the $K\alpha_1$ line as originating from an *electron* transition from the *L*III level to the *K* level; physicists regard it as originating from an *atom* transition from the *K* state (or *K* quantum state) to the *L*III state.

X-ray spectral lines are grouped in series (*K*, *L*, *M*, *N*); all lines in a series result from electron transitions from various levels to the same shell. (Physicists would say *atom* transitions from the same initial state to various final states.) Sometimes, series are further divided into subseries, such as *L*I, *L*II, *L*III; all lines in a subseries result from electron transitions from various levels to the same level.

1.6.2.2. *Band Spectra*

Electron transitions originating from sharply defined levels emit sharp spectral lines having breadth at half height ~0.001 Å. However, the outermost levels in atoms in solids may be broadened under the influence of neighboring atoms. Electron transitions originating from such levels give rise to broad band spectra. The *K* spectra of the lightest elements are of this type. These phenomena are shown in Figure 1.7.

1.6.2.3. *Selection Rules*

Electron transitions cannot occur from *any* higher to *any* lower level. Only certain transitions are "permitted" by the selection rules, which are

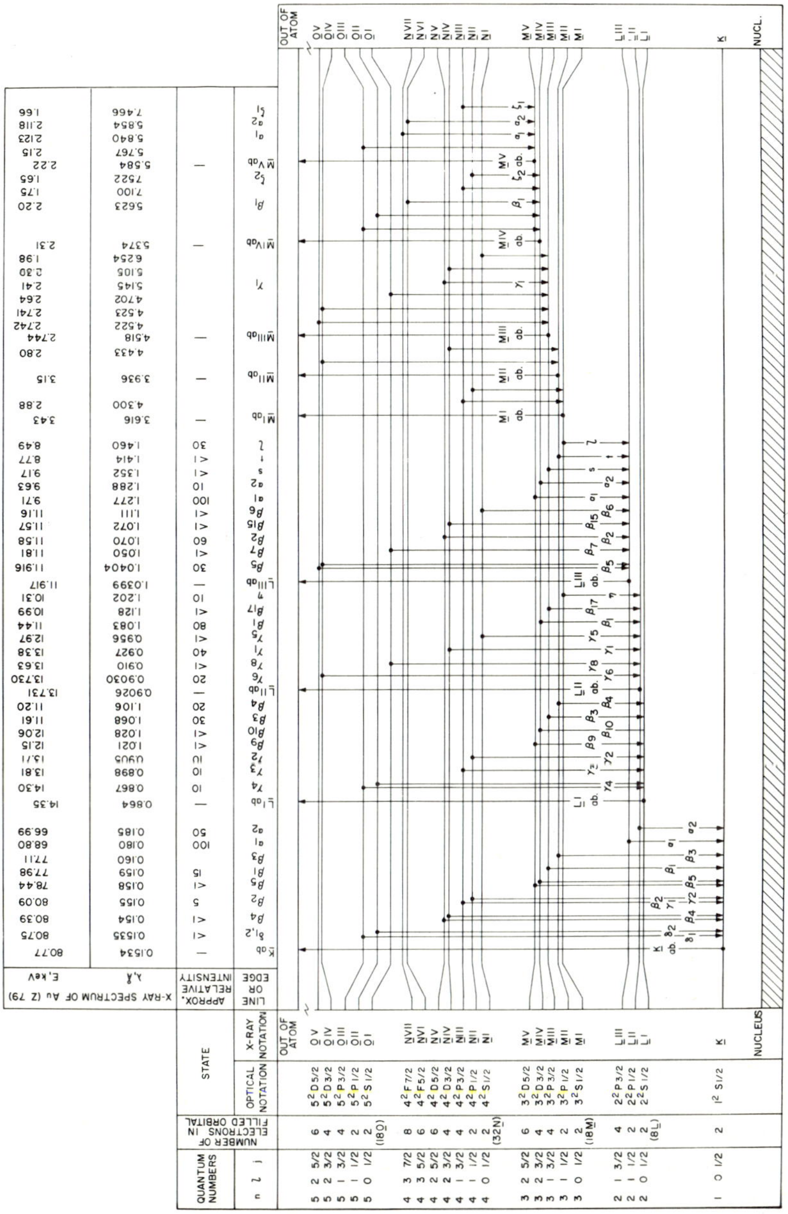

Fig. 1.6. Electron transitions producing the x-ray spectrum.

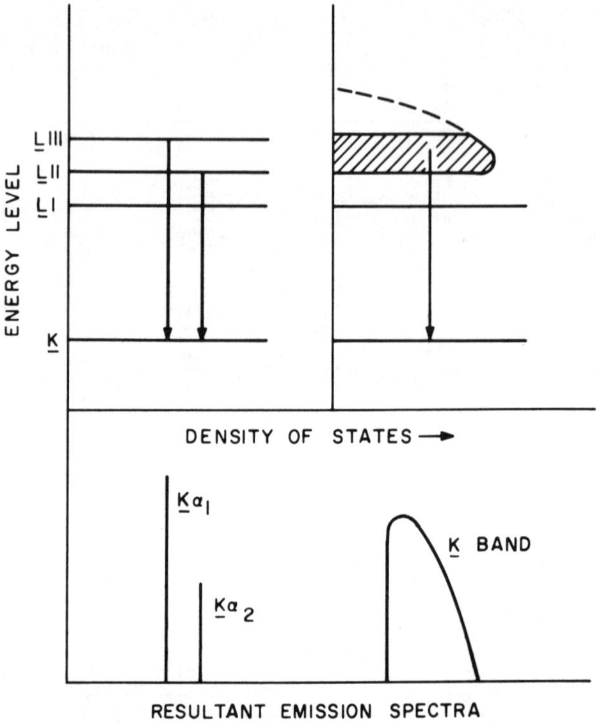

Fig. 1.7. Origin of the line and band spectra.

given in the last column of Table 1.3. For example, an $LIII \rightarrow LI$ transition violates the first rule, $\Delta n \neq 0$. An $NVI(l=3) \rightarrow LI(l=0)$ violates the second rule, $\Delta l = \pm 1$; $MV(j=5/2) \rightarrow LII(j=1/2)$ violates the third rule, $\Delta j = \pm 1$ or 0. Spectral lines that "obey" the selection rules are referred to as *diagram lines.*

1.6.2.4. *Notation*

The symbol of an x-ray spectral line (for example, Ni $K\alpha_1$, Au $L\beta_2$, W $L\eta$, U Ll) consists of: (1) the symbol of the chemical element; (2) the symbol of the series (K, L, M, N, etc.); the line originates from the filling of a vacancy in, or an electron transition to, the indicated shell; and (3) a lower-case Greek or italic letter, usually with a numerical subscript, denoting the particular line in the series.

The α_1 line is usually the strongest in the series, and α lines arise from $\Delta n = 1$ transitions. The β_1 line is usually the second strongest, except that $K\alpha_2$ is stronger than $K\beta_1$. The β and γ lines usually arise from $\Delta n = 1$ or

2 transitions. Aside from these conventions, the notation is not systematic. However, the line resulting from a given transition in any element is always given the same symbol.

The symbol $K\alpha$ indicates the $K\alpha_1\alpha_2$ pair and has wavelength equal to the average of the individual lines weighted for their relative intensities, $K\alpha_1:K\alpha_2 = 2:1$,

$$\lambda K\alpha = \frac{2\lambda K\alpha_1 + \lambda K\alpha_2}{3} \tag{1.18}$$

The symbol $K\beta$ indicates the $K\beta_1\beta_3$ pair.

1.6.2.5. *Wavelength*

The photon energy of a spectral line is the difference in energy ΔE between the initial and final levels involved in the electron transition. For example, in Figure 1.6, for gold,

$$\begin{array}{llll} K \text{ state} \rightarrow & L\text{III state} + & K\alpha_1 & \\ 80.77 & 11.92 & 68.80 & \text{eV} \end{array}$$

$$\begin{array}{llll} K \text{ state} \rightarrow & M\text{III state} + & K\beta_1 & \\ 80.77 & 2.74 & 77.98 & \text{eV} \end{array}$$

$$\begin{array}{llll} L\text{I state} \rightarrow & M\text{V state} + & L\beta_9 & \\ 14.35 & 2.22 & 12.15 & \text{eV} \end{array}$$

(The slight discrepancies in the potentials are not real and are due to small errors in the table values.)

Wavelength is inversely proportional to energy and is calculated from Equation (1.9). The energy differences among K, L, M, and N levels of atoms are such that the wavelengths of the K and L x-ray spectra lie in the region ~0.1–100 Å. Figure 1.8 shows the strongest K, L, and M lines of all the chemical elements. The absorption edges and higher orders also shown are discussed later.

For a given element, wavelengths of lines in a series decrease as ΔE of the energy transition increases; thus,

$$\lambda K\beta_2(N \rightarrow K) < \lambda K\beta_1(M \rightarrow K) < \lambda K\alpha(L \rightarrow K)$$

$$\lambda L\alpha_1(M\text{V} \rightarrow L\text{III}) < \lambda L\alpha_2(M\text{IV} \rightarrow L\text{III}) < \lambda Ll(M\text{I} \rightarrow L\text{III})$$

Also, for a given element, wavelengths of lines decrease from series M to L to K because electrons must "fall" farther to fill vacancies in levels succes-

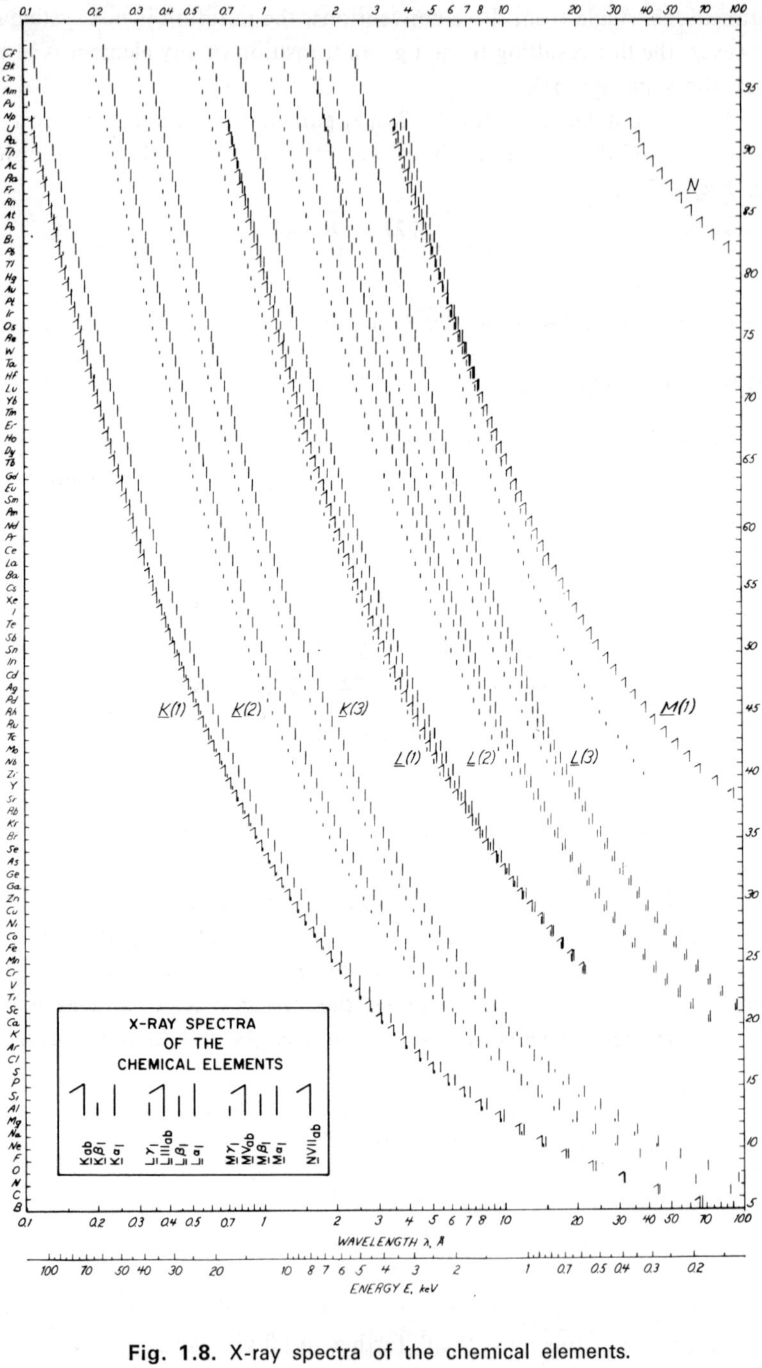

Fig. 1.8. X-ray spectra of the chemical elements.

sively closer to the nucleus. Figure 1.6 illustrates these considerations for gold.

For the same spectral line, wavelength decreases as atomic number increases because positive nuclear charge, binding of orbital electrons, and interorbital energy differences all increase. Figure 1.8 illustrates this property.

Because x-ray spectra originate in inner orbitals, which are substantially unaffected by valence, variation of line wavelength with Z is not periodic, as in optical spectroscopy, but uniformly progressive (Figure 1.8). Also, the wavelengths of x-ray lines are substantially—but not entirely—independent of the chemical state of the atom. The proximity of the nucleus to these inner orbitals explains the strong dependence of wavelength on Z, as defined by the Moseley law,

$$\sqrt{\nu} = (c/\lambda_{\text{cm}})^{1/2} = k_1(Z - k_2) \tag{1.19}$$

where ν, λ_{cm}, and Z are frequency, wavelength, and atomic number, respec-

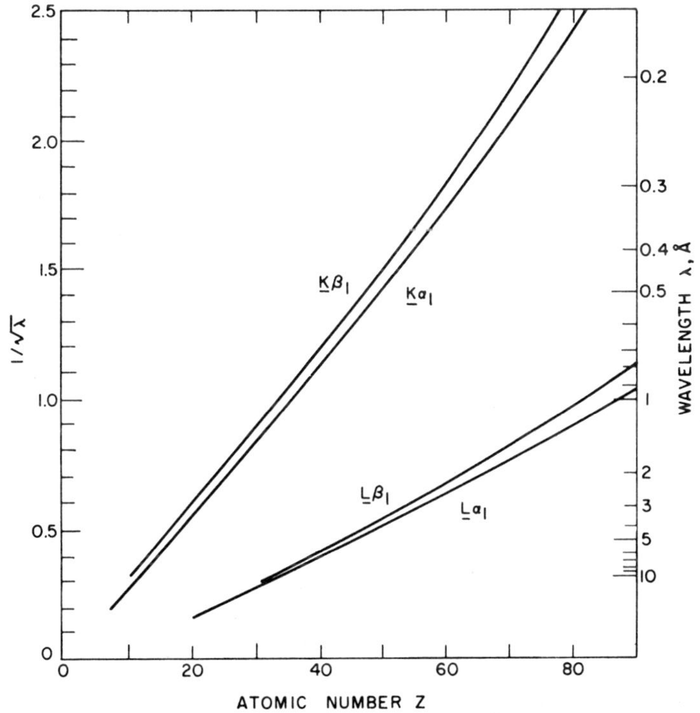

Fig. 1.9. Moseley-law diagram for selected x-ray spectral lines.

tively, and k_1 and k_2 are constants different for each line; k_2 is a screening constant to correct the effect of orbital electrons in reducing the effective nuclear charge, that is, the effective atomic number. Figure 1.9 is a Moseley-law plot for some prominent lines. The curves are linear except for a small deviation at high Z due to failure of k_2 to correct for the screening effect of the large number of electrons.

1.6.2.6. *Intensity*

The relative intensities of lines depend on the relative probabilities of their respective electron transitions. For example, in the K series, the relative intensities of $K\alpha_1(100) > K\alpha_2(50) > K\beta_1(15) > K\beta_2(5)$ are proportional to the relative probabilities of an electron transition to the K shell from LIII > LII > MIII > NIII. These probabilities change with Z. For example, the $K\alpha$:$K\beta$ intensity ratio is 25:1 for 13-Al, 5:1 for 29-Cu, and 3:1 for 50-Sn. This is because $K\beta$ originates from an $M \rightarrow K$ transition, and the M shell is not filled in the light elements. The L lines do not occur for the light elements for the same reason. However, among the medium and heavy elements, the relative intensities within the K and L series are substantially as shown in Figure 1.6. In general, in the same spectral region, first-order $K\alpha$ lines are seven to ten times as intense as first-order $L\alpha_1$ lines excited at the same conditions from elements having the same concentration.

1.6.3. Excitation—General

We have seen that x-ray spectra arise when vacancies occur in the inner orbitals of atoms. We now consider how these vacancies occur. Five processes are illustrated in Figure 1.10.

In primary, or direct, excitation (Figure 1.10A), the vacancies are created by bombarding matter with high-speed electrons. The deflected incident electron becomes a scattered electron, and the ejected electron is termed a secondary electron. High-speed protons, α-rays, or ions from radioisotopes or particle accelerators have the same effect. Primary excitation occurs in x-ray tubes, electron-probe x-ray microanalyzers, and certain instruments in which the specimen is bombarded by electrons in vacuum or outside a Lenard window, and on irradiation of matter with α or β radioisotopes.

In secondary, or fluorescence, excitation (Figure 1.10B), the vacancies are created by irradiating matter with high-energy photons, usually x-rays from an x-ray tube. The incident photon is absorbed, giving up its energy in expelling and imparting kinetic energy to the orbital electron. This process is termed photoelectric absorption, and the ejected electron is a

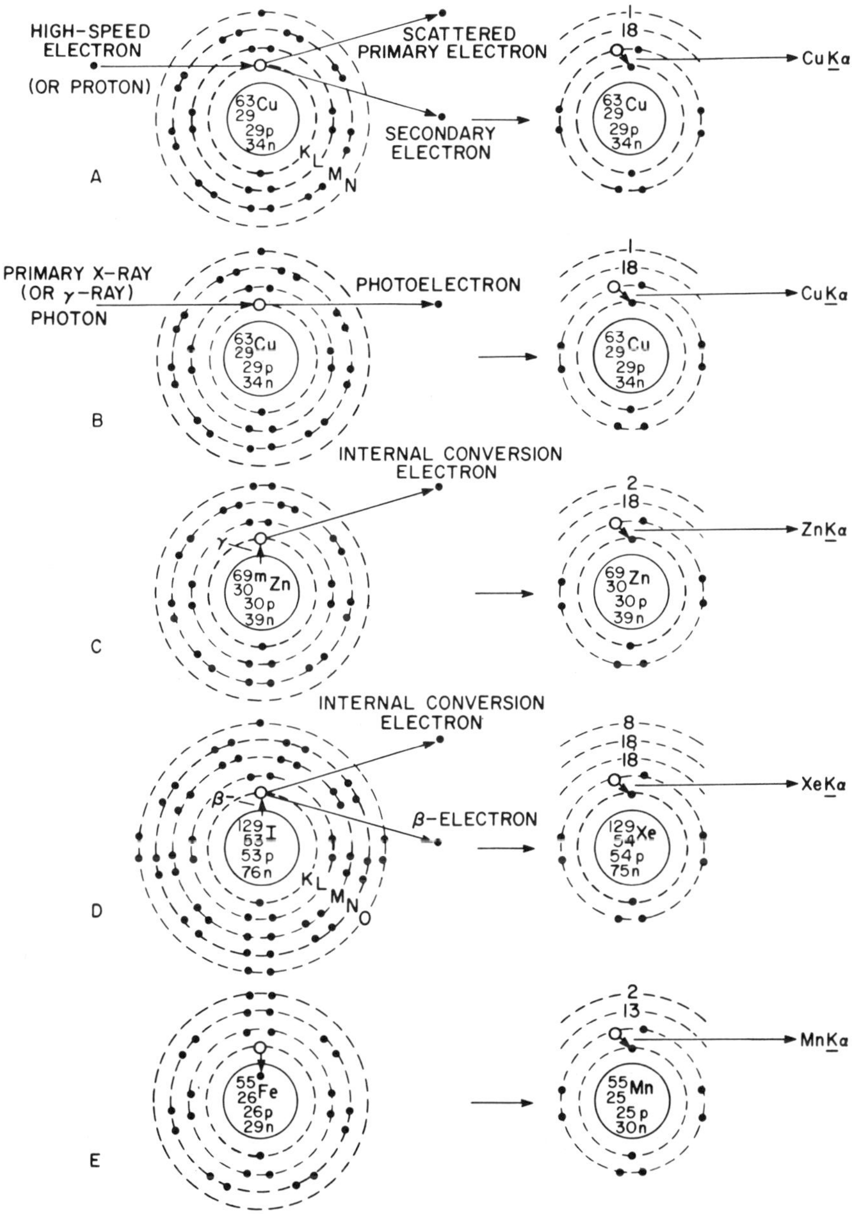

Fig. 1.10. Excitation of characteristic x-ray line spectra. A. Primary excitation. B. Secondary excitation (fluorescence). C. Internal γ conversion. D. Internal β conversion. E. Orbital-electron capture.

photoelectron. Gamma rays have the same effect. Secondary excitation occurs in x-ray fluorescence spectrometers and on irradiation of matter with γ radioisotopes.

Radioactive isotopes that decay by γ-emission may undergo internal conversion (Figure 1.10C). The γ photon is absorbed within the atom of its origin, giving up its energy in expelling and imparting kinetic energy to an orbital electron, which then becomes an internal-conversion electron. Internal γ-conversion may be regarded as secondary excitation in which the ionizing photon originates within the atom. The γ-ray may also escape its own atom and cause secondary excitation in another atom (Figure 1.10B).

Radioactive isotopes that decay by β-emission may also undergo internal conversion (Figure 1.10D) or cause primary excitation (Figure 1.10A) of other atoms. Internal β-conversion may be regarded as primary excitation in which the ionizing electron originates within the atom. However, the emission of a β-particle (electron, e^-) results from the conversion of a neutron to a proton ($n \rightarrow p^+ + e^-$), and thereby increases the atomic number by 1. Then the internal conversion results in the emission of x-rays characteristic of this new element. In the example in Figure 1.10D,

$$^{129}_{53}\text{I} \rightarrow {}^{129}_{54}\text{Xe} + \beta^-$$

Thus, an I-129 source emits the xenon x-ray spectrum.

In radioactive isotopes that decay by orbital-electron capture (Figure 1.10E), a K or L electron actually falls into the nucleus, thereby neutralizing a proton ($p^+ + e^- \rightarrow n$) and decreasing the atomic number by 1. In the example given,

$$^{55}_{26}\text{Fe} \xrightarrow{K\ \text{capture}} {}^{55}_{25}\text{Mn}$$

Thus, an Fe-55 source emits the manganese x-ray spectrum.

Radioactive isotopes are often used instead of x-ray tubes to excite the characteristic x-ray spectrum of the specimen. The excitation may be primary or secondary depending on whether the isotope is, respectively, a β-emitter or a γ- or x-ray emitter. Radioactive sources require no power supply and are relatively inexpensive and extremely compact and convenient. The disadvantages are that the intensity of the radioactive source is very low and decays gradually with time. However, the nondispersive optics, much useful work is done with such sources.

In primary, including β-ray, excitation, the characteristic line spectrum is superposed on strong continuous radiation generated simultaneously. In secondary, including γ-ray, excitation, only line x-ray spectra are *emitted.*

However, these line spectra are superposed on *scattered* primary continuum or the unabsorbed γ-ray spectrum. Radioisotopes that decay only by orbital-electron capture produce very nearly pure line spectra. In all the various modes of excitation of line spectra, some feeble continuum is generated by photo, internal-conversion, recoil, and Auger electrons.

Throughout this section, *K* excitation is used as the example, but the principles apply as well to *L*, *M*, and *N* excitation. Thus, there can be primary or secondary *L* excitation, internal conversion of *L* electrons, and *L*-orbital electron capture.

In any of the modes of excitation, the ionizing agent—incident electron or photon, β- or γ-ray, etc.—must be sufficiently energetic to expel the orbital electrons, and more energy is required the more tightly the electrons are bound. This means that electron expulsion from the same level—e.g., the *K*—requires more energy as *Z*, and therefore the positive nuclear charge, increases; and expulsion from any one atom requires more energy the closer the level lies to the nucleus. Of course, the line wavelengths of an element are the same regardless of how the vacancies originate.

We have seen that the orbital electron vacancies necessary for generation of characteristic line spectra arise principally in two ways—by primary or secondary (fluorescence) excitation, that is, by electron bombardment or x-irradiation. We now consider each of these modes in detail.

1.6.4. Primary Excitation

The minimum x-ray tube potential that can accelerate electrons from filament to target with enough energy to expel an electron from a given level in an atom of a given target element is known as the *excitation potential* of that level of that element. In fact, if the numerical value of the excitation potential is expressed in electron volts (or kiloelectron volts) instead of volts (or kilovolts), one gets the corresponding electron energy. Each element has many excitation potentials: one *K* (V_K), three *L* (V_{LI}, V_{LII}, V_{LIII}), five *M*, seven *N*, etc. For each element, the excitation potential increases for levels progressively closer to the nucleus: $V_K > V_{LI} > V_{LII} > V_{LIII} > V_M > \ldots$. For the same level, excitation potential increases with *Z*. The *K*, *L*, and *M* excitation potentials of gold are represented in Figure 1.6 by the arrows directed upward from each level, and the potentials are given at the top of the figure. The *K*, *L*, and *M* excitation potentials of all the elements are listed in Appendix 4.

Figure 1.11A illustrates primary excitation using the molybdenum *K* spectrum as an example, but the principles are general. (Figure 1.11B is

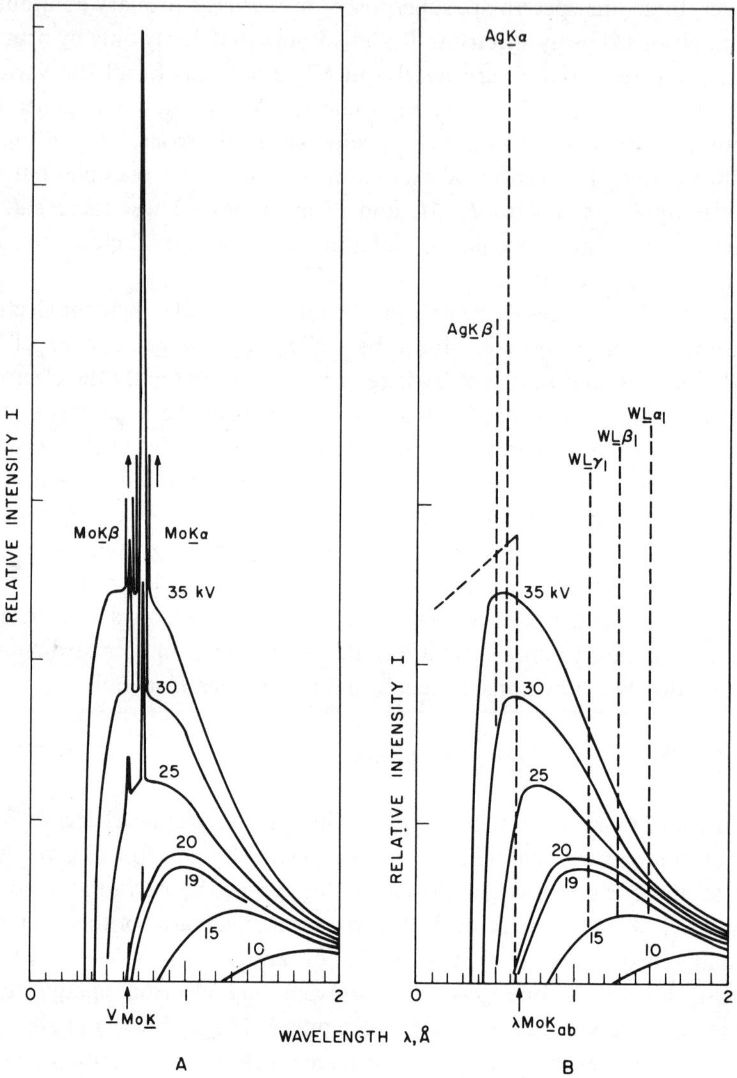

Fig. 1.11. (A) Primary and (B) secondary excitation of the molybdenum K spectrum.

discussed later.) The figure shows spectra from an x-ray tube having a molybdenum target and operating at various potentials. Below the molybdenum K excitation potential, $V_{\text{Mo}K} = 20$ kV, only continuum is generated. At 20 kV, very feeble molybdenum K lines appear superposed on the continuum. At higher potentials, the molybdenum line intensity increases strongly. In general,

$$I_K \propto i(V - V_K)^{\sim 1.7} \tag{1.20}$$

where I_K is the intensity of a specific K line; V_K is K excitation potential; and V and i are x-ray tube potential and current, respectively. For full-wave rectification at $V \geq 3V_K$, the exponent becomes ~ 1. Analogous equations apply to the L and M levels.

Some characteristic target line radiation originates from secondary excitation of the target element by its own continuum (*C35*).

An electron that has enough energy to expel a K electron obviously can also expel any L or M electron. However, an electron having just enough energy to expel, say, an LII electron can also expel LIII, M, and N electrons, but not LI or K electrons. It follows that all spectral lines appear simultaneously that result from electron transitions to the innermost level excited and to all levels farther out. Even if only, say, LII vacancies were created by the incident electrons, vacancies farther out would result from electron transitions into the LII level (Figures 1.5, 1.6), insofar as permitted by the selection rules.

In general, all lines of a given *level* appear at once. The exception is that lines arising from transitions to that level from levels very far from the nucleus do not appear if the atom is too small to have any electrons in that outer level.

1.6.5. Secondary Excitation

1.6.5.1. *X-Ray Absorption Edges*

Although x-ray absorption is discussed in detail in Chapter 2, some consideration of absorption edges is necessary here for an understanding of secondary excitation. X-ray absorption curves for several elements are shown in Figure 1.12. It is evident that in general the x-ray absorption coefficient of an element—that is, its x-ray "stopping power"—decreases with x-ray wavelength. This is to be expected because the shorter the wavelength, the greater the energy and penetrating power. However, there are abrupt discontinuities on these curves, and it is these *absorption edges* that are significant in secondary excitation.

The maximum wavelength (minimum photon energy) that can expel an electron from a given level in an atom of a given element is known as the *absorption edge* of that level of that element. Each element has as many absorption edges as it has excitation potentials: one K (K_{ab}), three L

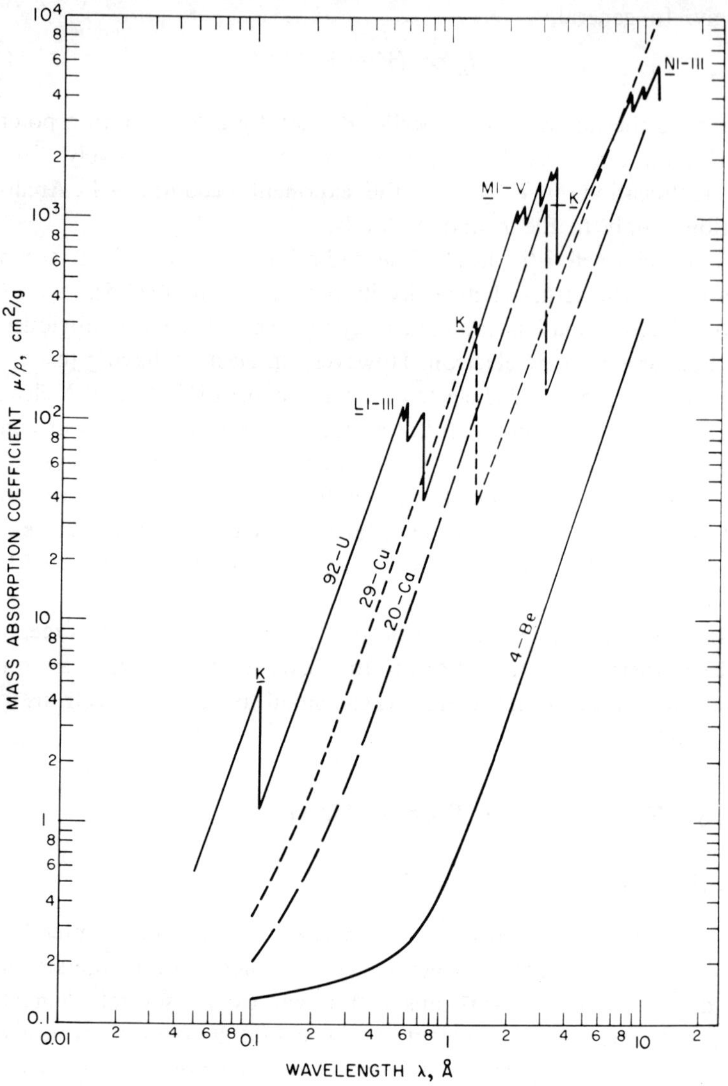

Fig. 1.12. Typical x-ray absorption curves.

(LI_{ab}, LII_{ab}, LIII_{ab}), five M, seven N, etc. The most prominent edges in the multiple series are the LIII, MV, and NVII. For each element, the absorption-edge wavelength decreases (photon energy increases) for levels progressively closer to the nucleus: $K_{ab} < LI_{ab} < LII_{ab} < LIII_{ab} < MI_{ab} < \ldots$. For a given level, this wavelength decreases as Z increases. The K, L, and M absorption edges of gold are represented in Figure 1.6 by the arrows directed upward from each level, and the wavelengths are given at the top of the figure. The *K*, *L*, and *M* absorption edges of all the elements are listed in Appendix 5, and absorption curves showing *K* and *L* edges of many elements are shown in Figures 2.5 and 2.6. Figure 1.8 shows, for all elements, the *K*, *L*III, *M*V, and *N*VII edges that occur in the wavelength region 0.1–100 Å.

1.6.5.2. *Principles.*

Referring to the absorption curves in Figure 1.12, consider a monochromatic primary x-ray beam of wavelength λ incident on a secondary target of atomic number *Z*. Consider what happens as λ is made progressively shorter (more energetic).

At $\lambda > ZK_{ab}$, the photons do not have enough energy to expel *ZK* electrons; consequently, no *ZK* lines appear. As λ decreases, the photons become more energetic, and the absorption coefficient decreases, that is, the secondary target becomes more transparent. At ZK_{ab}, the photons have exactly the energy required to expel *ZK* electrons, and the absorption increases abruptly. The absorbed photons expel *K* electrons, and *ZK* lines are emitted. This process is *photoelectric absorption*, and the expelled electrons are *photoelectrons*. The *ZK* electron expulsion and line emission are most efficient at λ just less than ZK_{ab}.

At $\lambda \ll ZK_{ab}$, the photons have much more than enough energy to expel *ZK* electrons. However, they are so energetic that they may not be absorbed, or else, before being absorbed, they may penetrate the target to a depth from which *ZK* radiation cannot emerge. Thus, as λ becomes progressively shorter, line emission decreases. Note the difference in this respect between primary and secondary excitation. In primary excitation, emitted line intensity *increases* as x-ray tube potential *V* is made greater than the target *K* excitation potential V_K: $I_K \propto i(V - V_K)^{\sim 1.7}$ [Eq. (1.20)]. In secondary excitation, the emitted line intensity decreases as primary wavelength is decreased below ZK_{ab}.

Even though at $\lambda < ZK_{ab}$ photons have more than enough energy to expel *ZK* electrons, each photon that is absorbed by an electron is wholly absorbed. The surplus energy imparts kinetic energy to the photoelectron, so that the shorter the wavelength, the higher the velocity of the photoelectrons.

The preceding discussion regarding wavelengths bracketing the *K* absorption edge applies as well to wavelengths bracketing the several *L* and *M* edges.

A photon that has enough energy to expel a K electron obviously can also expel any L or M electron. However, a photon having wavelength between LI_{ab} and LII_{ab} (see the absorption curve for uranium in Figure 1.12) can expel LII, LIII, M, and N electrons, but not LI or K electrons. The last two paragraphs of Section 1.6.4 regarding just which line series do and do not appear simultaneously apply here as well.

1.6.5.3. *Relationship of Absorption Edges and Spectral-Line Series*

Figure 1.13, which complements Figure 1.6, shows all K, L, and M absorption edges of gold and the principal lines associated with each. The figure is largely self-explanatory, but some features warrant discussion.

It must be reemphasized that only photons having wavelength equal to or less than that of an absorption edge can excite the associated lines. It is evident in Figures 1.6 and 1.13 that there is a very narrow wavelength interval between each absorption edge and the associated line of shortest wavelength. Even a wavelength in this interval—shorter than the shortest series line, but nevertheless slightly longer than the edge—cannot excite the spectrum of this series.

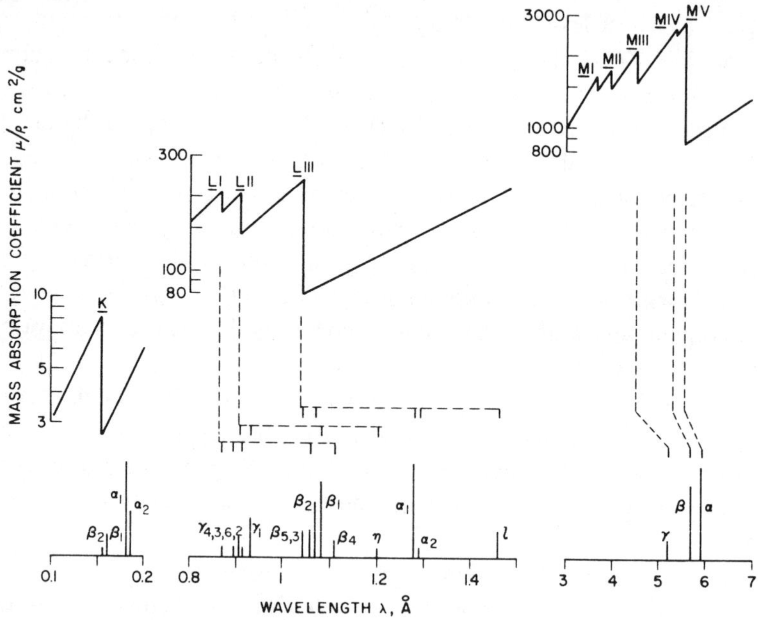

Fig. 1.13. Relationship between absorption edges and spectral-line series.

Figure 1.13 shows that the lines of each series of an element occur on the long-wavelength side of the corresponding absorption edge where absorption is relatively low. It follows that an element is relatively transparent to its own spectral lines. It also follows that lines of a given series of an element cannot excite that series of that element, although they can excite a series having an absorption edge at longer wavelength. Thus, barium *K* lines excite barium *L* lines, but not barium *K* lines.

No line in a series can have wavelength shorter than that of the series edge. For example, in Figure 1.13, although some *L*I and *L*II lines are shorter than the *L*III edge, no *L*III line is shorter than the *L*III edge. Also, in Figure 1.8, the $L\gamma_1$ lines on the short side of the *L*III edges might be construed as an error in the figure. However, the $L\gamma_1$ line is associated with the *L*II edge (not shown), not the *L*III edge.

The reason lines must have less energy—longer wavelength—than their absorption edges is as follows. The wavelength of an absorption edge is equivalent to the energy *absorbed* in creating an inner-orbital vacancy, that is, in lifting an electron from that level clear out of the atom—to the conduction band or free state. The wavelength of a spectral line is equivalent to the energy *emitted* on filling that vacancy; this is effected by an electron transition, not from outside the atom, but from an outer orbital. Thus, less energy is emitted than was absorbed. The farther out the orbital from which the electron falls, the more nearly equivalent the transition to the expulsion, and the closer the line lies to the absorption edge. This can be shown in Figure 1.6 by comparing in each series the transition for the shortest line with those for the other lines.

1.6.5.4. *Excitation with Polychromatic X-Rays*

In an x-ray fluorescence spectrometer, the primary beam is not monochromatic, but consists of the continuum and the line spectrum of the x-ray tube target element. Only that part of the primary beam having wavelength shorter than the absorption edge of an element contributes to the excitation of the spectrum of that element.

Figure 1.11B illustrates secondary excitation using the molybdenum *K* spectrum as an example, but the principles are general. The figure shows the spectra of the primary beam from the x-ray tube that irradiates an external molybdenum secondary target in the specimen compartment. The molybdenum *K* absorption edge Mo $K_{ab} = 0.62$ Å, is also shown.

If the molybdenum *K* spectrum is to be excited, the primary beam must contain x-rays having wavelength $\leq$ Mo K_{ab}. This means that the short-wavelength limit of the continuum $\lambda_{min} \leq$ Mo K_{ab}. Consequently, by Equation (1.14), the tube must be operated at a potential

$$V = \frac{12{,}400}{\lambda \text{Mo } K_{ab}} = V_{\text{Mo}K} \tag{1.21}$$

For Mo $K_{ab} = 0.62$ Å, $V = V_{\text{Mo}K} = 20{,}000$ V, or 20 kV. Figure 1.11B shows that at potentials <20 kV, the primary spectrum contains no x-radiation having wavelength $\leq$ Mo K_{ab}, and even at 20 kV, there is very little. However, as V is increased, the amount of useful primary radiation increases sharply, and the molybdenum K spectrum is excited with great efficiency. In general, it is desirable to operate the tube at a potential such that the "hump" of the continuum ($\lambda_{I_{max}}$) occurs at a wavelength shorter than that of the absorption edge to be excited.

So far, it has been tacitly assumed that only primary continuum excites the molybdenum K lines. However, the primary beam also contains target lines superposed on the continuum. In Figure 1.11B, if the target is tungsten, and the tube is operated $\geq V_{\text{W}L_{\text{II}}}$ (~11.5 kV), tungsten $L\alpha_1$, $L\beta_1$, and $L\gamma_1$ lines appear as shown. These lines, having wavelengths > Mo K_{ab}, cannot excite the Mo K spectrum. However, if the target is silver, and the tube is operated $\geq V_{\text{Ag}K}$ (~25.5 kV), the silver $K\alpha$ and $K\beta$ lines appear as shown. These lines have wavelengths < Mo K_{ab} and contribute to the excitation of the Mo K spectrum.

1.7. COMPARISON OF PRIMARY AND SECONDARY EXCITATION

1.7.1. X-Ray Tube Potential

It should be obvious that excitation potentials (which, as we have seen, are numerically equal to electron energies in electron volts or kilo-electron volts) and absorption edges are analogous quantities. Both are measures of the energy required to expel an electron from a specified level in an atom of a specified element, one applying to electrons in primary excitation, the other to photons in secondary excitation. One might say that the excitation potential is the potential or kilovolt equivalent of the absorption edge, and that the absorption edge is the wavelength or angstrom equivalent of the excitation potential. The two are related by equations having the form of Equations (1.9) and (1.14):

$$V_K = \frac{12{,}400}{\lambda K_{\text{ab}}}; \qquad \lambda K_{\text{ab}} = \frac{12{,}400}{V_K} \tag{1.22}$$

$$V_{L\text{III}} = \frac{12{,}400}{\lambda L\text{III}_{\text{ab}}}; \qquad \lambda L\text{III}_{\text{ab}} = \frac{12{,}400}{V_{L\text{III}}} \tag{1.23}$$

Thus, to excite, say, the molybdenum K spectrum, an x-ray tube must be operated at or above $V_{\text{Mo}K}$. This is true regardless of whether the molybdenum is made the tube target (primary excitation) or a tube having any arbitrarily chosen target is used to irradiate the molybdenum (secondary excitation). Below $V_{\text{Mo}K}$, in the former case, the electrons will not be accelerated sufficiently to expel K electrons from the molybdenum target, and in the latter case, there will be in the primary beam no x-rays of wavelength shorter than Mo K_{ab}.

Moreover, if the molybdenum spectrum is to be excited efficiently, the tube must be operated at $V \gg V_{\text{Mo}K}$—for either mode of excitation, as shown in Figure 1.11. High x-ray tube potential in primary excitation has the effect stated in Equation (1.20); in secondary excitation it enriches the primary beam in the wavelength region below Mo K_{ab}.

1.7.2. Features

Primary excitation gives a high-intensity line spectrum on a high-intensity continuous spectrum. The intensity ratio of target line and continuum radiation depends on x-ray tube potential, excitation potential, and whether full-wave or constant potential is used. Recent work indicates that this ratio is greater than previously believed (*G4*). Ratios of 1:3 and 3:1 have been observed at 45 kV CP for tungsten (L lines) and chromium (K lines) targets, respectively. Total intensity of the continuum increases in proportion to target atomic number and the square of the x-ray tube potential [Equation (1.16)]. Target-line intensity increases in proportion to the 1.7 power of the difference between operating and excitation potentials [Equation (1.20)].

Secondary excitation gives only a line emission spectrum because primary x-ray photons cannot lose energy incrementally, as electrons do. Most continuum in the secondary spectrum is scattered primary continuum; the remainder is generated by photo, recoil, and Auger electrons.

For a given x-ray tube power (kV, mA), the line spectrum is ~100 times as intense with primary excitation. Only a relatively small portion of absorbed primary x-radiation is reemitted as secondary line radiation, and this is emitted in all directions. Secondary radiation emitted too far below the emitter surface is wholly absorbed in the emitter.

The phenomenon of self-absorption directs attention to an interesting difference between primary and secondary x-ray sources. It is possible to obtain a small, intense *apparent* source of x-rays by viewing a primary-excited x-ray source at a small angle. The electrons penetrate the surface to a depth of only a fraction of a micrometer, so that the x-ray source is a layer so thin that it is substantially transparent, even to x-rays emitted at glancing angles. This phenomenon is applied to great practical advantage in x-ray diffraction. X-ray tubes for diffraction have linear focal spots. The takeoff angle of the emergent x-ray beam is made very small relative to the target surface to provide a very small, extremely intense line or point source, depending on whether the focal spot is viewed from the side or end. The phenomenon is also applied in the takeoff geometry of the x-ray spectrometer tube, as shown in Figure 4.2. However, it is not possible to obtain such a small apparent source by viewing a *secondary*-emitting source of x-rays at a small angle. The primary x-rays penetrate the emitting surface to a depth of 10–100 μm, and secondary x-rays emitted at small angles cannot emerge through the long path within the surface. Consequently, a loss, rather than a gain, in "brightness" occurs at small angles. Essentially, a secondary-emitting x-ray surface is Lambert-type emitter.

The advantages of primary excitation are: (1) better sensitivity, especially for light elements, because of the greater excitation efficiency; and (2) reduced absorption-enhancement effects, because the electrons penetrate only a thin surface layer. The disadvantages include: (1) higher background because of the continuum; (2) the requirement of high vacuum; (3) possible alteration of specimen composition by the electron beam; and (4) the specimen surface must be an electrical conductor, or it must be coated with a conductive film.

Instrumentation for primary excitation is described in Section 9.4.2.

Chapter 2

Properties of X-Rays

In this chapter, we consider several x-ray properties that are of particular significance in x-ray spectrometry.

2.1. ABSORPTION

2.1.1. X-Ray Absorption Coefficients

In Figure 2.1, a perfectly collinear monochromatic x-ray beam of intensity I_0 is directed on an absorber having uniform thickness t cm and density ϱ g/cm^3. X-ray-opaque masks having windows of the same area in perfect register are placed on both sides of the absorber. The emergent collinear beam has intensity I, which is always less than I_0 and is given by the Lambert law,

$$I = I_0 \exp(-\mu t) \tag{2.1}$$

where μ is the linear absorption coefficient in reciprocal centimeters (cm^{-1}). Four x-ray absorption coefficients are derived from this equation:

1. *Linear absorption coefficient* μ gives absorption per unit thickness. From Equation (2.1),

$$I/I_0 = \exp(-\mu t) \tag{2.2}$$

Taking natural logarithms, one gets

$$\log_e(I/I_0) = -\mu t; \qquad \log_e(I_0/I) = \mu t \tag{2.3}$$

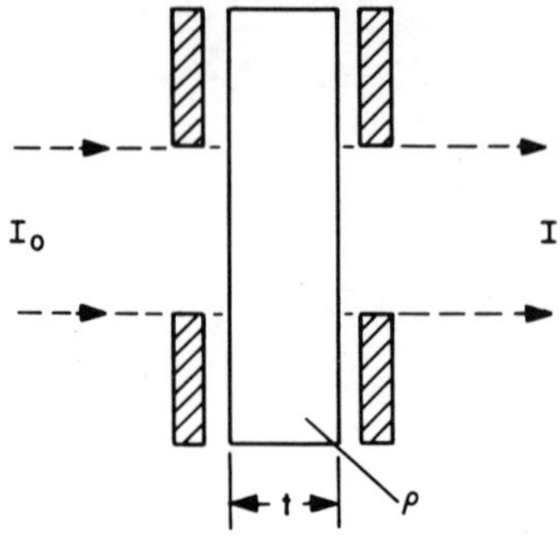

Fig. 2.1. Arrangement for x-ray absorption.

from which

$$\mu = \frac{\log_e(I_0/I)}{t} \quad \text{cm}^{-1} \tag{2.4}$$

2. *Mass-absorption coefficient* μ_m gives absorption per unit mass:

$$\mu_m = \mu/\varrho \quad \text{cm}^2/\text{g} \tag{2.5}$$

3. *Atomic absorption coefficient* μ_a gives absorption per atom:

$$\mu_a = \frac{\mu}{\varrho}\frac{A}{N_0} = \frac{\mu}{N} \quad \text{cm}^2/\text{atom} \tag{2.6}$$

4. *Molar absorption coefficient* μ_{mol} gives absorption per mole, that is, per gram-atom:

$$\mu_{\text{mol}} = (\mu/\varrho)A \quad \text{cm}^2/\text{mole} \tag{2.7}$$

The four coefficients are related as follows:

$$\mu = \mu_m\varrho = \mu_a\varrho(N_0/A) = \mu_{\text{mol}}\varrho/A \tag{2.8}$$

In Equations (2.2)–(2.8), ϱ is density (g/cm^3); A is atomic weight (g/mole); N_0 is the Avogadro number (6.02×10^{23} atoms/mole); N is number of atoms per cubic centimeter; and N_0/A is number of atoms per gram.

Of these four coefficients, mass-absorption coefficient μ/ϱ ("mu over rho") is by far the most useful. It is a property of each substance independent of state of physical aggregation. It is a function only of wavelength and atomic number, and the function for each is relatively simple. Values of μ/ϱ for different substances are directly comparable, and μ/ϱ for a compound, solution, or mixture is readily calculated from the values for the constituents. None of these advantages applies to the linear absorption coefficient, as is strikingly illustrated by a hypothetical experiment described by Sproull [p. 72 in (*21*)]: "A beam of x-rays passing from the ceiling to the floor of a chamber filled with hydrogen and oxygen may be [say] 10% absorbed, or 90% of it will reach the floor. If a spark explodes the hydrogen and oxygen, filling the chamber with steam, 90% of the x-rays will still reach the floor. Then if the chamber is chilled so that the steam condenses to a thin layer of water or ice on the floor, 90% of the x-rays will still reach the floor. This is not true for light or ultraviolet or infrared radiation, and it explains why the mass absorption coefficient of x-rays is commonly used, whereas the linear absorption coefficient is ordinarily used in optics."

Equation (2.1) may be rewritten in terms of μ/ϱ as follows:

$$I = I_0 \exp[-(\mu/\varrho)\varrho t] \tag{2.9}$$

where ϱ is density (g/cm^3), and t is thickness (cm); ϱt is area concentration (g/cm^2).

Throughout this book, the symbol μ/ϱ indicates mass-absorption coefficient in square centimeters per gram. A table of mass-absorption coefficients is given in Appendix 6, and references to other tables are given in the Bibliography (*41*, *43–45*, *47*, *48*). Methods for calculation of these coefficients have been published (*V7*). Absorption curves of most of the chemical elements are shown in Figures 2.5 and 2.6 (*M34*).

The mass-absorption coefficient of a compound, solution, or mixture of elements A, B, C, ... in weight fractions W_A, W_B, W_C, ... is given by

$$(\mu/\varrho)_{ABC\ldots} = \Sigma\, W_i(\mu/\varrho)_i = W_A(\mu/\varrho)_A + W_B(\mu/\varrho)_B + \ldots \tag{2.10}$$

In some cases, a more useful form may be the following:

$$(\mu/\varrho)_X = W_A(\mu/\varrho)_A + (1 - W_A)(\mu/\varrho)_{X-A} \tag{2.11}$$

where X is the specimen, A the analyte, and X − A the matrix.

2.1.2. X-Ray Absorption Phenomena

The linear and mass-absorption coefficients discussed in the preceding section are measures of that portion of the incident collinear beam that does not appear in the emergent collinear beam, regardless of reason (Figure 2.1). They are *total* absorption coefficients and are the result of three phenomena, each having its own linear and mass-absorption coefficients:

$$\mu = \tau + \sigma + \pi \tag{2.12}$$

$$\mu/\varrho = (\tau/\varrho) + (\sigma/\varrho) + (\pi/\varrho) \tag{2.13}$$

where τ and τ/ϱ are *true* or *photoelectric* absorption coefficients; σ and σ/ϱ are *scatter* coefficients; and π and π/ϱ are *pair-production* coefficients.

In true or photoelectric absorption, photons are really absorbed, giving their energy to expelling and imparting kinetic energy to orbital electrons with consequent emission of x-ray spectral lines. This process is discussed in Sections 1.6.3 and 1.6.5.1–3.

In the scatter process, photons are not really absorbed, but deflected from their collinear path in the absorber (Figure 2.1), in effect, disappearing from the emergent beam. This process is discussed in Section 2.2. Photoelectric absorption predominates over scatter except for the lightest elements at very short wavelengths. In the absorption curves of beryllium in Figure 1.12 and of all the light elements at short wavelengths in the curves in Figure 2.5, the marked departure from linearity is due to predominance of scatter over photoelectric absorption.

In pair production, photons passing close to atomic nuclei give their energy to creating and imparting kinetic energy to two charged particles, an electron (e^-) and a positron (e^+):

$$\text{X or } \gamma \text{ photon} \rightarrow e^- + e^+$$

This is an example of conversion of energy to matter in accordance with the Einstein equation,

$$E = mc^2 \tag{2.14}$$

where E is energy (ergs); m is mass (g); and c is velocity of light (cm/s). If we let m be the rest mass of the electron, substitute values for m (9.11×10^{-28} g) and c (3×10^{10} cm/s), and insert the conversion factor 1.6×10^{-6} erg/MeV, the equation gives the energy equivalent of the electron, 0.51 MeV. It follows that the threshold photon energy for pair production is 1.02 MeV, and that the phenomenon has no significance in x-ray spectrochemistry.

2.1.3. Relationship of μ/ρ, λ, and Z

Mass-absorption coefficient μ/ϱ is a property of a substance; it is different for every substance, and, for the same substance, it is different for every wavelength. For a given element Z, $(\mu/\varrho)_Z$ should increase with wavelength because as wavelength becomes longer, x-rays become less energetic and less penetrating. At a given wavelength, $(\mu/\varrho)_Z$ should increase with Z because the "heavier" the element, the greater is its stopping power. Figures 1.12 and 2.2 show x-ray absorption curves for several elements, and it is evident that these predictions are realized to a certain extent and

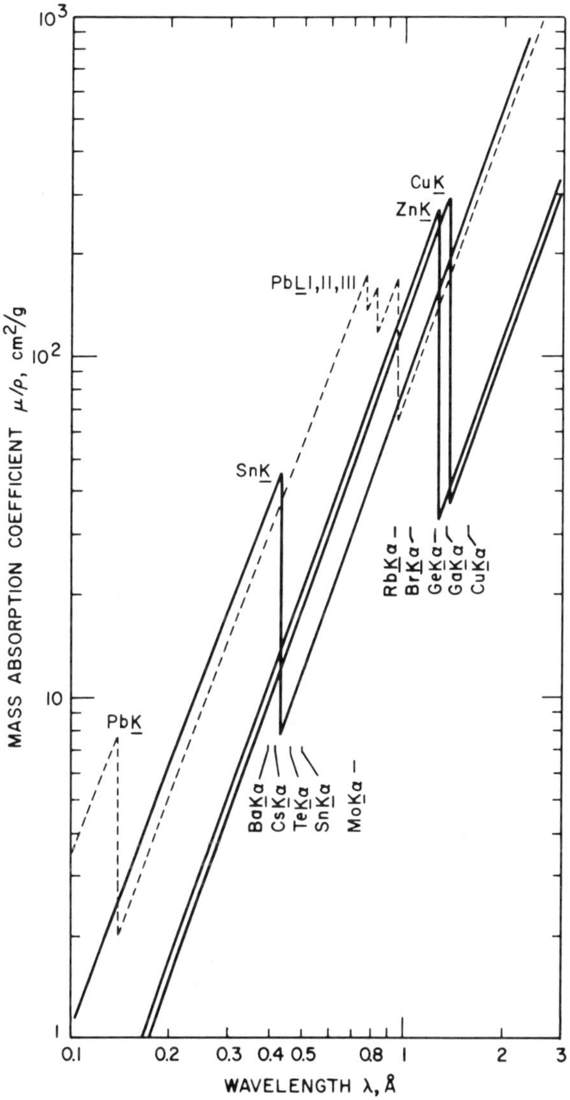

Fig. 2.2. X-ray absorption curves of copper, zinc, tin, and lead.

would be valid generally were it not for the absorption edges (Sections 1.6.5.1–3).

Many examples of departures from predicted regularity appear in Figure 2.2. The μ/ϱ of tin increases regularly with wavelength up to Sn K_{ab}, then falls abruptly and increases regularly again. Below Pb K_{ab}, μ/ϱ increases with Z, as it should; however, just above Pb LIII_{ab}, copper and zinc (Z 29 and 30) have higher absorption than tin and lead (Z 50 and 82), and tin and lead (ΔZ 32) differ in absorption no more than copper and zinc (ΔZ 1). The μ/ϱ of lead is greater than that of tin below Pb K_{ab}, less between Pb K_{ab} and Sn K_{ab}, greater between Sn K_{ab} and Pb LIII_{ab}, then slightly less at longer wavelengths. It is evident that absorption cannot be estimated simply on the basis of atomic number and wavelength.

We have seen that three phenomena—photoelectric absorption, scatter, and pair production—constitute total absorption. In the wavelength region of x-ray spectrochemistry, pair production does not occur, and photoelectric absorption predominates over scatter. Consequently, μ/ϱ is largely determined by τ/ϱ. The coefficient τ/ϱ is itself the sum of a series of coefficients representing photon absorption due to electron expulsion from each of the atomic levels:

$$(\tau/\varrho)_\lambda = \Sigma\,(\tau/\varrho)_{\lambda,i} = (\tau/\varrho)_{\lambda,K} + (\tau/\varrho)_{\lambda,LI} + (\tau/\varrho)_{\lambda,LII} + (\tau/\varrho)_{\lambda,LIII} + (\tau/\varrho)_{\lambda,MI} + \cdots \tag{2.15}$$

where $(\tau/\varrho)_\lambda$ is the photoelectric mass-absorption coefficient at λ, and $(\tau/\varrho)_{\lambda,i}$ is the photoelectric mass-absorption coefficient at λ for i-level electrons. As wavelength is increased, the absorption edge of each level is exceeded in turn and that term drops from the equation, causing an abrupt drop in τ/ϱ. Between absorption edges,

$$\tau/\varrho = K'(N_0/A)Z^4\lambda^3_{\mathrm{cm}} \tag{2.16}$$

Here, K' is a constant that changes at each absorption edge, N_0 is the Avogadro number (atoms/mole), and A is atomic weight (g/mole), so that N_0/A is number of atoms per gram. For any given substance, N_0/A is constant and may be combined with K', giving the Bragg–Pierce law:

$$\mu/\varrho = KZ^4\lambda^3_{\mathrm{cm}} \tag{2.17}$$

where K still changes at each absorption edge. Thus, in a log-log plot of μ/ϱ *vs.* λ, the segments of the curve between the edges are linear and substantially mutually parallel, as shown in Figures 1.12, 2.2, 2.5, and 2.6.

2.1.4. Absorption Edges

Absorption edges are discussed in Sections 1.6.5.1–3. Here, we need only add a few details. A multielement substance exhibits all the absorption edges of all its constituent elements. Figure 2.3 shows the wavelengths of the K, LIII, and MV edges as a function of atomic number. Figure 2.4 shows various absorption-edge jump ratios r and differences δ as functions of atomic number:

$$r = \frac{(\mu/\varrho)_S}{(\mu/\varrho)_L} \tag{2.18}$$

$$\delta = (\mu/\varrho)_S - (\mu/\varrho)_L \tag{2.19}$$

where S and L refer, respectively, to the short- and long-wavelength sides of the edge, that is, the "top" and "bottom," or maximum and minimum values of μ/ϱ. Use is made of these functions in Chapter 4.

The, K, L, and M absorption edges of all the elements are listed in Appendix 5, and absorption curves showing K and L edges of many elements are shown in Figures 2.5 and 2.6 (*M34*). Figure 1.8 shows, for all elements, the K, LIII, MV, and NVII edges that occur in the 0.1–100-Å region.

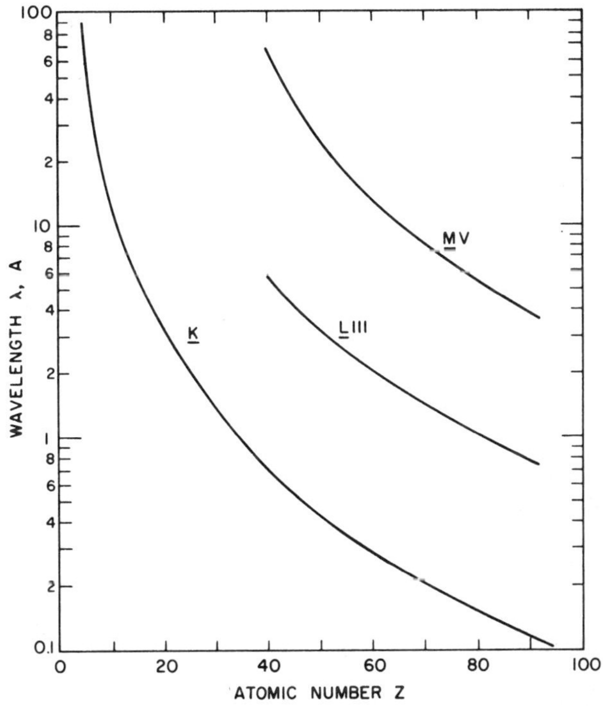

Fig. 2.3. Wavelengths of K, LIII, and MV absorption edges.

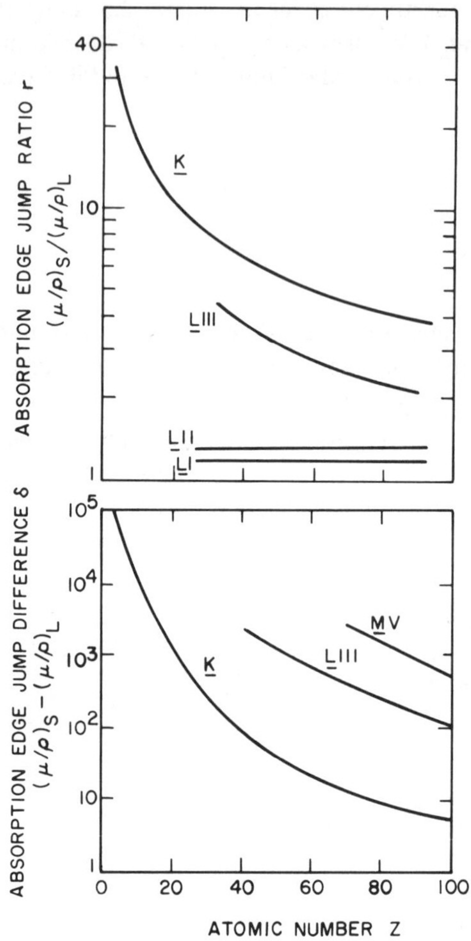

Fig. 2.4. Absorption-edge jump ratios and differences.

2.1.5. Comparison of X-Ray and Optical Absorption

X-ray and optical (visible and near-infrared and ultraviolet) absorption differ essentially only in that they occur in inner and outer orbitals of atoms, respectively. Nevertheless, their absorption spectra are strikingly dissimilar.

Optical absorption and emission spectra are complementary, the former consisting of dark lines (or bands), the latter bright lines (or bands) at

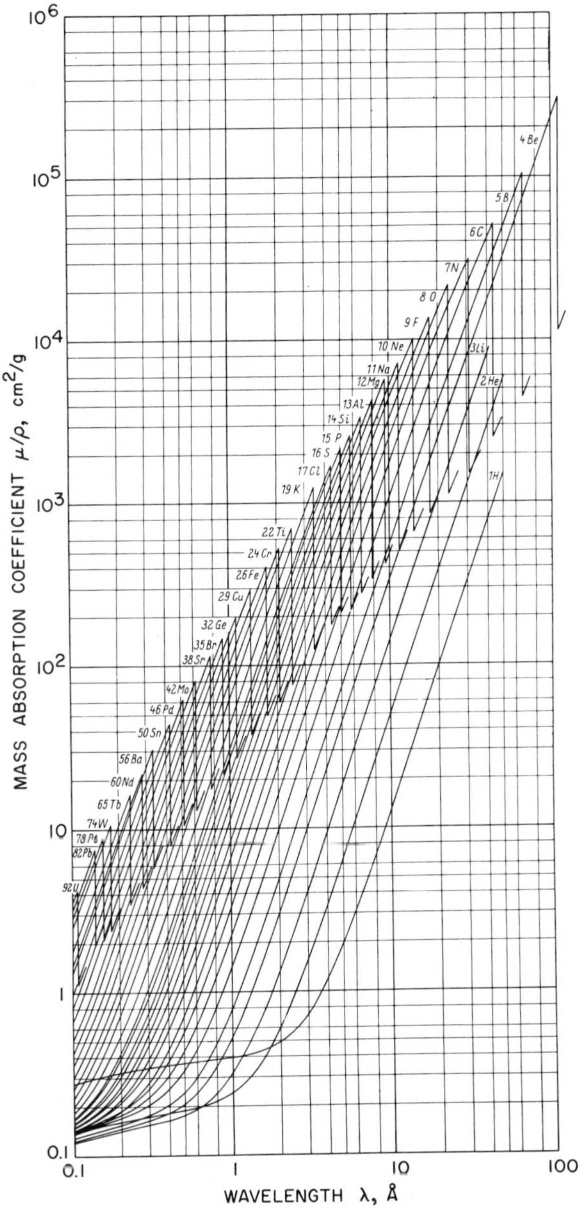

Fig. 2.5. X-ray absorption curves of the chemical elements showing *K* absorption edges. [E. A. W. Müller, *Archiv fuer Technisches Messen und Industrielle Messtechnik*, R. Oldenbourg, Munich (1956); courtesy of the author and publisher.]

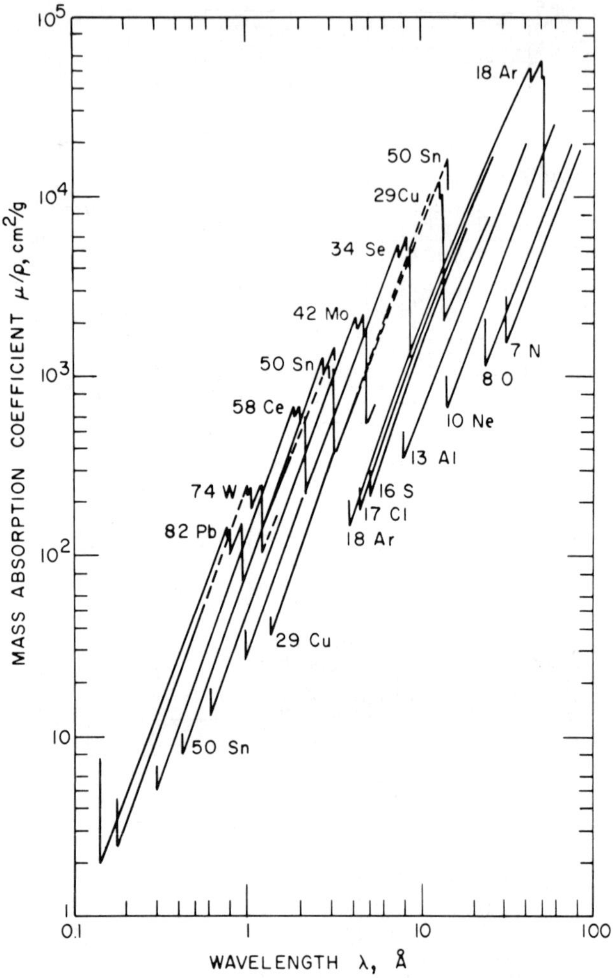

Fig. 2.6. X-ray absorption curves of selected chemical elements showing L absorption edges. [E. A. W. Müller, *Archiv fuer Technisches Messen und Industrielle Messtechnik,* R. Oldenbourg, Munich (1956); courtesy of the author and publisher.]

the same wavelengths. X-ray absorption increases gradually with wavelength, falls abruptly at a wavelength shorter than that of the emission lines, then gradually rises again. Figure 2.7 shows this difference. X-ray photoelectric absorption is explained in Sections 1.6.5.1–3. In optical absorption, outer orbital electrons are elevated to higher energy levels—not clear out of the atom, as in x-ray absorption. Such excitation requires absorption

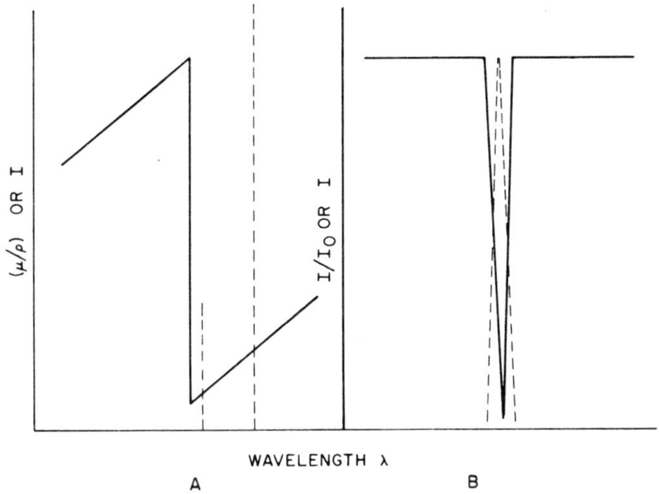

Fig. 2.7. Comparison of (A) x-ray and (B) optical absorption.

of photons having exactly the same energy as those emitted when the excited electrons return to their original levels. Thus, optical absorption and emission spectra both consist of the same discrete wavelengths.

With certain exceptions, x-ray absorption, unlike optical absorption, is substantially independent of chemical state. A striking example is provided by diamond and graphite, which are, respectively, transparent and opaque to light, but have the same x-ray mass absorption coefficient. Sproull's hypothetical experiment (Section 2.1.1) further illustrates this point. Equation (2.10) shows the simple additive nature of the mass absorption coefficient of an element in a compound, solution, or mixture. No such equation is possible for optical absorption.

2.1.6. Significance

Secondary excitation of characteristic x-ray spectra is itself based on photoelectric absorption, and in almost every component of the instrument, absorption is of great significance, sometimes beneficially, sometimes detrimentally.

Absorption of primary x-rays in the x-ray tube window becomes significant at the long wavelengths required to excite elements of low atomic number. In the specimen, absorption of primary and analyte-line radiation and absorption-enhancement effects determine sensitivity and dictate the analytical strategy. Absorption in liquid and powder cell windows becomes significant in determination of low-Z elements. Photoelectric absorption of secondary x-rays in some crystals results in line emission from the crystal. Absorption of secondary x-rays in the radiation path may require use of helium or vacuum, and for the lighter elements, even these may have significant absorption. Absorption of analyte-line radiation in the detector tube window may require use of windows of thin Mylar or polyurethane or ultrathin Formvar. The gas-ionization process in gas-filled detectors and the scintillation process in scintillation counters are absorption phenomena. All filters are based on absorption: primary-beam filters to remove target lines or improve peak-to-background ratio, and secondary-beam filters to reduce spectral-line interference or permit nondispersive analysis. Enhancement radiators are based on photoelectric absorption by the radiator and by the analyte.

Absorption is also the basis of several analytical methods—polychromatic and monochromatic absorption, absorption-edge spectrometry, and emission-absorption methods.

2.1.7. Half-Thickness

Another way to express x-ray attenuation is the *half-value thickness* or *half-thickness* $t_{1/2}$, the thickness required to reduce the incident intensity by one half. From Equations (2.1)–(2.3),

$$t = \frac{\log_e(I_0/I)}{\mu} \tag{2.20}$$

If $I = 0.5I_0$,

$$t = t_{1/2} = 0.693/\mu \tag{2.21}$$

TABLE 2.1
Half-Thicknesses

Wavelength λ, Å	Half-thickness $t_{1/2}$, cm				
	Air[a]	Cellophane	13-Al	29-Cu	82-Pb
0.1	—	4.3	1.6	0.21	0.016
0.7	410	0.4	0.050	0.0016	0.00044
1.5	62	0.11	0.0056	0.0016	—
2.0	26	0.049	0.0025	0.00071	—

[a] 0°C, 760 mm Hg.

where 0.693 is $\log_e 2$ and μ is linear absorption coefficient (cm^{-1}). Table 2.1 lists the half-thicknesses of some selected materials at several wavelengths. Half-thickness has little application in x-ray spectrometry, but is useful for expressing attenuation of polychromatic x-radiation.

2.1.8. Inverse-Square Law

Another form of x-ray attenuation is the reduction in intensity per unit area resulting from divergence and distance. X-rays emanating spherically from a point—or very small—source follow the inverse-square law,

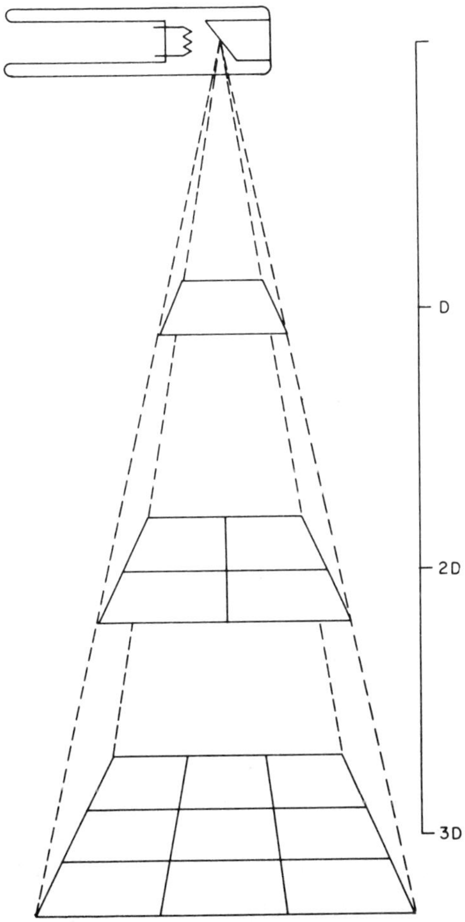

Fig. 2.8. The inverse-square law.

as shown in Figure 2.8 and the following equation:

$$I_2 = I_1(D_1^2/D_2^2) \tag{2.22}$$

where I_1 and I_2 are intensities per unit area at, respectively, distances D_1 and D_2 from the x-ray source. The inverse-square law does not apply to collinear beams. Soller collimators are an intermediate case, in that divergence occurs parallel to, but not perpendicular to, the foils.

2.2. SCATTER

2.2.1. General

Just as total absorption is the sum of three phenomena, scatter is the sum of two: (1) *coherent* scatter (also known as unmodified, elastic, and Rayleigh scatter), in which the x-ray photon is deflected without loss of energy, that is, without increase in wavelength; and (2) *incoherent* scatter (also known as modified, inelastic, and Compton scatter), in which the x-ray photon is deflected with loss of energy and increase in wavelength. The linear scatter coefficient is given by

$$\sigma = Zf^2 + (1 - f^2) \tag{2.23}$$

where the two right terms represent, respectively, the contributions of coherent and incoherent scatter, and f is electronic structure factor, that is, the scatter power of an electron.

In coherent scatter, the incident x-rays cause the electrons in the irradiated matter to oscillate at the same frequency as the x-rays. The oscillating electrons emit x-rays in all directions, again at the same frequency.

2.2.2. Incoherent Scatter

The mechanism of incoherent scatter (*C39*) is shown in Figure 2.9. The incident x-ray photon collides with a loosely bound electron in an outer orbit of an atom. The electron recoils under the impact, leaving the atom and carrying away some of the energy of the photon, which is deflected with corresponding loss of energy or increase in wavelength. The encounter is described by the laws of conservation of energy and momentum. The recoil electrons have the predicted direction and velocity, and the x-rays the predicted change in wavelength:

$$\lambda'_{\text{cm}} - \lambda_{\text{cm}} = \Delta\lambda_{\text{cm}} = (h/mc)(1 - \cos\phi) \tag{2.24}$$

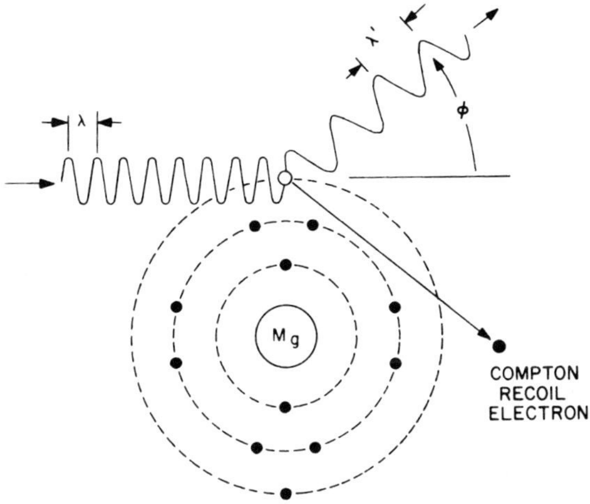

Fig. 2.9. Incoherent (Compton) scatter of an x-ray photon.

where λ and λ' are wavelengths (cm) of the incident and incoherently scattered x-rays, respectively; h is Planck's constant (6.6×10^{-27} erg · s); m is the rest mass of the electron (9.11×10^{-28} g); c is the velocity of light (3×10^{10} cm/s); and ϕ is the angle (deg) between unscattered and scattered x-rays. Substitution of these values and insertion of the conversion factor 10^8 Å/cm gives

$$\Delta\lambda = 0.0243(1 - \cos\phi) \tag{2.25}$$

where $\Delta\lambda$ is in angstroms.

The foregoing explanation is based on the corpuscular nature of both x-rays and electrons. Schrödinger explains the phenomenon on the basis of the wave nature of both x-rays and electrons. The initial and final states of the recoil electron mutually interfere to form a space grating of standing *matter* waves that reflects the incident x-radiation in much the same way as the space grating of atom layers in a crystal. The change in wavelength is attributed to Doppler shift caused by the recoil velocity of the reflecting space grating.

Every wavelength in the scattered primary continuum is subject to both coherent and incoherent scatter. However, in x-ray spectrometry, the principal manifestation of incoherent scatter is a broader peak on the long-wavelength side of each coherently scattered x-ray tube target line and separated from it by $\Delta\lambda$.

The $\Delta\lambda$ is independent of both x-ray wavelength and atomic number of the scatterer, and varies only with ϕ. In most commercial x-ray spectrometers, the axis of the primary beam is perpendicular to the axis of the source collimator. Thus, the collimator–crystal–detector system "sees" mostly scattered primary radiation that leaves the specimen at an angle about 90° to the primary beam. Then $\phi = 90°$, $\cos\phi = 0$, and, from Equation (2.25), $\Delta\lambda \sim 0.024$ Å.

In general, incoherently scattered target lines are broader than coherently scattered lines for two reasons. The primary beam has the form of a cone, and the specimen plane is usually inclined to the axis of this cone. So primary x-rays arrive at the specimen surface at various angles, and, since the collimator permits divergence in the direction parallel to the foils, the spectrometer accepts secondary rays over a substantial angle. Thus, ϕ, and therefore $\Delta\lambda$, may have a range of values. Moreover, the recoil electrons may have a range of velocities and recoil directions, taking a range of energies from the scattered photons. Unless the incident x-rays have extremely short wavelength, recoil electrons have much less energy than x-ray photoelectrons.

2.2.3. Relationship of Coherent and Incoherent Scatter

The proportion of incoherent scatter increases as: (1) the energy of the x-rays increases, that is, as wavelength decreases; (2) the binding of the orbital electrons decreases, that is, as atomic number of the scatterer decreases; and (3) ϕ increases. Incoherent scatter predominates when the x-ray energy greatly exceeds the binding energy of the orbital electrons. When this binding energy approaches or exceeds the x-ray energy, coherent scatter predominates. The variation of incoherent scatter with wavelength

TABLE 2.2

Ratio of Incoherent to Coherent Scatter[a]

Element	I_{inc}/I_{coh}
3-Li	All incoherent
6-C	5.5
16-S	1.9
26-Fe	0.5
29-Cu	0.2
82-Pb	All coherent

[a] Page 23 in (*1*).

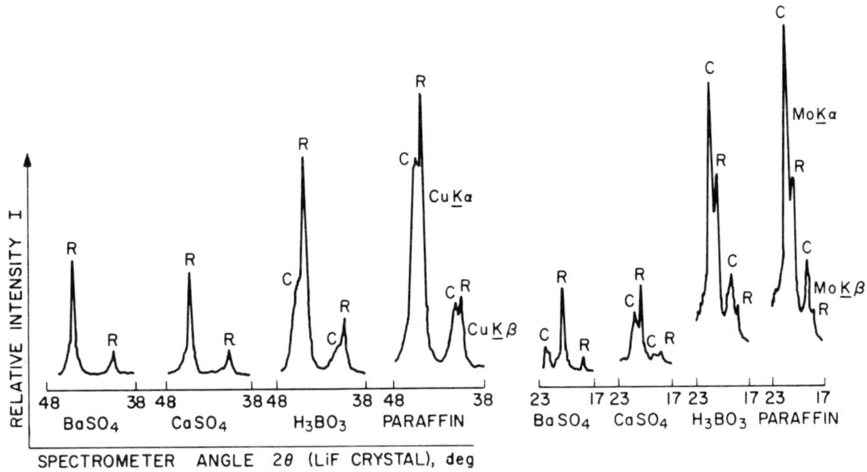

Fig. 2.10. Rayleigh (*R*) and Compton (*C*) scatter of copper and molybdenum *K* lines from heavy, medium, and light scatterers. [C. M. Johnson and P. R. Stout, *Analytical Chemistry* **30**, 1921 (1958); courtesy of the authors and the American Chemical Society.]

and atomic number is shown in Table 2.2 and Figure 2.10 (*J10*). The figure shows primary *K* lines from copper and molybdenum x-ray tube targets scattered from four scatter targets having various effective atomic numbers from heavy ($BaSO_4$) to light (paraffin). For the longer wavelengths, Cu $K\alpha, \beta$ (~1.5 Å), Compton scatter (*C*) is absent from the two heaviest scatterers and is substantially weaker than Rayleigh scatter (*R*) even from the lightest. For the shorter wavelengths, Mo $K\alpha, \beta$ (~0.7 Å), prominent Compton scatter is present from even the heaviest scatterer and is more than twice as intense as Rayleigh scatter from the two lightest scatterers.

An interesting example of incoherent scatter is the frequent appearance in scattered primary spectra from x-ray tubes of a weak line corresponding to Pd $K\alpha$ (0.587 Å). Actually, this line is incoherently scattered Ag $K\alpha$ (0.561 Å) from silver solder used in construction of the tube. Note that 0.587 Å − 0.561 Å = 0.026 Å, in agreement with the $\Delta\lambda$ predicted above. The longer-wavelength lines of iron, nickel, and copper target contaminants are mostly coherently scattered, except from very light specimens.

2.2.4. Significance

Formerly, scatter was regarded only as a nuisance (*J10*). Background is mostly scattered primary radiation, and both coherent and incoherent scattered target lines complicate the spectrum and increase the possibility

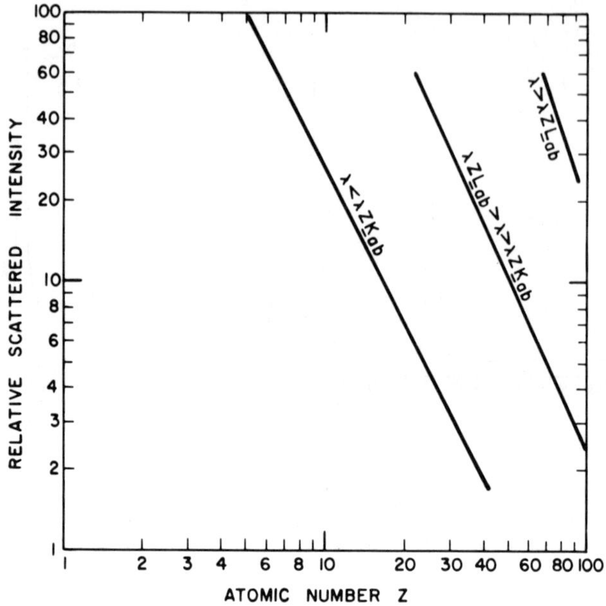

Fig. 2.11. Scattered x-ray intensity as a function of atomic number. [G. Andermann and J. W. Kemp, *Analytical Chemistry* **30**, 1306 (1958); courtesy of the authors and the American Chemical Society.]

of spectral-line interference. However, diffraction, a form of coherent scatter, is the basis of dispersion by crystals. More recently, several techniques have been developed for using scattered x-rays to correct absorption-enhancement effects, surface texture, and other difficulties. Finally, several analytical methods are based on x-ray scatter rather than emission, including a method for determining carbon in hydrocarbons. Figure 2.11 shows scattered x-ray intensity as a function of atomic number for wavelengths shorter than the K absorption edge, between the K and L absorption edges, and longer than the L edge (*A15*).

2.3. DIFFRACTION BY CRYSTALS

In the x-ray spectrometer, the analyzer crystal disperses the secondary characteristic x-ray beam so that each wavelength may be measured individually. An understanding of the function of the crystal requires an elementary knowledge of the diffraction of x-rays by crystals. Diffraction is also significant in that certain spurious effects are caused by diffraction of the primary beam by crystalline specimens and of the secondary beam by Soller collimator foils.

Diffraction arises from interference of coherently scattered x-rays. Figure 2.12 shows interference of two rays of the same wavelength for three phase relationships. In Figure 2.12A, the rays are one half-wavelength out of phase ("peak to trough and trough to peak") and undergo complete destructive interference or cancellation. In Figure 2.12B, the rays are in phase ("peak to peak and trough to trough") and undergo complete con-

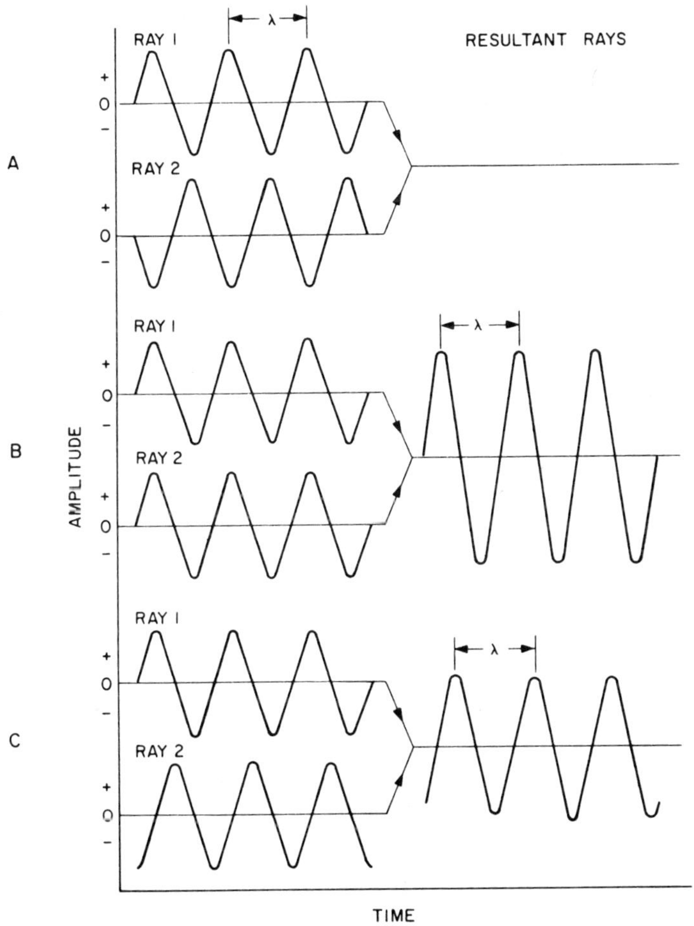

Fig. 2.12. (A) Destructive, (B) constructive, and (C) intermediate wave interference.

structive interference or reinforcement. The resultant ray has the same wavelength as the individual rays and the sum of their peak amplitudes. Figure 2.12C represents all intermediate phase relationships. The resultant ray has the same wavelength as the individual rays and the vector sum of their amplitudes—greater than zero, but less than the sum of the peak amplitudes. The more nearly in or out of phase the two interfering rays are, the more nearly they approach the extremes of Figures 2.12B and 2.12A, respectively.

One version of diffraction of x-rays by crystals is that of Laue (*F19*), who regards the phenomenon in terms of scatter from the individual lattice sites of the crystal. In Figure 2.13, a collimated monochromatic x-ray beam of wavelength λ and frequency ν is directed at angle θ on a set of crystal

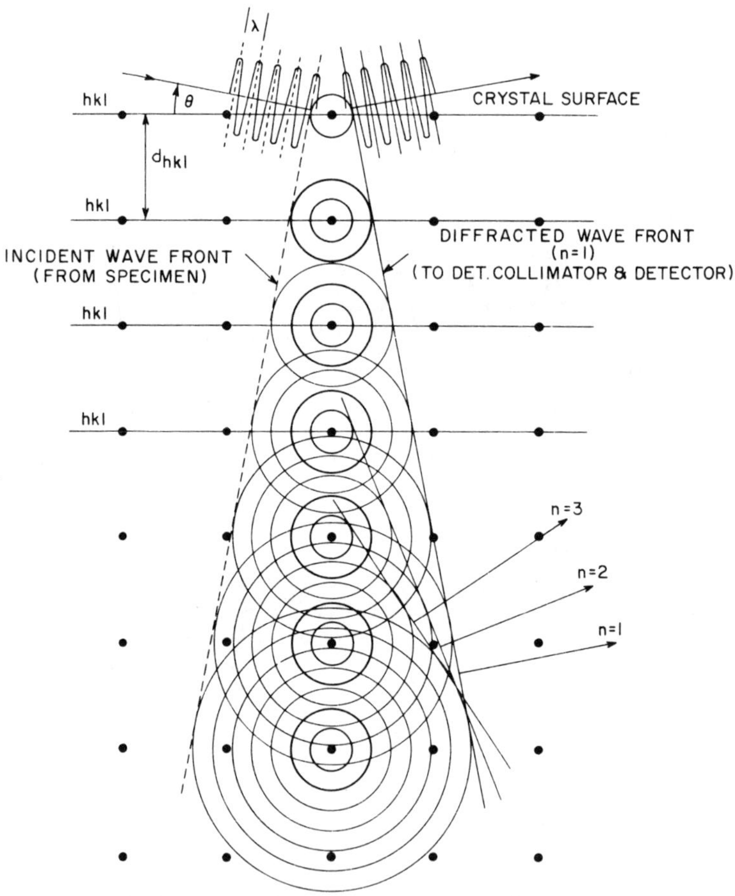

Fig. 2.13. Diffraction by crystals—Laue version.

planes (*hkl*) of interplanar spacing *d*. The atoms in the crystal lattice are excited to oscillate at frequency ν and act as spherical radiators of x-rays of the same wavelength as the incident x-rays. This is the coherent scatter process. In most directions, the scattered rays are out of phase, and undergo destructive interference. However, in certain directions, they are in phase, and mutually reinforce. The group of such reinforced rays in a given direction constitutes a diffracted x-ray beam. The conditions for diffraction are as follows: (1) the angles made by the diffracting planes (*hkl*) with the incident and diffracted beams are equal; (2) the directions of the incident and diffracted beams and the normal to the diffracting planes are coplanar; and (3) the waves emitted by individual atoms in the direction of a diffracted beam are in phase.

Figure 2.13 shows the condition for first-order diffraction of λ ($n = 1$). The first-order wave front consists of the first wave from the first (top) atom, the second wave from the second atom, etc.; that is, waves from successive atoms differ by 1. If θ is increased to appropriate discrete values, the crystal diffracts the second order ($n = 2$), where waves from successive atoms differ by 2, the third ($n = 3$), where they differ by 3, etc.

A more useful version of diffraction is that of Bragg (*B51*), who regards the phenomenon in terms of reflection from a stack of crystal planes. In Figure 2.14, a collimated monochromatic x-ray beam of wavelength λ is directed at angle θ on a set of crystal planes (*hkl*) of interplanar spacing *d*. Again the incident beam is scattered in all directions from each plane, but only in certain directions does the scatter reinforce. The conditions for diffraction are the same as for the Laue version, but the third condition might

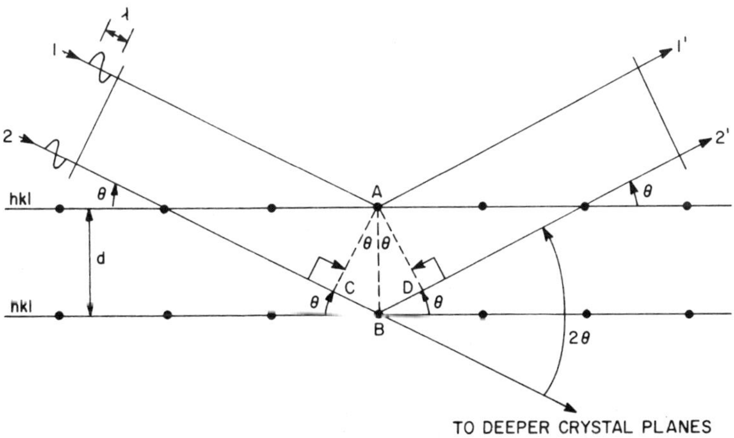

Fig. 2.14. Diffraction by crystals—Bragg version.

be more appropriately restated as follows: Reflected rays from successive *planes* differ in path length by an integral number of wavelengths.

In Figure 2.14, rays 1 and 2 are directed on two successive planes and scatter in all directions. Of the scattered rays, only 1′ and 2′ fill the conditions for diffraction. Rays 1 and 2 travel equal distances to AC, and rays 1′ and 2′ travel equal distances beyond AD. The path difference between rays 1A1′ and 2B2′ is then

$$CBD = CB + BD = 2AB \sin\theta = 2d \sin\theta \tag{2.26}$$

If these rays are to be in phase, this path difference must be an integral number of wavelengths, that is,

$$n\lambda = 2d \sin\theta \tag{2.27}$$

This is the Bragg law, where n is the order of the diffracted beam and is numerically equal to the path difference, in wavelengths, for successive planes; d is the interplanar spacing of the diffracting planes (Å); and θ, the Bragg angle, is the angle between the incident x-rays and the diffracting planes (deg). X-ray spectrometers are usually calibrated in terms of 2θ, the angle between the diffracted beam and the undeflected incident beam (Figures 2.14 and 2.15).

We have seen that complete destructive interference occurs only for two rays out of phase by exactly $\frac{1}{2}\lambda$. Two rays scattered from successive planes at an angle $\theta + \Delta\theta$ such that they differ in path length by only, say, $\frac{1}{4}\lambda$, do not cancel, but combine to a ray of smaller amplitude than if they were in phase (Figure 2.12). However, the ray scattered at $\theta + \Delta\theta$ from the third plane down in the crystal is $\frac{1}{4}\lambda$ out of phase with that from the second plane and $\frac{1}{2}\lambda$ out of phase with that from th first plane. Consequently, the $\theta + \Delta\theta$ rays from the first and third, second and fourth, fifth and seventh, etc. planes throughout the crystal cancel one another. Consider now rays scattered from successive planes at an angle $\theta + \Delta\theta$ such that they differ in path length by only, say, 0.005λ. The ray to cancel this ray from the first plane originates deep within the crystal, $0.5\lambda/0.005\lambda = 100$ planes down, that is, the 101st plane. Thus, the $\theta + \Delta\theta$ rays from planes 101 to 200 respectively cancel those from planes 1 to 100. Thus, diffraction occurs only at the Bragg angle.

The analogy of the Bragg version of diffraction with optical reflection is evident, but the two phenomena are quite different. Optical reflection is a wholly surface effect and occurs at all incident angles greater than a certain small critical glancing angle.

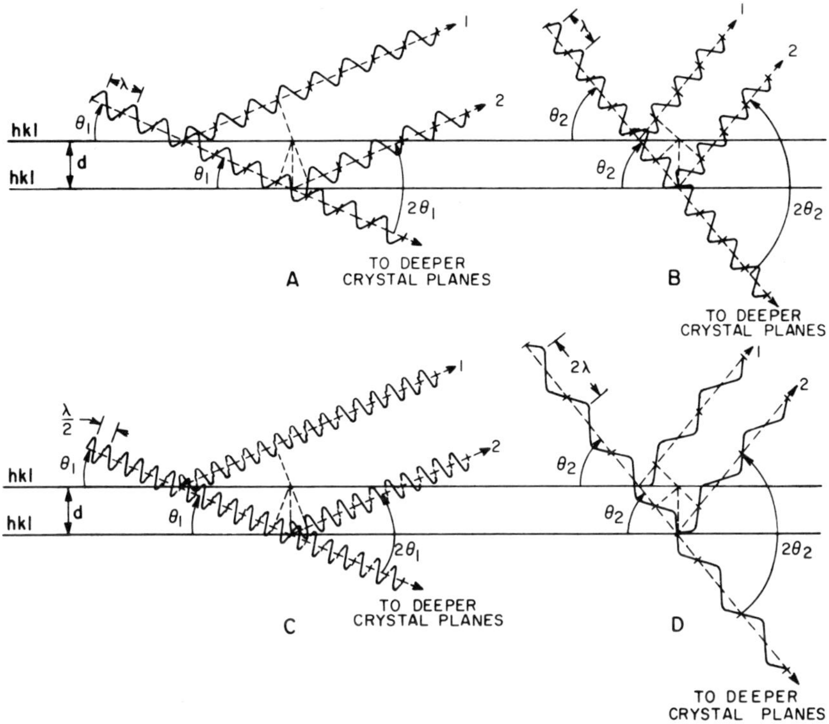

Fig. 2.15. Origin of diffraction orders.

Figure 2.15 further illustrates the phenomenon of orders: Figure 2.15A shows first-order diffraction of wavelength λ at θ_1 with path difference 1λ for rays 1 and 2. Figure 2.15B shows second-order diffraction of λ at θ_2 with path difference 2λ. Figure 2.15C shows second-order diffraction of $\lambda/2$ at θ_1, the same angle at which first-order diffraction of λ occurs. Figure 2.15D shows first-order diffraction of 2λ at θ_2, the same angle at which second-order diffraction of λ occurs. Thus, the same wavelength may be diffracted by the same planes d_{hkl} of the same crystal at more than one Bragg angle, and different wavelengths may be diffracted at the same angle.

If θ is increased progressively, the wavelength required to satisfy Equation (2.27) also increases progressively, and a polychromatic spectrum may be diffracted in sequence. At fixed θ, λ, $\lambda/2$, $\lambda/3$, etc. satisfy the equation in orders 1, 2, 3, etc.

The Bragg law disregards refraction of x-rays. The resultant error is usually negligible because the index of refraction of most substances for x-rays is of the order $1 - 10^{-6}$. The effect is greatest at the longer wavelengths, where the Bragg law may be in error by as much as 0.02° 2θ.

The application of crystals to dispersion is discussed in detail in Chapter 5.

2.4. SPECULAR REFLECTION; DIFFRACTION BY GRATINGS

2.4.1. Specular Reflection

The index of refraction of a substance for x-rays is given by

$$n_x = 1 - \frac{N_0 \varrho Z e^2 \lambda_{\text{cm}}^2}{2\pi A m c^2} = 1 - \delta \tag{2.28}$$

where N_0 is the Avogadro number (atoms/mole); ϱ is density (g/cm³); Z is atomic number; e is the charge of the electron (esu); λ is x-ray wavelength (cm); π is 3.1416...; A is atomic weight (g/mole); m is the rest mass of the electron (g); and c is the velocity of light (cm/s). In this expression,

$$N_0 \varrho Z / A = N \tag{2.29}$$

where N is the number of electrons per cubic centimeter of the refracting medium.

In Equation (2.28), δ is of the order 10^{-6}, so that the index of refraction for x-rays is of the order $1 - 10^{-6}$, and x-rays in air or vacuum are totally reflected from any medium if incident upon it at a very small glancing angle ϕ less than ϕ_C, the critical glancing angle. By the Snell law,

$$\cos \phi_C = 1 - \delta; \qquad 1 - \cos \phi_C = \delta \tag{2.30}$$

By trigonometric identity,

$$1 - \cos 2x = 2 \sin^2 x \tag{2.31}$$

from which,

$$1 - \cos \phi_C = 2 \sin^2(\tfrac{1}{2}\phi_C) = \delta \tag{2.32}$$

Since $\delta \approx 10^{-6}$, the sine may be replaced by its angle in radians, giving

$$\delta = 2(\tfrac{1}{2}\phi_C)^2 = \tfrac{1}{2}\phi_C{}^2 \tag{2.33}$$

from which

$$\phi_C = \sqrt{2\delta} \tag{2.34}$$

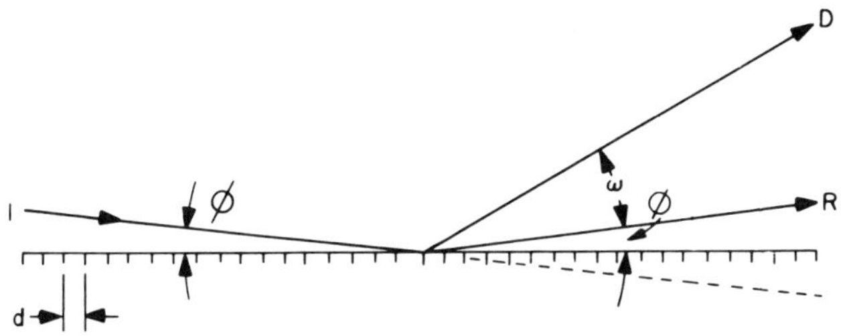

Fig. 2.16. Specular reflection from a smooth surface and diffraction by a grating.

In Figure 2.16, the highly collinear monochromatic x-ray beam I is directed at angle $\phi < \phi_C$ on a surface that for now may be regarded as plane and smooth, giving reflected ray R. The incident and reflected rays make the same angle ϕ with the reflecting surface. For a typical example, for Cu $K\alpha$ (λ 1.54 Å) reflected from aluminum, δ is 8.32×10^{-6}, and ϕ_C is 4.08 milliradians or 13.0 min of arc.

Reflection has been applied to wavelength dispersion in the ultralong-wavelength region, and it occurs on the foils of Soller collimators.

2.4.2. Diffraction by Gratings

In Figure 2.16, a highly collinear monochromatic x-ray beam I having wavelength λ is directed at angle ϕ less than the critical angle for reflection on a ruled reflection grating having line spacing d. Here, R is the specularly reflected beam, and D is the first-order diffracted beam. Diffraction occurs if

$$n\lambda = 2d \sin \tfrac{1}{2}(2\phi + \omega) \sin \tfrac{1}{2}\omega \qquad (2.35)$$

where n is the order; ϕ is the angle between the reflected (or incident) beam and the reflecting surface; and ω is the angle between the diffracted and reflected beams. For example, if carbon $K\alpha$ is diffracted by a grating having 12,000 lines/cm:

$$\lambda(CK\alpha) = 44.4 \text{ Å}, \qquad d = \frac{10^8 \text{ Å/cm}}{12{,}000 \text{ lines/cm}} = 8333 \text{ Å}$$

$$\phi = 0.75°, \qquad \omega + \phi = 6.3°$$

The angle between the direction of the incident x-rays and the first-order diffracted beam D corresponds to 2θ in crystal diffraction and is $\sim 7°$ [pp. 318–19 in (*16*)]. Reflection gratings have been applied to wavelength dispersion in the ultralong-wavelength region.

2.5. THE AUGER EFFECT; FLUORESCENT YIELD

2.5.1. Auger Effect

When an atom is ionized in one of its inner shells, it rearranges by filling the vacancy with an electron from a higher orbit and releasing the energy of the transition. The energy may be released as an x-ray photon; this process, of course, is x-ray emission. Alternatively, the energy may be released as an electron; this process is known variously as *radiationless transition*, *internal conversion*, or the *Auger effect*, and the electron is an *internal-conversion* or *Auger electron* (*A20–22*). Either way, the energy of the x-ray photon or of the Auger electron is characteristic of the emitting element.

The Auger effect may be regarded in either of two ways. Consider an atom in which an initial *K*-shell vacancy is filled by an *L* electron and which then undergoes the Auger effect. It may be assumed that the atom releases the energy of the electron transition from the *L* to the *K* shell by emission of an *L* or *M* electron. Or it may be assumed that the *L*-to-*K* shell electron transition results in the production of a $K\alpha$ photon in the usual way. However, in this case, the photon does not leave the atom of its origin, but rather is absorbed within the atom with consequent expulsion of an *L* or *M* electron. This process may be regarded as internal photoelectric absorption and is analogous to the internal γ-conversion described in Section 1.6.3. Either version results in a doubly ionized atom having two vacancies, the one created by filling of the initial vacancy and the one ejected by the Auger process. These atoms are said to be in, for example, the *LL* or *LM*

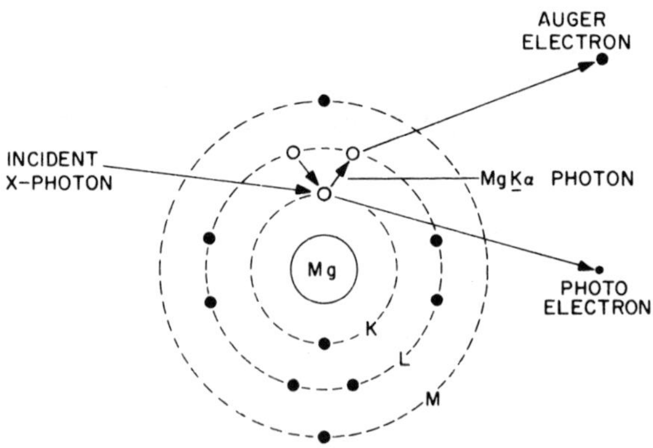

Fig. 2.17. The Auger effect.

state. The second version of the Auger effect and the resultant production of a doubly ionized (LL) atom are illustrated in Figure 2.17.

The Auger effect is more common in elements of low atomic number because the electrons are more loosely bound and the characteristic photons more readily absorbed. The effect is more marked for the L series than for the K series for the same reasons. In Figure 2.17, the initial vacancy is created by secondary excitation, but the Auger effect also occurs with primary excitation. In fact, the effect is more marked with primary excitation, but only because, as already explained, electrons are about 100 times as efficient as x-rays photons in exciting x-ray spectra. The more efficient the excitation, the greater is the opportunity for Auger absorption.

The Auger effect provides the basis for fluorescent yield, satellite lines, and the analytical method of Auger-electron spectrometry (Section 20.3.2).

2.5.2. Fluorescent Yield

One consequence of the Auger effect is that the lines in a given series are not as intense as would be predicted from the number of vacancies created in the associated orbital. The *K fluorescent yield* ω_K is the number of photons of all lines in the K series emitted in unit time divided by the number of K-shell vacancies formed during the same time:

$$\omega_K = \frac{\Sigma\,(n_K)_i}{N_K} = \frac{n_{K\alpha_1} + n_{K\alpha_2} + n_{K\beta_1} + \cdots}{N_K} \tag{2.36}$$

where ω_K is K-fluorescent yield; N_K is the rate at which K-shell vacancies

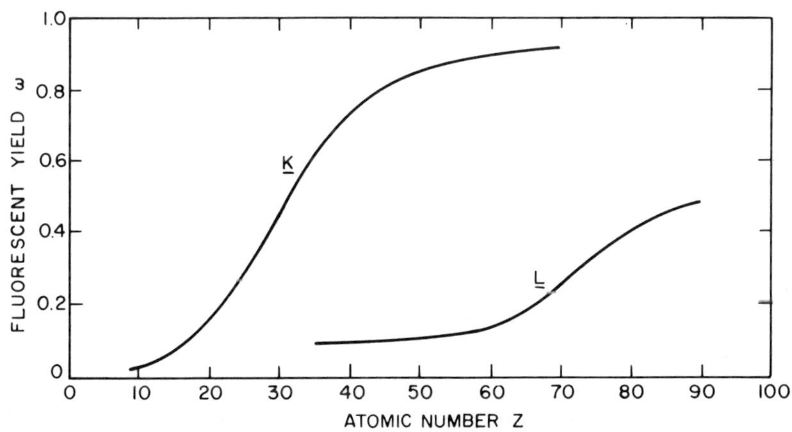

Fig. 2.18. *K* and *L* fluorescent yields.

TABLE 2.3

Fluorescent Yield[a]

Element	ω_K	ω_L
8-O	0.08	—
29-Cu	0.40	—
42-Mo	0.75	0.07
62-Sm	—	0.19
92-U	—	0.45

[a] Page 27 in (*1*).

are produced; and $(n_K)_i$ is the rate at which photons of a spectral line i are emitted. The L and M fluorescent yields ω_L and ω_M are defined similarly. Auger *yield* is defined as $1 - \omega$. Were it not for the Auger effect, ω would always be 1. Actually, ω varies with atomic number and series, as shown in Figure 2.18 and Table 2.3. Fluorescent yield is a major limitation of sensitivity, especially for elements of low atomic number. Note that the *fluorescent* yield is the probability that the filling of a vacancy in a specified shell will result in emission of a characteristic x-ray photon in the series associated with that shell—*regardless* of whether the vacancy arose from primary or secondary excitation.

2.5.3. Satellite Lines

Another consequence of the Auger effect is the production of doubly ionized atoms, that is, atoms with two inner-orbital vacancies. For example, suppose that (1) a K-shell vacancy is created by the primary x-rays; (2) an L-shell electron falls into this vacancy, producing a $K\alpha$ photon; and (3) this photon undergoes Auger absorption with consequent expulsion of another L electron. The result is an atom in the LL state, that is, with two L-shell vacancies, one created by the L–K electron transition, the other by the expulsion of the Auger electron. If an M electron falls into one of the L vacancies and an L-series photon is emitted, the atom is left in the LM state. Doubly ionized atoms can also be produced without the Auger effect if the incident electrons (primary excitation) or x-ray photons (secondary excitation) expel more than one orbital electron in one atom.

Electron transitions in such doubly ionized atoms result in emission of lines having wavelengths slightly different from those emitted from singly charged atoms. Such lines are known as *satellite lines* or *nondiagram lines.* For elements of high and intermediate atomic number, satellite lines usually have very low intensity and are of little significance in x-ray spectrochemistry. However, for low-Z elements, they may be relatively intense; for example, Al $K\alpha_3$, which results from an LK–LL transition, has ~10% of the intensity of Al $K\alpha_{1,2}$. The influence of chemical combination and electron distribution in the outermost orbits on the wavelength and intensity is more marked for satellite lines than for diagram lines. Satellite lines are listed by White *et al.* (*49*); typical pages of their work are reproduced in Figures 10.3 and 10.6; the satellite lines are those having symbols prefixed by the letter S.

Chapter 3

X-Ray Secondary-Emission (Fluorescence) Spectrometry; General Introduction

Having considered the excitation, nature, and properties of x-rays and x-ray spectra, we now begin our study of the application of x-ray spectra to chemical analysis. However, first we define some terms used throughout the book.

3.1. NOMENCLATURE

Imagine going to the laboratory chemical cabinet and being confronted with shelves of bottles labeled lye, caustic potash, lunar caustic, tartar emetic, cream of tartar, lime, calomel, alum, Epsom salt, Glauber's salt, Rochelle salt, blue vitriol, green vitriol, copperas, oil of vitriol, niter, saltpeter, sal ammoniac, corrosive sublimate, ... Fortunately, chemists have long since realized the vital importance of systematic nomenclature. More recently, units of physical measurement have been systematized. However, no such order exists in the nomenclature of analytical chemical instruments and methods.

The name of an instrument (or method) should: (1) classify it, showing its relationship to similar instruments; (2) distinguish it from all others in the class; (3) be appropriate and descriptive, briefly conveying its nature; and (4) be specific, that is, applicable only to that instrument. The term "atomic absorption spectrophotometer" is an example of very poor nomenclature. It obscures the relationship of the instrument to its complement —the optical flame emission spectrometer, and it does not indicate the applicable spectral region. "Atomic" absorption is certainly not limited to the optical region; as we have seen, x-ray absorption is "atomic" absorption in every sense of the word. The common name for the method—simply "atomic absorption"—even obscures the fact that the method is a spectrometric one. Another example of very poor nomenclature is found on many of the printed programs of technical meetings. The sessions on optical emission spectrography and x-ray secondary-emission (fluorescence) spectrometry are distinguished by the terms *emission spectrography* and *x-ray spectrography*—as if x-ray fluorescence spectrometry were not also an emission method!

Nomenclature for a spectrometric instrument should classify the instrument as a spectroscope (which provides for visual observation of the spectrum), spectrometer (which provides for measurement of wavelength), or spectrograph (which records the spectrum). The name should then distinguish the instrument with respect to type of spectrum (emission, absorption, Raman, nuclear magnetic resonance, etc.) and/or radiation or spectral region (microwave, infrared, visible, ultraviolet, optical, x-ray, γ-ray, β-ray, mass, etc.). In the present case, another adjective is required to distinguish primary or secondary excitation.

For the instrument dealt with in this book, the writer has always preferred the term *x-ray secondary-emission spectrometer*; the term *spectrograph* is also acceptable because of the ratemeter-recorder. The term *fluorescence* is unfortunate, but so widely used that its abandonment seems unlikely. However, it should be used only as a synonym for secondary characteristic spectral-line emission. Thus, the instrument should be referred to as an *x-ray secondary-emission spectrometer* or *x-ray fluorescence spectrometer*, and the method as *x-ray secondary-emission spectrometry* or *x-ray fluorescence spectrometry*—but never simply *x-ray fluorescence*. In no case should the term *fluorescence* be used as a synonym for primary-excited characteristic line spectra, such as the target line spectrum from an x-ray tube or the specimen line spectrum in an electron-probe microanalyzer.

The term *x-ray emission spectrometry* may be used to distinguish x-ray emission and absorption spectrometry. However, the term is indefinite in that it can refer to either primary or secondary emission. The term *x-ray spectrochemical analysis* is a good general term for all the various analytical methods based on x-ray emission and absorption spectra, even including electron-probe microanalysis. Throughout this book, whenever there is no question that x-ray secondary-emission spectrometry is referred to, the simplified terms *x-ray spectrometry*, *x-ray spectrometer*, and *x-ray spectrometric analysis* may be used for brevity.

While on the subject of nomenclature, we shall define some other terms used in the book. There has long been the need for a word to replace those cumbersome terms inevitable in all papers and books on chemical analysis—"element sought," "element to be determined," "element of interest," "analytical element," etc. In this book the term *analyte* is used. The term *specimen* indicates the object placed in the instrument specimen compartment and on which measurements are made. The specimen may be a *sample*—a material of unknown composition to be analyzed or a preparation derived from it, or a *standard*—a similar material or preparation having known composition for calibration purposes. The intensity measured at the line is the *peak* intensity I_P and consists of the *line* or *net* intensity I_L and background intensity I_B, contributed by the spectral line being measured and the background, respectively; that is, $I_L = I_P - I_B$. Some workers use the term *total* intensity I_T for peak intensity.

3.2. PRINCIPLE AND INSTRUMENT

3.2.1. Principle

X-ray secondary-emission spectrometry is a method of qualitative and quantitative analysis for chemical elements based on measurement of the wavelengths and intensities of their x-ray spectral lines emitted by secondary excitation. The primary beam from an x-ray tube irradiates the specimen (sample or standard), exciting each chemical element to emit secondary spectral lines having wavelengths characteristic of that element (basis of qualitative analysis) and intensities related to its concentration (basis of quantitative analysis). The principles and practice of the method have been summarized in several excellent reviews (*24–28*, *F18*, *K16*, *L4*, *P4*).

3.2.2. The X-Ray Spectrogoniometer

The essential features of the x-ray secondary-emission spectrometer are shown in Figure 3.1. The primary x-ray beam irradiates the specimen *A*

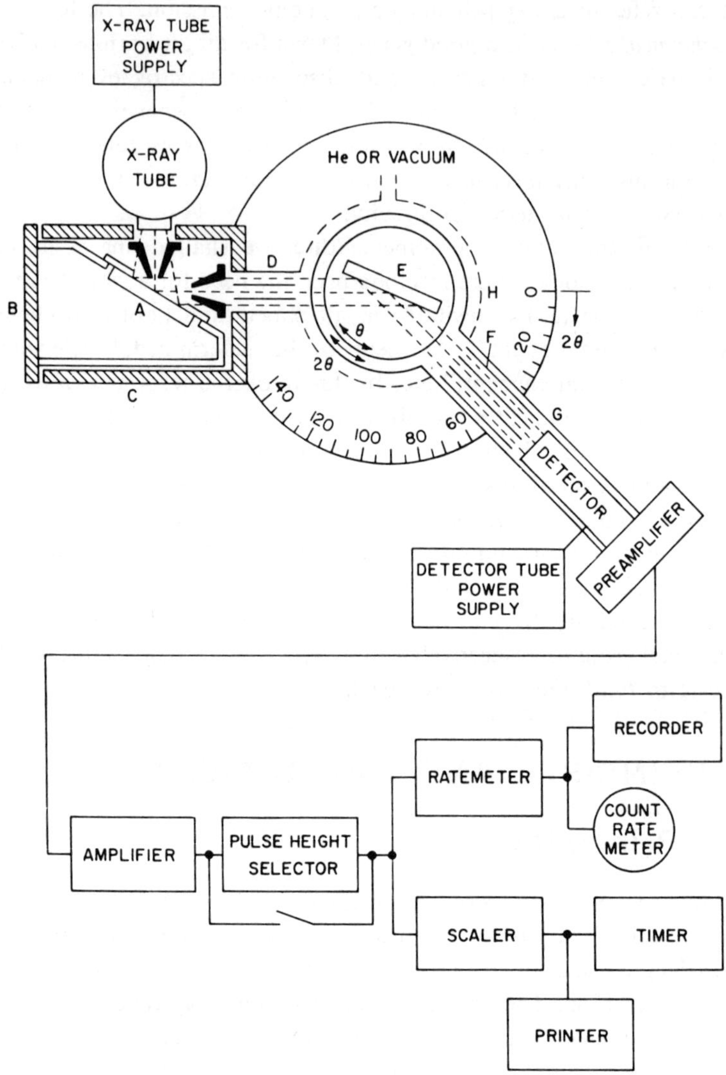

Fig. 3.1. X-ray secondary-emission (fluorescence) spectrometer.

held in some type of drawer or holder *B* in a specimen compartment *C*. A leaded shutter mechanism (not shown) is provided to block the primary x-ray beam at the x-ray tube window when the specimen is inserted and removed from its compartment. The secondary radiation thus excited contains many wavelengths and is emitted in all directions, but only those rays directed substantially parallel to the foils of the source Soller collimator *D* emerge from the specimen compartment and reach the analyzer crystal *E*. The crystal is mounted on a stage and may be rotated, the axis of rotation of the crystal being also the goniometer axis. As the crystal rotates, the angle θ it presents to the secondary beam varies, and each wavelength in the beam in turn satisfies the Bragg law and is diffracted. In this way, the crystal disperses the secondary spectrum, in effect causing each wavelength to go in a different direction.

A second detector collimator *F* and the detector are mounted on a common arm *G* that rotates on the same axis as the crystal and is geared to it with a "two-to-one" ratio, that is, as the crystal turns through an angle θ, the detector arm turns through twice that angle, 2θ. Thus, the detector and incident secondary beam always present equal angles to the crystal, and the detector is always in the correct position to receive any diffracted x-rays (Figures 2.14, 2.15). The detector converts each x-ray photon it absorbs into a pulse of electric current; in most detectors, these pulses have amplitudes proportional to the energy of the x-ray photons. These pulses are amplified by the preamplifier, which is mounted close to the detector on the detector arm.

When wavelengths greater than ~2 Å are measured, it is necessary to reduce the absorption of the radiation path (x-ray tube window to detector window) with a helium or vacuum enclosure *H*. When very small specimens or small selected areas on large specimens are measured, apertures *I*, *J* may be placed in the primary or secondary x-ray beam (Chapter 19).

The instrument just described—consisting of the specimen chamber, crystal, source and detector collimators, helium or vacuum enclosure, detector, preamplifier, and the drive mechanism for the crystal and detector arm—is known as the *goniometer* (angle-measuring device). When used for spectrometry in this way, the goniometer is referred to as a *spectrogoniometer* or *spectrometer*. The alternative mode of operation, described briefly below, is as a *diffractometer*. The crystal and detector arm may rotate in a horizontal or vertical plane—that is, the goniometer axis may be vertical or horizontal. The Picker goniometer includes the x-ray tube as an integral part and may be mounted in any position.

This type of spectrogoniometer is typical of the earlier commercial instruments, which were simply converted diffractometers. Today, all manufacturers offer instruments designed specifically for spectrometry. However, the older goniometers having specimen drawers are still available and very useful. In small laboratories and schools, they perform both spectrometric and diffractometric functions economically. They also provide

TABLE 3.1
Comparison of the X-Ray Spectrometer and Diffractometer

Bragg law	$n\lambda =$	$2d$	$\sin\theta$
Spectrometer	Calculated	Known	Measured
Diffractometer	Known	Calculated	Measured

versatility in choice of modes of operation, components, accessories, and specimen presentation not feasible with the newer versions.

Although both the spectrometer and diffractometer are based on the Bragg law, their modes of operation are entirely different. In a diffractometer, the specimen drawer *B* (Figure 3.1) and compartment *C* are omitted, and the x-ray tube is rotated about its axis so that the primary beam is directed through the collimator *D*. The single crystal *E* is replaced with the polycrystalline sample to be investigated. Thus, a known wavelength—the filtered $K\alpha$ line of the x-ray tube target—is directed on the sample, and as the sample is rotated through successive θ values, each set of crystal planes (hkl) having its own interplanar spacing d_{hkl} in turn satisfies the Bragg law and diffracts the incident wavelength. The functions of the spectrometer and diffractometer may be contrasted in terms of the Bragg law as shown in Table 3.1. Both instruments produce charts consisting of peaks on an intensity *versus* 2θ ("two theta") scale. On the spectrometer chart, each peak represents a different wavelength diffracted by the same set of planes in the analyzer crystal. On the diffractometer chart, each peak represents the same wavelength—the target $K\alpha$ line—but diffracted by a different set of planes in the crystalline sample.

3.2.3. Electronic Readout Components

The preamplifier is connected by cable to the amplifier, which further amplifies the detector output pulses. As we have seen, the crystal can diffract more than one wavelength at a given Bragg angle, and when this happens, each wavelength gives detector output pulses of different amplitude. The pulse-height selector sorts these pulses, passing only those having a certain preselected height. The pulse-height selector may be by-passed, as shown schematically in Figure 3.1. The amplified pulses—with or without selection—may be registered by the ratemeter, which integrates them and displays on a meter or strip-chart recorder the rate at which pulses enter the ratemeter, and thus the rate at which x-ray photons are absorbed in the detector. Alternatively, the pulses may be individually counted by the scaler–timer, the data from which may be recorded manually or by a printer. All these components and the detector-tube power supply are mounted in the electronic-circuit panel or console.

The x-ray tube and power supply, spectrogoniometer, and electronic-circuit panel collectively constitute the x-ray secondary-emission spectrometer. The particular instrument just described is a single-channel manual spectrometer—that is, it can measure only one wavelength at a time, and it is operated manually. Dual and multichannel and automatic instruments are described in Chapter 9.

3.2.4. Qualitative, Semiquantitative, and Quantitative Analysis

In qualitative analysis, the spectrometer is made to scan through the entire 2θ region as detected x-ray intensity is recorded by the ratemeter recorder. The goniometer motor drive rotates the crystal and detector arm at rates proportional to θ and 2θ, respectively, per unit time. The elements present are identified by the 2θ angles—or wavelengths—at which peaks occur on the chart, and their concentrations may be estimated from the peak heights—or intensities. Analyte peaks may be recorded from standards for comparison. Qualitative and semiquantitative analyses are described in Chapter 10.

In quantitative analysis, the intensities of analyte lines are measured from samples and standards. The measurements are almost always made with the scaler and timer, although the ratemeter may be used. Calibration curves or mathematical relationships are established from the standard data. Quantitative analysis is described in Chapter 14.

3.2.5. Phases of a Quantitative X-Ray Spectrometric Analysis

X-ray spectrochemical analyses involve four essential phases as follows:

1. Excitation of the characteristic line spectrum of the analyte involves the x-ray tube and power supply (Chapter 4) and the technique of specimen presentation (Chapters 16 and 17).

2. Dispersion of the secondary spectrum to isolate the analyte lines and permit their individual measurement is usually effected in either of two ways or by a combination of both: Crystal dispersion (wavelength or spatial dispersion) involves the crystal–collimator system (Chapter 5);

"nondispersive" analysis (or energy dispersion) involves the pulse-height analyzer (Chapter 8).

3. Detection, measurement, and readout of the analyte line intensity involves the detector, amplifiers, pulse-height analyzer, ratemeter and recorder, scaler and timer, and printer (Chapters 6 and 7).

4. Reduction of x-ray intensity data to analytical concentration involves the calibration or calculation technique (Chapter 14) and correction for absorption-enhancement effects (Chapter 15).

The first three of these phases are effected by the spectrometer, the last may be done manually or by a computer.

3.3. APPRAISAL

3.3.1. Advantages

3.3.1.1. *X-Ray Spectra*

Unlike optical emission spectra, x-ray spectra are simple and regular. In general, each element has substantially fewer x-ray lines than optical lines. The selection rules exclude many electron transitions, and many of the permitted transitions result in lines so weak as to be insignificant. However, weak lines from major constituents may interfere with determination of minor or trace constituents, especially in the electron-probe microanalyzer, where electron excitation and curved-crystal optics result in high efficiency. In general, each element has the same *K*, *L*, and *M* lines, and the wavelength of each line varies regularly with atomic number, regardless of the nature of the element—metal, nonmetal, liquid, or gas. The relative simplicity of x-ray spectra is illustrated by a comparison of the optical emission spectra of iron, copper, and an Alnico alloy shown in Figure 3.2 with the x-ray spectra shown in Figure 10.1.

Other advantages of x-ray spectra arise from their relative independence, compared with optical spectra, of chemical state. Although small displacements of wavelength of x-ray lines may occur, especially in light elements, the effect is much more severe in optical spectra. Several other undesirable effects present in optical spectra are completely absent in x-ray spectra: Differences in volatility of analyte elements affect relative intensities of optical lines. If the analyte volatilizes as a molecular, rather than atomic, species, the strength of the chemical bond affects the optical line intensity. Doppler line broadening arises from the rapid motion of the atoms in the arc or spark.

6Cu–0.5Ti–0.5Zr), and copper. Compare with x-ray spectrum of Alnico in Figure 10.1.

Internal standardization, almost always required in optical emission spectrography, is usually not necessary in x-ray spectrometry. However, when it is required, the simplicity of x-ray spectra and their relationship to atomic number make selection of an internal standard element much easier than it is in optical spectrography.

The simplicity and uniform regular relationship of x-ray spectra and atomic number are shown in Figure 1.8. No such simple diagram can be drawn for selected optical emission lines.

3.3.1.2. *Excitation and Absorption*

These properties vary with atomic number in the same uniform regular manner as x-ray spectra, and this regularity is also uninterrupted by the physical and chemical state of the element. Elements having about the same atomic number have very similar excitation and absorption properties.

3.3.1.3. *Absorption-Enhancement Effects*

The elements occurring with the analyte in the specimen may cause absorption-enhancement effects ("matrix" effects) in both x-ray and optical emission. In optical emission, these effects are largely unpredictable, except by experience. In x-ray emission, they are systematic, predictable, and readily evaluated. Moreover, there are many methods of dealing with them, including the following: (1) standard addition and dilution; (2) calibration standards; (3) internal standardization; (4) dilution; (5) thin-film methods; (6) methods based on x-ray scatter; (7) emission–absorption methods; (8) association (indirect) methods; and (9) mathematical correction. In addition, there are many special experimental x-ray spectrometric techniques, and the instrument can be arranged for x-ray absorption-edge spectrometry, in which absorption-enhancement effects largely disappear.

3.3.1.4. *Spectral-Line Interference*

Due to the simplicity of x-ray spectra, spectral-line interference is relatively infrequent. However, when it does occur, there are many ways to deal with it: (1) choice of an alternative spectral line or a higher order; (2) prevention or reduction of excitation of the interfering line by operating the x-ray tube below the interferant excitation potential, selection of the x-ray tube target, or monochromatic excitation; (3) use of a blank to measure the interfering line intensity; (4) increased collimation; (5) selection of the crystal for greater dispersion or resolution or to eliminate even-order interfering lines; (6) filtration; (7) use of air path to eliminate long-wavelength interfering lines; (8) selection of the detector for maximum efficiency for the analyte line or minimum efficiency for the interfering line; (9) pulse-height selection; (10) dilution; and (11) mathematical correction, including unfolding (stripping) of overlapped lines.

3.3.1.5. *Nondestruction of Specimen*

X-ray secondary-emission spectrometry is nondestructive in the sense that the specimen placed in the instrument remains substantially unchanged during the analysis. However, biological, organic, and liquid specimens may undergo radiolysis; minerals and other inorganic substances may become colored, usually temporarily; and plastics and electronically active materials may sustain radiation damage. The method is also nondestructive in the sense that, frequently, sampling is not required; the instrument can be arranged to accommodate very large objects, which then need not be damaged by cutting samples from them. These features are advantageous in many ways.

Specimen composition does not change during analysis, as it frequently does during arcing in optical spectrography. Since the instrument can accommodate the specimen in nearly any form, little or no specimen preparation is required in many cases. Other analyses and tests can be made, and questionable analyses can be repeated on the same sample. Samples can be subjected to additional treatment or processing, then reanalyzed. Standards, test specimens, expensive or precious samples, criminal evidence, etc. are not consumed. Fabricated parts, art objects, antiquities, etc. are not damaged (*C23*).

3.3.1.6. *Specimen Versatility*

No analytical method can deal with as wide a variety of specimen forms as the x-ray spectrometer. The physical form may be solid, briquet, powder, paste, liquid, or even gas. The material may be metal, mineral, ceramic, plastic, rubber, textile, paper, or practically anything else. The size may be anything from a barely visible speck or a film having less material than a monolayer, to a very large massive object. The shape may be plane, cylindrical, filamentary, or irregular, and the method is applicable to small fabricated parts of all shapes. The analyte may be distributed on filter paper, Mylar film, cellulose tape, ion-exchange resin, etc. The method is applicable to special conditions such as high and low temperatures, special atmospheres, and dynamic conditions.

3.3.1.7. *Operational Versatility*

The x-ray spectrometer can be fitted with a variety of components and accessories, and operated in a variety of modes under a variety of conditions, all selected to give optimum results in a given analysis. The selectable components include: (1) the x-ray tube target—about ten are available commercially, including a dual-target (tungsten–chromium); (2) collimators of various lengths and foil spacings; (3) crystals, selected for "reflectivity," dispersion, resolution, or elimination of orders; and (4) detectors—sealed and flow proportional counters having various gas fillings, scintillation counters, or Geiger counters. The selectable accessories include rotating specimen stages, inverted optics, bulk-specimen holders, selected-area apertures, curved-crystal accessories, and helium or vacuum enclosures. The modes of operation, in addition to x-ray secondary-emission spectrometry, include polychromatic and monochromatic absorption; absorption-edge spectrometry; methods based on scattered, rather than emitted, x-rays; nondispersive analysis; and even polychromatic and monochromatic contact microradiography.

The work of Cowgill (*C47*) with lake core samples provides an outstanding example of versatility of operation and analytical strategy (see below) in x-ray spectrometry. The material was dried and ground to -200 mesh, then briqueted at 15,000 lb/inch^2. The briquets were used to determine 41 elements ranging in atomic number from 11 to 82 (sodium to lead) and in concentration from <1 ppm to $\sim 50\%$. The work involved three x-ray tube targets (chromium, molybdenum, platinum), eight crystals (KHP, ADP, EDDT, germanium, silicon, NaCl, LiF, topaz) and two detectors (flow and scintillation). Three basic analytical methods were used—comparison with calibration standards and the methods of Mitchell (Section 15.3) and Reynolds and Ryland (Section 14.5.4).

3.3.1.8. *Versatility of Analytical Strategy*

X-ray spectrometry lends itself to a variety of analytical methods and techniques of specimen preparation and presentation so large that it would be futile to attempt to summarize them here. They are described in detail in Chapters 14, 16, and 17.

3.3.1.9. *Selected-Area Analysis*

Small selected areas can usually be analyzed in place so that contamination and loss by chemical or physical separation are avoided. Uniformity and homogeneity of large-area samples and uniformity of thickness and composition of films and platings can be evaluated.

3.3.1.10. *Semiquantitative Estimations*

These can be made from data taken without standards other than the pure analyte to within a factor of two or three in x-ray spectrometry, to within an order in optical spectrography. Simple calculations based on absorption coefficients can improve the x-ray estimates.

3.3.1.11. *Concentration Range*

X-ray spectrometry is applicable over an extremely wide concentration range, from 100% for any element above fluorine to 0.0001% for sensitive elements in favorable matrixes. Small differences in concentration of major constituents are readily measured. For most elements, the method is applicable to micro and trace analysis. In favorable cases, traces can be determined without separation, so that contamination or loss by chemical or physical separation is avoided. Quantities $<0.1\ \mu g$ and films "thinner" than one monolayer can be detected for most elements above magnesium on modern commercial flat-crystal spectrometers.

3.3.1.12. *Sensitivity*

The sensitivity, although in general not as good as that of optical emission spectrography, is still very high. Instrument components, accessories, and conditions may be chosen for optimal sensitivity for a given analysis. With a given x-ray tube target, crystal, collimator system, and detector, sensitivity of the pure analyte is a relatively simple function of atomic number. Elements having about the same atomic number are likely to have about the same sensitivity in a given system. No such simplicity exists in optical emission spectrography.

3.3.1.13. *Precision and Accuracy*

These qualities are generally very good. Precision may usually be made as great as desired by appropriate selection of the accumulated count. The accuracy compares favorably with that of other instrumental analytical methods. In conventional x-ray spectrometry, the specimen area is usually several square centimeters, so that inhomogeneity, although still a serious problem, is not as serious as in optical spectrography, where the sample area is only a very few square millimeters.

3.3.1.14. *Excitation*

Electronic and/or magnetic line, kV, and mA stabilization result in extremely well-regulated excitation conditions. Except for very small variations due to slight instability of x-ray tube operating conditions, x-ray emission intensity is constant with time. Optical emission intensity from an arc or spark may vary markedly with time.

3.3.1.15. *Speed and Convenience*

Because wavelength and intensity are measured directly on the instrument and because of the nondestructive feature, x-ray spectrometry is extremely rapid and convenient. In general, no analytical method can deal as quickly and conveniently with very large numbers of samples when they can be analyzed in the as-received form. Even when specimen preparation is required, the method is often more rapid and convenient than other methods. X-ray spectrometry permits extremely rapid confirmation, identification, or selection of known alternatives on the basis of the presence or absence, or high or low concentration, of some element. For example, Kovar and stainless steel parts can be sorted on the basis of presence or absence of cobalt; 52-alloy and Invar parts (48–52 and 65–35 iron–nickel alloys, respectively) can be distinguished on the basis of relative intensities of Fe $K\alpha$ and Ni $K\alpha$ lines. Errors, anomalous results, out-of-specification results, etc. may often be noted while data is being taken, and checked immediately.

3.3.1.16. *Operating Cost*

Although the initial cost of the equipment is substantial, and expensive x-ray tubes must be replaced from time to time, operating costs are low because there are no consumed supplies.

3.3.1.17. *Automation*

Great ingenuity has been shown in designing accessories to index large numbers of samples through the specimen compartment of standard commercial x-ray spectrometers. Other instruments are available commercially for automatic sequential or simultaneous determination of up to 22 elements in a sample. Some instruments may be programmed to index a large number of samples through the specimen compartment in turn and for each to perform a series of determinations by automatic selection of kV, mA, 2θ, collimator, crystal, detector, pulse-height selector settings, and the number of counts scaled. The intensity data is printed out, or intensity data from samples and standards may be processed by a computer and analyte concentrations printed out directly. Some of these instruments are described in Chapter 9.

3.3.1.18. *Process Control*

Application to on-line plating thickness gaging and chemical process control has been successful (Chapter 9).

3.3.1.19. *Use with Other Methods*

X-ray spectrometry has been used to complement x-ray diffraction and absorption, infrared absorption spectrometry, electrophoresis, chromatography, gas chromatography, radioactivity, and other analytical methods (Section 14.8.3).

3.3.2. Disadvantages

3.3.2.1. *Light Elements*

Standard commercial x-ray spectrometers are subject to certain difficulties and inconveniences for elements having atomic number 22 (titanium) or less; these difficulties become progressively more severe as atomic number decreases, and are quite serious for fluorine through silicon (atomic number 9 through 14). Present commercial instruments are substantially useless for elements of atomic number less than 9. However, specially built ultra-long-wavelength x-ray secondary-emission spectrometers and commercial electron-probe x-ray microanalyzers are capable of measuring K spectra of elements down to beryllium (Z 4).

3.3.2.2. *Penetration*

Only a relatively thin surface layer ($\leq$0.1 mm) contributes to the measured x-ray line intensity, so that the method is sensitive to surface texture and gives average bulk composition only for homogeneous substances. The effect becomes more severe as line wavelength increases.

3.3.2.3. *Absorption-Enhancement Effects*

The relationship between intensity and concentration may be seriously affected by absorption-enhancement effects caused by the matrix elements. However, these effects are systematic and predictable, and there are many ways of dealing with them.

3.3.2.4. *Sensitivity*

In general, x-ray spectrometry is less sensitive than optical spectrography by two or more orders. However, there are many exceptions to this, especially the nonmetallic elements.

3.3.2.5. *Qualitative Analysis*

Recording one x-ray spectrum with the ratemeter–recorder may require half an hour or more, during which time a whole plate of optical spectra may be recorded and developed.

3.3.2.6. *Standards*

In quantitative analysis, the standards must have as nearly as possible the same chemical composition as the samples in both x-ray and optical spectrochemical analysis. However, in x-ray spectrometry, standards must also have the same physical form as the samples: solid metal (blanks, sheet, and foil $\geq$0.1 mm thick) can be analyzed only with solid standards, powders packed in cells only with similarly packed powder standards, briquets only with briquets, etc. The same applies to analysis of analytes distributed on filter paper, Mylar film, ion-exchange resins, etc. Moreover, solid standards must have the same surface texture as the samples, powders the same particle size distribution and packing density, etc. No such restrictions apply in optical spectrography, where, for example, metal drillings may often be analyzed with synthetic standards consisting of mixed metal powders—or even metal oxide powders.

This limitation is particularly inconvenient when the same substance must be analyzed in a number of physical forms—for example, metal blanks, drillings, shot, filings, powder, thin rod, wire, and small fabricated parts. In optical spectrography, one set of standards would suffice—and each could be in a different form, but not in x-ray spectrometry.

When standards are not available in the same physical form as the samples, and when the same substance must be analyzed in several forms, all standards and samples must be reduced to a common form—powder, fusion product, solution, etc. In such cases, much of the advantage of speed and convenience is lost.

3.3.2.7. *Instrument Preparation*

The availability of a wide selection of x-ray tube targets, crystals, collimators, and detectors has already been cited as advantageous. However, it must be admitted that in a laboratory where the variety of samples requires frequent change and adjustment of components, much time can be lost. Of course, much of this time is regained by the rapidity of x-ray spectrometric analyses, and the problem is largely eliminated in automatic instruments.

3.3.2.8. *Components*

Frequently, all analytes of interest in a series of specimens cannot be measured optimally with the same x-ray tube target, crystal, collimators, and/or detector. In such cases, if all measurements are to be made with the same set of components, a compromise must be made in the interest of convenience and economy at the expense of intensity. Usually, loss of intensity can be afforded except for trace constituents. The alternative is to measure one or more analytes in all specimens with one set of components, then one or more others with another set of components, etc. The difficulty may be minimized by use of a dual-target (tungsten–chromium) tube, crystal changer, and dual detector, or by providing two spectrogoniometers, each fitted with a different set of components. The difficulty is almost completely eliminated in some of the modern automatic spectrometers described in Chapter 9.

3.3.2.9. *Instrument Settings*

Frequently, all analytes of interest in a series of specimens cannot be measured optimally with the same settings of x-ray tube potential and current, detector-tube potential, amplifier gain, and pulse-height selector baseline and window. Of course, the individual settings may be made for each analyte on each specimen; some instruments make provision to preset several values of kV and mA and to select them with selector switches. Or, here again, the analytes may all be determined at the same settings representing a compromise between convenience and intensity, or they may be determined in groups, each group at different settings. A manual single-channel instrument provides no other alternatives, but automatic instruments change these settings automatically for each analyte in each specimen.

3.3.2.10. *Error*

In the operation of single-channel manual instruments, there are many opportunities for error and reduction of precision when two or more analytes are to be measured in each of many specimens. The likelihood increases as the number of settings to be changed increases—2θ (line and background), kV, mA, detector tube potential, amplifier gain, pulse-height analyzer baseline and window, and preset time or count. Finally, one must read the scaler or timer at least once for each analyte in each specimen, and usually twice (line and background). A printer eliminates this last source of error, and automatic instruments eliminate many of the others.

3.3.2.11. *Tedium*

In the operation of a single-channel manual instrument, data-taking may be very tedious. A printer may record the data, but specimens must be loaded and settings made manually, so that constant attention is required throughout the analysis. Here again, automatic instruments minimize the problem.

3.3.2.12. *Cost*

Initial cost of an x-ray spectrometer—especially an automatic one—is high, and x-ray tubes must be replaced from time to time.

Chapter 4

Excitation

4.1. PRINCIPLES

The first phase of any x-ray spectrometric analysis is excitation of the characteristic line spectrum of the analyte. There are many ways to accomplish this: primary excitation by electrons, protons, or ions; excitation by α-, β-, γ-, or x-rays from radioisotopes; excitation by internal conversion of β- or γ-rays and by K- or L-orbital electron capture. There are also certain special secondary-excitation techniques—selective, preferential, and differential excitation—discussed in more-appropriate sections. However, this chapter deals only with conventional secondary excitation by primary radiation from x-ray tubes.

4.1.1. General

In Figure 4.1, the number of analyte-line photons of wavelength λ_L that enter the x-ray optical system (collimator–crystal–collimator–detector) is (*S28*):

$$N = \frac{\Gamma\Omega}{4\pi} \int_{\lambda_{\min}}^{\lambda_{\mathrm{ab}}} T(\lambda_P, \lambda_L) J(\lambda_P)\, d\lambda_P \qquad (4.1)$$

In the figure and equation:

Γ is the effective solid angle subtended at the x-ray tube target by the irradiated area of the specimen.

Ω is the effective solid angle subtended at the specimen by the x-ray optical system.

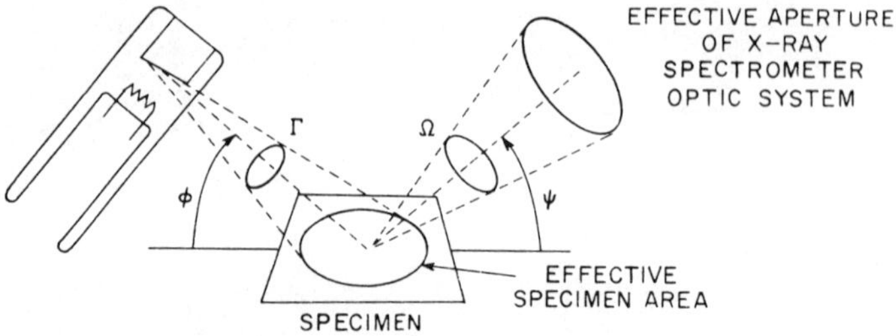

Fig. 4.1. Geometry of secondary excitation.

ϕ is the angle between the central ray of the primary cone and the specimen plane.

ψ is the angle between the central ray of the secondary cone and the specimen plane—the takeoff angle.

λ_P is the wavelength of the primary x-rays.

λ_L is the wavelength of the measured analyte line.

λ_{min} is the shortest wavelength present in the primary beam.

λ_{ab} is the absorption edge associated with the analyte line.

$T(\lambda_P, \lambda_L)$ is the conversion efficiency of primary wavelength λ_P to analyte line λ_L—the number of λ_L photons emitted per incident λ_P photon.

$J(\lambda_P)$ is the spectral distribution of the primary x-rays, that is, the number of photons of each primary wavelength λ_P per unit solid angle Γ.

The number N depends upon: (1) the geometry of the x-ray spectrogoniometer; (2) the primary beam intensity and spectral distribution, which, in turn, depend upon x-ray tube target, potential, and current; and (3) the conversion efficiency $T(\lambda_P, \lambda_L)$.

Conversion efficiency is given by (*V10*):

$$T(\lambda_P, \lambda_L) = \tau\omega g G/\mu_{\lambda_P} \tag{4.2}$$

in which

$$G = \frac{1}{1 + (\mu_{\lambda_L}/\mu_{\lambda_P})(\sin\phi/\sin\psi)} \tag{4.3}$$

where μ_{λ_P} and μ_{λ_L} are linear absorption coefficients of the specimen for the primary and analyte-line radiation, respectively; τ is that portion of μ_{λ_P} due to photoelectric absorption in the atom shell (K, L, M) associated with the analyte line; ω is the fluorescent yield of the analyte for the spectral-line series of the analyte line; and g is the relative intensity of the analyte line in its series, or the probability of the orbital electron transition causing the analyte line.

The next two sections discuss secondary excitation in somewhat more detail. The treatment follows closely that of Jenkins and de Vries [pp. 19–25 in (*14*)].

4.1.2. Excitation by Monochromatic X-Rays

The process of secondary excitation of an analyte line is most easily understood for the case of a monochromatic primary beam (*B31*). In Figure 4.1, consider an incremental layer of thickness Δt at depth t in a specimen of density ϱ. The incident angle of the primary beam is ϕ, and the takeoff angle of the secondary beam is ψ. Now consider a single wavelength λ_P having intensity I_{0,λ_P} in the incident primary beam. If the specimen has mass-absorption coefficient $(\mu/\varrho)_{\mathrm{M},\lambda_P}$ for λ_P, the fraction of incident photons penetrating to depth t and arriving at layer Δt is

$$\exp-[(\mu/\varrho)_{\mathrm{M},\lambda_P}\varrho t \csc \phi]$$

Of the photons arriving at layer Δt, the fraction absorbed in the layer is $(\mu/\varrho)_{\mathrm{M},\lambda_P}\varrho\ \Delta t \csc \phi$. Of the photons absorbed in layer Δt, the fraction absorbed by the analyte A having concentration C_A and mass-absorption coefficient $(\mu/\varrho)_{\mathrm{A},\lambda_P}$ for λ_P is

$$\frac{C_\mathrm{A}(\mu/\varrho)_{\mathrm{A},\lambda_P}}{(\mu/\varrho)_{\mathrm{M},\lambda_P}}$$

Of the photons absorbed by the analyte in layer Δt, the fraction that create electron vacancies in the atom shell corresponding to the series of the analyte line is related to the absorption-edge jump ratio of the analyte for that shell r_A; the fraction is $(r_\mathrm{A} - 1)/r_\mathrm{A}$. Of the photons causing such vacancies, the fraction that lead to actual emission of photons of the spectral lines in that series is given by the fluorescent yield of the analyte for that series ω_A. Of the primary photons causing such secondary photon emission, the fraction that lead to emission of the particular analyte line L to be measured is given by the probability of the orbital electron transition resulting in that line in analyte atoms g_L.

Of the analyte-line photons emitted by layer Δt, only that fraction directed toward the source collimator can contribute to the measured line intensity; this fraction is given by $d\Omega/4\pi$ (Figure 4.1). Of the analyte-line photons directed toward the collimator, the fraction that emerges from the overlying layer t having mass-absorption coefficient $(\mu/\varrho)_{\mathrm{M},\lambda_L}$ for the analyte line L is

$$\exp-[(\mu/\varrho)_{\mathrm{M},\lambda_L}\varrho t \csc \psi]$$

The foregoing processes may be grouped for convenience into five factors as follows:

1. Incident photon intensity: I_{0,λ_P}.
2. Attenuation of the primary beam in reaching layer Δt:

$$\exp-[(\mu/\varrho)_{\mathrm{M},\lambda_P}\varrho t \csc \phi]$$

3. Probability of excitation of analyte line L:

$$\frac{C_{\mathrm{A}}(\mu/\varrho)_{\mathrm{A},\lambda_P}}{(\mu/\varrho)_{\mathrm{M},\lambda_P}} \frac{r_{\mathrm{A}}-1}{r_{\mathrm{A}}} \omega_{\mathrm{A}} g_L \left[\left(\frac{\mu}{\varrho}\right)_{\mathrm{M},\lambda_P} \varrho\, \Delta t \csc \phi\right]$$

4. Fraction of analyte-line photons emitted toward source collimator: $d\Omega/4\pi$.

5. Attenuation of this analyte-line radiation in emerging from the specimen:

$$\exp-[(\mu/\varrho)_{\mathrm{M},\lambda_L}\varrho t \csc \psi]$$

Combination of all these expressions gives the analyte-line intensity ΔI_L emitted from layer Δt that actually enters the source collimator:

$$\Delta I_L = I_{0,\lambda_P}\left\{\exp-\left[\left(\frac{\mu}{\varrho}\right)_{\mathrm{M},\lambda_P} \varrho t \csc \phi\right]\right\}\left(\frac{\mu}{\varrho}\right)_{\mathrm{M},\lambda_P} \varrho\, \Delta t \csc \phi$$
$$\times \frac{C_{\mathrm{A}}(\mu/\varrho)_{\mathrm{A},\lambda_P}}{(\mu/\varrho)_{\mathrm{M},\lambda_P}} \frac{r_{\mathrm{A}}-1}{r_{\mathrm{A}}} \omega_{\mathrm{A}} g_L \frac{d\Omega}{4\pi} \exp-\left[\left(\frac{\mu}{\varrho}\right)_{\mathrm{M},\lambda_L} \varrho t \csc \psi\right] \quad (4.4)$$

Simplification of Equation (4.4) gives

$$\Delta I_L = I_{0,\lambda_P}\left\{\exp-\left[\left(\frac{\mu}{\varrho}\right)_{\mathrm{M},\lambda_P} \varrho t \csc \phi + \left(\frac{\mu}{\varrho}\right)_{\mathrm{M},\lambda_L} \varrho t \csc \psi\right]\right\} \varrho \Delta t \csc \phi$$
$$\times C_{\mathrm{A}}\left(\frac{\mu}{\varrho}\right)_{A,\lambda_P} \frac{r_{\mathrm{A}}-1}{r_{\mathrm{A}}} \omega_{\mathrm{A}} g_L \frac{d\Omega}{4\pi} \quad (4.5)$$

Integration of Equation (4.5) from $t = 0$ to $t = \infty$ gives

$$I_L = I_{0,\lambda_P} \omega_{\mathrm{A}} g_L \frac{r_{\mathrm{A}}-1}{r_{\mathrm{A}}} \frac{d\Omega}{4\pi} \frac{C_{\mathrm{A}}(\mu/\varrho)_{\mathrm{A},\lambda_P} \varrho \csc \phi}{(\mu/\varrho)_{\mathrm{M},\lambda_P} \varrho \csc \phi + (\mu/\varrho)_{\mathrm{M},\lambda_L} \varrho \csc \psi} \quad (4.6)$$

If we define

$$P_A = \omega_A g_L \frac{r_A - 1}{r_A} \frac{d\Omega}{4\pi} \tag{4.7}$$

and

$$A = \sin\phi/\sin\psi \tag{4.8}$$

Equation (4.6) becomes

$$I_L = P_A I_{0,\lambda_P} \frac{C_A(\mu/\varrho)_{A,\lambda_P}}{(\mu/\varrho)_{M,\lambda_P} + A(\mu/\varrho)_{M,\lambda_L}} \tag{4.9}$$

P_A is constant for a given analyte, matrix, and spectrometer (including a specified set of components). Equations (4.6) and (4.9) are valid only for homogeneous specimens, and do not allow for multiple scatter or enhancement effects.

4.1.3. Excitation by Continuous Spectra

If, now, instead of a monochromatic beam of wavelength λ_P, a continuous spectrum excites the analyte line, it becomes necessary to consider the following (*B14*, *G4*, *S27*):

1. Analyte-line radiation λ_L is excited by all primary wavelengths between the short-wavelength limit $\lambda_{\min}$ [Equation (1.14)] and the absorption edge λ_{ab} associated with λ_L. The total effective primary-beam intensity is then

$$I_\lambda - \int_{\lambda_{\min}}^{\lambda_{ab}} J(\lambda_P)\, d\lambda \tag{4.10}$$

$J(\lambda_P)$ represents the primary spectral distribution, which consists of the continuous spectrum and characteristic target lines, and depends on the x-ray tube target and potential. The distribution can be evaluated by recording the primary spectrum scattered by material of low atomic number, such as paraffin or carbon, and correcting for incoherent scatter.

2. Mass absorption coefficients must be considered for all elements in the specimen matrix for all effective primary wavelengths and for the analyte line. Expressions of the following form are obtained from Equation (2.10):

$$(\mu/\varrho)_{M,\lambda_P} = \Sigma\, C_i(\mu/\varrho)_{i,\lambda_P} \tag{4.11}$$

$$(\mu/\varrho)_{M,\lambda_L} = \Sigma\, C_i(\mu/\varrho)_{i,\lambda_L} \tag{4.12}$$

where $(\mu/\varrho)_{M,\lambda_P}$ and $(\mu/\varrho)_{M,\lambda_L}$ are mass absorption coefficients of the specimen for a specific primary wavelength λ_P and the analyte line λ_L; $(\mu/\varrho)_{i,\lambda_P}$ and $(\mu/\varrho)_{i,\lambda_L}$ are mass absorption coefficients of a specific element i for λ_P and λ_L; and C_i is concentration of element i.

Substitution of Equations (4.10)–(4.12) in Equation (4.9) gives a general expression for I_L:

$$I_L = P_A C_A \int_{\lambda_{\min}}^{\lambda_{ab}} J(\lambda_P)\left(\frac{\mu}{\varrho}\right)_{A,\lambda_P} \frac{1}{\sum C_i[(\mu/\varrho)_{i,\lambda_P} + A(\mu/\varrho)_{i,\lambda_L}]} \tag{4.13}$$

where P_A and A are defined by Equations (4.7) and (4.8), respectively.

The efficiency of a specified primary wavelength λ_P for exciting analyte line λ_L may be defined as follows:

$$\frac{I_L}{I_{0,\lambda_P}} = P_A \frac{C_A(\mu/\varrho)_{A,\lambda_P}}{\sum C_i[(\mu/\varrho)_{i,\lambda_P} + A(\mu/\varrho)_{i,\lambda_L}]} \tag{4.14}$$

Then, for a given analyte, matrix, and spectrometer arrangement, an efficiency factor T_{λ_P,λ_L} for primary wavelength λ_P in exciting analyte line λ_L may be defined as follows:

$$T_{\lambda_P,\lambda_L} = \frac{(\mu/\varrho)_{A,\lambda_P}}{\sum C_i[(\mu/\varrho)_{i,\lambda_P} + A(\mu/\varrho)_{i,\lambda_L}]} \tag{4.15}$$

For a pure element A, Equation (4.15) reduces to

$$T_{\lambda_P,\lambda_L} = \frac{(\mu/\varrho)_{A,\lambda_P}}{(\mu/\varrho)_{A,\lambda_P} + A(\mu/\varrho)_{A,\lambda_L}} \tag{4.16}$$

4.2. THE X-RAY TUBE

The x-ray generator, shown schematically in Figure 4.2, consists of the x-ray tube and its power supply. We now consider each of these in turn.

4.2.1. Function and Requirements

The x-ray tube generates the primary x-ray beam that irradiates the specimen and excites the characteristic line spectrum of its elements. The primary beam consists of the continuum and the spectral lines of the target and its contaminants. Formerly, x-ray tubes designed for other applications were used in spectrometers, but today tubes are designed specifically for spectrometers, where the following features are required: (1) high energy

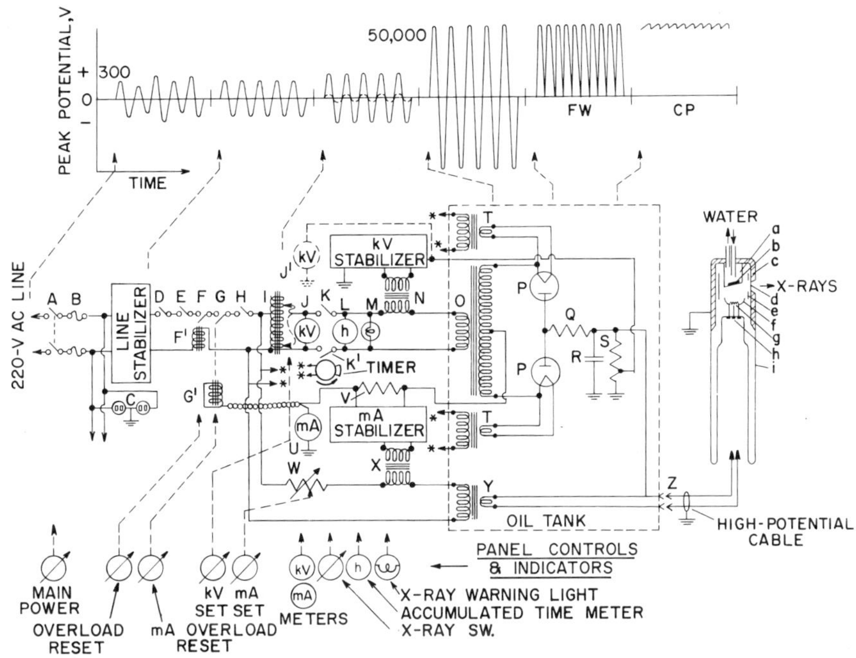

Fig. 4.2. The x-ray generator. Panel controls and indicators are shown along the bottom. Wave forms (potential *vs.* time) at selected points are shown along the top. *X-Ray Power Supply*: *A*. Main power switch. *B*. Fuses. *C*. Accessory outlets. *D*. High-pressure water switch. *E*. Low-pressure water switch. *F*. Line overload relay. *F'*. Solenoid coil for *F*. *G*. mA overload relay. *G'*. Solenoid coil for *G*. *H*. Door interlocks. *I*. Autotransformer—kV selector. *J*. kV meter. *J'*. kV meter, alternate position. *K*. X-ray switch. *K'*. Time-delay switch. *L*. Accumulated time meter. *M*. X-ray warning light. *N*. kV stabilizer reactor. *O*. High-potential transformer. *PP*. High-potential rectifiers. *Q*. Resistor. *R*. Capacitor. (*Q* & *R* make up constant-potential filter.) *S*. Potential-divider resistor. *TT*. Rectifier filament transformers. *U*. mA meter. *V*. Resistor for mA stabilizer. *W*. Filament-control rheostat. *X*. mA stabilizer reactor. *Y*. X-ray tube filament transformer. *Z*. High-potential socket. *X-Ray Tube*: *a*. Water-cooled target block. *b*. Target. *c*. Rayproof head. *d*. Beryllium window. *e*. Internal shield. *f*. Focusing electrode. *g*. Filament. *h*. Filament terminals. *i*. Glass envelope.

output, effected by high power loading, small target-to-window distance, and thin beryllium windows; (2) constancy of energy output; short-term constancy is effected by suitable design to minimize variation of internal geometry; long-term constancy is realized by prevention of target pitting and sublimation of metals onto the inside surface of the window; (3) cool operating temperature, especially important with liquid specimens; and (4) long life, with respect to filament burnout and to deterioration of energy output and spectral purity.

4.2.2. Construction

The construction of an x-ray spectrometer tube is shown in Figure 4.2 (*R9*). The external features consist of a heavy rayproof metal head *c* that confines the x-radiation, a thin beryllium window *d* that permits emergence of the useful beam, and a reentrant glass envelope *i* that retains the vacuum and provides a long electrical insulation path. This vacuum enclosure houses the x-ray generating structure.

The tungsten filament *g* is heated to incandescence by an electric current passed through the external filament terminals *h*. The hot filament emits electrons (thermionic emission) that are roughly focused by a concave focusing electrode *f* and attracted and accelerated to the target (anode) *b*, which is operated at a high positive potential. The target consists of a thin disk or plating of the target metal *b* imbedded in or plated on a heavy hollow copper block *a*, which serves to conduct heat away from the focal spot. The focal spot is the area bombarded by the filament electrons; it is the source of the primary x-rays, and "target-to-window" and "target-to-specimen" distances are measured from the center of the focal spot. The x-rays are emitted in all directions, but emerge from the metal head *c* only through the window *d*. The target surface is inclined $\sim 20°$ (target angle) toward the window. The internal structure is surrounded by a cylindrical metal shield *e* having an opening in register with the window. This shield intercepts improperly focused electrons from the filament and scattered electrons from the target, and minimizes metallization of the window by tungsten sublimed from the filament and by metal sputtered from the target.

In practice, the target, which is an integral part of the tube head, is operated at ground potential and the filament at up to 50 (or 75 or 100) kV negative. This mode of operation permits contact with the tube head and specimen compartment, and water-cooling without precaution against grounding through the water line. A rayproof housing must be provided to prevent access to the glass end of the operating tube, both because of the poor x-ray shielding at this end and because of the electric-shock hazard. The glass envelope may be encased in an oil-filled metal cylinder that seals to the adjoining rim of the tube head. This arrangement, known as shock-proof mounting, permits operation of the tube in any position and remote from the generator.

4.2:3. Special X-Ray Tubes

The x-ray tube just described is typical of conventional side-window tubes. Several other tubes are of interest.

4.2.3.1. *Dual-Target Tube*

The construction is essentially the same as that shown in Figure 4.2 except for two features: The target has two separate areas—one tungsten, the other chromium—and two filaments, each having its focal spot on one of the target areas. The appropriate filament is energized by an external selector switch. This tube permits high-efficiency excitation of both long- and short-wavelength spectra without the inconvenience of changing tubes.

4.2.3.2. *End-Window Tube*

X-ray tubes are available in which the window is in the end of the tube head, normal to the axis. These tubes give somewhat more uniform specimen irradiation (if the specimen is normal to the axis) and are especially advantageous for multichannel spectrometers. (See Figure 9.4).

4.2.3.3. *Demountable Tubes*

Continuously pumped demountable tubes are still used on special laboratory-built spectrometers and have many advantages: burned-out filaments are easily replaced; the target can be removed for cleaning of surface contamination or resurfacing; targets are easily interchanged, and there is no limit on the element that can be used, because gas-free operation and long life are not essential—thus, oxides and other reasonably stable compounds can be used when metals are unavailable; a rotable target wheel can be made having many sectors, each of a different element; provision can be made to position the wheel without dismantling the tube; very thin, or even open, windows can be used because of the continuous pumping and ease of replacement; and high anode power is permissible because the target can be replaced easily.

Three excitation modes are feasible with demountable tubes: (1) primary excitation with the specimen on the target (*C45*, *C46*, *H1*, *M32*), giving high analyte-line intensity, but with high background; (2) secondary excitation with the specimen external to the tube; and (3) secondary excitation with the specimen inside the tube and very near the target (*E4*, *E5*). This last mode is intermediate between the other two: it is not as efficient as primary excitation, but much more so than external secondary excitation because of the proximity to the primary source, and it still produces no continuum other than that scattered by the specimen.

In another form of demountable tube, a thin aluminum foil serves as both x-ray target and window. The specimen is placed just outside the window and is subjected to secondary excitation (*H25*).

Demountable tubes are especially advantageous for secondary excitation of trace analytes. A target can be used that has a strong line near the short-wavelength side of the analyte absorption edge. High anode power and thin windows can be used, and the sample can be placed inside the tube.

The principal disadvantages are the need for continuous pumping and, if specimens are placed inside the tube, the need for dismantling the tube for each specimen.

4.2.3.4. *Tubes for Ultralong Wavelength*

Figure 4.3 shows a cross section of the head of the Henke x-ray tube for excitation of ultralong-wavelength spectra (>15 Å) (*D23, H20*). Electrons are emitted by the thermionic cathode. An electrostatic field established by the focusing electrode and the positive anode constrains these electrons to move in the paths shown by the dotted lines until they strike the anode (target). The useful x-rays emerge through an ultrathin window. The position of the cathode behind the anode prevents sublimation of tungsten to the target and window and prevents direct thermal irradiation of the window. The anode has a large emitting area. It is easily replaced and may be selected to have characteristic lines at wavelengths just shorter than that of the analyte absorption edge. The tube is operated at several hundred volts to 15 kV and up to 300 mA, depending on the anode material. The anode is operated at positive potential with respect to the grounded cathode and tube head to prevent electron bombardment of the window. The window

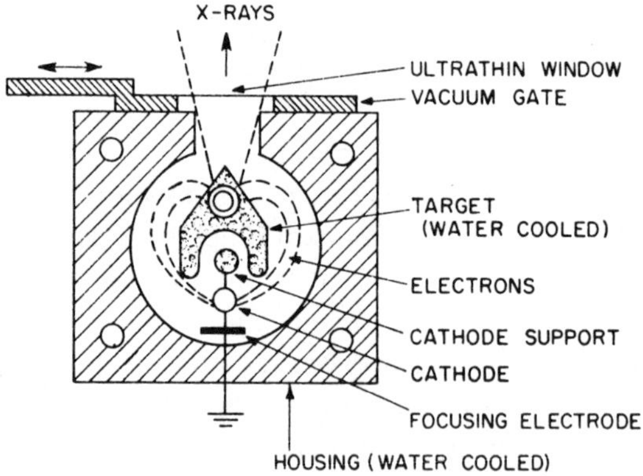

Fig. 4.3. Henke tube for excitation of ultralong-wavelength spectra. [B. L. Henke, *Advances in X-Ray Analysis* **5**, 285 (1962); courtesy of the author, University of Denver, and Plenum Press.]

isolates the high clean vacuum of the x-ray tube from the working vacuum of the spectrometer chamber, to which the x-ray tube is attached. The window consists of ultrathin Formvar (1000–2000 Å, 30–60 $\mu g/cm^2$) supported on a 200-mesh grid having 70% open area. The vacuum gate is advanced to close the exit port until both the x-ray tube and spectrometer chamber are evacuated by their separate pumps. The gate is then placed with the window over the port. This procedure prevents rupture of the window by pressure differential during evacuation. The anode and tube head are water-cooled. An ion pump and liquid-nitrogen cold finger are used to provide clean operating vacuum of 10^{-6} mm-Hg or less and to maintain a clean anode surface during many hours of operation under high-current electron bombardment. The Henke tube is available commercially (Section 9.4.4).

4.2.4. Design Considerations

The focal spot in an x-ray spectrometer tube, unlike that in a diffractometer tube, is not itself a part of the x-ray optical system, and, consequently, may be relatively large—10 × 5 mm, compared with 10 × 1 mm in a diffraction tube. The larger size permits greater power dissipation—2.5–3.5 kW (or kVA), compared with ~1 kW for a diffraction tube. However, the area and position of the focal spot must be constant throughout the analysis so that primary beam intensity and intensity distribution over the specimen area are constant.

An x-ray tube is extremely inefficient. Only ~1% of the power dissipated by the target—2.5 kW at 50 kV, 50 mA—is converted to x-rays, the other 99% to heat. Consequently, the target is supported on, and in intimate contact with, a block of copper, which is an excellent heat conductor. This block is water-cooled, as shown in Figure 4.2. Unless the target metal is itself a good heat conductor, it must be very thin, either a thin disk or plating. This presents no problem, because electrons penetrate only an extremely thin surface layer.

The focal spot is relatively small, so the primary beam is highly divergent. The useful cone subtends at the target an angle of ~20° from the central ray. Thus, the primary beam "obeys" the inverse-square law, and if the specimen is to receive high primary intensity, the target-to-specimen —and therefore the target-to-window—distance must be as small as possible. Different manufacturers accomplish this by reducing the diameter of the tube head, flattening the window side of the head, or placing the focal spot off axis in the direction of the window.

The window must be as transparent to x-rays as possible, especially at the longer wavelengths, which are generated only inefficiently to begin with, and which are most efficient in exciting long-wavelength spectra. Beryllium, the most transparent practical window material, is none too good: 0.010-inch beryllium transmits only 10% at 5 Å, only 0.1% at 8 Å; even 0.001-inch beryllium transmits only 20% at 10 Å. Beryllium thinner than 0.001 inch and free of pinholes is difficult to make. The window is heated by electrons scattered by the target, and the higher the atomic number of the target, the more efficient is the scatter. The temperature of the window may rise to several hundred degrees centigrade, and, since beryllium is a relatively poor heat conductor, large temperature gradients may develop across its diameter. If the window is too thin, it may rupture. The higher the atomic number of the target and the smaller the target-to-window distance, the thicker the window must be. In practice, this works out satisfactorily. The high-Z targets—molybdenum, tungsten, platinum, gold—have relatively thick (1 mm, 0.040 inch) windows. But these tubes are used to excite medium- and short-wavelength spectra, and the thick windows transmit these wavelengths efficiently. The chromium (Z 24) target is used to excite long-wavelength spectra, and a 0.010-inch beryllium window is feasible.

4.2.5. Practical Considerations

4.2.5.1. *Excitation Efficiency* (*B14*, *S28*)

X-ray spectral lines of a given series are excited most efficiently by primary radiation having wavelength near the short-wavelength side of the absorption edge associated with that series. If the x-ray tube target has intense lines near this wavelength, they make the dominant contribution to the excitation; otherwise, the continuum makes the dominant contribution. Target lines constitute a greater portion of the primary spectrum than was previously believed (*G4*). Recent measurements indicate that W L lines constitute ~25% of the spectrum of a tungsten target, Cr K lines ~75% of the spectrum of a chromium target. There are only about eight targets available commercially (although others may be ordered specially), so it is unlikely that a favorable target would be available for all analytes. Moreover, it is impractical to change targets too frequently, certainly for different analytes in the same specimen. Consequently, the continuum is frequently relied on for excitation, and, when this is the case, for given excitation conditions (kV, mA), the higher the atomic number, the more intense is the continuum.

All things considered, tungsten is the best general-purpose target. It has high atomic number (74), and thus gives intense continuous radiation; it is refractory (melting point 3410°C), so it withstands high target loading without pitting; and it is available in high purity. Moreover, it has intense *L* lines that contribute to excitation of *K* lines of elements below germanium (*Z* 32) and *L* lines of elements below iridium (*Z* 77).

Molybdenum ($K\beta$ 0.632 Å, $K\alpha$ 0.711 Å) has some advantage over tungsten for *K* lines of niobium (*Z* 41, K_{ab} 0.653 Å) to germanium (*Z* 32) and *L* lines of uranium (*Z* 92, $L\text{III}_{ab}$ 0.722 Å) to osmium (*Z* 76). Gold ($L\beta_1$ 1.084 Å, $L\alpha_1$ 1.276 Å) has some advantage for *L* lines of iridium (*Z* 77, $L\text{III}_{ab}$ 1.107 Å) to hafnium (*Z* 72). Platinum (*Z* 78) is substantially equivalent to tungsten (*Z* 74).

Chromium ($K\beta$ 2.085 Å, $K\alpha$ 2.291 Å) is advantageous for *K* lines of vanadium (*Z* 23, K_{ab} 2.269 Å) or titanium (*Z* 22, K_{ab} 2.497 Å) down, and for *L* lines of cerium (*Z* 58, $L\text{III}_{ab}$ 2.164 Å) down. Rhodium ($L\beta_1$ 4.374 Å, $L\alpha_1$ 4.597 Å) and silver ($L\beta_1$ 3.934 Å, $L\alpha_1$ 4.154 Å) are advantageous for chlorine (*Z* 17, K_{ab} 4.397 Å) or sulfur (*Z* 16, K_{ab} 5.018 Å) down. These tubes must have thin (0.010-inch) beryllium windows.

The almost ideal tube would seem to be the tungsten–chromium dual-target tube, tungsten for elements having atomic number greater than 22 (titanium) or 23 (vanadium), the chromium for the lighter elements. Figure 9.1 shows the useful spectral regions of several commercially available targets.

4.2.5.2. *Spectral-Line Interference* (*B27*, *L1*)

The primary spectrum contains the following spectral lines: (1) spectra of the target element; (2) spectra of impurities in the target; (3) spectra of contaminants sublimed or sputtered on the target during operation of the tube; these contaminants include tungsten from the filament; chromium, iron, and nickel from the internal shield and focusing electrode; copper from plated structures and the target block; and silver from silver solder; the intensities of these lines increase with life of the tube; (4) spectra arising from primary excitation of tube structures other than the target by scattered or improperly focused electrons; and (5) spectra arising from secondary excitation of these structures. Fortunately, modern tubes designed specifically for x-ray spectrometry have high spectral purity when new and maintain it with life. Any lines present in the primary beam are scattered by the specimen into the spectrometer. The lower the effective atomic number of the specimen, the more intense is the scatter.

In determination of major constituents, especially in high-Z matrixes (which scatter inefficiently), scattered primary lines may be tolerable if a suitable way of measuring background is used. However, for determining minor or trace concentrations of a given element, a target of that element should not be used, and tubes having extraneous lines of that element should be avoided. Another condition to be avoided is the possible preferential excitation by target lines of a matrix element having lines that interfere with analyte lines.

Recording of target spectra and use of filters in dealing with spectral-line interference are discussed in Sections 4.2.5.4 and 4.4, respectively.

4.2.5.3. *Temperature*

While in operation, the x-ray tube must be cooled by a continuous flow of water at a rate of approximately two liters per minute. Usually, tap water is passed through the tube and discharged into the drain. Alternatively, the cooling water may be circulated through a closed system including a radiator cooled by forced air or by a heat exchanger cooled by tap water. The last two methods have the advantage that distilled or deionized water—rather than tap water—passes through the tube, reducing the possibility of corrosion or obstruction. The radiator arrangement has the advantage of economy of tap water, the disadvantage that the target heat is discharged into the room by the radiator, rather than into the drain by the tap water.

Additional cooling may be provided by fitting the tube head with a jacket through which the target cooling water is passed. The jacket should be on the inlet side, that is, the water should pass through the jacket before the tube. This additional cooling is particularly desirable when liquid specimens are to be measured.

Operation of the tube at power exceeding its maximum ratings, or within its ratings with inadequate water supply, results in accelerated deterioration of the target. Operation without cooling water for even a fraction of a minute results in complete destruction of the target.

Slight changes occur in the position and dimensions of the filament and in other internal dimensions as the tube warms up to its equilibrium temperature. These changes may affect the intensity and intensity distribution of the primary beam. Consequently, it is necessary to operate the x-ray tube with both filament and target power for at least half an hour before making any quantitative measurements.

4.2.5.4. *Evaluation of the Condition of the X-Ray Tube*

The x-ray tube focal spot can be photographed at 10–20-fold magnification with a pinhole camera. The pinhole is made by piercing a hole ~0.005 inch in diameter in a piece of lead foil ~0.005 inch thick. A piece of x-ray film is wrapped in black photographic paper or aluminum foil. The pinhole is placed near and parallel to the window and the film some distance away in line with the pinhole and focal spot. The magnification is the ratio of the pinhole-to-film and pinhole-to-target distances. Such photographs taken from time to time show the condition of the focal spot.

The target spectrum can be observed as follows. A piece of paraffin or spectroscopically pure carbon or a briquet of starch is placed in the specimen compartment to scatter the primary beam into the spectrometer. A 2θ scan is made of the scattered x-rays and recorded on the ratemeter recorder. Such spectra should be recorded whenever a new tube is first used to identify the extraneous lines, and occasionally thereafter to observe increase of spectral contamination with life.

The specimen area irradiated by the target is observed by placing a piece of unexposed x-ray film or a glass microscope slide in the specimen plane without the specimen mask and irradiating it at full power for 3–5 min. The film or glass is colored in the irradiated area. This technique aids in aligning the x-ray tube for optimal irradiation of the specimen. A method of quantitatively mapping iso-intensity contour lines over the specimen plane is given in Section 19.2.1.

4.3. X-RAY POWER SUPPLY

4.3.1. Function and Requirements

The x-ray power supply provides filament power for the x-ray tube filament and high-potential power for the target. The filament power must be well regulated; the high-potential power must meet the following requirements: (1) constant-potential dc, although full-wave rectified dc is also satisfactory; (2) maximum potential at least 50 kV (peak or constant potential), although 75 or 100 kV may have some advantage; (3) maximum x-ray tube current at least 50 mA for a 50-kV supply; (4) maximum power output at least 2.5 kW (kVA) for a 50-kV supply; higher powers are preferable for 75- or 100-kV supplies; (5) regulation to ±0.05% or less provided by line, mA, and kV stabilization; (6) it must be possible to select any x-ray tube potential and current from the maximum values down to

10 kV or less and 5 mA or less; this selection should be continuous, rather than stepwise; provision should be made to preset several values of kV and mA and to select them repeatedly with selector switches when more than one analyte is to be measured in each specimen at more than one set of conditions; (7) it should be possible to operate two x-ray tubes, preferably, but not necessarily, simultaneously and at or near full power; this permits operation of two goniometers—both as spectrometers or one as a spectrometer, the other as a diffractometer.

The power supply, shown in Figure 4.2, consists of: (1) the high-potential supply; (2) the x-ray tube filament supply; (3) line, mA, and kV stabilizers; and (4) safety and protective devices. Along the bottom of the figure, the panel controls and indicators are shown. Along the top of the figure, the wave forms of the current at various key components are shown.

4.3.2. Components and Operation

In Figure 4.2, if the main power switch *A* is closed, the 220-V ac line passes through the switch and fuses *B* to the accessory outlets *C* and the line stabilizer. Auxiliary electrical equipment, such as a vacuum pump for a vacuum tunnel or a water pump and cooler, are also connected to the unstabilized ac line. The line stabilizer maintains a constant 220-V output for input fluctuations of 190–250 V.

If all the safety and protective devices *D–H* are closed, which they normally are, the stabilized 220-V line goes to: (1) the autotransformer *I* in the high-potential supply; (2) the x-ray tube filament transformer *Y*; (3) the rectifier filament transformers *TT*; (4) the mA and kV stabilizers; and (5) a 30–60-s time-delay switch *K*′.

4.3.2.1. *High-Potential Supply*

The continuously variable autotransformer *I* permits the line potential applied to the high-potential transformer to be varied from 0 to perhaps 250 V. The stabilized 220-V line is connected to the input of the autotransformer winding at the bottom and at a point a certain distance down from the top. The output is taken from the bottom and the movable contact. When the movable contact is set at the bottom of the winding (dotted arrow), the output is 0 V. As the contact is moved upward, the output potential increases until, at the point where the input is connected, output and input are equal at 220 V. If the contact is moved still higher, the output exceeds the input up to ~250 V; this is possible because of induction in the upper part of the winding.

The movable contact is operated by the kV-set knob, and the potential across the output of the autotransformer is indicated on the kV meter *J*, as shown. However, the scale of the meter is graduated 0–50 (or 75 or 100) kV, rather than 0–250 V as one would expect; how this is feasible is explained below. The output of the autotransformer is not applied to the high-potential transformer until the x-ray switch *K* and time-delay switch *K'* are closed, whereupon the accumulated time meter *L*, x-ray warning light *M*, and primary winding of the high-potential transformer *O* are energized.

Transformers are inductive devices for increasing or decreasing an ac potential without substantial loss of power. A transformer consists of two separate windings on a laminated iron core. If an ac potential V_{pri} is applied to the primary (input) winding, an induced ac potential V_{sec} of the same frequency appears across the secondary (output) winding:

$$V_{sec} = V_{pri}(N_{sec}/N_{pri}) \tag{4.17}$$

where N_{pri} and N_{sec} are the numbers of turns in the primary and secondary windings, respectively. If $N_{sec} > N_{pri}$, the device is a *step-up* transformer, and $V_{sec} > V_{pri}$; if $N_{sec} < N_{pri}$, the device is a *step-down* transformer. In the present case, if the output of the autotransformer *I*—that is, the input of transformer *O*—is 0–250 V, and the maximum output of the x-ray supply is to be 50 kV, the turns ratio must be

$$\frac{N_{sec}}{N_{pri}} = \frac{V_{sec}}{V_{pri}} \tag{4.18}$$

Then, 50,000/250 = 200/1. Thus, the secondary potential of transformer *O* is always 200 times the primary potential, that is, 200 times the output potential of the autotransformer *I*. Thus, for an autotransformer output of 50 V, the high-potential transformer output is $50 \times 200 = 10{,}000$ V (10 kV); for 125 V, it is 25 kV, etc., until, for 250 V, it is 50 kV.

It is now evident why the knob that operates the movable contact of the autotransformer is the kV-set knob. It is also evident why the panel meter *J* can be graduated in kV, since the high-potential transformer output is always 200 times the autotransformer output. Thus a small, rugged, relatively inexpensive, and easily insulated 0–250 ac voltmeter at the autotransformer output serves the same function as a bulky, expensive 0–50 ac kilovoltmeter at the output of transformer *O*. Another practical way of indicating x-ray tube potential *J'* is given below.

In Figure 4.2, transformer *O* actually has a turns ratio of 400/1 and delivers a total of 100 kV maximum. The winding is center-tapped and actually consists of two 50-kV windings. This arrangement is required for full-wave rectification, as explained below.

The transformer violates no conservation laws because, neglecting losses, the primary and secondary *powers* are the same, that is,

$$(Vi)_{\text{sec}} = (Vi)_{\text{pri}} \tag{4.19}$$

where V and i are potential and current, respectively. In the present case, if the x-ray supply is to provide 50 kV at 50 mA, the primary current, again neglecting power losses, is

$$i_{\text{pri}} = (Vi)_{\text{sec}}/V_{\text{pri}} \tag{4.20}$$

Then, $(50{,}000 \times 0.050)/250 = 10$ A.

The autotransformer I may be regarded as a transformer in which a single winding serves as both primary and secondary. The primary is the portion of the winding across which the 220-V line is connected. The secondary is the portion of the winding between the bottom and the movable contact.

The ac output of the high-potential transformer is rectified—that is, converted to dc—by the two rectifier tubes *PP*, each having its own filament transformer *TT*. Rectifiers are devices for converting alternating current to direct current and are available in many forms. The rectifiers in the figure are of the electron-tube type and consist of an evacuated glass tube having an electron-emissive tungsten filament surrounded by a cylindrical metal anode. Electrons can flow from the filament to the anode only when the anode is positive with respect to the filament, so the rectifier acts as a one-way electronic valve. Solid-state high-potential rectifiers are now also being used in x-ray power supplies. The rectifier filament transformers *TT* are step-down transformers and reduce the stabilized 220-V ac line to 5–10 V to heat the rectifier filaments. The ac input to the primary windings of these transformers is constant and in no way affected by the autotransformer *I* or the rheostat *W* for the x-ray tube filament transformer.

In full-wave supplies, the pulsating dc output of the rectifiers is connected directly to the high-potential socket *Z*. In constant-potential supplies, the full-wave rectified current is smoothed by an electric filter consisting of a resistor *Q* (or an inductor) and capacitor *R*. The potential divider *S* is a high resistance—10 MΩ or more—and allows the charge on the capacitor to leak off when the supply is turned off. Otherwise, the capacitor would retain its charge and present an electric-shock hazard, even when the supply is turned off. Other uses of the potential divider are given below.

The high-potential transformer *O*, rectifiers *PP*, constant-potential filter

QR, potential divider *S*, rectifier filament transformers *TT*, and x-ray tube filament transformer *Y* (see below) all require high electrical insulation—up to 100 kV for a 50-kV supply. Consequently, all these components are mounted in an oil-filled tank, as shown in Figure 4.2. Incidentally, even though the three filament transformers *TTY* have primary and secondary potentials of only 220 and ~10 V, respectively, the insulation between their windings must withstand the full 100-kV potential of transformer *O*. Otherwise, the high-potential could arc through the filament transformers to the ac line and ground.

4.3.2.2. *X-Ray Tube Filament Supply*

The x-ray tube filament transformer *Y* is a step-down transformer similar to the rectifier filament transformers *TT* and reduces the stabilized 220-V ac line to 5–10 V to heat the x-ray tube filament. The rheostat *W* varies the primary potential in much the same way as the autotransformer *I* varies the primary potential to transformer *O*. In fact, a second autotransformer can be used in place of the rheostat. The secondary of the filament transformer is connected to the high-potential socket *Z*, thence through the high-potential cable to the x-ray tube filament terminals *h*. If the resistance of the rheostat *W* is decreased, the potential applied to the primary winding, and therefore the output of the secondary winding, increases. As a result, more current flows through the x-ray tube filament, increasing its temperature and electron emission. Thus, the x-ray tube current and x-ray output increase. It is now evident why the knob that operates the rheostat is the mA-set knob, as shown.

4.3.2.3. *Operation*

Closing the main power switch *A* immediately energizes the following components: (1) any equipment connected directly to the ac line (such as a water-circulating system or vacuum pump) or plugged into the outlets *C* (such as a specimen spinner); of course, any of these devices may have their own switches; (2) the line stabilizer; (3) the mA and kV stabilizers; (4) the autotransformer *I*; (5) the x-ray tube filament transformer *Y*; (6) the rectifier filament transformers *TT*; and (7) the time-delay switch *K′*. However, in order to energize the high-potential transformer *O* and thereby generate x-rays, the x-ray switch *K* must be closed. Even then, the x-ray tube is not energized unless all the safety and protective devices *D–H*, *K′* are closed; these devices are discussed in Section 4.3.2.5.

We may now consider the operation of the high-potential system. Suppose that, at a particular instant, the phase of the ac at the output of the high-potential transformer O is such that the top of the winding is positive, the bottom negative. Electron current always flows from the negative terminal of a source, so current flows to the filament of the lower rectifier P, across the vacuum to the anode, through the resistor Q (if present) to the socket Z, thence through the cable to the x-ray tube filament g. The x-ray tube filament electrons flow across the vacuum to the target b, where some generate x-rays, others flow through the copper block a and tube head c to ground. The return current flows through the ground to the milliammeter U, through winding G' and resistance V (both considered later), to the center tap of the transformer O.

On the next half-cycle, the polarity of the secondary winding of O is reversed, with the top of the winding negative. Now, the same thing happens, except that the electron current flows from the top of the winding through the upper rectifier. Thus, successive half-cycles flow from alternate halves of the transformer and through alternate rectifiers, but the current applied to the x-ray tube is always in the same direction.

The electron current just described is the x-ray *tube current* and is responsible for the generation of the x-rays; it is indicated on the milliammeter U, and is usually of the order 50 mA or less. This current must be distinguished from the x-ray tube *filament current*, which flows in the circuit consisting of the x-ray tube filament and the secondary of the filament transformer Y, is responsible only for heating the x-ray tube filament, and is usually of the order 5 A or less.

The wave forms of the potential at key components in the system are shown at the top of Figure 4.2 on a scale of potential *vs.* time. The 220-V ac line is represented by the first wave form, where line fluctuations are indicated—rather naively—by the unequal amplitudes of the alternations. The stabilized 220-V ac is represented by the second wave form, where all the alternations are of equal amplitude. The 0–250-V ac output of the autotransformer I is represented by the third drawing. When the kV knob is set with the contact low on the autotransformer (dotted arrow), the output potential is low (dotted wave); when the knob is near the top of the winding (solid arrow), the output potential is high (solid wave) and may be made even higher than the 220-V input. The 50-kV ac output of the high-potential transformer O is shown next, followed by the 50-kV full-wave rectified pulsating output of the rectifiers. For 60-Hz ac, the full-wave pulse rate is 120 per second. In a full-wave generator, this is substantially the wave form applied to the x-ray tube. In a constant-potential generator, the full-wave pulsating current is smoothed by an electric filter QR, providing nearly pure direct current, as shown by the last wave form.

It is customary to state ac and pulsating dc potentials in volts *rms* (root mean square), also known as volts *effective*. This convention is followed in the x-ray generator except for the ac output of the high-potential transformer and the pulsating dc output of the rectifiers. These potentials are stated in kilovolts *peak*—kV(P)—for the following reasons. It is the peak x-ray tube potential that determines the wavelength profile of the continuous x-ray spectrum, and it is the peak potential that must exceed the excitation potential of an element if its characteristic spectrum is to be excited. For sine-wave form,

$$V_{\text{peak}} = 1.414 V_{\text{RMS}} \tag{4.21}$$

$$V_{\text{RMS}} = 0.707 V_{\text{peak}} \tag{4.22}$$

where 1.414 and 0.707 are $\sqrt{2}$ and $1/\sqrt{2}$, respectively. Constant-potential dc is stated in kilovolts constant potential—kV(CP).

In constant-potential generators, both the kilovoltmeter and milliammeter indicate substantially continuous dc, and x-ray tube power is given by

$$P_{\text{CP}} = Vi \tag{4.23}$$

where V is x-ray tube potential (V), i is x-ray tube current (A), and P_{CP} is power (W). In full-wave generators, the kilovoltmeter indicates peak kilovolts and the milliammeter rms current; x-ray tube power is given by

$$P_{\text{FW}} = 0.707 V_{\text{peak}} i_{\text{RMS}} \tag{4.24}$$

For example, an x-ray tube operating at 50 kV, 50 mA from a constant-potential generator dissipates $50{,}000 \times 0.050 = 2500$ W; the same tube operating at the same conditions from a full-wave generator dissipates $(0.707 \times 50{,}000)(0.050) = 1768$ W. About 1% of the power dissipated by the x-ray tube is converted to x-rays. It follows that, at the same nominal operating conditions (kV, mA), the x-ray output is substantially lower with full-wave than with constant-potential operation.

4.3.2.4. *Stabilization*

Because primary x-ray intensity is proportional to iV^2 [Equation (1.17)], it is essential that the x-ray tube current and, especially, potential be highly regulated, ±0.05% or less. All commercial x-ray spectrometers have an ac line stabilizer and an "mA stabilizer," and some also have a "kV stabilizer." These three types of stabilizer are shown schematically in Figure 4.2, and are considered below in order.

A simple "constant-potential" or "isolation" transformer may be used between the ac line and the input terminals of the x-ray generator in laboratories where extremely wide line fluctuations are encountered. However, these transformers used alone provide only about ±1% output regulation for ±15% input fluctuation. A good line stabilizer holds the output constant at 220 V ±0.1% or less for an input fluctuation of 190–250 V or more, and may be electronic, electromagnetic, or a combination of both. The electronic type usually have a temperature-controlled diode. The line fluctuations cause corresponding fluctuations in the electron emission in this tube. This varying emission current generates error potentials, which are amplified and used to correct the input fluctuations.

The "mA stabilizer" functions substantially as follows. The x-ray tube current returning to the center tap of the high-potential transformer *O* through the milliammeter *U* passes through the resistor *V* and develops a potentials across it. Fluctuations in the tube current cause corresponding fluctuations in this potential, which are amplified and, through transformer *X* (or a reactor), induce potentials in the primary circuit of the x-ray tube filament transformer *Y* in such phase as to counteract the fluctuations.

The "kV stabilizer" functions similarly. The high potential output of the power supply appears across the resistive potential divider *S*. A certain small fraction of this total potential is applied to the kV stabilizer, where it is compared with a fixed reference potential, thereby generating a difference potential. Fluctuations in the high potential across *S* cause corresponding fluctuations in the difference potential, which are amplified and, through transformer *N* (or a reactor), induce potentials in the primary circuit of the high-potential transformer in such phase as to counteract the fluctuations.

Incidentally, the potential taken from the divider *S* may also be indicated on a dc panel voltmeter *J′* graduated in kV, providing an alternative means of indicating x-ray tube potential. This alternative is advantageous in that the potential indicated on *J′* is derived directly from the potential actually applied to the x-ray tube.

4.3.2.5. *Safety and Protective Devices*

K′ is a 30–60-s time-delay switch that opens when the power supply is turned off and is closed by a bimetallic strip heated by a resistor or, preferably, by a small timer motor. When the main power switch *A* is turned on, the bimetallic strip heater or timer motor is energized and, after 30–60 s, the time-delay switch is closed. The time-delay switch prevents the application of power to the primary of the high-potential transformer *O* until the x-ray tube and rectifier filaments and the mA and kV stabilizers have had time to warm up. Closing switch *K* cannot energize transformer *O* unless switch *K′* has closed, and not even then unless all the other safety and protective devices are closed, which they normally are unless a hazardous condition exists.

Provision must be made to shut off the instrument if any difficulty develops that might present a hazard to equipment or personnel. Safety and protective devices are indicated schematically by switches *D–H*, shown open in Figure 4.2, but normally closed. Here, *D* and *E* represent high- and low-pressure switches in the cooling water line of the x-ray tube. *F* represents the line overload relay; if any of the circuits connected to the stabilized 220-V line draw excessive current, the solenoid *F′* is magnetized sufficiently to open *F*. *G* represents the x-ray tube current overload relay; if the x-ray tube current becomes excessive, the solenoid *G′* opens *G*. The overload relays *F* and *G* have reset buttons on the panel, as shown. *H* represents the several interlocks on cabinet doors and the x-ray tube housing; if any of these are disturbed while the equipment is in operation, *H* opens and turns off the equipment.

The fuses *B* will "blow" if excessive current is drawn by the line stabilizer or by any equipment connected directly to the 220-V ac line or to the outlets *C*, or if the protective devices fail.

4.3.3. Practical Considerations

4.3.3.1. *Constant Potential*

Figure 4.4 shows the wave forms of 50-kV ac, half-wave and full-wave rectified pulsating dc, and constant-potential dc. The *K*-excitation potentials of barium (37.4 kV), molybdenum (20.0 kV), and iron (7.1 kV) are indicated. The *K* lines of these elements are excited only when the x-ray tube potential exceeds their respective excitation potentials, and they are excited efficiently only when the excitation potentials are exceeded substantially. It is evident that the smaller the difference between operating and excitation potentials, the greater is the advantage of constant-potential (CP) compared with full-wave (FW) power. It is also evident that constant potential is particularly advantageous for determination of minor and trace constituents having high excitation potentials. Figure 4.5 shows the intensity

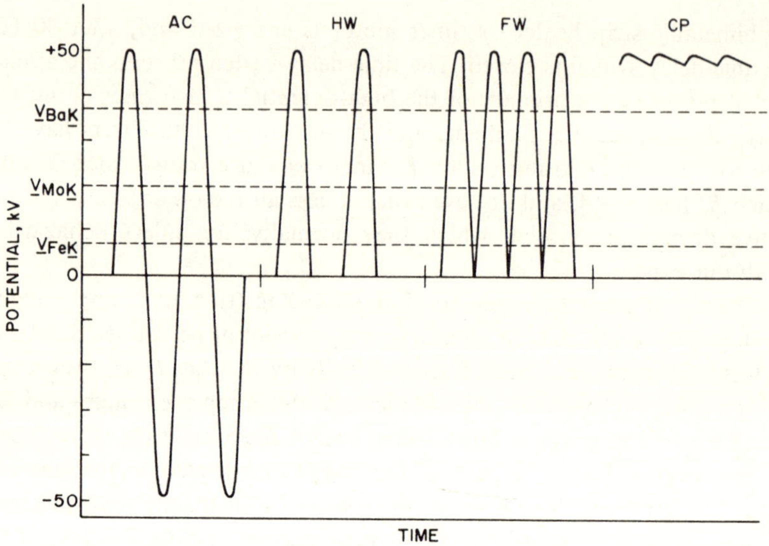

Fig. 4.4. Secondary excitation by half-wave, full-wave, and constant-potential supplies.

ratio for CP and FW supplies for excitation potentials up to 30 kV. The ~30% advantage of CP over FW at low excitation potentials (<10 kV) is due to the improved power factor (lead-lag phase relationship of potential and current) with CP. However, for these light elements, line and background increase at the same rate, so that line-to-background ratios with CP are no better than with FW.

If ac were applied to an x-ray tube, electrons could flow from filament to target only on the positive half-cycles when the target is positive with respect to the filament. Thus, in effect, the x-ray tube acts as its own rectifier, and its performance is substantially the same with ac and half-wave-rectified dc. In both cases, the tube is energized only half the time, and, for a given analyte, the secondary-emission intensity is about half that with FW. No commercial x-ray spectrometer operates on ac or half-wave-rectified dc, and they are of only academic interest.

Constant potential has another advantage when a Geiger counter is used. Suppose a Geiger counter, used on a spectrometer having a full-wave generator, gives a measured intensity at or near the upper limit of its linear region. Since the analyte is emitting photons only part of the time (Figure 4.4), it is likely that the linear limit is being exceeded during the effective part of the pulse. The error is more serious the greater the excitation potential for the measured line. No such problem arises with a CP supply.

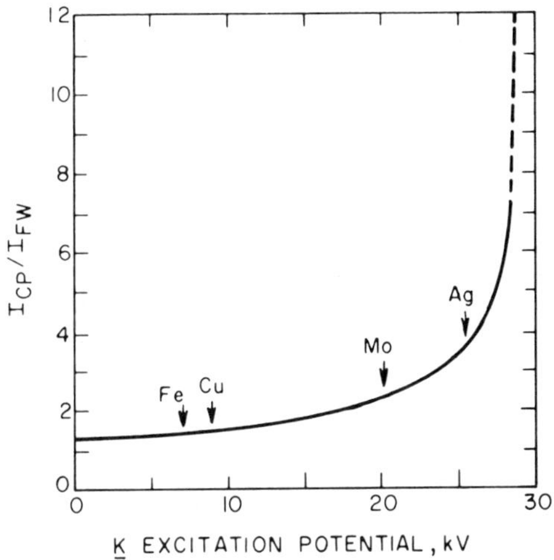

Fig. 4.5. Intensity ratio for constant-potential and full-wave excitation.

The problem is not particularly serious anyway because Geiger tubes are used relatively infrequently in x-ray spectrometers.

4.3.3.2. *Maximum Target Potential*

Commercial x-ray spectrometers are available with maximum high potentials of 50, 75, and 100 kV peak or CP. 50 kV excites *K* lines up to europium (*Z* 63), 75 kV up to osmium (*Z* 76), and 100 kV up to radon (*Z* 86). Of course, each of these potentials excites *K* lines efficiently only up to about five atomic numbers below those cited above. The advantage of the *K* lines of the heaviest elements is questionable. No crystal is available having high dispersion and "reflectivity" for these lines. However, *K* lines are fewer in number and more intense than *L* lines, and all *K* and *L* lines excited at 50 kV are excited more efficiently at 75 or 100 kV [Equation (1.20)]. Background is also higher, but line-to-background ratio stays about the same, so there is some net advantage. Electrical insulation and radiation protection problems are more severe at 75 and 100 kV.

4.3.3.3. *Operating Conditions*

Choice of x-ray tube current (mA) is less a problem than choice of potential (kV) because current affects both line and background intensity linearly and therefore does not affect the line-to-background ratio. However, potential affects the two intensities very differently.

Generally, it is preferable to select the optimum potential, then set the current to the maximum permitted by the target power rating (kW or kVA). Actually, it is prudent to set the current at ~5 mA less than this value: a few milliamperes costs very little intensity, but may substantially increase the tube life.

For qualitative analysis, it is prudent to operate at maximum potential so that lines having high excitation potentials are excited. This is particularly important when the spectrum is recorded in air path, and *K* lines of the heavy elements must be excited because the *L* lines may lie in the helium or vacuum region.

For quantitative analysis, maximum potential always gives maximum intensity for all lines, regardless of whether the line is excited primarily by continuum or by target lines. The intensities of all primary wavelengths increase with x-ray tube potential [Equations (1.17) and (1.20)]. Maximum potential is not used when excitation of a certain element is to be avoided (selective excitation), when long-wavelength spectra are to be excited, and when maximum line-to-background is to be realized. Also, it is inefficient and uneconomical to operate the tube at a potential so high that the primary x-rays are so short that they penetrate the specimen to a depth from which analyte-line radiation cannot emerge.

For long-wavelength lines, the line-to-background ratio is smaller at higher x-ray tube potentials. As potential is increased, intensity increases in the continuum hump more rapidly than at the longer wavelengths most efficient in exciting the long-wavelength spectra. The continuum radiation scatters from the specimen and increases the background. However, pulse-height selection can eliminate this background, so if line intensity cannot be spared, maximum x-ray tube potential can be used.

Optimal x-ray tube potential may be defined as that which gives the highest line-to-background ratio. For excitation of short-wavelength spectra by continuum, optimal potential is usually that which places $\lambda_{I_{max}}$ (Section 1.5.1)—the continuum hump—on the short-wavelength side of the analyte absorption edge.

4.4. FILTERS IN SECONDARY EXCITATION

4.4.1. Attenuation Filters

An attenuation filter is a thin foil or other layer placed in an x-ray beam to preferentially absorb a certain wavelength or band of wavelengths; it is based on the fact that an element absorbs wavelengths shorter than its absorption edge strongly, but is relatively transparent to longer wavelengths.

The practice of using simple (Hull) filters to remove the $K\beta$ lines from x-ray tube target spectra is well known in diffraction. The filter has an absorption edge between the $K\beta$ and $K\alpha$ lines and absorbs the $K\beta$ line very strongly. The best β filter for element Z is usually element $Z - 1$ or $Z - 2$. The filter thickness is chosen to reduce the $K\beta$:$K\alpha$ ratio from 1:5 to 1:500; this usually reduces the $K\alpha$ intensity to about half its original value.

Attenuation filters are used to remove target and other primary-beam lines when they might interfere with the analysis (*S4*). For example, Cr $K\beta$ (2.085 Å) and Cr $K\alpha$ (2.291 Å) are absorbed by a titanium filter (K_{ab} 2.497 Å); W $L\gamma_1$ (1.099 Å), W $L\beta_1$ (1.282 Å), and W $L\alpha_1$ (1.476 Å) are absorbed by nickel (K_{ab} 1.488 Å), or, better, by nickel and brass (Cu K_{ab} 1.380 Å, Zn K_{ab} 1.283 Å). If iron (K_{ab} 1.743 Å) is added to the filter, Cu $K\alpha$ (1.542 Å) and Ni $K\alpha$ (1.659 Å) extraneous lines are removed, and if aluminum is added, Fe $K\alpha$ and Cr $K\alpha$ extraneous lines are strongly attenuated nonspecifically.

4.4.2. Enhancement Filters

An enhancement filter is a thin foil or other layer placed in the primary x-ray beam to improve the analyte-line-to-background ratio. However, the analyte-line intensity is actually smaller than it would be without the filter.

Gunn (*G16*) determined traces of nickel in hydrocarbon oils using the intensity ratio of Ni $K\alpha$ to adjacent background as a function of nickel concentration. Zinc, gallium, germanium, selenium, bromine, and strontium, as powdered element or compound deposited on aluminum foil, were used as enhancement filters. Zinc ($K\alpha$ 1.436 Å), predictably the optimum enhancement filter for nickel (K_{ab} 1.488 Å), simply reduced Ni $K\alpha$ intensity without increasing the Ni $K\alpha$:background intensity ratio. Germanium increased the ratio by a factor of 1.6, but reduced the intensity to ~17% of its unfiltered value. Strontium was most effective, increasing the ratio by a factor of 1.5 while reducing the intensity to only ~40%.

The enhancement filter technique has been developed by Caticha-Ellis *et al.* (*C24*). Figure 4.6 shows the primary continuous spectrum with and without filtration, the analyte absorption edge and lines, and the absorption edge of an enhancement filter having wavelength slightly greater than the analyte lines. Such a filter absorbs that part of the primary spectrum that excites the analyte less than it does that part that only contributes to the

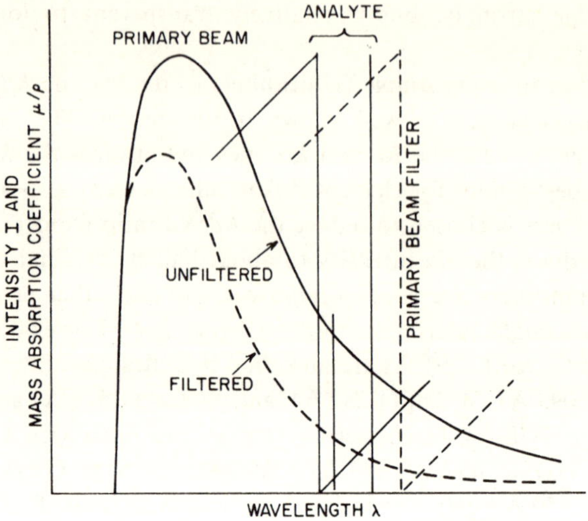

Fig. 4.6. Function of primary-beam enhancement filter.

scattered primary background at the analyte line. Consequently, higher line-to-background ratios are to be expected.

The enhancement-filter technique is advantageous only at intermediate wavelengths. At longer wavelengths: (1) the continuum has low intensity, and filtering of the useful part is less tolerable; and (2) the scatter of secondary radiation from the crystal contributes more to the background than scatter of primary radiation from the specimen. Short-wavelength spectra lie in the region of the continuum hump, where the short wavelength and high intensity necessitate relatively thick filters to suppress the background significantly. These thick filters also substantially suppress the part of the continuum that excites the analyte. Elements having short-wavelength spectra have high excitation potentials—approaching the x-ray tube potential—so their excitation efficiency is relatively low. Consequently, high primary filtration may prohibitively reduce the analyte-line intensity.

4.4.3. Enhancement Radiators

Many microchemical and thin-film specimens are substantially radiolucent. These specimens include small specimens—specks, flakes, fibers, etc.; films on radiolucent substrates; thin layers of solid, powder, or liquid; and milligram or microgram specimens supported on filter paper, Mylar film, ion-exchange membrane, etc. A substantial portion of the primary beam passes through such specimens and is lost. Some of this wasted radiation may be retrieved by backing the specimen with a secondary radiator having a strong spectral line near the short-wavelength side of the analyte absorption edge (*H35*). The primary x-rays transmitted by the specimen excite the radiator line, which, in turn, excites additional analyte-line radiation. Enhancements of 50% have been obtained in favorable cases. Alternatively, the excitant may be mixed with liquid and powder specimens. Some additional excitation may be provided by simply backing x-ray-transparent specimens with beryllium or a hydrocarbon to scatter some of the transmitted primary x-rays back through the specimen. However, this technique is likely to reduce the line-to-background ratio.

Other applications of filters are given in Section 8.2.4.

4.5. SPECIMEN PRESENTATION

Specimen presentation is discussed in detail in Section 16.1.1 and throughout Chapters 16 and 17. However, a few general comments are in order here.

The specimen plane may be horizontal below or above the x-ray tube window, inclined at 30–60° to the horizontal below or above the window, or vertical. In most x-ray spectrometers, the specimen plane is defined by a mask in the specimen chamber. The effective specimen area is the area on the specimen plane from which the x-ray optical system receives secondary emission. It is defined by the projection of the x-ray optical system on the specimen plane, or, in effect, by the projection of the source collimator (Section 5.2.2) on that plane. In general, the central rays of the primary and secondary beams meet at the center of the effective specimen area. The specimen plane is inclined at the incident angle ϕ to the central primary ray and at takeoff angle ψ to the central secondary ray. The takeoff angle is usually 30–45°, and the angle between the central primary and secondary rays is usually 90°.

The specimen mask material must have high mechanical rigidity and adequate absorption for the shortest wavelength emitted from the specimen area covered by the mask. Its spectral lines must lie in a region where they do not interfere with analyte lines. It must be available in high purity so that the spectrum is not complicated by spectral lines of trace constituents. High-purity aluminum ~1/8 inch thick is by far the most commonly used mask material. If short-wavelength lines penetrate the mask, 0.005-inch nickel or copper foil may be cemented to the under side of the mask. The foil absorbs the short-wavelength x-rays, and the aluminum prevents emission of the nickel or copper *K* lines. In the long-wavelength region even Lucite, nylon, or Teflon may be used; if higher-order short-wavelength lines penetrate these masks, they may be removed by pulse-height selection. Other mask materials may be used in special cases.

The specimen mask has a window of appropriate shape (usually rectangular) and dimensions, but not larger than the effective specimen area. With small windows, the area of the mask irradiated by the primary beam and, thereby, the intensity of scattered primary radiation increase. Consequently, for specimen areas of 1–50 mm^2 or so, a collimated specimen mask (*C31*, *G5*) may be used; the primary beam is conducted to the specimen and the secondary beam to the optical path in a continuous L-shaped tunnel in a block on the specimen mask. This arrangement reduces primary scatter from the mask and from air in the specimen compartment. For smaller specimen areas, selected-area techniques are used in which a pinhole or slit aperture is placed in the primary or secondary beam (Chapter 19).

Originally, all commercial x-ray spectrometers were constructed with the primary beam directed downward into a boxlike specimen compartment. Specimens were placed behind a mask in some type of drawer, which was then inserted in the compartment. When a wide variety of specimen forms must be dealt with, these drawer-type instruments are preferable because it is easy to add accessories to the drawer without altering the instrument itself. Such accessories include rotators for disk and cylindrical specimens, collimated specimen masks, selected-area accessories, and holders and manipulators for special specimen forms. For example, devices have been built to pass through the drawer continuous tapes of filter paper or other material having specimens spotted at intervals. Photographic slide changers have been arranged to index through the chamber specimens supported on 2 × 2 inch filter-paper squares. In more recent instruments, the specimen chamber has the form of a turret or carousel. Some can hold up to ten specimens. Others hold fewer specimens, but are fitted for continuous loading.

Chapter 5

Dispersion

5.1. INTRODUCTION

The second phase of the x-ray spectrometric analysis (Section 3.2.5) is dispersion of the secondary x-rays emitted by the specimen to permit isolation and measurement of the individual analyte lines. The modes of dispersion applied in most x-ray spectrometers may be broadly classified in two categories usually distinguished as "dispersive" and "nondispersive."

The dispersive mode is also referred to by the following terms: (1) *wavelength dispersion*, because the x-rays are "sorted" on the basis of their wavelengths, as indicated by the Bragg law; (2) *crystal dispersion* when a crystal is the dispersion device; (3) *diffractive dispersion* when x-ray diffraction is the dispersion phenomenon; and (4) *geometric, or spatial, dispersion*, because each wavelength leaves the crystal in a different direction (2θ). Optical gratings and specular reflectors are also used for dispersion of x-rays.

The nondispersive mode is also referred to by the following terms: (1) *energy dispersion*, because the x-rays are "sorted" on the basis of their photon energies; (2) *electronic dispersion* when the pulse-height analyzer is used; (3) *nondiffractive dispersion*; and (4) *nongeometric, or nonspatial, dispersion*, because there is no actual sorting of photons of different energies into different directions in space; this condition is the basis of the term *nondispersive*. Selective excitation, filtration, and detection are other forms of nondispersive analysis.

Both dispersion modes may be involved simultaneously. Energy dispersion may be applied to the output of a detector receiving a crystal-dispersed beam containing analyte line of wavelength λ, interfering higher-order lines $\lambda/2$, $\lambda/3$, etc., and spurious lines arising from any of several sources.

Nondispersive x-ray spectrometry is discussed in Section 8.2. This chapter is limited largely to crystal dispersion. The components of the x-ray optical system are considered first—the crystal itself, the collimator, and the optical path medium. Then, the basic flat- and curved-crystal dispersion arrangements are described. Finally, the most commonly used types of curved-crystal x-ray spectrometers are described.

5.2. COLLIMATORS

5.2.1. Function

Briefly, the function of the collimators (*W13*) is to intercept divergent x-rays so that a substantially parallel beam arrives at the crystal and detector window. The projection of the source collimator (see below) on the specimen plane defines the useful specimen area.

Each point on the specimen surface emits x-rays of many wavelengths—all the spectral lines of all the elements present at that point that can be excited by the primary beam. All these x-rays are emitted in all directions. In Figure 5.1A, consider such a point emitting a cone of x-rays containing wavelengths λ_1, λ_2, and λ_3. One ray of each wavelength arrives at the crystal at the correct Bragg angle and is diffracted, as shown. Thus, the crystal shown would diffract all wavelengths between λ_2 and λ_3. Similarly, the crystal would diffract a range of wavelengths from each point on the specimen. A pinhole or slit aperture at either of the positions shown collimates the beam and excludes all wavelengths other than λ_1. Of the x-rays emitted by the specimen, the aperture permits the crystal to receive only those that originate in the projection of the aperture on the specimen surface and that radiate in a direction θ_1 to the crystal surface. Of these x-rays, the crystal diffracts only those having wavelength λ_1 that satisfies the Bragg law for θ_1 and the *d*-spacing of the crystal. A pinhole and slit permit these conditions to exist at a spot and a line, respectively, on the specimen surface. Figure 5.1A is essentially the arrangement of the Bragg spectrometer (*B51*, *B52*). It is extremely wasteful, in that it excludes most of the x-rays

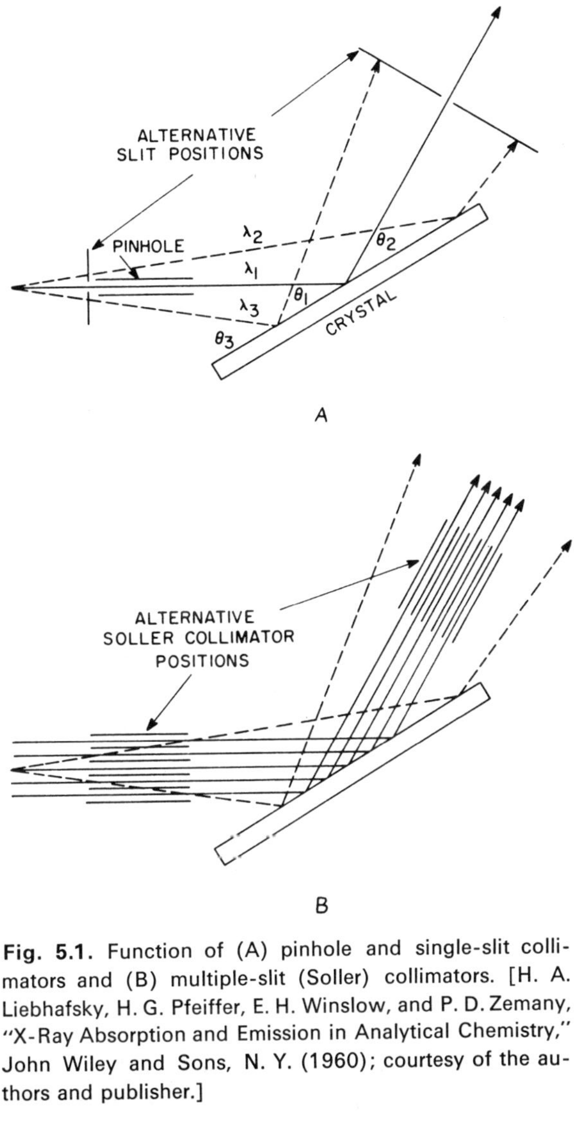

Fig. 5.1. Function of (A) pinhole and single-slit collimators and (B) multiple-slit (Soller) collimators. [H. A. Liebhafsky, H. G. Pfeiffer, E. H. Winslow, and P. D. Zemany, "X-Ray Absorption and Emission in Analytical Chemistry," John Wiley and Sons, N. Y. (1960); courtesy of the authors and publisher.]

emitted by the specimen at the wavelength for which the crystal is set; this is true even for small particulate or filamentary specimens.

The Soller arrangement (*S26*), shown in Figure 5.1B, is, in effect, a series of Bragg slit systems. The multiple-foil (Soller) collimator (Figure 5.2B) allows Bragg conditions to exist for each incremental strip of a large specimen area, thereby increasing the measured intensity and averaging

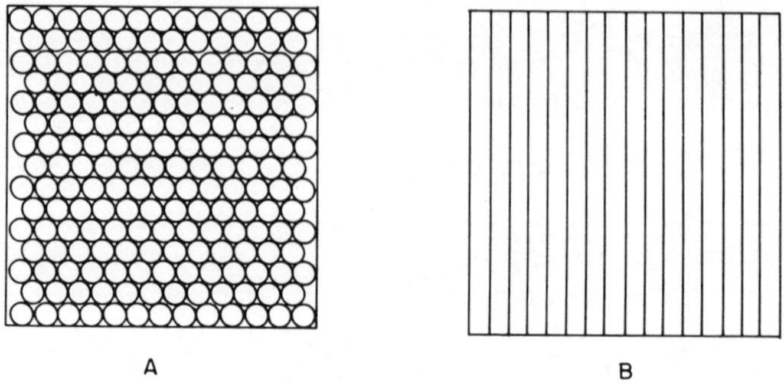

Fig. 5.2. End views of (A) multiple-tube and (B) multiple-slit (Soller) collimators.

out specimen inhomogeneity. The detector must have a window and active volume large enough for the large diffracted beam. All flat-crystal commercial x-ray spectrometers are now of the Soller type.

The earliest Philips spectrometer used multiple-tube collimators (Figure 5.2A) and was, in effect, a series of Bragg pinhole systems. The function is the same as for the Soller arrangement, except that Bragg conditions exist for each incremental spot on the specimen area.

5.2.2. Features and Considerations

It has already been mentioned that the collimators had the form of simple pinholes or slits in the Bragg spectrometer, and multiple-tubes in the first commercial Philips spectrometers. Today, all flat-crystal spectrometers use multiple-foil (multiple-blade, multiple-slit, Soller) collimators.

The multiple-tube collimator transmits much lower intensity than the multiple-foil collimator for two reasons. It has a smaller cross-sectional open area, and it limits divergence in the direction parallel to the crystal axis and thereby excludes many rays that would pass a Soller collimator. The divergence perpendicular to the crystal axis is about the same for both types of collimator. The divergence parallel to the crystal axis is proportional to the tube diameter or foil height, giving the Soller collimator a substantial advantage in transmitted intensity. The advantage decreases as θ approaches 90°.

Soller collimators are placed at either or both of two positions in the optical path (Figure 3.1). The *source* or *divergence* collimator is placed between the specimen and crystal, and is fixed. The *detector* or *receiving* collimator is placed between the crystal and detector, and moves with the detector. When collimators are used in both positions, one is usually coarse (short and/or wide-spaced foils), the other fine (long and/or close-spaced foils). When both collimators are fine, the spectrometer is difficult to align and to keep in alignment (*W13*, *W14*). Another difficulty encountered when two fine collimators are used is line splitting, or separation of a diffracted peak into two or more subpeaks. This phenomenon occurs when rays from the source collimator are blocked by foils in the detector collimator, or cross over into adjacent channels in that collimator.

In passing through the collimator, the x-rays may be specularly reflected, scattered, or diffracted by the foils, thereby producing extraneous lines. The foils may be treated to reduce these effects. In passing through the collimator, the x-rays may also excite the characteristic spectrum of the foil metal, thereby producing extraneous pulse-height distributions in proportional detectors.

Detector collimators may be tapered, with the height of the collimator aperture and foils increasing from the crystal end to the detector end. This feature permits better utilization of the vertical divergence mentioned above, with consequent increase in transmitted intensity, and is especially advantageous for measurement of the K lines of the low-Z elements. The detector window must be high enough to receive the divergent beam.

The x-ray beam transmitted by a Soller collimator has a triangular intensity *vs.* 2θ profile (Section 13.2.1.3, Figure 13.4). For a collimator of length L and foil spacing S, the width of this profile at half-maximum is

$$W_{1/2} = \tan^{-1}(S/L) \tag{5.1}$$

The spacing S is measured *betweeen* the foils, and is not the width of the collimator aperture divided by the number of foils. Collimators are specified either in terms of length and foil spacing (for example, 4×0.010 inch) or by the angular divergence in degrees as given by Equation (5.1). Divergence is discussed in more detail in Section 13.2.

The effects of increased collimation, that is, longer and/or closer-spaced collimators, are: (1) narrower peaks; (2) narrower "tailing" or "skirts" on the peaks; (3) increased resolution, as a consequence of (1) and (2); (4) decreased background; and (5) decreased measured analyte-line intensity. In short, the higher the collimation, the higher is the resolution and the lower is the intensity. As a rule, the shortest, most widely spaced collimator is used that gives resolution adequate for the measurements to be made. At wavelengths greater than 3 or 4 Å, the separation of the K lines of neighboring elements increases as atomic number decreases

(Figure 1.8). Consequently, low collimation is adequate unless *M* or higher-order *K* or *L* lines of high-*Z* elements are present. This applicability of low collimation is fortunate because it partially compensates the low fluorescent yield and high absorption losses in this spectral region. Further considerations regarding choice of the crystal–collimator system are given in Section 5.4.2.3, and Figure 5.7 shows the effect of collimation for a crowded spectral region.

5.3. RADIATION PATH

Air path is usually satisfactory for *K* lines of elements down to atomic number 23 (vanadium) and *L* lines of elements down to atomic number 56 (barium). Air path can be used for *K* lines from relatively high concentrations of elements down to atomic number 19 (potassium). In other cases, helium or vacuum path must be used (*H41*, *M19*, *Z10*).

The measured intensities in air and helium paths are equal at wavelengths less than ~1 Å (Br *Kα*, Ra *Lα*). Helium gives some advantage at all longer wavelengths, and should always be used at wavelengths >2.5 Å (Ti *Kα*). Figure 5.3A shows the intensity ratio for helium and air paths for the *Kα* lines of calcium to manganese.

The measured intensity for helium, hydrogen, and vacuum paths are

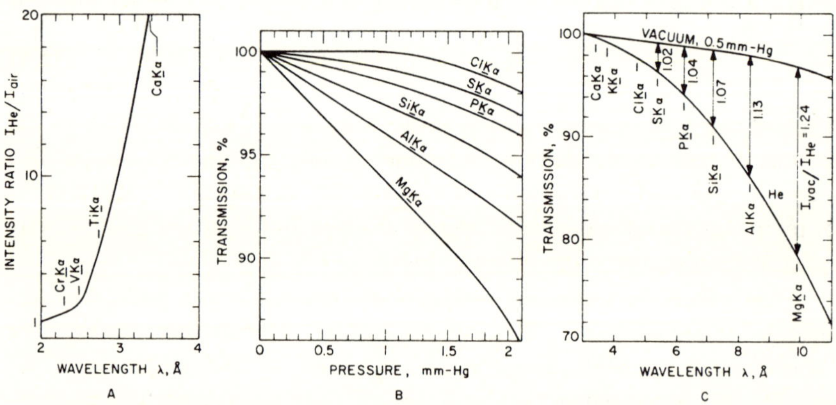

Fig. 5.3. Effect of radiation-path medium in the long-wavelength region. Data are for 25-cm path length. [J. H. Hoskins, *Norelco Reporter* **7**, 111 (1960); courtesy of the author and Philips Electronic Instruments, Inc.] A. Transmission ratio for helium and air paths. B. Transmission for *Kα* lines of elements 12–17 (Mg–Cl) as a function of pressure. C. Comparison of transmission for *Kα* lines of elements 12–20 (Mg–Ca) in 0.5-mm Hg vacuum and in helium.

equal at wavelengths less than ~6 Å (P $K\alpha$, Zr $L\alpha$). Vacuum gives some advantage at all longer wavelengths, and should be used at wavelengths >7 Å (Si $K\alpha$). Figure 5.3B shows transmission as a function of pressure for the $K\alpha$ lines of magnesium to chlorine. Figure 5.3C shows transmission as a function of wavelength for vacuum and helium and the intensity ratio for the two media for the $K\alpha$ lines of magnesium to calcium. Hydrogen (*B30*) is not commonly used because of the explosion hazard.

Increasing the transmission of the optical path is only one reason for using helium or vacuum path. Another is to reduce the scatter of primary x-rays in the specimen compartment and thereby to reduce background. A third application of vacuum path was reported by Ashby and Proctor (*A18*). With very high stabilization of the x-ray tube kV ($\pm$0.05%) and mA ($\pm$0.02%) and temperature control of the crystal ($\pm$0.5°C), changes in density of the air path may actually limit the instrumental stability and cause observable changes in the measured intensity. Thus, if it is instrumental stability that limits the overall precision, use of vacuum path may actually improve the precision.

As might be expected, there are problems peculiar to helium and vacuum paths. Distension of the windows of liquid-specimen cells and disruptive outgassing of powders and briquets in vacuum are discussed in Chapters 16 and 17. Thin windows on gas-flow detectors used in vacuum must be supported by nylon or metal grids. Some of the earlier vacuum enclosures had limited 2θ ranges and required use of less-favorable crystals.

Probably the greatest difficulties attend the use of rubber or plastic helium jackets. Following are some of the precautions that must be taken in their use. The connections of the jacket to the collimators and crystal stage must be secure and free of gaps. The folds or pleats may not protrude into the optical path. A 0–50-ft^3/h gas-flow meter must be provided in the helium line. A high helium flow rate (20–30 ft^3/h) for several minutes is used initially to flush the system. If the helium jacket is attached to the crystal end of the detector collimator, a short rubber tunnel or boot must be placed between the detector end of the collimator and the detector window. During the initial helium flush, this boot should be pried open slightly to allow the helium to sweep the air out of the collimator. Alternatively, a small hole or notch may be cut in the boot to permit a slow continuous outflow of helium. If these precautions are not taken, air may be permanently trapped in the collimator. After the initial flush, a somewhat lower flow rate (5–20 ft^3/h) may be used for normal operation. The optimal flow rate is established as follows. The intensity of a spectral line having wavelength longer than that of any line to be measured is observed on the ratemeter recorder as helium flow rate is increased incrementally until no further increase in intensity occurs. Each time the specimen compartment is opened to air, it should remain so far as short a time as possible, and a 15–60-s flush time must be allowed before resuming the measurements. The longer the wavelength to be measured, the higher will be the required operating flow rate and the longer the flush time after loading. When 2θ must be changed to measure background or other lines, the jacket may pull and leak air or protrude into the optical path. These conditions are particularly likely to occur at very high or very low 2θ angles.

5.4. ANALYZER CRYSTALS

5.4.1. Introduction

The analyzer crystal performs the same function in an x-ray emission spectrometer as the prism, or, more analogously, the grating, in an optical emission spectrograph. It spatially disperses the secondary x-ray beam into a wavelength spectrum. As the crystal is rotated from 0° 2θ, where it is parallel to the collimated secondary x-ray beam, through progressively larger angles, it passes successively through the Bragg angle for each wavelength present, up to a certain limit (see below) and diffracts each wavelength in turn. X-ray diffraction, the basic phenomenon involved in crystal dispersion, is discussed in Section 2.3. Table 5.1 gives data and Figure 9.1 shows the useful spectral regions for commonly used analyzer crystals. Table 5.2 lists some less commonly used crystals. The principles of flat-crystal (nonfocusing) optics are reviewed in (*15A*, *C16*, *S34*).

A good analyzer crystal should have the following features: (1) wavelength range appropriate for the analyte lines to be measured; (2) high diffracted intensity; (3) high resolution—high dispersion and narrow diffracted peak breadth; (4) high peak-to-background ratio; (5) absence of interfering elements; (6) low thermal coefficient of expansion; (7) high stability in air, with life, and on prolonged exposure to x-rays; (8) good mechanical strength; and (9) availability in suitable quality and size. Additional properties are required for crystals to be used in curved-crystal spectrometers: (1) availability in very thin plates; (2) ease of elastic or plastic curvature to specified radii; (3) ease of grinding to specified radii; and (4) ability to withstand repeated flexing. The following sections discuss most of these features in some detail.

TABLE 5.1

Data for Commonly Used Analyzer Crystals

Symbol (if any) Common and chemical names Chemical formula	(*hkl*), (*hkil*) 2*d*, Å Crystal system	Longest wavelength and lowest atomic no. diffracted		Practical useful region (10–145° 2θ)	Reflectivity Resolution Secondary emission ($K\alpha$ lines)	Optimal application (*) Remarks
α-Quartz Silicon dioxide SiO_2	$(50\bar{5}2)$ 1.624 Hexagonal	λ, Å $K\alpha$ $L\alpha_1$	1.624 29-Cu 71-Lu	0.142–1.55 83-Bi to 29-Cu ≥73-Ta	Very low Very high Si	—
Topaz $Al_2SiO_4(F,OH)_2$	(303) 2.712 Orthorhombic	λ, Å $K\alpha$ $L\alpha_1$	2.712 23-V 57-La	0.236–2.59 70-Yb to 23-V ≥58-Ce	Medium Very high F, Al, Si	* Improves dispersion in the region V *K*–Ni *K* where Z $K\beta$ overlaps $(Z + 1)\ K\alpha$ Intensity ~5–10% of LiF(200) Individual crystals may vary widely in reflectivity Inferior to LiF(220)
α-Quartz Silicon dioxide SiO_2	$(20\bar{2}3)$ 2.750 Hexagonal	λ, Å $K\alpha$ $L\alpha_1$	2.750 23-V 57-La	0.240–2.62 69-Tm to 23-V ≥58-Ce	Low Very high Si	* Same application as topaz Somewhat sharper lines, but lower intensity than topaz Inferior to LiF(220)
LiF(220) Lithium fluoride LiF	(220) 2.848 Cubic	λ, Å $K\alpha$ $L\alpha_1$	2.848 22-Ti 56-Ba	0.248–2.72 68-Er to 22-Ti ≥57-La	High Very high F	* Same application as topaz Diffracted intensity ~4× topaz, ~$\frac{1}{2}$× LiF(200)

TABLE 5.1 (*continued*)

Symbol (if any) Common and chemical names Chemical formula	(*hkl*), (*hkil*) 2*d*, Å Crystal system	Longest wavelength and lowest atomic no. diffracted		Practical useful region (10–145° 2θ)	Reflectivity Resolution Secondary emission ($K\alpha$ lines)	Optimal application (*) Remarks
Fluorite Calcium fluoride CaF_2	(220) 3.862 Cubic	λ, Å $K\alpha$ $L\alpha_1$	3.862 19-K 49-In	0.337–3.68 59-Pr to 20-Ca ≥50-Sn	Medium Medium F, Ca	—
LiF Lithium fluoride LiF	(200) 4.028 Cubic	λ, Å $K\alpha$ $L\alpha_1$	4.028 19-K 48-Cd	0.351–3.84 58-Ce to 19-K ≥49-In	Very high Very high F	* Best general crystal for K $K\alpha$ to Lw L Highest intensity for greatest number of elements of any crystal Combines high intensity and high dispersion
Aluminum Al	(200) 4.048 Cubic	λ, Å $K\alpha$ $L\alpha_1$	4.048 19-K 48-Cd	0.353–3.86 58-Ce to 19-K ≥49-In	High — Al	* Curved, especially doubly curved optics
Aluminum Al	(111) 4.676 Cubic	λ, Å $K\alpha$ $L\alpha_1$	4.676 18-Ar 45-Rh	0.408–4.46 54-Xe to 18-Ar ≥46-Pd	Very high — Al	* Curved, especially doubly curved optics
NaCl Rock salt Sodium chloride NaCl	(200) 5.641 Cubic	λ, Å $K\alpha$ $L\alpha_1$	5.641 16-S 42-Mo	0.492–5.38 50-Sn to 16-S ≥44-Ru	Very high High Na, Cl	* S $K\alpha$ and Cl $K\alpha$ in light matrixes Like LiF(200), good general crystal (S $K\alpha$–Lw L) May give slightly higher intensity than LiF(200) for Ge $K\alpha$–Ba $K\alpha$ Background may be higher than for LiF(200) due to Cl $K\alpha$ emission

TABLE 5.1 *(continued)*

Symbol (if any) Common and chemical names Chemical formula	(*hkl*), (*hkil*) 2*d*, Å Crystal system	Longest wavelength and lowest atomic no. diffracted		Practical useful region (10–145° 2θ)	Reflectivity Resolution Secondary emission ($K\alpha$ lines)	Optimal application (*) Remarks
Calcite Calcium carbonate $CaCO_3$	$(20\bar{2}0)$ 6.071 Hexagonal	λ, Å $K\alpha$ $L\alpha_1$	6.071 16-S 41-Nb	0.529–5.79 48-Cd to 16-S ≥41-Nb	Medium Very high Ca	* Very precise measurement of wavelength Extremely high degree of crystal perfection with resultant sharp lines
Silicon Si	(111) 6.276 Cubic	λ, Å $K\alpha$ $L\alpha_1$	6.276 15-P 40-Zr	0.547–5.98 47-Ag to 16-S ≥41-Nb	High — Si	* Eliminates even-order spectral-line interference in measurement of intermediate- and high-Z elements where Si $K\alpha$ emission from the crystal is absorbed by the air path
Fluorite Calcium fluoride CaF_2	(111) 6.306 Cubic	λ, Å $K\alpha$ $L\alpha_1$	6.306 15-P 40-Zr	0.550–6.01 47-Ag to 16-S ≥41-Nb	Medium Medium F, Ca	Very weak second order, strong third order
Germanium Ge	(111) 6.532 Cubic	λ, Å $K\alpha$ $L\alpha_1$	6.532 15-P 39-Y	0.569–6.23 46-Pd to 15-P ≥40-Zr	High Very high Ge	* Eliminates even-order spectral-line interference in measurement of intermediate- and low-Z elements where Ge $K\alpha$ emission from the crystal may be minimized by pulse-height selection

TABLE 5.1 *(continued)*

Symbol (if any) Common and chemical names Chemical formula	(*hkl*), (*hkil*) 2*d*, Å Crystal system	Longest wavelength and lowest atomic no. diffracted		Practical useful region (10–145° 2θ)	Reflectivity Resolution Secondary emission ($K\alpha$ lines)	Optimal application (*) Remarks
α-Quartz Silicon dioxide SiO_2	$(10\bar{1}1)$ 6.687 Hexagonal	λ, Å $K\alpha$ $L\alpha_1$	6.687 15-P 39-Y	0.583–6.38 46-Pd to 15-P ≥40-Zr	High Very high Si	* P $K\alpha$ in low-Z matrixes, especially in presence of Ca P $K\alpha$ intensity ~1.8× EDDT, but less than PET Intensity for P $K\alpha$ to K $K\alpha$ greater than for EDDT, but less than for PET
α-Quartz Silicon dioxide SiO_2	$(10\bar{1}0)$ 8.510 Hexagonal	λ, Å $K\alpha$ $L\alpha_1$	8.510 13-Al 33-Br	0.742–8.12 41-Nb to 14-Si 100-Fm to 36-Kr	Medium Very high Si	Same applications as EDDT and PET below; higher resolution, but lower intensity
PET Pentaerythritol $C(CH_2OH)_4$	(002) 8.742 Tetragonal	λ, Å $K\alpha$ $L\alpha_1$	8.742 13-Al 35-Br	0.762–8.34 41-Nb to 13-Al 99-Es to 36-Kr	High Low None	* Al, Si, P, S $K\alpha$, especially P and S $K\alpha$ in intermediate- and high-Z matrixes Good general crystal for Al–Sc $K\alpha$ Al, Si, P, S $K\alpha$ intensities ~1.5–2× EDDT Low background due to absence of any element to emit secondary spectral lines Crystal is soft, must be mounted on rigid support and handled carefully. Tends to deteriorate with age, especially on prolonged exposure to high-intensity x-rays Highest thermal-expansion coefficient of all common analyzer crystals

TABLE 5.1 (*continued*)

Symbol (if any) Common and chemical names Chemical formula	(*hkl*), (*hkil*) 2*d*, Å Crystal system	Longest wavelength and lowest atomic no. diffracted	Practical useful region (10–145° 2θ)	Reflectivity Resolution Secondary emission ($K\alpha$ lines)	Optimal application (*) Remarks
EDDT Ethylenediamine-*d*-tartrate $C_6H_{14}N_2O_6$, COOH—NH_2 HCOH CH_2 HCOH CH_2 COOH—NH_2	(020) 8.808 Monoclinic	λ, Å 8.808 $K\alpha$ 13-Al $L\alpha_1$ 35-Br	0.768–8.40 41-Nb to 13-Al 99-Es to 35-Br	Medium Low None	* Al and Si $K\alpha$ in any matrix * P, S, Cl $K\alpha$ in intermediate- and high-Z matrixes Same remarks as for PET above, except: Substantially lower intensity than PET Substantially lower thermal-expansion coefficient than PET Rugged, but not always available in full 1 × 3-inch blanks
ADP Ammonium dihydrogen phosphate $NH_4H_2PO_4$	(101) 10.640 Tetragonal	λ, Å 10.640 $K\alpha$ 12-Mg $L\alpha_1$ 32-Ge	0.927–10.15 37-Rb to 12-Mg 91-Pa to 33-As	Low — P	* Mg $K\alpha$ Background may be high due to P $K\alpha$ emission Can be used for Al and Si $K\alpha$ when Mg is also to be determined, but intensity is much lower than for EDDT and PET

TABLE 5.1 *(continued)*

Symbol (if any) Common and chemical names Chemical formula	(*hkl*), (*hkil*) 2*d*, Å Crystal system	Longest wavelength and lowest atomic no. diffracted	Practical useful region (10–145° 2θ)	Reflectivity Resolution Secondary emission (*K*α lines)	Optimal application (*) Remarks
Gypsum Calcium sulfate dihydrate $CaSO_4 \cdot 2H_2O$	(020) 15.185 Monoclinic	λ, Å 15.185 *K*α 10-Ne $L\alpha_1$ 28-Ni	1.32–14.48 31-Ga to 11-Na 77-Ir to 29-Cu	Medium — S, Ca	* Na *K*α Good general crystal for all low-*Z* elements when Na is also to be determined Background higher than EDDT or PET due to S *K*α and Ca *K*α emission from crystal May effloresce in low humidity or prolonged use in vacuum path
Mica (muscovite), cleavage plane $K_2O \cdot 3Al_2O_3 \cdot 6SiO_2 \cdot 2H_2O$	(002) 19.84 Monoclinic	λ, Å 19.84 *K*α 9-F $L\alpha_1$ 25-Mn	1.73-18.92 27-Co to 9-F 69-Tm to 26-Fe	Low — Al, Si, K	* Variable-radius curved crystal spectrometers for selected-area analysis and electron-probe microanalysis
Rock sugar Sucrose $C_{12}H_{22}O_{11}$	(100) 20.12 Monoclinic	λ, Å 20.12 *K*α 9-F $L\alpha_1$ 25-Mn	1.75–19.19 27-Co to 9-F 68-Er to 26-Fe	Low — None	—
KHP ("KAP"[a]) Potassium hydrogen phthalate $KOOC \cdot C_6H_4 \cdot COOH$, 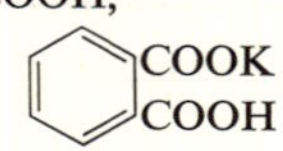	$(10\bar{1}0)$ 26.632 Hexagonal	λ, Å 26.632 *K*α 8-O $L\alpha_1$ 24-Cr	2.32-25.40 23-V to 8-O 60-Nd to 23-V	Medium — K	* F, Ne, Na, Mg *K*α Good general crystal for all low-*Z* elements when F and/or Na are also to be determined Background may be high due to K *K*α emission

TABLE 5.1 (*continued*)

Symbol (if any) Common and chemical names Chemical formula	(*hkl*), (*hkil*) 2*d*, Å Crystal system	Longest wavelength and lowest atomic no. diffracted		Practical useful region (10–145° 2θ)	Reflectivity Resolution Secondary emission (*Kα* lines)	Optimal application (*) Remarks
OHM						
Octadecyl hydrogen maleate		λ, Å	63.5	5.53–60.56	High	* Ultralong-wavelength region
Octadecyl hydrogen	63.5	*Kα*	6-C	15-P to 6-C	—	
cis-butanedioate		$L\alpha_1$	Any	41-Nb to 20-Ca	C	
$CH_3(CH_2)_{17}OOC\cdot CH{:}CH\cdot COOH$						
Lead myristate	80.5 Å	λ, Å	80.5	7.02–76.77	Medium	* Ultralong-wavelength region
Lead tetradecanoate	Smectic soap	*Kα*	5-B	14-Si to 5-B	—	
$[CH_3(CH_2)_{12}COO]_2Pb$	multilayer	$L\alpha_1$	Any	≤37-Rb	C, Pb[b]	
PbSt	100.4 Å	λ, Å	100.4	8.75–95.75	Medium	* Ultralong-wavelength region
Lead stearate[c]	Smectic soap	*Kα*	5-B	12-Mg to 5-B	—	
Lead octadecanoate	multilayer	$L\alpha_1$	Any	≤34-Se	C, Pb[b]	
$[CH_3(CH_2)_{16}COO]_2Pb$						
PbL	130 Å	λ, Å	130	11.35–124	—	* Ultralong-wavelength region
Lead lignocerate	Smectic soap	*Kα*	4-Be	11-Na to 4-Be		
Lead tetraeicosanoate	multilayer	$L\alpha_1$	Any	≤31-Ga		

[a] For potassium *acid* phthalate, which is not preferred nomenclature.
[b] Also the spectral lines of the substrate (mica or glass).
[c] Other smectic soap multilayers having $2d \approx 100$ Å: barium stearate and lead stearate decanoate (LSD, or LOD, for lead octadecanoate decanoate).

TABLE 5.2
Some Less Commonly Used Analyzer Crystals

Crystal	Chemical formula	(*hkil*)	2*d*, Å
β-Alumina	Al_2O_3	(0002)	23.8
Ammonium citrate	$NH_4OOC(OH)C(CH_2COONH_4)_2$	—	8.80
Ammonium dihydrogen phosphate	$NH_4H_2PO_4$	(200)	7.50
Ammonium tartrate	$NH_4OOC(CHOH)_2COONH_4$	—	8.80
Beryl	$3BeO \cdot Al_2O_3 \cdot 6SiO_2$	(100)	15.954
Bismuth titanate	$Bi_2(TiO_3)_3$	—	32.84
Clinochlore	Hydrated Mg–Al–Fe silicate aluminate	(001)	28.42
Dioctadecyl adipate (dioctadecyl hexanedioate)	$(CH_2)_4[COO(CH_2)_{17}CH_3]_2$	—	90
Germanium	Ge	(220)	4.000
Graphite	C	(002)	6.74
Gypsum (calcium sulfate dihydrate)	$CaSO_4 \cdot 2H_2O$	(002)	4.990
Itaconic acid (methylene butanedioic acid)	$CH_2{:}C(COOH)CH_2COOH$	—	18.50
Lead melissate (lead hentriacontanoate)	$[CH_3(CH_2)_{29}COO]_2Pb$	—	160
Lead stearate decanoate (lead octadecanoate decanoate)	$CH_3(CH_2)_{16}COO \cdot Pb \cdot OOC(CH_2)_8CH_3$	—	100
Mica (muscovite)	$K_2O \cdot 3Al_2O_3 \cdot 6SiO_2 \cdot 2H_2O$	$(33\bar{1})$[a]	3.00
Octadecyl hydrogen succinate (octadecyl hydrogen butanedioate)	$CH_3(CH_2)_{17}OOC(CH_2)_2COOH$	—	96.9
Oxalic acid dihydrate	$(COOH)_2 \cdot 2H_2O$	(001)	11.92
Potassium bromide	KBr	(200)	6.584
Potassium chloride	KCl	(200)	6.292
α-Quartz (silicon dioxide)	SiO_2	$(22\bar{4}3)$	2.030
		$(13\bar{4}0)$	2.355
		$(10\bar{1}4)$	2.576
		$(21\bar{3}1)$	3.082
		$(11\bar{2}2)$	3.634
		$(10\bar{1}2)$	4.564
		$(11\bar{2}0)$	4.916
		$(30\bar{3}8)$	?
		$(40\bar{4}3)$	?

TABLE 5.2 *(continued)*

Crystal	Chemical formula	(*hkil*)	2*d*, Å
Rubidium hydrogen phthalate	RbOOC · C_6H_4 · COOH	$(10\bar{1}0)$	26.12
	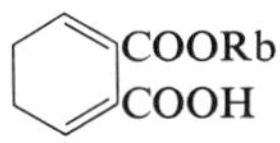		
Silicon	Si	(220)	3.840
Silver acetate	CH_3COOAg	(001)	20.0
Topaz		(006)	2.795
		(200)	4.64
		(002)	8.374

[a] Transmission: 2θ range, 10–145° 2θ; wavelength range, 0.26–2.9 Å.

5.4.2. Features

5.4.2.1. *Wavelength Range*

In principle, the longest wavelength that a crystal can diffract is equal to twice the interplanar spacing ($2d$) of the crystal planes parallel to its surface. From the Bragg law,

$$\sin\theta = n\lambda/2d \tag{5.2}$$

If $n\lambda > 2d$, $\sin\theta > 1$, which is not possible. It follows that for dispersion of long wavelengths, long d-spacing is required (*B40*). Incidentally, interplanar spacings for analyzer crystals are usually given in terms of $2d$, rather than d, because of this particular significance of $2d$.

Figure 5.4 shows some mechanical features that limit the useful 2θ range of the crystal (*B13*). For a collimator having breadth B in the direction perpendicular to the crystal axis, the crystal length L required to intercept the entire secondary beam at a specified diffraction angle θ is

$$L = B/\sin\theta \tag{5.3}$$

Alternatively, if a crystal of length L is used with a collimator of breadth B, the smallest diffraction angle at which the crystal can intercept the entire secondary beam is

$$\theta = \arcsin(B/L) \tag{5.4}$$

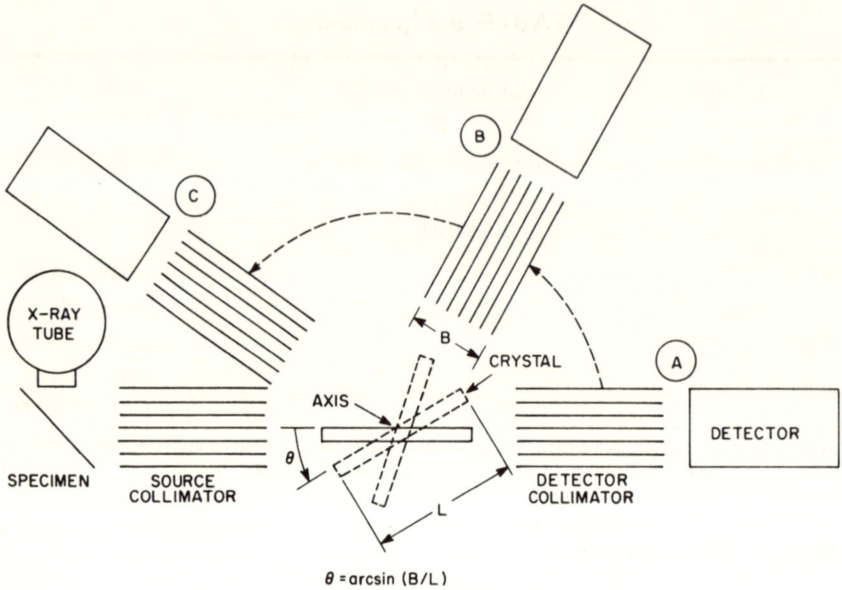

Fig. 5.4. Mechanical limitations on the 2θ range of the analyzer crystal. At the 2θ angle represented by position *B*, a crystal of length *L* just intercepts the entire usefu secondary beam. At smaller 2θ angles (position *A*), some secondary rays bypass the crystal. At a certain large 2θ angle (position *C*), the detector arm strikes the x-ray tube or specimen compartment.

These are the conditions for the spectrometer in position *B* in Figure 5.4. As 2θ is made progressively smaller, the crystal approaches parallelism with the Soller foils, and an increasing portion of the secondary beam passes by the crystal with resultant loss of diffracted intensity. When 2θ becomes very small, an increasing portion of this passed radiation enters the detector directly, until at 0° (position *A*), the entire secondary beam passes the crystal and enters the detector. Typically, *B* is ~0.5 inch and *L* is ~3 inches, giving $\theta \approx 10°$ ($2\theta \approx 20°$). The Bragg law [Equation (5.2)] shows that for a specified wavelength, the smaller the $2d$-spacing of the crystal, the larger is 2θ. If practicable, the crystal should be chosen to have $2d$ small enough to diffract the shortest wavelength to be measured at a minimum of 10–15° 2θ.

The high-2θ limit is set by contact of the detector arm with the tube head, specimen compartment, or some other structure, as shown in position *C* of Figure 5.4. However, peak breadth increases and peak intensity decreases at high 2θ angles, and, if practicable, a crystal should not be used at angles much above ~120° 2θ.

Rubber or plastic helium jackets may be difficult to use at very high and very low 2θ angles (see above), and, if practicable, a crystal should be chosen to permit use of intermediate 2θ angles.

A practical useful 2θ range for the crystal is 10–145° 2θ, and Table 5.1 gives the wavelengths and $K\alpha$ and $L\alpha_1$ lines corresponding to these limits.

Crystals good for the long-wavelength region are not also good for the short-wavelength region. Such crystals usually give low diffracted intensity at short wavelengths, and, because of their large $2d$, they diffract the short wavelengths in the low-2θ region, where background is higher and the resolution lower.

5.4.2.2. *Diffracted Intensity*

High diffracted intensity is desirable to give high statistical precision in relatively short counting time, to give low minimum detectable analyte concentration, and to permit the use of rapid 2θ scanning rates. Figure 5.5

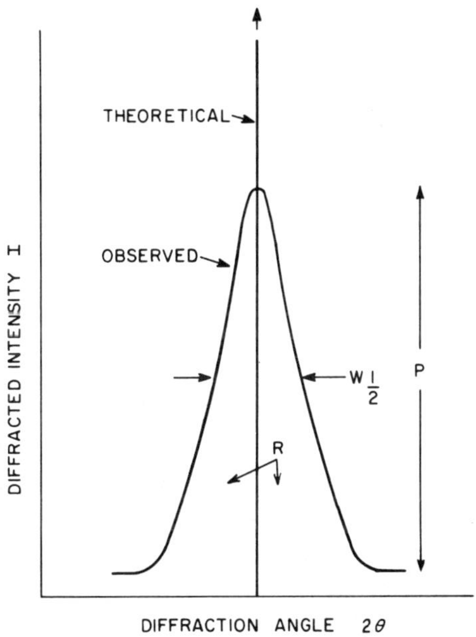

Fig. 5.5. Profile of the diffracted peak of a monochromatic x-ray beam, showing crystal parameters. [L. S. Birks, "Electron-Probe Microanalysis," Interscience Publishers, N. Y. (1963); courtesy of the author and publisher.] P: peak intensity; R: integrated intensity (area under the curve); $W_{1/2}$: rocking angle.

shows the substantially Gaussian profile of a diffracted peak and indicates peak P and integrated R intensities. These parameters may vary widely even among crystals of the same material cut on the same crystallographic plane. Diffracted intensity is affected by many factors, principally the following:

1. *Crystallographic perfection* (see below). A truly perfect crystal gives low diffracted intensity because of primary extinction. A certain amount of mosaic misorientation is desirable.

2. *Natural "reflectivity."* The coefficient of reflection for the first order depends in part on the structure factor, which, in turn, depends on atomic number.

3. *Crystallographic plane.* The reflectivity is different for different crystallographic planes in the same crystal.

4. *Order.* The second order is usually only 5–20% as intense as the first order, and is wholly or nearly absent in some crystals, such as Si(111), Ge(111), and CaF_2(111). Mica(002) has strong odd orders, weak even orders. A crystal may be selected purposely to eliminate an even-order spectral-line interference.

5. *Wavelength.* Diffracted intensity varies with wavelength, as is illustrated by the following diffracted intensity ratios for topaz and LiF: Cu $K\alpha$ (1.54 Å), 0.2; Mo $K\alpha$ (0.71 Å), 0.3; Ba $K\alpha$ (0.4 Å), 0.4.

6. *Surface.* Etched surfaces are more efficient for short wavelengths, polished surfaces for long wavelengths; the long wavelengths are absorbed by a rough surface, especially at low angles.

7. *Treatment.* By means of suitable treatment, the diffracted intensity from a crystal can be increased 2–20 times without prohibitive increase in peak breadth. Birks notes that whereas a twofold increase in primary intensity would be difficult and expensive to obtain, a twofold increase in diffracted intensity is relatively easy to obtain and is just as beneficial. The treatments (*B39*, *W8*) include polishing; etching; dry abrasion; wet abrasion with 15–20-μm Carborundum on glass; heating; heating to 80–300°C, then quenching in water; curving and flattening at ~300°C; and stroking with the tip of a cold ultrasonic soldering iron (*P12*). The treatment to be used depends on the nature of the crystal. Of course, there is the risk of ruining a crystal, and these techniques must be applied judiciously. Birks and Seal (*B39*) report that the diffracting power of LiF for Mo $K\alpha$ is increased four-, seven-, and tenfold by abrading, quenching, and bending, respectively; the diffracting power for Cu $K\alpha$ is increased threefold by all three techniques.

5.4.2.3. *Resolution*

High resolution is desirable to minimize spectral interference. The resolution of a crystal is its ability to distinguish or recognize as separate two spectral lines of nearly the same wavelength. Resolution is the combined effect of the dispersion—the separation of the two lines in 2θ—and the divergence—the breadth of the diffracted lines in 2θ.

Dispersion increases as $2d$ decreases, as indicated by differentiation of the Bragg equation:

$$\frac{d\theta}{d\lambda} = \frac{n}{2d \cos \theta} \tag{5.5}$$

As $2d$ decreases, $d\theta/d\lambda$ increases. For a specified wavelength, best dispersion occurs at a large θ angle with a crystal having $2d$ only slightly greater than the wavelength. Resolution is poorest at small θ angles. Dispersion is shown as a function of $2d$ in Figure 5.6 and in Table 5.3.

An increase in $\Delta\theta/\Delta\lambda$ does not necessarily result in good resolution if the diffracted peaks are too broad and have broad tailings. The line breadth is the combined effect of the natural breadth of the line and the rocking curve of the crystal. The greater the line breadth, the lower is the intensity at the peak and the more difficult is the resolution of the line from adjacent lines.

Natural line width is of the order 10^{-3}–10^{-4} Å for short-wavelength lines, but increases and becomes asymmetrical at long wavelengths, due to the increased effect of valence electrons.

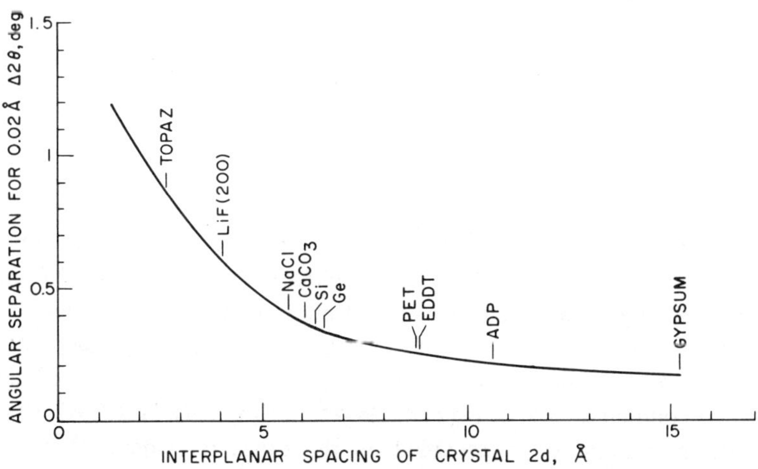

Fig. 5.6. Dispersion of common analyzer crystals as a function of interplanar spacing (2*d*).

TABLE 5.3

Crystal Dispersion as a Function of 2*d*
Dispersion of Mn Kα (2.103 Å) and Cr Kβ (2.085 Å), $\Delta\lambda$ 0.018 Å

Crystal	2*d*, Å	$\Delta 2\theta$, deg
Topaz	2.712	1.21
LiF	4.028	0.61
NaCl	5.641	0.40
EDDT	8.808	0.24
ADP	10.640	0.21

The rocking curve is shown in Figure 5.5, together with a line free of broadening. The rocking curve is a measure of the angular range over which the diffracted monochromatic wavelength is spread due to crystal imperfection. It is expressed numerically as the width in degrees 2θ of the diffracted peak at half-maximum—full width at half-maximum (FWHM), or half-width ($W_{1/2}$). The rocking curve is approximately Gaussian and is indicative of the degree of crystal imperfection. It is the combined effect of mosaic misalignment, imperfections, strain, surface texture, etc.

A crystal consists of small blocks ~500 Å on a side arranged nearly perfectly, but slightly misoriented with respect to one another by up to a few minutes of arc. Thus, there are regions in the crystal that diffract at angles slightly displaced from the Bragg angle. The greater the mosaic misorientation, the greater is the breadth of the diffracted peak. A truly perfect crystal would give extremely narrow diffracted lines, but very low intensity, because of primary extinction, and would be a very poor analyzer crystal.

The rocking angles of most common analyzer crystals are in the range 0.2–0.4° 2θ, but may be much less; for example, the rocking curve of quartz is ~0.05° 2θ. A crystal may be chosen to have a small 2*d* and/or a small rocking angle to improve resolution, minimize spectral interference, and permit accurate measurement of wavelength. Figure 5.7 shows 2θ scans of a crowded spectral region made with several combinations of coarse and fine collimators and LiF (2*d* 4.028 Å) and topaz (2*d* 2.712 Å) crystals. The improved resolution provided by the topaz is evident. Quartz and calcite crystals are often used for very accurate measurement of wavelength because of their narrow rocking curves.

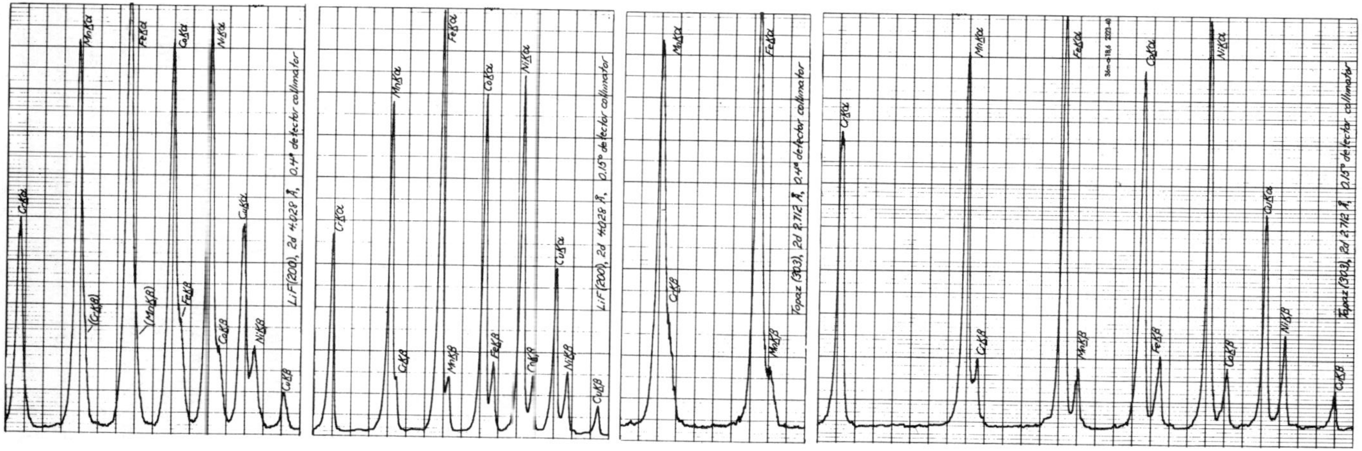

LiF(200), 2d 4.03 Å
0.4° COLL.

LiF(200)
0.15° COLL.

TOPAZ(303)
2d 2.71 Å
0.4° COLL.

TOPAZ(303)
0.15° COLL.

Fig. 5.7. Effect of $2d$-spacing of the analyzer crystal and of collimation on dispersion and resolution. The specimen was a mixture of oxides of Cr, Mn, Fe, Co, Ni, and Cu.

5.4.2.4. *Peak-to-Background Ratio; Crystal Emission*

The crystal scatters incident secondary x-rays, and the elements of which it is composed are excited by the incident x-rays to emit their own characteristic spectra. This scattered and emitted radiation emanates in all directions, contributes to the continuous background, reduces the peak-to-background ratios of the measured analyte lines, and may cause interference if the pulse-height analyzer is used. When the spectrometer is used in air path, *K* lines of all crystal elements up to chlorine are absorbed to insignificance, and *K* lines of potassium and calcium are reduced to very low intensity. However, when helium or vacuum path is used, this radiation reaches the detector, higher background results, and extraneous pulse-height distributions occur which may interfere directly with those of the analyte(s) if pulse-height analysis is used.

Organic crystals, such as EDDT and PET, are free from such emission. For example, NaCl, PET, and EDDT have diffracted intensities for S $K\alpha$ in ratio 4.7:2:1, respectively. In the determination of sulfur in hydrocarbons, the crystal receives, in addition to the spectra of sulfur and other impurities, only the scattered primary radiation, which would excite Cl $K\alpha$ in an NaCl crystal only inefficiently. In this case, the NaCl crystal would be chosen. However, in the determination of sulfur in iron, the iron *K* radiation would strongly excite Cl $K\alpha$, substantially increase the background, and create a pulse distribution close to that of S $K\alpha$. In this case, PET or EDDT would be preferable.

5.4.2.5. *Thermal Expansion*

A change in crystal $2d$-spacing resulting from thermal expansion or contraction may cause displacements of the measured analyte peak in 2θ, with consequent error in its measured wavelength or intensity (*L7*). The angular displacement is given by

$$-\frac{\Delta\theta}{\Delta t} = \alpha \tan\theta = \left(\frac{\Delta d}{d}\right)\left(\frac{1}{\Delta t}\right)\tan\theta \qquad (5.6)$$

where α is thermal coefficient of linear expansion (°C^{-1}) in the direction perpendicular to a specified set of planes (*hkl*); d is the interplanar spacing of those planes (Å); t is temperature (°C); and θ is diffraction angle.

Figure 5.8 shows $-\Delta 2\theta/°\text{C}$ as a function of 2θ for several common crystals. Table 5.4 gives the change in intensity due to thermal expansion of the crystal as a function of 2θ, collimator divergence $W_{1/2}$, and temperature

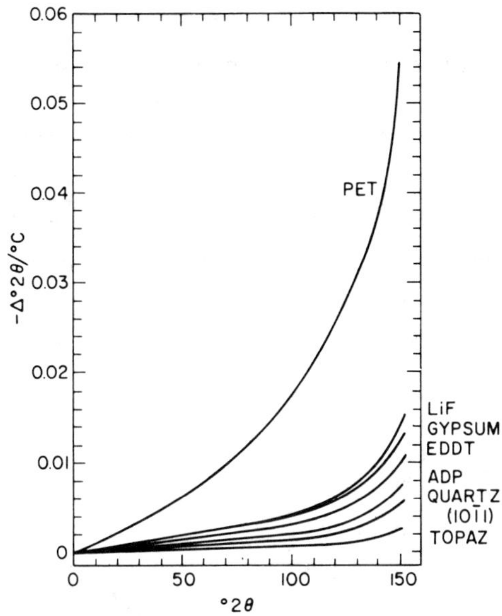

Fig. 5.8. Curves of temperature coefficient of linear expansion for some common crystals. [R. Jenkins and J. L. de Vries, "Practical X-Ray Spectrometry," Springer-Verlag N. Y. (1967); courtesy of the authors and publisher.]

for topaz, LiF, and ADP crystals. It is evident that the error increases as α, θ, and temperature increase, and as collimator divergence decreases.

Some modern instruments provide constant-temperature chambers for the crystal maintained at $\pm 0.5°C$. These accessories are most useful for precise, high-resolution work, such as measurement of line profiles, absorption-edge fine structure, and chemical effects.

5.4.2.6. *Miscellaneous Features*

Crystals should be stable in air and mechanically strong. Most of them are, but there are exceptions. For example, gypsum may effloresce in low humidity and in vacuum, and PET is very soft and fragile—about like paraffin. Some crystals, such as PET, tend to deteriorate with time and on prolonged exposure to x-rays. All crystals should be handled carefully and only by the edges. They should be stored in individual dust-free boxes when not in place on the instrument. Great care must be taken not to mar the diffracting surface and to keep it clean.

Crystals are available in several forms. Many instruments use un-

TABLE 5.4

Change in Intensity as a Function of Bragg Angle, Collimation, and Temperature

Crystal	α, °C^{-1}	2θ, deg 2θ	$W_{1/2}$, deg 2θ	ΔI, % ΔT, °C → 1	2	3	4	5
Topaz	7.5×10^{-6}	30	0.15	0.00	0.00	0.02	0.03	0.05
		60	0.15	0.00	0.03	0.08	0.15	0.24
		90	0.15	0.01	0.08	0.23	0.44	0.73
		120	0.15	0.03	0.24	0.68	1.32	2.18
		120	0.35	0.01	0.05	0.13	0.24	0.40
		120	0.55	0.00	0.02	0.05	0.10	0.16
		120	0.75	0.00	0.01	0.03	0.05	0.09
ADP	17×10^{-6}	30	0.15	0.00	0.03	0.08	0.17	0.27
		60	0.15	0.02	0.14	0.39	0.76	1.26
		90	0.15	0.05	0.42	1.17	2.27	3.73
		120	0.15	0.14	1.26	3.46	6.66	10.80
		120	0.35	0.03	0.23	0.64	1.26	2.07
		120	0.55	0.01	0.09	0.26	0.51	0.84
		120	0.75	0.01	0.05	0.14	0.28	0.46
LiF	34×10^{-6}	30	0.15	0.01	0.12	0.34	0.66	1.09
		60	0.15	0.06	0.56	1.55	3.02	4.94
		90	0.15	0.19	1.67	4.58	8.78	14.1
		120	0.15	0.56	4.94	13.1	24.1	36.6
		120	0.35	0.10	0.93	2.55	4.93	8.03
		120	0.55	0.04	0.38	1.04	2.03	3.33
		120	0.75	0.02	0.20	0.56	1.10	1.81

[a] From F. S. Lee and W. J. Campbell (*L7*).

mounted flat crystals in the form of 3 × 1 × 1/8-inch bars. Other instruments, particularly those with crystal changers, use mounted crystals of about the same length and width, but ~1 mm thick cemented in accurately machined metal holders, which are different for different instruments.

Crystals usually have one finished side that is clearly intended to be the diffracting surface. The other side is usually rougher, labeled, or cemented to a mount. When a crystal is first acquired, spectral lines should be recorded in the low-, medium-, and high-2θ regions with the crystal in both its end-for-end orientations to compare performance with respect to intensity and peak breadth. If either orientation appears to be superior, the good end should be indicated on the back. However, crystals are now of excellent quality, and it is unlikely that any significant difference will be found. The crystal surface usually lies within 5 min of arc of the nominal crystallographic plane.

In manual single-channel instruments, whenever the crystal or collimator is changed, it is usually necessary to adjust the crystal so that the spectral lines occur at or near their nominal 2θ angles. An adjustment is provided for this purpose on the crystal stage. When several analyte lines are to be measured, it may not be convenient—or even possible!—to adjust the instrument so that all lines are diffracted at their correct 2θ angles. In such cases, the instrument is adjusted for a line of intermediate wavelength, and the actual position of each peak is established prior to beginning the analysis. The simplest way to do this is to turn the 2θ control slowly through the peak while intensity is observed on the ratemeter. For greater accuracy, each peak may be recorded at a very slow scanning speed, or intensity may be scaled at intervals of 0.1–0.2° 2θ through the peak. In automatic instruments and manual instruments provided with crystal changers, the crystals are of the mounted type and are aligned when installed on the changer. Thereafter, the alignment requires only occasional checking.

Several curved crystals are available commercially for Applied Research Laboratories spectrometers. Several companies will, on special order, supply any specified crystal curved, or curved and ground, to any specified dimensions and radii—any crystal, that is, that can be so treated. Some crystals can be curved, but not ground. Some can be cleaved thin enough for use in spectrometers applying transmission or variable-radius modes. Some crystals can be curved plastically and retain their curvature; others curve elastically and must be held in curvature. Fixed-radius reflecting crystals, even plastically curved ones, are always mounted on a concave metal holder of the required radius. Many workers prefer to prepare their own curved crystals, and instructions for curving and grinding crystals have been published [pp. 127–31 in (*6*); pp. 185–94 in (*7*); (*B6*, *B39*, *P6*)]. Curved-crystal dispersion arrangements and complete curved-crystal spectrometers are described in Sections 5.5.2 and 5.6, respectively.

The analyzer crystal may be the source of extraneous spectral lines. The phenomenon is discussed in Section 13.3.2 and illustrated in Figure 13.6.

5.4.3. Other Dispersion Devices

For dispersion of very long wavelengths, analyzer crystals having very long interplanar spacing are required. Thus far, no such crystals having suitable properties have been developed, with the possible exception of octadecyl hydrogen maleate (OHM), which has $2d$ 63.5 Å. The next three sections describe other devices for dispersion of x-rays of long and ultralong wavelength.

5.4.3.1. *Gratings and Specular Reflectors*

Total specular reflectors (Section 2.4.1) (*F14*) and ruled optical gratings (Section 2.4.2) (*F13*, *S7*) have been applied to x-ray dispersion in the long- and ultralong-wavelength region. However, these devices require specially built, highly precise instruments because the critical angle for reflection and the incident and diffraction angles for gratings are very small. Moreover, the diffracted intensity for gratings is about an order lower than that from multilayer soap films (see below). Certainly, neither device can be used simply to replace the analyzer crystal on a conventional x-ray spectrometer.

5.4.3.2. *Multilayer Metal Films*

Dinklage and Frericks (*D9*) prepared a multilayer analyzer consisting of 140 layers of lead 2.5 Å thick separated by magnesium layers 26.6 Å thick, by alternate vacuum evaporation of the two metals on a glass plate. The "*d*-spacing" of this "crystal" was 26.6 Å, so that $2d$ was ~53 Å. The analyzer was very satisfactory when first prepared, but its performance was seriously impaired by diffusion, which reduced the diffracted intensity at the peak by half in two days at room temperature.

5.4.3.3. *Multilayer Soap Films*

Mesomorphic pseudocrystals consisting of multiple smectic layers of Langmuir–Blodgett (*B47*) heavy-metal soap films (*E6*, *E7*, *H22*) have been very successful for dispersion of ultralong-wavelength x-rays. The interplanar spacing and diffracted intensity can be "tailored" by selection of the length of the aliphatic acid molecule and the heavy-metal atom, respectively. The films can be prepared in any reasonable area and can be curved. High-quality pseudocrystals of several different soaps are available commercially (Table 5.1). A description of the pseudocrystals and their preparation is given below in terms of lead stearate as an example.

A solution of stearic acid (octadecanoic acid, $C_{17}H_{35}COOH$) in benzene is added dropwise to the surface of a dilute water solution of a lead salt. The benzene spreads over the water surface and evaporates, leaving a monolayer of lead stearate molecules (Figure 5.9) with their hydrophilic ends toward the water and their hydrophobic (hydrocarbon) ends upward.

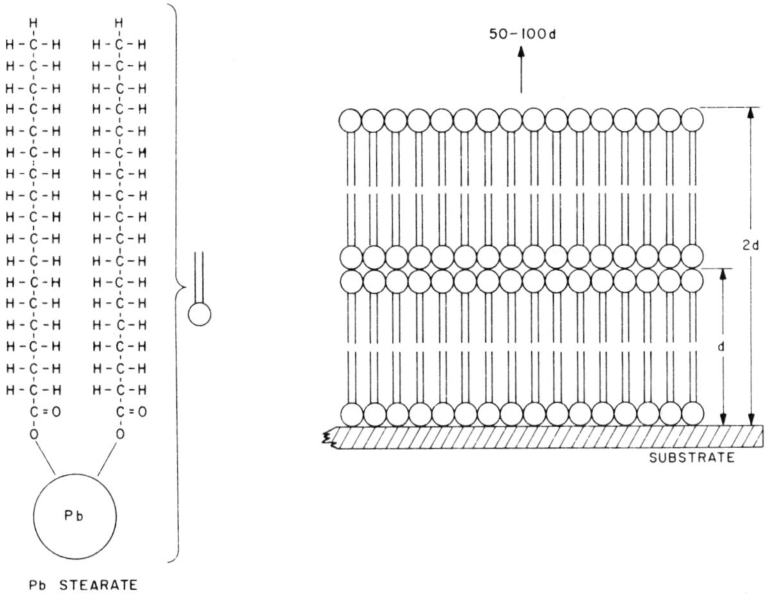

Fig. 5.9. Lead stearate molecule and the first four layers (first two repetition intervals *d*, first 2*d* interval) of a Langmuir–Blodgett, multilayer, smectic, lead stearate pseudocrystal.

The method of preparation of the multilayers is shown in Figure 5.10. A clean hydrophobic plate (glass or mica) is immersed in the solution at near normal incidence. The lead stearate does not adhere to the clean plate on immersion, but on withdrawal (pass 1), a monolayer of lead stearate adheres with the lead ends toward the plate, as shown. On reimmersion (pass 2), a second layer coheres to the first, but with opposite orientation. By successive immersion and withdrawal of the plate, a series of alternately oriented monolayers is built up, the first four of which are shown in Figure 5.9. If the substrate is first made hydrophobic, by application of a thin coat of stearic acid, for example, the film does adhere on the first downward pass, with the hydrocarbon ends toward the substrate and the lead atoms outward. On withdrawal, a second layer coheres with lead to lead, etc. The resulting multilayers are otherwise identical.

The stearate chain is ~25 Å long, so the *d*-spacing (repeat interval, Pb to Pb) is ~50 Å, giving $2d \sim 100$ Å. The analyzer pseudocrystals usually consist of 100–300 monolayers. Similar pseudocrystals having *d* 30–60 Å have been prepared from the 6- to 12-carbon saturated aliphatic acids. Table 5.1 gives data for several of these analyzers.

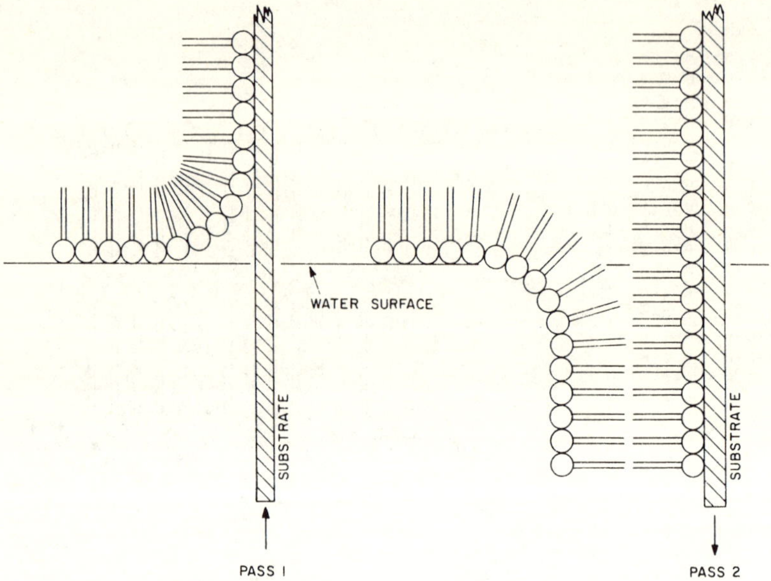

Fig. 5.10. Preparation of multilayer, smectic, heavy-element, soap-film pseudocrystals, showing the first two passes.

5.4.3.4. *Pyrolytic Graphite*

Polycrystalline analyzers of highly oriented pyrolytic graphite have been evaluated (*C19*). The analyzers are prepared by pyrolytic decomposition of hydrocarbon vapors on heated substrates, followed by annealing at high temperature (>3000°C) to effect crystallite growth and a high degree of preferred orientation. The (002) planes (*d* 3.35 Å) orient parallel to the surface. The crystallite size is ~1000 Å parallel to (002), ~400 Å perpendicular to (002). The analyzers are very rugged and can be prepared in any form, including cylindrically or toroidally curved. However, the diffracted intensity and resolution are not as good as for good conventional analyzer crystals of comparable 2*d*-spacing.

5.5. BASIC CRYSTAL-DISPERSION ARRANGEMENTS

Crystal-dispersion arrangements may be broadly classified as flat-crystal (nonfocusing) or curved crystal (focusing). Actually, curved crystals do not focus x-rays in the true optical sense, but diffract them in such a way that they converge to a line or point.

Most x-ray spectrometers are single-channel instruments in the sense that they have only one crystal–collimator–detector system and thus can measure only one wavelength at a time. However, the advantage of measuring more than one wavelength at a time was soon realized. Carson (*C22*) added a parallel flat-crystal–collimator–detector system to the Philips wide-range spectrogoniometer to permit simultaneous measurement of plate-metal and substrate lines. Adler and Axelrod built a two-channel curved-crystal instrument (*A9*) for simultaneous measurement of analyte and internal-standard lines, and later a four-channel instrument (*A23*) for simultaneous measurement of two analyte–internal standard line pairs (Rb/Sr and Cs/I $K\alpha$ line ratios). Birks and Brooks (*B35*) built a three-channel curved-crystal instrument for simultaneous measurement of chromium, nickel, and molybdenum lines from steel. Modern commercial multichannel instruments are described in Chapter 9.

5.5.1. Flat-Crystal Dispersion Arrangements

5.5.1.1. *Bragg and Soller*

The Bragg and Soller flat-crystal dispersion systems are described in Section 5.2.1 and shown in Figure 5.1. Another feature of these arrangements may now be considered.

Consider the spectrometer shown in Figure 5.11 with the crystal in

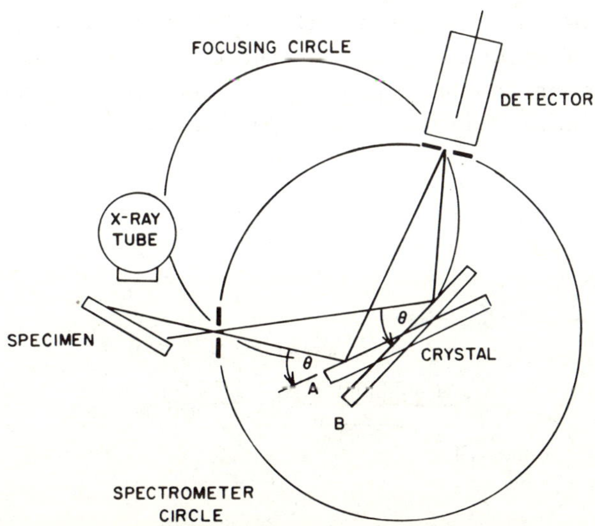

Fig. 5.11. Bragg–Brentano parafocusing.

position *A*. For every wavelength present, a cone of divergent rays from the x-ray point or line source intersects the crystal surface at a range of angles. Consider now the cone for a specific wavelength. One of the rays is likely to intersect the crystal at the Bragg angle and be diffracted, as shown for the lower ray in the figure. If now the crystal is rotated slightly ("rocked") to position *B*, some other ray forms the Bragg angle and is diffracted, as shown for the upper ray. Both these rays are diffracted in turn to the same point on the spectrometer circle. In fact, as the crystal is turned from *A* to *B*, a series of different rays in the cone is diffracted, all to the same point. Each wavelength present undergoes the same phenomenon at its own Bragg angle and is diffracted to a different point on the spectrometer circle, resulting in dispersion of the spectrum on the basis of wavelength. The diameter of the focusing circle is larger, the shorter is the wavelength. For large-area sources, a slit or pinhole must be placed at the source or detector for the purpose already explained in Section 5.2.1.

The phenomenon just described is not true focusing—even in the x-ray definition of the term—because (1) it results in *successive*, rather than *simultaneous*, diffraction of a number of rays to the same point or line, and (2) it does not increase instantaneous measured intensity. Brentano characterized this phenomenon and suggested a term for it; it is known as Bragg–Brentano *parafocusing* (*B53*).

5.5.1.2. *Edge-Crystal*

The edge-crystal arrangement is shown in Figure 5.12 (*B36*). For each secondary wavelength λ_1, λ_2, λ_3, ... emitted by the specimen, a certain series of rays lying in a line perpendicular to the figure and parallel to the crystal edge strikes the crystal edge at the Bragg angle and is diffracted

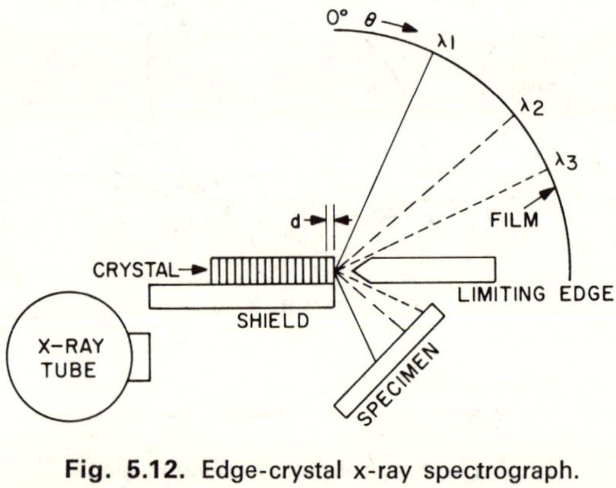

Fig. 5.12. Edge-crystal x-ray spectrograph.

to the film. Thus, a line spectrum is formed. The crystal is ~0.1 mm thick, and the breadth of each spectral line is the projection of the crystal on the film. Exposure time is ~1 h. Various crystals may be used to record all wavelengths of interest. It must be remembered that the effective $2d$-spacing is that of the edge, rather than the major face, of the crystal. The instrument is small and mounts on an x-ray diffraction unit. Having no moving parts, it is simple in structure and convenient to operate. An inexpensive version is available commercially. It may be flushed with helium or evacuated. Spectra may be recorded on the same film from series of samples and/or standards by displacing the film and changing the specimen after each exposure. Because each wavelength originates at a different place on the specimen surface (Figure 5.12), homogeneous specimens are required for semiquantitative and quantitative analysis. For accurate quantitative work, film calibration is required to convert photographic density to intensity.

5.5.1.3. *Laue*

Figure 5.13 shows the one-crystal multichannel Laue arrangement having a flat crystal placed so that several wavelengths are diffracted simultaneously, each from a different crystallographic plane (*L2*). The planes must all be of the same crystallographic zone having zone axis parallel to the crystal axis. The figure shows a quartz crystal cut parallel to the $(10\bar{1}\bar{2})$ planes reflecting five $K\alpha$ lines, each from a different plane ($hkil$) as indicated in the figure. Five detectors are arrayed about the crystal, each placed to receive one of the diffracted beams. Soller collimators are provided for the source beam and for each detector. The different planes have different "reflectivities"; the figure gives, for each $K\alpha$ line, the reflecting plane ($hkil$) and the ratio of the intensity from the indicated plane to the intensity of the same line from the $(10\bar{1}\bar{2})$ plane. Since $(10\bar{1}\bar{2})$ is the surface plane, this ratio is, in effect, the ratio of the intensities from the Laue and conventional Soller spectrometers, both using a quartz $(10\bar{1}\bar{2})$ crystal.

The Laue arrangement has many advantages. It provides many of the favorable features of a multichannel instrument with one crystal, while retaining many of the favorable features of a single-channel instrument. A spectrometer of the type described above can be built on a standard flat-crystal spectrogoniometer. The source geometry can be optimized because the same secondary beam from the specimen is used for all channels. Minimum distance between the x-ray tube target and specimen and optimum takeoff angle may be used. The cost of the collimators can be reduced by use of a fine source collimator and relatively short, wide-spaced detector collimators. Normalization of the measured intensities, if required, is

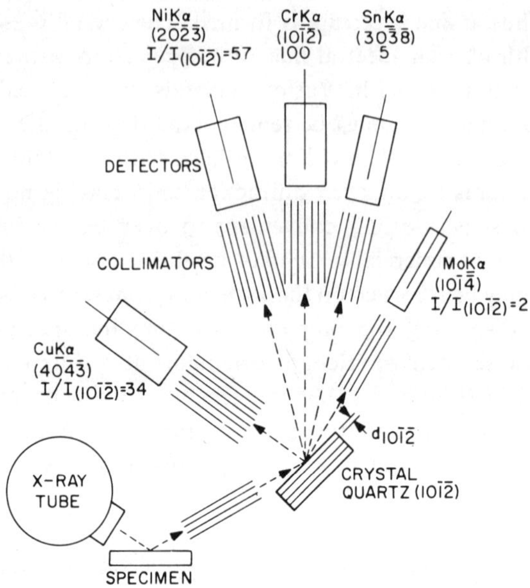

Fig. 5.13. Multichannel Laue spectrometer having five collimator–detector channels arranged to receive five different analyte lines, each from a different crystallographic plane (*hkil*) from the same quartz crystal cut parallel to the (10$\bar{1}\bar{2}$) planes. For each channel, the figure shows: (1) the measured spectral line; (2) the diffracting crystallographic plane; (3) the intensity ratio of that line measured from that plane and of the same line measured from the (10$\bar{1}\bar{2}$) plane.

easy because it involves the relative reflectivities of crystallographic planes in the same crystal. The measured intensity of a specified analyte line can be increased by measuring it with more than one channel, that is, from more than one crystallographic plane.

The unfavorable features include the following. The calculations required to establish the crystallographic planes and instrument arrangement are rather complex. The intensities from the individual planes are usually lower than those measured from the same crystal in Soller mode. High absorption losses occur in the beams diffracted from planes inclined at large angles to the crystal surface. Large-area detectors and collimators may be required to intercept the full useful beams at high 2θ angles.

The Laue spectrometer may be operated in other modes, as follows.

a. Sequential. The crystal is placed in the proper position with respect to the specimen, and the crystal θ-drive is disengaged so the crystal will remain fixed while leaving the detector free to move. The spectrum may be recorded or the individual analyte lines scaled in the usual manner. This mode of operation is readily applicable to a standard flat-crystal instrument.

b. Transmission. If thin crystals are used, the spectrometer can use transmitted diffracted beams or both transmitted and reflected beams.

c. Focusing. If a curved crystal is used, a one-crystal multichannel focusing spectrometer is feasible. Again, only crystallographic planes of the same zone having zone axis parallel to the crystal axis can be used. The instrument can be arranged so that the focal points of the diffracted beams lie on a focal circle (see below). A slit aperture is provided at the specimen.

d. Scanning. A one-crystal multichannel scanning spectrometer has been described in which simultaneous 2θ scans of the short-, intermediate-, and long-wavelength regions can be made (*S33*). The three wavelength regions are recorded from separate crystallographic planes having, respectively, small, intermediate, and long $2d$-spacings, and each channel is provided with accessories optimal for its wavelength region.

5.5.1.4. *Other Flat-Crystal Arrangements*

In the Suoinen *reflection–transmission spectrometer*, the same thin, flat LiF crystal is used to measure two wavelengths simultaneously, one by reflection, one by transmission (*S44*).

The use of pinhole or slit apertures to effect monochromatic diffraction from a flat crystal receiving a divergent beam from a point or line source is described in Section 5.2.1 and shown in Figure 5.1. The *two-crystal spectrometer* (*B41*) shown in Figure 5.14 achieves the same result without limiting apertures. In either arrangement in Figure 5.14, if the x-rays incident on crystal 1 are divergent, a range of wavelengths is diffracted, depending on the divergence between source and crystal, as shown in Figure 5.1A. However, of the wavelengths diffracted by crystal 1, only one satisfies Bragg conditions at any one setting of crystal 2. Thus, crystal 2, coupled in θ-2θ ratio with a detector, can disperse the x-rays from crystal 1 in the usual manner. In general, only a relatively small range of wavelengths is diffracted at any one setting of crystal 1, and, to diffract a new range, crystal 1 must be rotated slightly.

Several two-crystal arrangements are possible, but the three most commonly used are the following. The parallel arrangement of Figure 5.14A has the following features: (1) the two crystals are always parallel;

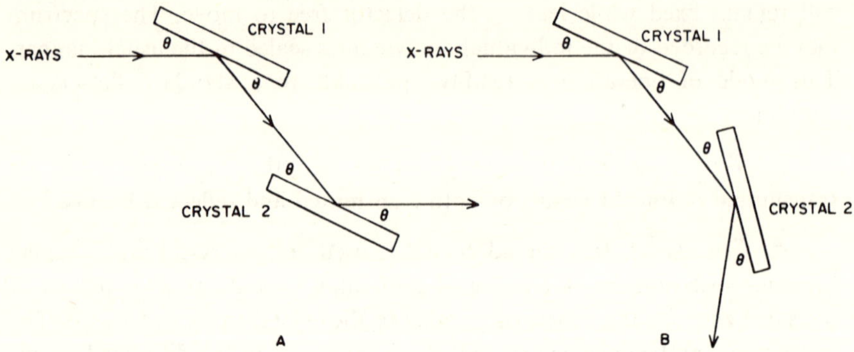

Fig. 5.14. (A) Parallel and (B) antiparallel two-crystal x-ray spectrometers.

(2) diffraction occurs from opposite faces of the two crystals; (3) both are inclined at the Bragg angle θ to the direction of the central ray of the incident beam; (4) the crystals are coupled and driven to rotate together at the same rate, but in opposite directions; and finally (5), the incident beam and the emergent, doubly diffracted beam are parallel. In a simpler arrangement, essentially similar to the one just described, crystal 1 is not coupled to crystal 2, but is mounted on an adjustable rotatable stage. At any specified setting of crystal 1, crystal 2 disperses the range of wavelengths diffracted by crystal 1. In the antiparallel arrangement of Figure 5.14B, the crystals are coupled to rotate together at the same rate and in the same direction, and diffraction occurs from the same faces of the two crystals. In all three arrangements, the detector is coupled in θ-2θ ratio with crystal 2. Similar instruments having curved crystals can also be built.

The *multicrystal spectrometer* (*D18*) achieves curved-crystal focusing with flat crystals. Fifty small, flat calcite crystals are arranged with their reflecting surfaces arrayed similarly to reflecting planes in the curved crystal of the Cauchois arrangement (Figure 5.15), except that the radius of the array is half the diameter of the focusing circle ($R/2$), rather than R as in the Cauchois instrument.

5.5.2. Curved-Crystal Dispersion Arrangements (*7, 15A*)

5.5.2.1. *General*

It has already been mentioned that curved crystals do not focus x-rays in the true optical sense, but diffract them in such a way that they converge to a line or point. Consider a flat-crystal (nonfocusing) spectrometer having single- or multiple-slit collimation and set at the Bragg angle for a specified wavelength. Of all the x-rays of that wavelength emitted by each point on the useful specimen surface, only a V-shaped, thin sheet of rays arrives at the crystal at the correct Bragg angle and is diffracted. The point of the V lies at the point on the specimen, its ends lie at the top and bottom of the crystal, and its plane is vertical, that is, parallel to the crystal axis. All other x-rays emitted from that point at the wavelength for which the crystal is set are wasted. Curvature of the crystal permits rays diverging laterally also to arrive at the crystal at the Bragg angle, so that more x-rays are "collected" and diffracted. The curvature also results in convergence of the diffracted rays toward the detector.

The basic curved-crystal arrangements are shown in Figure 5.15. In this figure and the remainder of this chapter, the following terminology and nomenclature are used.

The *focal circle*, or *focusing circle*, is analogous to the Rowland circle in optics. For x-ray focusing, it is necessary that the secondary-emissive x-ray source (real or virtual), its focused image (real or virtual), and the crystal all lie on this common circle having radius $R/2$ (see below).

The *caustic circle* has as its diameter the distance between the source (real or virtual) and its focused image (real or virtual). It has radius $R \sin \theta$.

R is the radius of cylindrical curvature (bending) of the crystal. For the doubly curved crystal, R_H and R_V are radii of curvature in the horizontal and vertical directions, respectively.

S is the optical source of the secondary characteristic x-rays, and may have one of three forms: (1) a small-area particulate or filamentary specimen on the focal circle; (2) a pinhole or slit aperture on the focal circle over an extensive specimen outside the circle; or (3) a virtual source, that is, the

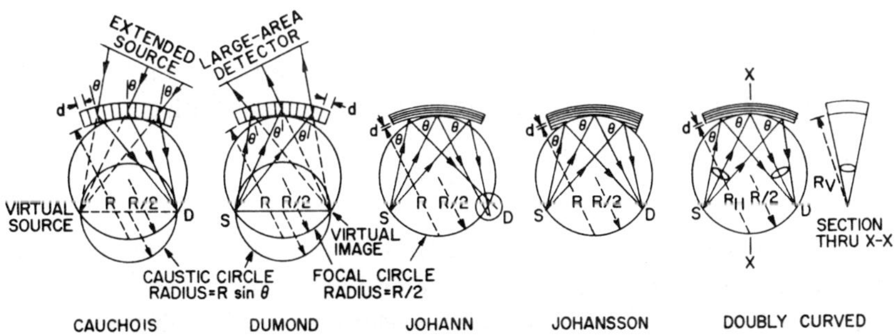

Fig. 5.15. Basic curved-crystal dispersion arrangements. Here, R is the radius of curvature of the crystal.

point on the focal circle at which the x-rays appear to originate, as shown in the Cauchois arrangement in Figure 5.15.

D is the optical focus of the source, and may have either of two forms: (1) a true line focus of the source at *S* (or point focus for the doubly curved crystal); or (2) a virtual focus, that is, the point on the focal circle at which the x-rays appear to focus, as shown in the DuMond arrangement in Figure 5.15. For true images, the detector slit (or circular aperture for the doubly curved crystal) is placed on the focal circle at *D*, behind which, outside the focal circle, lies the detector window.

θ is the Bragg angle, that is, the angle between the incident x-ray and the diffracting crystal plane in either reflection or transmission mode.

d is the interplanar spacing of the diffracting planes in either reflection or transmission mode.

L is the distance from the center of the crystal or from the crystal axis to the source *S* or detector slit *D*.

Slits at *S* and *D* and filamentary specimens at *S* are perpendicular to the plane of the figure. The relationships between the source and focal image for the various curved-crystal arrangements are as follows:

Cauchois: area to line.
DuMond: line or point to area.
Johann: line or point to line.
Johansson: line or point to line.
Doubly curved: point to point.
Von Hamos: area to area.

5.5.2.2. *Transmission*

In the Cauchois arrangement (*C25*), with a polychromatic extensive source outside the focal circle and a fixed-radius crystal, each wavelength is diffracted to a different point on the focal circle. Thus, the crystal–detector distance *L* is different for each wavelength. If *L* is to be constant, provision must be made to vary the crystal radius *R* continuously with θ:

$$R = L/\sin\theta \tag{5.7}$$

The DuMond arrangement (*D17*) is the reverse of the Cauchois arrangement. With a polychromatic line or point source on the focal circle and a fixed-radius crystal, only one wavelength is diffracted through the crystal with the source at a given point. Thus, the source–crystal distance *L* is different for each wavelength. Again, if *L* is to be constant, provision must be made to vary the crystal radius with θ. A large-area detector or two or three conventional detectors may be required to receive the entire useful transmitted divergent beam. The DuMond arrangement gives substantially greater measured intensity than the Cauchois, and it is sometimes used in electron-probe microanalyzers because of its applicability to point sources.

Cauchois has described a universal curved-crystal transmission-reflection spectrometer (*C26*) in which the secondary-emission specimen can be made to occupy the Cauchois or Johann position (see below).

Both the Cauchois and DuMond methods require crystals that can be cleaved very thin, yet remain strong enough to curve elastically; mica, aluminum, LiF, and NaCl are commonly used. A transmission crystal intercepts a greater solid angle of x-radiation than a reflecting crystal by a factor $(\cos\theta)/(\sin\theta)$. However, some of the advantage is lost by attenuation in the crystal, even at optimal thickness, which is that thickness that reduces the transmitted intensity to $1/e$ (1/2.72) of its incident value. Two or three crystals of different thickness are required for optimal performance over the full wavelength range. The crystals are usually cemented to thin spring-metal frames. Another advantage of the transmission arrangements is that measurements can be made at 2θ angles approaching 0°, where use of reflecting crystals is precluded by mechanical limitations (Section 5.4.2.1, Figure 5.4). Both transmission arrangements give excellent spectral resolution.

5.5.2.3. *Reflection*

In the Johann arrangement (*J8*), the crystal is simply cylindrically curved to radius R equal to the diameter of the focal circle. The crystal–source and crystal–detector distances L are equal,

$$L = n\lambda R/2d = R\sin\theta \tag{5.8}$$

where n, λ, and d are diffraction order, wavelength, and crystal interplanar spacing, respectively.

In semifocusing arrangements, L and R are constant at all values of θ, but for optimal performance, either R is kept constant and L varies with θ, or L is kept constant and R varies with θ. Since L increases as θ increases, it may become inconveniently large at large angles for a crystal of given radius. If L is to have convenient length at all angles, two or three crystals having different radii may be required.

The focusing is not sharp, as shown in exaggerated scale at D in Figure

5.15. The defect is most serious at very small 2θ angles, but decreases as 2θ increases and becomes insignificant at $2\theta > 90°$. The Johann arrangement is useful for crystals that are not readily ground for the Johansson arrangement (see below) and for spectrometers in which the crystal radius is varied continuously with θ.

The Johansson arrangement (*J9*) is substantially the same as the Johann except that the focusing defect is corrected by bringing the inner surface of the crystal into contact with the focal circle. This is done by cylindrically curving the crystal to radius R, then grinding it to radius $R/2$, as shown in Figure 5.15. Actually, it makes no difference whether the crystal is prepared this way or ground to R and then curved to $R/2$. The method is not applicable to crystals that cannot be ground, nor to spectrometers in which the crystal radius is varied continuously with θ. Equation (5.8) applies to the Johansson arrangement, as does the requirement for two or three crystals of different radii to keep L of convenient length at all angles.

In the doubly curved (toroidal) arrangement (*B6*, *H3*), the crystal is first cylindrically curved to radius R_H, the diameter of the focal circle. It is then curved to conform to the surface formed by rotation of the focal circle about the chord joining S and D, or to R_V. The surface of the resultant doubly curved crystal is a portion of a tore. Doubly curved crystals focus point sources to point images, but any one crystal gives point-to-point focusing for only one wavelength; at other wavelengths, the second (vertical) focus degenerates to a line. At its correct wavelength, a doubly curved crystal may give diffracted intensity up to 16 times that from a cylindrically curved crystal. Doubly curved crystals are applied mostly for x-ray diffraction, where monochromatic x-rays are commonly used.

5.5.2.4. *Von Hamos Image Spectrograph*

Von Hamos describes an *x-ray image spectrograph* or *x-ray emission micrograph* (*V9–11*) which permits photography of the area distribution of chemical elements over a flat specimen surface. The method would also be useful for mapping variations in composition and/or thickness of thin films and platings on flat substrates. The instrumental arrangement is shown in Figure 5.16. The specimen is irradiated by the primary beam from the x-ray tube, and each chemical element in the specimen surface is excited to emit its characteristic x-ray spectrum, just as in a conventional x-ray spectrometer. For each secondary wavelength $\lambda_1, \lambda_2, \lambda_3, \ldots$ emitted by the specimen, the intensity distribution over the specimen surface corresponds to the concentration distribution of the element producing it. In thin films and

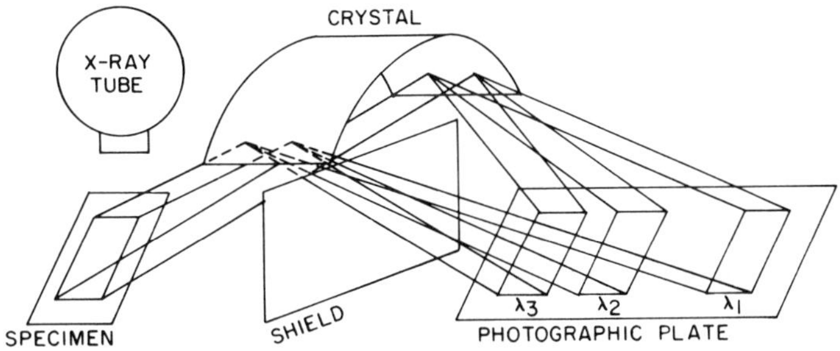

Fig. 5.16. Von Hamos image spectrograph.

platings, the intensity distribution would also correspond to the point-to-point thickness. A cylindrically curved crystal is arranged to form a series of separate true monochromatic images of the specimen surface on a plane fine-grain photographic plate. Each image shows the intensity distribution of a certain spectral line over the specimen surface—and thereby the concentration or thickness distribution of the element emitting that line. Figure 5.17A shows a conventional light photograph of the surface of a

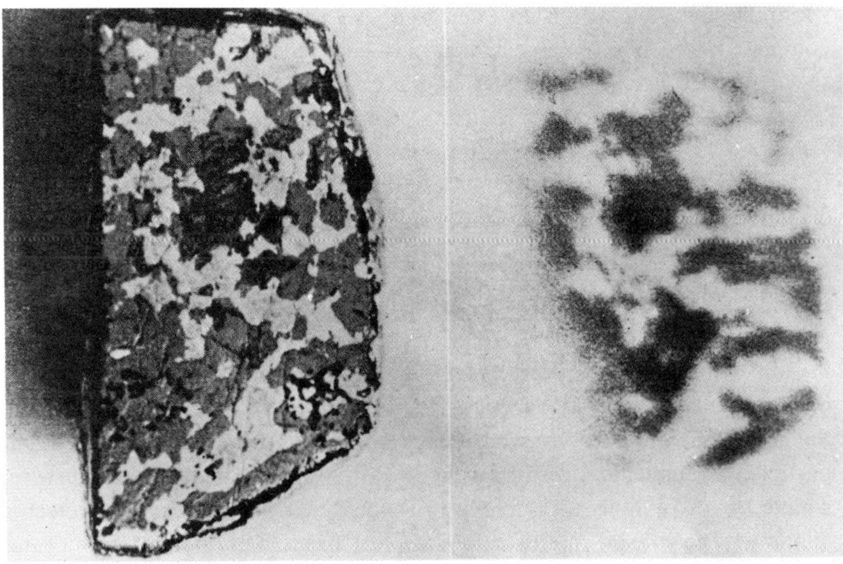

A B

Fig. 5.17. (A) Light photograph of a polished plane surface of a copper-bearing mineral and (B) Cu $K\alpha$ image of that surface. [H. Von Hamos, *Transactions of the Royal Institute of Technology* (Stockholm); courtesy of the author and Institute.]

copper-bearing mineral, and Figure 5.17B shows the corresponding Cu $K\alpha$ image spectrum.

Estimations of surface concentrations or film thickness are made by comparing the image spectrum of the sample with spectra of standards having known composition or thickness. Determinations of surface concentrations or film thickness are made by measuring photographic densities of the image spectra of samples and standards with a microdensitometer. A photographic emulsion calibration curve is prepared to permit conversion of densities to intensities.

The useful specimen area may be as much as $\sim$1 cm^2 and the spatial resolution $\leq$0.1 mm. An idea of the sensitivity may be derived from the following datum: 0.02 μg Co in 0.01 mm^2 gave a useful photographic density (0.12) in a 1-h exposure.

5.6. CURVED-CRYSTAL SPECTROMETERS

The crystal-dispersion arrangement is combined with the x-ray tube, specimen-presentation system, detector, and drive mechanism to form the complete x-ray spectrometer, which may take any of several forms. Crystal spectrometers may be classified as flat-crystal or curved-crystal. All flat-crystal instruments are nonfocusing. Curved-crystal instruments may be semifocusing or full-focusing. The term *focusing* is used in the sense given in Section 5.5.2.1. This section describes the principal types of curved-crystal spectrometers in common use.

5.6.1. Semifocusing Spectrometer

In semifocusing spectrometers, the flat crystal of the nonfocusing arrangement is simply replaced with a fixed-radius curved crystal, and the Soller collimators are removed. If the specimens have the form of small particles or filaments, no further modification may be required. Otherwise, simple slit or pinhole apertures are provided at the specimen and detector window. The crystal radius and axis are fixed. The crystal–source and crystal–detector distances are constant and should be equal, although this may not be true for a converted commercial flat-crystal instrument. Figure 5.18 can represent a semifocusing spectrometer at two wavelength settings if the crystal radius is assumed to remain the same as the diameter of the focal circle changes with θ.

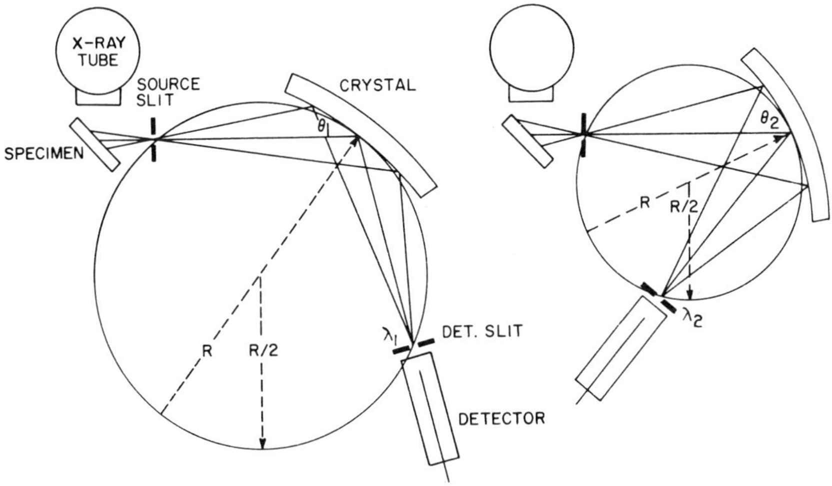

Fig. 5.18. Reflection curved-crystal x-ray spectrometer having continuously variable crystal radius, shown at two wavelength settings $\lambda_1 < \lambda_2$. The figure may also represent a semifocusing spectrometer if the crystal radius R is assumed to be the same at both settings.

The crystal and coupled detector arm rotate about the crystal axis at rates proportional to θ and 2θ, respectively, as in a conventional spectrometer. As the crystal rotates, the tangent to its center presents a series of Bragg angles to the incident secondary beam. The idea is that even though the crystal may not have the optimal radius, at a given Bragg angle, it collects and diffracts more of the x-rays having wavelength corresponding to that angle than a flat crystal does.

The fixed-radius crystal is likely to have the correct radius for the specimen–crystal–detector distances at a certain wavelength. As the instrument is set at wavelengths progressively farther from that wavelength, the sharpness of the focusing progressively deteriorates. Nevertheless, if the crystal radius is optimal near the center of the 2θ range, and if the spectrum is not crowded, one crystal can be used for the entire 2θ range. Otherwise, two or three crystals having different radii may be used to obtain good focusing in the low-, medium-, and high-2θ regions.

In full-focusing spectrometers, provision is made to vary the crystal radius or the specimen–crystal–detector distances continuously as a function of θ to maintain optimum focusing geometry over the full 2θ range. The following sections describe several full-focusing curved-crystal spectrometers.

5.6.2. Continuously Variable Crystal Radius

An instrument that maintains optimal focusing by continuous variation of the crystal radius with θ is shown in Figure 5.18 (*E9*). The specimen aperture and crystal axis are fixed. The distances L from the crystal axis to the specimen and detector apertures are equal and constant. The crystal and detector rotate about the crystal axis at rates proportional to θ and 2θ, respectively, as in a conventional spectrometer. Provision is made to vary the crystal radius R as a function of θ:

$$R = \frac{L}{n\lambda/2d} = \frac{L}{\sin \theta} \tag{5.9}$$

The takeoff angle, the angle between the central ray of the secondary beam and the specimen plane, is constant.

The arrangement is readily adaptable to standard spectrometers because of the constant L and fixed crystal axis. It is readily applicable to selected-area work because of the constant takeoff angle. One crystal serves for the full 2θ range. Because R varies with θ, this type of instrument is applicable only to Johann-type crystals (curved only).

5.6.3. Naval Research Laboratory Design

The essential features of this instrument are shown at two wavelength settings in Figure 5.19 (*B35*, *B42*). The specimen-presentation system may have any of three forms: (1) an extensive specimen lying outside the focal circle, excited by an x-ray tube in the position indicated by the solid outline, and having a pinhole or slit aperture (not shown) on the focal circle; (2) an extensive specimen lying on the focal circle, but receiving primary irradiation through an aperture from an x-ray tube in the alternate position indicated by the dotted outline; and (3) a small-area particulate or filamentary specimen lying on the focal circle and irradiated in the same way as (2) above, or even without the aperture.

The crystal radius R is constant. The diameter of the focal circle is constant, and its center is fixed and serves as the axis of the goniometer drive mechanism. Provision is made to vary the crystal–source and crystal–detector distances L in accordance with Equation (5.8). The crystal and detector are mounted on separate arms of equal length pivoted on concentric shafts at the center of the focal circle and rotated at rates proportional to θ and 2θ, respectively. Provision must also be made to keep the detector window directed toward the center of the crystal as θ varies, rather than

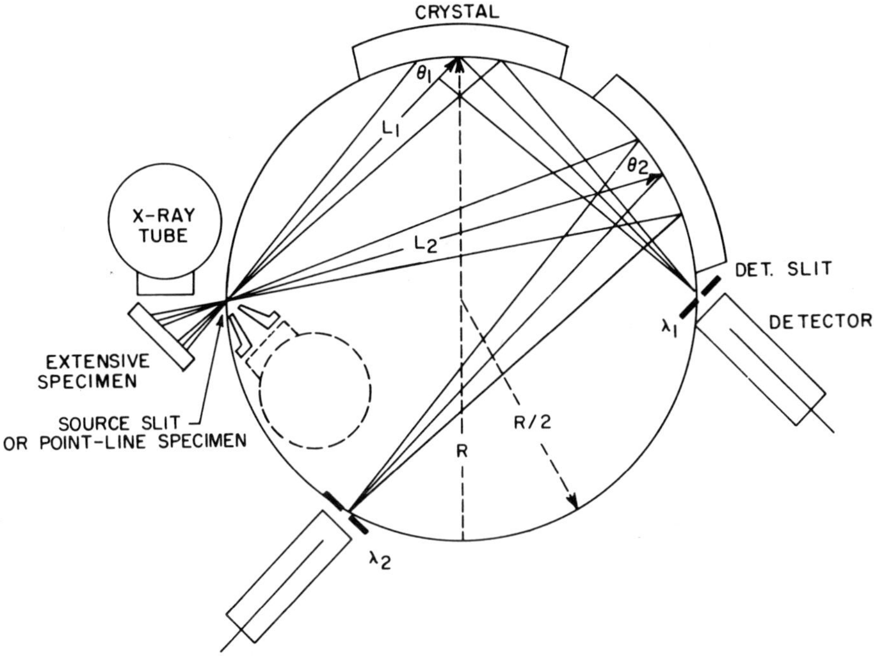

Fig. 5.19. Reflection curved-crystal x-ray spectrometer of the Naval Research Laboratory design, shown at two wavelength settings $\lambda_1 < \lambda_2$. Also shown are two alternative positions of the x-ray tube.

tangent to the focal circle. The instrument drive mechanism is very simple and compact. Since $L = R \sin \theta$, if L is to remain of convenient length as θ varies from 0 to 90°, two or three crystals having different radii must be used. Since R is constant at all values of θ, this type of instrument is applicable to both Johann and Johansson crystals.

For extensive specimens lying outside the focal circle, the takeoff angle varies with θ as shown, and homogeneous specimens are required for semiquantitative and quantitative analysis. However, the instrument is also applicable to small-area specimens and to small selected areas on extensive specimens lying on the circle and irradiated through an aperture.

5.6.4. Applied Research Laboratories Design

The essential features of this instrument are shown at three wavelength settings in Figure 5.20 (*K8*). The crystal radius is constant. Provision is made to effect the following movements as a function of θ: (1) The crystal

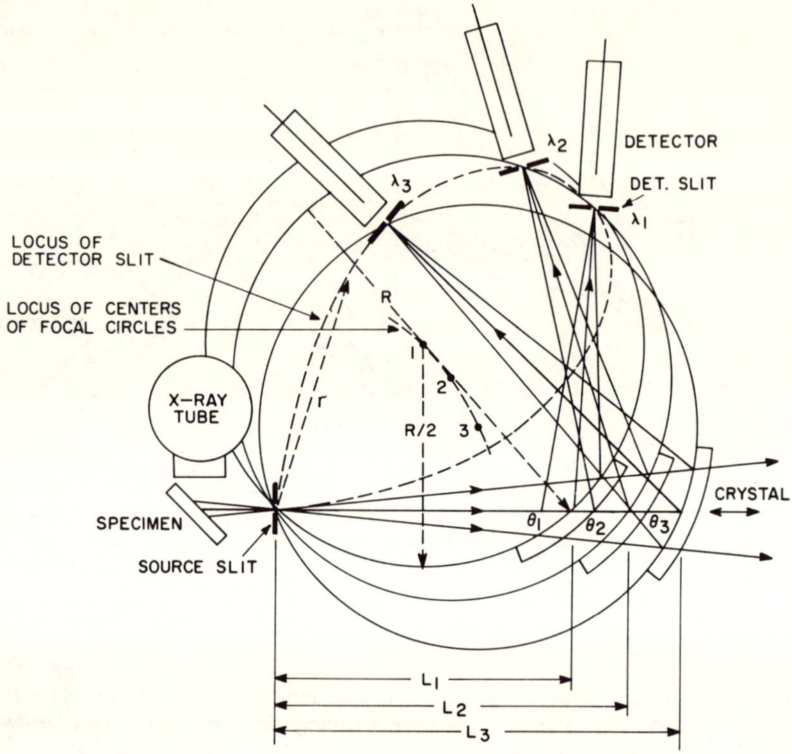

Fig. 5.20. Reflection curved-crystal x-ray spectrometer of the Applied Research Laboratories design, shown at three wavelength settings $\lambda_1 < \lambda_2 < \lambda_3$.

moves in a straight line away from the source aperture, thereby increasing the crystal–source distance in accordance with Equation (5.8). (2) The detector moves in a path constituting one loop (shown) of an analytic geometric curve known as a four-leaf rose and having equation

$$r = R \sin 2\theta \tag{5.10}$$

where r is the direct distance between the source and detector aperture, as shown in the figure. (3) Finally, the detector window is continuously directed toward the center of the crystal. As a result of these movements, the center of the focal circle moves with θ in a curved path, as shown in the figure. The takeoff angle is constant, so the instrument is applicable to selected-area work. Just as in the NRL instrument, crystals of both Johann and Johansson type are applicable, and if L is to have convenient length at all values of θ, two or three crystals of different radii are required. ARL

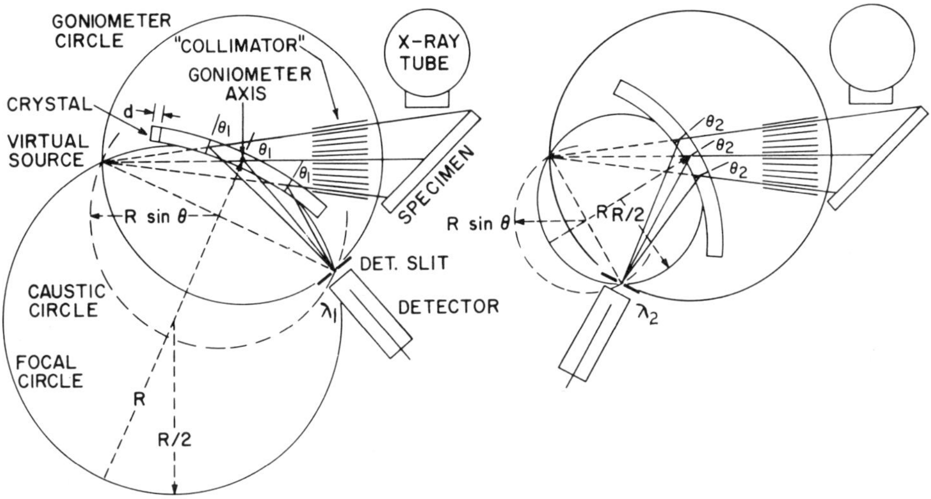

Fig. 5.21. Transmission curved-crystal x-ray spectrometer of the Cauchois type, shown at two wavelength settings $\lambda_1 > \lambda_2$.

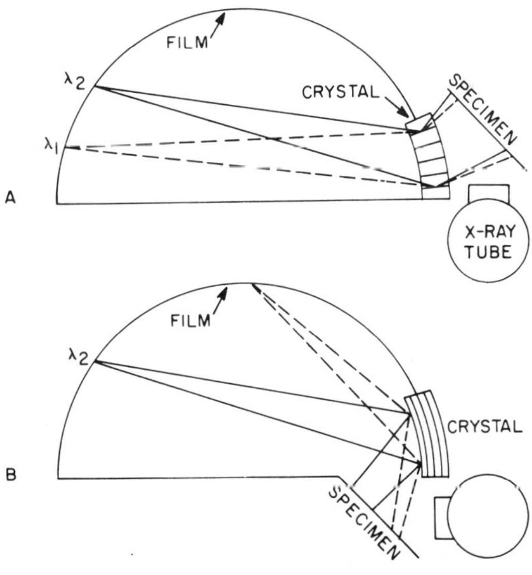

Fig. 5.22. Curved-crystal (A) transmission and (B) reflection x-ray spectrographs having photographic film registration. In both figures, $\lambda_1 < \lambda_2$.

spectrometers are used in the ARL line of automatic simultaneous instruments (Chapter 9).

5.6.5. Cauchois Spectrometer

A Cauchois spectrometer (*C25*) is shown at two wavelength settings in Figure 5.21. The crystal axis is fixed, and the distance L from the crystal axis to the detector aperture is constant. A selected wavelength from an extensive source outside the focal circle is focused to a line at the detector slit on the focal circle. If L is to be constant with θ, provision must be made to vary R in accordance with Equation (5.9). Although the crystal axis is fixed, the center of the focal circle moves with θ as shown. The convergent multiple-foil "collimator" shown excludes directly transmitted wavelengths other than that for which the instrument is set and prevents their reaching the detector. The original General Electric model XRD-3/SPG-1 spectrogoniometer was of the Cauchois type. The fixed crystal axis, constant crystal–detector distance, and specimen-presentation system permitted easy conversion to flat-crystal Soller optics.

5.6.6. Photographic Curved-Crystal Spectrographs

Figure 5.22 shows curved-crystal transmission and reflection spectrographs having photographic film registration (*P13*). It is evident that in both instruments, each recorded wavelength originates at a different area on the specimen surface, so homogeneous specimens are required for quantitative analysis.

Chapter 6

Detection

6.1. INTRODUCTION

The third phase of the x-ray spectrometric analysis (Section 3.2.5) is the detection and measurement of the characteristic secondary x-rays. Detection is the subject of this chapter; measurement is considered in Chapter 7. The detector transduces the x-rays into an electrical, luminous, or other analog that can be measured or otherwise used. Most commonly used x-ray detectors are based on the effect of x-rays on the electronic structure of matter. In *photographic emulsion*, x-rays render grains of silver halide developable—that is, labile to chemical reduction to silver—by an electronic process not fully understood. In *fluorescent screens and scintillation phosphors*, the x-rays elevate valence-band electrons to higher levels; when these electrons return to their unexcited state, the energy is reemitted in the visible and ultraviolet regions. In *gas-filled detectors*, the x-rays ionize gas atoms, producing ion–electron pairs. In *solid-state semiconductor detectors*, the x-rays elevate valence-band electrons to the conduction band, creating electron–hole pairs.

Detection may take place after crystal dispersion, or the undispersed secondary beam may be detected and electronic dispersion applied to the detector output. In the latter case, the detector itself plays an essential part in the electronic dispersion. Electronic detectors may be *continuous-current detectors*, producing a continuous output current proportional to the incident x-ray intensity, or *photon counters*, producing a discrete pulse of current for each detected photon. The detector–readout system may be *instantaneous*, indicating the instantaneous intensity directly, or *integrating* (or *accumulative*), accumulating the electrical analog of the x-rays for a selected time. Instantaneous and integrating detector–readout systems are functionally analogous to ammeters and coulombmeters or to speedometers and odometers, respectively. The respective readout components are the ratemeter and scaler. The common x-ray detectors—Geiger, proportional, and scintillation—may be operated in either mode by use with a ratemeter or scaler.

In analytical x-ray spectrometry, photographic film is sometimes used, but the detector is usually an electronic device—a gas-filled Geiger or proportional counter or a scintillation counter—in which the x-ray photons are converted to pulses of electric current. These detectors are discussed in Sections 6.2–6.4. Some less commonly used detectors are described in Section 6.5. Readout components are considered in Chapter 7. The subject of x-ray detection is reviewed in (*15D*, *K12*, *P3*, *P5*).

6.2. GAS-FILLED DETECTORS

6.2.1. Structure

6.2.1.1. *Components, Classifications*

The essential features of the gas-filled x-ray detector are shown in Figure 6.1. The active components are the cylindrical cathode, axial wire anode, and gas filling. In operation, the cathode and anode are operated at negative and positive potential, respectively, by an external dc power supply, as shown. Physically, the detector consists of a cylindrical metal envelope, which: (1) usually serves as the cathode, but, if not, houses, supports, and provides for external connection to an enclosed cylindrical cathode; (2) supports the insulator, which, in turn, supports and provides for external connection to the thin (0.010-inch or less) wire anode; (3) retains the gas filling and excludes air; (4) provides windows for entrance and, sometimes, exit of the x-rays; and (5) may provide an inlet and outlet for continuous flow of detector gas.

Gas-filled detectors are classified in terms of at least six criteria: (1) by mode of operation—ionization chamber, proportional counter, or Geiger counter; (2) as sealed or gas-flow (*H19*) counters; (3) by window position—end or side; (4) by window thickness or material; (5) by kind of gas filling; and (6) by certain special features. The mode of operation is

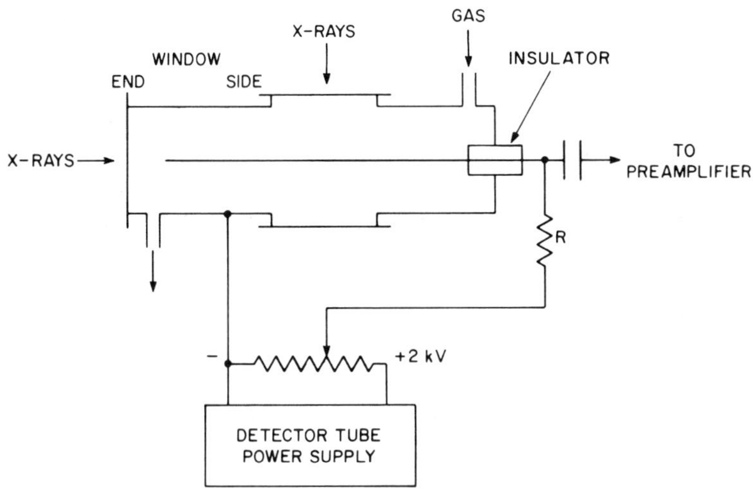

Fig. 6.1. Structure of the gas-filled x-ray detector. The tube is shown to have both end and side windows for illustration only.

determined essentially by the operating potential and is discussed in detail in Section 6.2.2. The other criteria are considered in the next two sections.

6.2.1.2. *Windows*

Sealed detectors have sealed-on gas-tight beryllium, mica, or Lindemann-glass windows that retain the gas filling and exclude air more or less permanently. Gas-flow detectors have thin ($\leq$0.00025-inch) windows of organic polymer film to permit measurement of x-rays of long wavelength. Such windows cannot be permanently sealed and are not impervious to air and the detector gas. Consequently, the gas must be passed through the detector continuously while it is in use. A flow meter, 0–0.5 ft^3/h or so, must be provided in the input gas line. Because pulse height may change substantially with detector-gas pressure, the outlet should discharge into a manostat to ensure constant pressure in the detector with fluctuating ambient barometric pressure. Some ionization-chamber detectors cannot be classified as either the sealed or gas-flow type. The thin window simply isolates a volume of air at atmospheric pressure.

The detector may have a single window in the end or side, or two diametrically opposite side windows, as shown in Figure 6.1. However, the detector never has both end and side windows, as shown in the figure for illustration only. The two-window arrangement serves two purposes. Unabsorbed x-rays that might excite secondary x-ray or photoelectron emission from the back wall of the detector are allowed to emerge from the tube. When a thin-window flow counter is used to detect long-wavelength x-rays, it may be backed by a second detector more suitable for short-wavelength x-rays. The disadvantage of the side window is that the x-ray path in the active detector volume is shorter, and a gas filling having higher absorption coefficient is required for adequate absorption at short wavelengths.

When the detector is to be operated as a proportional counter (*L19*) and its output is to be subjected to pulse-height analysis, uniformity of the electric field within the active detector volume is essential to: (1) improve the proportionality of the pulse height to the photon energy; (2) reduce the width of the pulse-height distribution; and (3) reduce the slope of the plateau. (All these features are described below.) Two structural features are essential to the realization of this uniformity: use of the side-window configuration and provision of a conductive coating on the window. The side-window configuration has a more uniform electric field because the active volume does not include the dead space (end effect) and distorted electric field (clamping effect) at the end of the anode wire. A two-window detector also permits emergence of unabsorbed x-rays that might otherwise excite secondary x-ray and photoelectron emission from the rear wall. The end and clamping effects may also be compensated by external electronic circuits (*C37*, *R6*). If the detector is to be used with pulse-height analysis, the inside surface of the window should be vacuum-coated with ~200 Å of aluminum, beryllium, or carbon to give a more uniform electric field than is possible with an insulating window. When coated thin-film windows are mounted, the coated side is identified by use of an ohmmeter. Of course, the coating, especially aluminum, reduces the window transmission at the long wavelengths. If pulse-height analysis is not to be used, end and side, and coated and uncoated windows perform about equally satisfactorily.

Sealed windows are usually of 0.5–1-mm beryllium, 12–15-μm mica, or Lindemann glass. Materials for flow-counter windows include the following: 0.00012- or 0.00025-inch (3- or 6-μm) Mylar, 0.00005–0.00025-inch (1–6-μm) polypropylene, 0.001-inch polyethylene stretched as thin as possible, and 100–500-Å Formvar. Windows 1–6 μm thick are regarded as "thin," those 100–1000 Å thick as "ultrathin."

Ultrathin collodion (nitrocellulose) films are prepared by placing a drop of a 2–5% solution of collodion in amyl or butyl acetate on a water surface, allowing it to spread and evaporate, and scooping up the film on a wire hoop. Much stronger Formvar films are prepared by dipping an extremely clean microscope slide in a 2–5% solution of Formvar in ethylene dichloride or dioxane, allowing the solution to drain and evaporate, scoring the film around the edge of the slide, then immersing the slide gently and obliquely into water; the film floats off and is scooped up as before.

Mylar (polyethylene terephthalate) contains carbon, hydrogen, and oxygen and has higher x-ray absorption coefficients than polyethylene and polypropylene, which are hydrocarbons. Consequently, for a given thickness, Mylar has lower transmission, or, for a given transmission, polyethylene and polypropylene may be thicker and therefore less fragile. Figure 6.2 shows transmission as a function of wavelength for several window materials. When a flow counter is to be backed by another detector for short wavelengths, the front window must be thin enough to have adequate transmission for the longest wavelength to be measured. However, the back window may be much thicker.

Since the gas-flow detector discharges into the atmosphere—either directly or through a manostat—the gas filling is at atmospheric pressure. Thus, the windows of flow counters used with helium path—also at atmospheric pressure—experience no pressure differential if care is taken not to use excessive helium flush rates. However, windows of flow counters used with vacuum path are subjected to an outward pressure differential of ~1 atm and must be supported. Nickel or nylon grids having a high proportion of open area are commonly used, but do reduce the window transmission.

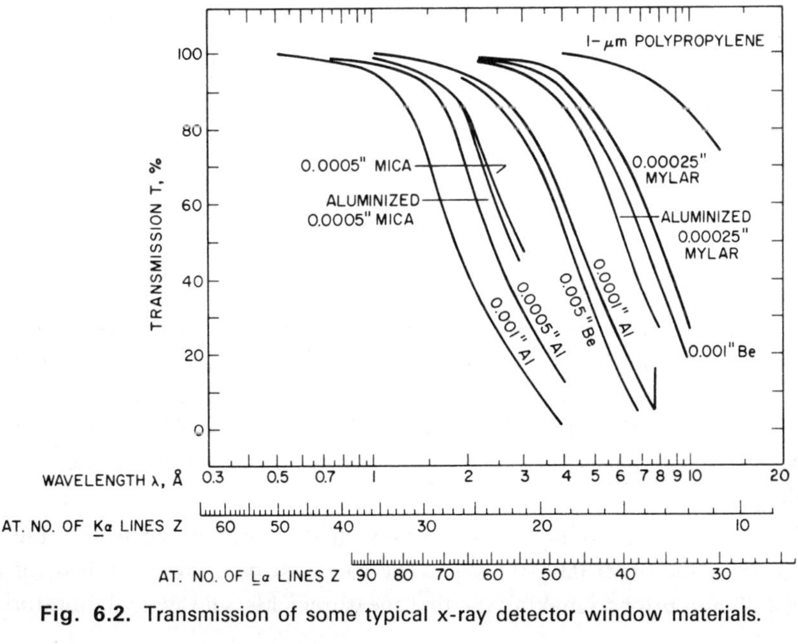

Fig. 6.2. Transmission of some typical x-ray detector window materials.

One commercial instrument is arranged so that the ends of the detector collimator foils are in contact with the window, thereby supporting it without the interposition of additional material.

6.2.1.3. *Gas Fillings*

The function of the gas filling is described in detail in Sections 6.2.2 and 6.2.3. Let it suffice here to state that the gas converts the energy of the x-ray photons to electric charge and contributes to the quenching process. (Section 6.2.2.4). The two functions may be performed by the same gas, but more often a mixture of two gases is used, usually a Group 0 gas—helium, neon, argon, krypton, or xenon—for detection, and an organic vapor or halogen for quenching. In the remainder of this chapter, the terms *detector gas* and *quench gas* are used to differentiate these two functions. Only those photons that are absorbed in the gas are "counted," so it is necessary for the detector gas to have absorption coefficient sufficiently high to absorb the incident x-rays in the path length of the active volume. In general, this means that the shorter the wavelength to be measured, the higher must be the atomic number, or effective atomic number, of the detector gas.

Sealed detectors usually have fillings of argon, krypton, or xenon at pressure 10–60 mm-Hg together with a halogen or organic quench gas. Probably the most common flow-counter gas in "P10" gas (90% argon + 10% methane), although many other fillings have been used, especially for the ultralong-wavelength region (*F8*). These gases include the following: 98% argon + 2% carbon dioxide; 25% argon + 75% methane; neon–methane mixtures; 95% helium + 5% methane; 96–97% helium + 4–3% butane or isobutane; and pure methane, propane, and butane. The helium and neon gases and the pure hydrocarbons are especially advantageous for the ultralong-wavelength region and for the *K* lines of elements 9–17 (fluorine through chlorine). They absorb the long-wavelength analyte lines strongly, but are highly transparent to the short-wavelength background and higher-order lines of high-*Z* elements. They give narrow pulse-height distributions and do not produce escape peaks (*J7*). Methane and propane also act, to a certain extent, as their own quench gas, but require higher operating potentials than the Group 0 gases—up to ~3 kV. Figures 6.12 and 9.1 show the useful ranges of detectors having several different gas fillings.

The optimal gas flow rate varies with the particular detector, but is usually of the order 0.05–0.5 ft^3/h. The gas must be allowed to flow for at least 30 min prior to making the first measurements, and many laboratories never interrupt the flow at all. The cathode cylinder, anode wire, and conductive window coating must be chemically inert to the gas filling; in general, this is a matter for concern only in the presence of halogen quench gases.

6.2.2. Operation

In this section, the operation of the gas detector is considered in general terms. Section 6.2.3 gives a more detailed discussion of the operation of the proportional counter.

6.2.2.1. *Phenomena in the Detector Gas Volume*

The basic initial process in a gas-filled x-ray detector is *gas ionization* or *ion–electron pair production* (not to be confused with electron–positron pair production, Section 2.1.2). The incident x-ray photon ionizes gas atoms along its path to form ion–electron pairs, each consisting of a positively charged gas atom (ion) and an electron. The process is photoionization, and the electrons are photoelectrons. With each ion pair produced, the x-ray photon loses energy about equivalent to the ionization potential of the detector gas, and its energy decreases and it becomes more prone to absorption as it progresses. Consequently, the density of ion pairs increases toward the end of the x-ray path and is very dense at the end. This process occurs whenever x-rays pass through any gas—in or out of an x-ray detector and with or without an electric field. This ionization involves only the outermost electrons in the detector-gas atoms and is not to be confused with expulsion of K and L electrons with resultant emission of characteristic x-ray spectral lines. Such K and L excitation of the gas atoms can and does occur, but for simplicity, the process is disregarded here and discussed in Section 6.2.3.

Gas ionization is the basic process of the gas detector in that it initiates the other phenomena that occur in the gas volume and provides the basis for proportionality of output pulse height and incident photon energy in proportional counters.

The electrons formed by the incident photon may undergo any of several processes. On production of an ion pair, rapid *recombination* occurs unless the electrons are removed from the proximity of the positive ions. This separation is effected by application of an electric potential between the cathode and anode, as shown in Figure 6.1, causing the electrons to undergo *migration* toward the positive anode wire. If the applied potential is high enough, the electrons undergo *acceleration* and may attain such velocities that they undergo *collision* with gas atoms along their path, causing ionization and additional *ion–pair production.* The additional electrons so formed may produce still more ion pairs, the electrons from which produce still more, and so on. Thus, a single electron may initiate a shower of hundreds to billions of electrons, a phenomenon known as *Townsend avalanching.* The electrons may also excite *ultraviolet emission* from the gas atoms, and, on finally striking the anode, they may cause *secondary-electron emission* and *x-ray emission* in the form of a long-wavelength continuous spectrum.

The heavier positive ions have mobility about 1000 times slower than the electrons. They undergo slow *migration* toward the cathode, eventually undergoing *recombination* or *neutralization*—that is, return to the uncharged state—by capturing a free electron or receiving an electron from a quench-gas molecule or from the negative cathode. The process of neutralization may result in *ultraviolet emission* from the detector-gas atom. The process of collision with the cathode wall may result in *electron emission by ion bombardment.* The ultraviolet radiation permeates the entire gas volume, and, by *photoionization* of gas atoms, initiates new *avalanching.* The electrons expelled from the cathode wall also initiate *avalanching.*

Quenching, another process that occurs, at least in part, in the gas, is discussed in Section 6.2.2.4.

6.2.2.2. *Proportionality in Gas Detectors*

If the energy of the incident x-ray photon is expended in producing ion pairs, and if each ion pair requires the same amount of energy, it follows that the number of ion pairs should be proportional to the photon energy. In fact, the number of ion pairs should be simply the x-ray photon energy divided by the ionization potential of the detector gas. This would probably be true if only singly charged ions were formed. However, more than one electron may be expelled from a gas atom, and the ionization potential increases for each successive ionization; for example, the first five ionization potentials for xenon are 12, 21, 32, 46, and 76 eV, respectively. Thus, each detector gas has an *effective* ionization potential somewhat larger than its first ionization potential, the relative difference between the two increasing as atomic number of the gas increases. The *average* number of ion pairs produced is given by the x-ray photon energy divided by the effective ionization potential. Table 6.1 shows, for each of the five stable Group 0 gases, the first ionization potential, the effective ionization potential, and the average number of ion pairs produced by Cu $K\alpha$ (1.54 Å, 8040 eV) and Mo $K\alpha$ (0.71 Å, 17,440 eV) photons.

It must be emphasized that even with a strictly monochromatic x-ray beam, the numbers of ion pairs produced by the individual photons have a

TABLE 6.1

Basis of Proportionality in Gas and Scintillation X-Ray Detectors

Detector gas	First ionization potential, eV	Effective ionization potential,[a] eV	Average number of ion pairs produced by each x-ray photon[b]	
			Cu $K\alpha$ 1.54 Å, 8040 eV	Mo $K\alpha$ 0.71 Å, 17,440 eV
2-He	24.5	27.8	289	628
10-Ne	21.5	27.4	293	637
18-Ar	15.7	26.4	304	660
36-Kr	13.9	22.8	352	765
54-Xe	12.1	20.8	386	838
Scintillator	**Light-photon energy, eV**	**Effective light-photon energy,[a] eV**	**Average number of light photons produced by each x-ray photon[b]**	
			Cu $K\alpha$	Mo $K\alpha$
NaI(Tl)	2	50	161	349

[a] Average energy actually required to produce one ion pair or one light photon.
[b] X-ray photon energy divided by the effective ionization potential or light-photon energy.

wide Gaussian distribution having a mean value $\bar{N}$ and a deviation $\sqrt{\bar{N}}$. However, Table 6.1 shows that Mo $K\alpha$ photons, having about twice the energy of Cu $K\alpha$ photons, do indeed produce about twice as many ion pairs. Thus, the gas ionization process provides the basis for proportionality of output pulse height and incident x-ray photon energy.

6.2.2.3. *Gas Amplification; Types of Gas Detectors*

Consider an x-ray photon entering the detector shown in Figure 6.1 and passing through the active gas volume. It produces ion pairs along its path until its energy is expended. The resulting electrons and ions may undergo various processes (Section 6.2.2.1). It is the potential applied to the detector that largely determines the specific behavior of the electrons and ions, and thereby determines whether the detector functions as an ionization chamber, proportional counter, or Geiger counter. Figure 6.3 shows the *gas amplification factor*—the number of electrons collected on the anode wire for each electron produced by the x-ray photon—as a function of detec-

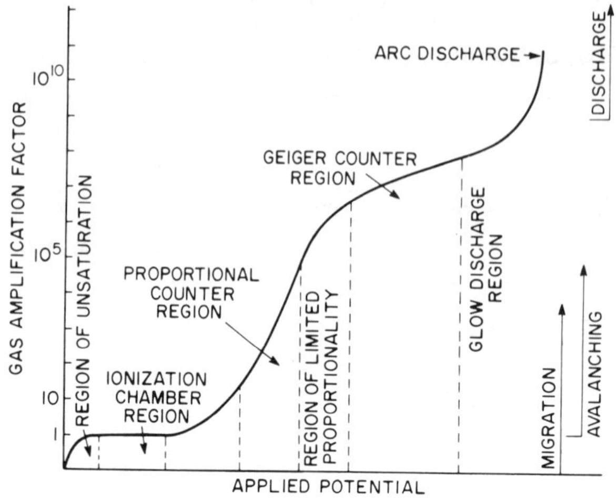

Fig. 6.3. Gas-amplification factor as a function of applied potential for the gas-filled x-ray detector.

tor potential. The curve is naturally divided by its inflections into a number of regions. Let us consider in detail the phenomena in the detector as the anode potential is increased from zero (potentiometer arm at extreme left in Figure 6.1) to progressively higher positive values.

a. Zero Applied Potential. At zero potential, the ion pairs quickly recombine, and no charge is collected on the anode. The gas amplification is zero.

b. Region of Unsaturation. The applied potential is so low that many electrons recombine with positive ions before they can reach the anode. The collected current is unsaturated, that is, fewer electrons are collected than were formed by the x-ray photon, and the gas amplification is <1. However, as the potential is increased, electron acceleration increases, and progressively more electrons escape recombination and reach the anode.

c. Ionization-Chamber Region. The potential is just sufficient to overcome recombination completely, all electrons are collected (saturation) and the gas amplification is 1 and remains constant over a substantial interval of applied potential. However, no secondary ionization occurs in the gas or at the electrodes. Ionization-chamber detectors operate in this region. In practice, ionization chambers are usually constructed with the two electrodes in the form of large-area parallel plates separated by the gas. The electrons and ions are all collected on the positive and negative plates, respectively. Ionization chambers have little application in x-ray spectrometry.

d. Intermediate Region. The potential is now high enough to initiate mild avalanching, and the gas amplification becomes >1. This region is of no value for x-ray detection.

e. Proportional-Counter Region. The potential is high enough for marked avalanching to occur, and the gas amplification is 10^2–10^5. However, this avalanching is limited in extent. It is localized in a region within a few wire diameters of the anode, and, in general, each electron produced by the x-ray photon initiates only one avalanche, and the avalanches are largely free of any interaction. The number of avalanches is substantially the same as the number of initial ionizations, and, since all electrons are collected, the total collected charge is proportional to the x-ray photon energy. The proportional region has another favorable feature. Since the electron mobilities are $\sim$1000 times faster than the ion mobilities, the electrons reach the anode and are collected before the ions move very far. As the electrons are collected, the region surrounding the wire becomes depleted of negative charge, and a positive-ion space-charge sheath rapidly froms concentric with the wire. The charge density and diameter of this sheath are very similar for each successive x-ray photon counted, and the sheath tends to limit the charge accumulated—and thereby the output pulse height—for successive photons. This effect results in greater uniformity of pulse height and narrower pulse-height distribution. The proportional counter is discussed in more detail in Section 6.2.3.

f. Region of Limited Proportionality. As potential is increased still further, avalanching becomes more generalized. The number of avalanches per unit volume increases, and they begin to interact. The same electron initiates multiple avalanches. The region in which avalanching occurs spreads outward from the wire. Proportionality deteriorates as potential is increased.

g. Geiger-Counter Region. The potential is now so high that the avalanching becomes generalized, and the other phenomena noted in Section 6.2.2.1 occur. The gas ions and atoms are excited in the gas volume and at the cathode wall to emit ultraviolet radiation. Secondary electrons and low-energy x-rays are emitted when the electrons strike the anode, and electrons are expelled by ion bombardment of the cathode. These emitted electrons initiate new avalanches. The ultraviolet radiation permeates the entire gas volume, causing photoionization, the photoelectrons from which also initiate avalanches. Thus, in the Geiger region, the initial ionizing event merely triggers the general discharge of the entire tube. All vestiges of proportionality disappear, and the output pulses are of substantially the same high amplitude—~1000 times the amplitude of pulses in the proportional region. The Geiger, or Geiger–Muller, counter operates in this region (*F16*, *G2*).

h. Glow-Discharge Region. The discharge is now so severe that it does not quench (Section 6.2.2.4) after each incident photon. The photon triggers not one output pulse, but a continuous, sustained discharge. If the tube window is transparent, the gas volume can be seen to glow. Operation in this condition for even a short time is likely to ruin the tube.

i. Arc Discharge. An arc discharge occurs between the cathode and anode, and the wire is likely to "burn out."

6.2.2.4. *Quenching*

The duration of the discharge initiated by the x-ray photon is *quenched*, that is, terminated, by the action of the quench gas, the positive ions, and the external resistor (*R* in Figure 6.1).

The quench gas is mixed in relatively low concentration with the detector gas and may be an organic vapor or a halogen. The organic gases used include methane (CH_4), ethane (CH_3CH_3), propane ($CH_3CH_2CH_3$), butane [$CH_3(CH_2)_2CH_3$], and isobutane [$(CH_3)_2CHCH_3$]; methyl chloride (CH_3Cl), methyl bromide (CH_3Br), and methyl iodide (CH_3I); and ethanol (CH_3CH_2OH). Molecules of these compounds undergo irreversible decomposition as a result of their quenching action and limit the useful life of the detector to ~10^9 counts. The halogens used include bromine (Br_2) and bromine chloride (BrCl). Molecules of these substances dissociate to halogen atoms, but recombine to restore the molecule. Detectors having halogen quench gas have substantially unlimited life (*L16*).

The quench gas functions in several ways to terminate the discharge. Quench-gas molecules neutralize some of the positive detector-gas ions by donating electrons to them. This is possible because the ionization potential of the quench gas is much lower than that of the detector gas. Thus, a slow, positive detector-gas ion is replaced by an even slower, positive quench-gas ion, which is less likely to expel electrons from the cathode wall. Moreover, the difference between the work functions of the detector and quench gases is much smaller than the difference between the work functions of the detector gas and cathode wall. Thus, no ultraviolet emission accompanies neutralization by the quench gas. The quench gas also absorbs secondary electrons and ultraviolet radiation produced by neutralization of detector-gas ions at the cathode.

The positive ions and the external resistor both tend to reduce the *effective* potential between cathode and anode to the extent that avalanching is no longer sustained. Since the positive-ion mobilities are ~1000 times slower than the electron mobilities, the electrons are all collected before the ions have migrated very far. Thus, the ions form a positive space-charge sheath around the anode wire. This sheath migrates outward toward the cathode and reduces the effective electric field potential near the anode. The external resistor (*R* in Figure 6.1) prevents the electrons arriving at the cathode from leaking off as fast as they are collected. This causes the anode to become more negative (more accurately, less positive) and reduces the effective potential across the tube.

Sometimes, an external electronic quenching circuit is provided. As already explained, during a pulse, the effective potential across the detector drops. Quenching circuits reduce the effective potential even further, hold it down until the detector is cleared of ions, then restore it to equivalence with the applied potential. Special quenching circuits for Geiger counters reduce or even reverse the applied potential soon after initiation of the discharge to reduce the number of ions reaching the cathode, and they do this so quickly that the discharge cannot spread throughout the tube.

6.2.3. Proportional Counters (*A17, L5, M35*)

6.2.3.1. *Phenomena in the Detector Gas Volume*

It has been explained that: (1) the output of a gas detector consists of a pulse of electric current for each x-ray photon absorbed; (2) the numbers of ion pairs produced by the individual photons of a strictly monochromatic x-ray beam are not the same, but have a Gaussian distribution about a mean value; and (3) in proportional and Geiger counters, these ion pairs initiate a complex gas-amplification process, resulting in an output pulse having 10^2–10^{10} times as many electrons as were produced by the x-ray photon. Consequently, it is not surprising that the amplitudes of the detector output pulses also have a Gaussian distribution. In a Geiger counter, the average pulse height is the same regardless of x-ray photon energy. In a proportional counter, the average pulse height is proportional to the photon energy.

We are now in a position to consider the operation of the proportional

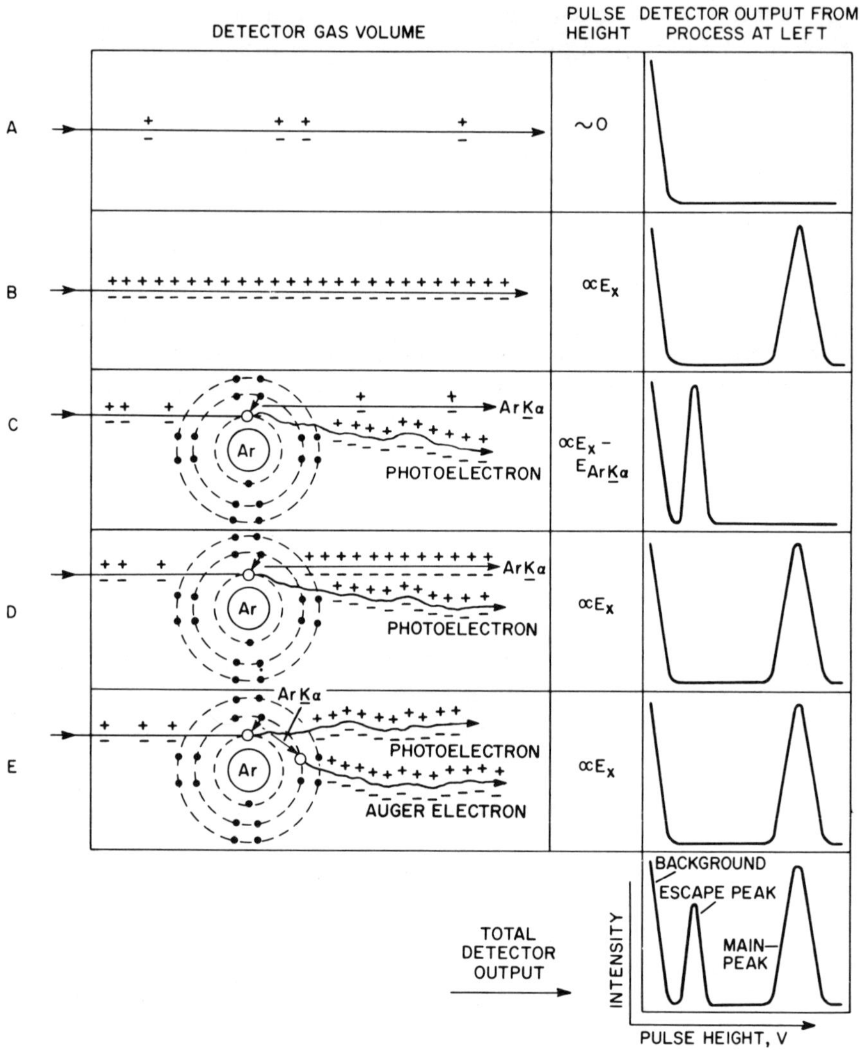

Fig. 6.4. Phenomena in the active gas volume of an argon-filled proportional counter, showing output pulse height and pulse-height distribution for each phenomenon. Here, E_X is the energy of the incident x-ray photon. A. X-ray photon largely unabsorbed. B. X-ray photon absorbed wholly by formation of Ar^+, e^- pairs. C. Ar *K* excitation followed by "escape" of Ar $K\alpha$ photon. D. Ar *K* excitation followed by absorption of Ar $K\alpha$ photon. E. Ar *K* excitation followed by Auger effect.

counter in more detail. Figure 6.4 shows five processes that an x-ray photon may undergo in the active volume of a proportional counter. The center column gives the output pulse height for each process in terms of incident x-ray photon energy E_x. The last column shows, for each process, the pulse-height distribution curve for a detector that is receiving a continuous monochromatic x-ray beam and in which only that process occurs. The "peaks" on these curves are *not* profiles of individual pulses; such profiles would be plotted on a grid of pulse height in volts *vs.* time. Nor are they to be confused with the familiar 2θ scan of intensity *vs.* 2θ angle. Rather, they show the numbers of pulses of each pulse height present in the detector output in unit time. It is also necessary to distinguish *pulse height*—the amplitude (in volts) of the individual pulse triggered by an individual x-ray photon—and *intensity*— the number of pulses per second. Pulse height for a detector is the maximum potential that appears across resistor R in Figure 6.1 during the life of that pulse. A careful study of Figure 8.3 may clarify these relationships.

Let us consider the five processes in detail. In Figure 6.4 and the following discussion, argon is assumed to be the detector gas, but the principles apply to any detector gas.

a. Transmission (Figure 6.4A). The x-ray photon may pass through the gas wholly or substantially unabsorbed, to be absorbed in the detector wall or to emerge through a rear window. Of course, no output pulse is produced. This process is more likely, the shorter is the x-ray wavelength and the lower is the absorption coefficient of the detector gas. In this way, short-wavelength x-rays pass through a two-window detector optimum for long-wavelength x-rays to a backing scintillation or sealed proportional counter.

b. Ion-Pair Production (Figure 6.4B). The x-ray photon may ionize the detector gas along its path to produce Ar^+, e^- pairs. Actually, the ion pairs are not uniformly distributed along the photon path, but become denser as the photon progressively loses energy. The output pulse height is proportional to the photon energy. If only this process occurred in a detector receiving a monochromatic beam, the output would consist of a single pulse-height distribution (peak) of average pulse height proportional to the photon energy.

It has been assumed so far that the ion-pair production process is the only process by which x-ray photons are converted to electric charge. In fact, it would be preferable if this were true. However, if the x-ray photon has wavelength shorter than that of the argon K-absorption edge, it may undergo photoelectric absorption in a single argon atom (Figure 6.4C,D,E), expelling an argon K electron and expending the remainder of its energy in imparting kinetic energy to the photoelectron. The photoelectron expends its energy in producing Ar^+, e^- pairs as shown. However, the Ar $K\alpha$ (or $K\beta$) photon resulting from the filling of the K-shell vacancy can undergo one of three processes.

c. Argon K Excitation Followed by Escape of the Ar $K\alpha$ *Photon* (Figure 6.4C). The Ar $K\alpha$ (or $K\beta$) photon may escape the active volume of the detector. This is not an unlikely process, because an element is relatively transparent to its own characteristic radiation. The output pulse height is proportional to the difference between the energies of the incident and Ar $K\alpha$ photons. If only this process occurred in a detector receiving a monochromatic beam, the output would consist of a single pulse-height distribution of correspondingly lower average pulse-height than that in Figure 6.4B. This distribution is known as an *escape peak* because it is produced when some of the energy carried into the detector by the x-ray photon "escapes" the active volume.

d. Argon K Excitation Followed by Absorption of the Ar $K\alpha$ *Photon* (Figure 6.4D). The Ar $K\alpha$ photon may expend its energy producing Ar^+, e^- pairs. To the extent that this occurs, the problems presented by the escape peak are obviated. The sum of the numbers of ion pairs formed by (1) the incident x-ray photon prior to its photoelectric absorption, (2) the photoelectron, and (3) the Ar $K\alpha$ photon equals the number formed in Figure 6.4B. The pulse height and pulse-height distribution are the same as in Figure 6.4B.

e. Argon K Excitation Followed by Auger Effect (Figure 6.4E). The Ar $K\alpha$ photon may undergo absorption within the atom of its origin (Auger effect) with resultant emission of an Auger electron (shown) and a satellite-line x-ray photon (not shown). The Auger electron and satellite photon expend their energies producing Ar^+, e^- pairs. The sum of the numbers of ion pairs formed by (1) the incident x-ray photon prior to its photoelectric absorption, (2) the photoelectron, (3) the Auger electron, and (4) the satellite-line photon equals the number formed in Figure 6.4B. The pulse height and pulse-height distribution are the same.

6.2.3.2. *Detector Output; Escape Peaks*

The content of the output of a proportional counter may be summarized as follows. For monochromatic incident x-rays having wavelength longer

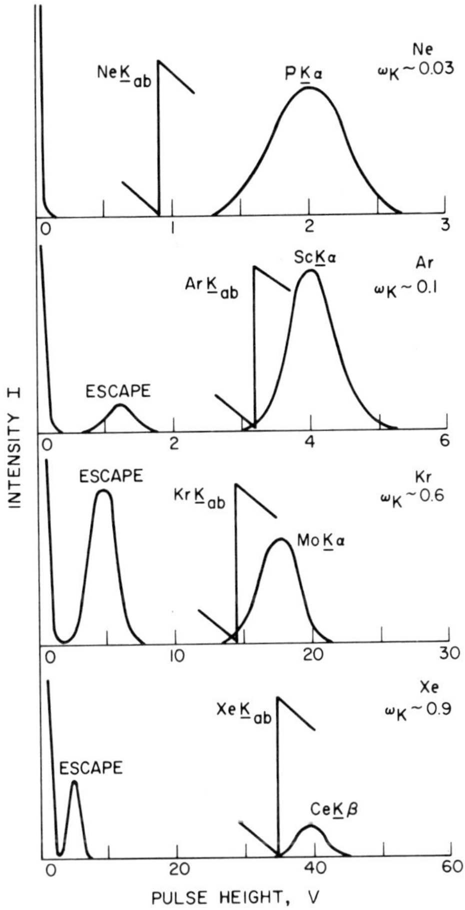

Fig. 6.5. Effect of K-fluorescent yield ω_K of the detector gas on the relative intensities of the main and escape peaks (pulse-height distributions) for detectors filled with neon, argon, krypton, and xenon, respectively. In each case, the incident x-rays have sufficient photon energy to expel K electrons from the detector gas atoms. As ω_K increases, the intensity of the escape peak relative to the main peak increases. [R. Jenkins and J. L. de Vries, "Practical X-Ray Spectrometry," Springer-Verlag N. Y. (1967); courtesy of the authors and publisher.]

than that of the absorption edge of the detector gas, the output contains a single pulse-height distribution (peak) having mean pulse height proportional to the x-ray photon energy. For monochromatic incident x-rays having wavelength shorter than that of the absorption edge of the detector gas, the output may contain two pulse-height distributions (peaks). The *main* (or *principal* or *natural*) peak has mean pulse height proportional to the x-ray photon energy. The *escape* peak has mean pulse height proportional to the difference between the photon energies of the incident x-rays and the spectral line of the detector gas. For incident x-rays containing more than one wavelength, the output contains a pulse-height distribution for each and an escape peak for each that can excite the detector gas. Since these peaks are relatively broad, they may overlap one another, and the main peak of the analyte line may be overlapped by a main or an escape peak of another element.

The intensity of the escape peak relative to the main peak is greater (1) the closer the wavelength of the x-rays lies to the short side of the absorption edge of the detector gas, and (2) the greater the fluorescent yield of the gas. This second condition is illustrated in Figure 6.5, which shows the output of a neon detector for P $K\alpha$, an argon detector for Sc $K\alpha$, a krypton detector for Mo $K\alpha$, and a xenon detector for Ce $K\beta$. Each spectral line lies close to the K-absorption edge of its detector gas and can excite its K lines very efficiently. It is evident that the intensity of the escape peak relative to the main peak increases substantially as the fluorescent yield (shown) of the detector gas increases. The escape peaks for the krypton and xenon detectors are more intense than their main peaks.

6.3. SCINTILLATION COUNTERS

6.3.1. Structure

The essential features of a scintillation counter are shown in Figure 6.6 (*H37*, *N11*, *W5*). The detector consists of a scintillation crystal (phosphor) affixed to the envelope of a multiplier phototube ("photomultiplier") just outside the photocathode. Each incident x-ray photon is converted into a burst of visible-light photons, which fall on the photocathode and emit photoelectrons, which, in turn, initiate the secondary-electron multiplication process.

6.3.1.1. *Scintillation Crystal*

The scintillation crystal (phosphor, scintillator) must have the following characteristics: (1) The absorption coefficient should be high enough to

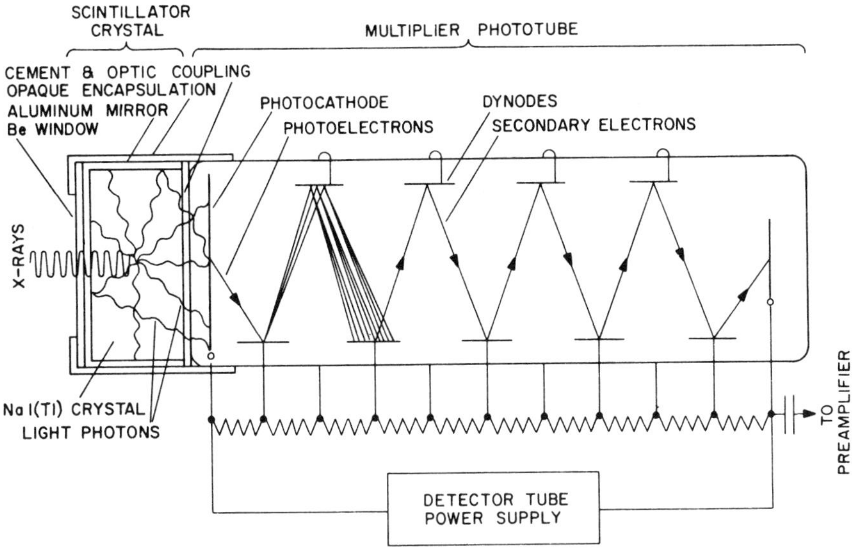

Fig. 6.6. Structure of the scintillation counter.

permit high absorption of the incident x-rays in a convenient thickness; (2) the ratio of visible-light photons to x-ray photon energy must be high; (3) the visible fluorescence spectrum should match the spectral sensitivity of the photocathode of the multiplier phototube; (4) the crystal should be optically clear and transparent to its own visible fluorescence; (5) the crystal should have index of refraction similar to that of the glass of the phototube envelope (or face plate) to ensure efficient transfer of light from the phosphor to the photocathode; and (6) finally, the rise and decay times of the scintillation should be short enough to permit high counting rates without serious coincidence loss.

Thallium-activated sodium iodide, NaI(Tl) or NaI:Tl, meets most of these requirements and is exclusively used in scintillation counters for analytical x-ray spectrometry. The heavy iodine atoms give high x-ray absorption over the entire useful spectral range of analytical x-ray spectrometry. A light photon is produced for each 50 eV of incident photon energy, and the resulting light is in the actinic blue spectral region. Sodium iodide can be grown synthetically in large, clear single crystals and cut to any required shape. The decay time is $\sim$0.25 μs. The principal disadvantages are the somewhat high index of refraction, $\sim$1.8, and, especially, the high hygroscopicity, which necessitates hermetic sealing.

The scintillator is a single-crystal disk of NaI(Tl) $\sim$2.5 cm in diameter and 2–5 mm thick. The disk is aluminized around the edge and on the side that is to be toward the x-rays to reflect scintillation light radiating away from the detector and direct it toward the phototube. The aluminized disk is sandwiched between a beryllium window ~0.2 mm thick on the x-ray side and a thin Lucite disk on the phototube side to act as a light coupler. The sandwich is then hermetically sealed and perhaps coated with opaque paint to exclude moisture and light, respectively. The packaged scintillator is affixed to the face of an end-window multiplier phototube just outside the photocathode. A layer of silicone is applied between the Lucite and phototube to improve optical coupling.

6.3.1.2. *Multiplier Phototube*

The multiplier phototube ("photomultiplier") consists of a photocathode, a series of dynodes, and a collector sealed in an evacuated glass envelope and provided with a base having prongs to permit external connection of the electrodes. The tube is coated with optically opaque black paint to exclude external light and may be enclosed in a mu-metal shield to exclude stray magnetic fields. A regulated dc potential of 600–1000 V is applied to the tube with the photocathode negative. Each successive dynode is maintained at a higher positive potential by means of a potential divider.

When illuminated by the light from the scintillation, the photocathode emits photoelectrons, which are attracted to the first dynode, where each expels two to four secondary electrons. These secondary electrons are attracted to the second dynode, where more secondary electrons are emitted, etc. Thus, a progressively increasing shower of secondary electrons cascades through the multiplier. The collector receives the secondary-electron current, which constitutes the output pulse from the scintillation counter. The gain G of the electron multiplier is

$$G = k\sigma^n \tag{6.1}$$

where k is the efficiency of collection of photoelectrons on the first dynode (~0.9); σ is the secondary emission ratio, that is, the number of secondary electrons emitted for each electron incident on the dynode; and n is the number of dynodes, usually 9–14.

6.3.2. Operation

6.3.2.1. *Proportionality in Scintillation Counters*

The production of light photons in the scintillation crystal is analogous to the production of ion–electron pairs in the gas detector. If the energy of the incident x-ray photon is expended in producing light photons, and if each light photon requires the same amount of energy, it follows that the number of light photons should be proportional to the x-ray photon energy. In fact, the number of light photons should be simply the x-ray photon energy divided by the light-photon energy (~2 eV). However, the scintillation process is much less efficient than the ion-pair process, and the *effective* light-photon energy, that is, the energy actually required to produce a light photon, is ~50 eV. Table 6.1 shows the average number of light photons produced by Cu $K\alpha$ (1.54 Å, 8040 eV) and Mo $K\alpha$ (0.71 Å, 17,440 eV) photons.

Again it must be emphasized that even with a strictly monochromatic x-ray beam, the numbers of light photons produced by individual x-ray photons have a wide Gaussian distribution having a mean value $\bar{N}$ and a deviation $\sqrt{\bar{N}}$. However, Table 6.1 shows that Mo $K\alpha$ photons, having about twice the energy of Cu $K\alpha$ photons, do indeed produce about twice as many light photons. Thus, the scintillation process provides the basis for proportionality of output pulse height and incident photon energy.

6.3.2.2. *Phenomena in the Scintillation Counter*

To continue the analogy of light-photon (scintillation) and ion–electron pair production, the x-ray photon in the scintillation crystal may undergo one of five processes directly analogous to those that occur in the gas volume of a proportional counter (Section 6.2.3.1, Figure 6.4):

1. The x-ray photon may pass through the crystal wholly or substantially unabsorbed, to be absorbed in the glass envelope or internal structure of the phototube. This process is unlikely except at very short wavelengths.
2. It may produce a burst of light photons (a scintillation), as already explained.

If the x-ray photon has wavelength shorter than that of the iodine K-absorption edge, it may undergo photoelectric absorption in a single iodine atom, expelling an iodine K electron and expending the remainder of its energy in imparting kinetic energy to the photoelectron. The photoelectron expends its energy in producing light photons. However, the I $K\alpha$ (or $K\beta$) photon resulting from the filling of the K-shell vacancy can undergo one of three processes:

3. The I $K\alpha$ (or $K\beta$) photon may escape the scintillation crystal. This is not an unlikely process, because an element is relatively transparent to its own characteristic radiation.

4. It may expend its energy producing light photons.

5. It may undergo absorption within the atom of its origin (Auger effect) with resultant emission of an Auger electron and a satellite-line x-ray photon.

The first of the foregoing processes produces substantially no light photons. The second process produces a number of photons proportional to the x-ray photon energy. In the fourth process, the sum of the photons produced by the x-ray photon prior to its photoelectric absorption, the photoelectron, and the I $K\alpha$ photon equals the number produced by the second process. In the fifth process, the sum of the photons produced by the x-ray photon, the photoelectron, the Auger electron, and the satellite-line photon equals the number produced by the second process. Thus, the second, fourth, and fifth processes produce a number of light photons proportional to the x-ray photon energy. However, in the third process, the number of light photons is proportional to the difference between the energies of the incident and I $K\alpha$ photons. This is, of course, the escape-peak phenomenon.

A certain fraction ($\sim$0.9) of the scintillation photons find their way to the photocathode. A certain fraction (0.1–0.2) of these result in emission of photoelectrons, a certain fraction ($\sim$0.9) of which fall upon the first dynode and initiate the secondary-electron multiplication process. If there are n dynodes having a secondary-emission ratio of σ, the output pulse contains $0.9 \times 0.1 \times 0.9 \times \sigma^n$ electrons for each photon in the scintillation.

Thus, the output of the scintillation counter, like that of the proportional counter, contains a pulse-height distribution (peak) for each incident x-ray wavelength and may contain an escape peak for each incident wavelength shorter than the iodine K-absorption edge. The closer the incident wavelength lies to the K-absorption edge, the greater is the intensity of the escape peak relative to the main peak. Escape peaks are not nearly as troublesome with scintillation counters as with argon- and krypton-filled proportional counters. The iodine LIII absorption edge lies at 2.7 Å and the principal iodine L lines at $\sim$3 Å or more. These wavelengths are at the long-wavelength limit of the usual working range of the scintillation counter. The iodine K-absorption edge lies at 0.37 Å. Thus, escape peaks occur only for $K\alpha$ lines of elements having atomic number 57 (lanthanum) or higher, and these elements are usually determined by measurement of their L lines with gas-flow proportional counters.

6.4. EVALUATION OF X-RAY DETECTORS

6.4.1. Detector Characteristics

This section describes some parameters that may apply to the detector or to the detector–readout system. However, modern x-ray spectrometers are usually equipped with readout components having very fast count-rate capability, and it is usually the detector that limits the performance with respect to these characteristics.

6.4.1.1. *Rise Time* (Figure 6.7).

This is the time interval, after initiation of a pulse, for the pulse to attain 90% of its full amplitude.

6.4.1.2. *Dead Time* (Figure 6.7).

This is the time interval, after initiation of a pulse, during which the detector is completely insensitive and gives no response whatever to a second incident x-ray photon. In a gas detector, dead time is determined by the time required for the positive-ion sheath to decay to the extent that some avalanching can occur again. In proportional counters, the avalanching is localized very near the anode wire, whereas, in Geiger counters, it is generalized throughout the gas volume. Thus, the positive-ion sheath decays much more rapidly in proportional counters than in Geiger counters. In a scintillation counter, dead time is determined by the time required for the scintillation to decay.

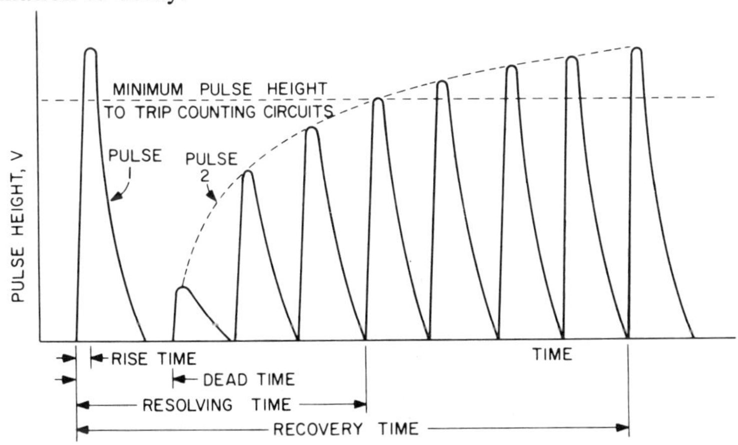

Fig. 6.7. Amplitude of a detector output pulse as a function of the time interval after initiation of the preceding pulse.

Normally, dead time is ~200 μs for Geiger counters, ~0.2 μs for proportional counters, and ~0.25 μs for scintillation counters. However, dead time is affected by the type of x-ray generator, and by the intensity and energy of the measured x-rays. Dead time is longer if the measured x-rays are excited by a full-wave generator than if by a constant-potential generator. If the second x-ray photon follows closely after expiration of the dead time for pulse 1, pulse 2 has an abnormally small amplitude and is followed by an abnormally short dead time. Finally, dead time is a function of, and in fact varies about linearly with, pulse height.

The detector responds to an x-ray photon after the expiration of the dead time, but the pulse height is abnormally low. The longer after the expiration the photon enters the detector, the more nearly normal is the pulse height.

6.4.1.3. *Resolving Time* (Figure 6.7).

This is the time interval, after initiation of a pulse, after which the detector can produce a second pulse high enough to actuate the counting circuits, or, more simply, it is the minimum time interval between pulses that can be counted. In calculation of coincidence loss (see below), resolving time is often incorrectly referred to as dead time because, in effect, the detector is "dead" during the resolving time, in the sense that it cannot count a photon.

6.4.1.4. *Recovery Time* (Figure 6.7).

This is the time interval, after initiation of a pulse, after which the detector can produce a second pulse of substantially full pulse height. To summarize, after the dead time, a second pulse can occur; after the resolving time, a second pulse can be counted; after the recovery time, a second pulse has normal amplitude.

6.4.1.5. *Linear Counting Range* (Figure 6.8).

This is the incident x-ray intensity range over which the number of countable pulses per second from the detector is directly proportional to the number of x-ray photons arriving at the detector per second. The linear ranges of Geiger, proportional, and scintillation counters are given in Table 6.3. A detector capable of very high count rate without substantial departure from linearity is known as a *fast* counter.

Figure 6.9 shows the multiple-foil method for establishing the linear

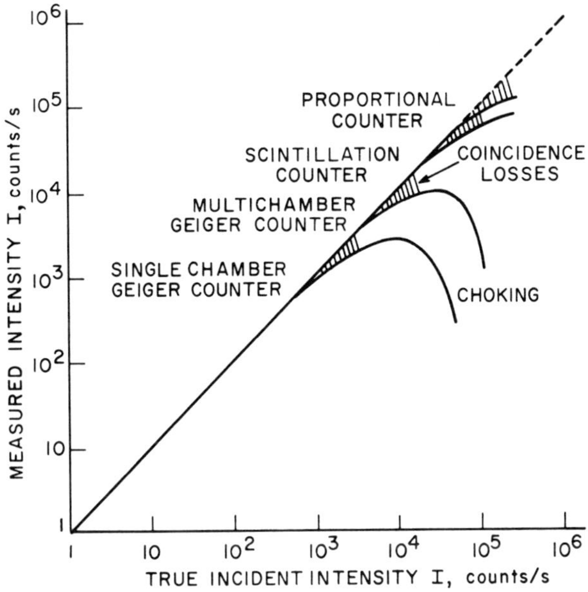

Fig. 6.8. Linear counting range, coincidence loss, and choking in commonly used x-ray detectors.

range of a detector (or detector–readout system) and for measuring the true incident intensity when it lies beyond the linear region. A stack of metal foils is placed in the x-ray path, preferably at the detector window. The foils are removed one at a time, and x-ray intensity is scaled after each removal. In the figure, the plotted data show noticeable departure from linearity at ~7000 counts/s and true incident intensity ~150,000 counts/s.

6.4.1.6. *Coincidence Loss* (Figure 6.8)

This term refers to the x-ray photons that are "lost," that is, not counted, because they arrive at the detector too soon after the preceding photon. The nonlinear portions of Figure 6.8 are the result of coincidence loss and occur at intensities so high that the rate at which x-ray photons enter the detector exceeds its resolving time. The following equation may be used to correct for coincidence loss or, with the multiple-foil method, to calculate resolving time:

$$I_{\text{true}} = I_{\text{obs}}(1 - I_{\text{obs}}T_R) \tag{6.2}$$

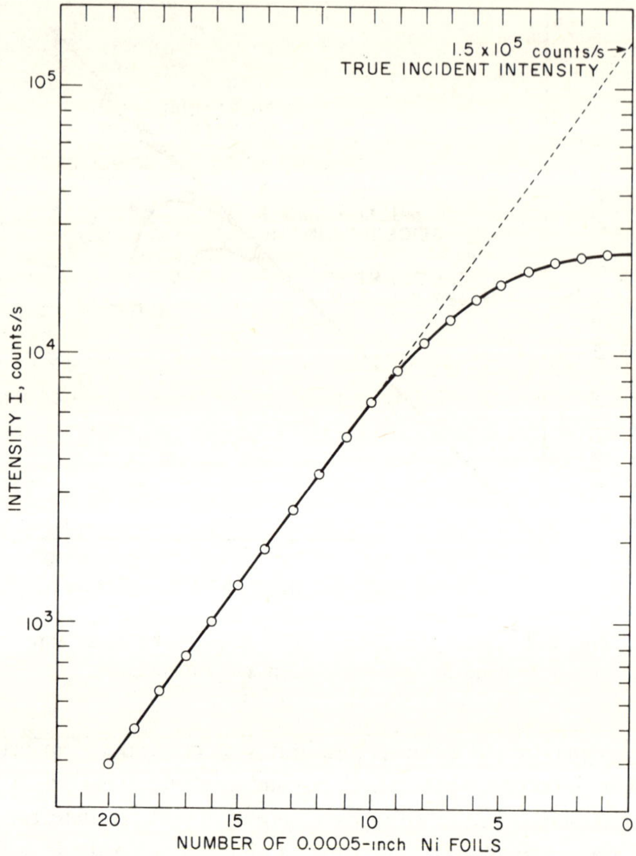

Fig. 6.9. The multiple-foil method for measurement of true intensity.

where I_{true} and I_{obs} are true and observed intensity, respectively (counts/s), and T_R is resolving time (s). This equation is applicable to intensities measured at peaks, but not to integrated intensities, because the observed intensity, and therefore the coincidence loss, varies continuously during the integration. The equation is also less satisfactory with full-wave x-ray excitation than with constant-potential.

Coincidence losses in the fixed-count method (Section 7.2.2.2) are calculated with the equation

$$I_{\text{true}} = N/(T - NT_R) \tag{6.3}$$

where N is the accumulated count, T the counting time (s), and T_R the resolving time.

Finally, suppose that a Geiger counter having a resolving time of 200 μs gives a measured intensity of 1000 counts/s. The detector does not count photons for 200 μs after the initiation of each counted pulse. Thus, at 1000 counts/s, the detector is inactive $(1000)(200 \times 10^{-6}) = 0.2$ s/s, or 20% of the time, and 20% of the incident photons are lost. The true intensity is then 1250 counts/s.

6.4.1.7. *Choking* (Figure 6.8)

Choking or blocking is the condition whereby an increase in incident intensity actually results in a decrease in measured intensity. In Figure 6.7, if pulse 2 occurs shortly after the dead time of pulse 1, it produces a pulse too small to be counted, and, in effect, extends the dead time following pulse 1. The higher the incident intensity, the more such small pulses occur, until, finally, all pulses are too small to be counted, that is, the dead time is extended indefinitely, and the counter completely chokes.

6.4.1.8. *Plateau* (Figures 6.10 and 6.11)

This is the substantially straight, horizontal portion of the intensity *vs.* potential characteristic of the detector tube. Suppose that monochromatic x-ray photons are arriving at the detector at some constant intensity, say 1000/s, and that amplifier gain and pulse-height analyzer settings (Chapter 7) are constant. If, now, the detector tube potential is increased from zero, the observed detector output intensity follows the curve shown in Figure 6.10. The measured intensity just perceptibly rises above zero at the *starting potential*, becomes substantially constant at the *threshold potential*, remains substantially constant over a range termed the *plateau*, and finally rises

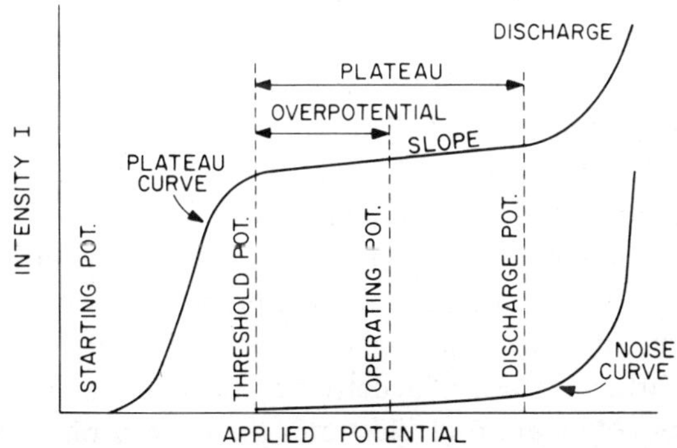

Fig. 6.10. Plateau and noise curves of an x-ray detector.

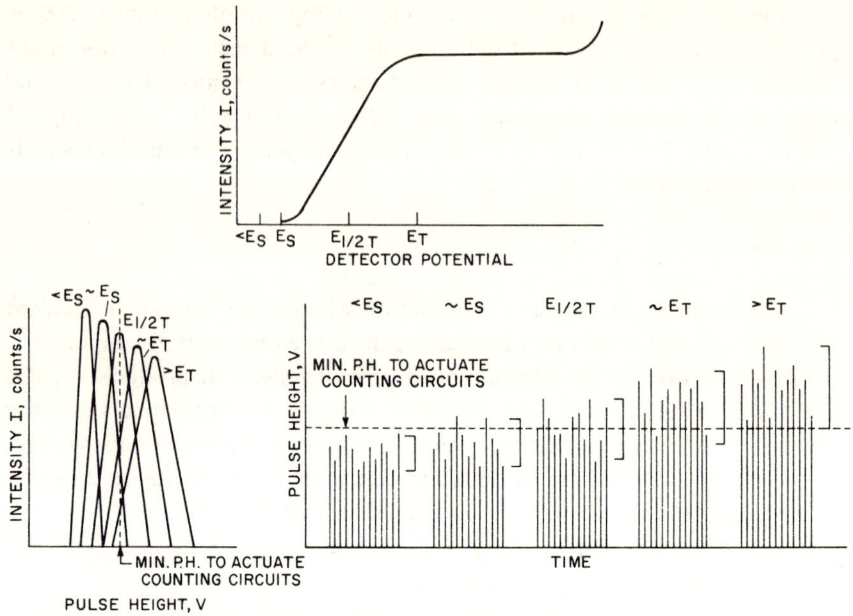

Fig. 6.11. Origin of the plateau of an x-ray detector. As detector potential increases, average pulse-height increases and progressively more—and finally all—of the pulses exceed the minimum pulse height required to actuate the counting circuits.

again at the *discharge potential.* The *operating potential* is usually set near the center of the plateau. The interval between the threshold and operating potentials is termed the *overpotential.* The plateau curve is displaced to progressively lower potential the higher the amplifier gain, the lower the discriminator setting (that is, the lower the minimum countable pulse height), and the shorter the wavelength of the incident x-rays.

Figure 6.11 shows the origin of the plateau phenomenon. It has already been explained that the output of a detector receiving a monochromatic x-ray beam is a pulse-height distribution. In Chapter 8, it is explained that the average amplitude of this distribution can be varied, among other ways, by varying the detector potential. Figure 6.11 shows that at detector potentials below the starting potential E_S, none of the pulses in the distribution exceeds the minimum pulse height required to actuate the counting circuits. As detector potential is increased, progressively more of the pulses exceed this minimum, until, at the threshold potential E_T, all pulses are counted and further increase in detector potential has no effect on the measured intensity. Figure 6.11 illustrates the plateau phenomenon in two ways, with plots of intensity *vs.* pulse height and pulse-height *vs.* time.

It is also evident in Figure 6.11 why the plateau curve shifts toward lower potentials with increased amplifier gain, decreased discriminator setting, and shorter wavelength. Increased amplifier gain increases all pulses proportionally so they exceed the minimum detectable height at lower detector potential. Decreased discriminator setting lowers this minimum detectable height. Shorter wavelength means higher photon energy and proportionally higher pulse amplitude.

6.4.1.9. *Slope* (Figure 6.10)

Slope is the inclination of the straight portion of the plateau curve expressed as the ratio of the change in measured intensity to the change in applied potential near the midpoint of the plateau. It may be expressed in counts per second per volt, or it may be expressed as per cent change in measured intensity per volt. The slope increases and the plateau shortens as the detector deteriorates with life, and these are good criteria of the condition of the detector.

6.4.1.10. *Inherent Noise and Background* (Figure 6.10)

These two, taken together, constitute the intensity measured from the detector without incident x-rays, and they are evaluated by running a plateau curve without x-rays.

6.4.1.11. *Quantum Efficiency* (Figure 6.12)

Quantum efficiency (*T4*) is the fraction, usually expressed as per cent, of incident x-ray photons that produce pulses of countable amplitude, or the number of countable output pulses from the detector divided by the number of incident x-ray photons, both in the same time interval. Figure 6.12 shows typical quantum efficiency curves for some common detectors. However, these curves are only typical, and the same kind of detector—say, a gas-flow proportional counter having 0.00025-inch Mylar window and argon–methane gas—from different manufacturers may give somewhat different curves.

6.4.1.12. *Resolution*

Resolution of a detector is given in terms of the width of its pulse-height distribution at half maximum (FWHM) relative to the average pulse height, which is the pulse height corresponding to the peak of the distribu-

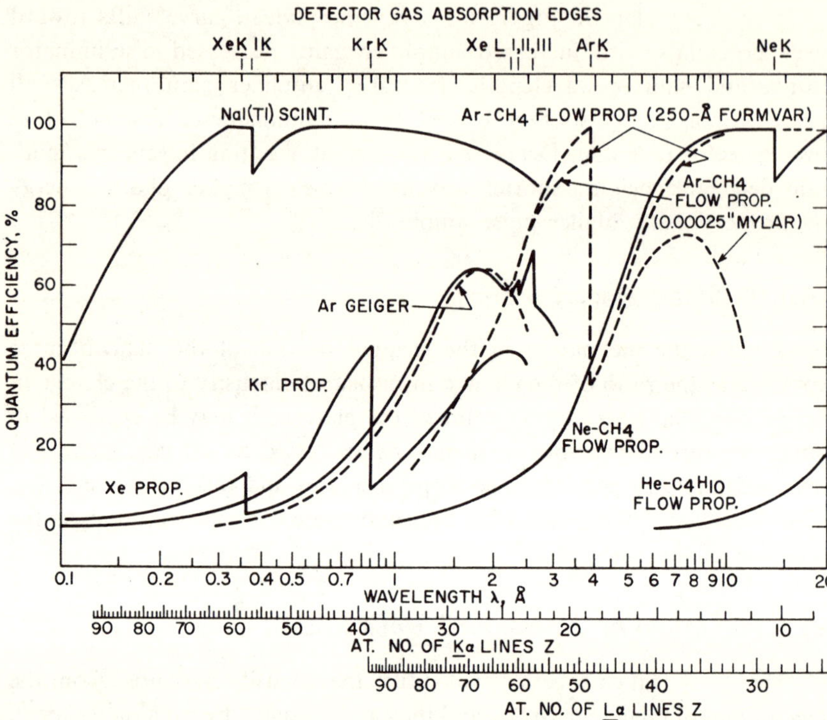

Fig. 6.12. Typical quantum efficiencies of several x-ray detectors.

tion. Assuming a Gaussian distribution, the root-mean-square deviation is

$$\sigma = \sqrt{\bar{N}} \tag{6.4}$$

where $\bar{N}$ is the average number of electrons in the statistically limiting population. In a proportional counter, $\bar{N}$ is the average number of ion pairs produced by the x-ray photon. In a scintillation counter, $\bar{N}$ is the average number of useful photoelectrons collected by the first dynode of the multiplier phototube for each x-ray photon (not the number of scintillation photons or the number of photoelectrons they produce). The width of the pulse-height distribution at half-maximum (full width at half-maximum, FWHM) is

$$W_{1/2} = 2\sigma \tag{6.5}$$

The resolution is then

$$R = 100(W_{1/2}/\bar{N}) \tag{6.6}$$

TABLE 6.2

Width and Energy Resolution of Pulse-Height Distributions of Proportional, Scintillation, and Semiconductor X-Ray Counters

Cu $K\alpha$ photon energy		8040 eV
Ar-filled proportional counter:		
Effective ionization potential of Ar		26.4 eV per Ar^+,e^- pair
Number of Ar^+,e^- pairs per Cu $K\alpha$ photon,[a] $\bar{N}$	8040/26.4	305 pairs
Deviation for Gaussian distribution, σ	$\sqrt{305}$	17.5 pairs
Half-width of distribution, $W_{1/2}$	$2\sqrt{305}$	35 pairs
Increase in half-width due to fluctuations in gas gain in the avalanching process	$\sim 1.3\times$	45.5 pairs
Resolution, R	$(45.5/305)\times 100$	15%
NaI(Tl) scintillation counter:		
NaI(Tl) phosphor efficiency		50 eV/(light) photon
Number of light photons per Cu $K\alpha$ photon	8040/50	161 photons
Number of light photons arriving at photocathode (efficiency of optical coupling between phosphor and photocathode $\sim$0.9)	0.9×161	145 photons
Number of photoelectrons emitted (efficiency of photocathode $\sim$0.1)	0.1×145	14.5 photoelectrons
Number of photoelectrons collected by the first dynode[a] (efficiency of collection $\sim$0.9), $\bar{N}$	0.9×14.5	13 photoelectrons
Deviation for Gaussian distribution, σ	$\sqrt{13}$	3.6 photoelectrons
Half-width of distribution, $W_{1/2}$	$2\sqrt{13}$	7.2 photoelectrons
Increase in half-width due to fluctuations in the secondary-electron multiplication process	$\sim 1.2\times$	8.6 photoelectrons
Resolution, R	$(8.6/13)\times 100$	66%
Si(Li) semiconductor detector:		
Energy to produce one electron–hole pair		$\sim$3 eV per (e^-, hole) pair
Number of (e^-, hole) pairs per Cu $K\alpha$ photon,[a] $\bar{N}$	8040/3	2680 pairs
Deviation for Gaussian distribution, σ	$\sqrt{2680}$	52 pairs
Half-width of distribution, $W_{1/2}$	$2\sqrt{2680}$	104 pairs
Resolution, R	$(104/2680)\times 100$	4%

[a] Statistically limiting process.

Mean pulse amplitude, half-width, and energy resolution are all related to photon energy E; they are proportional to E, $E^{1/2}$, and $E^{-1/2}$, respectively.

Table 6.2 gives calculations of the energy resolution for an argon proportional counter, NaI(Tl) scintillation counter, and lithium-drifted silicon semiconductor detector. The table shows that the width of the pulse-height distribution, and thereby the energy resolution, of the scintillation counter is limited by the collection of photoelectrons on the first dynode, but is also affected by statistical fluctuations in the scintillation process, light-photon collection at the photocathode, photoelectron emission, and secondary-electron multiplication. Conversely, the width of the pulse-height distribution of the proportional counter is limited by the deviation of the initial ionizing event. The only other contributing factor is statistical fluctuation in the gas amplification.

6.4.2. Comparison of Detectors *(D14)*

Geiger, proportional, and scintillation counters are compared in detail in Table 6.3, and the useful spectral regions of the scintillation counter and of neon, argon, krypton, and xenon proportional counters are shown in Figure 9.1. Probably the three most commonly used detectors on standard commercial spectrometers are the argon–methane (P10 gas) gas-flow proportional counter for the longer wavelengths, and the NaI(Tl) scintillation counter or sealed xenon proportional counter for the shorter wavelengths.

Because no one detector is suitable for the entire spectral region, combination (double, tandem, universal) detectors are used to facilitate selection of the optimal detector for each region. Combination detectors consist of a two-window flow-proportional counter for long wavelengths backed by a sealed krypton or xenon proportional counter or a scintillation counter for short wavelengths. The output may be taken from either detector, or the combined output from both may be used. Figure 6.13 shows the performance of a combination flow–scintillation counter. It is evident that over half the spectral region, the combined outputs exceed the output of either detector alone.

In combination detectors, the front window of the flow counter may be thin (1–6 μm) Mylar, polyethylene, or polypropylene, or ultrathin (100–500 μÅ) Formvar to pass the long wavelengths. However, the rear window, which passes the short-wavelength x-rays to the rear detector, may be much heavier, so it need not be supported in vacuum nor be replaced with prolonged use. In some instruments, a short collimator is placed between

TABLE 6.3

Comparison of Common X-Ray Detectors

Geiger counter	Proportional counter	Scintillation counter
Detection process		
1) Gas ionization	1) Gas ionization	1) Scintillation production in NaI:Tl crystal
2) Townsend avalanching (generalized throughout tube)	2) Townsend avalanching (localized near anode wire)	2) Photoelectron emission from photocathode
3) Ultraviolet excitation		3) Secondary electron multiplication in dynodes
Autoamplification, output electrons/incident x-ray photon		
$\sim 10^9$	$\sim 10^6$	$\sim 10^6$
Output pulse amplitude, V		
~ 1	$\sim 10^{-3}$	$\sim 10^{-3}$
Useful wavelength region, Å		
0.3–4	Sealed 0.3–4 Flow (0.00025-inch Mylar) 0.7–10 Flow (ultrathin Formvar) 0.7–50	0.1–3 or 4
Dead time, μs		
~ 200	≥ 0.2	≥ 0.1
Maximum useful count rate without substantial coincidence loss, counts/s		
Single channel ~ 1000 Multichannel $\sim 10^4$	10^5	10^6
Background, counts/s		
~ 1	Sealed ~ 0.5 Flow ~ 0.2	~ 10

TABLE 6.3 (*continued*)

Geiger counter	Proportional counter	Scintillation counter
Energy resolution (width of pulse-height distribution at half height ÷ mean pulse amplitude), % (for Cu $K\alpha$)		
—	~15	~45
Life		
Practically unlimited (if halogen quench used)	Sealed: ~10^9 counts Flow: Unlimited, but windows and anode wire require occasional replacement	May be relatively short (see limitations below)
Advantages		
1) Output pulse amplitude ~1000× higher than proportional or scintillation counter; requires little amplification or associated circuitry 2) Construction simple, since no need to realize narrow pulse-height distribution 3) Plateau wide, precludes need for high regulation of detector power supply 4) Stability high 5) Quantum efficiency high 6) Background low 7) Life substantially unlimited for tubes having halogen quench gas 8) Since pulse amplitude is independent of wavelength, best detector for counting method that uses capacitor to integrate detector output	1) Pulse height proportional to x-ray photon energy; pulse-height selection can be used 2) Energy resolution better (2–3×) than scintillation counter: a) Pulse-height distribution for monochromatic x-rays is narrower b) Difference in pulse height for a given difference in x-ray photon energy is greater 3) Resolving time fast; useful at intensities up to 10^5–10^6 counts/s 4) Useful wavelength region very wide if suitable construction, window, and gas filling are used	1) Pulse-height proportional to x-ray photon energy; pulse-height selection can be used 2) Output pulse amplitude somewhat greater than proportional counter 3) Resolving time somewhat faster than proportional counter; useful at intensities up to 10^6–10^7 counts/s 4) Quantum efficiency substantially independent of wavelength and very high (~100%) over wide useful wavelength range 5) Escape peak phenomena not serious with NaI(Tl) scintillator 6) Pulse-height-distribution drift and distortion relatively uncommon

TABLE 6.3 *(continued)*

Geiger counter	Proportional counter	Scintillation counter
Limitations		
1) Pulse height independent of x-ray photon energy; pulse-height selection cannot be used 2) Resolving time very long; maximum useful count rate much lower than proportional or scintillation counter 3) Dead time dependent on wavelength (for full-wave generator): efficiency of excitation of a spectral line depends on excitation potential and varies continuously with time 4) Quantum efficiency highly wavelength dependent: a) For a given gas filling, efficiency varies widely with wavelength b) Different gas fillings required for maximum efficiency at different wavelengths 5) Since maximum useful count rate is very low, poorly suited for semi-quantitative scanning of spectra where strong and weak lines occur	1) Cannot resolve close wavelengths: a) With crystal resolves ΔZ 1 (low Z) to 3 (high Z) b) Without crystal (nondispersive) resolves ΔZ 3 (low Z) to 6 (high Z) 2) Output pulse amplitude only a few mV, requires high amplification 3) Detector-tube potential must be highly stabilized ($\pm 0.1\%$), since gas amplification factor varies rapidly with detector potential 4) Quantum efficiency highly wavelength dependent: a) For a given gas filling, efficiency varies widely with wavelength b) Different gas fillings required for maximum efficiency at different wavelengths 5) Thin (0.00025-inch Mylar) or ultrathin (100-200-Å Formvar) windows and continuous gas flow required to realize full useful wavelength region	1) Cannot resolve close wavelengths: a) With crystal resolves ΔZ 1–3 b) Without crystal (nondispersive) resolves ΔZ 6–10 2) Energy resolution poorer (2–3 ×) than proportional counter: a) Pulse-height distribution for monochromatic x-rays is wider b) Difference in pulse height for a given difference in x-ray photon energy is smaller 3) Wavelength region limited to maximum of ~4 Å 4) Background high due to multiplier phototube noise; increases with age and detector-tube potential 5) Life may be limited: a) Not as rugged as gas-filled detector to shock: encapsulation may fail, crystal may crack or deteriorate, multiplier phototube may break or its electrodes become misaligned

TABLE 6.3 (*continued*)

Geiger counter	Proportional counter	Scintillation counter
6) For same reason, poorly suited for integrated counts (intensity "under a peak") where intensity is very high at the peak and falls off on either side	6) Life of sealed tubes limited to $\sim 10^9$ counts: a) Width of pulse-height distribution increases progressively with life b) Require replacement or refilling after $\sim 10^9$ counts 7) Flow counter requires occasional cleaning and replacement of windows and anode wire 8) Escape peaks more troublesome than for scintillation counter (with Ar and Kr gas fillings) 9) Pulse-height-distribution drift and distortion more prevalent than for scintillation counter; especially so for gas-flow counter 10) Flow counter requires gas cylinder and lines, may require pressure and temperature control	b) Phototube noise increases with age

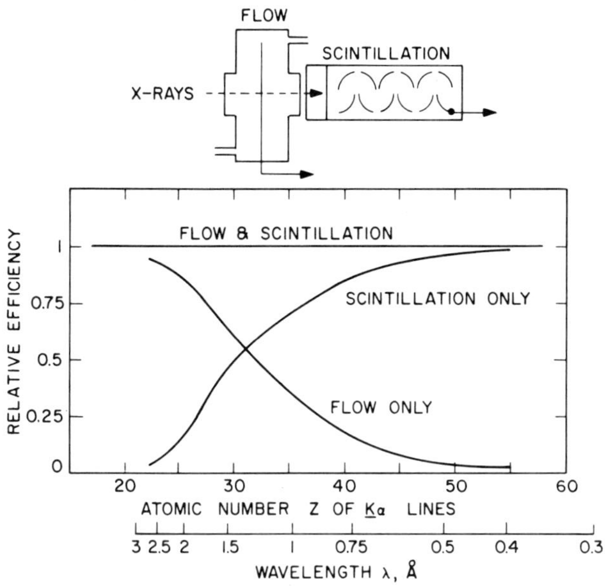

Fig. 6.13. Performance of the combination gas-flow and scintillation counter. [R. Jenkins and J. L. de Vries, "Practical X-Ray Spectrometry," Springer-Verlag N. Y. (1967); courtesy of the authors and publisher.]

the two detectors. In some vacuum instruments, the flow counter is inside the vacuum chamber, the scintillation counter outside, with the vacuum window of the chamber between.

Efforts have been made to extend the limited linear intensity range of the Geiger counter to take advantage of its high sensitivity and high pulse amplitude. In the *multichamber Geiger tube*, the active volume is partitioned into two or more separate compartments having a common anode wire and gas. Each chamber counts independently of the other(s), giving increased resolving time and substantially linear intensity range up to 10^4 counts/s. Use of a *mesh cathode cylinder*, rather than sheet, reduces the cathode surface and thereby the electron emission by ion bombardment. Use of a *beaded anode* (*C56*), having small glass beads spaced along the wire, limits the discharge to a small fraction of the total volume, giving a linear intensity range up to several thousand counts per second. However, the reduced sensitivity necessitates higher applied potential. Thin-window *gas-flow Geiger counters* have been used for long wavelengths; these tubes have shorter length to reduce absorption of short wavelengths. Special quenching circuits for Geiger counters have already been mentioned (Section 6.2.2.4).

Efforts have also been made to extend the useful range of the scintillation counter farther into the long-wavelength region, beyond 3 Å. This limit is imposed partly by the absorption of the beryllium window and aluminum reflecting layer, but principally by the inherent noise pulses of the multiplier phototube, which have the same amplitude as pulses from x-rays having wavelength ~3 Å or more. The noise originates from: (1) thermionic emission from the photocathode; (2) field emission from points, corners, and edges, where high potential gradients occur; (3) ionization of residual gas in the phototube; and (4) electrical leakage over insulating surfaces. The noise may be reduced in several ways, including the following: The multiplier phototube may be selected for low noise level. Thermionic emission may be reduced by cooling. The socket may be water-cooled to reduce heating by the potential divider, which usually consists of a bank of resistors mounted in the socket. The tube itself may be cooled with liquid nitrogen. The outputs of two multiplier phototubes sharing a common phosphor may be connected to a coincidence circuit, which rejects all pulses that do not come from both tubes simultaneously. Of these techniques, only the selection and water-cooling are practical on an analytical spectrometer.

6.5. OTHER X-RAY DETECTORS

6.5.1. Photographic Film

Instruments using photographic-film registration have been described in Sections 5.1 (the earliest Bragg spectrograph), 5.5.1.2 (edge crystal), 5.5.2.4 (Von Hamos image spectrograph), and 5.6.6 (curved-crystal film spectrographs). In the image spectrograph, the focal surface is plane, and photographic plate can be used. In the other instruments, the focal surface is curved, and x-ray diffraction film is used.

For quantitative analysis, a film calibration curve must be established to permit correlation of photographic density and x-ray intensity. A uniform monochromatic x-ray beam is photographed through a stack of strips of thin metal foil having progressively shorter length from bottom to top, exposing a stepped wedge of graded thickness. The intensity transmitted by each step is calculated from the wavelength and intensity of the incident x-rays, the absorption coefficient of the metal, and the thickness of the step. The photographic density of the image of the step is measured with a microdensitometer. A plot of photographic density *vs.* logarithm of exposure (*IT*, intensity × time) is plotted. The photographic densities of the spectral lines are measured and applied to the curve to derive *IT*, from which intensity is calculated from the exposure time.

6.5.2. Photoelectric Detectors

6.5.2.1. *The Phosphor-Phototube Detector*

This detector is physically similar to a scintillation counter, in that it consists of a phototube or multiplier phototube on the envelope of which is mounted a piece of fluorescent screen with the fluorescent side facing the photocathode. Alternatively, a layer of phosphor may be applied directly to the bulb just outside the photocathode. This detector is used principally for continuous-current instantaneous measurements, especially in industrial x-ray thickness and filling gages. It has no application in spectrometry.

6.5.2.2. *Photoelectric Detectors for the Ultralong-Wavelength Region.*

This section describes four detectors that are used principally in the long- and ultralong-wavelength region (*G10*). These detectors all require vacuum, but may be operated without windows if mounted in or attached to a vacuum spectrometer. The detectors are usually prepared so that their active surfaces can withstand repeated exposure to air.

The *windowless multiplier phototube* (*B11*, *F8*, *G10*, *J1*) is simply an open electron multiplier, the first electrode of which is made of beryllium–copper alloy and serves as an x-ray-sensitive photocathode. The x-rays to be measured irradiate this photocathode, thereby expelling photoelectrons, which initiate the secondary-electron multiplication process.

In the *photoelectron scintillation counter* (*G10*, *S11*) (Figure 6.14C), a series of up to six parallel, spaced, thin metal foils are arranged just above and perpendicular to a plastic or crystalline phosphor cemented to a multiplier phototube. Foils successively farther from the x-ray source are maintained at successively higher potential by a potential divider. The incident x-rays expel photoelectrons from the front and back surfaces of the foils. Some of these photoelectrons strike the phosphor and excite scintillations, which are counted by the phototube.

In the *resistance-strip magnetic photoelectron multiplier* (*G10*) (Figure 6.14A), two plane semiconductor strips, a field strip and a dynode strip, are arranged mutually parallel and a short distance apart. The dynode strip

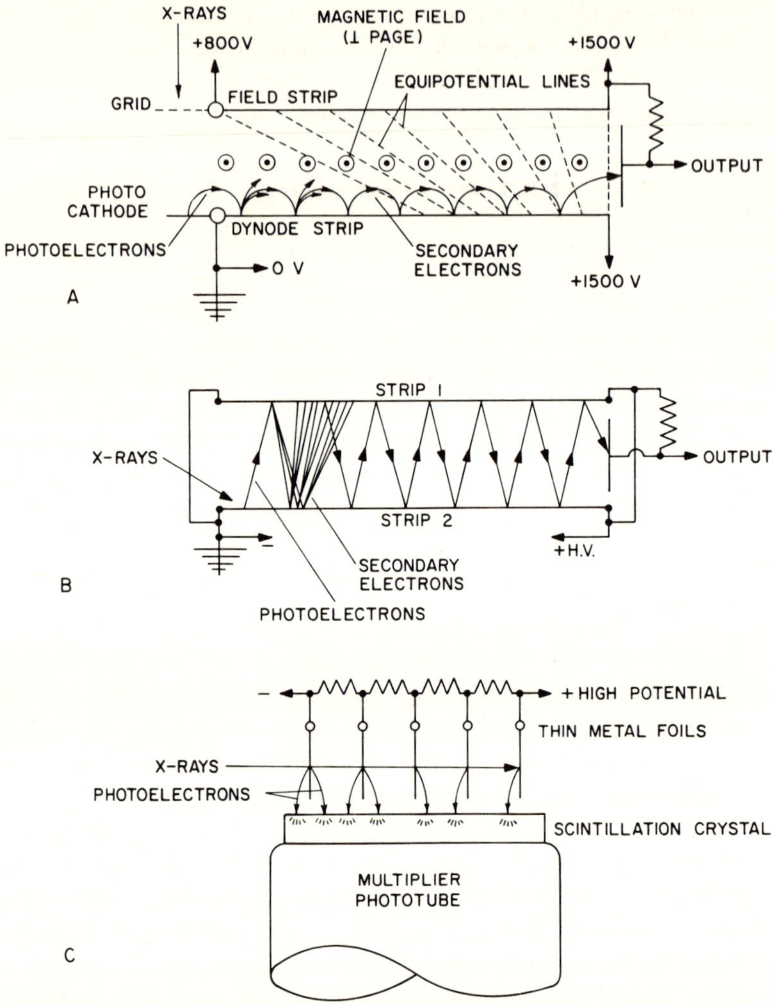

Fig. 6.14. Some photoelectric x-ray detectors for use in the ultralong-wavelength region. A. Resistance-strip magnetic photoelectron multiplier. B. Continuous-strip photoelectron multiplier. C. Photoelectron scintillation counter.

acts as a secondary-electron emitter. The electric potentials applied to the strips cause uniform potential drops along the strips in such a way as to initiate an electric field having inclined equipotential lines as shown. A uniform magnetic field is applied parallel to the planes of the strips by means of small, external, permanent magnets. The incident x-rays irradiate the photocathode, emitting photoelectrons, which initiate the secondary-electron multiplication process, as shown. In the crossed electric and magnetic fields, the motion of a photo or secondary electron is a cycloid directed perpendicular to the magnetic field and along the equipotential line on which the electron originated. Thus, each electron travels a short distance, ~1 mm, then falls back on the dynode surface, emitting additional secondary electrons.

In the *continuous-strip photoelectron multiplier* (*G10*) (Figure 6.14B), two secondary-emissive strips are deposited longitudinally and diametrically opposite one another on the inside surface of a glass tube. Alternatively, two such parallel strips may be separated and supported by insulating side walls. A potential of 500–2000 V is applied to the ends of the two strips, causing an electric field having equipotential lines perpendicular to the strips. The incident x-rays expel photoelectrons, which initiate the secondary-electron multiplication process, as shown.

6.5.3. Solid-State Semiconductor Detectors

6.5.3.1. *Crystal Counters*

If metal electrodes are affixed to the ends of a small single crystal of a suitable material and a potential is applied to the electrodes, the device functions as a photon counter (*F15*). Cadmium sulfide, zinc sulfide, germanium, and diamond act as crystal counters at room temperature, silver chloride and silver bromide at very low temperature. Crystal counters are very small and require relatively simple electronic equipment. They are applicable to pulse-height selection, perform well at high intensity, and have high sensitivity and signal-to-noise ratio. The disadvantages include the following: They are sensitive to the purity, crystallographic perfection, and heat treatment of the crystal; the pulse height changes with time, due to accumulation of charge in the crystal; dark current may be high, and signal current persistent. Crystal counters have found little application in analytical x-ray spectrometry.

6.5.3.2. *Lithium-Drifted Silicon and Germanium*

Probably the most significant new development in x-ray intensity measurement is the lithium-drifted semiconductor detector, principally lithium-drifted silicon, Si(Li) or Si:Li, and germanium, Ge(Li) or Ge:Li (*B49*, *B58*, *S19*). A lithium-drifted detector consists of a silicon (or germanium) single-crystal semiconductor having a compensated region sandwiched between *p*- and *n*-type regions. The compensated region is formed by diffusing ("drifting") lithium into *p*-type silicon (or germanium) under carefully controlled conditions. The lithium compensates impurities and dopants already present. Lithium has an extremely high diffusion rate even at room temperature, so lithium-drifted detectors must be maintained at liquid-nitrogen temperature at all times, even when not in use.

In operation, a reverse bias potential is applied across the detector, and the resulting electric field depletes the compensated region of electron–hole free-charge carriers. The depleted region constitutes the radiation-sensitive detector volume and typically has an area of 30–100 mm^2 and a depth of 2–5 mm. Because of the carrier-depleted region, the lithium-drifted detector is a high-impedance device.

Each incident x-ray photon absorbed produces a large number of electron–hole pairs in the semiconductor—just as it produces ion–electron pairs in a gas detector or light photons in a scintillation counter. However, the semiconductor detector does not have the inherent high multiplication provided by gas amplification and secondary-electron multiplication. Consequently, a low-noise, high-gain preamplifier is required. Usually, a field-effect transistor is used, and this must also be maintained at liquid-nitrogen temperature to reduce the noise level. A lithium-drifted detector, then,

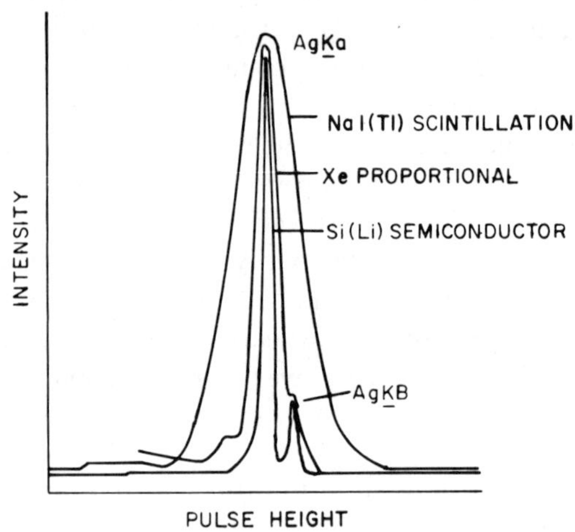

Fig. 6.15. Comparison of the Ag $K\alpha,\beta$ pulse-height distributions for the NaI(Tl) scintillation counter, Xe proportional counter, and Si(Li) semiconductor detector. [P. G. Burkhalter and W. J. Campbell, *Oak Ridge National Laboratory Report* **ORNL-IIC-10**, vol. 1 (1967); courtesy of the authors.]

actually consists of the Si(Li) or Ge(Li) detector itself, the field-effect transistor preamplifier, and the liquid-nitrogen cryostat and reservoir.

Detection efficiency is a function of the atomic number of the semiconductor and the depth of the depletion layer. A 3-mm layer of silicon absorbs 50% of incident 30-kV (0.4-Å) photons, but photoelectric absorption decreases rapidly with increasing photon energy. Thus, Si(Li) detectors are optimal at wavelengths >0.4 Å (Cs $K\alpha$), Ge(Li) detectors are preferable at shorter wavelengths. It is evident that the Si(Li) detector is suitable for substantially the entire spectral range of a 50-kV x-ray spectrometer.

The lithium-drifted semiconductor detector has three principal advantages:

1. The energy conversion is very efficient. Only ~3 eV is required to produce one electron–hole pair, compared with 20–26 eV to produce an

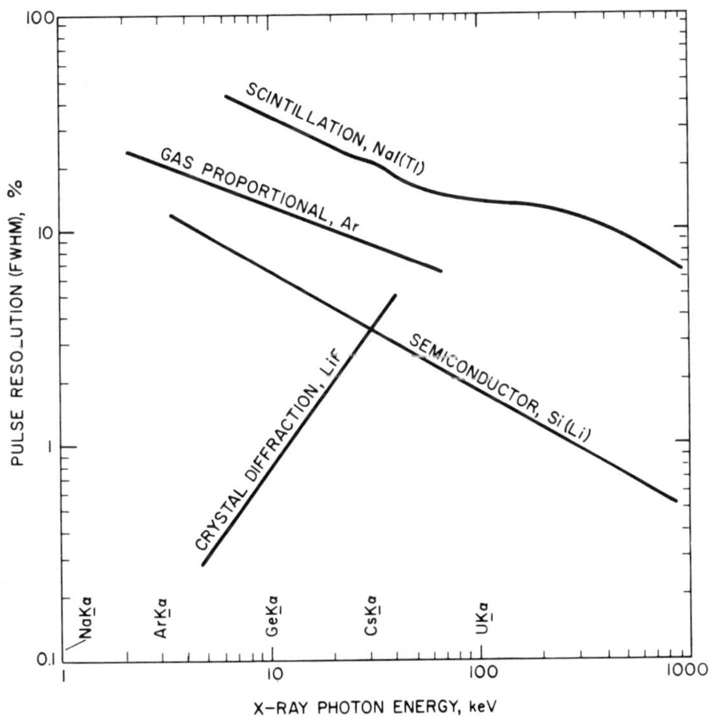

Fig. 6.16. Comparison of resolutions of the NaI(Tl) scintillation counter, Ar proportional counter, and Si(Li) semiconductor detectors, and LiF analyzer crystal. [P. G. Burkhalter and W. J. Campbell, *Oak Ridge National Laboratory Report* **ORNL-IIC-10**, vol. 1 (1967); courtesy of the authors.]

ion–electron pair in a gas detector and 50 eV to produce a light photon in a scintillation detector.

2. It follows that the resolution is excellent; $W_{1/2}$ of the pulse-height distribution may be as small as 1.1 keV for 6–8-keV photon energy, and spectral lines of the same series for adjacent atomic numbers can be resolved. Figure 16.15 shows resolution of Ag $K\alpha$ and Ag $K\beta$. Table 6.2 gives the calculation of resolution of a Si(Li) detector for Cu $K\alpha$ photons.

Figure 16.15 compares the Ag $K\alpha, \beta$ pulse-height distribution for the NaI(Tl) scintillation, xenon proportional, and Si(Li) semiconductor detectors. Figure 16.16 compares the resolution of the NaI(Tl) scintillation, argon proportional, and Si(Li) semiconductor detectors and the LiF analyzer crystal as a function of photon energy. Figure 6.17 shows the performance of the Si(Li) detector used with a 400-channel analyzer for the K lines of some elements of high atomic number.

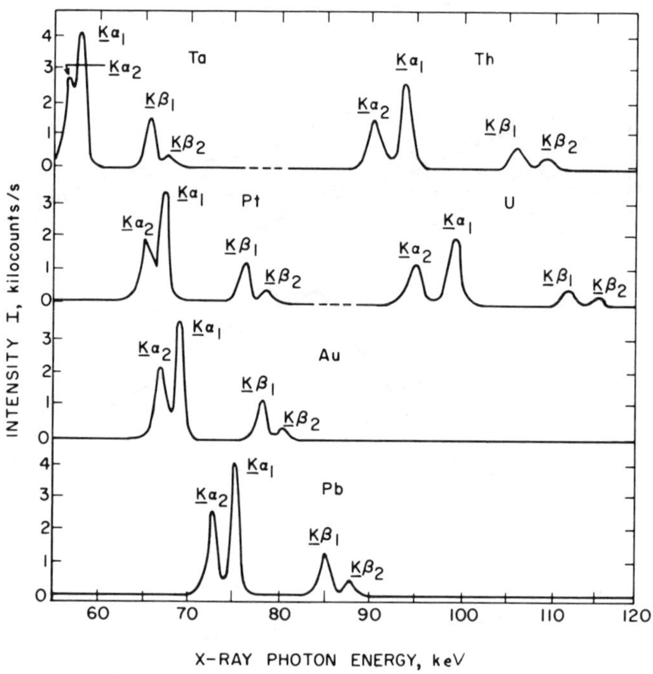

Fig. 6.17. Pulse-height distribution curves from a Si(Li) semiconductor detector used with a 400-channel analyzer. [H. R. Bowman, E. K. Hyde, S. G. Thompson, and R. C. Jared, *Science* **151**, 562 (1966); courtesy of the authors and the American Association for the Advancement of Science.]

3. A third advantage of the Si(Li) detector is its freedom from pulse-height drift and broadening at intensities up to 15,000 counts/s.

The principal disadvantage is that the high mobility of the lithium causes substantial deterioration of detector characteristics, even at room temperature. Consequently, the detector must be maintained perpetually at liquid-nitrogen temperature.

Chapter 7

Measurement

7.1. INSTRUMENTATION

7.1.1. Introduction

In Chapter 6, it is shown that the outputs of x-ray detectors of the types used on analytical x-ray spectrometers consist of series of pulses of electric current, one pulse for each photon detected. The pulses are randomly distributed in time. In Geiger counters, all these pulses have about the same amplitude (height). In proportional and scintillation counters, the pulses have average amplitude proportional to the photon energy.

Even for strictly monochromatic x-rays, the pulses are not all of the same height, but have a Gaussian distribution about a *mean* height proportional to the photon energy. For incident x-ray beams containing more than one wavelength, that is, more than one spectral line, the detector output contains (1) a pulse-height distribution peak for each wavelength, and (2) another escape peak for each wavelength shorter than the absorption edge of the detector gas or of the iodine in the NaI(Tl) scintillator.

The electronic readout components count these pulses or measure their rate of production for analytical purposes. The electronic components must have a response time at least as fast as the detector resolving time, and preferably as fast as the recovery time. Otherwise, the linear counting range is limited by the counting circuits, rather than the detector. However, clearly, the counting circuits cannot improve the linear range of the detector.

In addition to the pulses that arise from the discrete incident spectral wavelengths, the detector output also contains pulses arising from back-

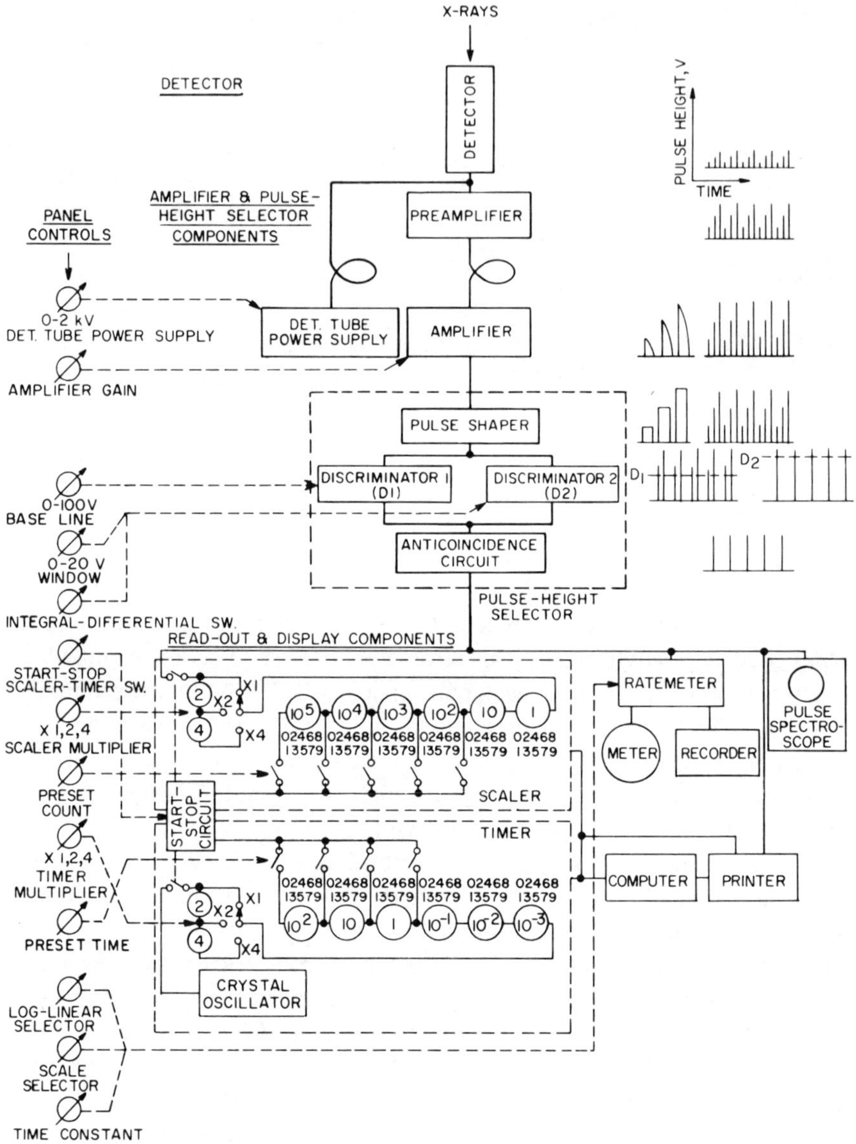

Fig. 7.1. Electronic detection and readout components of an x-ray spectrometer. Panel controls are shown along the left. Simplified wave forms (pulse height *vs.* time) are shown in the upper right.

ground and, especially in scintillation counters, pulses arising in the detector itself.

The detector and electronic components of an x-ray spectrometer are shown schematically in Figure 7.1. The principal panel controls are shown along the left edge of the figure. Highly simplified diagrams of pulse height *vs.* time are shown at the upper right for pulses of three amplitudes. The low pulses represent low-energy background and detector and amplifier noise. The medium and high pulses originate from two incident x-ray spectral lines having relatively low and high photon energies, respectively.

The components are divided into three groups: the detector itself, the amplifier and pulse-height analysis components, and the readout and display components. The detector and preamplifier are mounted on the spectrogoniometer. All the other components are mounted in some sort of relay rack or console, with the possible exception of the computer (if any), which may be separate, even in a location remote from the spectrometer. The detector power supply is included with the electronic components. It is a well-regulated ($\pm 0.05\%$ or so) 0–2 kV dc supply. Formerly, all the electronic components used electron tubes, and many still do. However, today, the components have been largely converted to solid-state semiconductor devices—transistors, diodes, rectifiers, even integrated circuits—with consequent reduction in size and power consumption.

7.1.2. Preamplifier

The preamplifier is mounted on the detector arm of the spectrogoniometer as an almost integral part of the detector itself. In fact, some instruments provide water-cooling of the detector mounting plate to reduce heat conduction from the preamplifier to the detector. The preamplifier output is connected by a relatively long coaxial cable to the input of the main amplifier in the electronics console. The preamplifier usually consists of one or two linear amplifier stages and a cathode-follower stage to provide optimal coupling to and minimum attenuation along the cable. The preamplifier also prevents amplifier feedback from affecting the detector. "Sine-function" preamplifiers are one means of providing automatic pulse-height analysis and of correcting for pulse-height shift (Section 8.1.7).

7.1.3. Amplifier

It is the basic function of the main amplifier to linearly amplify the detector–preamplifier output pulses to potentials high enough to actuate the pulse-height analyzer, ratemeter, and scaler. The amplifier is a linear, low-noise, fast-response, video-type amplifier having gain that can be varied up to ~10,000. Alternatively, many of these amplifiers have a fixed gain of, say, 10,000, and provide for *attenuation* to lower gain. The amplifier is designed to amplify preamplifier output pulses to a maximum of 5–100 V. Geiger, proportional, and scintillation counters require low, intermediate, and high amplification, respectively.

7.1.4. Pulse-Height Selectors

7.1.4.1. *Pulse-Height Analyzer; Discriminator*

A *discriminator* passes all pulses having height greater than some preset minimum, and rejects all lower pulses. A *pulse-height analyzer*, or *pulse-height selector*, passes on to the ratemeter and scaler all pulses having height greater than one preset level, but lower than a second preset level, and rejects all pulses outside this *window* or *channel.* All x-ray spectrometers have a discriminator to reject pulses having height, after amplification, of ~5 V or less. This discriminator eliminates some background and all of the detector and amplifier noise. Although most modern x-ray spectrometers have pulse-height analyzers, this component may be omitted, and, even if present, it may be by-passed.

The pulse-height analyzer is shown in Figure 7.1 enclosed in a dash-line box. The particular type shown consists of four components, the functions of which are illustrated by the pulse diagrams at the right of each component. It will be recalled that the low pulses represent low-energy background and detector and amplifier noise, and the medium and high pulses represent two x-ray spectral lines having medium and high photon energy, respectively. The four components are as follows.

The *pulse shaper* forms rectangular or square-wave pulses from the amplified pulse forms, as shown by the two groups of enlarged pulses, but does not alter their relative amplitudes. The shaped pulses are passed on to the inputs of both of the discriminators.

Discriminator 1 passes all pulses having amplitude greater than level D_1. Thus, it passes the pulses of both spectral lines, but rejects the low-amplitude noise. A discriminator functioning in this way is always required, even when the pulse-height analyzer is absent or by-passed.

Discriminator 2 passes all pulses having amplitude greater than level D_2. Thus, it passes only the pulses of the high-energy spectral line.

The *anticoincidence circuit* cancels all pulses that arrive simultaneously from both discriminators. Thus, it rejects the high-amplitude pulses, and passes on to the ratemeter and scaler only the pulses of intermediate amplitude.

To summarize, pulses of amplitude smaller than a selected minimum D_1 do not pass the first discriminator. Pulses of amplitude greater than a selected maximum D_2 pass both discriminators and are eliminated by the anticoincidence circuit. Thus, only pulses having amplitude between the two levels pass on to be counted.

The potential V corresponding to the lower discriminator level is usually referred to as the *baseline*, sometimes as the *window level* or *channel level.* It represents the minimum countable pulse height referred to so frequently in Chapter 6 and shown in Figures 6.7 and 6.11. The potential interval ΔV between the two levels is usually referred to as the *window* or *channel,* sometimes as the *window width* or *channel width.* The baseline may be varied from 0–100 V in most instruments; the window may be varied from 0 to 10, 20, or 30 V, depending on the instrument.

The circuit may be operated as a discriminator, passing only pulses having amplitude greater than the baseline V, or as a pulse-height analyzer, passing only pulses having amplitude greater than the baseline V, but less than the potential of the baseline plus window, $V + \Delta V$.

The instrument described above is a single-channel pulse-height analyzer. The multichannel analyzer is described briefly in Section 8.1.2. Pulse-height analysis is described in detail in Section 8.1.

7.1.4.2. *Reverter*

The pulse-selection function performed by the pulse shaper, two discriminators, and anticoincidence circuit can also be performed by a single circuit known as a *pulse reverter* or, simply, *reverter*—but not as well. The reverter was used in early General Electric x-ray spectrometers. The function of the reverter is shown in Figure 7.2.

The curves in Figure 7.2A show preamplifier output pulse height as a function of detector potential for Cr $K\alpha$, Cu $K\alpha$, and Ag $K\alpha$, relatively long-, intermediate-, and short-wavelength x-rays, respectively. The curve is shown in an unusual orientation to permit graphic correlation with the other parts of the figure. The arrows on the axes indicate directions of increasing pulse height and detector potential.

Figure 7.2B shows the reverter response characteristic and reverter output pulse height as a function of input pulse height. Because the reverter follows the preamplifier, the preamplifier output is the reverter input.

The curves in Figure 7.2C show reverter output pulse height as a function of detector potential for Cr $K\alpha$, Cu $K\alpha$, and Ag $K\alpha$. The discriminator

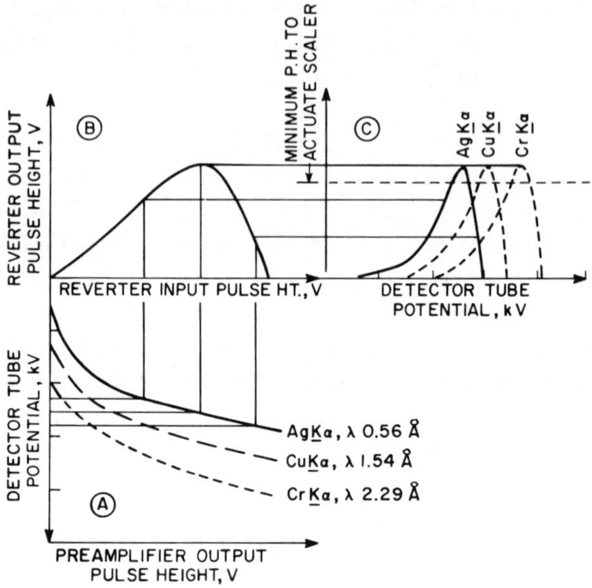

Fig. 7.2. Function of the reverter. [Courtesy of General Electric Co., X-Ray Dept.] A. Preamplifier output pulse height *vs.* detector-tube potential for a short, medium, and long wavelength. B. Reverter output pulse height *vs.* input pulse height. C. Reverter output pulse height *vs.* detector potential for a short, medium, and long wavelength.

baseline, or the minimum pulse height that will actuate the scaler, is also indicated.

It is evident that as detector potential is increased, the height of the Ag $K\alpha$ pulses increases progressively at the preamplifier output (Figure 7.2A), but increases to a maximum, then decreases at the reverter output (Figure 7.2C). Thus, only for a relatively narrow range of detector potential do Ag $K\alpha$ pulses exceed the discriminator baseline. The Cu $K\alpha$ and Cr $K\alpha$ pulses undergo the same phenomenon at successively higher detector potentials. Thus, by suitable choice of detector potential, the chromium, copper, or silver pulses may be passed individually.

7.1.5. Ratemeter and Recorder

The ratemeter indicates the mean value of the current consisting of the individual pulses from the discriminator or pulse-height analyzer accumulated over a certain selected short time interval determined by the time constant. This mean current is proportional to the instantaneous x-ray intensity incident on the detector and is displayed as intensity or count rate on a panel meter or a strip-chart recorder. Thus, the ratemeter provides instantaneous intensity readout. The ratemeter may have linear or logarithmic response. Linear response reduces tailing of the peaks and is preferable for estimation of intensity. Logarithmic response may be advantageous for recording spectra having both intense and weak peaks. The ratemeter–recorder combination is used principally to record the x-ray spectrum for qualitative or semiquantitative analysis, or preparatory to quantitative analysis. Its use and modification for these purposes are described in Section 10.3.

7.1.6. Scaler and Timer

It has already been mentioned that the ratemeter is analogous to a speedometer, whereas the scaler is analogous to an odometer. The *scaler* or *count register* actually counts the individual pulses passed by the discriminator or pulse-height analyzer during a certain time interval. Their rate of arrival is determined by the x-ray intensity at the detector, and their time distribution is random. The scaler thus provides integrated or accumulative intensity readout.

The *timer* or *time register* is a clock, but it keeps time by counting the positive pulses from the sine-wave output of a high-frequency crystal oscillator. Their rate of arrival is determined by the frequency of the oscillator and is, say, 1000 Hz, that is, 1000 positive pulses per second. Their time distribution is perfectly uniform, and, at 1000 Hz, they are spaced at 1-ms intervals.

Both the scaler and timer, then, are electronic binary or decimal counters, and if there is any essential difference between them, it is that the scaler must be capable of counting at extremely high rates, whereas the timer need not count faster than $\sim$1000/s. Thus, the scaler and timer are "fast" and "slow" counters, respectively.

Figure 7.1 shows, very schematically, a scaler and timer, each having a six-decade decimal counter and provision for two-fold or four-fold binary multiplication.

A *binary circuit* may be regarded as one that passes every second pulse it receives; that is, it blocks pulses 1, 3, 5, . . . and passes pulses 2, 4, 6, ... Usually, provision is made to indicate which condition the circuit is in. When it has received and blocked a pulse, a small neon lamp glows until the next pulse is received and passed.

A *decimal circuit* may be regarded as one that passes every 10th pulse it receives, blocking—or rather, storing—pulses 1–9. Provision is made to indicate how many pulses are accumulated in the circuit. In the E1T decimal indicator tube, a fluorescent spot advances sequentially from 0 to 9 on a numerical scale. In the Nixie indicator, wire numerals 0 to 9, mounted in a neon-filled bulb, glow in sequence. Other indicators are also in use.

In Figure 7.1, suppose that both the scaler and timer multiplier switches are at $\times 1$, that all of the preset-count and preset-time switches are open, and that both start–stop switches are simultaneously closed (start). The analyte-line pulses from the pulse-height analyzer pass directly to the first scaler decade, and the pulses from the crystal oscillator pass directly to the first timer decade.

The first nine analyte-line pulses are accumulated in and indicated by the first scaler decade. Pulse 10 passes to the second decade, causing it to indicate 1 and returning the first decade to zero. Pulses 11–19 accumulate in the first decade, and pulse 20 passes to the second. This sequence continues until both decades are filled by pulse 99. Pulse 100 then passes to the third decade, returning the first two to zero. Thus, the individual decades indicate, respectively, individual pulses, 10's, 100's, 1000's, etc. as shown.

Meanwhile, the timer is counting positive pulses from the crystal oscillator in the same way, but since these pulses arrive at exact 1-ms intervals, the timer is, in effect, counting milliseconds of elapsed time. However, the decades are labeled 0.001, 0.01, 0.1, 1, . . . *seconds*, rather than 1,10, 100, 1000, . . . *milliseconds*.

Since the scaler and timer were started together, if they are now stopped together at any arbitrary time, the accumulated count and counting time may be read on the scaler and timer, respectively. The two instruments are stopped by opening the start–stop switches.

If it is intended to make a preset-count measurement of, say, 10^4 counts, the switch following the 10^3 scaler decade, that is, between the 10^3 and 10^4 decades, is closed. The scaler and timer are started simultaneously. The action is the same as before up to the 9999th pulse. However, the 10,000th pulse, instead of passing to the 10^4 decade, passes through the switch and actuates the stop circuit, opening both start–stop switches and stopping the scaler and timer. The counting time is then read on the timer.

Similarly, if it is intended to make a preset-time measurement of, say, 100 s, the switch following the 10 timer decade is closed. The first 99,999 millisecond pulses fill the first five timer decades. The 100,000th pulse (100.000 s) stops the scaler and timer. The accumulated count is then read on the scaler.

Now, suppose that the scaler multiplier switch is set at $\times$ 2. The analyte pulses no longer pass directly to the decimal counter, but first pass through a binary stage (marked "2" in Figure 7.1), which passes only every second pulse. The action is the same as before, but the accumulated count in the scaler must be multiplied by 2. Similarly, if the multiplier switch is set at $\times$ 4, the incoming pulses pass through two binary stages in series, so that only every fourth pulse reaches the counter, and the accumulated count must be multiplied by 4. Binary multipliers are also shown on the timer. Thus, the scaler and timer in Figure 7.1 have capacities of 1, 2, and 4×10^6 counts and 1, 2, and 4×10^3 s, respectively.

Provision is also made to clear the scaler and timer, that is, reduce all decades to zero, prior to starting a new measurement.

Another way to measure intensity is to allow the pulses to accumulate in a capacitor for a certain time interval, then measure the accumulated charge. This method is used in Applied Research Laboratories automatic x-ray spectrometers (Section 9.3.3).

A printer may be provided and programmed to print out the analytical data—sample number; symbol of analyte; and line, background, and net intensities—on a paper strip.

7.2. MEASUREMENT OF INTENSITY

7.2.1. Ratemeter Methods

Intensity can be measured with the ratemeter panel meter when only very rough estimates are required—for example, in distinguishing 35–65 and 50–50 nickel–iron alloys.

Intensity is frequently measured with the ratemeter–recorder for semiquantitative analysis (Section 10.5) and for quantitative analysis when highest accuracy is not required. The ratemeter should have linear, rather than logarithmic, response for this application and should be carefully calibrated. In general, the accuracy of the intensity measurement is greater the more closely the following conditions are met: The 2θ scanning rate should be slow; the time constant should be fast enough to permit the recorder to indicate the true peak intensity, but slow enough to dampen excessive recorder fluctuations; insofar as practicable, peaks from all samples and standards should be recorded at the same full-scale deflection and other instrument settings; the sample and standard peaks should have as nearly as possible the same deflection.

Four ratemeter–recorder techniques are used for measurement of x-ray intensity, of which three involve recording of the entire analyte peak for each sample and standard. The least accurate technique is the measurement of peak height above background. The intensity measured this way is usually less than the scaled intensity, and the discrepancy is greater for peaks having different intensity and for different time constants and other instrument settings. A more accurate method is the measurement of peak height above background at half-width, in the manner commonly used in x-ray diffraction. An even more accurate method is the measurement of the area under the peak, but above the background, by counting squares or with a planimeter. The fourth, and possibly most accurate method, does not require recording of the full peak. The goniometer is set at the 2θ angle for the analyte line and left undisturbed. The recorder traces analyte-line intensity for several centimeters from each sample and standard in turn. A mean line is drawn through the recorder noise of each tracing. Background may be recorded the same way. In all four methods, measurements are made above background, and calibration curves or mathematical relationships are established from the standard data.

7.2.2. Scaler-Timer Methods

7.2.2.1. *Preset-Time Method*

In the preset-time (fixed-time) method, the timer is set for a specified time T long enough to permit accumulation of enough counts for the required statistical precision. The timer and scaler are started simultaneously, operate until the preset time has elapsed, then stop simultaneously. The number of counts accumulated N is read on the scaler. Intensity $I = N/T$. If the same preset time is used for all samples and standards, the accumulated counts N may be used in calculations and calibrations instead of intensities. If this is done, and background corrections are made, the same preset time must be used for them also.

7.2.2.2. *Preset-Count Method*

In the preset-count (fixed-count) method, the scaler is set for a specified number of counts N large enough to give the required statistical precision. The timer and scaler are started simultaneously, operate until the preset count has accumulated, then stop simultaneously. The elapsed time T is read on the timer. As before, intensity $I = N/T$. If the same preset count is used for all samples and standards, the elapsed times may be used in calculations and calibrations instead of intensities. If this is done, and background corrections are made, the same preset count must be used for them also. The preset-time and preset-count methods are compared statistically in Sections 11.2.2 and 11.2.3.1.

7.2.2.3. *Integrated-Count Method*

The spectrometer is set at a 2θ angle on one side of the peak to be measured (*a* in Figure 14.4). The 2θ drive, timer, and scaler are started simultaneously, operate until the entire peak has been scanned (*a–b* in Figure 14.4), then are stopped simultaneously. The accumulated count and lapsed time are read. The integrated count or intensity may be used. Alternatively, the integrated count may be made by moving the spectrometer manually or with a step-scanner in small, equal-2θ increments and making a fixed-time or fixed-count measurement at each position. This method is used when the peak profile is required—to deal with spectral overlap, for example (*B33*).

7.2.2.4. *Monitor and Ratio Methods*

It is evident that the preset-time, preset-count, and integrated-count methods are simply ways of defining the duration of the counting time. In the three techniques, the count is continued for the duration of, respectively, a preset time, the time to accumulate a preset count, and the time to scan through the analyte peak. Two other ways to define the counting time are the monitor and ratio methods.

When a line having very low intensity is to be measured, a very long counting time is required to accumulate enough counts to give reasonable statistical precision. During such long times, the count is also affected by fluctuations in excitation and other instrument conditions. Thus, correlation of the measurements from a series of specimens may be meaningless. A monitor provides one way to deal with this condition. A small portion of the primary x-ray beam is diverted, usually by a small scatter target, directly into a second detector–scaler system. The spectrometer and monitor counts are started simultaneously and terminated when the monitor has accumulated a preset count. All samples and standards are counted for the time for the monitor to accumulate this same preset count.

In the ratio method, the time to accumulate a specified count from the standard defines the time during which the analyte-line count is accumulated from the sample. The standard count may be: (1) the analyte line measured from a calibration standard or from pure analyte; (2) an internal-standard line measured from the sample itself; (3) a scattered target line or primary-continuum wavelength measured from the sample; or (4) a scattered target line or continuum wavelength measured from a separate scatter specimen. These methods are widely used on automatic spectrometers.

The monitor and various ratio techniques are not equally effective in compensating fluctuations in x-ray tube potential V. The integrated intensity of the primary continuum is proportional to V^2 [Equation (1.17)]. The intensity of a specific wavelength in the primary continuum is proportional to $(V_\lambda - V)$, where V_λ is the potential equivalent to the wavelength. The intensity of a secondary analyte line is proportional to $(V - V_A)^2$, where V_A is the analyte-line excitation potential [Equation (1.20)]. Thus, fluctuations in x-ray tube potential are compensated most effectively by those ratio techniques in which the count is terminated by a preset count of the analyte line from a standard or from pure analyte.

7.2.3. X-Ray Dose and Dose Rate

For purposes of radiation health and safety, x- and γ-radiation are measured in terms of *dose* or *dose rate*, which have units roentgens (R) and roentgens per unit time, respectively (Section 9.5). Equations for conversion of x-ray measurements in counts or counts per second into roentgens or roentgens per hour have been derived (*C2*, *K18*). Kohler and Parrish (*K18*) show that in the wavelength region 0.2–3 Å, the dose produced by one photon/cm² in air is approximated by

$$d \approx (4.8 \times 10^{-10})\lambda^2 \quad \text{R per photon/cm}^2 \tag{7.1}$$

and the number of photons per square centimeter in air required to produce a dose of 1 R is approximated by

$$n_R \approx 1/d \approx (2.1 \times 10^9)\lambda^{-2} \quad \text{photons/cm}^2 \text{ per R} \tag{7.2}$$

In both approximations, λ is in angstroms. The lower curve in Figure 7.3 is a plot of n_R as a function of wavelength.

For a total measured count N of x-rays of wavelength λ in a photon detector having effective area A cm² and quantum efficiency Q, the dose D in roentgens in given by

$$D = k_1 N = \frac{1}{n_R A Q} N \tag{7.3}$$

$$= \frac{(4.8 \times 10^{-10})\lambda^2}{AQ} N \tag{7.4}$$

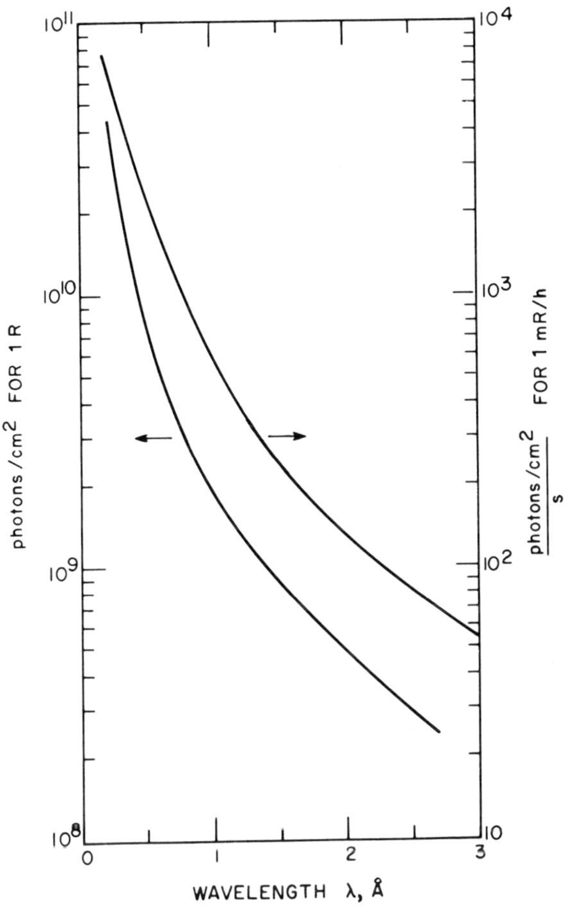

Fig. 7.3. Conversion of accumulated counts into dose in roentgens, and conversion of intensity in counts per second into dose rate in milliroentgens per hour.

For a measured intensity I counts/s, the dose rate D/T in roentgens per hour is given by

$$\frac{D}{T} = k_2 I = \frac{3600}{n_R A Q} I \tag{7.5}$$

$$= \frac{(1.7 \times 10^{-6})\lambda^2}{AQ} I \tag{7.6}$$

where 3600 is the number of seconds per hour. The upper curve in Figure 7.3 is a plot of intensity in photons per square centimeter per second required to give a dose rate of 1 mR/h as a function of wavelength. The constants

k_1 and k_2 apply to a particular detector at a specified wavelength, and have units R/count and R/h per count/s, respectively.

7.3. BACKGROUND

7.3.1. Definition and Significance

In its broadest sense, background may be defined as the intensity that would be measured at the 2θ angle of the analyte line if the analyte were absent. In this sense, the background consists of a more or less continuous spectrum and the contributions of spectral lines at or near the 2θ angle of the analyte line.

Background is significant in x-ray spectrometric analysis for three reasons: (1) frequently, it must be subtracted from the measured analyte and internal-standard peak intensities; (2) in the background-ratio method, it may compensate absorption errors and certain specimen and instrument errors; and (3) it limits the minimum detectable amount, which is that amount that gives a net intensity of $3\sqrt{I_B}$.

7.3.2. Origin and Nature

Background radiation originates from: (1) cosmic radiation; (2) radioactive emission from the environment; (3) spurious pulses and noise arising in the detector and in the electronic amplification components; (4) radioactive emission from the specimen, if it is radioactive; (5) primary x-ray emission from the x-ray tube; and (6) secondary x-ray emission from the specimen, crystal, and other components in the x-ray path. The first two sources constitute the irreducible minimum background and contribute an intensity of about two counts/s. The detector and amplifier noise pulses have very low energy and are substantially eliminated by the discriminator. The background, then, largely results from the collective transmission, scatter, diffraction, or reflection of x-rays from the last two sources, and each component in the spectrometer contributes to the background.

Air in the specimen compartment scatters the primary continuous spectrum and the target-line and other spectral-line radiation originating in the x-ray tube. The *specimen mask* and other structures in the specimen compartment also scatter the primary spectrum and may contribute additional emission lines. The *specimen* scatters the primary spectrum and emits spectral lines that may wholly or partially overlap the analyte line (spectral-line interference); crystalline specimens may diffract specific wavelengths from the primary continuum, or even one or more spectral lines in the primary spectrum. The *crystal*, aside from its intended function of diffracting the analyte-line radiation corresponding to its θ setting: (1) diffracts any incident x-radition having product $n\lambda$ that satisfies the Bragg law for its θ setting; (2) scatters all incident x-rays; and (3) emits the characteristic line spectrum of its own elements. The *collimators* scatter, diffract, or specularly reflect x-rays passing through them and emit their own characteristic spectrum. The *detector*, especially the multiplier phototube of the scintillation counter, contributes spurious pulses (noise) (Section 6.4.2). The *amplifier* and other electronic components also contribute noise.

In the foregoing discussion, scatter may be coherent or incoherent, diffraction may be in first and higher orders. The primary spectral lines scattered coherently and incoherently by the air, specimen compartment, and specimen are diffracted by the crystal as spectral lines, as, of course, are the interfering lines emitted by the specimen. Diffraction from the specimen, and diffraction and specular reflection from the collimator foils may have the form of very sharp "pseudolines" or "spikes." All other emission and scatter appear as continuous background unless the pulse-height analyzer is used, in which case each wavelength entering the detector produces its own pulse-height distribution. For example, without the pulse-height analyzer, emission from the analyzer crystal causes only continuous background, because the emission radiates in all directions. With the pulse-height analyzer, the pulse distributions of the spectral lines of the elements in the crystal may interfere with the distribution of the analyte line.

The continuous component of the background may have minor discontinuities at 2θ angles corresponding to the absorption edges of the x-ray tube target element, the major constituents of the specimen, and the detector gas or scintillation crystal. The pulse-height analyzer may also introduce anomalous discontinuities in the background. An example of this phenomenon is given in Section 8.1.8.4 in the paragraph immediately preceding the discussion of double or sum peaks.

Neglecting the contributions of spectral-line interference, for practical purposes, the background intensity I_B measured at the 2θ angle of the analyte line consists principally of the sum of three contributions:

$$I_B = I_{\text{pri,sc}} + I_{C,\text{sc}} + I_{C,\text{em}} \tag{7.7}$$

where $I_{\text{pri,sc}}$ is intensity of coherent and incoherent scattered primary x-rays; $I_{C,\text{sc}}$ is intensity of secondary x-rays scattered from the crystal; and $I_{C,\text{em}}$ is intensity of x-rays emitted from the crystal.

7.3.3. Measurement

Figure 14.3 shows one way to evaluate background intensity. Analyte-line peak intensities are measured and plotted against analyte concentration. The curve, extrapolated to zero concentration, intercepts the intensity axis at the background intensity. Figure 14.4 shows a composite typical x-ray spectrum (2θ scan) having emitted and scattered x-rays that may be used for, or interfere with, an analysis. The significance of each feature is given in the caption. The analytical features of the figure are discussed in Section 14.3.1. Here, the figure serves to illustrate several ways to measure background. In the following discussion, the notation has the significance given in the caption of the figure.

If the analyte line A is superposed on uniform continuous background, background is measured adjacent to the peak at B_{adj}. If the analyte line A' is superposed on nonuniform continuous background, background is measured at equal 2θ intervals on the low-2θ B_L' and high-2θ B_H' sides of the peak. Then,

$$B_P' = (B_L' + B_H')/2 \tag{7.8}$$

If the analyte line A'' is superposed on a scattered target line or some other scattered spectral line T, two alternatives are available: B_P'' may be measured from a blank at the 2θ angle for the analyte; alternatively, a substantially constant relationship usually exists between adjacent background B''_{adj} and the background at the peak B_P'', which, in this case, includes T. Then B_P'' may be calculated from measured B''_{adj}:

$$B_P'' = kB''_{adj} \tag{7.9}$$

The constant k is evaluated from measurements at the 2θ angles of B_P'' and B''_{adj} on suitable blanks containing no analyte. A variation of this technique has been used for measurement of uniform background in determination of trace elements in hydrocarbons (*D25*). From measurements on pure hydrocarbons having various hydrogen/carbon ratios, a relationship is established between background at the analyte peak and the intensity ratio of a coherently and incoherently scattered target line. Thereafter, the ratio is measured from each sample and standard and used to derive the background at the peak. The technique has the advantage that the scattered target lines from a low-Z matrix are measured more rapidly and more precisely than the less-intense, uniform, scattered background.

If the analyte line A''' is superposed on a line of a matrix constituent M, background, including M, is usually measured from a blank.

If the peak intensity is measured by a continuous or stepwise integrated count (a–b), background may be measured by an integrated count on an equal-2θ interval of adjacent uniform background (a'–b'), or by a count at B_{adj} for a time interval equal to that required for the integrated peak count.

Reference is made above to the use of blanks. A blank is usually defined as a specimen identical with those to be measured in composition and physical form, except that the analyte is absent. In practice, if the analyte is a trace constituent, this definition may apply rigorously. However, if the analyte is a minor or major constituent, its omission may result in a change in scattering power of the remaining matrix. The change may be to higher or lower intensity, depending on the relative values of the atomic number of the analyte and effective atomic number of the matrix. In such cases, a correction factor may be applied, or the analyte may be replaced with an equal concentration of an element having about the same scattering power, but no interfering spectral lines. An element having atomic number one or two higher or lower than the analyte may be suitable.

In general, at the peak of element Z, the background intensity due to scattered primary x-rays is substantially the same as the intensity measured at 2θ for the Z peak from the same concentration of element $Z \pm 1$ or $Z \pm 2$. Usually element Z *plus* one or two is preferable for this purpose. There is a discontinuity in the intensity of the scattered spectrum at the absorption edge of the scattering element, the intensity being slightly less on the short-wavelength side. The phenomenon is analogous to the discontinuity in the primary *emission* spectrum, as described in Section 1.5.1. If element $Z - 1$ or $Z - 2$ is used, this low-intensity region is certain to extend to the wavelength of the Z line, giving an abnormally low scattered intensity. However, the absorption edges of elements of higher Z occur at wavelengths shorter than the Z line, and the scattered intensity at the Z line will be substantially normal.

In a multielement system, the background intensity $(I_B)_i$ at the peak of analyte element i may be closely approximated by

$$(I_B)_i = W_i I_i + \Sigma\, W_j I_j \tag{7.10}$$

where W is weight fraction, and I_j is the intensity measured at the analyte i peak from pure specimens of each of the other elements j in the specimen. I_i may be measured from pure analyte i adjacent to the i peak, or from element $i \pm Z$ or $i \pm 2Z$ at the i peak (see above). Of course, if for any element j, I_j has a component—such as a spectral line—other than scattered primary x-rays, an adjacent element must be used for that element. Finally, for highest accuracy, the measured I_j should be multiplied by a factor to compensate the slight increase in scattered intensity with decreasing Z or *vice versa*, that is,

$$I_Z = I_{Z'}(Z'/Z) \tag{7.11}$$

where I is the scattered primary intensity at the analyte peak, and Z and Z' are the atomic numbers of the analyte and matrix element, respectively.

7.3.4. Reduction

In general, background intensity is reduced by the following conditions:

1. X-ray tube target atomic number is as low as possible.
2. Excitation conditions (kV, mA) are as low as possible.
3. Filters are used to remove target lines or to reduce continuum intensity in the region of a specific analyte line.
4. The specimen compartment is evacuated or flushed with helium to eliminate air scatter.
5. The specimen compartment is fitted with baffles, primary and/or secondary apertures, or a collimated specimen mask to minimize background from air in the specimen compartment, the surface outside the specimen area of interest, or other structures in the specimen compartment.
6. Matrixes of low atomic number are avoided, such as solutions, fusion products, and glass disks.
7. Very small specimens are supported on fine fibers, and microgram and submicrogram quantities are supported on very thin films—Mylar 0.00025 inch thick or less, or ultrathin Formvar.
8. Low-Z binders or diluents and plastic cell windows are avoided for powder specimens.
9. Liquid specimens are frozen or held in open-top cells, again to avoid cell-window covering.
10. High-order lines are used; these lines lie at larger 2θ angles where background is lower.
11. The crystal contains no element(s) of relatively high atomic number to produce secondary emission.
12. Collimator foils are treated to minimize reflection and scatter.
13. Fine collimation is used.
14. The crystal axis is not parallel to the specimen plane (see below).
15. A gas-filled detector is used, rather than a scintillation counter, to avoid multiplier phototube noise.

16. The detector is operated at relatively low potential.
17. The amplifier is operated at relatively low gain.
18. Pulse-height analysis is used.

It is evident that some of the techniques that reduce background also reduce analyte-line intensity, and therefore may give little or no increase in peak-to-background ratio, and may even reduce it. Peak-to-background ratio is mostly a matter of excitation conditions and is discussed in Section 4.3.3.3.

The pulse-height analyzer discriminates against only that portion of the background that has wavelength substantially different from that of the analyte line. At a specified spectral line (2θ), that component of the background originating from primary radiation scattered by the specimen (or by the mask or other structure in the specimen compartment) *and* diffracted by the crystal has the same *product* of order and wavelength $n\lambda$ as the analyte line. However, the pulse-height analyzer is still applicable if the *wavelengths* are not the same. Most of that component of the background scattered by the crystal is likely to be rejected by pulse selection. The crystal emission can be eliminated by pulse selection unless the emitting element of the crystal lies within about three atomic numbers of the analyte.

The effectiveness of the pulse-height analyzer is reducing background varies with the spectral region, as shown in Figure 7.4. The effectiveness is least in the region of the scattered primary continuum hump. Birks [pp. 102–3 in (*5*)] points out a precaution to be taken in measurement of background with the pulse-height analyzer. If the baseline and window are set for the analyte-line distribution and left at these settings when the goniometer is set to a different 2θ angle to measure adjacent background, the measured background intensity will be erroneously low. Rather, the baseline must be reset to move the window into the adjacent background region. Of course, automatic pulse-height selection (Section 8.1.7) effects this function automatically.

Champion and Whittem (*C27*) have pointed out that certain instrument geometries are more prone to high background than others. Background consists mostly of scattered primary x-rays. In most instruments, the scatter angle intercepted by the collimator–crystal–detector system is $\sim 90°$, so the primary x-rays scattered by the specimen are highly plane-polarized. Conversely, the characteristic x-rays emitted by the specimen are completely random in direction and unpolarized. Consequently, a substantial increase in peak-to-background ratio would be effected by a polarization analyzer to discriminate against the polarized component of the x-radiation leaving

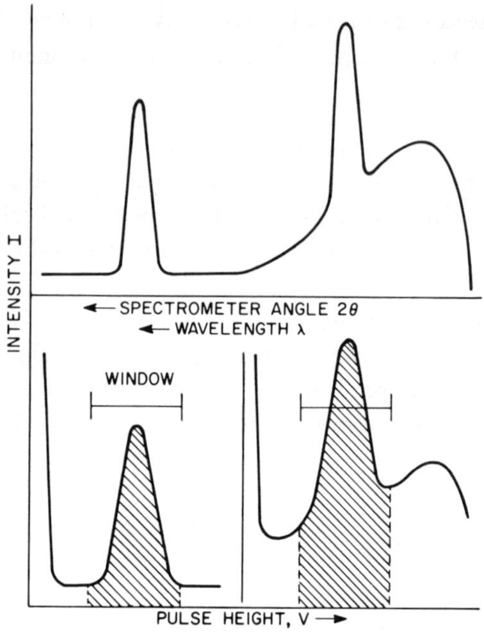

Fig. 7.4. Effectiveness of the pulse-height analyzer in reducing background. The upper curve is an idealized 2θ scan showing peaks in the high-2θ region, where background is low, and in the low-2θ region, where it is high. The lower curve shows the corresponding pulse-height distributions, and the shaded areas are the pulses passed by the windows shown.

the specimen. The analyzer crystal may constitute just such a device. In spectrometers having a horizontal axis (vertical plane) (Section 9.2.2, Figure 9.3), the axis of rotation of the crystal is parallel to the specimen plane, and at 2θ angles near 90°, the crystal and specimen planes themselves are substantially parallel. This arrangement transmits the polarized component efficiently, especially near 90° 2θ, as shown in Figure 7.5A. In spectrometers having a vertical axis (horizontal plane) (Section 9.2.2, Figure 9.2), the crystal and specimen axes and planes are more or less mutually perpendicular, and the instrument discriminates against the polarized x-rays, as shown in Figure 7.5B. Figure 7.5C shows the ratio of the background intensities scattered from a cell of pure water for arrangements B (crystal and specimen planes perpendicular) and A (planes parallel).

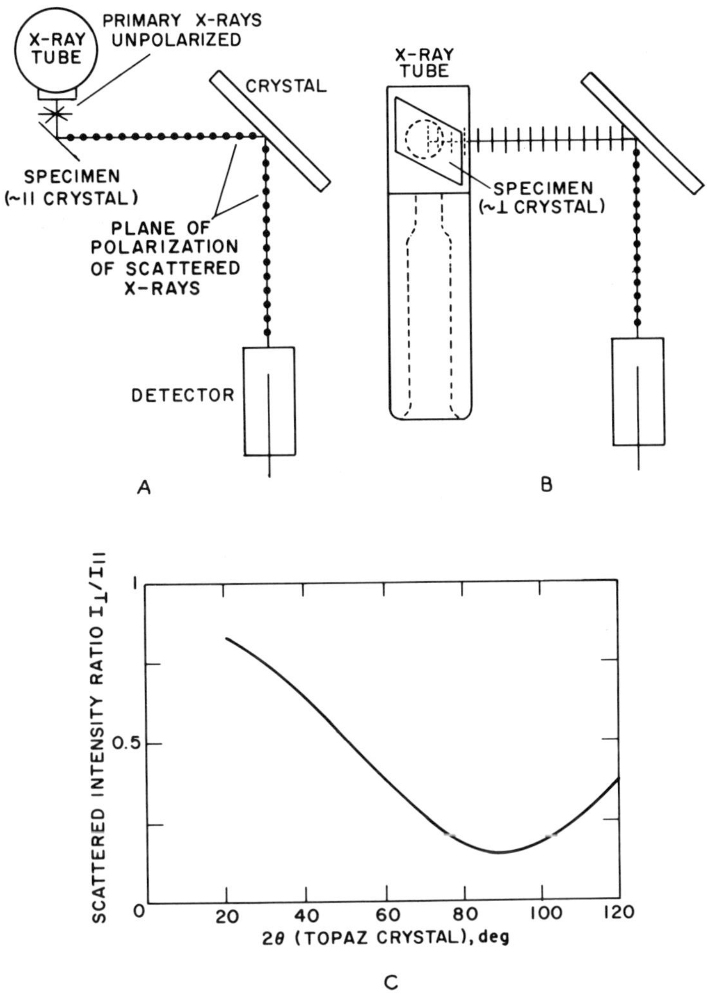

Fig. 7.5. Reduction of background by discrimination against polarized scattered primary x-rays. [K. P. Champion and R. N. Whittem, *Nature* **199**, 1082 (1963); courtesy of the authors and the Macmillan Co.] A. Crystal and specimen planes essentially parallel, especially at 90° 2θ. B. Crystal and specimen planes essentially perpendicular. C. Ratio of scattered primary intensity from water for arrangements B and A.

7.3.5. Considerations

It is by no means always necessary to correct analyte-line intensity data for background, and there are actually disadvantages to doing so. Unless background intensity can be derived from a blank or by extrapolation, at least one additional measurement is required from each specimen, with consequent increase in analysis time and cost. Moreover, background correction increases counting error because the statistical error in $I_P - I_B$ is greater than that in I_P alone.

The advantages of background correction include the following. The background measurement may serve to correct absorption-enhancement effects (Section 14.5.1) and as an indication of improper or abnormal operation of the instrument or of error in procedure. When working without a calibration curve, for example, with a single-standard or standard-addition or standard-dilution method, the net intensity is preferable.

Liebhafsky and his colleagues [pp. 212–15 in (*16*)] classify background problems in six categories as follows. In the discussion, I_P and I_B are analyte-line peak and background intensities, respectively, and C_A is analyte concentration.

1. The background is negligible, that is, I_B is so small that $(I_P - I_B) \approx I_P$, and background need not be measured.

2. The background is substantial, but need not be corrected. If I_B is constant over the entire analytical range of interest, and if the analyte-line *vs.* concentration curve is linear, the function $I_P - I_B$ *vs.* C_A is linear and intercepts the origin, whereas I_P *vs.* C_A is linear and intercepts the intensity axis at I_B. In such cases, there is no advantage in correcting background.

3. Background correction is required, but simple; for example, if I_B can be measured from a single measurement on a blank.

4. Background correction is required, and there are no complications, but it must be measured from each specimen on one or both sides of the analyte peak.

5. Background correction is required and complex; for example, when there are interfering lines superposed on the continuous background near the analyte peak and a blank is not feasible, or the background is affected by absorption edges.

6. Background is so large with respect to the analyte line that it may seriously interfere with the analysis. In trace analysis, $I_P \approx I_B$, and statistical fluctuations in both may lead to reporting of an analyte as present when it is absent, and *vice versa*.

In general, background correction should be made only when it leads to greater analytical reliability. However, it is always prudent to exercise all practical means of reducing background.

Chapter 8

Pulse-Height Analysis; Nondispersive Analysis

8.1. PULSE-HEIGHT ANALYSIS

The instrumentation and general function of the pulse-height analyzer (or pulse-height selector) are described in Section 7.1.4.1. This chapter considers the method of pulse-height analysis in greater detail.

8.1.1. Principle of Pulse-Height Selection

(*F12, H16, M18, V3, V4*)

The height of an individual pulse and the mean height and width of a pulse-height distribution are influenced by many parameters, principally the following: (1) photon energy; (2) the number of useful electrons produced by the initial ionizing event; (3) the internal electron multiplication (gas or secondary) in the detector; (4) the relationship of incident x-ray intensity and detector resolving and recovery times; (5) detector potential; and (6) amplifier gain. One might be inclined to wonder how proportionality can have any significance with so many variables. Of course, the answer is that if all the variables except photon energy are kept constant, the photon energy becomes the determining parameter.

Mean pulse height is given by

$$V = \frac{E}{E_D} A_D \frac{A_E}{L} \qquad (8.1)$$

where V is mean pulse height for a specified spectral line (V); E is x-ray photon energy for the line (eV); E_D is the average energy to produce one useful electron in the detector (eV); this would be one ion pair in a proportional counter, and one photoelectron collected at the first dynode in a scintillation counter; A_D is internal detector amplification—gas amplification or secondary-electron multiplication; A_E is external amplifier gain; and L is amplifier attenuation.

In general, if all other parameters are constant, the pulse-height distribution has lower mean amplitude V, narrower half-width, $W_{1/2}$ or ΔV, and higher energy resolution $\Delta V/V$: (1) for proportional counters than for scintillation counters; (2) the longer the x-ray wavelength (the lower the photon energy), or, for a specified spectral line, such as $K\alpha$ or $L\alpha_1$, the lower the atomic number; (3) the lower the applied detector potential; and (4) the lower the amplification (gain). The relationship of mean pulse height, energy resolution, and atomic number for $K\alpha$ lines is shown in Figure 8.1. The last three conditions above can all be illustrated by Figure 8.2, which is explained in its caption and should be studied carefully. In all three cases, although the peak intensities decrease progressively from A to C, the integrated intensities, the areas under the peaks, remain substantially constant.

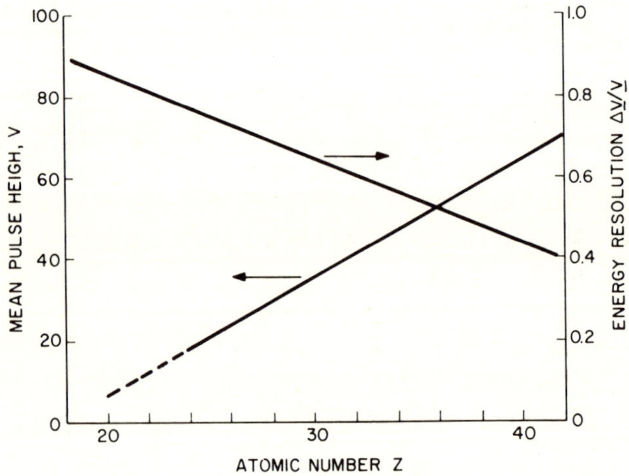

Fig. 8.1. Mean pulse height, and ratio of width of pulse-height distribution and mean pulse height as functions of atomic number for $K\alpha$ lines.

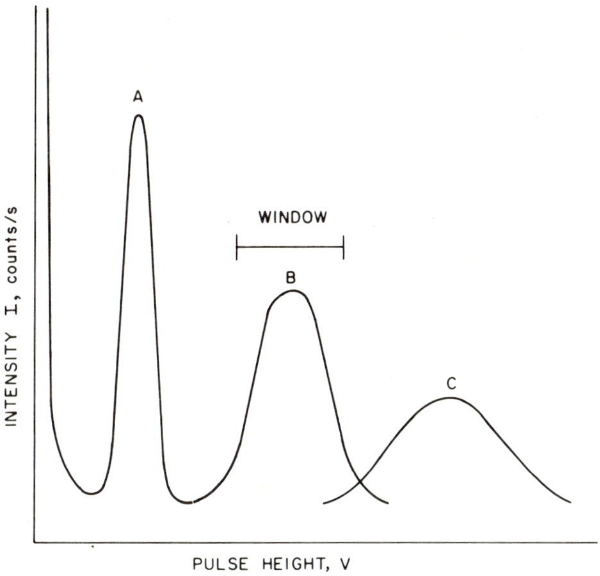

Fig. 8.2. Effect of incident x-ray photon energy, detector potential, and amplifier gain on mean pulse height, width, and peak intensity of the pulse-height distribution. The areas under the three peaks are equal. The figure represents any of three cases. I. Detector potential and amplifier gain constant: peaks *A*, *B*, and *C* represent pulse-height distributions for low, medium, and high photon energy, respectively. II. Photon energy and amplifier gain constant: peaks *A*, *B*, and *C* represent successively higher detector potentials. III. Photon energy and detector potential constant: peaks *A*, *B*, and *C* represent successively higher amplifier gain.

Figure 8.2 can also be used to illustrate the principle of pulse-height selection. Assume that the figure shows three pulse-height distributions of x-rays having low, medium, and high photon energy (long, medium, and short wavelength), respectively (case I in the caption). It is evident that with the window set as shown, the *B* peak is passed to the exclusion of *A* and *C*. If the baseline potential is varied, the pulse-height distributions remain fixed, but the window is moved and can be set to admit any one of them. If the detector potential and/or amplifier gain are varied, the window remains fixed, but the pulse-height distributions are moved, and any one can be moved into the window. Finally, the window width can be varied to pass as many as possible of the pulses in the analyte-line distribution while excluding as many as possible of the pulses from neighboring peaks, which may, unlike the peaks shown here, overlap. For example, it is evident that the window width shown in Figure 8.2 just admits the entire *B* peak. If the window were moved to the *A* peak, it could be made narrower without loss of *A* pulses, but if it were moved to the *C* peak, many *C* pulses would be excluded unless the window were made wider. In practice, it is customary to move the pulse-height distribution to the window, rather than the window to the distribution.

Incidentally, although emphasis is placed on proportionality of mean pulse height and photon energy—and therefore atomic number—the half-width is also proportional to photon energy and atomic number. Consequently, half-width can be used for identification of elements if the $W_{1/2}$ *vs.* E relationship is previously established (*H18*).

The *analyzer transmission* is the fraction of the pulses originating from a monochromatic x-ray beam that pass the pulse-height analyzer and are counted. The *detector efficiency* is the fraction of the photons of a monochromatic x-ray beam arriving at the detector window that are counted. It is equal to the product of the detector quantum efficiency (fraction of photons detected) and the analyzer transmission. Both these terms are usually expressed as per cent.

8.1.2. Pulse-Height Distribution Curves

Pulse-height distribution curves are best described in specific, rather than general, terms. A good illustrative example is the determination of silicon at concentration ~1% in iron with an EDDT crystal (*K10*). Fourth-order Fe $K\beta$ slightly overlaps Si $K\alpha$. Table 8.1 gives the pertinent data.

TABLE 8.1

Data for Si $K\alpha$ and Fourth-Order Fe $K\beta$[a]

Spectral line	Si $K\alpha$	Fe $K\beta(4)$
Wavelength λ, Å	7.13	1.76
$n\lambda$ ($n = 4$)	—	7.04
Energy E, eV	1739	7045
Diffraction angle (EDDT) 2θ, deg	108	106
Peak intensity I, counts/s	~160	~1600

[a] Data apply to Figures 8.3–8.6.

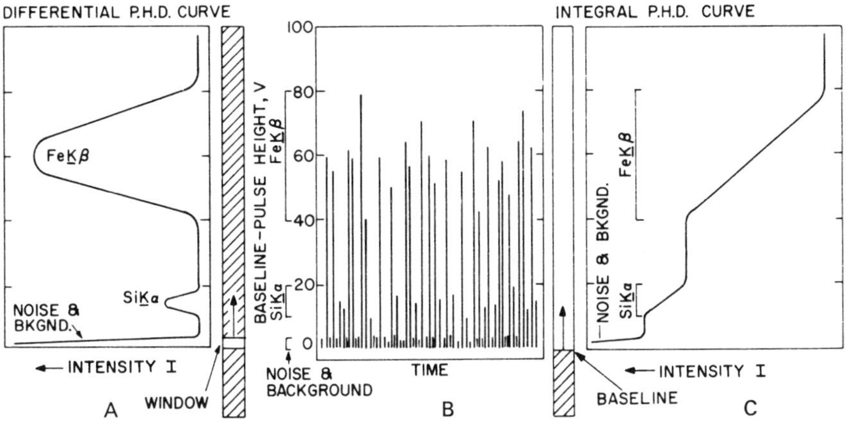

Fig. 8.3. Function of the pulse-height analyzer. The center figure (B) represents the amplifier output, pulse height *vs.* time, and shows three groups of pulses from background, Si $K\alpha$, and Fe $K\beta$. The left figure (A) shows the differential pulse-height distribution curve. The right figure (C) shows the integral pulse-height distribution curve.

Actually, the two lines are separated by 2° 2θ, but the Fe $K\beta$(4) is so intense relative to the Si $K\alpha$ that the tail of the iron line contributes intensity at the 2θ angle of the silicon line. Thus, with the spectrogoniometer at 108° 2θ (Si $K\alpha$), the detector output, and therefore the amplifier output, contains three groups of pulses: very low-amplitude detector noise, low-amplitude Si $K\alpha$, and relatively high-amplitude Fe $K\beta$. There are several ways to display pulse-high distribution data, three of which are shown in Figure 8.3.

In Figure 8.3B, individual pulses are displayed on a grid of pulse height *vs.* time. The three pulse-height distributions are distinguishable and are marked at the left of the drawing. As expected, the high-energy Fe $K\beta$ distribution is wider than the low-energy Si $K\alpha$ distribution. Such a display is obtained by connecting the vertical input of a cathode-ray oscilloscope to the output of the detector, preamplifier, or amplifier. The three displays would differ only in linear amplification. The x-ray photons incident on the detector, and therefore the output pulses, have a random distribution in time.

The pulse-height analyzer and ratemeter recorder may be used to display the pulse-height distribution in two ways.

A *differential pulse-height distribution curve* (Figure 8.3A) is recorded by: (1) setting the spectrogoniometer at 108° 2θ (Si $K\alpha$); (2) setting the window at an arbitrary small width ΔV, say 1 V; and (3) "scanning the baseline," that is, moving the baseline discriminator level V—and with it the window—from 0 to 100 V as intensity is recorded. At any instantaneous baseline setting, the window passes to the ratemeter only those pulses having height between V and $V + \Delta V$. The resulting curve of intensity *vs.* pulse height, Figure 8.3A, shows two pulse-height distribution peaks and the detector noise. The orientation is unconventional, to permit correlation with Figure 8.3B. At very low baseline potential, the high-intensity, low-energy noise passes the window, giving high response, which then falls to zero until the window passes through the Si $K\alpha$ distribution, giving a small peak. The response then again falls to zero until the window passes the intense Fe $K\beta$ distribution.

An *integral pulse-height distribution curve* (Figure 8.3C) is recorded by: (1) setting the spectrogoniometer at 108° 2θ as before; (2) making the window inoperative or, in effect, "wide open"; and (3) scanning the baseline V from 0 to 100 V as intensity is recorded. At any instantaneous baseline setting, *all* pulses having height greater than V pass to the ratemeter. The resulting curve of intensity *vs.* pulse height, Figure 8.3C, shows three *steps*, instead of the *peaks* of Figure 8.3A—the noise, Si $K\alpha$, and Fe $K\beta$. Again, the orientation is unconventional, to permit correlation. At very low baseline potential, *all* pulses—noise, Si $K\alpha$, Fe $K\beta$, and background—pass to the recorder, and the response is very high. At 3–5 V, the noise is discriminated out, and the intensity decreases. At a somewhat higher potential, the Si $K\alpha$ distribution is excluded, and at still higher potential, the Fe $K\beta$.

Both types of pulse-height distribution curve convey the same information, and the choice is largely a matter of personal preference. Such curves are made prior to undertaking analysis of a new sample type. They reveal escape peaks and overlapping of neighboring pulse-height distributions with that of the analyte, and permit the choice of baseline and window settings for the analysis. The window settings for Si $K\alpha$ and Fe $K\beta$ (of academic interest only) are shown in Figure 8.3B and 8.3C.

The baseline can be scanned in any of four ways: (1) The baseline control may be motor-driven; (2) the baseline may be advanced manually in equal, small increments at equal, short time intervals—say, 0.5 V every other chart division—as the recorder continuously registers intensity; (3) the baseline may be advanced incrementally, as above, and the intensity scaled at each setting (step scanners are available to perform this function automatically); and (4) the baseline may be scanned by electronic means.

In the *multichannel analyzer* (*V5*), the baseline in Figure 8.3 is, in effect, divided into 200, 400, 800, 1200, or even more individual equal, narrow, fixed windows, each connected to its own accumulator. The analyzer receives the amplified detector output and sorts the incoming pulses by height, directing each to its proper "box" along the baseline.

The process continues long enough to permit accumulation of a suitably large count—say 10^5—in the channel corresponding to the highest peak. Then the number of counts in each channel is rapidly read out electronically and indicated on an *X–Y* recorder or oscilloscope. A 400-channel analyzer, in effect, step-scans the baseline in 400 incremental steps and scales the intensity at each step—but it does it "all at once," rather than sequentially.

8.1.3. Pulse-Height Analyzer Displays

It is informative at this point to summarize and compare the various ways of displaying x-ray intensity data, particularly with respect to the pulse-height analyzer. The silicon–iron example is used again for illustration. Figure 8.4 shows four types of ratemeter–recorder display.

Figure 8.4B is a 2θ scan of the spectral region near the Si $K\alpha$ line recorded without pulse-height selection, that is, with only a discriminator set to reject detector noise. The scan shows intensity as a function of 2θ (wavelength). The continuous background is very high, and the tail of the Fe $K\beta(4)$ peak contributes intensity at the 2θ angle for Si $K\alpha$. The Si $K\alpha$ line-to-background ratio, that is, the intensity ratio of Si $K\alpha$ to background plus Fe $K\beta$, is only ~1:1.

Figure 8.4A is the same 2θ scan, but with pulse-height selection. The baseline and window are set to pass the Si $K\alpha$ distribution, as shown in Figure 8.3. The continuous background has been substantially eliminated. Only a vestigial Fe $K\beta(4)$ peak remains, and it no longer contributes intensity at the Si $K\alpha$ peak. The Si $K\alpha$ net intensity is substantially the same as before, but the line-to-background ratio is now ~37:1.

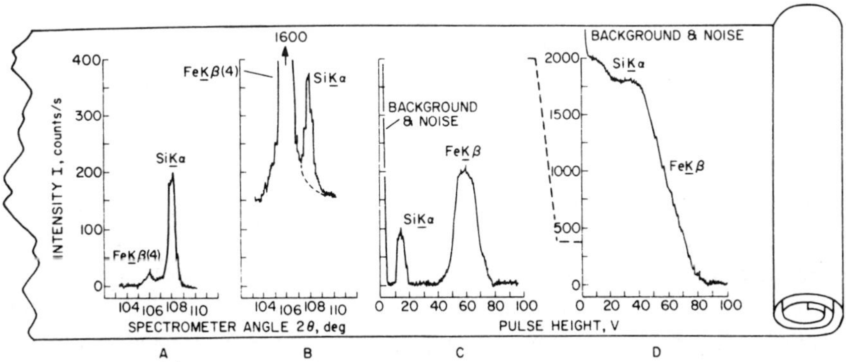

Fig. 8.4. Ratemeter–recorder readout of intensity data. A. 2θ scan (EDDT crystal) with pulse-height selection. B. 2θ scan without pulse-height selection. C. Differential pulse-height distribution curve. D. Integral pulse-height distribution curve.

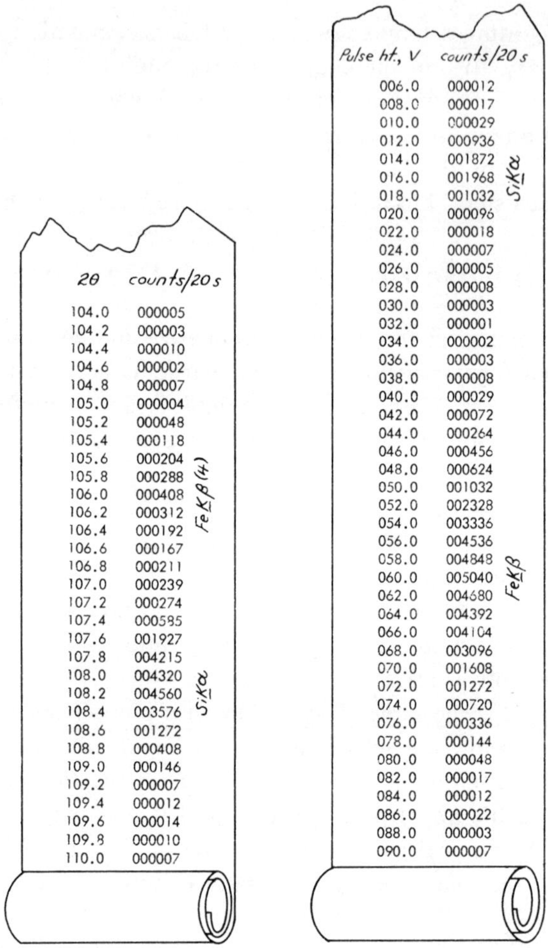

A B

Fig. 8.5. Printer readout of intensity data. A. 2θ scan with pulse-height selection (Fig. 8.4A). B. Differential pulse-height distribution curve (Figure 8.4C).

Figure 8.4C is a differential pulse-height distribution curve showing intensity as a function of pulse height. It is equivalent to Figure 8.3A. The spectrogoniometer is fixed at 108° 2θ (Si $K\alpha$). The window is set at 1 V. The baseline is scanned continuously from 0 to 100 V. The curve shows the number of pulses per second passing through the narrow window at each instantaneous baseline potential. The differential pulse-height distribution curve is an x-ray *energy* spectrum directly analogous to the x-ray *wavelength* spectrum represented by a 2θ scan.

Figure 8.4D is an integral pulse-height distribution curve, also showing intensity as a function of pulse height. It is equivalent to Figure 8.3C. The spectrogoniometer is fixed at 108° 2θ as before. The window is inoperative ("wide open"). The baseline is scanned continuously from 0 to 100 V. The curve shows the number of pulses per second passing over the baseline at each instantaneous baseline potential.

Figure 8.5 shows digital printer readouts of two series of intensity measurements. Figure 8.5A shows data for a 2θ scan of the spectral region near the Si $K\alpha$ line with pulse-height selection. The data plotted as intensity *vs.* 2θ would resemble Figure 8.4A. The spectrogoniometer is manually indexed or automatically step-scanned in increments of 0.2° 2θ, and, at each position, the intensity is scaled for a preset time of 20 s. Figure 8.5B shows data for a differential pulse-height distribution curve. The data plotted as intensity *vs.* pulse height would resemble Figure 8.4C. The spectrogoniometer is fixed at 108° 2θ. The window is set at 1 V. The baseline is manually or automatically scanned in increments of 2 V, and, at each potential, the intensity is scaled for a preset time of 20 s.

Figures 8.3B and 8.6 show cathode-ray oscilloscope displays, the former as a drawing. Figure 8.3B, as already mentioned, shows the individual pulses in the output of the detector, preamplifier, or amplifier on a scale of pulse height *vs.* time. The spectrogoniometer is fixed at 108° 2θ, and the display shows the pulses due to detector noise, Si $K\alpha$, and Fe $K\beta$. Figure

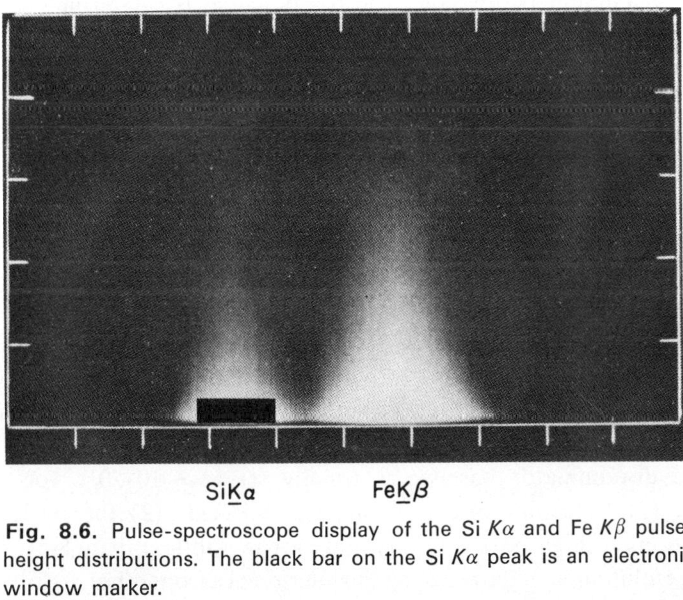

Fig. 8.6. Pulse-spectroscope display of the Si $K\alpha$ and Fe $K\beta$ pulse-height distributions. The black bar on the Si $K\alpha$ peak is an electronic window marker.

8.6 shows a differential pulse-height distribution curve photographed from a Siemens pulse spectroscope (*B54*). The conditions are similar to those for Figure 8.4C, except that the scanning is effected electronically. The black bar on the Si $K\alpha$ peak is an electronic marker showing the window position. The marker lengthens and moves as the window is widened and shifted along the baseline.

8.1.4. Pulse-Height Analyzer Operating Controls

The operating controls involved in the use of the pulse-height analyzer are the detector potential (up to ~2 kV), amplifier gain (up to ~10,000), baseline (0–100 V), window (0–10, 20, or 30 V), and differentiate–integrate mode switch.

The detector potential and amplifier gain both affect the mean pulse height, peak intensity, and half-width of the pulse-height distribution. Decreasing of either control increases the peak intensity and decreases the width; increasing of either control has the reverse effect (Figure 8.2). The detector potential and amplifier gain controls are used together to: (1) give optimum signal-to-noise, or peak-to-background, ratio; and (2) to move the pulse-height distributions along the baseline, usually to a relatively low potential.

The maximum permissible detector potential is established from the detector noise curve, and the useful potential range from the plateau curve (Section 8.1.5.1).

With the inherently noisy scintillation counter, low detector potential and high amplifier gain must be used. With the relatively noise-free proportional counter, relatively high detector potential and intermediate amplifier gain may be used. The higher detector potential permits advantageous use of the relatively noise-free gas amplification of the detector. With the Geiger counter (which, of course, is not applicable to pulse selection), the high output pulse amplitude permits use of low amplifier gain.

Usually, the amplifier gain is set so that pulses of the lowest amplitude to be measured (that is, those originating from the longest wavelength) fall at a selected baseline setting. Then the gain is attenuated to bring higher-amplitude (shorter-wavelength) pulses to the selected working range.

The discriminator baseline is usually set at ~10–20 V for several reasons: (1) Detector noise is completely rejected; (2) the pulse-height distribution has high peak intensity and narrow width (Figure 8.2); (3) the energy resolution is high, reducing the interference from other distributions;

(4) a narrow window can be used without substantial loss of analyte pulses; and finally (5), the line-to-background ratio is increased because the narrow window admits fewer continuous background pulses.

In general, the window is set wide enough to pass the entire analyte pulse-height distribution and its escape peak, if any. If the detection process is preceded by crystal dispersion, the window may usually be much wider than the distribution because the crystal eliminates neighboring wavelengths, and distributions from higher orders are well separated in energy. Thus, the analyte distribution is likely to be free from neighboring distributions. A wide window is advantageous to allow for shift and distortion effects. If the optics are nondispersive, neighboring distributions are likely to be present, and it may be necessary to narrow the window, sacrificing analyte pulses to exclude pulses from adjacent peaks.

The baseline and window controls are usually 10-turn helical potentiometers. Provision may be made to motor-drive the baseline control continuously or stepwise. Provision may be made to couple the baseline alone or the baseline and window to the spectrogoniometer θ-2θ drive (Section 8.1.7). The window always "tracks" the baseline, that is, the window may be regarded as resting on the baseline.

In some instruments, provision is made to operate the window in either of two modes: In *fixed-width* mode, the window has the same preset width in volts at all baseline settings; in *relative-width* mode, the window width varies in proportion to the baseline setting, always maintaining a preset relative value. In either case, the width may be selected: 0–10, 20, or 30 V in fixed width, 0–20% of baseline in relative width.

The differential–integral control determines the mode of operation of the pulse-height analyzer. In differential mode, both discriminator levels are activated, and the analyzer functions as a pulse selector, passing only pulses having heights between the two levels. In integral mode, the second discriminator, and therefore the upper level, is inactivated, making the window inoperative (wide open). The analyzer now functions simply as a discriminator, passing all pulses having height greater than the baseline.

8.1.5. Use of the Pulse-Height Analyzer

8.1.5.1. *Evaluation of Detector Characteristics*

The amplifier gain is set at low, medium, or high for Geiger, proportional, or scintillation counters, respectively. The Geiger counter is not applicable to pulse selection, but does require the evaluation described in this section.

The *detector noise level* is established as follows. With no x-rays, the baseline discriminator setting is reduced to very *low* values until a sharp increase in response is observed on the ratemeter or recorder. This is the detector noise level. The minimum baseline setting should be 2–5 V above this level to ensure adequate discrimination against the noise. Usually, a minimum baseline of 5 V is satisfactory. For pulse-height analysis, a baseline of 10–20 V is typical.

The *detector noise curve* is established as follows. Again, with no x-rays, the detector potential is increased incrementally from zero, and intensity is read on the recorder at each point. At some relatively high potential, the response increases, gradually at first, then sharply. The plot of intensity *vs.* detector potential without x-rays is the detector noise curve and indicates the maximum permissible operating potential (Section 6.4.1.10, Figure 6.10).

The *detector plateau curve* is established as follows. A specimen of pure element (or its oxide) having a major spectral line in the region of interest is excited to give an intensity of ~1000 counts/s for Geiger counters, 2000–5000 counts/s for proportional and scintillation counters. The baseline is set at 5–10 V. Detector potential is increased incrementally from zero, and intensity is read on the recorder at each point. The response rises, remains constant over a substantial potential range, then rises again. The plot of intensity *vs.* detector potential, with x-rays, is the detector plateau curve. The linear plateau itself is the useful range of detector potential. Geiger counters are operated near the center of the plateau. So are proportional and scintillation counters, but their potential may be varied somewhat as an operating variable to displace pulse-height distributions along the baseline. The detector potential should never be moved off the plateau. It must be recalled that the plateau curve is shifted to lower detector potential at shorter wavelengths, higher amplifier gain, and lower discriminator baseline (Section 6.4.1.8, Figure 6.10).

The foregoing evaluation need be made carefully for each specific detector only at intervals of several months. However, rapid checks at more frequent intervals monitor the condition of the detector and give warning of onset of deterioration.

8.1.5.2. *Establishment of Pulse-Height Analyzer Settings*

The pulse-height analyzer operating conditions for determination of a specified analyte in a specified type of sample are established by the following—or a similar—procedure.

First, the peak of the analyte pulse-height distribution is moved along the baseline to a potential of 10–20 V, and the detector potential and amplifier gain settings to be used for recording the pulse-height distribution curves are established. The detector potential and amplifier gain are set temporarily in accordance with the guides outlined in Section 8.1.5.1, or on the basis of previous experience. A specimen of pure analyte is placed in the specimen chamber. The oxide or another compound may be used, but no element should be present having an observable pulse-height distribution that may be confused with that of the analyte. The spectrometer is set at the analyte line, and excitation conditions are selected to give an intensity of several thousand counts per second as observed on the recorder in the usual way. The differential–integral switch is set for differential mode. The baseline is set at 10–20 V. The window is set at 1 V. The detector potential and/or amplifier gain are now adjusted to give maximum intensity, indicating that the peak of the analyte-line distribution has been moved to the selected baseline potential. These detector and amplifier settings are left undisturbed, for they are to be used for recording the pulse-height distribution curves.

The pure element is now replaced by a specimen of the type to be analyzed. Excitation conditions are readjusted as required to give an analyte-line intensity of several thousand counts per second, or, if they have been established, to the conditions to be used for the analysis.

A differential pulse-height distribution curve is recorded as follows. The differential–integral switch is set for differential mode. The window is set at 1 V. The baseline is turned back to zero, then manually turned, fairly rapidly, through its full 0–100-V range while intensity is observed on the recorder. In this way, the maximum intensity to be encountered is established, and the recorder scale set accordingly. The baseline is now returned to zero and motor-driven from 0 to 100 V as intensity is recorded.

An integral pulse-height distribution curve is recorded as follows. The switch is set for integral mode, making the window inoperative (wide open). The baseline is set at zero. Since maximum intensity occurs at zero, the recorder scale can be set now. The baseline is motor driven from 0 to 100 V as intensity is recorded.

In the absence of a motor drive, the baseline can be scanned manually in equal, small increments at equal, short time intervals—say, 0.5 V every other chart division. Electronic scanning may also be used.

An examination of either pulse-height distribution curve reveals escape peaks and possible overlapping peaks. The curves permit selection of the optimal baseline and window settings to admit as many pulses as possible from the analyte peak, as few as possible from peaks of other elements.

Finally, all settings—kV, mA, detector potential, amplifier gain, baseline, and window—are recorded and always used thereafter to determine the specified analyte in the specified type of sample. The foregoing procedure must be followed for each other analyte to be determined with the aid of the pulse-height analyzer. If the analyte having spectral line of longest wavelength is evaluated first, it may be that the other analyte lines simply require progressively lower amplifier gains to move their pulse distributions to the window. This preliminary work is done only once for a given sample type. However, it is prudent to check conditions occasionally, especially if the detector plateau curve shows a change. The great convenience of the Siemens pulse spectroscope (Figure 8.6) for this preliminary work is obvious.

8.1.6. Applications and Limitations

Certainly the pulse-height analyzer is not required, perhaps not even beneficial, in many analyses. When sharp, intense spectral lines occur in a sparse spectrum having low, continuous background, there is little, if any, advantage in use of the analyzer. However, in other situations, it is extremely beneficial, even indispensable.

The principal applications of the pulse-height analyzer are the following: (1) elimination of higher-order spectral-line interference; (2) reduction of background by elimination of that portion having wavelength different from that of the measured line; (3) increase of detection limit in trace and micro analysis because of the reduced background; (4) higher resolution of lines that are relatively close in wavelength; (5) higher sensitivity at long wavelengths because crystal dispersion need no longer resolve higher-order interfering lines and coarser collimators may be used; (6) elimination of radioactive background from radioactive specimens; and (7) feasibility of nondispersive operation (Section 8.2) with resultant ~100-fold increase in measured intensity.

Probably the principal applications of pulse-height analysis are the first two listed above and the last. It is very significant that two spectral lines that diffract at the same—or nearly the same—2θ angle in crystal dispersion may have widely separated pulse-height distributions. For example, Cu $K\alpha$ (1.542 Å) and third-order In $K\alpha$ (λ 0.513 Å, 3λ 1.539 Å) are inseparable by crystal dispersion, but have wide energy dispersion (photon energies 8040 and 24,140 eV, respectively). However, the pulse-height analyzer cannot separate pulse distributions from similar wavelengths.

The resolution of a pulse-height analyzer used with a proportional counter is far superior to that with a scintillation counter—but is still none too good. Pulse distributions for the same lines of elements differing in atomic number by ± 1 are not resolved; those differing by ± 2 or 3 are resolved only by sacrifice of a substantial fraction of the total intensity, with consequent loss in counting efficiency; those differing by ± 4 are resolved with little loss of counting efficiency. The ability of a counter (proportional or scintillation) to resolve pulse distributions of neighboring elements is substantially constant throughout the periodic table (*H18*). With the advent of the lithium-drifted silicon and germanium detectors (Section 6.5.3.2), resolution of lines of adjacent elements has been accomplished.

The application of the pulse-height analyzer in reduction of background is discussed in Section 7.3.4 and illustrated in Figures 7.4 and 8.4A,B. Again, the analyzer removes only background having wavelength substantially different from that of the analyte line.

The effectiveness of the pulse-height analyzer is limited by: (1) the difference in mean pulse height of the pulse-height distributions to be resolved; (2) their half-widths; (3) their profiles, that is, whether Gaussian or distorted; (4) their relative intensities; (5) the presence or absence of escape peaks and their character with respect to the first four properties; and (6) the continuous background pulses. The pulse-height analyzer is more effective: (1) the more widely separated, narrower, and more nearly Gaussian the distributions; (2) the more intense the analyte distribution and the less intense the interfering one; (3) the fewer, weaker, narrower, and more separated the escape peaks; and (4) the lower the background.

8.1.7. Automatic Pulse-Height Selection

Because mean pulse height is proportional to x-ray photon energy, for each wavelength to be measured, different settings are required for the baseline and window, or for the detector potential and/or amplifier gain. Such resetting is very inconvenient when several analyte lines are to be measured on a number of specimens, and is intolerable in automatic instruments and in recording of 2θ scans. Accordingly, methods have been devised to couple the pulse-selection function to the θ-2θ drive of the spectrogoniometer so as to vary the pulse-selection conditions automatically and synchronously with 2θ, that is, with wavelength.

The methods for effecting this function are of two basic types—those that vary the pulse height and those that vary the baseline, or, in effect, those that move the pulses to the window and those that move the window to the pulses. The two approaches are best discussed in terms of Equation (8.1), $V = (E/E_D)A_D(A_E/L)$, where V is mean pulse height; E is x-ray photon energy; E_D is energy to produce one useful electron in the detector (Section 8.1.1); A_D is internal electron multiplication in the detector; A_E is external amplifier gain; and L is amplifier attenuation.

In the first method, the baseline and window are constant, and, as 2θ varies, the amplifier gain and/or attenuation A_E/L is varied to amplify the pulses to different degrees and move them into the preset window. The same effect would result from variation of detector potential, and, thereby, of A_D. However, A_D is highly dependent on the specific detector, and this method is avoided. For a specified detector, E_D is constant if the detector potential, and therefore A_D, are constant. Then, from Equation (8.1) and the Bragg law,

$$V \propto E(A_E/L), \qquad E \propto 1/\lambda, \qquad \lambda \propto (2d \sin\theta)/n$$

For a specified analyzer crystal and order n of λ, $2d/n$ is constant. Then,

$$A_E/L \propto V \sin\theta$$

Consequently, if A_E/L is varied sinusoidally with θ, all pulse-height distributions occur at V. Amplifiers designed to effect this function are known as *sine-function amplifiers* (*W14*).

In the other method, as 2θ varies, the pulse-height distributions of the spectral lines are allowed to undergo their natural variation in mean height and in width. The baseline (*R6*, *S5*), and perhaps also the window (*W3*), are moved to receive them. Again, for a specified detector and potential, E_D and A_D are constant. Then, if A_E/L is also constant,

$$V \propto E \propto 1/\lambda \propto 1/\sin\theta$$

Consequently, if, as θ varies, the baseline is varied in proportion to $1/\sin\theta$, the baseline follows the pulse-height distributions. An added refinement is provision for variation of the window width to correspond to the variation in width of the pulse-height distributions. The window should vary as $1/(\sin\theta)^{1/2}$, but, in practice, a linear variation with θ is satisfactory.

8.1.8. Problems with Pulse-Height Selection

8.1.8.1. *General*

If the detector is receiving a monochromatic x-ray beam and the pulse-height analyzer is not used, each pulse resulting from an incident x-ray photon is counted, provided only that it is higher than the discriminator baseline. It does not matter whether the pulse is a natural or escape pulse, or whether the pulse distribution has shifted or become distorted. However, when the pulse-height analyzer is used, many conditions become troublesome that would otherwise be insignificant. Jenkins and de Vries have given a particularly good review of these problems, and much of the treatment below is derived from theirs. They classify pulse-height analysis problems in four categories: (1) pulse-amplitude shift; (2) pulse-amplitude distortion; (3) additional pulse-height distributions arising from the measured wavelength; and (4) additional pulse-height distributions arising from other wavelengths.

The fourth category comprises distributions arising from: (1) true spectral interference, where a second line is present having wavelength very close to the measured wavelength; (2) higher-order lines, where a second line is present having $n\lambda$ very close to the measured wavelength; (3) secondary characteristic emission from the analyzer crystal; (4) spurious lines diffracted from crystal planes inclined slightly to the surface planes (Figure 13.6); and (5) spurious lines that enter the detector because of its relatively large angle of acceptance. These phenomena are discussed in Section 13.3.2.

The other three classes of problems listed above are discussed in this section, where the treatment is based on that of Jenkins and de Vries [pp. 73–85 in (*14*)].

8.1.8.2. *Shift of Pulse-Height Distribution*

Consider a pulse-height analyzer set to pass the distribution outlined by the solid outline in Figure 8.7A. Now suppose that the distribution shifts as shown by the dotted outline. Those pulses in the shaded area now lie outside the window and are not counted, so the measured intensity decreases proportionally. Such shifts may occur during the course of an analysis, or on a long-term basis, affecting analyses in which the same long-established analyzer settings are used without occasional verification.

Proportional counters are more prone to pulse-amplitude shift than scintillation counters, and, in proportional counters, the effect is greater for: (1) flow counters than for sealed counters; (2) high applied potential, that is, high gas amplification; (3) high-atomic-number detector gas; (4) small anode wire diameter, that is, small detector capacitance; and (5) stainless steel wire than for tungsten wire of the same diameter.

Pulse-amplitude shifts usually result from four causes:

a. Drift in Applied Detector Potential. A change of 1.5% in applied potential may cause a 25% change in average pulse height in a

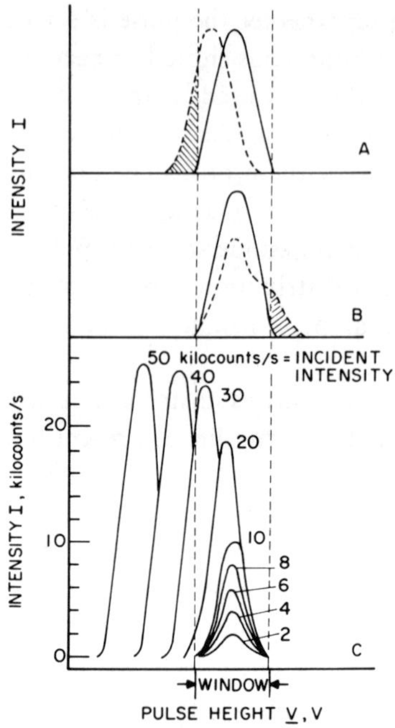

Fig. 8.7. Effect of shift and distortion of the pulse-height distribution. A. Shift. B. Distortion. C. Intensity effect—shift caused by dead-time effect. The detector is assumed to be free from dead-time losses up to 10,000 counts/s. At lower intensities, the detector output intensity decreases proportionally and without change in pulse height. At higher incident intentities, the detector output intensity increases at a progressively slower rate than the incident intensity, and the entire pulse-height distribution shifts to smaller average pulse height.

flow counter. The obvious remedy is high regulation ($\pm$0.05% or less) of the detector power supply.

b. Gas Density Effect. Increase in pressure or decrease in temperature increase the density of the detector gas, thereby decreasing the mean free path of electrons, the gas amplification, and the average pulse height. Decrease in pressure or increase in temperature have the reverse effect. Typically, a 1% change in density may cause ~5% change in pulse height. The effect is dealt with in several ways. The detector may be mounted inside a temperature-controlled spectrometer chamber. The detector mount may be water-cooled to prevent thermal conduction from the preamplifier. The gas

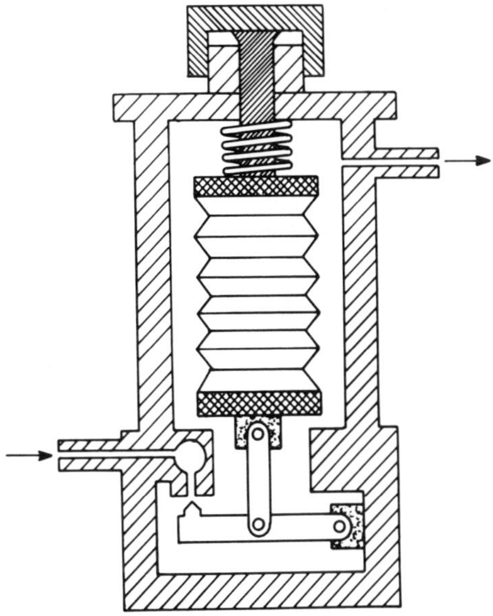

Fig. 8.8. Manostat for gas-flow proportional counter. [R. Jenkins and J. L. de Vries, "Practical X-Ray Spectrometry," Springer-Verlag N. Y. (1967); courtesy of the authors and publisher.]

outlet of a flow counter should not discharge directly into the atmosphere, but into a gas-density compensator or manostat to maintain constant pressure in the detector with fluctuating barometric pressure and ambient temperature. A typical manostat is shown in Figure 8.8. The ambient pressure and temperature act on the fixed mass of gas inside the bellows, causing expansion or contraction against the adjustable spring load at the top. The movement actuates the needle valve and regulates the effluent gas flow to maintain constant pressure in the detector. Another solution is to make the analyzer window wide enough to accommodate a reasonable amount of shift.

c. Gas Composition Effect. The ratio of detector and quench gas may change when the gas cylinder is nearly empty (*J6*). A change of 0.5% in the methane concentration in P10 gas (90% argon, 10% methane) may cause a 10–20% shift in pulse height. The obvious remedy is to replace the cylinder when it is 90% empty or so.

d. Intensity Effect. When the incident intensity is such that photons enter the detector between the dead and recovery times, the pulses are higher than the baseline, but not of normal height. The effect is serious with a proportional counter at 50,000 counts/s, but may not be serious with a scintillation counter even at 100,000 counts/s. The remedies include the following. The most obvious remedy is to reduce the analyte-line intensity by reducing the x-ray tube operating potential or current or by using a higher-order line. If a combination detector is being used, the signal may be taken from the *less* favorable one for the wavelength being measured. A more risky remedy is to use standards having analyte concentrations very close to those of the samples in the hope that the intensity effect will be similar for all specimens. To shorten the dead time, all dirt and corrosion must be cleaned from the tube, especially from the anode wire. If a sine-function amplifier is provided for automatic pulse-height analysis, it has the added benefit of correcting the intensity effect because it automatically varies the amplification and keeps the distribution in the window.

The intensity effect may be visualized by reference to Figure 8.7C. Let the outline marked 10 represent the pulse-height distribution for the maximum intensity at which no significant coincidence loss occurs—say, 10,000 counts/s. If now the incident intensity is increased incrementally to 20, 30, 40, 50, etc. kilocounts/s, the pulse-height distribution undergoes two changes: (1) The intensity of the entire distribution increases, but each 10-kilocount increase gives a progressively smaller increase in measured intensity; and (2) the entire distribution shifts to progressively *smaller* average pulse height—that is, to the *left* in the figure. Conversely, if the incident intensity is decreased incrementally to 8, 6, 4, 2, etc. kilocounts/s, the distribution remains fixed in position along the pulse-height axis, but decreases in intensity in proportion to the incident intensity.

Secondary characteristic emission from the analyzer crystal may sometimes cause an *apparent* pulse-height shift. Jenkins and de Vries give this example: If P $K\alpha$ is measured with a gypsum ($CaSO_4 \cdot 2H_2O$) crystal, a S $K\alpha$ (2.31 eV) distribution may occur with the *P* $K\alpha$ (2.02 eV) distribution. The two are not resolved, so the observed distribution is a composite of the two. At low P $K\alpha$ intensities, the S $K\alpha$ peak predominates, and the peak of the composite distribution lies near the peak of the sulfur distribution. As P $K\alpha$ intensity increases, its peak assumes dominance, with a resultant shift of the peak of the composite distribution.

8.1.8.3. *Distortion of Pulse-Height Distribution*

The distortion effect is shown in Figure 8.7B, where the pulses in the shaded area lie outside the window if the pulse-height distribution (solid outline) becomes distorted (dotted outline), that is, departs from its Gaussian profile. Distortion, like shift, is most serious with flow-proportional counters. It is not uncommon for the resolution of a new counter, or a counter that has been cleaned and fitted with a new anode wire, to deteriorate by a factor of two or three within a month. Distortion results from an inhomogeneous electric field caused by window effects; anode-wire end and clamping effects; and dirt, corrosion, pitting, and variation in diameter of the anode wire. Side-window counters are much less prone to distortion than end-window counters, and the inside surfaces of windows must be metallized or carbonized. When resolution has deteriorated to an unacceptable extent, the anode wire may be cleaned by gentle brushing with a camel-hair brush dipped in benzene, or replaced with a new, similarly cleaned wire.

8.1.8.4. *Additional Pulse-Height Distributions Arising from the Measured Wavelength*

In addition to its own natural pulse-height distribution peak, the measured wavelength may initiate *escape peaks* and *double* or *sum peaks.*

a. Escape Peaks. The origin of escape peaks (and of the term) is explained for proportional counters in Sections 6.2.3.1 and 6.2.3.2 and for scintillation counters in Section 6.3.2.2. The escape peak as a source of interference is discussed in Section 13.3.2.

The ratio of the mean pulse heights of the escape and main peaks is

$$\frac{V_{\text{esc}}}{V_{\text{main}}} = \frac{E - E_{\text{det}}}{E_x} \tag{8.2}$$

where V_{esc} and V_{main} are mean pulse heights of the escape and main peaks, respectively; E_x is the energy of the incident x-ray photon; and E_{det} is the energy equivalent of the absorption edge of the detector gas or of the iodine in the NaI(Tl) scintillator. The intensity of the escape peak relative to the main peak increases with (1) proximity of the incident wavelength to the absorption edge of the detector gas or iodine, and (2) the fluorescent yield of the detector gas or iodine. The effect of fluorescent yield is shown in Figure 6.5.

In general, escape peaks are not a serious problem in analytical x-ray spectrometry. In argon proportional counters, elements of atomic number 18 (argon) and below do not excite the argon *K*-absorption edge, and for elements of atomic number 28 (nickel) and above, the escape and main peaks are not resolved. Thus, escape peaks may be troublesome only for elements 19–27 (potassium through cobalt). Even for these, the low argon *K* fluorescent yield (~0.1) results in relatively weak escape peaks.

In xenon proportional counters, the *L*III absorption edge (2.6 Å) is outside the usual spectral region of the detector. In NaI(Tl) scintillation counters, the I *L*III absorption edge (2.7 Å) is at the long-wavelength limit of the useful range. Thus, x-ray lines most efficient in exciting these edges give very low-energy escape peaks that are lost in the detector noise. Although such escape peaks are lost for counting purposes, at least they cannot interfere with other useful pulse distributions. In both the xenon and NaI(Tl) scintillation detectors, the *K* edges (both ~0.36 Å) are excited only by the *K* lines of elements of atomic number 57 (lanthanum) or above. The *K* lines of these elements are excited relatively inefficiently by a 50-kV generator, and the elements are usually determined by use of their *L* lines. However, the *K* fluorescent yields of xenon and iodine are high (~0.9), and, when escape peaks do occur, they have two to three times the intensity of the main peak.

In the krypton proportional counter, the *K* edge (0.86 Å) is excited by elements of atomic number 38 (strontium) and above, and the *K* fluorescent yield is relatively high (~0.6). Thus, the krypton detector is the most prone to escape peaks of all the common detectors, but even it has a wide useful range.

Escape peaks, like main peaks, arise from the x-ray spectral line being measured, so, in general, the pulse-height analyzer window should be set to admit both. However, this is not feasible if: (1) the escape peak has such low pulse height that it is lost in the detector noise; (2) the two peaks are too widely separated; (3) too many background pulses are admitted; or (4) a main or escape peak of another element is wholly or partially admitted. Sometimes, even the last three of these difficulties can be surmounted if two pulse-height analyzers are available and set for the main and escape peaks, respectively. In this way, background and interfering peaks between the two peaks of interest are rejected.

Escape peaks may be troublesome in two ways. If, for any of the reasons given above, it is not feasible to include both the main and escape peaks in the window of a single-channel analyzer, only one can be measured. Then the measured intensity is reduced by the intensity represented by the rejected peak. Also, escape peaks complicate the energy spectrum and increase the probability of overlapping (Section 13.3.2).

An interesting escape-peak phenomenon occurs when a NaI(Tl) scintillation counter is used with automatic pulse-height selection to record a 2θ scan from a low-*Z* specimen having high scattered background (*P2*). A discontinuity occurs in the background at ~15.8° 2θ (topaz crystal), the intensity being lower at lower 2θ (shorter wavelength). The explanation follows. For topaz, 15.8° 2θ corresponds to the iodine *K* absorption edge. Below 15.8°, the scattered continuum is sufficiently energetic to excite the I $K\alpha$ line in the scintillator. An escape peak forms, but its amplitude lies outside the window automatically set to receive natural pulses corresponding to each 2θ angle, and it is not counted. Thus, measured background is abnormally low below 15.8° 2θ, but is normal at higher angles.

b. Double or Sum Peaks. If two x-ray photons enter the detector simultaneously, or very nearly so, a pulse of twice normal amplitude occurs. If this happens often enough, a second pulse distribution appears in the detector output at mean amplitude about twice that of the main peak. This new peak is known as a *double* or *sum* peak. Such a peak sometimes appears in pulse-height distribution curves recorded with fast counters at very high intensities. Triple and even higher sum pulses may also occur. If a pulse-height analyzer is used and the window set to admit the main peak, the number of photons that are detected but not counted is *twice* the number of double pulses. Even without the analyzer, there is a loss *equal* to the number of double pulses.

Sum pulses may be difficult to understand in view of the dead-time phenomenon. However, the problem is resolved if one considers that, in effect, the two concurrent photons are equivalent to a single photon of twice the energy. A photon entering the detector after the quench process has begun is, of course, subjected to the dead-time effect.

8.1.9. Unfolding of Overlapping Pulse-Height Distributions

Pulse-height distributions from neighboring elements are very poorly resolved. Thus, total measured intensity at the mean pulse-height potential for element i is the sum of the intensity from i and the fractional contributions from the pulse-height distributions of other neighboring elements j at the position of the i peak.

Dolby (*D11*) has shown that a set of linear simultaneous equations may be written for the intensities at the mean pulse-height positions—that is, the pulse-height distribution peaks—corresponding to each element. These equations are solved algebraically by use of the intensities measured at the position of the i peak from the sample, pure i, and pure specimens of each other element j contributing to the total intensity at the i peak. These equations have the form

$$I_i = I_{ii}R_i + \sum_{1}^{j} (R_j I_{jj} P_{ji}) \tag{8.3}$$

where I_i and I_{ii} are net intensities measured at the position of the i peak from the sample and pure i, respectively; I_{jj} is net intensity measured at the position of the j peak from pure j; P_{ji} is the fractional net intensity measured at the position of the i peak from element j; and R_i, R_j, . . . are true relative net intensities of elements i, j, . . . in the sample. These R's are the values required for quantitative analysis and are found by solution of the simultaneous equations.

This technique of deriving the individual contributions in overlapping pulse-height distribution curves is known variously as *unfolding*, *convolution sorting*, or *stripping*. The following examples are taken from work by Birks and Batt (*B32*) [p. 98–102 in (*5*)].

The basic approach is illustrated in Figure 8.9. The three solid curves represent pulse-height distributions from three consecutive adjacent elements A, B, and C in a hypothetical specimen. The curves were drawn with Gaussian distributions and arbitrarily assigned relative peak intensities of 4.5, 2.0, and 2.5, respectively. The dotted curve is the summation of the three individual curves and represents the single unresolved pulse-height distribution that would be observed from the hypothetical specimen. The composite curve has relative *measured* intensities at the positions of A, B, and C peaks of 6.06, 6.25, and 4.30, respectively.

The simplest approach to the unfolding is to assume that the individual distributions are Gaussian (this would not be known if only the composite curve were available), and have standard deviation σ equal to the separation

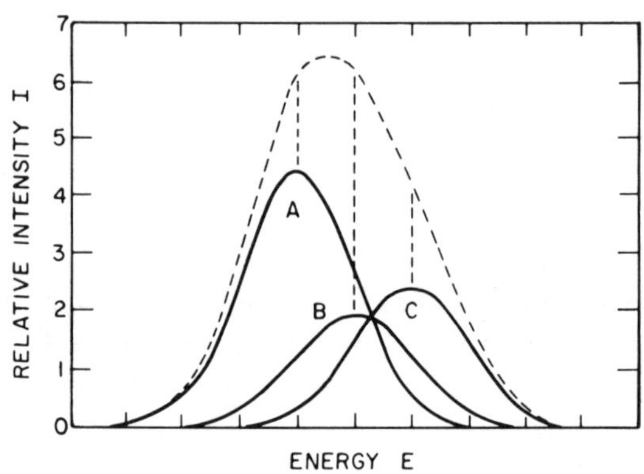

Fig. 8.9. Idealized pulse-height distribution curves of arbitrary intensities for spectral lines of adjacent elements A, B, and C in a hypothetical specimen. The dashed curve is the summation of the three individual curves. [L. S. Birks, "Electron-Probe Microanalysis," Interscience Publishers, N. Y. (1963); courtesy of the author and publisher.]

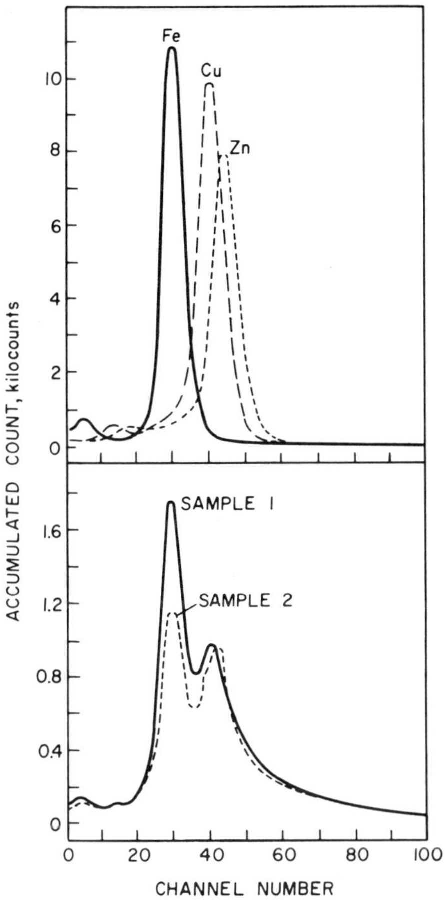

Fig. 8.10. Pulse-height distribution curves of *K* lines from iron, copper, and zinc standards (above) and from two samples (below). [L. S. Birks, "Electron-Probe Microanalysis," Interscience Publishers, N. Y. (1963); courtesy of the author and publisher.]

between peaks for adjacent atomic numbers. For a Gaussian distribution, the relative intensities at the peak, 1σ, and 2σ are 1.00, 0.607, and 0.135, respectively. Then the relative intensities of element A at its own peak and at the peaks of its first and second consecutive adjacent neighbors are 1.00, 0.607, and 0.135, respectively. These assumptions and the data for Figure 8.9 may be used to establish three simultaneous equations:

$$I_A = R_A + 0.607R_B + 0.135R_C = 6.06 \quad (8.4a)$$

$$I_B = 0.607R_A + R_B + 0.607R_C = 6.25 \quad (8.4b)$$

$$I_C = 0.135R_A + 0.607R_B + R_C = 4.30 \quad (8.4c)$$

where I indicates intensities measured from the specimen (dotted curve in Figure 8.9), each I being the sum of contributions from all three elements; and R indicates true relative intensity in the specimen.

If only the composite curve is known, and it is required to calculate R_A, R_B, and R_C, the three equations are solved by use of determinants. The solution gives 4.56, 2.06, and 2.50, respectively, in excellent agreement with the starting values given above.

In practice, the pulse-height distributions are not exactly Gaussian, and the peak intensity varies with atomic number due to dependence of detector sensitivity on wavelength. Consequently, the multipliers in Equations (8.4) must be determined by actual measurements on the pulse-height distributions of the pure elements. Figure 8.10 shows the pulse-height distributions for pure iron, copper, and zinc and for two samples containing

TABLE 8.2

Data for Unfolding of Pulse-Height Distribution Curves Shown in Figure 8.10[a]

Specimen	$K\alpha$ net intensity, counts/s	Pulse-height distribution peak position		
		Fe	Cu	Zn
		Fractional $K\alpha$ net intensity at peak position indicated above		
Fe	10,765	1.0	0.017	0.003
Cu	9,730	0.077	1.0	0.559
Zn	7,845	0.069	0.506	1.0
		Measured $K\alpha$ net intensity at peak position indicated above, counts/s		
Sample 1	—	1015	659	680
Sample 2	—	1610	740	630

[a] From Birks and Batt (*B32*).

these elements as minor (4–14%) constituents in a low-Z matrix. Table 8.2 gives the data derived from these curves at the positions of the iron, copper, and zinc peaks. The three simultaneous equations established from the data for sample 1 are as follows:

$$\text{Fe:}\quad 1015 = 10765R_{\text{Fe}} + (9730\times 0.077R_{\text{Cu}}) + (7845\times 0.069R_{\text{Zn}}) \tag{8.5a}$$

$$\text{Cu:}\quad 659 = (10765\times 0.017R_{\text{Fe}}) + 9730R_{\text{Cu}} + (7845\times 0.506R_{\text{Zn}}) \tag{8.5b}$$

$$\text{Zn:}\quad 680 = (10765\times 0.003R_{\text{Fe}}) + (9730\times 0.559R_{\text{Cu}}) + 7845R_{\text{Zn}} \tag{8.5c}$$

The simultaneous equations for sample 2 are identical except for the quantities to the left of the equality signs; these values for sample 2 are given in Table 8.2. The set of equations for each sample is solved for the three R values.

These unfolding techniques are made feasible by the speed and convenience of the multichannel analyzer (Section 8.1.2, last paragraph). Collection of the data in Table 8.2 requires ~1 h with a single-channel pulse-height analyzer, but only ~5 min with the multichannel analyzer. Incidentally, there appears to be no reason why unfolding techniques cannot be applied to overlapped wavelength spectra (2θ scans).

8.2. NONDISPERSIVE ANALYSIS

8.2.1. Introduction

In any method of chemical analysis based on measurement of characteristic x-ray spectral lines, some means must be provided to permit the lines of each analyte element in the specimen to be measured individually. There are many ways to do this. The elements may be excited individually (selective excitation) by (1) potential discrimination, (2) monochromatic excitation, or (3) differential excitation. Alternatively, the elements may be excited all at once and some means provided to isolate the lines from each element so they can be measured individually; this can be done by (1) selective diffraction (crystal or wavelength dispersion), (2) selective filtration, (3) selective detection (gas discrimination), or selective counting (electronic or pulse-height discrimination, energy dispersion).

Of these methods for isolation of the line to be measured, the two most commonly applied are, in order, wavelength (crystal) dispersion and energy dispersion (pulse-height analysis). However, two or more methods are often used in combination. Pulse-height analysis is very commonly used to supplement crystal dispersion. Selective, monochromatic, and differential excitation, selective filtration, and selective detection may be used to supplement either wavelength or energy dispersion or their combination.

Only in crystal dispersion are the various wavelengths actually dispersed spatially, and only in this method is it necessary to move the detector for each wavelength. Hence, all the methods other than crystal dispersion are regarded as "nondispersive." They are also referred to as nondiffractive.

8.2.2. Instrumentation

The various nondiffractive methods are conveniently applied on a standard commercial x-ray spectrometer (*J7*). The instrument arrangement is the same as for secondary-emission spectrometry, except that the goniometer is fixed at $0°\ 2\theta$, and the crystal—preferably also the crystal stage—is removed. Coarse collimators may be used in both positions. The distance from specimen to detector may be reduced by mounting the detector directly after the source collimator or even at the exit port of the specimen compartment. In instruments having specimen drawers, a special drawer may be provided in which the specimen plane is inclined in the opposite direction, away from the crystal, toward the front of the drawer. A window may be cut in the front of the drawer and the detector mounted at this window, preferably after a short, wide-spaced collimator.

Nondiffractive methods are also applied on a standard diffractometer. The specimen is placed in the diffractometer specimen position on the axis of the goniometer, which is then set at $\sim 90°\ 2\theta$. Care is taken to avoid any lines diffracted by the specimen. Wide slits and coarse collimators may be used. The specimen is irradiated by the direct primary beam, preferably from a diffraction tube having a tungsten target.

Many specially designed nondispersive spectrometers have been built. Many of these consist basically of a specimen chamber, x-ray tube, detector, and pulse-height analyzer. A more sophisticated instrument is shown in Figure 19.10.

The elimination of crystal dispersion and reduction of optical path length result in extremely low analyte-photon loss between specimen and detector. This economy of analyte radiation makes excitation by radioactive isotopes feasible (*C8*, *K5*, *M12*, *R3*, *R4*, *S12*). Figure 8.11 shows two arrangements using sealed radioisotope sources having annular and button forms. The source–specimen–detector arrangement provides extremely short paths for both source and specimen emission. The button and ring provide their own shielding to exclude source radiation from the detector.

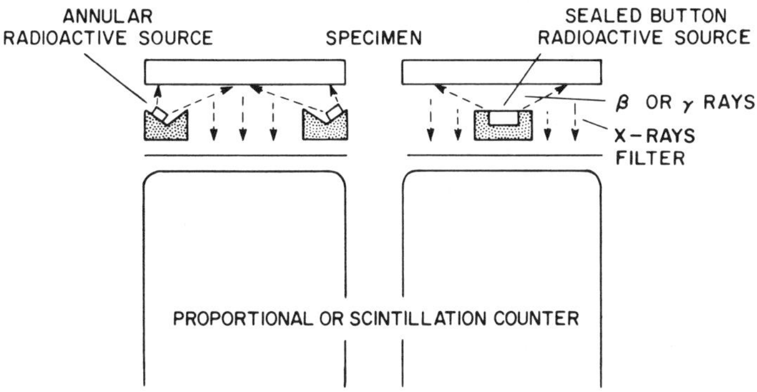

Fig. 8.11. Annular (left) and button (right) arrangements for radioactive excitation and nondispersive analysis.

In some applications, selective excitation, filtration, or detection may permit measurement of analyte-line intensity simply by counting the detector output directly. This may be the case when the specimen contains only two or a very few elements having relatively widely spaced absorption edges and spectral lines, or when differential excitation or balanced filters are used (see below). Otherwise, the detector output must be subjected to pulse-height analysis.

When pulse selection is required, the resolution is poorer with scintillation counters than with proportional counters, and the best resolution is realized with lithium-drifted silicon semiconductor detectors, which permit resolution of lines of adjacent elements. A single-channel pulse-height analyzer may be used, but a multichannel analyzer is much more rapid and convenient (*L6*); 200 or 400 channels are sufficient for analytical x-ray spectrometry.

8.2.3. Selective Excitation

The analyte line may be excited selectively—that is, to the exclusion, or substantially so, of the lines of other elements in the specimen—by potential discrimination, monochromatic excitation, or differential excitation.

In *potential discrimination*, the x-ray tube is operated at a potential below that at which the interfering line is excited. The details of this technique are given in Section 13.3.3.2. The method is applicable to simple systems having relatively few elements, and in which the analyte excitation potential is relatively low; it may be used as a supplement to crystal dispersion. Loss of analyte-line intensity results from the reduced operating potential. An example of an ideal application is selective excitation of Fe $K\alpha$ in tin-plated steel.

In *monochromatic excitation*, the primary beam may be used to irradiate a secondary target or radiator having strong spectral lines on the short-wavelength side of the analyte absorption edge. The resulting spectral emission excites the analyte line. The method may be useful when intensity can be sacrificed for a "clean" spectrum. The analyte-line intensity is very low, but its measurement by the nondispersive method is feasible. Birks [p. 16 in (*6*)] cites the following typical application: In measurement of Cr $K\alpha$ in the presence of cobalt, Fe $K\alpha$ from an iron radiator excites the chromium, but not the cobalt. In instruments having drawer-type specimen presentation, a special drawer can be provided to accommodate the radiator and specimen in proper geometry.

In *differential excitation*, the specimen is irradiated successively by two wavelengths λ_S and λ_L bracketing the analyte absorption edge on the short- and long-wavelength sides, respectively. A detector is placed to receive x-rays emitted and scattered by the specimen. The detector receives analyte-line emission at λ_S, but not at λ_L. The difference in measured intensity at the two wavelengths gives analyte-line intensity—without pulse-height analysis. More details of the method, especially as applied to selected-area analysis, are given in Section 19.3.

8.2.4. Selective Filtration

One of the ways to isolate a specified spectral line from a beam containing many such lines is by use of filters (*S13*) (Figure 8.12).

The principle of the *simple filter* (*S4*) is shown in Figure 8.12A. The filter is a thin foil or other layer of an element having a *K* or *L*III absorption edge at a wavelength just shorter than that of the analyte line λ_2. Thus, the filter has low absorption for the analyte line. All lines of shorter wavelength, such as λ_1, are strongly absorbed. The method is very simple, but gives only rough discrimination, and none for interfering lines on the long-wavelength side of the filter edge. The intensity of the passed wavelength is lower the more nearly complete the discrimination—that is, the thicker the filter. The method is most useful for removing the shorter of two interfering lines bracketing the filter absorption edge.

The principle of the *balanced* (*Ross*) filter (*K14*, *R13*) is shown in Figure 8.12B. A balanced filter consists of two separately mounted thin foils or other layers of two elements of adjacent atomic number selected so that the

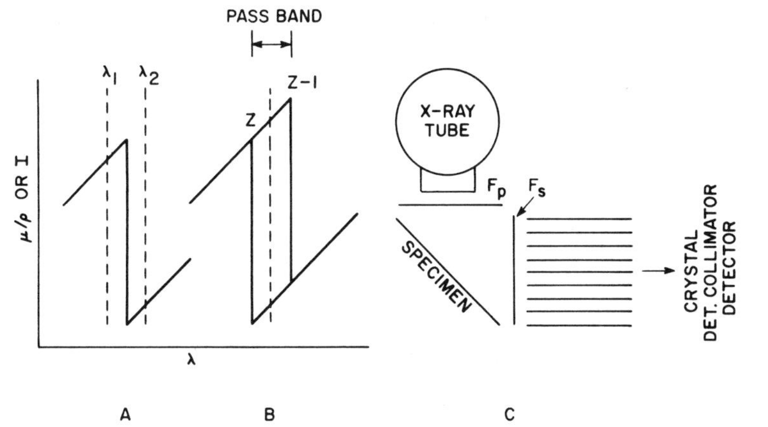

Fig. 8.12. Filter arrangements. A. Simple. B. Balanced (Ross). C. Kustner.

wavelength to be measured lies between their absorption edges. The thicknesses of the filters are adjusted so that they transmit equal intensities at all wavelengths except those between their absorption edges. This is done by actual adjustment of their thickness or area concentration (mg/cm²), or by stretching or tilting the filters in their frames. Intensity is measured first with one filter in the x-ray beam, then the other. The difference between the two intensities is due entirely to x-rays having wavelengths between the two absorption edges, principally the analyte line. The balanced-filter method can be substantially specific for any wavelength, but the filters are difficult to balance, and a different pair of filters is required for each analyte line (*A1, M11, T2*).

The principle of the *Kustner* filter method (*K23, K24*) is shown in Figure 8.13C. The filter is a foil or other layer of an element having a *K* or *L*III absorption edge at a wavelength just longer than that of the analyte line. Thus, the filter has high absorption for the analyte line. Intensity is measured first with the filter in the primary beam F_P, then in the secondary beam F_S. The intensity of the scattered x-rays reaching the detector is the same with the filter in either position, whereas the analyte-line intensity is much lower with the filter in the secondary beam. Thus, the difference between the two intensities is due to analyte-line emission from the specimen.

The *Dothie "Filterscan"* method (*D12*) is based on the fact that a spectral line of an element of atomic number *Z* is transmitted by a filter of element $Z - 1$ more efficiently than by one of element $Z - 2$, etc. A series of filters of successive atomic numbers are placed in turn in the secondary beam, and the intensity transmitted through each is measured and plotted.

The resulting plot is, in effect, a "filter scan" of the secondary spectrum. The method can resolve lines of elements of adjacent atomic number. The filters must have about the same area concentration (mg/cm^2) of element.

In all filter methods, *K*-absorption edges are used whenever practical. The *L*III edges of elements of higher atomic number may be used, but, in a specified spectral region, elements with *L*III edges have higher absorption than elements with *K* edges.

Filters are prepared in several ways. Ductile metals may be rolled to thin foil. Nonductile metals may be ground and polished with abrasive, electropolished, or chemically etched. Metals may be electroplated or vacuum-evaporated on thin, x-ray-transparent substrates. Metal or oxide powders may be slurried and spread on similar substrates. Filter paper may be impregnated with solutions.

Other uses of filters in x-ray spectrometry include elimination of x-ray tube target lines; enhancement of the emission of thin-film and micro specimens; and improvement of line-to-background ratio. Filters are also used in combination emission–absorption methods. All these techniques are described elsewhere in this book.

8.2.5. Selective Detection

Selective detection, or "gas discrimination," takes advantage of the dependence of quantum efficiency of the detector (Figure 6.12) on wavelength to realize high detection efficiency for the analyte line—or low efficiency for interfering lines. The method is somewhat inconvenient to apply to specific spectral lines, but may be used in special cases. Gas discrimination is most effective in the long-wavelength region. Helium, neon, methane, and propane detector gases are relatively efficient at these wavelengths, but have very low absorption for higher-order, shorter-wavelength lines of elements of higher atomic number.

8.2.6. Selective Counting

In the method of selective counting (*B29*), the detector receives the "undispersed" secondary beam comprising all excited lines of all specimen elements. The amplified detector output is subjected to energy dispersion by pulse-height selection. By appropriate setting of the baseline and window, the pulse-height distribution of each analyte line can be counted individually. Alternatively, by appropriate setting of the detector potential and/or amplifier gain, the analyte distributions can be moved individually into a fixed window. The method of pulse-height analysis is the most widely applied "nondispersive" method, and is treated in detail in Section 8.1.

The loss of analyte-line photons between specimen and detector can be made extremely small because (1) the crystal is eliminated, (2) collimation may be eliminated, or at least made very coarse, (3) the distance between specimen and detector may be greatly reduced, and (4) the detector can be made to intercept a greater solid angle of specimen emission. The resulting inherently high sensitivity permits use of monochromatic and radioactive excitation, which give analyte-line intensities too low for crystal dispersion. The increased sensitivity is also advantageous when selective excitation and filtration are used, and for trace and micro specimens. The method is very convenient for relatively simple systems having well-separated spectral lines. It is easily applied to standard commercial x-ray spectrometers, or to very simple, specially built instruments.

The principal limitation of selective counting is the poorer resolution of energy dispersion with proportional and, especially, scintillation counters, compared with wavelength dispersion by crystals. There is also the inconvenient and time-consuming procedure of establishing the single-channel pulse-height analyzer settings for each analyte. The former limitation is removed by use of a lithium-drifted silicon detector, the latter by use of a multichannel analyzer.

Chapter 9

Laboratory, Automated, and Special X-Ray Spectrometers

9.1. INTRODUCTION

In earlier chapters, we have considered the principles of x-ray spectrometry and the individual components that comprise the instrument. By way of a summary, Figure 9.1 shows the useful ranges of the principal components. In this chapter, we consider the integration of these components into complete x-ray spectrometer instrument systems.

An x-ray spectrometer may be a standard or custom-built commercial instrument, or it may be wholly designed and built in the laboratory, or assembled from commercial mechanical and electronic components. X-ray spectrometers may be classified in many ways, principally the following.

The *intended application* may be for the physics laboratory, analytical chemistry laboratory, production control, process control, or portable/mobile field use.

The *mode of operation* may be manual, semiautomatic, or automatic; and automatic instruments may be further classified as sequential or simultaneous.

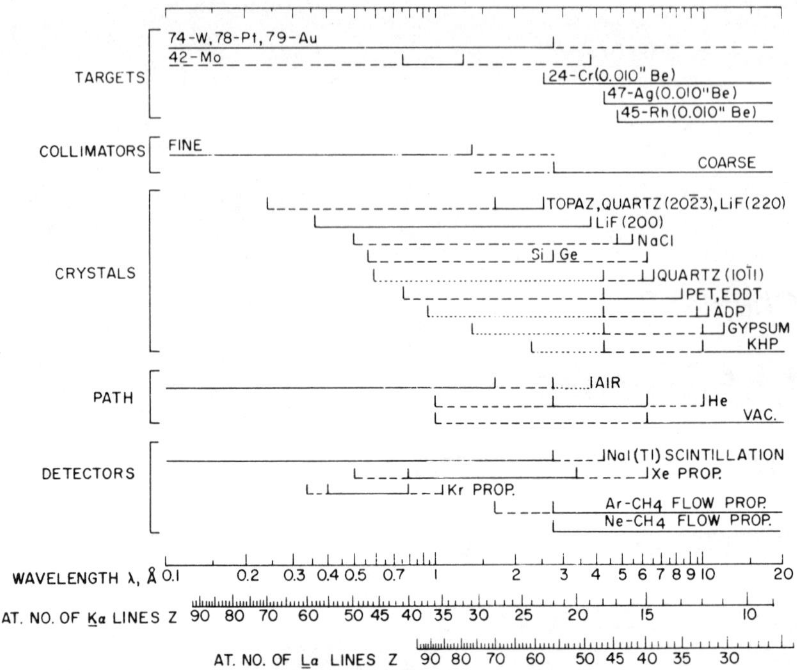

Fig. 9.1. Useful spectral regions of the most commonly used x-ray tube targets, collimators, crystals, detectors, and radiation-path media. Solid, dashed, and dotted lines indicate optimal and progressively lower degrees of applicability, respectively.

The *excitation source* may be an x-ray tube, electron gun, or radioactive isotope; and the x-ray tube power may be full-wave or constant-potential and have maximum potential of 50, 60, 75, or 100 kV.

The *optics* may be wavelength-dispersive or energy-dispersive ("nondispersive"), and the former may be further classified as nonfocusing (flat-crystal), semi- or full-focusing (curved-crystal), or double-crystal. The instrument may have single-channel, two-channel, or multichannel optical systems. The goniometer may have a horizontal or vertical plane, and may or may not have "inverted" optics; or it may be possible to orient the goniometer in any plane.

The *readout* may have many forms, including panel meter, strip chart, electronic digital display, printed paper tape or typewriter copy, punched paper tape, and photographic film. The data itself may be analyte-line intensity, or the intensity data for samples and standards may be entered into a computer and analyte concentration read out directly.

Many of these criteria and instrument types are considered in other chapters. The others are defined and discussed in this chapter, which is devoted to commercially available standard, automated, and special x-ray emission spectrometers.

9.2. LABORATORY X-RAY SPECTROMETERS

9.2.1. General

Laboratory x-ray spectrometers are usually designed for application in physics or analytical chemistry. Physics laboratory instruments are designed for the greatest attainable precision and accuracy in measurement of x-ray wavelengths for applications such as cataloging of x-ray spectra and evaluation of atomic energy levels. Speed and convenience of operation are of secondary importance.

Analytical laboratory instruments are designed for use in chemical analysis and materials characterization in their broader senses. They have the following features: (1) speed and convenience of operation; (2) rugged construction to withstand continuous operation at the hands of relatively unskilled personnel; (3) favorable balance of intensity and resolution; (4) high precision and accuracy in setting wavelength (2θ) and measuring intensity; and (5) versatility of specimen form and presentation to permit application to a wide variety of analytical problems arising in science, engineering, technology, geology, medicine, etc. It is these analytical laboratory instruments that are the principal concern of this book.

All the original commercial analytical laboratory x-ray spectrometers were basically modified diffractometers. The goniometer was fitted with collimators, analyzer crystal, and a boxlike specimen compartment, which received a specimen holder having the general form of a drawer. Most manufacturers still offer instruments of this type. However, the trend is toward instrumentation designed specifically for x-ray spectrometry, in which the basic goniometer is barely recognizable and the specimen chamber has the form of a turret or carousel (*T8*). An intermediate phase is provided by the Philips Universal Vacuum Spectrometer, in which the x-ray tube, four-position specimen turret, collimators, crystal, dual flow-proportional and scintillation detector, and vacuum enclosure all form a self-contained unit. The standard Philips upright goniometer provides only the θ-2θ drive. The General Electric SPG-3 and SPG-5 vacuum spectrometers are examples of laboratory instruments that represent a complete departure from the former basic design.

9.2.2. Instrument Arrangements

The newer instrument designs are advantageous for laboratories where the principal work load is analysis of solids, briquets, powders, and liquids. However, for laboratories required to deal with a wide variety of specimens of all forms, quantities, sizes, and shapes, the older design permits a greater versatility. Moreover, these instruments are more conveniently arranged for modes of operation other than conventional secondary-emission spectrometry. These instruments have two principal forms—those having a horizontal goniometer plane (vertical axis), and those having a vertical goniometer plane (horizontal axis).

Figures 9.2 and 9.3 show various arrangements of instruments having horizontal and vertical goniometer planes, respectively. In either figure, arrangement *A* represents conventional geometry for x-ray secondary-emission spectrometry. The x-ray tube is rotated to place the window in position *W*1, where the primary beam is directed downward into a specimen compartment having the specimen plane in position 2. It is evident that horizontal-plane goniometers (Figure 9.2) have inclined specimen planes, whereas vertical-plane goniometers (Figure 9.3) have horizontal specimen planes. This arrangement can also be used for other x-ray methods, as described in the following paragraphs.

Selected-area analysis (Chapter 19) is done by placing apertures *PA* and *SA* in the primary and secondary beams, respectively.

All three of the principal absorption methods (Section 20.1) are feasible;

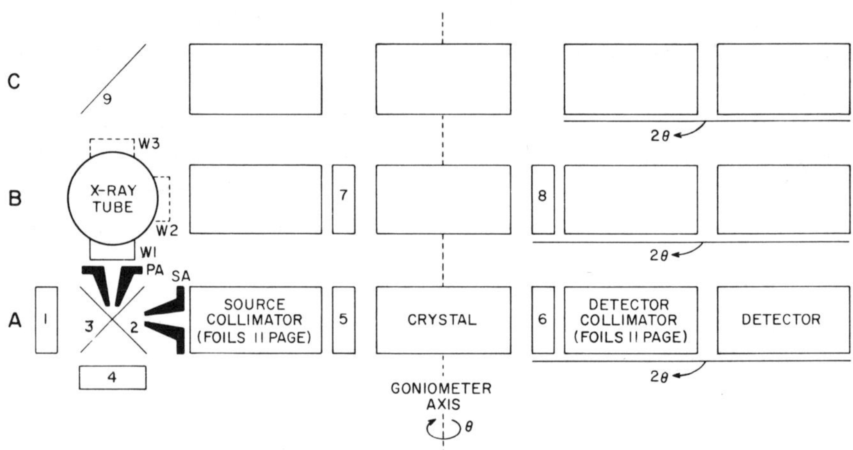

Fig. 9.2. Various arrangements of an x-ray spectrogoniometer having a horizontal plane (vertical axis).

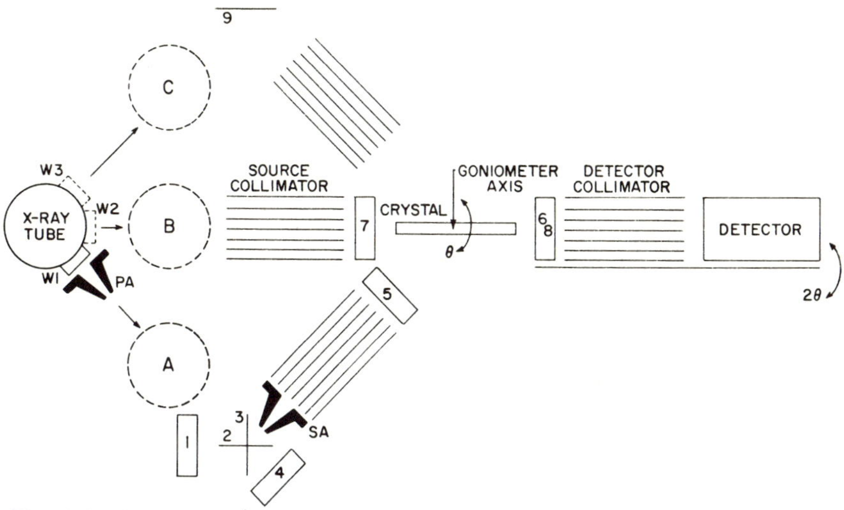

Fig. 9.3. Various arrangements of an x-ray spectrogoniometer having a vertical plane (horizontal axis).

absorption specimens are placed at position 5 or 6. Polychromatic absorptiometry is done by placing a very low-Z scatter target in position 2 and setting the goniometer at $0° \, 2\theta$. Monochromatic absorptiometry is done by placing at position 2 a secondary-emission target having a strong spectral line at the required wavelength and setting 2θ to diffract and detect that wavelength. Absorption-edge spectrometry is done the same way, except that the secondary target is selected to provide two wavelengths bracketing the analyte absorption edge—or else two targets are used in turn—and 2θ is set at each wavelength alternately.

Polychromatic contact microradiography is done by exposing the camera to the primary beam at position 2 or 4. Monochromatic contact microradiography is done by placing the camera at 5 and exposing it to the spectrum of a secondary target at 2, or the specimen drawer can be modified and the camera and secondary target placed at 1 and 3, respectively. Contact microradiography is discussed in more detail in Section 20.1.5.

Arrangement C (Figures 9.2 and 9.3) represents "inverted" geometry for x-ray secondary-emission spectrometry. The x-ray tube is rotated to place the window in position $W3$, where the primary beam is directed upward into a specimen compartment having the specimen plane in position 9. Here again, as in arrangement A, horizontal- and vertical-plane goniometers have inclined and horizontal specimen planes, respectively. The inverted horizontal plane is advantageous for liquid and loose-powder specimens. The geometry in *C* permits analysis of very large specimens (see below). The selected-area, absorption, and microradiography techniques feasible with arrangement *A* are also feasible with *C*.

Arrangement *B* represents the geometry of a diffractometer, although the components may also be those of a spectrometer. The x-ray tube window, at *W2*, is directed into the optical system. This arrangement can be used for absorption methods that use the primary beam as the source of incident x-rays. The absorption specimens are placed at position 7 or 8. For polychromatic absorptiometry, the goniometer is set at $0° \ 2\theta$, for monochromatic absorptiometry at 2θ to diffract and detect the appropriate wavelength. For absorption-edge spectrometry, the goniometer is set alternately at the 2θ angles for the wavelengths bracketing the analyte absorption edge.

Some other nonconventional modes of operation are discussed in Chapters 8 and 20.

9.2.3. Accessories

Instruments having the geometry shown in *A* of Figures 9.2 and 9.3 are converted to "inverted" geometry (*C* in the figures) by rotation of the x-ray tube about its axis to move the window from position *W*1 to *W*3 and elevation of the collimators, crystal stage, and detector on spacers to align them with the new specimen plane. On some instruments, the original specimen compartment can also be mounted on a bracket at an elevated inverted position above the x-ray tube window. However, more often, a multiple-specimen turret or large-specimen holder is provided.

A large-specimen holder consists of a stage having a specimen-mask window and hinged lead-lined lid; it accommodates specimens up to perhaps 10 inches square and 4 inches thick. If the lid is removed and other provision made for x-ray shielding, even larger specimens and parts can be accommodated.

A flat-crystal spectrometer can be converted to a semifocusing curved-crystal instrument without any modification other than replacement of the flat crystal with a fixed-radius curved crystal and removal of the Soller collimators. It is usually preferable to place at the detector window a simple slit parallel to the goniometer axis. Such an arrangement is optimal at only one 2θ angle, but if this optimal angle is $\sim 50° \ 2\theta$, and if some line broadening can be tolerated at high and low 2θ angles, one crystal may suffice for the entire 2θ range. However, for near-optimal performance over the full 2θ range, two or three crystals having different radii are required. Full-focusing curved-crystal accessories provide for continuous variation of crystal radius or of specimen–crystal–detector distances to maintain optimal focusing as 2θ is varied. Curved-crystal accessories of either type require primary or secondary apertures (*PA* and *SA* in Figures 9.2 and 9.3) when used with extensive specimens. They may be used with or without apertures for very small-area or filamentary specimens. Curved-crystal accessories are discussed more fully in Chapter 5.

Helium and vacuum accessories are described in Chapter 5. Specimen rotators for drawer-type specimen holders are described in Chapter 16 (*B21*, *T1*). The newer turret or carousel holders have built-in rotators, which may be energized by a panel switch. Various specimen cells and specimen-presentation systems are described in Chapters 16 and 17. Selected-area accessories are described in Chapter 19.

9.3. AUTOMATED X-RAY SPECTROMETERS

9.3.1. General

Suppose that a manual, single-channel laboratory spectrometer of the type considered so far throughout this book is used for production control. It would be required to analyze large numbers of similar multicomponent samples submitted continuously throughout the day, and, in such cases, it is common for results to be required very soon after samples are submitted. Even instruments used for incoming inspection and lot approval are required to make large numbers of such analyses from time to time, perhaps with somewhat less urgency. In such analyses, if the analytes have a substantial range of atomic number, no one set of components or conditions is likely to give optimal—or even acceptable—results for all of them. Sometimes, the same components may be used for the entire multicomponent analysis with a certain amount of sacrifice of sensitivity and statistical precision; otherwise, one or more components must be changed during the course of the analysis, with substantial loss of time. In multicomponent analyses, many individual settings may be required for each analyte and perhaps for one or more internal standards—kV, mA, 2θ, preset count or time, pulse-height selector, etc.—and the accumulated count or lapsed time must be read and recorded. For large numbers of multicomponent samples, the number of settings and readings becomes very large, and, aside from the tedium, the opportunity for error increases. If the specimen compartment can accommodate only one specimen and there is no provision for continuous loading, the instrument is completely idle while one sample is removed and the next one loaded. For large numbers of samples, the cumulative loss of time can be substantial.

Consequently, it is not surprising that ever since the first commercial instruments were introduced, modifications have been made to facilitate all these operations. The first innovations were the multiple-specimen changer, which permits indexing a series of preloaded specimens into the measuring position, and the continuous specimen loader, which permits loading the next specimen while one is being measured. The dual detector, crystal changer, and dual-target tube permit convenient interchange of these components without substantial interruption of a series of multielement analyses, and the printer automatically reads out scaler–timer data. An entirely different approach was the introduction of multichannel instruments having two or more complete optical systems. These were particularly useful for simultaneous measurement of analyte and internal-standard lines. These modifications are the direct antecedents of the sophisticated semiautomatic and fully automatic instruments available today. The components of automated instruments are essentially the same as those in the laboratory instruments, except, of course, for the electronic and mechanical automatic program and control components.

Kemp (*15L*) classifies automated x-ray spectrometers as production-control or process-control instruments. Production-control instruments are designed for rapid, accurate, routine, repetitive analysis of the same types of samples for the same elements day after day, and for reduced dependence on the attendance and training of the operator. The precision and accuracy are comparable with those of a laboratory instrument. The equipment is usually installed in a central laboratory.

Process-control instruments (*K11*, *M10*, *M36*) are designed for continuous, on-line, largely unattended analysis of a product during some phase of its processing, or of a by-product, residual or waste material, or other material that can indicate the condition of the process. The output is read out under continuous observation, and any changes may be used to guide adjustment of the process. The ultimate objective of the process-control instrument is to use the output to automatically control the process, and this has been realized for some simple processes, such as plating of metal sheet or recovery of a single element from a slurry of powdered ore. If no material in the process has suitable form, sample preparation is required, such as finishing, grinding, drying, briqueting, dilution, fluxing, or addition of an internal standard; this preparation must also be automated. It is fair to state that the principal problem in design of process-control equipment is the design of reliable dynamic sample-presentation systems.

Process-control instruments may be classified in four groups: (1) converted laboratory or production-control instruments fitted with special dynamic sample-presentation systems for continuous strip, slurries, liquids, etc.; (2) commercial instruments designed specifically for a single, widely-used industrial process; the Philips plating thickness gage is of this type; (3) commercial basic instruments designed for process control, but having a built-in flexibility for relatively convenient adaptability to a variety of specific processes; the Applied Research Laboratories Quantrol, General Electric XEG, and Philips Continuous On-Stream Process Analyzer are of this type; and (4) instruments custom-designed for the customer's application. Most manufacturers of x-ray spectrometers will undertake the design, construction, and instllation of such instruments, and probably most of the process-control instruments in use today are of this category. Although many process-control x-ray spectrometers are in use, the field must still be regarded as being in an early phase of its development. No further consideration is given to process-control instruments in this book.

A situation intermediate between production and process control is provided by processes that do not require continuous analysis. In such cases, an automated process for sampling at suitable intervals may be established and a production-control instrument used.

9.3.2. Sequential Automatic Spectrometers

A sequential automatic x-ray spectrometer is basically an automated single-channel instrument programmed to index automatically to a preselected series of analyte lines, stopping at each to accumulate a preset count from the sample and from a standard. The instrument is programmed to automatically select components and conditions optimal for each analyte line. This type of instrument greatly expedites routine analyses. The operator is relieved of the tedium of making the many instrument settings and readings required for multielement analyses of large numbers of routine samples, and the opportunity for error is minimized.

The sequential x-ray spectrometers presently available commercially are the Applied Research Laboratories MXQ, General Electric XRD-410, Hilger Fluoroprint, Philips PW-1212 (*W14*) (fully automatic) and PW-1220 (semiautomatic), Siemens SRS-1, and Solartron XZ-1030 (*L29*). The Philips instruments are the successors to the first commercial sequential automatic spectrometer, the Philips Autrometer (*C49*).

All these instruments provide for automatic selection of a number of the following components and conditions for each programed analyte, but none provides for all. Of course, any component or parameter for which selection is not provided is the same for all analytes.

1. *Sample*: Various instruments accommodate 2–10 samples, which are automatically indexed into position in turn by means of a carousel arrangement.

2. *Element*: The goniometer automatically indexes to up to 15–24 2θ angles, depending on the instrument.

3. *Sample motion*: Fixed or rotated.

4. *X-ray tube target*: Tungsten or chromium (only with a dual-target tube in the G. E. instrument).

5. *X-ray tube potential*: A selection of preset values up to 50, 60, 75, or 100 kV, depending on the instrument.

6. *X-ray tube current*: A selection of preset values up to 50, 75, or 100 mA, depending on the instrument.

7. *Collimator*: Coarse or fine.

8. *Crystal*: Choice of three or four.

9. *Diffraction order*: The instrument can be instructed to measure the first- or second-order line.

10. *Path*: Air, helium, or vacuum.

11. *Detector*: Most instruments have a dual scintillation and gas-flow proportional counter, and the output can be taken from either or both.

12. *Delay time*: To ensure adequate vacuum or helium flushing, the instrument may be set to delay a preset time before making a measurement each time a new specimen is indexed into position.

13. *Amplifier gain*: A selection of several preset values.

14. *Automatic pulse-height selection*: In or out, and window width.

15. *Preset count or preset time*: Which mode, and a selection of values. For each analyte, that count (or time) is programmed that gives the required statistical precision.

16. *Repetition*: The instrument can be set to repeat a measurement a specified number of times.

The spectrometer can be operated in any of three modes, as follows.

a. Manual. In this mode, the instrument becomes substantially a standard single-channel scanning spectrometer useful for qualitative analysis and for quantitative analysis of individual samples.

b. Semiautomatic. For each sample, the analytes are selected in turn, and for each analyte, the components and conditions are selected—all by manual operation of pushbuttons and multiposition selector switches on the control panel. As the operator makes each setting, the instrument indicates that it has executed its "instructions" correctly. This mode is used for analysis of large numbers of nonroutine samples for which no program has been established.

c. Fully Automatic. The samples are indexed into position and, on each sample, the specified analytes are measured, each with its specified components and conditions—all automatically. This mode is used for analysis of large numbers of routine samples that warrant establishment of a program. The program may be set manually, or by means of a patchboard, punched card, or punched tape specific for a certain series of analytes in a certain type of sample. Alternatively, the program can be stored in the computer memory (see below) and selected by addressing it on the computer input–output typewriter.

The spectrometer output may be "interfaced" with a computer, which may be "dedicated," that is, a permanent complement to the spectrometer, or a central "time-shared" computer. The computer converts intensity data measured from the samples directly to concentration by means of calibration functions automatically established from intensity data measured from the standards. The calibration may be based on a single standard, a linear function derived from two or more standards, or even a nonlinear function divided into regions, each considered to be linear.

The equipment is usually designed so that the basic manual instrument can be purchased first, then later expanded to a semiautomatic, then to a fully automatic, instrument by addition of controller and programing units, respectively. Finally, the computer may be added.

The output may be read out on a printer, a digital display, or the computer typewriter. A strip-chart recorder may be provided for ratemeter readout with manual operation. The output data for each output in each sample may be read out in any of three forms: (1) absolute accumulated count (or counting time); (2) ratio of accumulated count (or time) for the sample and standard; and (3) concentration.

The specimens may be solids, briquets, powders, slurries, or liquids. Provision is usually made for continuous loading—through an air lock if helium or vacuum path is used. Most sequential spectrometers are useful for all elements down to atomic number 9 (fluorine).

Several specific instruments have noteworthy features.

The *ARL model MXQ* has two full-focusing curved-crystal scanning channels, each with a two-crystal changer and a nondispersive ratio channel, which monitors the primary x-ray beam. The instrument can be operated manually, semiautomatically, or automatically, and, in automatic operation, the two channels can be operated serially or in parallel. In serial operation, one channel can be fitted and programmed for several lines in the short-wavelength region, the other for several lines in the long-wavelength region; then the first channel sequentially measures several analytes down to potassium, whereupon the other channel continues the analysis down to fluorine. In parallel operation, the two channels simultaneously measure two wavelengths from the specimen; the second wavelength may be a second analyte, an internal-standard line, or a scattered continuum wavelength or target line. The MXQ has features of both simultaneous and sequential instruments. The instrument can be purchased with a single channel and the other added later.

The *General Electric XRD-6SP* (*D4*), no longer available, operated unattended while up to 11 analytes were automatically measured on up to 90 samples 1/8–3/16 inch thick, or 45 samples 1/4–3/8 inch thick. The samples were placed in a vertical tower, which indexed downward, one sample at a time, discharging each sample onto an inclined belt, which, in turn, conveyed it to the specimen compartment. All measurements were made in air at a constant preset potential up to 75 kV, but any of the following parameters could be programmed automatically for each analyte line: (1) tungsten or chromium target; (2) any of four preset values of x-ray tube current up to 100 mA; (3) any of four crystals; (4) output from scintillation or flow-proportional counter or both; (5) any of six preset counting times; (6) any of six preset values of amplifier gain; (7) pulse-height selector in or out; (8) repetition of the measurement—single, duplicate, or quadruplicate. A digital printer printed sample number, channel (2θ) number, counting time; and accumulated count for each of the one, two, or four measurements. If the sample mechanism jammed, a loud alarm sounded, and if the instrument received no attention within a preset time, it shut off. The instrument permitted unattended overnight operation. The XRD-6SP could also function with helium path, but only four samples could be programmed.

The *Hilger Fluoroprint* has a nondispersive primary-beam monitor channel, against which all analyte-line measurements are ratioed.

The *Solartron XZ-1030* (*L29*), like the ARL MXQ, has two identical dispersive channels for sample and standard, respectively. It also provides for setting of pulse-height selector window width for each analyte line.

The *Siemens SRS-1* has several notable features:

1. The instrument can be operated in two sequential modes. In the

sample-sequential (conventional) mode, the instrument measures all analytes on one sample before automatically indexing to the next sample. In the element-sequential mode, the instrument measures one analyte on all samples before automatically indexing to the next analyte.

2. The system can be programmed to automatically periodically calibrate against a standard—or two or more. In this way, the instrument periodically checks and revises its calibration data, and uses the revised data to derive analyte concentration.

3. While a programmed analysis is in progress, if an urgent sample of a different type should require analysis, the instrument can be switched to manual or semiautomatic mode, the special sample analyzed, then the automatic program resumed.

4. Several features are included to ensure maximum precision. Precise specimen positioning is provided. The spectrometer compartment temperature is controlled to $\pm 0.5°C$ to keep the crystal temperature constant, to reduce scintillation counter noise, and to reduce change in gas amplification due to temperature. The gas-flow counter discharges into a constant-pressure manostat to reduce change in gas amplification due to pressure.

9.3.3. Simultaneous Automatic Spectrometers

A simultaneous x-ray spectrometer is basically a number of single-channel instruments, each having its own collimators, crystal, detector, and count scaler or integrator, but having a common specimen and readout. The simultaneous x-ray spectrometers presently available commercially are the Applied Research Laboratories VXQ and VPXQ X-Ray Quantometers, the Philips PW-1250, and the Siemens MRS-1.

In the ARL X-Ray Quantometers, most of the channels (monochromators) are fixed, that is, fitted, usually at the factory, with optimal components and set at 2θ for a specified analyte line, and thereafter left undisturbed. However, one or more of the channels may be scanning channels. These are conveniently set to measure elements determined less frequently than those for which the fixed channels are set, thereby precluding the necessity to disturb the fixed channels. The scanning channels can also be used to make 2θ scans for qualitative analysis. Finally, a nondispersive channel is provided to monitor the primary beam. In the ARL instruments, this channel receives x-rays from a separate semipermanent monitor specimen and has a phosphor-phototube detector. The VXQ can have up to nine analytical channels, not counting the nondispersive channel, of which three may be scanners; the VPXQ can have up to 22 channels, of which five may be scanning.

The original ARL X-Ray Quantometers (XIQ and XRQ) used flat crystals and Soller collimators (*H15*, *K8*), but more recent instruments use full-focusing curved-crystal optics and adjustable bilateral source and detector slits. The ARL curved-crystal spectrometer is shown in Figure 5.20 and described in Section 5.6.4. The crystals in these instruments are of the Johansson curved-and-ground type, and eight are available—LiF, NaCl, Ge, SiO_2, PET, EDDT, ADP, and KHP; a curved lead stearate decanoate crystal is also available. The slits are set individually for optimal balance of resolution and sensitivity for the analyte to be determined by that channel. The ARL spectrometers use end-window x-ray tubes, which are particularly well suited to instruments of this type because they occupy no space radial to the specimen plane and therefore permit arrangement of more channels about the sample. Figure 9.4 shows an end-window tube and two optical channels of the flat-crystal type used previously. The specimens may be solids, briquets, powders, slurries, or liquids, and may have dimensions up to 4-inch diameter and 2-inch thickness. An air lock permits loading of the next sample while one is being measured. Air, helium, and vacuum

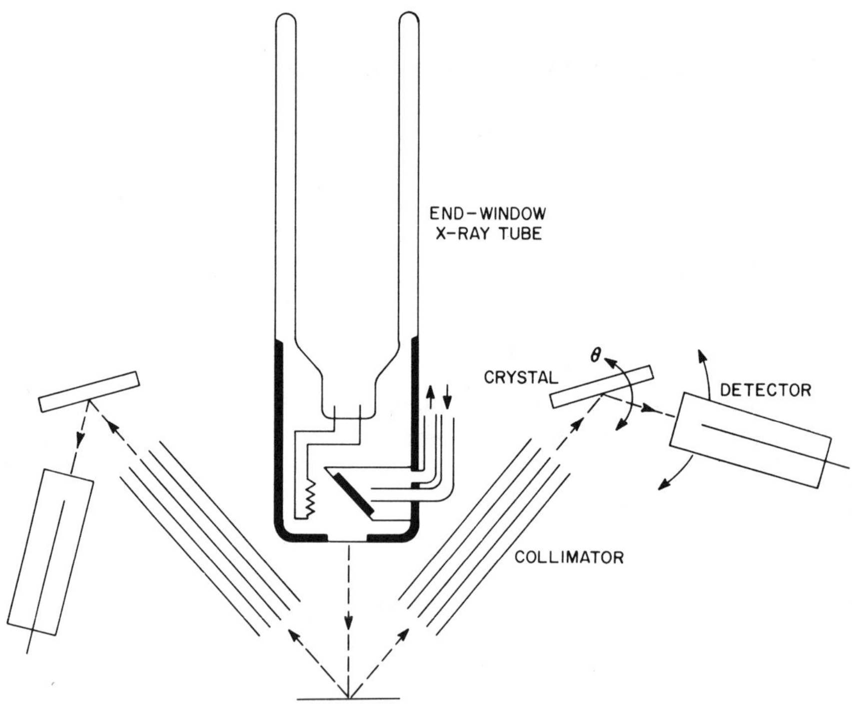

Fig. 9.4. End-window x-ray tube with two x-ray optical channels.

path can be used. The detectors are sealed proportional counters having aluminum or beryllium windows and xenon, krypton, argon, and neon gas fillings; a gas-flow counter having an ultrathin window is also available.

In operation, the output of the detector in every channel, including the nondispersive one, charges a separate integrating capacitor until the charge in the capacitor in the standard channel (see below) reaches a predetermined value, whereupon all detectors are disconnected from their capacitors. Then the accumulated charge in each analytical capacitor is subjected to high-speed sequential readout at a rate ~2.5 s for each. The standard channel may be the nondispersive one, which, as mentioned earlier, in effect monitors the primary beam. Alternatively, any one of the analytical channels can serve as the standard channel, and the crystal may be set to diffract a wavelength in the continuum, a scattered target line, or the line of an internal-standard element.

A computer can be used with a simultaneous spectrometer, and the output can be read out by the same components and in the same forms as for a sequential spectrometer. A significant feature of the ARL X-Ray Quantometers is that their output can be read out on the consoles of ARL Optical Emission Quantometers.

The ARL VPXQ has a unique modular construction. The basic unit can accommodate up to six channel positions arranged radially about the specimen compartment. Each position receives a metal plate on which can be placed one or two curved-crystal channels. Thus, the basic unit can accommodate up to 12 analytical channels. The single unit can be expanded to a dual unit by addition of a semicircular annex, not unlike a bay window, that can accommodate five more positions—10 more channels. Thus, the VPXQ can measure up to 22 elements simultaneously.

The Philips PW-1250 is similar in principle, but with some distinctive features. The maximum excitation power is 100 kV, 36 mA. Side-window x-ray tubes are used, and gold, tungsten, silver, molybdenum, and chromium targets are available. The analytical channels are of the flat-crystal type. The sample carousel holds up to 15 samples. In operation, a scaler in each channel accumulates counts, instead of the capacitor integration in the ARL instruments. The instrument can be operated in any of three modes as follows: (1) absolute—all channels accumulate counts for a preset time; (2) monitor—the analytical channels accumulate counts until the monitor channel has accumulated a preset count; and (3) ratio—analyte-line counts are accumulated from a standard and sample in turn; the standard is measured for the time required to accumulate a preset count; the sample is then measured for the same time, and the sample and standard counts are ratioed.

The simultaneous technique has the same advantages as the sequential, and some additional ones. The simultaneous measurement of all analytes permits greater economy of analysis time and cost. Analysis time is limited by the counting or integrating time required to give acceptable statistical precision for the analyte having lowest intensity, plus short loading and readout times. If no element requires more than 1-min counting time, an analysis for 22 elements can require as little as 2 min for both counting and readout. A sequential instrument requires the sum of the individual counting times plus the operation time. Another advantage is that many instrument and specimen variables are compensated by simultaneous measurement of all intensities. Finally, in trace analysis, higher sensitivity and statistical precision can be realized in relatively short time by setting two channels at the same trace analyte line and combining their outputs.

9.4. SPECIAL X-RAY SPECTROMETERS

9.4.1. Portable Spectrometer

The Philips Portaspec (*M16*) is a portable x-ray secondary-emission spectrometer consisting of two units. The power unit weighs 48 lb and houses a 30-kV, 4-mA constant-potential supply that operates from the 110-V ac line and uses 125 W at full power. The x-ray unit weighs 18 lb and houses the x-ray tube, specimen compartment, goniometer, detector, ratemeter, and detector-tube power supply. The x-ray tube has a beryllium end window, the inside of which is coated with gold to act as the target. The goniometer consists of source and detector collimators, LiF crystal, and an argon-filled detector; the 2θ angle is varied by means of an external knob in the range 14–78°, thereby permitting measurement of all elements down to atomic number 23 (vanadium). The ratemeter output is indicated on a panel meter. A specimen holder is available for powders, liquids, and small solids, another for bar, rod, and sheet stock. The instrument can be used in the plant and warehouse for identification and sorting of metal stock, chemicals, leaded fuels, etc. With a mobile generator, it can be used in the field for geological and mining applications.

9.4.2. Primary Excitation

An example of a commercial instrument having electron excitation is the "Betaprobe" manufactured by TELSEC Instruments, Oxford, England,

and available from Jarrell-Ash. A stabilized electron gun and an electron-beam scanning system provide a 1-mA electron beam covering an area $\sim 1.5 \times 1.5$ cm at the specimen plane. The emitted primary x-rays are intercepted by either of two interchangeable spectrometer units—an eight-channel spectrometer system or a single-channel scanning spectrometer. A metal plate is fitted with up to eight fixed x-ray optical channels, each consisting of a source collimator, flat crystal, and ultrathin-window gas-flow proportional counter. Each detector has its own preamplifier, and the output of each preamplifier is selected in turn and applied to an amplifier and pulse-height selector. Alternatively, the eight-channel assembly can be replaced with a single-channel, flat-crystal, 2θ-scanning spectrometer. For either arrangement, the readout can be by ratemeter and recorder or by scaler and printer or digital display.

The crystals available are LiF, NaCl, KBr, PET, EDDT, and KHP. The instrument is useful for 0.005–100% concentrations of elements down to atomic number 6 (carbon). The advantages derive from the high excitation efficiency and low penetration of electrons, compared with primary x-rays. The efficiency results in high analyte-line intensity, which is especially beneficial for the low-Z elements, and short counting times, typically ~ 10 s, so that eight elements can be read out in <1 min. The low penetration reduces absorption-enhancement effects.

Other instruments having direct excitation by an electron gun are described by Fox (*F11*) and by Kimoto and his co-workers (*K13*). Lenard windows provide another approach to primary excitation. The electron beam from an electron gun passes through a thin aluminum or glass window into the air (*F10*, *S23*). The specimen is placed just outside the window and is subjected to primary excitation.

9.4.3. Radioisotope Excitation

An example of a commercial instrument having a radioactive isotope source is the X-Spec x-ray secondary-emission spectrometer manufactured by Reactor Experiments, Inc., San Carlos, Calif. It consists of a radioisotope source of x- or γ-rays, specimen chamber, lithium-drifted silicon semiconductor detector, field-effect transistor (FET), preamplifier, amplifier, pulse shaper, and detector power supply. The detector, FET, and preamplifier are housed in a cryostat vacuum chamber evacuated by an ion pump and cooled by liquid nitrogen from a 10-liter reservoir. The detector gives 425-eV resolution (full width at half-maximum). With a suitable radioactive source, the instrument is applicable to microgram amounts or less of elements down to atomic number 15 (phosphorus). The instrument is used with a single or, preferably, multichannel pulse-height analyzer, not supplied.

9.4.4. Ultralong-Wavelength Spectrometry
(*151, F8, H20-24, H39, H40, P8*)

In x-ray spectrometry, the ultralong-wavelength (ultrasoft) x-ray region extends from 15 to 150 Å, or over about 1–0.1 keV. Its short-wavelength limit lies at the long-wavelength limit of standard commercial x-ray spectrometers. The region contains the $K\alpha$ lines of elements 9 (fluorine, 18.3 Å) to 4 (beryllium, 114 Å); the $L\alpha_1$ lines of elements 26 (iron, 17.6 Å) to 14 (silicon, 136 Å); and the M spectra of elements 57 (lanthanum, $M\alpha$ 14.9 Å) to 37 (rubidium, $M\zeta$ 129 Å).

Formerly, work in the ultralong region, at least beyond ~25 Å, was done on laboratory-designed and built optical-grating spectrometers. These instruments require highly collimated beams from pinhole or slit apertures and very small incidence and takeoff angles. They require critical adjustment and are generally inconvenient to operate. However, ultralong-wavelength accessories are now available commercially for the Philips Universal-Vacuum x-ray spectrometer (*D22, D23*) to permit measurement of K spectra down to beryllium and L spectra down to silicon. The accessories include a special x-ray tube power supply, x-ray tube, analyzer crystal, and detector.

The power supply delivers relatively low potential, but high current—up to 20 kV, 500 mA. The x-ray tube is the demountable, continuously pumped Henke tube described in Section 4.2.3.4. The tube mounts in the same position as the conventional tube, and the eight-position specimen turret may be used.

KHP crystals permit measurement of K spectra of elements down to oxygen, but Langmuir–Blodgett multilayer smectic crystals of heavy-element soaps are required for the longer wavelengths. Compared with gratings, these crystals have several advantages. They can be used with large-area specimens and the Soller-collimator and flat-crystal optics of standard commercial instruments, and they are readily interchanged with conventional flat crystals. They can also be used with line or point sources and curved-crystal optics. The solid angle of acceptance of a grating is small, and the diffracted intensity is comparable with that from a crystal only for a line or point source. For a large-area specimen, the crystal is much more efficient because Soller-collimated x-rays are accepted by a crystal at a Bragg angle that is large compared with the glancing angle of a grating. The principal disadvantage of the crystal is that its resolving power is only 0.2–0.5 that of a grating. However, it must be remembered that the grating requires a special spectrometer, critical alignment, and less-convenient operating procedures. Data for KHP and the commonly used multilayer crystals are given in Table 6.1.

The detector is a side-window gas-flow proportional counter. The window consists of ultrathin Formvar supported on a 200-mesh nickel grid having 70% open area and sandwiched between apertured metal plates. The gas may be P10 (90% argon, 10% methane) or neon, but is frequently pure methane or propane. The pressure is 1 atm and the flow rate 0.5 ft^3/h. The gas discharges into a manostat to ensure constant pressure in the active volume. Another type of detector suitable for ultralong-wavelength work is the open multiplier phototube having beryllium–copper photocathode and dynodes (Section 6.5.2.2).

Windows are required on the x-ray tube to isolate the clean high vacuum (10^{-6} mm-Hg or less) from the working vacuum in the spectrometer, and on the detector to isolate the detector gas from the spectrometer. Window materials include aluminum leaf dipped in Formvar solution in ethylene dichloride, stretched polypropylene film, and ultrathin (1000–2000 Å, 30–60 $\mu g/cm^2$) collodion or Formvar. The two ultrathin films are usually used as double layers, each having half the indicated thickness, to eliminate pinholes. Ultrathin films are supported on some kind of grid, such as 200-mesh nickel having 70% open area.

Table 9.1 gives some typical performance data for the modified Philips Universal-Vacuum spectrometer.

Alternatively, the eight-position specimen turret may be replaced with a continuously pumped, self-contained, primary-excitation accessory consisting of an electron gun and target. The x-rays from the target are directed into the optical system of the Universal-Vacuum spectrometer. This accessory permits analysis of specimens deposited on the target and provides a source of high-intensity ultralong-wavelength x-rays for absorption measurements or for irradiation of materials. This accessory is not available commercially.

Ultralong-wavelength x-ray spectrometry extends the applicability of x-ray spectrometric analysis from fluorine to beryllium. Because of the long wavelengths, the method is particularly applicable to relatively thin surface layers. The method is extremely valuable for investigation of chemical state and valence-electron band structure because the spectra in the ultralong region are relatively highly influenced by chemical state (Section 11.1.4). The chemical effects cause the spectral lines or bands to shift in wavelength, change in profile, or split into multiple bands. Such investigations are of

TABLE 9.1

Performance of an Ultralong-Wavelength X-Ray Spectrometer[a]

Element	Specimen	Analyte concentration, wt %	Spectral line and wavelength, Å	X-ray tube anode	Crystal[b]	Intensity, counts/s		Minimum detectable concentration,[c] wt %
						Peak	Background	
8-O	SiO_2	53.3% O	O *K*, 23.6	Cu	PbSt	13,297	759	0.03
7-N	BN	56.4% N	N *K*, 31.6	Cu	PbSt	3,840	462	0.11
6-C	Graphite	100% C	C *K*, 44.6	Cu	PbSt	21,681	696	0.04
5-B	BN	43.6% B	B *K*, 67.8	C	PbSt	12,444	108	0.01
4-Be	Be foil	100% Be	Be *K*, 114	C	PbL	1,855	132	0.2

[a] From B. L. Henke (*H24*). Philips Universal-Vacuum X-Ray Spectrometer; Henke demountable x-ray tube: 6 kV, 330 mA; Soller collimators: source, 3.7°, detector, 2.7°; detector: gas-flow proportional counter, methane gas at 1 atm, 0.5 ft^3/h; x-ray tube and detector windows: Formvar, 60 $\mu g/cm^2$, supported on 200-mesh nickel grid, 70% transmission.

[b] PbSt: lead stearate (octadecanoate) multilayer, $2d$ 100.6 Å. PbL: lead lignocerate (tetraeicosanoate) multilayer, $2d$ 130 Å.

[c] $(0.3C\sqrt{I_B})/(I_P - I_B)$, where I_P and I_B are peak and background intensities, respectively (columns 7 and 8), and C is analyte concentration in the specimen (column 3).

great significance in chemistry, solid-state physics, and semiconductor electronics.

More recently, ultralong-wavelength x-ray spectrometry has found application in high-temperature plasma physics and in the study of solar x-rays. In matter at temperatures of the order of millions of degrees, electron and atom energies are of the order of 0.1–1 keV, and the resulting electron accelerations and atom ionizations produce x-rays in the ultralong-wavelength region. Such phenomena occur in stars and in high-temperature plasmas, which are being investigated as a possible source of energy from controlled thermonuclear fusion. Ultralong-wavelength spectrometers are used to evaluate or characterize phenomena and performance in plasmas, and they have been sent up in rockets to measure solar x-rays.

9.5. X-RAY SAFETY AND PROTECTION

It has come to be almost obligatory to include in any book on x-ray practice a section on x-ray safety and protection. Actually, a careful, alert, competent worker who is aware of the potential hazards of x-rays need have little apprehension for his personal safety in working with commercial x-ray spectrometers. These instruments are adequately shielded to confine the x-rays and prevent irradiation of the environment.

The principal sources of possible extraneous x-rays from an x-ray spectrometer are: (1) the interface between the x-ray tube head and the specimen chamber; (2) the specimen chamber itself when specimens are inserted and removed, especially if the x-ray shutter assembly is faulty; (3) the high-potential rectifiers, which may radiate through the relatively thin steel of the oil tank on the peak-inverse half-cycles; and (4) the crystal, which diffracts, scatters, and emits x-rays in directions other than that intercepted by the detector collimator. Of course, the extraneous intensity from any source increases with increasing x-ray tube operating potential and current, and is substantially greater in 100-kV instruments than in 50-kV instruments.

The greatest possible radiation hazard in any x-ray generator is the primary beam, particularly close to the x-ray tube window. In even a 50-kV x-ray spectrometer, the primary intensity at the x-ray tube window may be of the order megaroentgens per minute (see below), and even momentary exposure to it is almost certain to result in a lesion at the irradiated site. The sources of extraneous radiation listed above are likely to have very low intensity. However, an operator taking data for several hours each day may well receive a substantial fraction of his maximum permissible dose (see below) from this source.

The hazard from ionizing radiation in general (alpha, beta, gamma, x) is compounded by the fact that such radiation is completely insidious. It is quite possible that one could be receiving a lethal dose of ionizing radiation and be wholly unaware of any sensation. Further, the physiological effects are delayed, and symptoms may not appear for days or even weeks after the exposure. Finally, the physiological effects are cumulative if one receives radiation at a rate faster than the body can repair the physiological damage, that is, faster than the maximum permissible dose rate (see below).

For purposes of radiation health and safety, x- and γ-radiation are measured in terms of *dose* or *dose rate*. The unit of dose is the *roentgen*, R, which is the quantity of x- or γ-radiation that produces in 1 cm^3 of air at 0°C and 760 mm-Hg, ions carrying 1 esu of charge, regardless of sign. Dose rate is given in units of roentgens per unit time, usually milliroentgens per hour, mR/h.

The maximum permissible whole-body dose rate is 5 R/year or 100 mR/week; in a 40-h work week, this is 2.5 mR/h. All evidence indicates that if a healthy normal adult receives x-radiation at a dose rate no greater than this, his body can easily repair the minor physiological damage. However, every effort should be made to reduce the dose rate to as little as possible, and, pursuant to this objective, certain simple precautions are prudent.

The x-ray spectrometer should be thoroughly monitored with a radiation-survey meter when it is installed and at routine 3–6-month intervals thereafter. Particular attention should be given to the regions of the sources of extraneous radiation given above and to the place at which the operator sits when taking data. The region of the x-ray tube head and specimen compartment should also be monitored each time the x-ray tube is changed or the compartment disturbed. All monitoring should be done with the x-ray tube operating at maximum potential (50 or 100 kV) and maximum permissible current, and with an emission specimen having a strong spectral line of short wavelength—molybdenum, for example. The survey meter should be a model having good response for x-rays having energy up to 50 keV (or 100 keV) and should be properly calibrated to read milliroentgens per hour.

In general, the survey meter should indicate not more than 0.5 mR/h at 5 cm from any accessible surface or part of the instrument. Actually, the 5 cm should be measured to the center of the active volume of the detector in the survey meter. Thus, in practice, the survey meter should be held as close to the spectrometer as possible without danger of damaging the meter. If the reading exceeds 0.5 mR/h, the source of the "leak" should be examined to ascertain that the components involved are properly assembled and that any shielding provided by the manufacturer is securely in place. If all is proper, and a radiation leak still exists, additional shielding must be provided.

Particularly hazardous conditions may exist when the x-ray tube is operated without its housing in place, that is, with its glass end exposed, or without its shutter, or without the specimen compartment in place. Laboratory-designed and built instruments and accessories and, especially, temporary arrangements should be carefully monitored.

All x-ray laboratory personnel should be provided with film badges and/or pocket dosimeters. The person responsible for the x-ray laboratory must be familiar with all federal, state, and municipal regulations pertaining to x-ray safety and must take all necessary measures to comply with them.

Another hazard is that of electric shock. Contact with the high-potential supply while in operation may well result in electrocution. Fortunately, when the high-potential cable is secured at both the oil-tank and x-ray-tube ends, contact with the high potential is unlikely. However, even when the power is off, there may be danger of shock from charge stored in capacitors in the high-potential circuit if there is no "bleeder" resistor or if the resistor is open. Also, great care must be taken when it is necessary to work on the equipment with the cabinet door interlocks thwarted. Contact with the 220-V line power is very hazardous.

Chapter 10

Qualitative and Semiquantitative Analysis

10.1. GENERAL

Many analytical problems require only qualitative or semiquantitative, rather than quantitative, analysis of the sample(s)—that is, *detection* or *identification* of the elements present, or *estimation*, rather than *determination*, of their concentrations. A quantitative x-ray spectrometric analysis usually gives analyte concentration within a relatively narrow known range of accuracy as derived from precisely measured and carefully corrected intensity data, and from calibration data from standards very similar to the samples in analyte concentration and matrix composition. A semiquantitative x-ray spectrometric analysis usually gives analyte concentration as derived from intensities measured with speed and convenience, rather than precision, as primary objectives, and from standards applicable to a variety of related sample types. If samples and standards are very similar, a semiquantitative analysis may give high accuracy, but the more they differ, the poorer is the accuracy.

X-ray secondary-emission spectrometry is extremely well suited to qualitative and semiquantitative analysis. It is convenient, rapid, nondestructive, and applicable to major, minor, and trace constituents in any sample that can be placed in the instrument. When practiced on commercial instruments, it is applicable to all elements down to magnesium, sodium, or fluorine, although sensitivity falls off for elements below phosphorus. Identification of the elements is greatly facilitated by: (1) synchronism of the goniometer and chart drives so that chart divisions are directly convertible to 2θ; (2) the simplicity of x-ray spectra; and (3) the availability of excellent x-ray spectrometer tables. Intensities are easily measured with reasonable accuracy either on the recorder charts or on the scaler. Semiquantitative analyte concentrations sufficiently accurate for many purposes are readily derived from these intensities, and often these concentrations are remarkably accurate.

10.2. RECORDING THE SPECTRUM

A qualitative x-ray spectrometric analysis consists of recording of the x-ray spectrum of the sample, identification of the peaks, and, perhaps, classification of each element identified as a major, minor, or trace constituent on the basis of the intensity of its peaks.

The spectrum is recorded as follows. The sample is placed in the x-ray spectrometer (Figure 3.1), where it is irradiated by the primary beam and emits secondary x-rays consisting of the wavelengths of the spectral lines of its constituent elements. This secondary beam is collimated and subjected to a 2θ scan (or θ scan) as follows. The analyzer crystal and detector arm are driven about a common axis at relative speed ratio 1:2. Thus, the crystal and the normal to the detector window always present angles θ and 2θ, respectively, to the direction of the secondary beam, and the detector is always in position to receive any x-rays diffracted by the crystal. As the crystal rotates, it presents a continuous succession of θ angles to the secondary beam and, one by one, in accordance with the Bragg law, diffracts each wavelength in the beam to the detector. The detector output is amplified and applied to the ratemeter, the output of which displaces the recorder pen. The recorder chart drive is synchronized with the goniometer drive. The result is a chart having the form of those shown in Figure 10.1. The chart is a plot of intensity *vs.* 2θ and is, in fact, an x-ray spectrum.

10.3. INSTRUMENT CONDITIONS

Of the many instrument conditions to be considered in recording the spectrum, the following are the most important. *Spectrogoniometer conditions*: x-ray tube target, potential (kV), and current (mA); collimators;

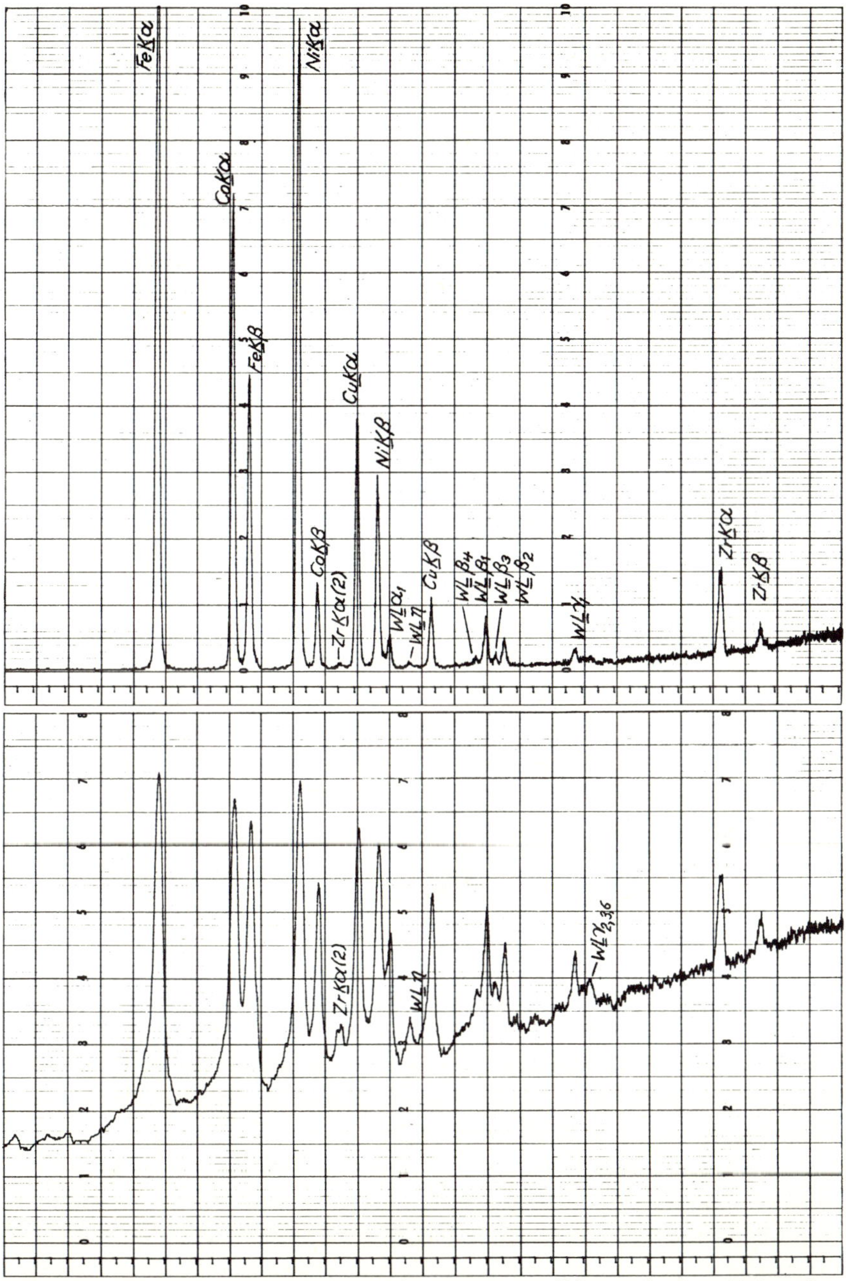

Fig. 10.1. Ratemeter–recorder charts of x-ray spectra (2θ scans) of an Alnico alloy (53Fe–17Ni–13Co–9.5Al–6Cu–0.5Ti–0.5Zr). The upper chart was recorded with linear, the lower with logarithmic ratemeter mode.

crystal; optical-path medium; detector; and goniometer scanning speed. *Readout conditions*: pulse-height analysis; ratemeter time constant (response time); ratemeter sensitivity (intensity required for full-scale recorder deflection); ratemeter mode (linear or logarithmic); recorder response time (time to reach full scale); and recorder chart speed. If there is interest in only one or a few elements, the instrument is arranged for optimal recording of these elements. If practically nothing is known of the specimen and maximum information is required, the following general considerations may guide the choice of conditions for standard commercial instruments.

The x-ray tube should be operated at maximum potential to ensure excitation of spectra of elements having high excitation potentials, and at maximum current to ensure maximum sensitivity for minor and trace elements. In general, a tungsten or platinum target is preferable because the high atomic number ensures maximum x-ray output for a given input power. If interest is limited to elements below vanadium, a chromium target is preferable. A dual tungsten/chromium target permits convenient changeover during the recording of the spectrum. A primary beam filter to remove target lines may be beneficial if the sample has low effective atomic number and therefore scatters primary radiation efficiently.

The collimators are selected to provide a balance between high intensity and high resolution. Certainly no crystal can exceed the combination of high diffracted intensity, high resolution, and wide useful range of elements (down to potassium) provided by lithium fluoride. A crystal changer permits convenient extension of the element range down to sodium or perhaps fluorine. Helium or vacuum is of some benefit for trace concentrations of elements having line wavelengths >1 Å, and is required if spectra of elements below vanadium are to be recorded. A combination detector consisting of a scintillation counter and a flow-proportional counter having an ultrathin window gives satisfactory response down to fluorine.

The goniometer 2θ scanning speed should be slow enough to permit response to weak peaks, yet fast enough to permit the spectrum to be recorded in reasonable time. A commonly used speed is $2°\ 2\theta$/min. Some form of automatic pulse-height selection may be beneficial to exclude certain types of spectral-line interference.

If there are several major constituents, the full-scale deflection is usually set so that the strongest peak is just on scale. If there is only one major constituent, or even two, it may be preferable to set the scale so that the strongest minor constituent peak is just on scale and let the major peak(s) run off scale. In either case, the peak to be used to set the scale may be known or found by making a rapid manual 2θ scan. It must be remembered that response is smaller when the goniometer scans through a peak than when it is stationary at the peak, and response decreases the faster the scanning speed, the longer the ratemeter time constant, and the slower the recorder response time. If the scale is made too small, background "noise" fluctuations become severe, and weak peaks may be obscured.

The ratemeter time constant should be short enough to permit response to weak peaks, yet long enough to damp background noise. Typical values are 1–4 s, but, in general, the time constant is selected for compatibility with the 2θ scanning speed and the full-scale deflection. If minor and trace as well as major constituents are of interest, that is, if very weak as well as very strong peaks are to be recorded, operation of the ratemeter in logarithmic, rather than linear, mode may be preferable. Recorder chart speed is selected principally to give a chart of convenient length. The charts in Figure 10.1 were recorded at a chart speed of 1 cm/min and a goniometer speed of $2° \ 2\theta$/min, giving a $52° \ 2\theta$ spectrum on a 26-cm chart.

An inspection of the recorded spectrum may reveal cases where the presence of a peak is questionable because of poor resolution from an adjacent peak or from background noise. These parts of the spectrum may be rerun at conditions to accentuate the questionable peak. Techniques for improving resolution are given in Section 13.2.2. The most convenient ones are use of higher collimation, a crystal having a smaller *d*-spacing, and higher-order spectra.

Figure 10.2 (*15P*) shows the effect of several instrument conditions on the recorded 2θ scans. Each pair of spectra was recorded at linear mode and identical conditions except for the one being evaluated. Figure 10.2A compares 4×0.020 inch and 4×0.005 inch source collimators. The fine collimator gives only $\sim$25% as much net peak intensity, but resolves many new peaks. Figure 10.2B compares $1\frac{1}{2} \times 0.023$ inch and $1\frac{1}{2} \times 0.005$ inch detector collimators. Here, the coarse collimator doubles the intensity with little loss of resolution and line/background ratio. Figure 10.2C shows the effect of continuous variation of pulse-height analyzer baseline in synchronism with 2θ. Second-order zirconium, lead, and strontium peaks are eliminated, leaving the first-order hafnium, nickel, and iron peaks clear. Figure 10.2D shows the combined effect of continuous pulse-height selection and an Al–Fe–Cu–Zn primary-beam filter to clear the germanium and selenium *K* lines from overlapping scattered tungsten *L* lines. Figure 10.2E shows that high intensity and resolution can be obtained at very high 2θ-scanning rates. The first spectrum was recorded at $2° \ 2\theta$/min and 1-s time constant on a conventional recorder having 1-s full-scale response. The other spectrum was recorded at $32° \ 2\theta$/min and 0.14-s time constant on a recorder having

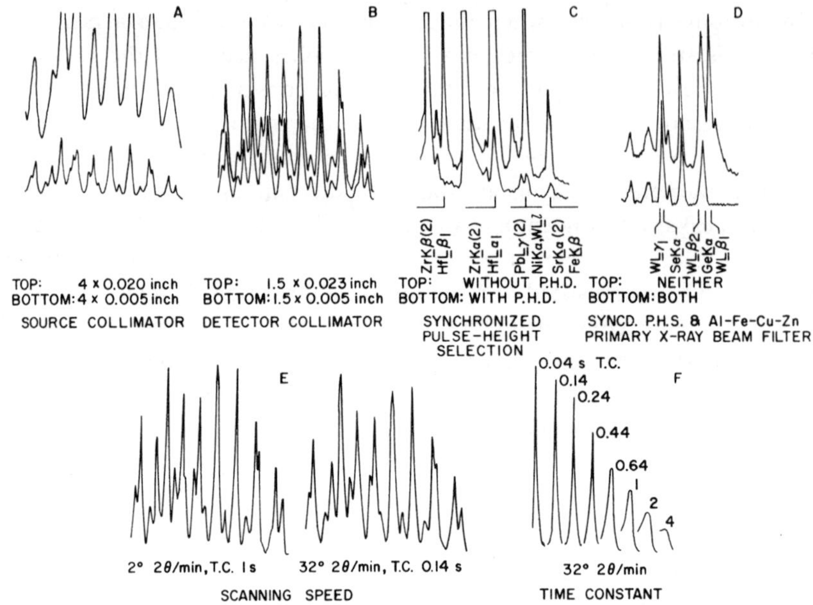

Fig. 10.2. Effect of certain instrument conditions on intensity and resolution in 2θ scans. [M. L. Salmon, in "Handbook of X-Rays," E. F. Kaelble, ed.; McGraw-Hill Book Co. (1967); courtesy of the author and publisher.]

0.25-s full-scale response, and has substantially the same intensity and resolution as the slow scan. Figure 10.2F shows the effect of time constant on the same peak scanned at 32° 2θ/min. Commercial spectrometers must be modified to have such rapid 2θ scan rates, ratemeter time constants, and recorder response times.

10.4. IDENTIFICATION OF THE PEAKS

Figure 10.1 shows the x-ray spectrum of an Alnico alloy (53Fe–17Ni–13Co–9.5Al–6Cu–0.5Ti–0.5Zr) recorded with linear and logarithmic ratemeter modes. Certainly the linear spectrum is preferable for general purposes, and is always used when intensities are to be estimated from the chart. However, the very weak Zr $K\alpha(2)$, W $L\eta$, and W $L\gamma$ peaks are accentuated on the logarithmic spectrum.

The peaks may be identified by use of transparent overlay scales (*15P*, *G8*, *S2*) showing the positions of all but the weakest spectral lines. A different overlay is required for each combination of crystal, 2θ-scanning speed, and chart speed. However, peaks are usually identified by use of tables.

The 2θ angle at which each peak on the chart occurs is obtained from the known 2θ scan and chart speeds or from a scale previously prepared from these parameters. Also, most recorders in x-ray spectrometers have a degree-marker pen that is actuated by a micro switch in the goniometer and makes a tick along the edge of the chart at 1° or 0.5° intervals. Degree-marker traces are shown along the bottom edges of the charts in Figure 10.1. The peaks are then identified by reference to x-ray spectrometer tables, several of which are listed in the Bibliography. These tables are of two forms: The *line-to-2θ* table enables one to find at what 2θ angle a specified order of a specified line is diffracted by a specified crystal; the *2θ-to-line* table enables one to identify a peak from the 2θ angle at which it is diffracted, knowing the crystal that was used. The 2θ-to-line table is also useful for searching for possible spectral-line interference. The analyte line is located in the table, and a search is made for 1° or 2° 2θ on either side of the line for lines of other elements known to be present in the specimen.

A portion of a page of a line-to-2θ table is shown in Figure 10.3 (*49*). Elements are listed in order of increasing atomic number, and, for each element, spectral lines are listed in order of wavelength. Many tables include absorption edges, even though these do not appear on the ratemeter-recorder charts. The table shown in Figure 10.3 includes satellite lines (those having line symbols prefixed by S) and some lines to which no symbols have been assigned (those indicated by initial and final atom state, for example, M3–M5). The column headings signify, in order, the element (EL), spectral line (LINE), order (N), relative intensity (I), atomic number (Z), reference to source of data (R), photon energy (KEV), and product of order and wavelength in angstroms $n\lambda$ (LAMBDA). The wavelength of a line (1λ) is the value given for the first order of that line. The last 11 columns give the 2θ angle at which the line is diffracted by each of 11 crystals. This table lists only three orders for each line, and is a little difficult to use because it contains so many lines of no significance in conventional x-ray spectrometric analysis.

The form of table shown in Figure 10.3 lists 2θ values for each line for all crystals together, that is, on the same page. Other line-to-2θ tables have the forms shown in Figures 10.4 and 10.5, where the data for each crystal is separate. Figure 10.4 (*46*) shows part of a page of such a table for a LiF crystal in which each spectral line has its own column and each of the five orders its own row. Figure 10.5 (*40*) shows part of a page of a table in which each spectral line has its own row and each of the five orders its own column.

PAGE 27.0

EL	LINE	N	I	Z	R	KEV	LAMBDA	LSC	KAP	MICA	GYPM	ADP	EDDT	PET	SIO2	CACO3	NACL	LIF
RU	KA2	3	50	44	1	19.15	1.9422	2.22	8.36	11.24	14.70	21.04	25.48	25.67	33.77	37.31	40.28	57.67
RU	KA1,2	1	150	44	1	19.23	0.6445	.74	2.77	3.72	4.87	6.95	8.39	8.46	11.06	12.19	13.12	18.42
RU	KA1,2	2	150	44	1	19.23	1.2890	1.47	5.55	7.45	9.74	13.92	16.83	16.96	22.23	24.52	26.42	37.33
RU	KA1,2	3	150	44	1	19.23	1.9335	2.21	8.33	11.19	14.63	20.94	25.36	25.56	33.61	37.14	40.C9	57.39
RU	KA1	1	100	44	1	19.28	0.6430	.73	2.77	3.71	4.85	6.93	8.37	8.44	11.04	12.16	13.09	18.37
RU	KA1	2	100	44	1	19.28	1.2860	1.47	5.54	7.43	9.72	13.88	16.79	16.92	22.17	24.46	26.36	37.24
RU	KA1	3	100	44	1	19.28	1.9290	2.2C	8.31	11.16	14.60	20.89	25.30	25.50	33.53	37.05	40.00	57.24
RU	SKBN	1	1	44	1	21.53	0.5757	.66	2.48	3.33	4.35	6.20	7.49	7.55	9.88	10.88	11.72	16.44
RU	SKBN	2	1	44	1	21.53	1.1513	1.32	4.96	6.65	8.70	12.42	15.02	15.14	19.83	21.86	23.56	33.22
RU	SKBN	3	1	44	1	21.53	1.7270	1.97	7.44	9.99	13.06	18.68	22.61	22.79	29.93	33.05	35.66	50.79
RU	KB3	1	15	44	1	21.63	0.5731	.65	2.47	3.31	4.33	6.17	7.46	7.52	9.83	1C.83	11.66	16.36
RU	KB3	2	15	44	1	21.63	1.1461	1.31	4.93	6.62	8.66	12.37	14.95	15.07	19.74	21.76	23.45	33.07
RU	KB3	3	15	44	1	21.63	1.7192	1.96	7.40	9.94	13.00	18.60	22.51	22.68	29.79	32.90	35.49	50.54
RU	KB1	1	15	44	1	21.65	0.5725	.65	2.46	3.31	4.32	6.17	7.45	7.51	9.82	10.82	11.65	16.34
RU	KB1	2	15	44	1	21.65	1.1449	1.31	4.93	6.62	8.65	12.35	14.94	15.05	19.72	21.74	23.42	33.03
RU	KB1	3	15	44	1	21.65	1.7174	1.96	7.39	9.93	12.99	18.58	22.49	22.66	29.76	32.86	35.45	50.48
RU	SKB7	1	1	44	1	21.72	0.5707	.65	2.46	3.30	4.31	6.15	7.43	7.49	9.79	10.79	11.61	16.29
RU	SKB7	2	1	44	1	21.72	1.1413	1.3C	4.91	6.60	8.62	12.32	14.89	15.00	19.65	21.67	23.35	32.93
RU	SKB7	3	1	44	1	21.72	1.7120	1.96	7.37	9.90	12.95	18.52	22.42	22.59	29.67	32.76	35.34	50.31
RU	SKB6	1	1	44	1	21.78	0.5691	.65	2.45	3.29	4.30	6.13	7.41	7.47	9.76	10.76	11.58	16.25
RU	SKB6	2	1	44	1	21.78	1.1383	1.3C	4.90	6.58	8.60	12.28	14.85	14.96	19.60	21.61	23.28	32.84
RU	SKB6	3	1	44	1	21.78	1.7074	1.95	7.35	9.87	12.91	18.47	22.36	22.53	29.59	32.67	35.24	50.17
RU	KB5	1	1	44	1	21.83	0.5678	.65	2.44	3.28	4.29	6.12	7.39	7.45	9.74	10.73	11.56	16.21
RU	KB5	2	1	44	1	21.83	1.1357	1.3C	4.89	6.56	8.58	12.25	14.82	14.93	19.56	21.56	23.23	32.76
RU	KB5	3	1	44	1	21.83	1.7035	1.95	7.33	9.85	12.88	18.43	22.30	22.47	29.52	32.59	35.16	50.05
RU	SKB8	1	1	44	1	21.98	0.5639	.64	2.43	3.26	4.26	6.08	7.34	7.40	9.68	10.66	11.48	16.10
RU	SKB8	2	1	44	1	21.98	1.1279	1.29	4.85	6.52	8.52	12.17	14.71	14.83	19.42	21.41	23.07	32.53
RU	SKB8	3	1	44	1	21.98	1.6918	1.93	7.28	9.78	12.79	18.30	22.15	22.32	29.31	32.36	34.91	49.68
RU	KB2	1	5	44	1	22.07	C.5616	.64	2.42	3.24	4.24	6.05	7.31	7.37	9.64	10.62	11.43	16.03
RU	KB2	2	5	44	1	22.07	1.1233	1.28	4.83	6.49	8.48	12.12	14.65	14.76	19.34	21.32	22.97	32.39
RU	KB2	3	5	44	1	22.07	1.6849	1.93	7.25	9.74	12.74	18.22	22.06	22.22	29.19	32.22	34.76	49.46
RU	KB4,	1	1	44	1	22.10	0.5608	.64	2.41	3.24	4.23	6.04	7.30	7.36	9.62	10.60	11.41	16.01
RU	KB4,	2	1	44	1	22.10	1.1217	1.28	4.83	6.48	8.47	12.10	14.63	14.74	19.31	21.29	22.94	32.34
RU	KB4,	3	1	44	1	22.10	1.6825	1.92	7.24	9.73	12.72	18.20	22.02	22.19	29.14	32.18	34.71	49.39
RU	SKB+2	1	1	44	1	22.18	0.5589	.64	2.41	3.23	4.22	6.02	7.28	7.33	9.59	10.56	11.37	15.96
RU	SKB+2	2	1	44	1	22.18	1.1179	1.28	4.81	6.46	8.44	12.06	14.58	14.69	19.25	21.22	22.86	32.23
RU	SKB+2	3	1	44	1	22.18	1.6768	1.92	7.22	9.70	12.68	18.13	21.95	22.12	29.04	32.07	34.59	49.21
RH	M3-M5	1	1	45	0	0.19	65.4500	81.47										
RH	M2-M4	1	1	45	0	C.21	59.5400	72.83										
RH	MZ1,2	1	1	45	0	C.26	47.5900	56.65										

Fig. 10.3. Portion of a page of a line-to-2θ x-ray spectrometer table giving data for all crystals together. [E. W. White, G. V. Gibbs, G. C. Johnson, Jr., and G. R. Zechman, Jr., "X-Ray Emission Line Wavelength and Two-Theta Tables," *ASTM Data Series* **DS-37**, American Society for Testing and Materials (1965); courtesy of the authors and A.S.T.M.]

ASCENDING ATOMIC NUMBER DATA FOR LIF 2D 4.028

ELEM	ORD	KA1	KA2	KB1	KB3	LA1	LB1	LB2	LY1	LB3	LB5	LL	LB4	LY6
81 TL	1	4.84	4.98	4.27	4.30	34.87	29.19	29.04	24.89	28.78	28.19	40.22	29.90	24.22
	2	9.68	9.97	8.54	8.60	73.64	60.53	60.20	51.06	59.61	58.30	86.89	62.11	49.61
	3	14.55	14.98	12.83	12.91	128.04	98.22	97.57	80.55	96.41	93.88		101.40	78.00
	4	19.44	20.02	17.13	17.25				119.08	167.48	153.90			114.10
	5	24.36	25.09	21.46	21.61									
82 PB	1	4.70	4.84	4.15		33.92	28.22	28.25	24.07	27.84	27.37	39.16	28.96	23.41
	2	9.40	9.68	8.31		71.38	58.36	58.43	49.30	57.52	56.48	84.18	60.00	47.87
	3	14.12	14.55	12.49		122.12	94.00	94.13	77.46	92.39	90.43		97.18	74.96
	4	18.86	19.44	16.67			154.41	154.93	113.06	148.42	142.30		180.00	108.45
	5	23.64	24.36	20.88										
83 BI	1	4.58	4.70	4.04	4.07	33.00	27.34	27.43	23.32	26.96	26.58	38.17	28.07	22.65
	2	9.17	9.40	8.09	8.14	69.23	56.42	56.61	47.68	55.58	54.75	81.68	58.04	46.25
	3	13.77	14.12	12.14	12.23	116.87	90.31	90.68	74.64	88.75	87.21	157.56	93.38	72.19
	4	18.40	18.86	16.21	16.33		141.95	143.01	107.87	137.65	133.72		151.96	103.53
	5	23.06	23.64	20.30	20.45									158.15

Fig. 10.4. Portion of a page of a line-to-2θ x-ray spectrometer table giving data for each crystal separately and having spectral lines in separate columns. [M. C. Powers, "X-Ray Fluorescent Spectrometer Conversion Tables," Philips Electronic Instruments, Inc. (1957); courtesy of the author and Philips Electronic Instruments, Inc.]

LIF 2D 4.0267

ELEMENT	LAMBDA	ORDER	1	2	3	4	5
82 PB	.14077	K	ABS	4.01			
82 PB	.14155	KB4	4.03	8.06	12.11	16.17	20.25
82 PB	.14191	KB2	4.04	8.08	12.14	16.21	20.30
82 PB	.14494	KB5	4.13	8.26	12.40	16.56	20.74
82 PB	.14596	KB1	4.15	8.31	12.49	16.67	20.88
82 PB	.14680	KB3	4.18	8.36	12.56	16.77	21.01
82 PB	.16536	KA1	4.71	9.42	14.15	18.91	23.70
82 PB	.16701	KA	4.75	9.52	14.29	19.10	23.94
82 PB	.17029	KA2	4.85	9.70	14.58	19.48	24.41
82 PB	.78152	L I	ABS	22.38			
82 PB	.78588	LY4	22.51	45.95	71.68	102.64	154.76
82 PB	.78710	LYP4	22.54	46.03	71.81	102.87	155.56
82 PB	.80231	LY11	22.99	46.97	73.42	105.69	170.05
82 PB	.81484	LY3	23.35	47.75	74.76	108.08	
82 PB	.81508	L II	ABS	23.36			
82 PB	.81686	LY6	23.41	47.87	74.97	108.47	
82 PB	.82127	LY2	23.54	48.15	75.45	109.34	
82 PB	.82327	LYP8	23.59	48.27	75.67	109.73	
82 PB	.82368	LY8	23.61	48.30	75.71	109.81	
82 PB	.83971	LY1	24.07	49.30	77.45	113.05	
82 PB	.86645	LY5	24.85	50.98	80.41	118.79	
82 PB	.92677	LB9	26.61	54.81	87.33	134.03	
82 PB	.93383	LB10	26.82	55.27	88.17	136.14	
82 PB	.95029	LIII	ABS	27.30			

Fig. 10.5. Portion of a page of a line-to-2θ x-ray spectrometer table giving data for each crystal separately and having diffraction orders in separate columns. ["X-Ray Wavelengths for Spectrometer," General Electric Co., X-Ray Dept.]

A portion of a page of a 2θ-to-line table is shown in Figure 10.6 (*49*), where the column headings have the significance given to Figure 10.3 above. The lines are listed in order of increasing $n\lambda$ or 2θ.

Another type of table is of incidental interest (*42*, *50*). Some x-ray spectrometers have 2θ-drive drums (odometers) calibrated to indicate wavelength in angstroms directly when used with a certain crystal, usually LiF or ADP. Such spectrometers are frequently used on electron-probe microanalyzers (Chapter 21). If a crystal other than the one for which it is calibrated is used on such an instrument, the odometer no longer indicates wavelength, and the indicated reading must be corrected as follows:

$$R_X = (d_C/d_X)\lambda_L \tag{9.1}$$

where R_X is odometer reading for the interchanged crystal X; d_C and d_X are interplanar spacings (Å) for the diffracting planes of, respectively, the crystal for which the odometer was calibrated C and the new crystal X; and λ_L is the wavelength of the line—or the odometer reading for that line with the original crystal. Tables have been compiled that give odometer settings for various crystals used on a spectrometer calibrated for a specified crystal. These tables have line-to-setting and setting-to-line forms analogous to the forms of coventional x-ray spectrometer tables. Figure 10.7 (*50*)

PAGE 141.0

EL	LINE	N	I	Z	R	KEV	LAMBDA	LSD	KAP	MICA	GYPM	ADP	EDDT	PET	SIO2	CACO3	NACL	LIF
PT	KA2	8	50	78	1	65.11	1.5231	1.74	6.56	8.81	11.51	16.46	19.92	20.07	26.33	29.06	31.33	44.44
HF	LN	1	1	72	1	8.14	1.5232	1.74	6.56	8.81	11.51	16.46	19.92	20.07	26.33	29.06	31.33	44.45
BI	LG4+	2	1	83	1	16.27	1.5239	1.74	6.56	8.81	11.52	16.47	19.93	20.08	26.34	29.07	31.35	44.47
IR	KB3	9	15	77	1	73.20	1.5241	1.74	6.56	8.81	11.52	16.47	19.93	20.08	26.35	29.08	31.35	44.47
ER	KA1,2	6	150	68	1	48.80	1.5241	1.74	6.56	8.81	11.52	16.47	19.93	20.08	26.35	29.08	31.35	44.47
PR	KB3	5	15	59	1	40.65	1.5246	1.74	6.56	8.81	11.52	16.48	19.94	20.09	26.36	29.09	31.36	44.49
TA	SLAS	1	1	73	1	8.13	1.5252	1.74	6.57	8.82	11.53	16.48	19.94	20.10	26.37	29.10	31.38	44.51
TH	L3-P2,	2	1	90	1	16.25	1.5253	1.74	6.57	8.82	11.53	16.48	19.94	20.10	26.37	29.10	31.38	44.51
AT	LG1	2	10	85	2	16.25	1.5258	1.74	6.57	8.82	11.53	16.49	19.95	20.10	26.38	29.11	31.39	44.53
HF	KB2	8	5	72	1	64.97	1.5263	1.74	6.57	8.82	11.54	16.49	19.96	20.11	26.39	29.12	31.40	44.54
GD	KB1	6	15	64	1	48.72	1.5265	1.74	6.57	8.83	11.54	16.50	19.96	20.11	26.39	29.12	31.40	44.55
TH	L3-P1	2	1	90	1	16.24	1.5267	1.74	6.57	8.83	11.54	16.50	19.96	20.12	26.39	29.13	31.41	44.55
TA	KB3	8	15	73	1	64.94	1.5271	1.74	6.57	8.83	11.54	16.50	19.97	20.12	26.40	29.14	31.42	44.57
IR	KA1	8	100	77	1	64.91	1.5279	1.75	6.58	8.83	11.55	16.51	19.98	20.13	26.42	29.15	31.43	44.59
PD	KB4	3	1	46	2	24.34	1.5279	1.75	6.58	8.83	11.55	16.51	19.98	20.13	26.42	29.15	31.43	44.59
CU	SKA+3	1	1	29	1	8.11	1.5280	1.75	6.58	8.83	11.55	16.51	19.98	20.13	26.42	29.15	31.43	44.59
BI	L1-N6,	2	1	83	1	16.22	1.5281	1.75	6.58	8.83	11.55	16.51	19.98	20.13	26.42	29.16	31.44	44.60
PO	LG3,6	2	3	84	1	16.22	1.5289	1.75	6.58	8.84	11.56	16.52	19.99	20.14	26.43	29.17	31.45	44.62
GD	LG3	1	2	64	1	8.11	1.5290	1.75	6.58	8.84	11.56	16.52	19.99	20.15	26.43	29.17	31.46	44.63
RN	KA2	10	50	86	2	81.07	1.5290	1.75	6.58	8.84	11.56	16.52	19.99	20.15	26.43	29.17	31.46	44.63
TH	LB5	2	1	90	1	16.21	1.5295	1.75	6.58	8.84	11.56	16.53	20.00	20.15	26.44	29.18	31.47	44.64
TB	LG1	1	10	65	1	8.10	1.5297	1.75	6.59	8.84	11.56	16.53	20.00	20.16	26.45	29.19	31.47	44.65
TM	LB1	1	50	69	1	8.10	1.5299	1.75	6.59	8.85	11.56	16.53	20.01	20.16	26.45	29.19	31.47	44.65
TH	LB1	2	50	90	1	16.20	1.5303	1.75	6.59	8.85	11.57	16.54	20.01	20.16	26.46	29.20	31.48	44.67
PD	KB2	3	5	46	1	24.29	1.5307	1.75	6.59	8.85	11.57	16.54	20.02	20.17	26.46	29.21	31.49	44.68
TL	KA1	9	100	81	1	72.86	1.5313	1.75	6.59	8.85	11.58	16.55	20.02	20.18	26.47	29.22	31.50	44.70
GD	KB3	6	15	64	1	48.57	1.5313	1.75	6.59	8.85	11.58	16.55	20.02	20.18	26.47	29.22	31.50	44.70
BI	L1-O1	2	1	83	1	16.18	1.5321	1.75	6.60	8.86	11.58	16.56	20.03	20.19	26.49	29.23	31.52	44.72
PB	KA2	9	50	82	1	72.81	1.5322	1.75	6.60	8.86	11.58	16.56	20.04	20.19	26.49	29.24	31.52	44.72
TA	LA2	1	10	73	1	8.09	1.5328	1.75	6.60	8.86	11.59	16.57	20.04	20.20	26.50	29.25	31.54	44.74
CU	SKA4	1	1	29	1	8.09	1.5330	1.75	6.60	8.86	11.59	16.57	20.05	20.20	26.50	29.25	31.54	44.75
LU	L2-M2	1	1	71	1	8.08	1.5333	1.75	6.60	8.86	11.59	16.57	20.05	20.20	26.51	29.26	31.55	44.76
CU	SKA3+	1	1	29	1	8.08	1.5339	1.75	6.60	8.87	11.60	16.58	20.06	20.21	26.52	29.27	31.56	44.78
GD	LG2	1	1	64	1	8.08	1.5341	1.75	6.60	8.87	11.60	16.58	20.06	20.21	26.52	29.27	31.56	44.78
TB	SLG9	1	1	65	1	8.08	1.5345	1.75	6.61	8.87	11.60	16.58	20.07	20.22	26.53	29.28	31.57	44.80
CU	SKA3	1	1	29	1	8.08	1.5347	1.75	6.61	8.87	11.60	16.59	20.07	20.22	26.53	29.28	31.58	44.80
OS	L3-M2	1	1	76	2	8.08	1.5347	1.75	6.61	8.87	11.60	16.59	20.07	20.22	26.54	29.28	31.58	44.80
TH	SLB2•7	2	1	90	1	16.15	1.5355	1.75	6.61	8.88	11.61	16.60	20.08	20.23	26.55	29.30	31.59	44.83
I	KB1	4	15	53	1	32.29	1.5355	1.75	6.61	8.88	11.61	16.60	20.08	20.23	26.55	29.30	31.59	44.83
CU	SKA+	1	1	29	1	8.07	1.5361	1.76	6.61	8.88	11.61	16.60	20.09	20.24	26.56	29.31	31.61	44.84

Fig. 10.6. Portion of a page of a 2θ-to-line x-ray spectrometer table. [E. W. White, G. V. Gibbs, G. G. Johnson, Jr., and G. R. Zechman, Jr., "X-Ray Emission Line Wavelength and Two-Theta Tables," *ASTM Data Series* **DS-37**, American Society for Testing and Materials (1965); courtesy of the authors and A.S.T.M.]

PAGE 84.0

EL	LINE	N	I	Z	KEV	LAMBDA	ACP LSD	ADP KAP	ALP EDDT	ACP SIO2	ADP NACL	ADP LIF	LIF LSD	LIF KAP	LIF ADP	LIF EDDT	LIF SIO2	LIF NACL
TL	L3-M3	3	1	81	12.05	3.0860		1.2329	3.7278	4.9101	5.8212	8.1533		0.4666	1.1680	1.4109	1.8584	2.2033
W	LG4+	3	1	74	12.05	3.0860		1.2329	3.7278	4.9101	5.8212	8.1533		0.4666	1.1680	1.4109	1.8584	2.2033
CA	SKB•5	1	1	20	4.02	3.0864		1.2330	3.7283	4.9108	5.8219	8.1544		0.4667	1.1681	1.4111	1.8587	2.2036
EU	LG4,	2	1	63	8.03	3.0876		1.2335	3.7298	4.9127	5.8242	8.1576		0.4669	1.1686	1.4117	1.8594	2.2044
RH	KA2	5	50	45	20.07	3.0882		1.2338	3.7305	4.9136	5.8253	8.1591		0.4669	1.1688	1.4120	1.8598	2.2048
CU	KA2	2	50	29	8.03	3.0886		1.2339	3.7310	4.9143	5.8261	8.1602		0.4670	1.1690	1.4121	1.8600	2.2051
TM	LB4	2	4	69	8.03	3.0886		1.2339	3.7310	4.9143	5.8261	8.1602		0.4670	1.1690	1.4121	1.8600	2.2051
RE	SLB2+	3	1	75	12.04	3.0890		1.2341	3.7315	4.9149	5.8269	8.1613		0.4671	1.1691	1.4123	1.8602	2.2054
SM	KA1	10	100	62	40.13	3.0892		1.2342	3.7317	4.9152	5.8272	8.1618		0.4671	1.1692	1.4124	1.8604	2.2056
CA	KB1	1	15	20	4.01	3.0896		1.2343	3.7322	4.9159	5.8280	8.1629		0.4672	1.1694	1.4126	1.8606	2.2058
CF	LG1	6	10	98	24.07	3.0900		1.2345	3.7326	4.9164	5.8287	8.1638		0.4672	1.1695	1.4128	1.8608	2.2061
FR	LA1	3	100	87	12.03	3.0902		1.2346	3.7329	4.9168	5.8291	8.1644		0.4672	1.1696	1.4129	1.8610	2.2063
BA	KA1,2	8	150	56	32.07	3.0926		1.2355	3.7358	4.9206	5.8337	8.1708		0.4676	1.1705	1.4140	1.8624	2.2080
PA	LB2	4	20	91	16.02	3.0946		1.2363	3.7382	4.9238	5.8374	8.1761		0.4679	1.1713	1.4149	1.8636	2.2094
W	L1-O1	3	1	74	12.02	3.0951		1.2365	3.7388	4.9246	5.8383	8.1773		0.4680	1.1714	1.4151	1.8639	2.2097
AU	SLB2•,	3	1	79	12.01	3.0959		1.2368	3.7398	4.9259	5.8399	8.1795		0.4681	1.1718	1.4155	1.8644	2.2103
RE	LG2	3	1	75	12.01	3.0968		1.2372	3.7409	4.9273	5.8416	8.1819		0.4682	1.1721	1.4159	1.8649	2.2110
PR	KA1	9	100	59	36.03	3.0968		1.2372	3.7409	4.9273	5.8416	8.1819		0.4682	1.1721	1.4159	1.8649	2.2110
PB	L1-M1	3	1	82	12.01	3.0969		1.2372	3.7410	4.9274	5.8417	8.1821		0.4683	1.1721	1.4159	1.8650	2.2110
MO	LB10	2	1	67	8.01	3.0970		1.2373	3.7411	4.9277	5.8420	8.1824		0.4683	1.1722	1.4160	1.8651	2.2111
IN	KA2	6	50	49	24.00	3.0992		1.2382	3.7438	4.9312	5.8461	8.1883		0.4686	1.1730	1.4170	1.8664	2.2127
HG	LB3	3	6	80	11.99	3.1004		1.2386	3.7453	4.9331	5.8484	8.1914		0.4688	1.1735	1.4175	1.8671	2.2136
CA	SKB+	1	1	20	4.00	3.1011		1.2389	3.7461	4.9342	5.8497	8.1933		0.4689	1.1737	1.4179	1.8675	2.2141
AU	SLB5•,	3	1	79	11.99	3.1022		1.2394	3.7474	4.9359	5.8518	8.1962		0.4691	1.1741	1.4184	1.8682	2.2149
U	L3-M2	3	1	92	11.98	3.1037		1.2400	3.7493	4.9383	5.8546	8.2002		0.4693	1.1747	1.4191	1.8691	2.2159
BI	L2-M2	3	1	83	11.98	3.1038		1.2400	3.7493	4.9384	5.8547	8.2003		0.4693	1.1747	1.4191	1.8691	2.2160
MO	KB2	5	5	42	19.96	3.1047		1.2404	3.7505	4.9399	5.8565	8.2028		0.4694	1.1751	1.4195	1.8697	2.2166
BK	LB1	5	50	97	19.96	3.1050		1.2405	3.7508	4.9403	5.8570	8.2035		0.4695	1.1752	1.4196	1.8699	2.2168
SM	KA1,2	10	150	62	39.92	3.1052		1.2406	3.7511	4.9407	5.8575	8.2042		0.4695	1.1753	1.4197	1.8700	2.2170
I	SLA9	1	1	53	3.99	3.1055		1.2407	3.7514	4.9412	5.8580	8.2049		0.4696	1.1754	1.4199	1.8702	2.2172
BR	SKA4	3	1	35	11.97	3.1058		1.2408	3.7518	4.9417	5.8586	8.2057		0.4696	1.1755	1.4200	1.8704	2.2174
BR	SKA3,	3	1	35	11.96	3.1091		1.2421	3.7558	4.9469	5.8648	8.2145		0.4701	1.1768	1.4215	1.8724	2.2198
RE	LG6	3	1	75	11.95	3.1109		1.2428	3.7580	4.9498	5.8683	8.2192		0.4704	1.1774	1.4224	1.8735	2.2211
PR	KA1,2	9	150	59	35.86	3.1112		1.2430	3.7583	4.9503	5.8688	8.2200		0.4704	1.1776	1.4225	1.8736	2.2213
AU	SLB2•5	3	1	79	11.95	3.1124		1.2434	3.7598	4.9522	5.8711	8.2232		0.4706	1.1780	1.4230	1.8744	2.2222
AC	LB3	4	6	89	15.93	3.1128		1.2436	3.7602	4.9527	5.8717	8.2241		0.4707	1.1781	1.4232	1.8746	2.2224
CS	KB2	9	5	55	35.82	3.1148		1.2444	3.7627	4.9560	5.8756	8.2296		0.4710	1.1789	1.4241	1.8758	2.2239
IN	KB2	7	5	49	27.86	3.1149		1.2444	3.7628	4.9562	5.8758	8.2298		0.4710	1.1790	1.4242	1.8759	2.2239
SN	LB9	1	1	50	3.98	3.1150		1.2445	3.7629	4.9564	5.8760	8.2301		0.4710	1.1790	1.4242	1.8759	2.2240
I	SLA8	1	1	53	3.98	3.1150		1.2445	3.7629	4.9564	5.8760	8.2301		0.4710	1.1790	1.4242	1.8759	2.2240

Fig. 10.7. Portion of a page of an x-ray spectrometer table for use when other crystals are used on spectrometers calibrated directly in angstroms for LiF or ADP crystals. [E. W. White, G. V. Gibbs, G. G. Johnson, Jr., and G. R. Zechman, Jr., "X-Ray Wavelengths and Crystal Interchange Settings for Wavelength-Geared Curved-Crystal Spectrometers," *Mineral Industries Experiment Station Special Publication* **3-64**, Pennsylvania State University (1965); courtesy of the authors and Pennsylvania State University.]

shows a portion of a page of the setting-to-line form of such a table. The first seven column headings have the same significance as in Figure 10.3. The six columns ADP/LSD, ADP/KAP, etc. give odometer readings for LSD, KAP, etc. used on a spectrometer calibrated for ADP. The six columns LIF/LSD, LIF/KAP, etc. have the same significance for a spectrometer calibrated for LiF.

There is a systematic procedure for identification of the peaks: First, by use of a line-to-2θ table, all peaks known to be present are identified. These include peaks of elements known to be present in the specimen and coherently and incoherently scattered lines of the target and its contaminants. The incoherently scattered peaks appear as relatively broad peaks on the high-2θ (long-wavelength) sides of the coherently scattered peaks (Figure 2.10). All scattered peaks are more intense the lower the effective atomic number of the specimen. Then, by use of a 2θ-to-line table, all remaining peaks are identified.

Peaks may be as much as $0.5° \ 2\theta$ away from their tabulated values because of slight misalignment of the goniometer or lag in ratemeter–recorder response, and this displacement may vary progressively from start to end of the spectrum. Moreover, it may be difficult to read the chart more precisely than $\pm 0.2° \ 2\theta$.

Insofar as possible, more than one of its lines should be identified for an element to be regarded as present. If a peak is believed to be Fe $K\alpha$, the weaker ($\sim$1/5) Fe $K\beta$ should be present. If a peak is believed to be Pb $L\beta_1$, the slightly stronger Pb $L\alpha_1$ and weaker Pb $L\gamma_1$ peaks should be present. If relatively strong Mo $K\alpha$–$K\beta$ peaks are present, their second- and perhaps third-order peaks should be present. If the associated peaks cannot be found, their absence must be accounted for. Some examples follow: If a $K\alpha$ peak is very weak, the $K\beta$ peak may not be distinguishable from the background; $K\beta$ lines of elements 9–17 (fluorine–chlorine) are very weak relative to the $K\alpha$ lines; a peak may be lost in an interfering peak; and second-order peaks may undergo extinction by the crystal.

Relative intensities of peaks must be considered and anomalies accounted for. For example, suppose a strong peak is believed to be Cu $K\alpha$, but the Cu $K\beta$ peak is very weak, when it should be $\sim$1/5 as intense as Cu $K\alpha$. Perhaps another line is superposed on Cu $K\alpha$ giving it an anomalously high apparent intensity, or perhaps a matrix element is preferentially absorbing the Cu $K\beta$—nickel, for example. Then again, perhaps the strong line is not Cu $K\alpha$ after all. In considering relative intensities, one can be guided by certain general relationships: the relative intensities of the lines within a series (K, L, etc.) are given in the tables and in Figure 1.6; in the same 2θ region and for the same concentrations, first-order $K\alpha$ lines are 7–10 times as intense as first-order $L\alpha_1$ lines; and for LiF crystals, second-order lines are ~1/10 as intense as first-order lines.

10.5. A GENERAL PROCEDURE FOR QUALITATIVE AND SEMIQUANTITATIVE ANALYSIS

The techniques developed by Salmon (*S2*) for qualitative and semiquantitative x-ray spectrometric analysis of minerals and ores permit large numbers of accurate analyses to be performed in a remarkably short time. These techniques provide an excellent example of the speed and accuracy attainable by x-ray spectrometric techniques and well warrant detailed consideration.

All samples are reduced to at least −200 mesh by a standardized grinding procedure, loosely packed in CaPlugs, sealed with 0.00025-inch Mylar, and placed in the inverted position in a Philips 100-kV spectrometer. The CaPlugs are available in a range of sizes. The Mylar-sealed samples minimize contamination of the samples, instrument, and laboratory; reduce loading time; and can be stored for future reexamination.

Six superposed 2θ scans are made. For each, the chart is rolled back to the appropriate starting place, and the recorder pen is replaced with another having ink of a different color. All scans are made at the following conditions, except as noted below:

W target
50 kV, 33 mA
LiF crystal
4 × 0.005 inch source collimator
1 1/2 × 0.023 inch detector collimator
Air path
Scintillation counter
Scanned region and rate: 90°–3° 2θ, 8° 2θ/min
Ratemeter time constant: 0.44 s
Full-scale recorder deflection (switched automatically):
  90°–60° 2θ, 800 counts/s
  60°–31° 2θ, 1600 counts/s
  31°–17° 2θ, 3200 counts/s
  17°–3° 2θ, 6400 counts/s
Pulse-height selection synchronized with 2θ drive

Scan 1 is made at the indicated conditions and provides complete qualitative data and intensity data for semiquantitative analysis for minor and trace constituents, except where scattered target lines interfere.

Scan 2 is made at the same conditions, but without pulse-height selection, and the goniometer, full-scale deflection, and chart position are adjusted to rerun 2θ regions where peaks are off-scale. This scan shows gross intensities at all wavelengths without discrimination against higher-order peaks. The data are used for estimation of major constituents and to supplement the data in scan 1 for minor and trace constituents.

Scan 3 is limited to the region 40°–28° 2θ with full-scale deflection 400 counts/s, and a filter is placed over the x-ray tube window to supress tungsten *L* lines. The filter consists of three foils: 0.002-inch aluminum, 0.001-inch brass (Cu–Zn), and 0.0005-inch iron. This scan permits *K* lines of gallium, germanium, arsenic, and selenium, and *L* lines of elements 73–81 (tantalum–thallium) to be observed without interference from scattered tungsten *L* lines.

Scan 4 is made only if in scan 1 there are instances of overlapping first-order peaks. Such peaks are rerun at the same conditions except that a filter having an absorption edge between the peaks is placed in the secondary beam. In this way, the peak having the shorter wavelength is absorbed.

Scan 5 is limited to the region 26°–3° 2θ with excitation power 100 kV, 16.5 mA. This scan gives improved data for estimation of minor and trace concentrations of elements 38–74 (strontium–tungsten).

Scan 6 is made with helium path. This scan gives improved intensity data for wavelengths >1 Å and permits more accurate estimation of minor and trace concentrations of titanium, vanadium, chromium, and manganese.

The peaks are identified by placing the chart over a transparent template mounted on an illuminated cylindrical drum. The template is prepared from the tables for the 2θ-scan and chart speeds used and shows the angles at which all orders of all lines of any interest appear. The marks on the template show through the illuminated chart.

The semiquantitative analyses are based on intensities measured from the charts. Net intensity of a peak is measured directly if interfering peaks are absent. If an analyte peak includes contributions from more than one element, the contributions of the interfering elements are estimated from their unobscured peaks, and the analyte peak is corrected.

Standards are obtained by chemical analysis of selected samples or by synthesis. Oxide powders may be mixed, or they may be added to a base mineral of known composition to adjust the concentration of the analytes. If the powders to be mixed vary substantially in density, a solution of the denser constituent is added to the lighter powder; the paste is then thoroughly mixed and dried. The standards are subjected to the same procedure as the samples to obtain intensity data from which to prepare calibration curves.

Several techniques are used to convert the intensity data to semiquantitative analytical data. The calibration curve may simply have the function intensity *vs.* concentration. Other semiquantitative techniques are based on the fact that scattered-background and analyte-line intensities decrease in constant ratio as the absorption coefficient of the sample increases. Sets of calibration curves of *net* intensity *vs.* concentration are prepared, each curve for a different background level. Each sample is then analyzed by use of the curve corresponding to its background level. The background-ratio method may also be used, in which the calibration curve has the form analyte line/background intensity ratio *vs.* concentration. The ratioed background intensity may be measured adjacent to the peak or in the scattered continuum hump. In all these methods, intensities are in terms of chart divisions.

Intensity ratios of lines of the same series for elements of nearly the same—preferably adjacent—atomic number are usually substantially proportional to their concentration ratios, unless an absorption edge intervenes.

In the emission-absorption technique, the analyte-line intensity is measured from each sample directly and through a thin filter of the same sample placed in the secondary beam. The filter consists of a 0.001–0.020-inch layer of the finely powdered sample in a thin cell having two Mylar windows or spread with binder on a radiolucent support. A rough matrix-correction factor is calculated from the transmittance measurement and used to correct the intensities derived from the direct measurement. The corrected intensities are then applied to the calibration curves. The absorption measurement can be made by recording another 2θ scan of the peaks to be corrected, this time with the sample filter in the secondary beam. Alternatively, the scaler can be used for both the direct and the filtered measurements.

Chapter 11

Precision and Error; Counting Statistics

11.1. ERROR IN X-RAY SPECTROMETRIC ANALYSIS

The principal objective of the analytical method or "strategy" (Chapters 14 and 15) is to eliminate, minimize, circumvent, or correct for absorption-enhancement effects. The principal objective of the specimen preparation and presentation technique (Chapters 16 and 17) is to eliminate or minimize specimen errors. It is appropriate that all these sources of error be discussed first. This chapter defines and discusses all the various sources of error that beset a quantitative x-ray spectrometric analysis and considers counting statistics and analytical precision in detail. Absorption-enhancement effects and spectral-line interference are discussed in Chapters 12 and 13, respectively.

11.1.1. Nature and Sources of Error

In a quantitative x-ray spectrometric analysis, analyte concentration is derived from analyte-line intensities measured from the sample and from one or more standards. The *precision* of the analytical result is the degree of agreement among replicate determinations made under conditions as nearly identical as possible; it is usually expressed as a relative error, and the greater the precision, the smaller is its numerical value. Analytical precision depends on the magnitude of the errors affecting the x-ray intensity measurements and on the effectiveness of the techniques used to minimize them; it may be evaluated experimentally or calculated by summation of the individual contributing errors. The *accuracy* of the analytical result is the degree of agreement with the "true," accepted, or most reliably known value; like precision, it is usually expressed as a relative error having numerical value inversely proportional to the degree of accuracy. Analytical accuracy depends on the accuracy and precision of the x-ray intensity measurements and on the validity of the calibration. In x-ray spectrometric analysis, the accuracy is usually as good as the precision because it can usually be assumed that: (1) the measured x-ray intensity can be adjusted to agree with the true analytical value by the calibration; and (2) the analyte concentration in the standards is accurately known, or accepted as true. Otherwise, the analytical result may be very precise, but very inaccurate. The *reliability* of the analytical result is the degree to which it possesses precision and accuracy.

The errors affecting an x-ray spectrometric analysis may be classified as random or systematic. *Random errors* consist of small differences in successive values of a measurement made repetitively with great care by the same competent person under conditions as nearly constant as possible. The magnitude of random errors can be evaluated, and the errors can be minimized, but not corrected for. *Systematic errors* are those that can be avoided, or at least evaluated and corrected for; they may be constant, in which case they account for the deviation or bias of the experimental result from the true value; or they may fluctuate about a mean value, in which case they contribute to the precision. To a first approximation, random and systematic errors limit the precision and accuracy, respectively.

The errors may also be classified by source as follows:

1. *Statistical counting error* (Section 11.2) constitutes the best possible attainable precision and depends only on the total accumulated count.

2. *Instrumental errors* (Section 11.1.2) consist of short-term and long-term variation, instability, and drift in instrumental components, conditions, and parameters, principally the following: (a) X-ray tube potential (kV) and current (mA). (b) Intensity and distribution of the primary x-ray beam (caused by changes in dimension and position of internal components of the tube). (c) Crystal interplanar spacing (caused by changes in temperature; such changes cause slight displacement in 2θ for the spectral line). (d) Gas-amplification of the proportional counter (caused by changes in ambient temperature and pressure, and by drift in detector-tube potential).

(e) Secondary-emission ratio of the scintillation counter (caused by drift in detector-tube potential). (f) Coincidence (dead time) losses in the detector and electronic circuitry. (g) Shift and distortion of pulse-height distributions (only when pulse-height analysis is used). (h) Electronic circuitry.

3. *Operational errors* (manipulative or resetting errors) (Section 11.1.2) consist of slight nonreproducibility in settings of instrument conditions, principally the following: (a) x-ray tube potential (kV); (b) x-ray tube current (mA); (c) goniometer angle (2θ); (d) amplifier gain; (e) pulse-height analyzer baseline and window; and (f) detector-tube potential.

4. *Specimen errors* (Section 11.1.3) arise in the specimen itself. However, specimen errors do not include *sampling errors*, that is, errors arising from failure of the sample to be representative of the bulk of the material to be analyzed. Specimen errors may be classified as follows. (a) Absorption-enhancement ("matrix") effects (Chapter 12). (b) Chemical effects (Section 11.1.4). (c) Position effects, including variations in specimen plane, takeoff angle, position, orientation, and flatness (warp, ripple, wrinkle, etc.); position effects include the specimen-insertion effect. (d) Physical attributes, including thickness (if less than infinite); heterogeneity of composition; heterogeneity of density (porosity, voids, cracks, etc.); surface texture (roughness, grind marks, etc.); particle size (average and distribution); packing density of loose and briqueted powders; and expansion, volatility, bubble formation, radiolysis, and window distension of liquid specimens.

5. *Error in estimation of concentration from the calibration curve.*

6. *Spectral-line interference.* Inasmuch as a nearby spectral line can contribute to an intensity measurement of an analyte line, spectral-line interference may constitute a source of error (Section 13.3).

Counting error, most instrumental errors, operational errors, and errors in the estimation of analyte concentration from the calibration curve are random. Specimen errors (including absorption-enhancement effects) and spectral-line interference are systematic. Each of the foregoing groups of errors, or each individual source of error, may be expressed in terms of the standard deviation (see below) that it contributes to the precision of the measured intensity or of the analytical result. Of all the errors listed above, only the statistical counting error is easily and accurately calculated. The others must be evaluated experimentally.

11.1.2. Instrumental and Operational Error

Instrumental instability is usually limited by x-ray generator stability, but may increase with the complexity of the electronic readout components.

Long-term instability usually has several times the magnitude of short-term instability. There are two approaches to reduction of error due to instrumental drift: one is to reduce the instability itself; the other is to choose a counting strategy that reduces the effect of any remaining instability on the intensity measurement.

Instrumental drift can be reduced in several ways, principally the following. Efficient kV and mA stabilization reduce drift in these two parameters. Efficient line stabilization ($\pm 0.05\%$ or less) and adequate warmup time (at least 30 min) reduce all forms of instrument instability except those caused by the effects of temperature and pressure on the crystal and detector.

Drift in 2θ due to change in crystal *d*-spacing with change in ambient temperature is minimized by avoidance of crystals with high thermal expansion coefficients (such as PET) and by temperature control of the spectrometer chamber. Temperature control also minimizes the effect of ambient temperature on the gas amplification of sealed and flow-proportional counters. The effect of ambient pressure on gas amplification in flow counters is minimized by discharging the effluent gas into a manostat. Shift and distortion of pulse-height distribution is discussed in Section 8.1.8.

Having reduced instrumental instability as much as possible, the operator can select a counting strategy to minimize the effects of the residual instability on the intensity measurement. These effects are minimized in ratio techniques (Chapter 14), and substantially eliminated when the two ratioed intensities are measured simultaneously. The instrumental effects are also substantially eliminated by monitor techniques (Section 7.2.2.4).

If measurements are made of a large number of specimens, total measurement time may be very long, and if maximum precision is required, it may be advisable to measure the samples and standards in random sequence to minimize the effect of any long-term drift. Adler proposes that the samples and standards be assigned serial numbers, then placed in random sequence by use of a table of random numbers. If replicate measurements or multielement analyses are to be made, the entire series of measurements should be thus randomized.

If the analyte-line intensity is very low, very long counting time is required to realize acceptable precision. In such cases, the effect of instrument drift may be minimized by dividing the line and background counting times into intervals and alternating them until the total times have been counted.

Operational error is substantially dependent on "slop" and backlash in the controls and on the attentiveness and care of the operator. It is a problem only when settings must be changed during an analysis, and is far less serious with stepwise controls than with continuous controls. The error is minimized by always approaching a setting from the same direction. Operational errors are substantially absent in modern semiautomatic and automatic instruments (Chapter 9).

It has been proposed that in a multielement analysis, maximum precision is realized by first measuring a certain analyte-line intensity on all specimens at $2\theta_L$, then background intensity on all specimens at $2\theta_B$. Then kV, mA, and the readout parameters are changed as required and another analyte line and background measured the same way. However, this procedure is very prone to specimen errors, and, because of the repeated reloading, briquet, powder, and liquid specimens must be handled with great care.

11.1.3. Specimen Error

Instrumental and operational errors may be very small in modern commercial instruments, and the counting error usually can be made very small by accumulation of large counts. Therefore, specimen errors may limit the precision of a quantitative x-ray spectrometric analysis. Techniques for reducing specimen error are discussed in detail in Chapters 16 and 17. Only general considerations are given here.

The measured surfaces of all specimens in an analytical series must lie precisely in the plane defined by the specimen mask. If the target-to-specimen distance is 25 mm along the central ray of the primary beam, a 0.5-mm shift along this line changes the analyte-line intensity by ~4%. If the takeoff angle is ~45°, a change of 1° changes analyte-line intensity ~3.5%. Solids must have plane surfaces and be seated firmly against the specimen mask. Rippling or wrinkling of sheet and foil specimens cannot be tolerated. Precautions must be taken to prevent warping of sheets, foils, and filter disks, and distension of liquid-cell windows due to heating by the primary x-rays. These and other precautions are discussed in Chapters 16 and 17. The specimen compartment and drawer must be replaced when they become worn, and even when they are in good condition, a leaf spring may be required to hold the drawer firmly in a reproducible position.

The concept of infinite thickness is discussed in Section 14.7.2. Let it suffice here to state that if the specimens are not infinitely thick, or of identical thickness, variations in thickness constitute a source of error. Moreover, in a given specimen, infinite thickness occurs at different depths for different analyte lines. The shorter the wavelength of the line, or the smaller the absorption coefficient of the specimen for the line, the deeper is the infinite thickness. It follows that a specimen may be infinitely thick for one wavelength, but not for another.

In most x-ray spectrometers, the central rays of the primary and secondary beams are mutually perpendicular, so there is the possibility of absorption and shielding of the two beams by surface texture. The surface need not have a high polish, but it must be plane and smooth, and the longer the wavelength of the analyte line or the higher the absorption coefficient of the specimen for the line, the finer must the specimen finish be. In general, surface roughness should be less than the infinite thickness for the analyte line, and must be the same for all samples and standards. If the specimens have substantially parallel grind marks, and are not rotated during measurement, the grind marks should be oriented in the same direction for all specimens. To reduce shielding, this direction should be parallel to the plane defined by the central rays of the primary and secondary beams.

Powder specimen errors are reduced by grinding the powders to 20-μm particle size or smaller. Specimen rotation reduces most specimen errors arising from inhomogeneities of one kind or another (composition, porosity, particle size, etc.) and from surface texture. Liquid specimens are discussed in Chapter 17.

A source of long-term specimen error often overlooked is the deterioration of some types of standards with repeated use. The principal causes of the deterioration are: (1) mechanical abrasion from repeated handling and insertion; (2) oxidation and corrosion, especially from perspiration and from ozone produced in the specimen chamber by the primary x-rays; and (3) radiation damage by the high primary-beam intensity. The types of standards subject to such deterioration include plated and coated metals and ceramics, and heterogeneous materials in which the phases are unequally prone to chemical or mechanical damage.

11.1.4. Chemical Effects *(15E, F2, H38, W7, Z3)*

In general, x-ray spectral lines of analytical interest in x-ray secondary-emission spectrometry arise from electron transitions to the *K* or *L* shell from the *L* or *M* shell. In most elements to which the method is applicable, the *L* and *M* shells are not the outermost shells and are therefore not involved in chemical phenomena. Thus, influence of chemical state of the atom on the wavelength of the analyte line is relatively uncommon and very small at most. However, for atomic numbers below 18 (argon), the $3p$ (M_{II}, M_{III}) orbital is unfilled. The electron transition giving the $K\beta$ line originates in this orbital, and, consequently, the $K\beta$ line of elements

9–17 (fluorine to chlorine) may show slight wavelength dependence on chemical state. The same consideration applies to the $K\beta_5$ line of elements 21–28 (scandium to nickel), which have only partially filled $3d$ orbitals, and to the $L\beta_2$ lines of elements 40–45 (zirconium to rhodium), which have unfilled $4d$ (NIV, NV) orbitals. In general, if a spectral line arises from an electron transition *from* the outermost orbital or *from* an unfilled orbital, there is probability of chemical effect.

Oxidation number is the most important chemical feature that contributes to wavelength shift of an x-ray spectral line; others include coordination number, type and electronegativity of bonds, induction effects, and the nature of the atom or radical to which the emitting atom is bonded. The magnitude of the wavelength displacement varies from 0.005 to 0.25 Å.

The chemical effect provides an excellent means for investigating the various chemical phenomena listed above. Such studies require an x-ray spectrometer capable of extremely high resolution and having a temperature-controlled crystal chamber. Although some of these studies have been done on standard commercial instruments, usually the small wavelength displacements are masked by other effects, such as 2θ shift due to thermal expansion of the crystal. However, chemical effects in the lighter elements and for certain lines of the heavier elements may constitute a source of error. If the goniometer 2θ angle for the analyte line is peaked for, say, the pure analyte, the line may occur at a slightly different 2θ for standards prepared from pure oxides, and at still different 2θ for complex mineral samples. Moreover, in complex substances, 2θ may vary from sample to sample.

11.2. COUNTING STATISTICS

It is the purpose of this section not to give a detailed discussion of statistics, but rather to present practical equations for calculating counting error for various x-ray spectrometric measurement techniques. The errors considered in this section are those arising *only* from the random arrival of x-ray photons and do not include possible errors arising in the x-ray source (x-ray tube and power supply) or detector (counter tube, power supply, and electronic readout components). The evaluation of other errors and of the overall analytical precision are discussed in Section 11.3.

The counting error always represents the maximum precision attainable in an x-ray spectrometric analysis. In modern x-ray spectrometers, instrumental errors are very small, and in semiautomatic and automatic instruments, operational errors are similarly reduced. Consequently, in analyses where specimen errors are absent or successfully dealt with, counting error may substantially constitute the analytical precision.

A particularly good discussion of counting statistics in x-ray spectrometric analysis is given by Jenkins and de Vries [Ch. 5 in (*14*)], and a substantial portion of the following treatment is based on theirs. Briefer reviews of counting statistics are given in (*M1, P10*).

11.2.1. Nature of the Counting Error

Consider a perfectly smooth, plane, homogeneous specimen in an x-ray spectrometer, all components of which, from the x-ray generator to the readout circuitry, are perfectly stable. Suppose that a series of equal preset-time intensity measurements 1, 2, 3, . . ., n are made of an x-ray spectral line of an element in the specimen without disturbing the specimen or any instrument setting during the entire series. Even under these ideal—and unattainable—conditions, the numbers of counts accumulated N_1, N_2, N_3, . . ., N_n during successive identical time intervals would not be equal. The reason is that emission of x-ray photons is a random process, so that the photons in an x-ray beam have a random time distribution. If a large number n of measurements are taken, the individual accumulated counts N always have a Poisson distribution:

$$P(N) = \frac{\bar{N}^N \exp -\bar{N}}{N!} \tag{11.1}$$

where $P(N)$ is the probability of occurrence of a specified accumulated count N; $\bar{N}$ is the mean of a very large number of measurements; and $N!$ is N-factorial, that is, $N \times (N-1) \times (N-2) \times \ldots \times 1$.

As $\bar{N}$ becomes progressively larger, the distribution approaches a Gaussian (normal) distribution, and for $N > 100$, the distribution may be regarded as Gaussian:

$$P(N) = \frac{1}{(2\pi\bar{N})^{1/2}} \exp - \frac{(N-\bar{N})^2}{2\bar{N}} \tag{11.2}$$

The Poisson distribution is represented by a histogram (bar graph), the Gaussian distribution by a smooth continuous curve. A plot of a Gaussian distribution is shown in Figure 11.1, where frequency of occurrence of a given value of N is plotted as a function of the value of N. In statistics, *standard deviation* σ is defined as the square root of the mean of the squares (rms) of the deviations of an infinite set of measurements, or at least of all possible measurements, from their arithmetic mean μ. In practice, the *estimate of the standard deviation* s is the rms deviation of a limited set of measurements from their mean $\bar{x}$, which is itself the *estimate of the mean.* However, the practice of using the symbol σ instead of s in the x-ray literature is very common, as is the practice of referring to precision in terms of one, two, or three "sigma." Therefore, the symbol σ is used here. Also in deference to common practice in the x-ray literature, the symbol ε (a form of epsilon), rather than ν (nu), is used for coefficient of variation (see below).

To return to our hypothetical experiment, the rms deviation of the set of n measurements of N from their mean $\bar{N}$ is the standard deviation and is given by

$$\sigma = \left[\frac{\Sigma (N - \bar{N})^2}{n - 1}\right]^{1/2} \tag{11.3}$$

in which

$$\bar{N} = \frac{\Sigma N_i}{n} \tag{11.4}$$

The practical significance of the standard deviation is illustrated in Figure 11.1 and may be stated in various ways, including the following:

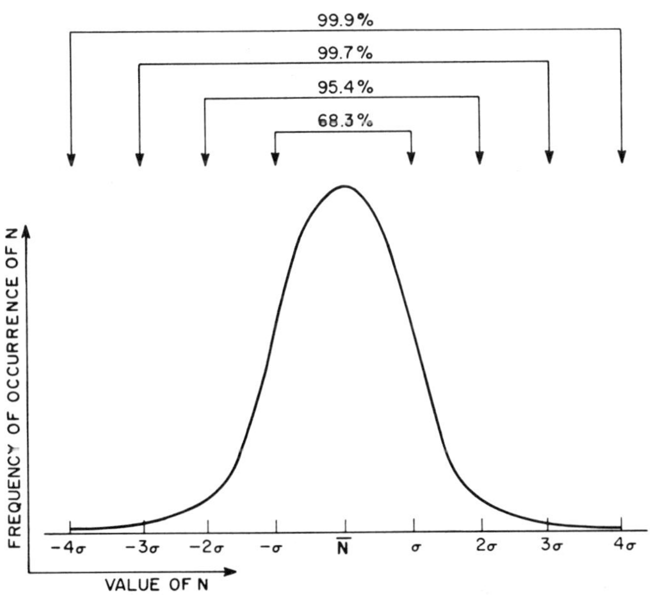

Fig. 11.1. The Gaussian (normal) distribution.

1. The probability is 68.3% (~7/10) that any individual value of N will deviate from the average $\bar{N}$ of a very large number of values by $\leq\sigma$.

2. The probability is 68.3% that any N will have a value between $\bar{N} \pm \sigma$.

3. Of a very large number of measurements of N, 68.3% will have values of N lying between $\bar{N} \pm \sigma$.

4. It follows that 68.3% of the area under the Gaussian distribution curve lies between $\bar{N} \pm \sigma$.

Similarly, the probability is 95.4% (19/20) than an individual N will lie between $\bar{N} \pm 2\sigma$, 99.7% (997/1000) that it will lie between $\bar{N} \pm 3\sigma$, and 99.9% (999/1000) that it will lie between $\bar{N} \pm 4\sigma$. Finally, the probability is 50% (1/2) that an individual N will lie between $\bar{N} \pm 0.67\sigma$; 0.67σ is known as the *probable error* or *probable fractional error*.

Figure 14.3 shows a typical calibration curve on which the associated error is indicated on each calibration point in the conventional manner. The ranges may indicate $\pm 1\sigma$ or some other specified precision. For example, if $\pm 1\sigma$ is indicated, the probability is 68.3% that the true intensity for each point is neither greater nor less than the limits shown. If the fixed-count method is used, the range is the same for each point, as shown, but if different numbers of counts are accumulated for each point, the range is also different. In addition to the random error shown for all the points, the point for the standard having second highest concentration also shows a substantial systematic error.

The larger σ is, the flatter is the Gaussian distribution curve; the width of the distribution at half its maximum height is 2.35 σ intervals. The flatter the curve, the greater is the spread in replicate intensity measurements or analytical results, and the lower is the precision in the measurements or results. The larger the number n of replicate measurements, the more nearly their distribution corresponds to the Gaussian curve and the more reliable is the calculated value of σ.

The standard deviation expressed as a per cent of the mean value is the *relative standard deviation* or *coefficient of variation* ε:

$$\varepsilon = 100(\sigma/\bar{N}) \tag{11.5}$$

In this section, for simplicity, ε is expressed as a fraction:

$$\varepsilon = \sigma/\bar{N} \tag{11.6}$$

TABLE 11.1
Statistical Precision

Symbol	Term	Confidence level	Probability
0.67σ	Probable error	50.0%	1/2
σ	Standard deviation, sigma	68.3%	7/10
2σ	Two sigma	95.4%	19/20
3σ	Three sigma	99.7%	997/1000
4σ	Four sigma	99.9%	999/1000

The square of the standard deviation is the *variance* V:

$$V = \sigma^2 \tag{11.7}$$

Variance is a very useful quantity. The standard deviation of a sum or difference of two or more measurements, each having its individual standard deviation σ_i, is given by combining variances. Relative standard deviations are combined in an analogous way:

$$\sigma = [\sum \sigma_i^2]^{1/2} \tag{11.8}$$

$$\varepsilon = [\sum \varepsilon_i^2]^{1/2} \tag{11.9}$$

Equations (11.8) and (11.9) are used to calculate σ and ε for net intensity measurements $(I_P - I_B)$ or net count measurements $(N_P - N_B)$, where P and B refer to the peak and background, respectively.

One final statistical term: The highest and lowest values of N between which the actual measurement is likely to fall with a specified probability are known as the *confidence limits* or *confidence levels.* For example, one may predict that a measured N will lie between $\bar{N} \pm \sigma$ at the 68.3% confidence level, and $\bar{N} + \sigma$ and $\bar{N} - \sigma$ are the 68.3% confidence limits. The most frequently cited confidence limits are given in Table 11.1.

11.2.2. Calculation of Counting Error

11.2.2.1. *Counting Error for Accumulated Counts*

Following are equations for calculation of the standard deviation σ_N and relative standard deviation ε_N in terms of accumulated count N measured by various counting techniques. In this application, some workers prefer the terms *standard counting error* to standard deviation and *relative fractional counting error* to relative standard deviation. The use of these terms should be encouraged, to avoid confusion with standard deviation and relative standard deviation of other errors and of the overall precision (Section 11.3).

a. Single Measurement. For a single measurement of N counts,

$$\sigma_N = \sqrt{N} \tag{11.10}$$

$$\varepsilon_N = \frac{\sigma_N}{N} = \frac{\sqrt{N}}{N} = \frac{1}{\sqrt{N}} \tag{11.11}$$

b. Net Intensity (*Z7*). For a measurement of N_P counts at the analyte peak corrected for a background measurement of N_B counts, both measurements taken for the same time interval, from Equation (11.8),

$$\sigma_N = (\sigma_P{}^2 + \sigma_B{}^2)^{1/2} \tag{11.12}$$

$$= (N_P + N_B)^{1/2} \tag{11.13}$$

$$\varepsilon_N = \frac{(N_P + N_B)^{1/2}}{N_P - N_B} \tag{11.14}$$

c. Ratio Method (Uncorrected). In this technique, the time T_S to accumulate N_S counts from a standard S is also used to accumulate N_X counts from the sample X. The N_X is measured at the analyte line; N_S may be measured at the analyte line from a standard, or it may be measured from the sample at an internal-standard line, or a scattered target line, or a scattered wavelength in the continuum. The method is very widely used in automatic spectrometers. If no background corrections are made,

$$\varepsilon_N = \left[\left(\frac{\sqrt{N}}{N}\right)_X^2 + \left(\frac{\sqrt{N}}{N}\right)_S^2\right]^{1/2} \tag{11.15}$$

$$= \left[\frac{1}{N_X} + \frac{1}{N_S}\right]^{1/2} \tag{11.16}$$

d. Ratio Method (Corrected). If the corresponding backgrounds $(N_B)_X$ and $(N_B)_S$ are measured for the same times T_S,

$$\varepsilon_N = \left[\frac{(N_P + N_B)_X}{(N_P - N_B)_X{}^2} + \frac{(N_P + N_B)_S}{(N_P - N_B)_S{}^2}\right]^{1/2} \tag{11.17}$$

e. Replicate Measurements. The foregoing equations apply to single measurements of N, $N_P - N_B$, or N_X/N_S, as the case may be. If n replicate determinations are made of each of these quantities, the standard counting error and relative fractional counting error are given by

$$(\sigma_N)_n = \frac{1}{\sqrt{n}}\,\sigma_N \tag{11.18}$$

$$(\varepsilon_N)_n = \frac{1}{\sqrt{n}}\,\varepsilon_N \tag{11.19}$$

It is evident that standard counting error decreases as N increases, and, in principle, may be made as small as required if a sufficiently long counting time is permissible. However, accumulation of a large N at low intensity requires a very long counting time; this may be disadvantageous for several reasons. Analysis time is increased. Extremely high stability is required in the x-ray tube potential and current and in the electronic

TABLE 11.2

Standard and Relative Counting Error

Number of counts N	σ,[a] counts/s	$\varepsilon_{68.3}$,[b] %	2σ,[a] counts/s	$\varepsilon_{95.4}$,[b] %	3σ,[a] counts/s	$\varepsilon_{99.7}$,[b] %
100	10	10.0	20	20.0	30	30.0
200	14	7.1	28	14.1	42	21.2
500	22	4.5	44	9.0	66	13.4
1,000	32	3.2	64	6.3	96	9.5
2,000	45	2.2	90	4.5	135	6.7
5,000	71	1.4	142	2.8	213	4.2
10,000	100	1.0	200	2.0	300	3.0
20,000	141	0.70	282	1.4	423	2.1
50,000	224	0.45	448	0.90	672	1.3
100,000	316	0.32	632	0.63	948	0.95
200,000	447	0.22	894	0.45	1341	0.67
500,000	707	0.14	1414	0.28	2121	0.42
1,000,000	1000	0.10	2000	0.20	3000	0.30
2,000,000	1414	0.07	2828	0.14	4242	0.21
5,000,000	2236	0.05	4472	0.09	6708	0.13
10,000,000	3162	0.03	6324	0.06	9486	0.10

[a] $\sigma = \sqrt{N}$.
[b] $\varepsilon = 100(n\sigma/N)$; the subscript of ε is the confidence level; n is 1, 2, or 3.

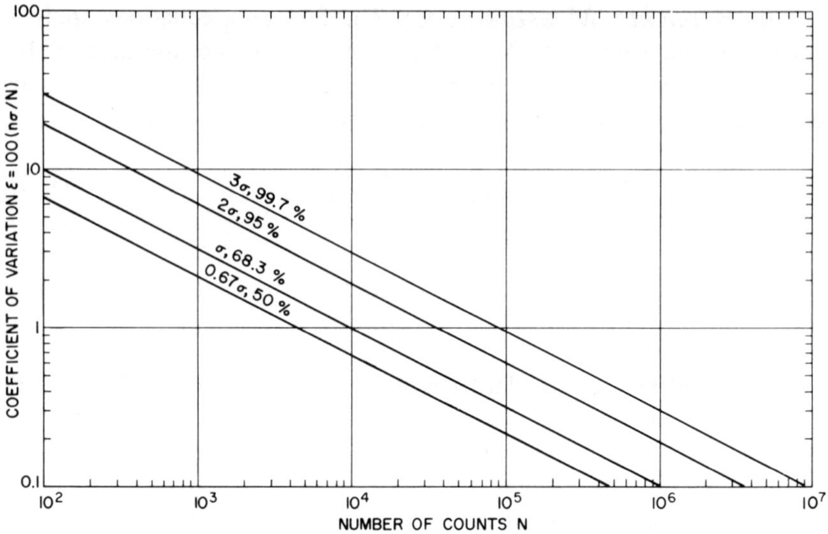

Fig. 11.2. Correlation of coefficient of variation and the accumulated count for various standard deviations and confidence levels.

detector and readout components. Finally, difficulties arise on prolonged exposure of liquid specimens to the x-ray beam.

Table 11.2 gives values of σ, 2σ, and 3σ and ε, 2ε, and 3ε for a range of values of N. Figure 11.2 shows the correlation of ε for $N = 10^2$–10^7 counts at 0.67σ, σ, 2σ, and 3σ.

11.2.2.2. *Counting Error for Intensities*

This section presents substantially the same equations as Section 11.2.2.1, but in terms of intensity I rather than accumulated count N. However, from Equation (11.6),

$$\varepsilon_N = \sigma_N/N \qquad \text{and} \qquad \varepsilon_I = \sigma_I/I \tag{11.20}$$

Therefore, for values of intensity and accumulated count related to the same counting time T,

$$N = IT \tag{11.21}$$

Then standard and relative counting errors in terms of N and I are equal, that is,

$$\varepsilon_I = \varepsilon_N; \qquad \sigma_I/I = \sigma_N/N \tag{11.22}$$

Equation (11.22) applies only if I and T are related by Equation (11.21) and if the error associated with the measurement of T is negligible and has no distribution of its own. When high precision is required and T is small, the error in T may require consideration. The equations for standard counting error σ_I and relative fractional counting error ε_I in terms of intensity follow.

a. Uncorrected Intensity. For a single measured intensity,

$$\sigma_I = I\varepsilon_N = \frac{I}{\sqrt{N}} = \frac{I}{(IT)^{1/2}} = \left(\frac{I}{T}\right)^{1/2} \tag{11.23}$$

$$\varepsilon_I = \frac{(I/T)^{1/2}}{I} = \frac{1}{(IT)^{1/2}} = \frac{1}{\sqrt{N}} = \varepsilon_N \tag{11.24}$$

Equation (11.24) confirms Equation (11.22).

b. Net Intensity. For a measurement of I_P counts/s at the analyte peak corrected for I_B background counts,

$$(\sigma_I)_P = \left(\frac{I_P}{T_P}\right)^{1/2}; \qquad (\sigma_I)_B = \left(\frac{I_B}{T_B}\right)^{1/2} \tag{11.25}$$

From Equation (11.8),

$$\sigma_I = (\sigma_P{}^2 + \sigma_B{}^2)^{1/2} \tag{11.26}$$

$$= \left[\frac{I_P}{T_P} + \frac{I_B}{T_B}\right]^{1/2} \tag{11.27}$$

$$\varepsilon_I = \frac{[(I_P/T_P) + (I_B/T_B)]^{1/2}}{I_P - I_B} \tag{11.28}$$

where, of course, $I_P - I_B$ is net intensity.

c. Ratio Method (Uncorrected). For I_X/I_S, both measured at T_S (see above),

$$\varepsilon_I = \frac{1}{\sqrt{T_S}}\left[\frac{1}{I_X} + \frac{1}{I_S}\right]^{1/2} \tag{11.29}$$

$$= \frac{1}{\sqrt{N_S}}\left[1 + \frac{I_S}{I_X}\right]^{1/2} \tag{11.30}$$

Equation (11.29) can be used to calculate the standard and relative fractional counting errors for the two commonly used counting methods, fixed count and fixed time, and for the optimal-fixed-time method (see below).

d. Fixed Count. This method is characterized by the following conditions: $I_P T_P = I_B T_B = N$, and $T_P + T_B = T$ (total counting time); then,

$$\sigma_{\text{FC}} = \left(\frac{1}{\sqrt{T}}\right)(I_P + I_B)^{1/2}\left[\frac{I_P}{I_B} + \frac{I_B}{I_P}\right]^{1/2} \tag{11.31}$$

e. Fixed Time. This method is characterized by the following conditions: $T_P = T_B = T/2$, and $T_P + T_B = T$; then,

$$\sigma_{\text{FT}} = \left(\frac{2}{T}\right)^{1/2}(I_P + I_B)^{1/2} \tag{11.32}$$

$$\varepsilon_{\text{FT}} = \left(\frac{2}{T}\right)^{1/2}\left[\frac{I_P + I_B}{I_P - I_B}\right]^{1/2} \tag{11.33}$$

f. Optimal Fixed Time. When the difference of two intensities is to be measured, as in a net intensity measurement, the two intensities should be known with about the same absolute error. The optimal-fixed-time method is essentially the same as the fixed-time method except that the apportionment of the total counting time T between the peak and the background is calculated to give minimum error in the measured net intensity. This condition exists when $T_P + T_B = T$ and

$$\frac{T_P}{T_B} = \left(\frac{I_P}{I_B}\right)^{1/2} \tag{11.34}$$

Then,

$$\sigma_{\text{FTO}} = \frac{1}{\sqrt{T}}(\sqrt{I_P} + \sqrt{I_B}) \tag{11.35}$$

$$\varepsilon_{\text{FTO}} = \frac{1}{\sqrt{T}}\left(\frac{\sqrt{I_P} + \sqrt{I_B}}{I_P - I_B}\right) \tag{11.36}$$

$$= \frac{1}{\sqrt{T}}\left(\frac{1}{\sqrt{I_P} - \sqrt{I_B}}\right) \tag{11.37}$$

Equations (11.18) and (11.19) for replicate measurements apply to the equations for σ_I and ε_I as well.

g. Ratemeter–Recorder. The precision of intensity measurements made with the ratemeter–recorder is also subject to statistical fluctuation. Measurement of an intensity with a ratemeter having a time constant of RC

seconds is equivalent to accumulation of counts during a time interval of $2RC$ seconds. Then the relative fractional counting error is

$$\varepsilon_I = 1/[2I(RC)]^{1/2} \tag{11.38}$$

Actually, R signifies resistance in ohms (Ω), and C signifies capacitance in farads (F). Dimensionally, $1\ \Omega = 1$ s/esu-cm, and $1\ \text{F} = 1$ esu cm. Then $RC\ [=]\ \Omega\text{F}\ [=]$ (s/esu-cm)(esu cm) $[=]$ s.

11.2.3. Counting Strategy

Counting strategy may be defined, perhaps with some oversimplification, as the selection of counting technique, counting time, and/or accumulated count to make a specified intensity measurement or series of measurements with a specified precision in the shortest time.

11.2.3.1. *Measurement of Net Intensity*

The choice of counting strategy for the measurement of a net intensity involves the choice of the fixed-time or fixed-count method, apportionment of the total counting time between measurement of peak and background intensities, and certain practical considerations. The counting errors for the fixed-count, fixed-time, and optimal-fixed-time methods may be compared by solving Equations (11.31) (σ_{FC}), (11.32) (σ_{FT}), and (11.35) (σ_{FTO}) with the following typical values: $I_P = 10{,}000$ counts/s; $I_B = 100$ counts/s, and $T = 100$ s. The result is: $(\sigma_{\text{FTO}} = 11) < (\sigma_{\text{FT}} = 14.2) < (\sigma_{\text{FC}} = 100)$, where the values are in counts per second. It can be shown that this sequence is generally valid. The proof follows.

a. Fixed Count vs. Fixed Time (*B37, G1*). It is always true that I is positive and $I_P > I_B$; then it follows that

(1) $(I_P - I_B) > 0$

(2) $(I_P - I_B)^2 > 0$

(3) $\dfrac{I_P^2 + I_B^2}{I_P I_B} > 2$

(4) $\left(\dfrac{I_P}{I_B} + \dfrac{I_B}{I_P}\right) > 2$

(5) $\left[\dfrac{I_P}{I_B} + \dfrac{I_B}{I_P}\right]^{1/2} > \sqrt{2}$

Hence,

$$\left[\frac{I_P}{I_B}+\frac{I_B}{I_P}\right]^{1/2}\left(\frac{1}{T}\right)^{1/2}(I_P+I_B)^{1/2}>\sqrt{2}\left(\frac{1}{T}\right)^{1/2}(I_P+I_B)^{1/2}$$

and, from Equations (11.31) and (11.32), $\sigma_{\text{FC}} > \sigma_{\text{FT}}$.

b. Fixed Time vs. Optimal Fixed Time. Similarly, for all values of I_P and I_B except for $I_P = I_B$,

(1) $(I_P - I_B)^2 > 0$

(2) $I_P^2 - 2I_PI_B + I_B^2 > 0$

(3) $I_P^2 + 2I_PI_B + I_B^2 > 4I_PI_B$

(4) $(I_P + I_B)^2 > [2(I_PI_B)^{1/2}]^2$

(5) $(I_P + I_B) > 2(I_PI_B)^{1/2}$

(6) $2(I_P + I_B) > I_P + 2(I_PI_B)^{1/2} + I_B$

Hence,

$$\sqrt{2}\,(I_P+I_B)^{1/2} > \sqrt{I_P}+\sqrt{I_B}$$

and, from Equations (11.32) and (11.35), $\sigma_{\text{FT}} > \sigma_{\text{FTO}}$. Then $\sigma_{\text{FC}} > \sigma_{\text{FT}} > \sigma_{\text{FTO}}$ for all values of I_P and I_B, provided $I_P \neq I_B$.

It has already been shown that in the measurement of a net intensity, the precision of a fixed-time measurement is always higher (σ lower) than that of a fixed-count measurement when the peak and background counting times are the same. However, these counting times need not be the same. Gaylor (*G1*) has shown that (1) for a specified total counting time for peak and background, the variance of the fixed-time method is always less than or equal to that of the fixed-count method, and (2) for a specified precision, the total counting time required for the fixed-time method is always less than or equal to that for the fixed-count method. Equations (11.31), (11.32), and (11.35) can be used to calculate the total counting time T required to obtain a specified standard counting error σ for the peak and background intensities actually encountered from the specimens. Then the optimal apportionment of T to be spent counting peak I_P and background I_B intensities can be calculated from the following two equations:

$$T_P + T_B = T \tag{11.39}$$

$$T_P/T_B = (I_P/I_B)^{1/2}$$

Of course, the instrument may not permit use of the exact times given by the calculations. In such cases, the closest preset times provided on the instrument are used.

Aside from the improved precision of the fixed-time method, established above, there are also practical advantages. Some examples follow.

Suppose that a method has been developed for the determination of a specified analyte in a specified type of sample in concentration range, say, 25–30%. For samples in which the analyte concentration falls in or near this range, the two counting techniques are equally *convenient*. However, for occasional samples in which the analyte is present in very low concentration, analysis time is lost in accumulating the same fixed count to determine a concentration outside the calibration range.

Now suppose that a calibration curve has been established for a specified analyte in concentration range 0–1%, that the slope of the curve is 100 counts/s per per cent analyte, and that the required precision is ±1% relative for 1σ at the 1% concentration level. Table 11.3 summarizes data for fixed-count and fixed-time measurements on two samples having analyte concentrations 1% and 0.1%, respectively. In this case, the analyst would usually prefer the fixed-time method at the sacrifice of some relative precision at low concentrations, rather than count ten times as long to achieve a precision (±0.001%) that, relatively, is ten times that at the 1% concentration level (±0.01%).

TABLE 11.3

Comparison of Fixed-Count and Fixed-Time Methods

	Fixed-count method $N = 10{,}000$ counts	Fixed-time method $T = 100$ s
Analyte concentration 1%, $I = 100$ counts/s		
N, counts	10,000	10,000
$\%\varepsilon = 100(\sqrt{N}/N)$	1%	1%
T, s	100	100
Analytical result	$1 \pm 0.01\%$ (σ)	$1 \pm 0.01\%$ (σ)
Analyte concentration 0.1%, $I = 10$ counts/s		
N, counts	10,000	1000
$\%\varepsilon = 100(\sqrt{N}/N)$	1%	3.2%
T, s	1000	100
Analytical result	$0.1 \pm 0.001\%$ (σ)	$0.1 \pm 0.003\%$ (σ)

The final example is the most significant in that it concerns the background measurement. In principle, background should be measured with the same precision as the analyte peak, that is, the same number of counts should be accumulated for each. However, when $I_P \gg I_B$, the absolute magnitude of the fluctuations at the peak is much greater than at the background, and high precision is not required for the background measurement. For example, a background of, say, 25 counts/s measured with a precision of $\pm 5\%$ would contribute an error of only ± 1 count/s. Thus, a fixed-count measurement may require a very long time to count the background, and much time is wasted in reducing the smaller component of the variance in the net intensity [Equation (11.26)], time better spent in reducing the larger component.

11.2.3.2. *The Ratio Method*

It can be shown that when applied to the measurement of intensity ratios, the fixed-time and fixed-count methods yield exactly the same precision in the same measuring time. However, when intensity ratios are measured, the numerator and denominator should have substantially the same relative error. Decreasing the relative counting error ε in only one of the values does not improve the precision of the ratio significantly. For such applications, maximum precision is realized by use of a counting strategy such that a constant relative error is realized at all intensities; this is the case for the fixed-count method.

In automatic instruments, the time T_S to accumulate N_S counts from a standard S is often used as the measuring time to accumulate N_X counts from the sample X. The relative fractional counting error is given by Equation (11.30),

$$\varepsilon = \frac{1}{\sqrt{N_S}}\left[1 + \frac{I_S}{I_X}\right]^{1/2}$$

If I_S is of the same order as I_X,

$$\varepsilon \approx (2/N_S)^{1/2} \tag{11.40}$$

It is evident that this value is $\sqrt{2}$ greater than the relative fractional counting error associated with a simple measurement of N counts, $(1/N)^{1/2}$ [Equation (11.11)]. However, although the counting error is greater in the ratio method, the instrumental errors are substantially smaller. The choice between absolute or ratio method must be based on the value of ε for the two methods in the following equation:

$$\varepsilon = (\varepsilon_N{}^2 + \varepsilon_{\text{instr}}^2)^{1/2} \tag{11.41}$$

If $\varepsilon_{\text{instr}}$ is, say, 0.3%, the corresponding N is 10^5 counts (Figure 11.2); then it is a waste of time to accumulate more than 10^5 counts in a single measurement because above this number, instrumental errors limit the precision. If the required precision is greater than that corresponding to the instrumental error, replicate measurements must be made. For example, for a single measurement of $N = 10^6$ counts, $\sigma_N = \sqrt{10^6} = 10^3$, or $\varepsilon = 0.1\%$; for 10 replicate measurements of $N = 10^6$, from Equation (11.18), $\sigma_N = (1/\sqrt{10})\sqrt{10^6} \approx 300$, or $\varepsilon = 0.03\%$.

Some other counting strategies are discussed briefly in Section 11.1.2: counting of samples and standards in random sequence and alternation of portions of very long peak and background counts.

11.2.4. Figure of Merit

Peak-to-background ratio $(I_P - I_B)/I_B$ is a commonly recognized figure of merit to be attained in setting instrument operating conditions. Jenkins and de Vries [p. 100 in (*14*)] propose other figures of merit based on Equation (11.37). For major and minor constituents (relatively high analyte-line intensity) and a properly apportioned total counting time T, if ε is to be minimum, $\sqrt{I_P} - \sqrt{I_B}$ should be maximum. For constituents having very low concentrations (low analyte-line intensity), I_P is not much greater than I_B and, in fact, approaches it at or near the detection limit. Then $(I_P I_B)^{1/2} \approx I_P$. In such cases, if ε is to be minimum, $(I_P - I_B)/\sqrt{I_B}$ should be maximum.

All three criteria can be used to compare the precision of intensity measurements on different instruments at substantially the same conditions, and of the same instrument at different conditions. Spielberg and Brandenstein propose still another figure of merit (*S31*).

11.3. ANALYTICAL PRECISION

11.3.1. Nature of Analytical Precision

Table 11.4 gives data for 12 (n) determinations of germanium made on the same sample by timing (T) 12 preset counts of 51,200 (N) of the Ge $K\beta$

TABLE 11.4

Precision of an X-Ray Spectrometric Analysis

i	Counting time T, s	Ge $K\beta$ net intensity I, counts/s	Ge concentration C, at % x_i	Deviation $d = x_i - \bar{x}$	Deviation squared d^2
1	26.1	1962	34.7	+0.3	0.09
2	26.0	1969	35.1	+0.7	0.49
3	26.3	1947	33.9	—0.5	0.25
4	26.2	1954	34.3	—0.1	0.01
5	26.1	1962	34.7	+0.3	0.09
6	26.2	1954	34.3	—0.1	0.01
7	26.2	1954	34.3	—0.1	0.01
8	26.2	1954	34.3	—0.1	0.01
9	26.4	1939	33.4	—1.0	1.00
10	26.0	1969	35.1	+0.7	0.49
11	26.2	1954	34.3	—0.1	0.01
12	26.2	1954	34.3	—0.1	0.01
		$\sum I = 23472$	$\sum x = 412.7$		$\sum d^2 = 2.47$
		$\sum I/n = 1956$	$\sum x/n = 34.4 = \bar{x}$		$\sum d^2/(n-1) = 0.22$

$$\sigma = [\sum d^2/(n-1)]^{1/2} = 0.47 \text{ at \% Ge}$$

$$\varepsilon = 100(\sigma/\bar{x}) = 1.4\%$$

Standard counting error for $N = 51200$

$$\sigma_N = \sqrt{N} = 226$$

$$\varepsilon_N = 100(\sigma_N/N) = 0.44\%$$

line, calculating the intensities ($I = N/T$), and applying them to a calibration curve. The curve was prepared previously from measurements on standards regarded as accurate. The specimen was reloaded without regard to orientation for each measurement, but no instrument setting was disturbed. The standard deviation σ of the set of n results x from their mean $\bar{x}$ is calculated with an equation similar to Equation (11.3),

$$\sigma = \left[\frac{\sum d^2}{n - 1}\right]^{1/2} \tag{11.42}$$

where d is the deviation of an individual determination x_i from the average $\bar{x}$ of the n replicate determinations. The coefficient of variation ε is then calculated with an equation similar to Equation (11.5):

$$\varepsilon = 100(\sigma/\bar{x}) \tag{11.43}$$

The standard deviation is a measure of absolute precision; the coefficient of variation is a measure of relative precision, that is, precision relative to the "amount" (concentration) of analyte present.

In Table 11.4, the analytical result of, say, the first determination ($i = 1$) may be reported by any combination of the following statements: 34.7% Ge ±0.5% absolute, or 34.7% Ge ±1.4% relative, or 34.7% Ge ±1.4% of the amount present; followed by: for one standard deviation, or for one sigma (1σ), or at the 68.3% confidence level. This means that the probability is 68.3% (7/10) that the "true" germanium concentration, that is, the mean of a very large number of determinations, lies within ±0.5% absolute of 34.7%, or that it lies within (0.014 × 34.7) of 34.7%. Similarly, the result may be reported as 34.7% Ge ±0.9% absolute or ±2.8% relative for 2σ or at the 95.4% confidence level. Table 11.1 gives other degrees of precision that are used in reporting analytical results. If 12 replicate determinations are made, as shown in Table 11.4, and their mean (34.4% Ge) is reported as the analytical result, the standard deviation of this mean result is gives by Equation (11.18): $\sigma = (1/\sqrt{12})(0.47) = 0.13$.

Table 11.4 also gives the standard counting error σ_N and relative counting error ε_N for the preset count N of 51,200. It is evident that the coefficient of variation of a single determination of germanium (1.4%) is nearly three times the relative counting error for a count of 51,200 (0.44%). This discrepancy shows that serious errors other than the counting error are impairing the precision of the determination. Standard counting error is—or is derived from—the square root of the mean. Standard deviation is a measure of the dispersion of a group of individual results about their mean. The two are identical in an x-ray spectrometric analysis only if the analytical process is truly random and subject to no error other than that arising from the random time distribution of the x-ray photons.

11.3.2. Evaluation of Precision

11.3.2.1. *General Considerations*

The standard counting error σ_N has twofold significance in x-ray spectrometric analysis: (1) It represents the maximum precision attainable, and cannot be eliminated from a process based on counting photons; and (2) it can be accurately calculated and used as a standard against which to evaluate the precision actually attained.

The precision of an x-ray spectrometric analysis is obtained by combining the variance for the individual sources of error [Equation (11.8)], so that the precision may be represented by

$$\sigma_T^2 = \sigma_C^2 + \sigma_I^2 + \sigma_O^2 + \sigma_S^2 + \sigma_M^2 \tag{11.44}$$

where the subscripts represent, in order, total, counting, instrumental, operational, specimen, and miscellaneous errors. A similar equation can be written for each of the individual errors, expressing the variance for each as the sum of the variances of several specific sources of error, as outlined in Section 11.1.1. The following paragraphs describe the general procedure for revealing, identifying, and evaluating errors.

If the standard deviation of the result of a series of replicate determinations of an analyte does not differ significantly from the standard counting error, the analysis is free of significant error, and the precision is limited by the counting statistics. If the standard deviation and standard counting error do differ significantly, some error(s) other than the counting error is limiting the precision. If the magnitude of the error is not acceptable, the source of the error must be found.

11.3.2.2. *Instrumental Instability*

A plane, smooth, thick ($\geq$1 mm), rigid piece of homogeneous metal—preferably a pure element—is placed in the specimen chamber. The x-ray generator, detector, and all electronic components are allowed to warm up for 30–60 min. Then, without disturbing the specimen or any instrument settings, 30-s to 1-min preset-time measurements are made of the line intensity of a spectral line of an element in the specimen for several hours. The standard deviation σ_1 is calculated for the entire set of measurements, and the counting error σ_N is calculated for the average number of counts accumulated during the preset time. Any significant difference between the two values may be attributed to some kind of long-term instrument instability. The variance for this instability is

$$\sigma^2 = \sigma_1{}^2 - \sigma_N{}^2 \tag{11.45}$$

Short-term instability can also be evaluated by calculation of the standard deviation σ_2 for sets of 10–20 consecutive measurements from the above set and comparison of these values with σ_N in the same way. If the sets of 10–20 consecutive measurements are distributed throughout the full series—say, a set every 30 min—a comparison of their σ_2 values reveals whether the short-term instability is constant or varies during prolonged operation of the instrument.

11.3.2.3. *Operational Error*

A series of measurements is made without disturbing the specimen, but this time, after each measurement, all instrument controls are reset—kV, mA, 2θ, and perhaps detector and pulse-height analyzer settings. Moreover, if a crystal changer is used, this, too, should be reset. If the standard deviation σ_3 for this series is significantly larger than that for short-term instability σ_2 (see above), some kind of operational error is indicated. If the magnitude of the error is unacceptable, each setting must be investigated individually. Settings made by stepwise selector switches are less likely to be sources of error than settings made by continuous controls.

If helium or vacuum path is to be used, the effect on precision must be evaluated, preferably with a spectral line that, although usually measured in helium or vacuum path, can also be measured in air—Ti $K\alpha$ or Ca $K\alpha$, for example. Several series of measurements may be required: Two series made without disturbing the specimen or settings, one in air, the other in helium or vacuum, permit evaluation of the constancy of the helium or vacuum path. Another series without disturbing the settings but with specimen reinsertion after each measurement permits evaluation of the effectiveness of the helium flush period or of the air lock. A final series should be made without disturbing the specimen and without disturbing any setting except the goniometer angle. After each measurement, the goniometer is displaced to a very high and a very low 2θ. This series reveals whether movement of the goniometer permits air to leak through the sliding vacuum seal (if any), or whether it pulls the helium jacket, allowing air to enter or pleats to protrude into the x-ray path.

11.3.2.4. *Specimen Error*

Specimen error is best evaluated with a specimen of the type to be analyzed. A series of measurements is made without disturbing specimen or settings. If the standard deviation σ_4 for this series is significantly larger than that for short-term instability σ_2 (see above) or than the counting error σ_N, deterioration of the specimen in the primary beam or wrinkling of foil specimens are indicated. Next, a series of measurements is made without disturbing any setting, but this time the specimen is reloaded, and therefore also reinserted, after each measurement. If the standard deviation σ_5 for this series is significantly larger than σ_4, some type of specimen error is indicated. The series is repeated, but this time, the specimen is only reinserted—not reloaded—care being taken to insert the specimen in the same orientation each time. A comparison of the standard deviation σ_6 of this series with σ_5 indicates how much, if any, of σ_5 is reinsertion error. If a substantial portion of σ_5 remains, other experiments are made to identify the source of the error.

For liquid specimens, a series of measurements on identical cells filled with the same liquid in the same way permits evaluation of the effect of slight differences in the degree of filling or in stretching of the Mylar cover. A series of measurements on the same undisturbed cell permits evaluation of the effect of temperature rise, bubble formation, and radiolysis on prolonged irradiation. If the cell is removed, jarred to release bubbles, then reinserted after each measurement, the effect of bubble formation can be evaluated.

Powders loosely packed in Mylar-covered cups may be shaken or jarred between measurements, and powders packed into cells may be dumped out and repacked.

Solids, glass disks, briquets, etc. may be marked at the edge and measurements made with the marks up, down, right, and left in the specimen holder. Measurements can be made on different areas of the sample, or the sample may be resurfaced between measurements. These specimen types, as well as filter disks, are best evaluated by use of a specimen rotator. Three series of measurements are made: without rotation, with rotation only between measurements, and with continuous rotation.

Finally, if specimen treatment or preparation is involved, a carefully selected homogeneous specimen should be divided into a number of portions, and each portion subjected to the preparation process. A measurement is made on each specimen, and a standard deviation σ_7 is calculated and compared with σ_5 for the reloaded single specimen (see above). A significant difference indicates that specimen preparation is contributing to the error.

The foregoing discussion is not intended to be a complete outline of techniques for evaluation of precision, nor is it to be inferred that such a comprehensive procedure must be followed for every analysis. Instrumental and operational error require only occasional evaluation—a few times a year at most. However, a suitable evaluation of specimen error is required for each combination of analyte and sample type. Once the individual errors have been identified and evaluated, measures can be taken to reduce those that contribute error of unacceptable magnitude. These errors are then evaluated again, and the overall analytical precision is calculated with an equation similar to Equation (11.44).

Chapter 12

Absorption-Enhancement Effects

12.1. INTRODUCTION

Consider a thick specimen free from all sources of positional, physical, and chemical specimen error in an x-ray spectrometer free from all sources of instrumental and operational error (Section 11.1.1). One would expect the intensity $I_{A,M}$ of a spectral line of element A in matrix M (see below) to be given by

$$I_{A,M} = W_{A,M} I_{A,A} \tag{12.1}$$

where $W_{A,M}$ is weight fraction of analyte A in matrix M, and $I_{A,A}$ is analyte-line intensity from pure A. Unfortunately, even under these ideal—and unattainable—conditions, this simple relationship rarely applies. Instead, $I_{A,M}$ is a function of the matrix M as well as the two parameters given in Equation (12.1), that is,

$$I_{A,M} = f(W_{A,M}, I_{A,A}, M) \tag{12.2}$$

This chapter considers the effect of the matrix on analyte-line intensity.

The *matrix* consists of the entire specimen except the particular analyte under consideration. Thus, in a multielement system, the matrix of the same specimen is different for each analyte in the specimen, and each analyte constitutes a part of the matrix of every other analyte. The term *matrix* applies to the specimen as measured in the spectrometer. For example, if an alloy is to be dissolved and measured as a solution, one may consider the alloy itself as consisting of certain analytes in a certain basic matrix, but it is the solution matrix that is of analytical significance. In such cases, the alloy and solution matrixes may be referred to as the *original* and *specimen* matrixes, respectively. The same applies if an internal standard is added; the treated matrix is the significant one. Some confusion may arise in the calculation of absorption coefficients for evaluation and correction of "matrix" effects; the absorption coefficient of the entire specimen—including the analyte—must be considered.

The effects of the matrix on the measured analyte-line intensity are known variously as *matrix*, *interelement*, *self-absorption*, and *absorption-enhancement* effects. Liebhafsky and his co-authors [p. 172 in (*16*)] raise valid objections to the first three of these terms: *Matrix effect* excludes the analyte itself, which contributes to the absorption coefficient of the specimen just as any matrix element does. *Interelement effect* excludes the self-absorption effect when only one element is present, or when only one element having a short-wavelength line is present in a very light matrix. *Self-absorption* excludes enhancement effects. These authors prefer the term *absorption-enhancement* effect, and this term is used here.

Absorption-enhancement effects arise from the following phenomena:

1. The matrix absorbs primary x-rays; it may have a larger or smaller absorption coefficient than the analyte for primary x-rays, and it may preferentially absorb or transmit those wavelengths that excite the analyte line most efficiently, that is, those near the short-wavelength side of the analyte absorption edge.
2. The matrix absorbs the secondary analyte-line radiation; it may have a larger or smaller absorption coefficient than the analyte for the analyte-line radiation, and it may preferentially absorb or transmit this wavelength.
3. The matrix elements emit their own characteristic lines, which may lie on the short-wavelength side of the analyte absorption edge, and thereby excite the analyte to emit additional line radiation.

The absorption-enhancement effects may be classified in two ways: On the basis of their effect on the analyte-line intensity, they may be positive or negative absorption effects, or true or apparent enhancement effects. On the basis of their origin or general nature, they may be nonspecific (general), specific, secondary (second order), or unusual (special).

In the *positive absorption effect*, the matrix has a smaller absorption coefficient than the analyte for the primary and analyte-line radiation, and the analyte-line intensity for a specified analyte concentration is greater than would be predicted from Equation (12.1). In the *negative absorption effect*, the matrix has a larger absorption coefficient than the analyte, and the analyte-line intensity is smaller than would be predicted. In the *true enhancement effect*, one or more spectral lines of matrix elements have wavelengths shorter than the analyte absorption edge. Thus, the matrix actually excites analyte-line radiation in addition to that excited by the primary beam, and the intensity is greater than that predicted by Equation (12.1). The *apparent enhancement effect* is simply the positive absorption effect. The analyte-line intensity is greater, but only because matrix absorption is lower; the matrix excites no additional analyte-line radiation.

Nonspecific or *general* absorption-enhancement effects result simply from differences in the absorption coefficients of analyte and matrix elements for the primary and, especially, analyte-line radiation. Specific absorption edges are not involved. Nonspecific effects involve only absorption—unless one considers apparent enhancement. *Specific* absorption-enhancement effects result from interactions of analyte and matrix spectral lines and absorption edges in close proximity. *Secondary* or *second-order* absorption-enhancement effects arise from the influence of the overall matrix on a nonspecific or specific effect on a certain analyte–matrix element pair. They take the form of departures from the effects predicted from the absorption coefficients and the wavelengths of spectral lines and absorption edges of the specified analyte–matrix pair. *Unusual* or *special* absorption-enhancement effects include cases in which analyte-line intensity remains substantially constant, or even decreases, as analyte concentration increases.

In the derivation of equations for mathematical correction of absorption-enhancement effects (Chapter 15) and in the interpretation of intensity *vs.* concentration calibration functions influenced by enhancement effects, it is sometimes expedient to regard enhancement as the opposite of absorption, that is, as "negative absorption, $-\mu/\varrho$." However, the concept of negative absorption is not to be confused with the negative absorption *effect*, in which analyte-line intensity is reduced. If analyte-line intensity measured in a specific matrix is greater than predicted by Equation (12.1), the cause may be either of the following: (1) The matrix may have a lower absorption coefficient than the analyte for analyte-line and/or primary radiation; this is the positive absorption effect; or (2) the spectral lines of the matrix element(s) may be enhancing the analyte line; this is the true enhancement effect, but it has the same effect on analyte-line intensity as the positive absorption effect. In other words, true enhancement, regarded as negative absorption, acts as an *apparent* positive absorption effect, just as true positive absorption acts as an apparent enhancement effect, as already mentioned.

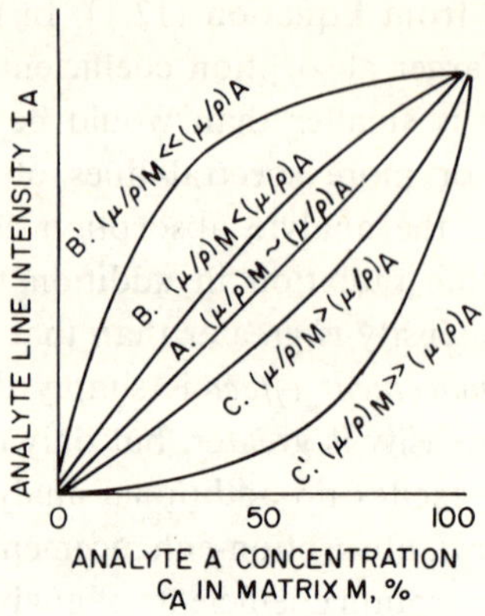

Fig. 12.1. Absorption-enhancement effects.

Figure 12.1 shows a series of calibration curves that illustrate absorption-enhancement effects. If only absorption effects are present, the curves represent various combinations of analyte and matrix absorption coefficient for the analyte line and for the primary radiation that excites it most efficiently. In the figure, μ/ϱ represents the total mass-absorption coefficient for both primary and analyte-line radiation. However, similar curves are obtained when enhancement effects are present, the enhancement appearing as a positive absorption effect (curves B and B'), as explained above.

Nonspecific effects can often be evaluated by simple calculations based on the absorption coefficients of analyte and matrix elements. Specific effects are evaluated by measurement of analyte-line intensity from mixtures of the analyte and a single matrix element in various proportions. The more complex secondary effects are evaluated by measurement of analyte-line intensity from a specific analyte–matrix element pair as the remainder of the matrix is varied. Chapters 14 and 15 outline several experimental and mathematical methods for evaluation and correction of absorption-enhancement effects.

12.2. NONSPECIFIC ABSORPTION EFFECTS

The relative intensities I_1 and I_2 measured under identical conditions from two samples having concentrations C_1 and C_2 of the same analyte in the same matrix are given by

$$\frac{I_1}{I_2} \approx \left(\frac{C_1}{C_2}\right) \frac{(\mu/\varrho)_2'}{(\mu/\varrho)_1'} \tag{12.3}$$

where $(\mu/\varrho)'$ is mass-absorption coefficient of the sample for both primary and secondary x-rays. The ratio of the slopes m_1 and m_2 of the calibration curve (Figure 14.3) at C_1 and C_2 is

$$\frac{m_1}{m_2} = \frac{(\mu/\varrho)_2'}{(\mu/\varrho)_1'} \tag{12.4}$$

Equations (12.3) and (12.4) show that as the ratio C_1/C_2 changes, I_1/I_2 also changes, but not necessarily at the same rate, as determined by $(\mu/\varrho)_2'/(\mu/\varrho)_1'$. This difference in rate may lead to positive and negative deviations from linearity of the calibration function. In special cases, when $(\mu/\varrho)'$ increases at a greater rate than C, even a negative slope may occur, as shown in Section 12.5.

It follows from Equation (12.3) that for the same analyte concentration in two samples 1 and 2 having different matrixes with mass absorption coefficients $(\mu/\varrho)_1'$ and $(\mu/\varrho)_2'$ for primary and secondary x-rays, the relative analyte-line intensities I_1 and I_2 are

$$\frac{I_1}{I_2} = \frac{(\mu/\varrho)_2'}{(\mu/\varrho)_1'} \tag{12.5}$$

In Equations (12.3) and (12.5), for practical purposes, $(\mu/\varrho)'$ may be replaced with the corresponding mass-absorption coefficient for the analyte line. Then, to a first approximation, analyte-line intensity may be corrected for matrix absorption simply by use of mass-absorption coefficients of the samples. If these are neither known nor readily calculated, they may be measured; this is the basis of emission–absorption methods (Sections 10.5 and 14.8.1).

Figure 12.1 shows a series of calibration curves that illustrate nonspecific absorption effects and apparent enhancement effects.

In a *neutral matrix* (curve A), the absorption coefficient of the matrix $(\mu/\varrho)_M$ and analyte $(\mu/\varrho)_A$ are substantially the same for both the analyte-line and the primary radiation that excites it most efficiently. The primary x-rays reach each incremental specimen volume, and the excited analyte-line radiation emerges from the specimen with about the same attenuation in the matrix as in pure analyte. Thus, analyte-line intensity I_A increases at about the same rate as C_A, the I_A–C_A curve is substantially linear, and the absorption effect is very small.

Elements within one or two atomic numbers of the analyte usually constitute neutral matrixes for that analyte, at least down to atomic number 22 (titanium). For these medium-Z and high-Z elements, absorption-enhancement properties of adjacent elements are so similar that they can serve as internal standards for one another (Section 14.4.2). However, at lower atomic numbers, the differences in wavelength (Figure 1.8) and absorption coefficient (Figure 2.5) for adjacent elements have increased to a degree such that the $Z\ K\alpha$ line is very strongly absorbed by element $Z - 1$. For example, one might expect curve A in Figure 12.1 to represent the curves of $K\alpha$ intensity *vs.* concentration for aluminum (*Z 13*) and silicon (*Z* 14) in their mixed oxides. The actual curves and the relationships among their K lines and absorption edges are shown in Figure 12.2. The Al $K\alpha$ curve shows marked enhancement even though Al $K\alpha$ is absorbed more strongly by silicon ($\mu/\varrho \sim 500$) than by aluminum ($\mu/\varrho \sim 385$). The reason is that Si $K\alpha$ is absorbed very strongly by aluminum ($\mu/\varrho \sim 3500$), thereby causing true enhancement of Al $K\alpha$ sufficient to more than balance the absorption. The Si $K\alpha$ curve shows very strong absorption because of the high absorption coefficient of aluminum just mentioned. Thus aluminum, although an adjacent element, actually constitutes a very heavy matrix for silicon. It is interesting to compare the aluminum–silicon

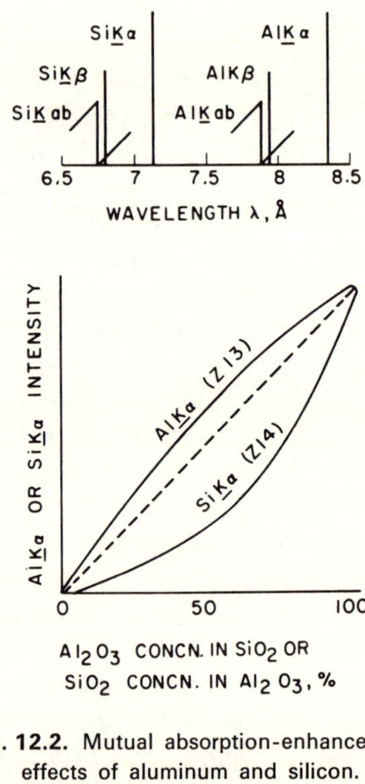

Fig. 12.2. Mutual absorption-enhancement effects of aluminum and silicon.

system with the copper–zinc system, where the absorption coefficients of both copper and zinc for both Cu $K\alpha$ and Zn $K\alpha$ are all substantially the same, ranging from ~44 to 60. Only a slight enhancement of Cu $K\alpha$ by Zn $K\beta$ occurs.

In a *light matrix* (curves B and B' in Figure 12.1), $(\mu/\varrho)_M$ is less than $(\mu/\varrho)_A$ for the analyte-line and primary x-rays. In such a matrix, the primary x-rays undergo less attenuation in reaching any incremental specimen volume and analyte-line radiation undergoes less attenuation in emerging from the specimen in the matrix than in pure analyte. Thus, I_A increases at a faster rate than C_A. At higher concentrations, the specimen composition approaches pure analyte, and the difference in rates of increase for I_A and C_A decreases. The I_A–C_A curves are nonlinear and show the positive absorption effect; the greater the difference between $(\mu/\varrho)_M$ and $(\mu/\varrho)_A$, the more distorted is the curve, as is evident from a comparison of curves B and B'. This positive absorption effect constitutes an apparent or pseudo enhancement effect. Examples of light matrixes are usually provided by analytes having relatively short-wavelength lines in relatively low-Z matrixes. Lead as tetraethyllead in gasoline is an example of an extreme case (curve B').

In a *heavy matrix* (curves C and C'), $(\mu/\varrho)_M$ is greater than $(\mu/\varrho)_A$, primary and analyte-line radiation undergo greater attenuation in the matrix than in pure analyte, I_A increases at a slower rate than C_A, and the curves show the negative absorption effect. Again, at higher analyte concentration, the difference in the two rates decreases, and the greater the difference in $(\mu/\varrho)_M$ and $(\mu/\varrho)_A$, the more distorted is the curve. Examples of heavy matrixes are usually provided by analytes having relatively long-wavelength lines in relatively high-Z matrixes. Determinations of magnesium, aluminum, and silicon in molybdenum alloys provide examples of extreme cases (curve C').

12.3. SPECIFIC ABSORPTION-ENHANCEMENT EFFECTS

If the analyte line occurs at a wavelength just less than that of an absorption edge of a particular matrix element B, the A line is highly absorbed by B, and A-line intensity is reduced in proportion to B concentration. This is the case for Fe $K\alpha$ (1.94 Å) in the presence of chromium (K_{ab} 2.07 Å), and chromium has a similar, but progressively weaker, negative absorption effect on the $K\alpha$ lines of elements of successively higher atomic number—cobalt, nickel, copper, etc.

Conversely, if the analyte absorption edge occurs at a wavelength just greater than that of a line of a particular matrix element B, the B line is absorbed by A, and A-line intensity is enhanced in proportion to B concentration. This is the case for Fe K_{ab} (1.74 Å) and Ni $K\alpha$ (1.66 Å), and nickel has a similar, but progressively weaker, enhancement effect on the $K\alpha$ lines of elements of successively lower atomic number—manganese, chromium, vanadium, etc.

The effects just described are specific, in that they depend on the proximity of spectral lines and absorption edges of analyte and matrix elements. These specific absorption-enhancement effects are now considered in detail, with Fe $K\alpha$ serving as an example of a typical analyte line [pp. 165–6 in (*16*)]. Ideally, in the determination of iron in various matrixes, Fe $K\alpha$ intensity should be given by Equation (12.1):

$$I_{\mathrm{Fe}K\alpha,\mathrm{M}} = W_{\mathrm{Fe,M}} I_{\mathrm{Fe}K\alpha,\mathrm{Fe}} \tag{12.6}$$

That is, Fe $K\alpha$ intensity in matrix M should be given by the product of the weight fraction of iron in matrix M and the Fe $K\alpha$ intensity from pure iron. Let us consider specific absorption-enhancement effects on the Fe $K\alpha$ intensity measured from binary alloys or mixtures consisting of iron (atomic number 26) in matrixes of aluminum (13), chromium (24), manganese (25), cobalt (27), nickel (28), cerium (58), and lead (82). Figure 12.3 shows the spectral lines, absorption edge, and mass-absorption curve for each of these elements in the spectral region of the iron K absorption edge and lines.

The effect of each matrix element on Fe $K\alpha$ intensity must be considered in terms of three phenomena: (1) absorption of that portion of the primary beam that excites Fe $K\alpha$ most efficiently, that is, having wavelength near the short-wavelength side of the iron K absorption edge; (2) absorption of the Fe $K\alpha$ line; and (3) enhancement of the Fe $K\alpha$ line by matrix lines having wavelengths on the short side of the iron K edge.

With respect to absorption of the primary x-rays, at all wavelengths on the short side of the iron K edge, manganese absorption is about the same as that of iron, cerium absorption is substantially higher, chromium and lead absorption are substantially lower, and aluminum absorption is very low. Cobalt and nickel absorption are very low in the extremely critical region within, respectively, ~0.15 Å and ~0.25 Å of the iron K edge, but they rise to about the same value as iron at shorter wavelengths.

With respect to the much more significant absorption of Fe $K\alpha$ radiation, aluminum, manganese, iron, cobalt, and nickel all have about the same very low absorption. Lead, chromium, and, especially, cerium have much higher absorption.

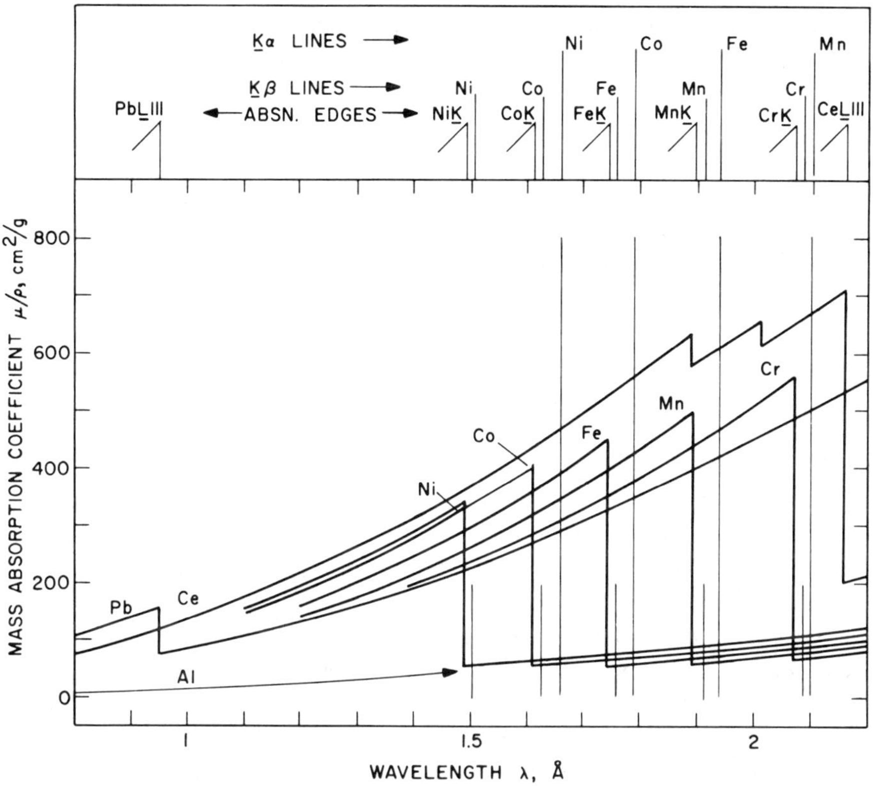

Fig. 12.3. Specific absorption-enhancement effects of several matrix elements on Fe $K\alpha$ intensity.

With respect to enhancement, aluminum, chromium, manganese, cerium, and lead have no spectral lines at wavelengths anywhere near the short side of the iron K edge, and therefore do not enhance Fe $K\alpha$. The Co $K\beta$ and Co $K\alpha$ lines bracket the iron K edge, with Co $K\beta$ on the short side; thus, cobalt mildly enhances Fe $K\alpha$. Both Ni $K\alpha$ and Ni $K\beta$ lie on the short side of the iron K edge, and nickel strongly enhances Fe $K\alpha$.

The net effect of each of these matrix elements on Fe $K\alpha$ intensity may be presented in terms of its predicted effect based on the foregoing discussion as compared with the intensity calculated from Equation (12.6); that is,

$$\frac{(I_{\text{Fe}K\alpha,\text{M}})_{\text{predicted}}}{(I_{\text{Fe}K\alpha,\text{M}})_{\text{Eq. (12.6)}}} = R \tag{12.7}$$

Aluminum. $R \gg 1$ because of the very low absorption of primary and

Fe $K\alpha$ radiation and despite the absence of any enhancing spectral lines (strong positive absorption or apparent enhancement effect).

Chromium. $R \ll 1$ because the high absorption of Fe $K\alpha$ outweighs the somewhat lower absorption of the primary x-rays (strong negative absorption effect).

Manganese. $R \approx 1$ because manganese has substantially the same absorption coefficient as iron for primary and Fe $K\alpha$ radiation and causes no enhancement (absorption-enhancement effects largely absent; neutral matrix).

Iron. $R = 1$.

Cobalt. R is somewhat greater than 1. The absorption of primary radiation within ~0.15 Å of the short side of the iron K edge is very low, and Co $K\beta$ enhances Fe $K\alpha$ (combination of positive absorption and mild enhancement).

Nickel. $R \gg 1$ because the absorption of primary radiation within ~0.25 Å of the short side of the iron K edge is very low, and both Ni $K\alpha$ and Ni $K\beta$ enhance Fe $K\alpha$ (combination of positive absorption and strong enhancement).

Cerium. $R \ll 1$ because of the very high absorption of both primary and Fe $K\alpha$ radiation and the absence of enhancement (strong negative absorption).

Lead. $R < 1$. The effect of lead is similar to that of cerium, except that the absorption coefficient of lead is not as high in this spectral region (relatively strong negative absorption).

An analysis of the possible absorption-enhancement effects should be made whenever an analytical method is to be developed for a new combination of analyte(s) and matrix. Often, it is sufficient to plot on a wavelength scale the principal spectral lines and absorption edges of all elements present, in the manner shown at the top of Figure 12.3.

12.4. SECONDARY ABSORPTION-ENHANCEMENT EFFECTS

12.4.1. General

Mitchell and Kellam (*M24*) have carried the characterization of absorption-enhancement effects a significant step farther by consideration of the overall matrix composition on the absorption-enhancement effect of a particular matrix element on a specified analyte line. They describe five secondary effects that may markedly influence the effect predicted for an analyte–

matrix element pair on the basis of the nonspecific and specific effects just described. The discussion that follows is largely summarized from their excellent paper.

Figure 12.4 illustrates the secondary effects, using Fe *Kα* as the analyte line. For all curves, Fe *Kα* is measured from synthetic specimens having the same iron concentration (10%) in a matrix M consisting of various proportions of a specific matrix element B and one other matrix element (M − B). Figure 12.4A shows secondary absorption effects on the absorption effect of chromium (element B) on Fe *Kα*. Figure 12.4B shows secondary enhancement effects on the enhancement effect of copper (element B) on Fe *Kα* Table 12.1 gives the data necessary for consideration of the figure. Following is a glossary of the symbols and terms used in the discussion:

A is the analyte element.

B is a specific matrix element, the effect of which on the analyte is being considered as a function of the overall matrix composition.

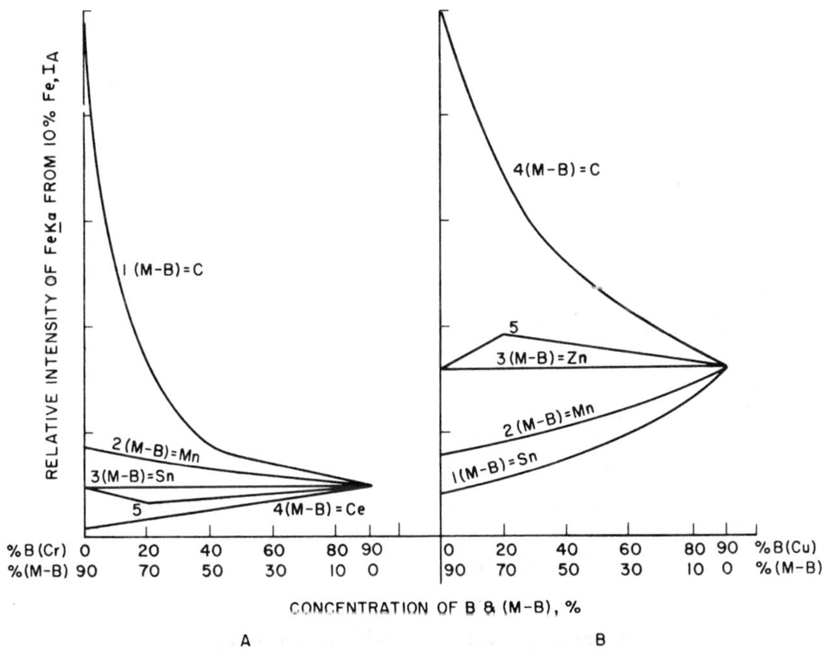

Fig. 12.4. Secondary absorption-enhancement effects. A. Secondary absorption effects of chromium on Fe *Kα* intensity in the presence of several other matrix elements. B. Secondary enhancement effects of copper on Fe *Kα* intensity in the presence of several other matrix elements. [B. J. Mitchell and J. E. Kellam, *Appl. Spectrosc.* **22**, 742 (1968).]

TABLE 12.1
Secondary Absorption-Enhancement Effects[a]

Curve[b]	Matrix type[b]	Absorption effects[c] (Figure 12.4A)		Enhancement effects[d] (Figure 12.4B)	
		(M — B)[e]	$(\mu/\varrho)_{M-B,FeK\alpha}$	(M — B)[e]	$(\mu/\varrho)_{M-B,FeK\alpha}$
1	Very light	6-C	11	6-C	11
2	Neutral	25-Mn	64	25-Mn	64
3	Equivalent	50-Sn	457	30-Zn	109
4	Heavy	58-Ce	636	50-Sn	457
5	Medium heavy	53-I	527	27-Co	81

[a] From Mitchell and Kellam (*M24*). Refer to Figure 12.4.
[b] Curve numbers and matrix types apply to both Figures 12.4A and 12.4B.
[c] For absorption effects: analyte A = Fe, Fe $K\alpha$ = 1.94 Å, $(\mu/\varrho)_{Fe,FeK\alpha}$ = 73 cm²/g; matrix element B = Cr, Cr K_{ab} = 2.07 Å, $(\mu/\varrho)_{Cr,FeK\alpha}$ = 490 cm²/g.
[d] For enhancement effects: analyte A = Fe, Fe K_{ab} = 1.74 Å, $(\mu/\varrho)_{Fe,FeK\alpha}$ = 73 cm²/g; matrix element B = Cu, Cu $K\alpha$ = 1.54 Å, $(\mu/\varrho)_{Cu,FeK\alpha}$ = 99 cm²/g.
[e] The entire matrix M consists of elements B and (M − B).

M is the whole matrix—all specimen constituents except the analyte; in the examples given below, the matrix consists of only two elements, B and "(M − B)."

(M − B) is the matrix excluding element B; it is the element, the effect of which on the absorption-enhancement effect of B on A is being evaluated.

μ/ϱ is the mass-absorption coefficient for the analyte line, unless specified otherwise.

A *light matrix* is one in which $(\mu/\varrho)_{M-B} < (\mu/\varrho)_A$.

A *neutral matrix* is one in which $(\mu/\varrho)_{M-B} \approx (\mu/\varrho)_A$.

A *heavy matrix* is one in which $(\mu/\varrho)_{M-B} > (\mu/\varrho)_A$.

An *equivalent matrix* is one in which $(\mu/\varrho)_{M-B} \approx (\mu/\varrho)_B$, that is, in which element B and the remainder of the matrix have about the same absorption coefficient for the analyte line.

12.4.2. Secondary Absorption Effects

In Section 12.3, it is shown that chromium has a strong negative absorption effect on Fe $K\alpha$ intensity, that is, strongly depresses it. However, when other elements are present in the matrix, the absorption effect becomes much more complex. Figure 12.4A shows the effect of various elements

(M − B) on the negative absorption effect of chromium (B) on Fe $K\alpha$ intensity. Table 12.1 gives the supplementary data. Mitchell and Kellam define five secondary-absorption effects as follows:

1. *Exaggerated absorption* (curve 1) occurs in a very light matrix, where $(\mu/\varrho)_{M-B} \ll (\mu/\varrho)_B > (\mu/\varrho)_A$. Absorption of Fe $K\alpha$ by carbon (μ/ϱ 11) is so much less than that by chromium (μ/ϱ 490) that, as chromium progressively replaces carbon in the matrix, the absorption of Fe $K\alpha$ by chromium appears to be strongly exaggerated; that is, Fe $K\alpha$ intensity is depressed at a much faster rate than would be expected.
2. *Normal absorption* (curve 2) occurs in a neutral matrix, where $(\mu/\varrho)_{M-B} \approx (\mu/\varrho)_A$. Chromium has about the same absorption effect on Fe $K\alpha$ in the presence of manganese (μ/ϱ 64) as in the presence of iron alone (μ/ϱ 73).
3. *Cancellation of absorption* (curve 3) occurs in an equivalent matrix, where $(\mu/\varrho)_{M-B} \approx (\mu/\varrho)_B$. Replacement of tin ($\mu/\varrho$ 457) with chromium (μ/ϱ 490) has relatively little effect on Fe $K\alpha$ intensity. The expected decrease in Fe $K\alpha$ intensity on increase in chromium concentration does not occur.
4. *Reversal of absorption* (curve 4) occurs in a heavy matrix, where $(\mu/\varrho)_{M-B} \gg (\mu/\varrho)_B > (\mu/\varrho)_A$. An expected reduction in analyte-line intensity due to the presence of a specific matrix element B having a high absorption coefficient for the analyte line may actually become an apparent enhancement effect upon change in the overall matrix composition. This is the case when, as B concentration increases, it replaces a matrix having much higher absorption than itself. As chromium (μ/ϱ 490) replaces cerium (μ/ϱ 636), the matrix absorption of Fe $K\alpha$ actually decreases.
5. *Absorption-enhancement inflection* (curve 5) occurs in a medium-heavy matrix, where $(\mu/\varrho)_{M-B} > (\mu/\varrho)_B > (\mu/\varrho)_A$. At low B concentrations, B absorbs analyte-line radiation. As B concentration increases, it replaces a matrix having somewhat higher absorption than itself, causing a slight apparent enhancement. This phenomenon might be expected for Fe $K\alpha$ in a matrix of chromium (element B, μ/ϱ 490) and iodine (element M − B, μ/ϱ 527).

It is evident that chromium may be said to decrease measured Fe $K\alpha$ intensity only under certain conditions, and even then to a varying extent.

12.4.3. Secondary Enhancement Effects

In Section 12.3, it was shown that nickel has a strong enhancement effect on Fe $K\alpha$ intensity. The same applies to copper; both Cu $K\alpha$ and

Cu $K\beta$ lie on and near the short side of the iron K edge. However, here again, when other elements are present in the matrix, the enhancement effect becomes much more complex. Figure 12.4B shows the effect of various elements (M − B) on the enhancement effect of copper (B) on Fe $K\alpha$ intensity. Table 12.1 gives the supplementary data. Mitchell and Kellam define five secondary enhancement effects analogous to the five secondary absorption effects described in Section 12.4.2:

1. *Exaggerated enhancement* (curve 1) occurs in a heavy matrix, where $(\mu/\varrho)_{\mathrm{M-B}} \gg (\mu/\varrho)_{\mathrm{B}} \approx (\mu/\varrho)_{\mathrm{A}}$. Replacement of tin ($\mu/\varrho$ 457) with copper (μ/ϱ 99) results not only in true enhancement by excitation of Fe $K\alpha$ by Cu $K\alpha$ and Cu $K\beta$, but additional apparent enhancement by reduction of the overall matrix absorption for Fe $K\alpha$.

2. *Normal enhancement* (curve 2) occurs in a neutral matrix, where $(\mu/\varrho)_{\mathrm{M-B}} \approx (\mu/\varrho)_{\mathrm{A}}$. The absorption of the analyte line (Fe $K\alpha$) is about the same for both copper (μ/ϱ 99) and manganese (μ/ϱ 64). Also, the absorption of the enhancing lines (Cu $K\alpha$ and Cu $K\beta$) is about the same for both iron (μ/ϱ for Cu $K\alpha$ is 324) and manganese (μ/ϱ for Cu $K\alpha$ is 284). Thus, copper has about the same enhancement effect on Fe $K\alpha$ in the presence of manganese as in the presence of iron alone.

3. *Cancellation of enhancement* (curve 3) occurs in an equivalent matrix, where $(\mu/\varrho)_{\mathrm{M-B}} \approx (\mu/\varrho)_{\mathrm{B}}$. Replacement of zinc ($\mu/\varrho$ 109) with copper (μ/ϱ 99) has relatively little effect on absorption of Fe $K\alpha$ intensity, and Cu $K\alpha$, Cu $K\beta$, Zn $K\alpha$, and Zn $K\beta$ all excite Fe $K\alpha$ with about the same efficiency. The expected increase in Fe $K\alpha$ intensity on increase of copper concentration does not occur.

4. *Reversal of enhancement* (curve 4) occurs in a very light matrix, where $(\mu/\varrho)_{\mathrm{M-B}} \ll (\mu/\varrho)_{\mathrm{B}} \approx (\mu/\varrho)_{\mathrm{A}}$. An expected enhancement of analyte-line intensity due to the presence of a specific matrix element B having a strong spectral line near the short-wavelength side of the analyte absorption edge may actually become an absorption effect upon change in the overall matrix composition. This is the case when, as B concentration increases, it replaces a matrix having much lower absorption than itself. As copper (μ/ϱ 99) replaces carbon (μ/ϱ 11), the matrix absorption of Fe $K\alpha$ increases so much that the enhancement by Cu $K\alpha$ is outweighed.

5. *Enhancement-absorption inflection* (curve 5) occurs in a medium-light matrix, where $(\mu/\varrho)_{\mathrm{M-B}} < (\mu/\varrho)_{\mathrm{B}} \approx (\mu/\varrho)_{\mathrm{A}}$. At low B concentrations, B enhances analyte-line radiation. As B concentration increases, it replaces a matrix having somewhat lower absorption than itself, causing a slight negative absorption effect. This phenomenon might be expected for Fe $K\alpha$

in a matrix of copper (element B, μ/ϱ 99) and cobalt (element M − B, μ/ϱ 81).

It is evident that copper may be said to enhance measured Fe $K\alpha$ intensity only under certain conditions, and even then to a varying extent.

12.5. UNUSUAL ABSORPTION-ENHANCEMENT EFFECTS

Some of the absorption-enhancement phenomena described in the preceding section may be considered unusual. However, the term is used here for cases in which analyte-line intensity remains substantially constant or even actually decreases with increasing analyte concentration. Two examples have been reported.

The x-ray secondary-emission spectrometer is generally unable to distinguish various concentrations of an analyte having a relatively short-wavelength line in an extremely light matrix. For example, it is, in general, not possible to distinguish the various oxides of an element of medium or high atomic number simply by comparing the intensities of its spectral line measured from the untreated oxides. Table 12.2 gives data for iron and its oxides [pp. 182–4 in (*16*)]. The oxygen is highly transparent to Fe $K\alpha$ (μ/ϱ 22) and to the primary x-rays, so the iron (μ/ϱ 73) largely determines the absorption coefficient of the oxide and thereby the critical thickness or depth from which Fe $K\alpha$ can emerge from the specimen. Thus, substantially the same number of iron atoms contribute to the measured Fe $K\alpha$ intensity for each oxide, and the relative intensity increases only from 95.0% to 95.7% while iron concentration increases from 70.0% to 77.8%. The difficulty is easily remedied by quantitative addition of a heavy element to increase matrix absorption or, for some elements, by measurement of an *L* or *M* line.

TABLE 12.2

Unusual Absorption Effect for Oxides of Iron[a]

	Fe	FeO	Fe_3O_4	Fe_2O_3
Fe concentration, wt %	100	77.8	72.5	70.0
Fe $K\alpha$ intensity, relative	100	95.7	95.5	95.0
Critical thickness for Fe $K\alpha$, cm $\times 10^{-3}$	1.36	2.31	2.72	2.77

[a] From Liebhafsky *et al.* [pp. 182–4 in (*16*)].

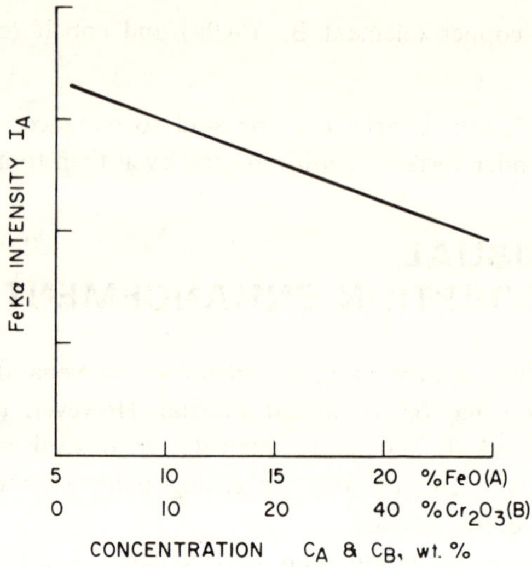

Fig. 12.5. Unusual matrix effect of chromium on Fe *Kα* intensity. [B. J. Mitchell and J. E. Kellam, *Appl. Spectrosc.* **22**, 742 (1968).]

It is also possible for analyte-line intensity to actually decrease as analyte concentration increases. This phenomenon occurs when the following two conditions occur simultaneously: (1) a specific matrix element B has a high absorption coefficient for the analyte line, and (2) the concentration of element B increases at a faster rate than analyte concentration. Figure 12.5 shows an example. Chromium has a high absorption coefficient for Fe *Kα* (μ/ϱ 490) compared with that of iron (μ/ϱ 73). The figure shows Fe *Kα* intensity as a function of composition for a substance in which FeO concentration increases from 5 to 25% as Cr_2O_3 concentration increases from 0 to 50%. The remainder of the specimen consists of light elements, such as magnesium and aluminum. The high absorption coefficient and the disproportionate rates of increase of chromium and iron concentration result in the negative absorption effect of chromium outweighing the increased Fe *Kα* emission from the iron. The phenomenon was first noted in mixture of the minerals chromite [$FeO \cdot (Cr,Al)_2O_3$] and olivine [$2(Mg,Fe)O \cdot SiO_2$] in the concentration range 8–17% iron (*B38*, *M24*).

Chapter 13

Sensitivity and Resolution; Spectral-Line Interference

13.1. SENSITIVITY

13.1.1. Definitions

Sensitivity in x-ray spectrometric analysis may be defined in either of two ways—in terms of minimum detectable amount of analyte, or rate of change of analyte-line intensity with change in amount of analyte.

The minimum detectable amount, or limit of detectability, may be defined in several ways, but the definition given by Birks [p. 54 in (*6*)] is probably the most widely accepted: that amount of analyte that gives a net line intensity equal to three times the square root of the background intensity, or, in statistical terms, that amount that gives a net intensity equal to three times the standard counting error of the background intensity. "Amount" signifies concentration (%, mg/ml, etc.) for infinitely thick specimens, area density (mg/cm^2) for specimens less than infinitely thick, or mass (μg) for extremely small particulate or filamentary specimens. Figure 13.1 shows a ratemeter–recorder 2θ scan of a weak peak on a uniform background. Suppose that in a specified preset time, the accumulated count for the background is 100 counts. The standard counting error for the background count σ_N [Equation (11.10)] is then $\sqrt{100}$, or 10 counts.

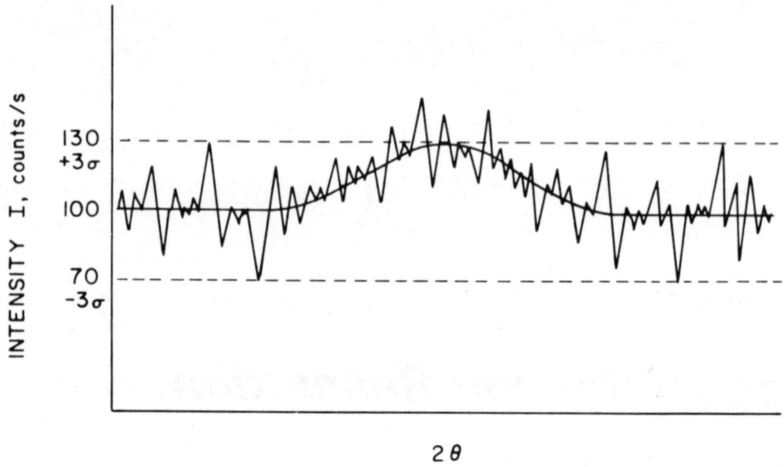

Fig. 13.1. Detection limit.

Then, if there is to be 99.9% certainty of detection of the peak, its accumulated count in the same preset time must exceed the background by $3\sigma_N$, or 30 counts, that is, the accumulated peak count must be 130. Whatever concentration, area density, or mass is required to give analyte-line intensity corresponding to this count is the minimum detectable amount, or limit of detectability.

Alternatively, sensitivity may be stated in terms of intensity per unit analyte concentration (counts/s per %, counts/s per mg/ml, etc.) for infinitely thick specimens; intensity per unit area density (counts/s per mg/cm^2) for thin specimens; or intensity per unit mass (counts/s per μg) for very small specimens. Sometimes, sensitivity is given per unit x-ray tube current —for example, counts/s per % per mA. When the calibration function—intensity *vs.* amount of analyte—is nonlinear, the sensitivity varies with the concentration or amount of analyte. Although this may not be the case with thin-film and very small specimens, it is often the case with the more common solid, briquet, powder, and liquid specimens. In such cases, it is necessary to express sensitivity at a specified concentration level, for example, counts/s per % "at the 10% concentration level."

13.1.2. Factors Affecting Sensitivity (*150*)

Regardless of the terms used to express sensitivity, the value must be rigorously qualified. So many factors affect the sensitivity that a given figure applies only for a specified list of specimen and instrument conditions.

This section lists the specimen and instrument conditions that result in increased analyte-line intensity. Sensitivity is usually, although not necessarily, improved by an increase in analyte-line intensity. Analyte-line intensity is increased by the following conditions.

13.1.2.1. *Excitation Conditions*

1. Constant-potential, rather than full-wave, generator.
2. X-ray tube potential and current as high as possible; however, background also increases with x-ray tube potential and current, and it may be preferable to reduce excitation to increase line-to-background ratio.
3. X-ray tube target lines close to the short-wavelength side of the analyte absorption edge; when inconvenience, limited choice of targets, or spectral interference by target lines precludes this condition, excitation is effected principally by the continuum, and the following condition assumes added significance:
4. X-ray tube target atomic number as high as possible to give high continuum intensity.
5. X-ray tube window of beryllium and as thin as possible, especially when low-Z analytes are to be determined.
6. X-ray tube target-to-window distance as small as possible so that the target-to-specimen distance can be made small.
7. X-ray tube target-to-specimen distance as small as possible.
8. X-ray tube target surface free of pitting from long use at high power dissipation, and x-ray tube window free of high-Z sublimate.
9. Primary-beam filters avoided; such filters always reduce primary-beam intensity, and therefore analyte-line intensity; however, such filters may improve line-to-background ratio, and thereby decrease the detection limit.
10. Irradiated area on the specimen plane as large as possible.

13.1.2.2. *Specimen Conditions*

1. Useful specimen area equal to or greater than the area of the projection of the spectrometer optical system on the specimen plane (for flat-crystal instruments).
2. Enhancement radiator (Section 4.4.3) used behind x-ray-transparent specimens.
3. Analyte absorption-edge wavelength just longer than the wavelengths of the x-ray tube target lines or just longer than the wavelength of the continuous radiation hump.

4. Matrix absorption coefficient low.

5. Matrix spectral lines enhance the analyte line.

6. Dilution used for a matrix having high absorption coefficient.

7. Preconcentration or separation of analytes of low concentration.

8. Particle size of powder or briquet or surface texture of solid specimens <50–$100\ \mu m$.

9. Grind marks parallel to the plane defined by the central rays of the primary and secondary beams.

10. Optimal analyte line used (*C7*): For $Z < 55$ (Cs), the $K\alpha$ line is most intense. For Z 55–65 (Tb), the $L\beta_1$ line is most intense. For $Z > 65$, the $L\alpha_1$ line is most intense. However, sometimes an L line, although less intense than a K line, may give a more favorable line-to-background ratio if it occurs in a less intense region of the scattered continuum. With a 50-kV generator, in practice, K lines are always used for elements up to atomic number 40 (zirconium), and L lines are always used above atomic number 60 (neodymium); K or L lines may be used for the intermediate elements. First-order lines have about 10 times the intensity of second-order lines with a LiF crystal.

13.1.2.3. *Optical System*

1. Source geometry favorable: specimen plane inclined at 60° or more to the primary beam, takeoff angle 30–60°.

2. Useful aperture as large as possible—that is, useful area of collimators, crystal, and detector window (*S21*).

3. Collimators short, with wide-spaced foils.

4. Axial divergence permitted, that is, divergence in the plane of, or parallel to, the goniometer axis.

5. High-reflectivity crystal used.

6. Crystal length and width sufficient to intercept all useful secondary radiation.

7. Crystal permits measurement of the analyte line at a 2θ angle at which the entire useful secondary beam is intercepted by the crystal.

8. Curved crystal used for small particulate or filamentary specimens.

9. Optical path as short as possible.

10. Optimal optical path used: helium or vacuum is beneficial for all wavelengths greater than 1 Å; vacuum is preferable to helium for the K lines of elements 9–20 (fluorine to calcium).

11. In multichannel instruments, two channels may be set to the same analyte line and their outputs combined.

13.1.2.4. *Detector and Readout Conditions*

1. Detector quantum efficiency for the analyte line as high as possible.
2. Detector window as thin as possible, especially for the lightest elements.
3. Pulse-height analysis avoided if possible so that all analyte pulses are counted.

13.1.3. Photon Losses in the X-Ray Spectrometer

At this point, it is very informative to calculate the number of x-ray photons generated at the x-ray tube target and follow their progress to the specimen, then to calculate the number of spectral-line photons generated in the specimen and follow their progress to the detector. The calculations reveal which components or processes in the spectrometer are responsible for the greatest losses in the measured intensity.

Consider a specimen consisting of a monatomic layer of copper having area 1 cm^2. For copper, the atomic diameter is 2.55 Å (2.55 × 10^{-8} cm), and the density is 8.92 g/cm^3. Then the mass of the monolayer is (1 cm^2) (2.55 × 10^{-8} cm) (8.92 g/cm^3) = 2.3 × 10^{-7} g, or 0.23 μg. Now consider the monolayer to be placed in an x-ray spectrometer and irradiated by the primary beam from an x-ray tube having a tungsten target and operating at 50 kV, 50 mA constant potential. The minimum wavelength in the primary spectrum from this tube is given by Equation (1.14):

$$\lambda_{\text{min}} = 12.4/50 \text{ kV} = 0.25 \text{ Å}$$

The effective wavelength (Section 1.5.1) is then

$$\lambda_{\text{eff}} \approx \lambda_{I_{\text{max}}} \approx 1.5\lambda_{\text{min}} \approx 0.375 \text{ Å}$$

However, it is the longer primary wavelengths nearer the copper K-absorption edge (1.38 Å) that excite Cu $K\alpha$ most efficiently. A wavelength of 1 Å represents a compromise between the most intense and the most efficient primary wavelengths, and, for simplicity, in the calculations it is assumed that the primary beam is monochromatic and has wavelength 1 Å. Table 13.1 gives the data required for the calculations, including specimen and instrument parameters and, for beryllium, copper, and air, densities and mass-absorption coefficients for primary (1 Å) and spectral-line (Cu $K\alpha$) x-rays. The calculations follow.

TABLE 13.1

Data for Calculation of Photon Losses in the Flat-Crystal X-Ray Spectrometer

Specimen data:	
Element	Cu
Area, a_X	1 cm^2
Thickness, t	2.55 Å, 2.55×10^{-8} cm
Spectral line	Cu $K\alpha$, 1.54 Å, 1.54×10^{-8} cm
K absorption edge jump ratio, r_K	~8
K fluorescent yield, ω_K	~0.4
Instrument data:	
X-ray tube target, Z	W, Z 74
X-ray tube potential, V	50 kV, 5×10^4 V
X-ray tube current, i	50 mA, 0.05 A
Target to specimen distance, d_1	4 cm
X-ray tube window to specimen distance, d_2	2 cm
Specimen to detector distance, d_3	32 cm
Soller collimator, $L \times S$	3.5×0.010 inch
Crystal (flat)	LiF (200)
Detector	NaI(Tl) scintillation counter

Mass-absorption coefficients and densities:

Material	Density ϱ, g/cm³	Mass-absorption coefficient μ/ϱ, cm²/g	
		1 Å (primary)	1.54 Å (Cu $K\alpha$)
Be	1.85	0.57	1.35
Cu	8.92	130	53
Air	0.0012	2.45	9.56

1. Number of primary photons (1 Å) produced at the target focal spot: The x-ray tube power expended in producing x-rays is given by Equation (1.17):

$$\begin{aligned} I_W &= (1.4 \times 10^{-9})ZiV^2 \\ &= (1.4 \times 10^{-9})(74)(0.05\ \text{A})(5 \times 10^4\ \text{V})^2 \\ &= 13\ \text{W} \end{aligned}$$

Then,

$$I_{\text{erg/s}} = (13\ \text{W})(10^7\ \text{ergs/W-s}) = 1.3 \times 10^8\ \text{ergs/s}$$

Photon energy for 1-Å x-rays is given by Equation (1.7):

$$\begin{aligned} E &= h\nu = h(c/\lambda) \\ &= (6.6 \times 10^{-27}\ \text{erg-s})\,\frac{3 \times 10^{10}\ \text{cm/s}}{1.54 \times 10^{-8}\ \text{cm}} \\ &= 1.3 \times 10^{-8}\ \text{erg} \end{aligned}$$

The number of 1-Å photons per second is then

$$I = \frac{1.3 \times 10^{8}\ \text{ergs/s}}{1.3 \times 10^{-8}\ \text{erg/photon}} = 10^{16}\ \text{photons/s}$$

2. Fraction of primary photons directed through the solid angle subtended by the specimen: If uniform spherical propagation of the primary x-rays is assumed, the fraction of the spherical propagation surface a_S intercepted by the specimen area a_X (1 cm²) at the target-to-specimen distance d_1 (4 cm) is

$$\frac{a_X}{a_S} = \frac{a_X}{4\pi d_1{}^2} = \frac{1\ \text{cm}^2}{(4)(3.14)(4\ \text{cm})^2} = 0.005$$

where $4\pi d_1{}^2$ is the area of a sphere having radius d_1.

3. Fraction of primary photons transmitted by the x-ray tube window: For a beryllium window having thickness t_{Be} 0.040 inch (0.1 cm), Equation (2.9) gives

$$\begin{aligned} I/I_0 &= \exp[-(\mu/\rho)_{\text{Be},1\text{Å}}\rho_{\text{Be}}t_{\text{Be}}] \\ &= \exp[-(0.57\ \text{cm}^2/\text{g})(1.85\ \text{g/cm}^3)(0.1\ \text{cm})] = \exp -0.105 \\ &= 0.90 \end{aligned}$$

4. Fraction of primary photons transmitted by the air path from x-ray tube window to specimen:

$$\begin{aligned} I/I_0 &= \exp[-(\mu/\rho)_{\text{air},1\text{Å}}\rho_{\text{air}}d_2] \\ &= \exp[(2.45\ \text{cm}^2/\text{g})(0.0012\ \text{g/cm}^3)(2\ \text{cm})] = \exp -0.0059 \\ &= 0.994 \end{aligned}$$

5. Fraction of primary photons absorbed in the specimen: The fraction transmitted is

$$\begin{aligned} I/I_0 &= \exp[-(\mu/\rho)_{\text{Cu},1\text{Å}}\rho_{\text{Cu}}t] \\ &= \exp[-(130\ \text{cm}^2/\text{g})(8.92\ \text{g/cm}^3)(2.55\times 10^{-8}\ \text{cm})] = \exp -(3\times 10^{-5}) \\ &= 0.99997 \end{aligned}$$

The fraction absorbed is $1 - 0.99997 = 3 \times 10^{-5}$.

6. Fraction of absorbed primary photons that create copper K-shell vacancies: This is given by the copper K-absorption edge jump ratio r_K (Section 4.1.2, Figure 2.4): $1 - (1/r_K) = 1 - (1/8) = 0.875$.

7. Fraction of copper K-shell vacancies that result in copper K photons: This is given by the copper K fluorescent yield ω_K (Section 2.5.2, Figure 2.18): $\omega_K \approx 0.4$.

8. Fraction of copper K photons consisting of Cu $K\alpha_1$ and Cu $K\alpha_2$ photons: The relative intensities of the principal Cu K lines are as follows: $K\alpha_1{:}K\alpha_2{:}K\beta = 100{:}50{:}20$; then,

$$\frac{K\alpha_1 + K\alpha_2}{K\alpha_1 + K\alpha_2 + K\beta} = \frac{150}{170} = 0.88$$

9. Fraction of Cu $K\alpha$ photons transmitted by the collimator: For a 3.5×0.010 inch Soller collimator having 2.5° opening in the plane of the crystal axis and ~80% open cross-sectional area, divergence is given by Equation (13.2):

$$\begin{aligned} B &= \tan^{-1}(S/L) = \tan^{-1}(0.010/3.5) = \tan^{-1} 0.003 \\ &= 0.17° \end{aligned}$$

Then,

$$\left(\frac{0.17°}{360°}\right)\left(\frac{2.5°}{360°}\right)(\pi)(0.8) = 9 \times 10^{-6}$$

10. Fraction of Cu $K\alpha$ photons transmitted by the air path from specimen to detector:

$$\begin{aligned} I/I_0 &= \exp[-(\mu/\varrho)_{\text{air},\text{Cu}K\alpha}\varrho_{\text{air}} d_3] \\ &= \exp[-(9.56\ \text{cm}^2/\text{g})(0.0012\ \text{g/cm}^3)(32\ \text{cm})] = \exp -0.367 \\ &= 0.692 \end{aligned}$$

11. Fraction of Cu $K\alpha$ photons diffracted by the crystal: For Cu $K\alpha$ and a LiF crystal, ~0.1.

12. Fraction of Cu $K\alpha$ photons counted by the detector: The quantum efficiency of a scintillation counter for Cu $K\alpha$ (Figure 6.12) is ~0.95.

Table 13.2 summarizes the calculations and permits convenient appraisal of the photon losses. In a monolayer specimen, absorption of primary x-rays is very low, and the loss due to transmitted primary photons is correspondingly great. Self-absorption of emergent Cu $K\alpha$ photons is absent in a monolayer. Coincidence losses are absent, because of the low measured intensity. Loss from rejection of pulses by the pulse-height analyzer is disregarded.

Table 13.2

Summary of Photon Losses in the Flat-Crystal X-Ray Spectrometer

Calculation	Process	Fraction	Cumulative fraction	Number of photons remaining, s^{-1}
1	Primary photons emitted by target	1.0	1.0	10^{16}
2	Primary photons directed toward specimen	0.005	5×10^{-3}	5×10^{13}
3	Primary photons transmitted by x-ray tube window	0.90	4.5×10^{-3}	4.5×10^{13}
4	Primary photons transmitted by window-specimen air path	0.994	4.47×10^{-3}	4.47×10^{13}
5	Primary photons absorbed in specimen	3×10^{-5}	1.34×10^{-7}	1.34×10^{9}
6	Cu *K*-shell vacancies created	0.875	1.17×10^{-7}	1.17×10^{9}
7	Cu *K* photons emitted	0.4	4.68×10^{-8}	4.68×10^{8}
8	Cu $K\alpha$ photons emitted	0.88	4.12×10^{-8}	4.12×10^{8}
9	Cu $K\alpha$ photons transmitted by collimator	9×10^{-6}	3.71×10^{-13}	3.71×10^{3}
10	Cu $K\alpha$ photons transmitted by specimen-detector air path	0.692	2.57×10^{-13}	2.57×10^{3}
11	Cu $K\alpha$ photons diffracted by crystal	0.1	2.57×10^{-14}	2.57×10^{2}
12	Cu $K\alpha$ photons counted by detector	0.95	2.44×10^{-14}	244

13.1.4. Sensitivity Performance

Because of the multiplicity of factors that affect sensitivity, it is difficult to cite "typical" performance. For extremely small particulate, filamentary, or residue specimens, a flat-crystal spectrometer can measure as little as 10^{-6}–10^{-7} g, a curved-crystal spectrometer as little as 10^{-8}–10^{-9} g. For comparison, the electron-probe microanalyzer can detect as little as 10^{-12}–10^{-15} g. For thin-film specimens having area 1 cm^2 or more, a flat-crystal instrument can measure a monolayer of any element having atomic number 12 (magnesium) or more, and perhaps as little as 0.01 monolayer of the more sensitive elements.

Table 13.3 gives what may be near-optimal sensitivity for x-ray spectrometric analyses. The measurements in columns 3, 4, and 5 were made on standard commercial flat-crystal spectrometers, those in column 6 on a full-focusing curved-crystal spectrometer (Figure 19.9). Column 3 gives sensi-

TABLE 13.3
Optimal Sensitivity of X-Ray Spectrometric Analysis

Atomic number	Element and line	Direct determination in solid Fe (*L24*), %	Borax-fusion method 1-g Fe sample (*L24*), %	Water solution (*C5*), μg/ml	Preconcentrated in 1/8-inch ion-exchange membrane disks (*L24*), μg
13	Al *Kα*	0.012[a]	—	—	—
14	Si *Kα*	0.007[a]	—	—	—
15	P *Kα*	0.003[a]	—	—	—
16	S *Kα*	—	—	140	—
19	K *Kα*	—	—	3	0.06
20	Ca *Kα*	—	—	2	—
22	Ti *Kα*	0.0001[b]	0.0001	—	0.008
25	Mn *Kα*	0.0007[c]	0.0001	—	—
26	Fe *Kα*	—	—	2	0.005
28	Ni *Kα*	0.0009[d]	0.0001	—	0.002
29	Cu *Kα*	0.0015[d]	—	2	—
31	Ga *Kα*	—	—	—	0.01
32	Ge *Kα*	—	0.0002	—	—
33	As *Kα*	0.001[e]	—	2	—
37	Rb *Kα*	—	—	—	0.03
38	Sr *Kα*	—	0.0004	1.5	—
42	Mo *Kα*	0.001[e]	0.0009	2	—
50	Sn *Kα*	0.002[e]	—	10	—
56	Ba $L\alpha_1$	—	0.0003	5	—
74	W $L\alpha_1$	0.006[e]	—	—	—
82	Pb $L\alpha_1$	—	0.0006	5.5	—

Footnote	Target	kV	mA	Crystal	Collimator, inches	Path	Detector
(a)	Ti	50	60	PET	3.5 × 0.020	He	Flow (Ar)
(b)	Cr	50	60	LiF	3.5 × 0.010	He	Flow (Ar)
(c)	W	50	75	LiF	3.5 × 0.010	He	Flow (Ar)
(d)	W	50	75	LiF	3.5 × 0.010	Air	Prop. (Xe)
(e)	W	50	75	LiF	3.5 × 0.010	Air	Scint.

Counting times: columns 3, 4, 6: 100 s; column 5: 2 min.

tivity for direct determination in solid iron at the conditions given in the table. Column 4 gives sensitivities for the borax-fusion method for 1-g portions of the iron samples in column 3; the analyte is preconcentrated before being subjected to the fusion (Section 16.4.3, last paragraph). Column 5 is taken from Table 13.4 (see below). For column 6, the analyte is separated from the matrix by microchemical techniques and taken up in ion-exchange membrane disks 1/8 inch in diameter, which are then measured in a full-focusing curved-crystal spectrometer.

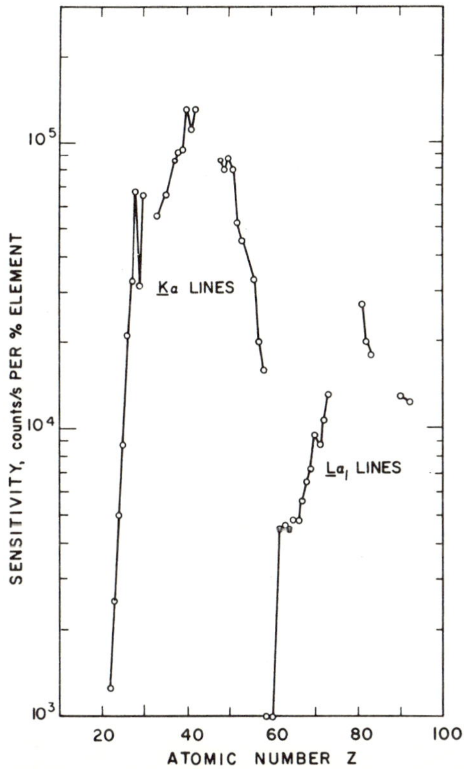

Fig. 13.2. Typical sensitivity performance of x-ray spectrometric analysis of powders. [Data from M. L. Salmon, in "*Handbook of X-Rays,*" McGraw-Hill Book Co., p. **32**-7 (1967).] Powders containing up to 1% analyte in silica matrix loosely packed in cells having 0.0014-inch Mylar windows. Philips 100-kV spectrometer; W target, 50 kV (CP), 50 mA; 4 × 0.005 inch collimator; LiF crystal; air path; scintillation detector.

TABLE 13.4
Typical Sensitivity of X-Ray Spectrometric Analysis in Water Solution[a]

Atomic number	Element and line	X-ray tube current,[b] mA	Crystal	Path	Detector[c]	Intensity for 1 mg/ml, counts/s: Line (net)	Intensity for 1 mg/ml, counts/s: Bkgnd.	Minimum detectable concentration,[d] μg/ml
16	S $K\alpha$	25	EDDT	He	Flow	3	2	140
17	Cl $K\alpha$	,,	EDDT	,,	,,	7	2	62
19	K $K\alpha$	,,	LiF	,,	,,	320	14	3.2
20	Ca $K\alpha$	,,	,,	,,	,,	311	7	2.3
23	V $K\alpha$	,,	,,	,,	,,	772	19	1.5
24	Cr $K\alpha$	,,	,,	,,	,,	1030	370	5.1
26	Fe $K\alpha$	,,	,,	Air	Scint.	2000	170	1.8
29	Cu $K\alpha$	10	,,	,,	,,	2100	350	2.4
33	As $K\alpha$	,,	,,	,,	,,	3140	460	1.8
38	Sr $K\alpha$	,,	,,	,,	,,	6400	1300	1.5
42	Mo $K\alpha$	,,	,,	,,	,,	8000	3300	2.0
48	Cd $K\alpha$	,,	,,	,,	,,	3200	4000	5.4
53	I $K\alpha$	,,	,,	,,	,,	1200	2800	12
56	Ba $K\alpha$	,,	,,	,,	,,	1070	3200	14
57	La $K\alpha$	,,	,,	,,	,,	635	2540	20
42	Mo $L\beta_1$	25	EDDT	He	Flow	18	11	49
48	Cd $L\beta_1$	,,	LiF	,,	,,	75	8	10
53	I $L\alpha_1$	,,	,,	,,	,,	232	13	4.2
56	Ba $L\alpha_1$	,,	,,	,,	,,	275	23	4.8
57	La $L\alpha_1$	,,	,,	,,	,,	275	29	5.3
62	Sm $L\alpha_1$	,,	,,	,,	,,	440	45	4.1
70	Yb $L\alpha_1$	,,	,,	Air	Scint.	675	285	6.8
79	Au $L\alpha_1$	10	,,	,,	,,	690	2400	19
82	Pb $L\alpha_1$	,,	,,	,,	,,	1060	460	5.5
90	Th $L\alpha_1$	,,	,,	,,	,,	1220	850	6.5

[a] From Campbell and Leon (*C5*, *C17*). The instrument was a Philips Inverted Three-Position x-ray spectrometer; a 4 × 0.020 inch collimator and pulse-height discrimination were used for all measurements.

[b] Tungsten target, 50 kV peak.

[c] *Flow* signifies a gas-flow proportional counter having a 0.00025-inch Mylar window and using P10 gas (90% argon, 10% methane); *Scint.* signifies a NaI(Tl) scintillation counter.

[d] Defined as that concentration that gives a net line intensity equal to three times the square root of the background intensity for a 2-min counting time.

Hale and King (*H8*) determined nickel in concentration 0.07–0.6 μg/ml with an absolute precision (1σ) of ± 0.035 μg/ml. The Ni $K\alpha$ intensity and background intensity 0.01 Å from the Ni $K\alpha$ peak were measured simultaneously on an Applied Research Laboratories multichannel spectrometer. A very narrow detector slit was used, and the detector output was integrated, rather than scaled, for up to 30 min.

Figure 13.2 (*15O*) gives sensitivity data for loosely packed powders. Table 13.4 (*C5, C17*) gives typical intensity and sensitivity data for water solutions, and figure 13.3 (*L4*) gives the concentration in water solution required to give a net intensity of 10 counts/s at various conditions given in the caption. Table 9.1 gives typical performance data for the ultralong-wavelength region, showing *K*-line intensities for elements 4–8 (beryllium to oxygen) measured on a Philips Universal Vacuum Spectrometer fitted with accessories for this region.

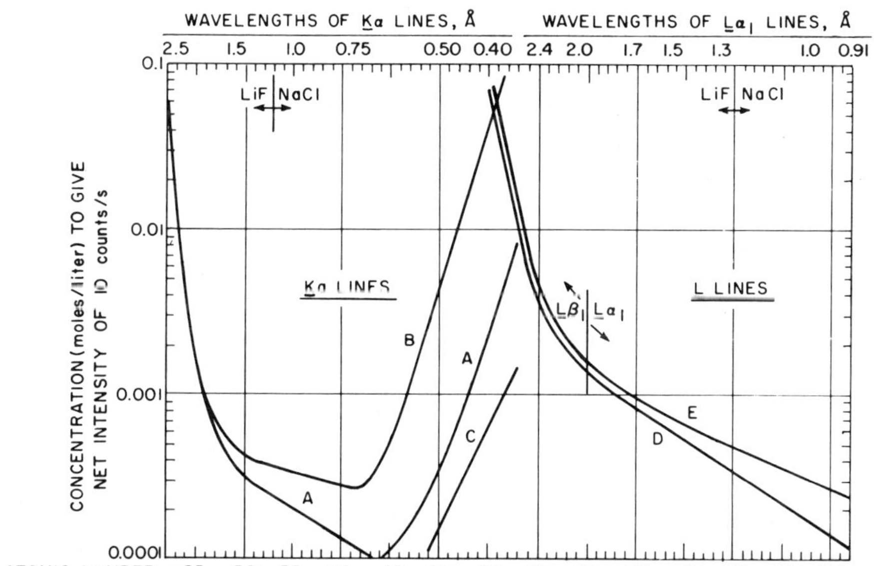

Fig. 13.3. Typical sensitivity performance of x-ray spectrometric analysis of water solutions. [Data from M. C. Lambert, *Norelco Reporter* **6**, 37 (1959).] Philips Inverted Three-Position spectrometer; W target, 50 kV (peak), 40 mA, except curve *C*, 56 kV (peak); collimation not specified; LiF or NaCl crystal, as indicated; scintillation counter for curves *A, B, D*; Ar-filled Geiger counter for curves *B, E*.

13.2. RESOLUTION

13.2.1. Definitions

13.2.1.1. *Resolution*

Resolution, or resolving power, is a measure of the ability of the spectrometer to distinguish, or recognize as separate, two closely spaced spectral lines, and is a function of their dispersion and divergence—that is, their separation and breadth. Depending on whether wavelength (2θ) or energy dispersion is used, x-ray spectrometer resolution is expressed in terms of *wavelength resolution*—separation of the 2θ peaks of the two lines —or *energy resolution*—separation of their pulse-height distributions. Wavelength resolution may be expressed in terms of the difference in wavelength $\Delta\lambda$ of two spectral lines of average wavelength λ that are just resolved, that is, $\lambda/\Delta\lambda$; $\Delta\lambda$ is placed in the denominator so that the higher the resolution, the higher is the number expressing it. Energy resolution may be expressed analogously in terms of the difference in photon energy ΔE of two spectral lines of average photon energy E that are just resolved, that is, $E/\Delta E$.

Perhaps more practically, x-ray spectrometric wavelength resolution is usually expressed as the quotient of the dispersion D of two peaks that are just resolved, by their divergence or breadth B, that is, D/B, where both terms are in degrees 2θ. Dispersion and divergence are discussed below.

Energy resolution may be expressed analogously as the quotient of the separation $V_1 - V_2$ of the peaks of two pulse-height distributions that are just resolved, by their half-widths $W_{1/2}$ or FWHM (Section 13.2.1.3), that is, $(V_1 - V_2)/W_{1/2}$, where all terms are in volts. However, energy resolution of an x-ray detector or spectrometer is usually expressed as the quotient of the half-width of an *individual* pulse-height distribution by the *average* pulse height in that distribution V_{av}, that is, $W_{1/2}/V_{av}$, usually as per cent. Again, both terms are in volts. It is evident that by this definition, the higher the resolution, the lower is the number expressing it.

13.2.1.2. *Dispersion*

Dispersion is the separation of the two lines in 2θ or of their pulse-height distributions in average pulse height, that is, $\Delta 2\theta/\Delta\lambda$ (degrees 2θ per angstrom) or $\Delta V/\Delta E$ (volts per electron volt). The dispersion of an analyzer crystal is given by differentiation of the Bragg law,

$$\frac{d\theta}{d\lambda} = \frac{n}{2d\cos\theta} \tag{13.1}$$

Crystal dispersion is greater, the smaller is the d-spacing of the crystal. As the d-spacing of the crystal decreases, $d\theta/d\lambda$ must decrease, because $\cos\theta$ is always <1.

Energy dispersion is more complex, and is discussed in Section 8.1. Let it suffice here to say that energy dispersion is increased by use of a proportional, rather than a scintillation, counter, and by low detector potential and amplifier gain.

13.2.1.3. *Divergence*

Divergence is a measure of the intensity distribution of the diffracted peak in 2θ (wavelength divergence), or of its pulse-height distribution in baseline potential (energy divergence). It is expressed as the half-width $W_{1/2}$ or breadth B of the diffracted peak in degrees 2θ, or of the pulse-height distribution in volts, measured midway between the peak maximum and the background. The choice of symbol, $W_{1/2}$ or B, is largely a matter of personal choice. Still another term having the same significance is *full width at half-maximum*, or FWHM.

The intensity distribution of the x-ray beam diffracted by the analyzer crystal is known as its rocking curve. The distribution is Gaussian and has divergence B_D, as shown in Figure 13.4A. The divergence B_D is smaller, the more nearly perfect is the mosaic structure of the crystal. The intensity distribution of the x-ray beam transmitted by a single collimator C_1 is triangular, as shown in Figure 13.4B. If the collimator has length L and foil spacing S, the divergence is given by

$$B_{C_1} - \tan^{-1}(S/L) \qquad (13.2)$$

When a crystal is used with a single collimator, the resultant intensity

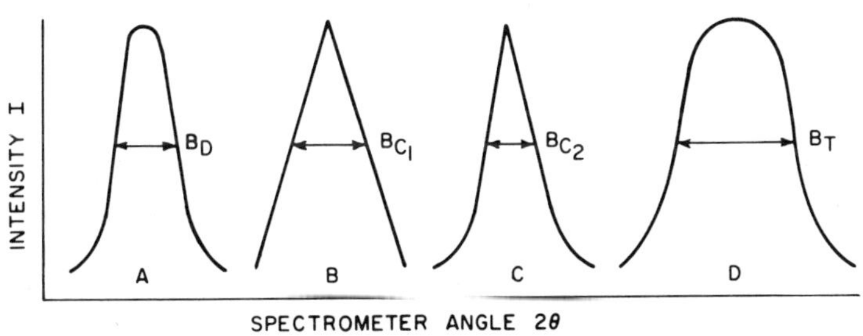

Fig. 13.4. Intensity distribution and divergence of diffracted and collimated beams. [L. S. Birks, "X-Ray Spectrochemical Analysis," Interscience Publishers, Inc., p. 22 (1959); courtesy of the author and publisher.] A. Crystal. B. One collimator. C. Two collimators having the same divergence. D. Resultant distribution of A and B.

distribution, shown in Figure 13.4D, is obtained from the convolution of the crystal rocking curve (Figure 13.4A) and the intensity distribution of the collimator (Figure 13.4B). The total divergence B_T is calculated by assuming the collimator curve to approximate a Gaussian distribution and applying the rule for combining variance [Equation (11.8)],

$$B_T{}^2 = (B_D{}^2 + B_{C_1}^2)^{1/2} \tag{13.3}$$

If a second collimator having the same divergence as C_1 is added, the intensity distribution for the two collimators, shown in Figure 13.4C, is obtained by squaring each ordinate of the curve for a single collimator (Figure 13.4B). The divergence of the squared curve $B_{C_2} \approx 0.6B_{C_1}$. In general, for two collimators having foil spacings S_1 and S_2, respectively, and distance L' from the specimen end of the source collimator to the detector end of the detector collimator, the divergence is (*S34*)

$$B_{C_2} = (S_1 + S_2)/L' \tag{13.4}$$

In practice, the crystal rocking curve has little effect on the total divergence, even with relatively imperfect crystals, and the divergence of the crystal–collimator system is substantially that of the collimators [Equations (13.2) and (13.4)].

13.2.2. Factors Affecting Resolution

Figure 13.5 shows two closely spaced peaks having divergence B and separation $\Delta 2\theta$. Resolution may be regarded as adequate when $\Delta 2\theta = 2B$, although two peaks may often be distinguished at much smaller separation, as shown in Figure 13.5. Combining Equation (13.1) and the Bragg law, one gets

$$\frac{\lambda}{\Delta\lambda} = \frac{2\tan\theta}{\Delta 2\theta} \tag{13.5}$$

Substitution of the minimum acceptable value of $\Delta 2\theta$—$2B$—gives

$$\frac{\lambda}{\Delta\lambda} = \frac{\tan\theta}{B} \tag{13.6}$$

The left side of Equation (13.6) may be regarded as the required resolution, the right side as the attainable resolution, which involves θ and B. The equation shows that resolution increases rapidly with θ (2θ) because of the tangent function. From the Bragg law $n\lambda = 2d\sin\theta$ [Equation (2.27)],

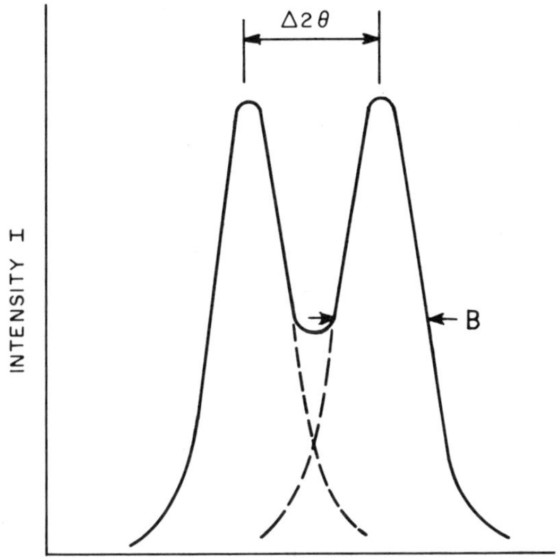

Fig. 13.5. Resolution of adjacent peaks.

it is evident that the smaller the *d*-spacing of the crystal, the larger is θ, and, thereby, the higher is the resolution. Also, for a specified crystal, successively higher orders occur at successively higher 2θ angles, and thus have successively higher resolution. The second-order spectrum of an element has approximately twice the resolution of the first-order spectrum.

For a flat crystal, in principle, the divergence depends on collimation and the degree of crystal perfection. However, as already mentioned, in practice, the divergence is determined principally by the collimation.

For a transmission curved crystal, divergence depends on the precision of the focusing and can be made very small. For a reflecting curved crystal, divergence is subject to many influences, and only the Johansson crystal has even theoretically perfect focusing; these influencing parameters include the following: (1) width of the source slit; (2) accuracy of curving and grinding; (3) mechanical tolerances in the instrument movements (these are especially serious when source-to-crystal and crystal-to-detector distances are not constant); (4) aging of the cement holding the crystal to its curved support block; (5) accentuation of the crystal mosaic structure due to the curving and grinding; and (6) penetration of the crystal by x-rays of short wavelength, leading to diffraction by planes lying off the focusing circle.

The divergence of a transmission curved crystal is smaller than that of a reflecting curved crystal, which, in turn, is substantially larger than that of a flat crystal.

When the pulse-height analyzer is used, it is the breadth and separation of the pulse-height distributions of the two lines that must be considered. Figure 13.5 could also illustrate two pulse-height distributions on intensity *vs.* pulse height axes. Birks [p. 47 in (*6*)] has pointed out an interesting way to improve energy resolution in cases where the escape peaks are intense. For neighboring elements, the main pulse-height distributions overlap. The escape peaks are separated by the same number of volts (energy dispersion), but their breadths are smaller because they result from smaller numbers of initial ionizations (Section 6.2.3.2, Figure 6.4). Thus, their resolution is better. However, it is still true that usually escape peaks only interfere with the measurements, and they should be avoided whenever possible.

In general, the resolution of an analyte line from an adjacent, potentially interfering line will be greater under the following conditions:

1. A second- or higher-order analyte line is measured.
2. The intensity of the interfering line is small.
3. Long, close-spaced collimators are used. The longer and the closer the foil spacing of the collimator, the more nearly parallel are the emergent x-rays. However, the measured intensity is proportional to the open cross-sectional area and decreases as foil spacing decreases.
4. A high-resolution crystal is used, that is, one having a small *d*-spacing and a narrow rocking angle (high mosaic perfection).
5. The crystal diffracts the analyte line at a high 2θ angle. The separation of lines of neighboring elements increases with increasing 2θ angle for a specified crystal.
6. A proportional counter is used, rather than a scintillation counter. The pulse-height distribution is narrower, and there is a greater difference in average pulse height for a given difference in photon energy.
7. Low detector-tube potential and amplifier gain are used to reduce the width of the pulse-height distributions.
8. The escape peak can be measured instead of the main peak (see above).

It is evident that the conditions favorable to high resolution are unfavorable to high intensity. In practice, only enough resolution is required to give sufficient separation of adjacent lines to permit measurement of their line and background intensities. It is rarely necessary for the spectrometer to have resolution greater than that required to partially resolve the

TABLE 13.5

Effect of Collimation[a]

Soller foil spacing,[b] S, inch		Equatorial divergence[c] α, deg	Peak half-width $W_{1/2}$ deg 2θ	Relative peak intensity I_P	Relative peak/bkgnd. intensity ratio I_P/I_B
Source collimator	Detector collimator				
0.005	—	0.14	0.24	53	8
0.010	—	0.29	0.42	91	5
0.020	—	0.58	0.74	100	3
0.005	0.005	0.04	0.15	27	100
0.005	0.010	0.06	0.20	37	45
0.005	0.020	0.11	0.24	40	35
0.010	0.010	0.09	0.28	57	35

[a] From Spielberg *et al.* (*S34*). All measurements made with a quartz $(10\bar{1}1)$ crystal and Cu $K\alpha$ x-rays.
[b] Length L of all collimators is 4 inches, foil thickness 0.002 inch.
[c] For one collimator, $\alpha = 2S/L$. For two collimators, $\alpha = (S_1 + S_2)/L'$; L' is 13.4 inches, the distance from the specimen end of the source collimator to the detector end of the detector collimator.

$K\alpha$ doublet. The degree of resolution required can often be established from a preliminary ratemeter–recorder 2θ scan.

Table 13.5 gives typical performance data for several collimator dimensions and arrangements.

13.3. SPECTRAL-LINE INTERFERENCE (*B27*)

13.3.1. Definition

It has already been pointed out that due to the relative simplicity of x-ray spectra, spectral-line interference is relatively infrequent, and that when it does occur, there are many ways to deal with it. Two types of x-ray spectral interference occur: (1) *wavelength* (2θ) *interference*, in which the interfering line overlaps the 2θ interval occupied by the analyte line; the interfering and analyte lines may have the same or nearly the same wavelength (λ) or product of order and wavelength ($n\lambda$); and (2) *energy interference*, in which the interfering pulse-height distribution overlaps the analyte pulse-height distribution; energy interference, as such, is observed only when pulse-height analysis is used. In either type of interference, the overlap may have any degree from slight to complete superposition.

Spectral interference, then, may be defined as that condition whereby (1) photons of a line other than the measured line enter the detector at the 2θ angle of the measured line, and/or (2) pulses caused by a line other than the measured line enter the window of the pulse-height analyzer at the settings for the measured pulse-height distribution. The measured line is usually the analyte line, but it may also be an internal-standard or internal intensity reference standard line, or a scattered target line to be used for standardization.

13.3.2. Origin of Interfering Spectral Lines

13.3.2.1. *Wavelength Interference*

The x-ray spectrum (2θ scan) recorded without the use of the pulse-height analyzer contains "lines" of various origins as listed below. Basically, the x-ray tube emits the primary continuous and line spectra, and any object in the x-ray path, including the path medium itself, can then act as a scatterer or secondary emitter.

1. Spectral lines originating in the x-ray tube and scattered by the specimen (*L1*): (a) Primary spectrum of the target element. (b) Primary spectra of impurities in the target. (c) Primary spectra of contaminants sublimed, sputtered, or otherwise deposited on the target during operation of the tube, principally the following: tungsten from the filament; copper from the anode block and from plated components; chromium, iron, and nickel from the stainless steel focusing electrode and filament-to-anode tunnel; and silver from silver solder. (d) Primary and secondary spectra of internal structures other than the target excited by scattered or improperly focused electrons or by the primary x-rays. (e) Primary and secondary spectra of contaminants on the tube window excited by the electrons that bombard the window and/or by the primary x-rays.

These x-ray tube lines may be scattered coherently or incoherently. The intensity of the scattered spectrum and the ratio of incoherent to coherent scatter become greater as the effective atomic number of the specimen decreases. The intensity of the spectra of the contaminants increases with the age of the tube and with the power at which it is usually operated. The tubes of each manufacturer have a more or less characteristic extraneous-line spectrum. Modern x-ray spectrometer tubes are designed specifically for this application and are much "cleaner" than their early predecessors. The scattered spectrum is identified by recording a 2θ scan of the x-rays scattered from a piece of paraffin or spectrographically pure graphite, or from a cell of pure water in the specimen compartment. The incoherent peaks are recognized as somewhat broader peaks on the high-2θ (long-wavelength) side of each coherently scattered peak. If reduction of total primary intensity can be afforded, any or all lines originating in the x-ray tube can be eliminated by filters placed at the tube window.

2. Secondary spectra of the specimen mask and associated structures.

3. Secondary spectra of trace impurities in the specimen support (Mylar, filter paper, etc.) or cell.

The spectra of the specimen mask and support are identified by comparison of the recorded 2θ scans of the unloaded mask and the mask loaded with the support or cell containing no specimen material.

4. Secondary spectra of the specimen itself, including feeble contributions of primary spectra excited by photo, recoil, and Auger electrons: (a) Diagram lines; of course, among these lines are the analyte lines and the lines of the internal standard or internal intensity reference standard, if any. (b) Nondiagram (satellite) lines; these lines result from electron transitions in doubly charged (*KL*, *LL*, *LM*, etc.) atoms. (c) Forbidden lines; these lines result from electron transitions forbidden by the selection rules; for example, in Figure 10.3, the second and third lines from the bottom (*M3*—*M5* and *M2*—*M4*), if correctly identified, are forbidden because $\Delta n = 0$ (Section 1.6.2.3).

K, *L*, *M*, or *N* spectra may be excited and recorded, depending on the x-ray tube potential, atomic numbers of the specimen elements, and spectral region recorded. The lines may appear at their "table" values, or they may be displaced as much as 0.25° 2θ by chemical effects (Section 11.1.4), or as much as 0.5° 2θ by positional effects (Section 19.2.1). Misalignment of the goniometer also results in displacement of line position.

5. Diffraction lines from highly crystalline specimens (*B59*, *E3*): Single-crystalline specimens may give Laue diffraction, and highly oriented polycrystalline specimens may give Debye–Scherrer diffraction of discrete primary wavelengths that meet the Bragg requirements for their *d*-spacings and inclination to the primary beam. The diffracted beams usually appear as series of close-spaced, extremely sharp peaks ("spiking"). They are further identified by comparison of spectra from the same specimen in various orientations about its own plane normal, or fixed and rotated about that normal, or in solid and powdered form. If the peaks are emission lines, all these spectra are substantially the same, but diffraction lines are highly

sensitive to orientation, and Laue peaks disappear on powdering the specimen. The phenomenon is eliminated by pulse-height selection because the spikes are diffracted primary wavelengths almost certainly different from the analyte-line wavelength. The phenomenon is minimized or eliminated by continuous rotation of the specimen.

6. Additional lines introduced by the dispersion system. (a) Higher-order diffracted lines. (b) Spurious diffracted lines (see below). (c) Split lines, that is, individual peaks split into two or more peaks because of interaction of long, close-spaced source and detector collimators. (d) Line displacement by thermal expansion of the crystal (Section 5.4.2.5).

7. Additional lines introduced by the relatively large angle of acceptance of the detector. For example, second-order Ca $K\beta$ (2λ 6.18 Å, 2θ 89.08° for EDDT) may interfere with P $K\alpha$ (λ 6.15 Å, 2θ 88.66°) when a short, wide-spaced detector collimator is used.

Normally, the detector receives only radiation diffracted from the planes parallel to the reflecting surface of the analyzer crystal. However, there may be a second set of planes only slightly inclined to the surface planes and arrayed longitudinally (as in topaz) or crosswise (as in EDDT). This condition is shown in Figure 13.6. Under certain circumstances, the detector set to receive the analyte peak from the surface planes may also receive x-rays of a different wavelength from these other planes. This phenomenon is unlikely with crystals of high symmetry, such as cubic crystals, but may occur with crystals of low symmetry, such as topaz (orthorhombic) and EDDT (monoclinic), especially when used with a short, wide-spaced detector collimator and a large-area detector window (*S29*, *S32*).

The spurious peaks are recognized by their greater width. They are further identified by use of a tandem detector, especially one having a

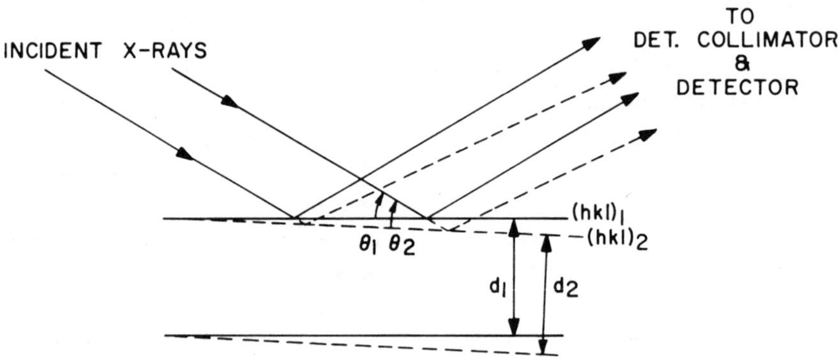

Fig. 13.6. Origin of spurious reflections from the analyzer crystal; $(hkl)_2$ is a second set of planes slightly inclined to $(hkl)_1$, on which the crystal is cut.

collimator between the front and back detectors. The spurious peaks are reduced in intensity or eliminated when measured from the back detector alone because of the increased distance and collimation. The spurious peaks may also be minimized or better resolved from the analyte line by increased detector collimation, by turning the crystal end for end or top for bottom, or by pulse-height selection if the spurious and analyte-line wavelengths are substantially different.

13.3.2.2. *Energy Interference*

If the pulse-height analyzer is not used, photons of any of the seven groups of lines listed above that enter the detector at the 2θ setting for the measured line constitute wavelength spectral interference. If the pulse-height analyzer is used, any such photons give rise to pulse-height distributions having average pulse height proportional to the photon energy. These distributions do not necessarily overlap the window set for the measured distribution. When they do not, the pulse-height analyzer eliminates a source of wavelength spectral interference, but when they do, spectral energy interference occurs. In addition, the pulse-height analyzer may introduce forms of energy spectral interference that have no analog in wavelength interference. The two principal forms are crystal emission and escape peaks.

a. Crystal Emission. The incident secondary x-rays excite the elements in the analyzer crystal to emit their characteristic spectra. If the crystal is purely organic (EDDT, PET, etc.), these spectra are of little significance in conventional x-ray spectrometry. If the crystal contains a heavier element, such as aluminum (topaz), silicon (quartz, silicon), phosphorus (ADP), sulfur (gypsum), chlorine (NaCl), or potassium (KHP), the spectra still do not interfere if high-Z elements are being determined in air path, or if low-Z elements are being determined in helium or vacuum *without* pulse-height analysis. In the former case, the air absorbs the crystal emission before it reaches the detector. In the latter case, since the crystal emission radiates in all directions, it merely contributes to the continuous background. However, if the pulse-height analyzer is used, that portion of the crystal emission that does enter the detector gives rise to a pulse-height distribution that may interfere with the measured distribution.

b. Escape Peaks. When the pulse-height analyzer is used, one must consider the overlap of the pulse-height distributions not only of the measured line and interfering line, but also of the measured line and the escape peak of the interfering line. In general, a pulse-height analyzer set to pass pulses from x-rays of energy E also passes pulses from x-rays of energy $E' \approx E + E_{\text{det}}$, where E_{det} is the energy of the principal spectral line of the detector gas or of the iodine in the scintillator. The reason is that the escape peak of x-rays of energy E' has the same average pulse height as the main peak of the measured line of energy E, that is, $E' - E_{\text{det}} \approx E$. For example, if a pulse-height analyzer is used with a krypton-filled proportional counter (Kr $K\alpha$: λ 0.98 Å, E 12.6 keV) to pass Cu $K\alpha$ (λ 1.54 Å, E 8.0 keV), it will also pass Rh $K\alpha$ (λ 0.62 Å, E 20.2 keV): $20.2 - 12.6 = 7.6$ keV. Partial overlap will occur for energies substantially higher and lower than 20 keV.

A particularly interesting example of this phenomenon occurs when a flow-proportional counter with P10 gas (90% argon, 10% methane) is used to measure Ca $K\alpha$ from specimens containing tungsten (*C14*). Second-order W Ll (2λ 3.356 Å, E 7.39 keV) gives an almost exact wavelength interference with Ca $K\alpha$ (λ 3.360 Å, E 3.69 keV). The pulse-height analyzer easily rejects the main W Ll pulse distribution, but with the argon detector gas (Ar $K\alpha$: λ 4.193 Å, E 2.96 keV), the W Ll escape peak ($7.39 - 2.96 = 4.43$ keV) partially overlaps the Ca $K\alpha$ peak.

13.3.2.3. *Common Sources of Spectral Interference*

Of course, among the lines listed above are the lines to be measured, that is, the analyte line and perhaps a line for use as an internal standard or an internal intensity reference standard. The reference line may come from the specimen (internal standard or internal intensity reference standard), the specimen mask (external intensity reference standard), or the x-ray tube (scattered target line). All other lines present in the x-rays reaching the detector are potential sources of interference.

Obviously, the higher the intensity of the measured line, and the lower the intensity and greater the separation of the interfering line, the less serious is the interference. Orders higher than the fifth, satellite lines, and M and N lines are usually troublesome in x-ray secondary-emission spectrometry only for weak measured lines, although they may be more troublesome in the electron-probe microanalyzer (Chapter 21). In general, the great majority of instances of spectral-line interference may be classified in one of the following four categories, regardless of the source of the interfering line:

1. Superposition of first-order lines of the same series (K, L, M) from adjacent or neighboring elements in the periodic table. Some examples

follow. The $K\alpha$ lines of adjacent elements above technetium (Z 43) lie within 0.03 Å of one another, as do the $L\alpha_1$ lines of adjacent elements above mercury (Z 80). For titanium through cobalt (Z 22–27), the Z $K\alpha$ line lies within 0.03 Å of the ($Z - 1$) $K\beta$ line, and for rhodium through indium (Z 45-49), the Z $L\alpha_1$ line lies within 0.03 Å of the ($Z - 1$) $L\beta_1$ line.

2. Superposition of first-order lines of different series; for example, As $K\alpha$ (λ 1.177 Å) and Pb $L\alpha_1$ (λ 1.175 Å).

3. Superposition of first-order K lines and higher-order lines from elements of higher atomic number; for example, Ni $K\alpha$ (λ 1.659 Å) and Y $K\alpha$ (2λ 1.660 Å), P $K\alpha$ (λ 6.155 Å) and Cu $K\alpha$ (4λ 6.168 Å), Al $K\alpha$ (λ 8.337 Å) and Cr $K\beta$ (4λ 8.340 Å) or Ba $L\alpha_1$ (3λ 8.325 Å). This type of interference is particularly troublesome in the determination of low-Z elements in multicomponent samples having many medium-Z and high-Z elements. Higher orders of scattered target lines also cause this type of interference with low-Z elements.

4. Superposition of first-order L lines and higher-order K lines from elements of lower atomic number. The classic examples are Hf $L\alpha_1$ (λ 1.570 Å) and Zr $K\alpha$ (2λ 1.574 Å), and Ta $L\alpha_1$ (λ 1.522 Å) and Nb $K\alpha$ (2λ 1.494 Å). Similar interferences also exist between $L\beta_1$ and second-order $K\beta$ lines of these elements.

13.3.3. Reduction of Spectral Interference

13.3.3.1. *General Methods*

Methods for elimination of certain specific types of spectral interference have already been cited in Section 13.3.2. The methods for dealing with spectral interference may be classified in nine categories as follows:

1. Selection of an alternative analyte line that is free from interference.
2. Removal or reduction of the concentration of the interfering element (Chapters 16 and 17).
3. Prevention or reduction of the efficiency of excitation of the interfering element or increase of the efficiency of excitation of the analyte.
4. Increase of the resolution of the spectrometer (Section 13.2).
5. Prevention or reduction of the transmission of the interfering line through the spectrometer, or increase of the transmission of the analyte line.
6. Prevention or reduction of the efficiency of detection of the interfering line, or increase of the efficiency of detection of the analyte line.
7. Prevention of the counting of detector pulses originating from the interfering line.

8. Correction for the interfering line by experimental technique.
9. Correction for the interfering line by mathematical means.

The practicability of any approach to the reduction of spectral interference is determined by the specific features of the particular case. The relative intensities of the analyte and interfering lines, the degree of their overlap, and the difference in their true wavelengths (λ, not $n\lambda$) determine the seriousness of the interference. The intensity of the analyte line determines how much intensity may be sacrificed to reduce spectral interference. If several elements are to be determined in a sample and only one or two are subject to interference, the "economy" of the overall analysis must be considered. The optimal instrument arrangement to reduce the interference may well be unfavorable for the other analyses.

In the first method listed above, the alternative spectral line may be in the same series ($K\beta$ instead of $K\alpha$) or in a different series ($L\alpha_1$ instead of $K\alpha$), or it may be a higher order of the same line. Of course, if it is the optimal line that is interfered with, the alternative line will necessarily have lower intensity or line-to-background ratio. The seventh method, the prevention of counting of output pulses originating from the interfering line, is the function of the pulse-height analyzer, which is discussed in detail in Chapter 8. It must be emphasized that the pulse-height analyzer is useless for discrimination against lines of the same or nearly the same wavelength. Nevertheless, resolution is improved by use of a proportional counter with low applied potential and low amplifier gain.

13.3.3.2. *Excitation of Analyte and Interferant Lines*

There are many ways to apply the third method given in the preceding section.

a. Choice of X-Ray Tube Target. The target may be selected to: (1) eliminate a scattered interfering target line; (2) excite the analyte line at maximum efficiency; or (3) excite the interfering line at minimum efficiency. The analyte line is excited most efficiently by a target having lines of wavelength just shorter than its absorption edge. The interfering line is excited least efficiently by a target having lines of wavelength just longer than its absorption edge. However, if the interferant is, say, chromium, a chromium target is a poor choice even though it excites chromium very inefficiently. The target-line radiation scattered by the specimen may more than compensate the reduced emission from the specimen, especially for low chromium concentration in a low-Z matrix.

b. Choice of X-Ray Tube Potential. The potential may be selected to: (1) prevent excitation of an interfering line (selective excitation); (2) excite the analyte line at maximum efficiency; or (3) excite the interfering line at minimum efficiency. Selective excitation is feasible if the excitation potential of the interfering line is much higher than that of the analyte line, *and* is high enough so that if the x-ray tube is operated just below this potential, the primary-beam intensity is still high enough for efficient analyte-line excitation. For example, if third-order In $K\alpha$ (3λ 1.542 Å) interferes with Cu $K\alpha$ (λ 1.542 Å), the x-ray tube can be operated below the indium K-excitation potential (28.0 kV); the x-ray tube potential is still well above the copper K-excitation potential (9.0 kV) and high enough to give high primary continuum intensity. However, the interference of fourth-order Cu $K\alpha$ with P $K\alpha$ could not be dealt with this way, because operation of the x-ray tube below 9 kV would prohibitively reduce P $K\alpha$ intensity. Of course, both these examples are largely academic, because both are best dealt with by pulse-height analysis.

If the analyte and interferant lines have high and low excitation potentials, respectively, analyte-line intensity is increased more than interferant-line intensity by increase of operating potential and by use of constant potential. If an L line of a high-Z element is subject to interference, use of a 75- or 100-kV generator may permit use of a K line.

c. Choice of X-Ray Tube Current. If the overlap of analyte and interfering lines is only partial and if some analyte-line intensity can be sacrificed, the overlap can be reduced, perhaps to insignificance, by reduction of the x-ray tube current.

d. Monochromatic Excitation. The primary beam may be used to irradiate a secondary target having strong spectral lines on the short-wavelength side of the analyte absorption edge and the resulting spectral emission used to excite the analyte line. This method may be useful when intensity can be sacrificed for a "clean" spectrum. The analyte-line intensity would be very low, but nondispersive measurement would probably be feasible.

e. Specimen Mask, Support, and Matrix. If the interfering line arises from one of these sources, judicious choice of materials is indicated (Chapters 16 and 17).

f. Shielding. The intensity of interfering lines may be reduced by appropriate use of apertures, shields, and baffles in the specimen compartment to confine the primary beam to the specimen area and to reduce scatter. With specimens of small area, selected-area techniques may be used (Chapter 19), or a collimated specimen mask having a continuous L-shaped tunnel to convey the primary beam to the specimen and the secondary beam to the optical path.

13.3.3.3. *Transmission and Detection of Analyte and Interferant Lines*

The fifth and sixth methods given in Section 13.3.3.1 may be applied as follows.

a. Filters. Filters are useful for reduction of spectral interference in three cases: (1) If the interfering line is a scattered target line, a filter at the x-ray tube window may reduce it to insignificance—at the expense of primary intensity; for example, a filter consisting of 0.001-inch foils of brass and nickel at the window of a tungsten-target tube effectively removes the tungsten L lines (*S4*). (2) If the superposed lines have very different wavelengths—as is the case for superposition of λ and $n(\lambda/n)$—a filter having an absorption edge of wavelength just longer than either line may be used, but the technique is likely to be more effective when the interfering line is the line of longer wavelength. (3) Finally, if the superposed lines have nearly the same wavelength, a filter having an absorption edge between them absorbs the line of shorter wavelength.

b. Shielding. If the analyte-line intensity is very low, air scatter of target lines may become significant and may be reduced by appropriate use of apertures, shields, baffles, and tunnels in the specimen compartment.

c. Radiation Path. The path may be used to deal with spectral interference in two ways. Use of helium or vacuum in the specimen chamber reduces target-line scatter. The other way is best explained with an example: If phosphorus is to be determined in the presence of copper, fourth-order Cu $K\alpha$ (4λ 6.16 Å) interferes with P $K\alpha$ (λ 6.16 Å). In the absence of a pulse-height analyzer, the intensity at the 2θ angle for P $K\alpha$ may be measured first in helium or vacuum path, then in air. The contribution of Cu $K\alpha$ is substantially the same in both paths, but P $K\alpha$ is completely absorbed in air. Thus, the difference between the two measured intensities is P $K\alpha$ intensity.

d. Crystal. The crystal may be selected to have high extinction for an interfering higher-order line (*L27*).

e. Detector. By use of Figure 6.12, a detector can be chosen having high quantum efficiency for the analyte line or low quantum efficiency for the interfering line. The use of gas-filled detectors in this way is known as *gas discrimination*, and is particularly effective in discriminating against high-order lines of high-Z elements when the long-wavelength lines of low-Z elements are being measured. For example, neon is preferable to argon for discrimination against wavelengths shorter than the argon K-absorption edge (3.9 Å).

13.3.3.4. *Experimental and Mathematical Correction of Spectral Interference*

Frequently, the net intensity of the interfering line can be measured from a blank at the 2θ angle of the analyte line, then subtracted from the net intensities of the analyte line measured from the specimens. Blanks are discussed in Section 7.3.3.

Another technique is to establish a ratio between the intensity of the interfering line at the 2θ angle for the analyte line and the intensity of another line of the interfering element that can be measured on the specimens without interference. The ratio is established from measurements on the pure interfering element or its oxide. Then, for each specimen, the interfering-line intensity can be calculated from the measured intensity of the second interferant line. The method has been used to correct U $L\alpha_1$ net intensity (λ 0.911 Å) for spectral interference by second-order In $K\beta$ (2λ 0.910 Å) (*B44*):

$$I_{\mathrm{net,U}L\alpha_1} = I_{\mathrm{peak,U}L\alpha_1} - I_B - \left(\frac{I_{\mathrm{net,In}K\beta(2)}}{I_{\mathrm{net,In}K\alpha}}\right) I_{\mathrm{net,In}K\alpha} \tag{13.7}$$

where I_B is adjacent continuous scattered background. The net intensity ratio of In $K\beta(2)$/In $K\alpha$, measured from pure indium, is 0.0451.

Birks and Brooks (*B33*) devised a correction technique for the determination of niobium and tantalum in their binary mixtures. Niobium was determined from the Nb $K\alpha$ line (λ 0.747 Å), which is free from interference. Tantalum may be determined from the Ta $L\alpha_1$ line (λ 1.522 Å), which is subject to interference from second-order Nb $K\alpha$ (2λ 1.494 Å), or from the Ta $L\beta_1$ line (λ 1.327 Å), which is subject to interference from second-order Nb $K\beta$ (2λ 1.332 Å). The integrated intensities of the Nb $K\alpha$ and unresolved Ta $L\beta_1$–Nb $K\beta(2)$ peaks were measured from each of several standards consisting of the mixed oxides in various proportions. A curve was plotted of the integrated intensity ratio of the Nb $K\alpha$ and Ta $L\beta_1$–Nb $K\beta(2)$ peaks *vs.* tantalum concentration. Another curve was plotted of Nb $K\alpha$ integrated intensity *vs.* niobium concentration. For each sample, the integrated intensities of the two peaks were measured and applied to the curves. The same technique was applied to mixtures of zirconium and hafnium oxides.

Zemany (*Z5*) determined vanadium, chromium, and manganese in steel by a correction method. The V $K\alpha$ is free of interference in steels that contain no titanium; V $K\beta$ interferes with Cr $K\alpha$, and Cr $K\beta$ interferes with Mn $K\alpha$. The procedure follows:

1. From each of several standards, the integrated intensity was measured for each of the three peaks of interest: V $K\alpha$ (I_V), unresolved Cr $K\alpha$–V $K\beta$ (I_{Cr}), and unresolved Mn $K\alpha$–Cr $K\beta$ (I_{Mn}). The measurements were made by making 100-s scans of the peaks at 2° 2θ/min with the scaler operating.
2. The $K\beta/K\alpha$ intensity ratio was established for vanadium and for chromium by measurement of the integrated intensities of V $K\alpha$ and V $K\beta$ from pure vanadium oxide (V_2O_5), and of Cr $K\alpha$ and Cr $K\beta$ from pure potassium dichromate ($K_2Cr_2O_7$). The values were V $K\beta$/V $K\alpha = 0.278$ and Cr $K\beta$/Cr $K\alpha = 0.247$.
3. Three curves were plotted from the standard data: I_V *vs.* %V, I_{Cr} *vs.* %Cr, and I_{Mn} *vs.* %Mn.
4. The intercept of each curve with the intensity axis at 0% element was read and designated as the background: $I_{B,V}$, $I_{B,Cr}$, and $I_{B,Mn}$.
5. The integrated intensity corresponding to a concentration of 1% element was read from each curve: $I_{1\%V}$, $I_{1\%Cr}$, and $I_{1\%Mn}$.
6. For each sample, the three integrated intensities I_V, I_{Cr}, and I_{Mn} were measured in the same way as for the standards and used in the following equations to derive analytical concentrations, C_V, C_{Cr}, and C_{Mn}:

$$C_V = \frac{I_V - I_{B,V}}{I_{1\%V}} \tag{13.8}$$

$$C_{Cr} = \frac{I_{Cr} - I_{B,Cr} - (\text{V } K\beta/\text{V } K\alpha)(I_V - I_{B,V})}{I_{1\%Cr}} \tag{13.9}$$

$$C_{Mn} = \frac{I_{Mn} - I_{B,Mn} - (\text{Cr } K\beta/\text{Cr } K\alpha)(I_{Cr} - I_{B,Cr})}{I_{1\%Mn}} \tag{13.10}$$

The various techniques used by Birks and Brooks (*B33*) and Campbell and Carl (*C11*) in the determination of hafnium in the presence of zirconium and tantalum in the presence of niobium provide an example of how a number of techniques, none entirely satisfactory by itself, can be combined to reduce spectral interference. In these analyses, Zr $K\alpha(2)$ (2λ 1.58 Å) interferes with Hf $L\alpha_1$ (λ 1.57 Å), and Nb $K\alpha(2)$ (2λ 1.49 Å) interferes with Ta $L\alpha_1$ (λ 1.52 Å). Similarly, Zr $K\beta(2)$ and Nb $K\beta(2)$ interfere with Hf $L\beta_1$ and Ta $L\beta_1$, respectively. The following techniques were applied:

1. The hafnium and tantalum $L\beta_1$ lines were used instead of the $L\alpha_1$ lines. The $L\beta_1$ lines are almost as intense as the $L\alpha_1$ lines, but the $K\beta(2)$ lines that interfere with $L\beta_1$ are only ~1/5 as intense as the $K\alpha(2)$ lines that interfere with $L\alpha_1$.

2. A molybdenum target (Mo $K\alpha$ 0.71 Å) was used to excite the hafnium and tantalum L lines more efficiently (LIII absorption edges 1.30 and 1.25 Å), eliminate tungsten target-line interference, and reduce the continuous background.

3. The x-ray tube potential was decreased (in some experiments) for selective excitation of the hafnium and tantalum L lines (LIII excitation potentials 9.5 and 9.9 kV) without excitation of the zirconium or niobium K lines (excitation potentials 18 and 19 kV).

4. Alternatively, the x-ray tube potential was increased to permit excitation of the hafnium and tantalum K lines (excitation potentials 65 and 67 kV), which are free of interference from zirconium and niobium.

5. High collimation, 4×0.005 inch, was used before and after the crystal.

6. A quartz ($11\bar{2}2$) crystal ($2d$ 3.636 Å) was used to provide good dispersion and narrow peak width.

7. An argon-filled detector was used to provide high quantum efficiency for the hafnium and tantalum L lines.

8. Integrated intensities of the unresolved Hf $L\beta_1$–Zr $K\beta(2)$ and Ta $L\beta_1$–Nb $K\beta(2)$ peaks were corrected by comparison with the integrated intensities of the fully resolved Zr $K\beta$ or Nb $K\beta$ peaks (see above).

Chapter 14

Methods of Quantitative Analysis

14.1. INTRODUCTION

It has already been pointed out that were it not for the effect of the matrix, the analyte-line intensity $I_{A,M}$ from analyte A in a thick specimen having matrix M would be simply a function of weight fraction $W_{A,M}$ of A in M and the analyte-line intensity $I_{A,A}$ from pure analyte, that is,

$$I_{A,M} = W_{A,M} I_{A,A} \tag{14.1}$$

It has also been pointed out that the principal objective of the analytical "method" or "strategy" is to eliminate, minimize, circumvent, or correct for the effect of the matrix, that is, for the absorption-enhancement effects. However, this is not necessarily the only objective of the analytical method.

Attempts have been made at absolute x-ray spectrometric analysis by conversion of intensity data to analytical data by wholly mathematical means. The calculations require knowledge of: (1) the intensity–wavelength distribution of the primary x-ray beam; (2) conversion efficiency of each primary wavelength to the analyte line; (3) absorption coefficient of the specimen for each primary wavelength and for the analyte line; (4) conversion efficiency of each primary wavelength to each matrix element line that can excite the analyte line; (5) absorption coefficient of the specimen for each of these lines; (6) conversion efficiencies of each of these lines to the analyte line; and (7) efficiency of the spectrometer (collimators, crystal, path, detector) for transmission, diffraction, and detection of the analyte line. Obviously, the calculations are extremely complex, and absolute methods have had only limited success. Even this has been accomplished only with simple systems having relatively few components, and then only with simplifying assumptions.

Other approaches to the absorption-enhancement problem have been more successful. Most of these methods involve the use of calibration standards. X-ray spectrometric analysis is still largely a calibration method, and intensity data are converted to analytical concentration by use of calibration curves or mathematical relationships derived from measurements on standards.

The many x-ray spectrometric analytical methods may be classified in eight categories with respect to their basic approach to the reduction of absorption-enhancement effects.

1. *Standard addition and dilution methods.* The analyte concentration is altered quantitatively in the sample itself. In a certain sense, the sample, subjected to one or more quantitative incremental concentrations or dilutions of the analyte, actually provides its own standard(s) in its own matrix.

2. *Thin-film methods.* The specimens are made so thin that absorption-enhancement effects substantially disappear.

3. *Dilution methods.* The matrix of all specimens is leveled or diluted to a composition such that the effect of the matrix is determined by the diluent.

4. *Comparison-standard methods.* The analyte-line intensity from the samples is compared with that from standards having as nearly as possible the same analyte concentration and matrix.

5. *Internal standardization.* The comparison-standard method is improved by quantitative addition to all specimens of an internal-standard element having excitation, absorption, and enhancement characteristics similar to those of the analyte in the particular matrix. The calibration function involves the intensity ratio of the analyte and internal-standard lines.

6. *Standardization with scattered x-rays.* The intensity of the primary x-rays scattered by the specimen is used to correct the absorption-enhancement effects.

7. *Experimental correction.* Various special experimental techniques have been devised to minimize or compensate absorption-enhancement effects.

8. *Mathematical correction.* Absorption-enhancement effects are corrected mathematically by use of experimentally derived parameters.

Effectiveness of Analytical Methods in Compensating Errors

Analytical method	Sources of error[a]										
	Instrumental		Operational	Specimen							
	Short-term	Long-term		Absorption	Enhancement	Inhomogeneity	Position	Surface	Particle size	Density	Bubbles
Standard addition and dilution	—	C	—	C	C	—	C	—	—	—	—
Thin-film method	—	—	—	C	C	—	—	C	C	—	—
Dilution											
Dry (low absorber)	—	—	—	C	C	—	—	P	P	C	—
Fusion (low absorber)	—	—	—	C	C	C	—	C	C	C	—
Solution (low absorber)	—	—	—	C	C	C	—	C	C	C	—
High absorber (dry)	—	—	—	C	C	—	—	P	P	C	—
Calibration standardization	—	—	—	X	X	—	—	X	X	X	—
Internal standardization	S	C	P	C	C	—	P	P	P	P	P
Other standardization methods											
Internal control standardization	S	C	P	C	P	—	—	—	—	—	—
Internal intensity-reference standardization	S	C	P	C	P	—	C	—	—	C	C
External standardization	S	C	P	C	P	—	—	—	C	C	—
Scattered x-ray methods											
Background-ratio method	S	C	P	C	—	—	C	C	C	C	P
Scattered target-line ratio method	S	C	P	C	—	—	C	C	C	C	P
Reynolds–Ryland method	S	C	P	C	—	—	C	C	P	C	—
Briqueting	—	—	—	—	—	—	—	—	P	C	—
Mathematical methods (geometrical, empirical, influence coefficient methods)	—	—	—	C	C	—	—	—	—	—	—

[a] C: complete or substantial correction; P: partial correction; S: correction only if both the analyte-line and standard intensities are measured simultaneously; X: correction only if standards and samples are very similar with respect to the specified property.

Most of these methods also reduce error from other sources. Table 14.1 summarizes the effectiveness of each method in compensating the principal sources of error. Mathematical methods are discussed in Chapter 15. In this chapter, each of the other methods is discussed with respect to principles, variations, advantages and disadvantages, limitations, and general applicability. Chapters 16 and 17 discuss specific techniques of specimen preparation and presentation.

14.2. STANDARD ADDITION AND DILUTION METHODS

14.2.1. Principles and Considerations

If the calibration function—analyte-line intensity I_A *vs.* analyte concentration C_A—is linear in the region of analytical interest, I_A is proportional to C_A, and an incremental or decremental change in C_A gives a proportional change in I_A. If such a linear relationship is questionable, a second change and new measurements may be made to ascertain linearity. If the calibration curve is nonlinear, the slope of the curve is established by repetitive additions. The terms *doping* or *spiking* are sometimes used for addition, and *dopant* for additive.

Standard addition and dilution methods usually require no standards or calibration curves, so the methods are useful for samples analyzed too infrequently to warrant preparation of standards and calibration curves, and for samples for which there is inadequate knowledge about the matrix. The methods are applicable only to trace and minor analytes (less than 5–10%) where the I_A–C_A calibration function is likely to remain linear for suitable incremental additions or dilutions. In standard-addition methods, an addition must be made for each analyte appropriate for its estimated concentration. In standard-dilution methods, analytes having the same order of concentration may be determined from the same diluted sample. However, a portion of the sample must be treated at a different level for each analyte or group of analytes having a different order of concentration. The methods are applicable only to samples to which additions can be made—principally powders, fusion products, and solutions.

The addition of analyte increases the analyte concentration, but, in effect, also reduces the original concentration in proportion to the dilution factor. Thus, to simplify the calculations, the actual weight of substance added should be small compared with the total sample weight. In other words, the analyte concentration in the additive should be high, and the pure analyte itself is preferable whenever its use is feasible. Intimate homogeneous admixture of the additive is required. This condition is easily attained for solutions and melts. For powders, the particle size and density of the additive must be similar to those of the samples.

The methods are most accurate when used with monochromatic primary x-rays, or with polychromatic x-rays and analytes having absorption edges at wavelengths just longer than those of the target lines.

As already mentioned, standard addition and dilution methods may be regarded as self-calibration or self-standardization methods in that the incrementally concentrated or diluted sample actually constitutes its own standard. Jenkins and de Vries regard standard-addition methods as internal standardization with the same (analyte) element.

14.2.2. Methods

14.2.2.1. *Standard Addition*

Essentially, a weighed portion of the sample X is treated with analyte to increase the analyte concentration by ΔX, say 1% absolute, from C_X to $C_{X+\Delta X}$. The analyte-line net intensity is then measured from the untreated I_X and treated $I_{X+\Delta X}$ samples. Then,

$$\frac{I_X}{I_{X+\Delta X}} = \frac{C_X}{C_{X+\Delta X}} \tag{14.2}$$

In practice, a weighed portion of the sample X is treated with a standard material S having known analyte concentration C_S to form a mixture XS containing concentration $C_{(X)}$ of *sample*. In the *standard-addition method* (*C9*, *C12*), analyte-line intensity is measured from the untreated sample I_X and mixture I_{XS}. Analyte concentration in the original sample is given by

$$C_X = C_S \frac{I_X/I_{XS}}{1 - [(I_X/I_{XS})C_{(X)}]} \tag{14.3}$$

In the *active-dilution method* (*L4*), analyte-line intensity is measured from the untreated sample I_X, mixture I_{XS}, and standard material (see above) I_S. Analyte concentration in the original sample is given by

$$C_X = C_S\left(\frac{1 - C_{(X)}}{C_{(X)}}\right)\left(\frac{I_{XS} - I_S}{I_{XS} - I_X}\right) \tag{14.4}$$

If the standard material and sample are mixed in ratio 1:1, $C_{(X)}$ is 0.5, the first parenthetic term in Equation (14.4) becomes unity, and the equation reduces to

$$C_X = C_S \frac{I_{XS} - I_S}{I_{XS} - I_X} \tag{14.5}$$

The standard material may have a matrix different from that of the sample and may be used for the same analyte in different kinds of samples. It may be a stoichiometric oxide or other simple compound of the analyte, or it may be chosen to have relatively high absorption for the analyte line. (See the discussion of high-absorption dilution in Section 14.6.1).

14.2.2.2. *Standard Dilution* (*W2*)

Weighed portions of the sample X and of a standard material S having known analyte concentration C_S of the same order as that of the sample are separately mixed in the same known proportion with an inert diluent containing no analyte(s). Analyte-line intensity is measured from the untreated sample I_X and standard I_S and from the diluted sample $I_{X'}$ and standard $I_{S'}$. Analyte concentration in the untreated sample is given by

$$C_X = C_S \frac{I_S - I_{S'}}{I_X - I_{X'}} \tag{14.6}$$

14.2.2.3. *Multiple Standard Addition* (*R10*)

The weighed sample is dissolved, treated as required, and adjusted in volume to an analyte concentration of, say, ~50 μg/ml. To five 1-ml portions of this solution, aliquots of a pure standard analyte solution are added containing increasing amounts of analyte—20, 40, 60, 80, 100 μg—bracketing the amount expected to be present in each portion of sample solution. Each of the five treated solutions, a sixth untreated 1-ml portion of sample solution, and a reagent blank are individually absorbed in 500-mg portions of powdered chromatographic paper. The specimens are then mixed, dried, ground, and briqueted against a backing of pure chromatographic paper to form a disk having 1-inch diameter. Analyte-line intensity is measured from all disks, corrected for background measured from the blank, and plotted as a function of analyte mass, as shown in Figure 14.1. The curve is extrapolated to the analyte-mass axis to derive analyte mass in the 1-ml portions of sample solution. The two curves in the figure are for 50-mg and 100-mg sample weights and give analytical results of 33.5 and 66.5 μg (per 1-ml portion of sample solution), respectively.

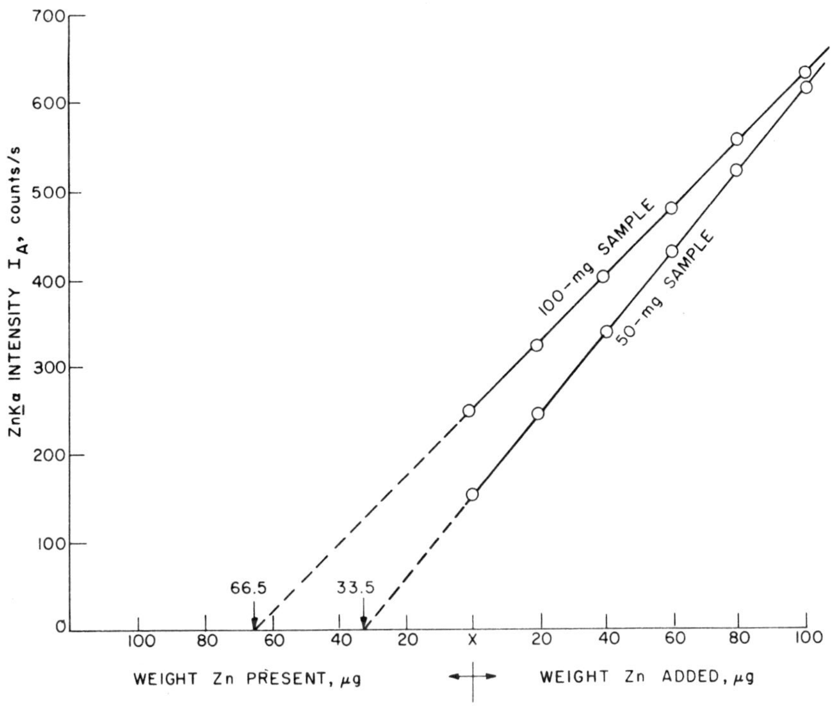

Fig. 14.1. Multiple standard-addition method with graphic solution. [H. J. Rose, Jr., and F. Cuttitta, *Advances in X-Ray Analysis* **11**, 23 (1968); courtesy of the authors, University of Denver, and Plenum Press.]

14.2.2.4. *Slope-Ratio Addition* (*R10*)

This method is a refinement of the multiple standard-addition method. It is used when the variation in density and composition of the samples is such that on addition of 1 ml of sample solution to 500 mg of chromatographic paper, the sample matrix would still dominate and override the dilution effect of the paper. The sample solution volume is adjusted to an analyte concentration of, say, 500 μg/ml. To five 1-ml portions of this sample solution, aliquots of a pure standard analyte solution are added containing increasing amounts of analyte—200, 400, 600, 800, 1000 μg—bracketing the amount expected to be present in each portion of sample solution. A sixth 1-ml portion of sample solution is left untreated. So far, the procedure is the same as that for the multiple standard-addition method. Now five more identical aliquots of pure analyte solution are added to 1-ml portions of water. Each of these 11 solutions and a reagent blank are taken up in chromatographic paper and briqueted as before. Analyte-

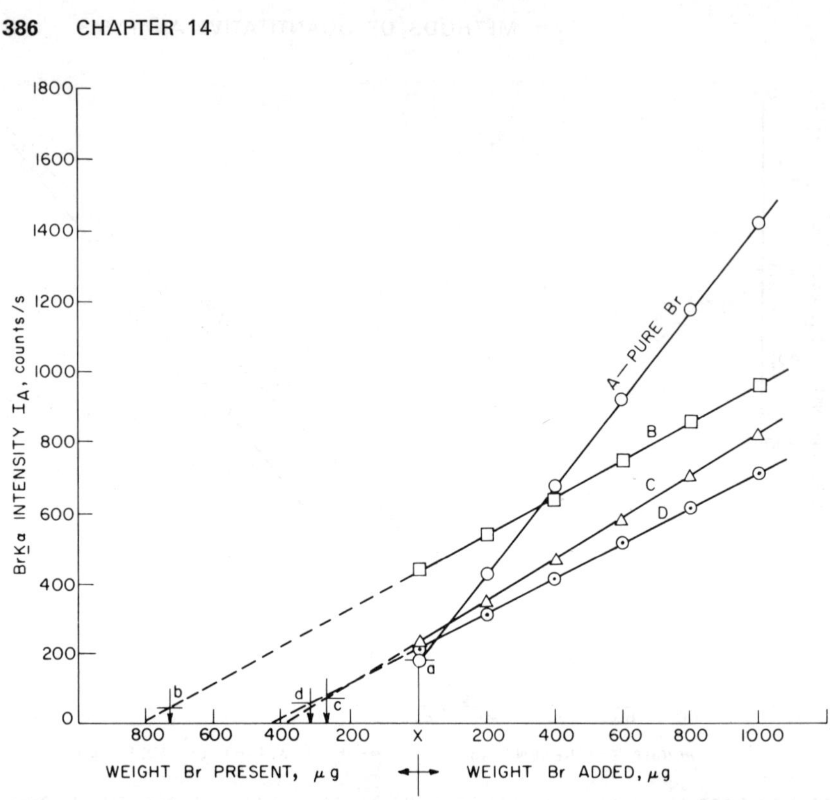

Fig. 14.2. Slope-ratio addition method applied to determination of bromine in brines. Curve *A* represents pure bromine (NaBr), curves *B*, *C*, and *D*, three brines. [H. J. Rose, Jr. and F. Cuttitta, *Advances in X-Ray Analysis* **11**, 23 (1968); courtesy of the authors, University of Denver, and Plenum Press.]

line intensity is measured from all disks and plotted as a function of analyte mass. Figure 14.2 shows such plots for determination of bromine in brines. Curve *A* represents the pure bromine (as sodium bromide, NaBr) disks. Curves *B*, *C*, and *D* represent sets of disks for three different brine samples.

For each sample, a corrected background is calculated from the slope ratios of the sample and pure-analyte curves:

$$I_{B,\mathrm{X}} = I_{B,\mathrm{S}}(m_\mathrm{X}/m_\mathrm{S}) \tag{14.7}$$

where $I_{B,\mathrm{X}}$ is corrected background intensity for a sample; $I_{B,\mathrm{S}}$ is background intensity measured from the blank (*a* in Figure 14.2); and m_X and m_S are the slopes of the sample and pure-analyte curves, respectively. The corrected backgrounds for curves *B*, *C*, and *D* are indicated as *b*, *c*, and *d*, respectively, in the figure. Normals dropped from these points to the analyte-mass axis give the analyte masses in the 1-ml portions of the sample solutions.

The multiple-addition and slope-ratio addition methods are discussed above in terms of the solution and paper-disk techniques used by their originators (*R10*). However, the methods are by no means limited to these techniques, and are applicable to any sample form suitable for any standard-addition method.

14.3. CALIBRATION STANDARDIZATION

14.3.1. Principles

With the exception of the standard addition and dilution methods, all x-ray secondary-emission spectrometric analytical methods now in common use are based on comparison of analyte-line intensities measured from the sample(s) and one or more standards. This applies even to the standard addition and dilution methods if one regards the concentrated or diluted samples as standards. The standards must be similar to the samples with regard to: (1) physical form—solid, briquet, powder, fusion product, solution; or specimen supported on Mylar, filter paper, ion-exchange membrane, etc.; (2) analyte concentration; (3) matrix composition; and

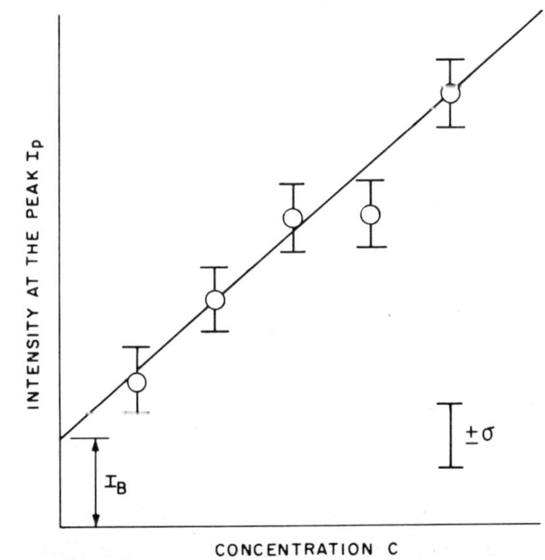

Fig. 14.3. Typical x-ray spectrometric calibration curve.

(4) physical features, such as surface finish, particle size, and packing density.

Accurate analysis is feasible with a simple calibration method—analyte-line intensity *vs.* concentration—if the effect of the matrix is not serious, or, if it is serious, if the analyte concentration range is small and the standards are very similar to the samples. Otherwise, the calibration method must be supplemented by one of the other basic approaches—internal standardization, thin-film, dilution, scatter, mathematical, or special experimental methods.

Figure 14.3 shows a typical x-ray spectrometric calibration curve. If the measured intensities are not corrected for background, the curve does not pass through the origin (zero intensity and concentration), but intercepts the intensity axis at a value corresponding to the background intensity, as shown. The figure illustrates one method of measuring background—by extrapolation. Alternatively, background can be measured for each standard and net intensity plotted, in which case the curve should pass through the origin. The figure also indicates the precision of the measured intensities and shows a data point subject to strong systematic error. The analysis may be done by measurement of the analyte-line intensity from the sample and application of this intensity to the calibration curve to derive analytical concentration graphically. Alternatively, a mathematical calibration factor may be derived from the curve or from the data itself. For example, in Figure 14.3,

$$I_P = mC + I_B \tag{14.8}$$

where m is the slope of the curve and may serve as a calibration factor,

$$m = \frac{I_P - I_B}{C} \tag{14.9}$$

The unit for m is counts per second per unit concentration (%, mg/ml, $\mu g/cm^2$, etc.). Then analytical concentration is given from the net intensity measured from the sample:

$$C = \frac{I_P - I_B}{m} \tag{14.10}$$

Figure 14.4 shows a composite typical x-ray spectrum (2θ scan) showing emitted and scattered radiation that may be used for, or interfere with, an analysis. The significance of each feature of the spectrum is given in the caption. Some features of the figure are discussed in the section on background measurement (Section 7.3). Here, the figure illustrates many of the various calibration functions that may serve as the basis of an x-ray

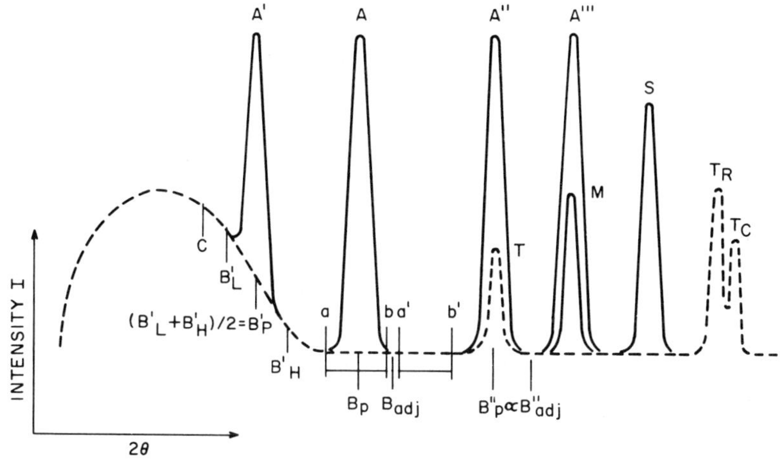

Fig. 14.4. Composite typical x-ray spectrum (2θ scan) showing emitted and scattered radiation that may be used for, or interfere with, an analysis. Dashed curve: scattered target spectrum. Solid curve: emitted specimen spectrum. Analytical lines: A—on uniform continuous background; A′— on nonuniform continuous background; A″—on scattered target line *T*; A‴—on emitted matrix line *M*. Background: B_{adj}—adjacent to peak; B_P—at peak; B_L', B_H'—adjacent to peak at equal 2θ intervals on the low *L* and high *H* sides. *C*—intensity at a certain 2θ angle in the "hump" of the scattered continuous spectrum. *M*—matrix line. *S*—internal standard, internal control-standard, internal intensity-reference standard, or external-standard line. Scattered target lines: *T*—scattered target line; T_C—Compton-scattered target line; T_R—Rayleigh-scattered target line. *a–b*: limits of integrated-count scan of peak. *a′–b′*: limits of integrated-count scan of background.

spectrometric analysis. The intensities or intensity ratios that may be plotted as a function of analyte concentration include the following:

Intensities:

1. Analyte-line peak intensity *A*.
2. Analyte-line net intensity $A - B$.
3. Intensity of a line of an element stoichiometrically combined with the analyte (association analysis, Section 14.8.2).

Intensity Ratios:

4. Analyte line/internal-standard line A/S.
5. Analyte line/internal control-standard line A/S (*M25*).
6. Analyte line/internal intensity-reference standard line A/S (*J14*).
7. Analyte line/adjacent background A/B_{adj} (*A15*).
8. Analyte line/background at the peak A/B_P or A/B_P'.
9. Analyte line/background measured in the scattered continuum "hump" A/C (*A15*).

10. Analyte line/coherently scattered target line A/T_R (*C52*).
11. Analyte line/incoherently scattered target line A/T_C (*B50*).
12. Minor or trace analyte line/line of major matrix element of substantially constant concentration A/A' (*H29*).
13. Line of analyte A/line of analyte B (binary-ratio method) A/A' (*B19*).
14. Analyte line from specimen/analyte line from standard.
15. Analyte line from specimen/analyte line from pure analyte.
16. Analyte line from specimen/radiation from external standard.

The intensities may be peak or net, depending on the particular system. For the last three ratios, the two ratioed intensities are not necessarily present in the spectrum simultaneously. The first two functions listed represent simple calibration methods. The other functions involve other methods of dealing with matrix effects and are discussed in subsequent sections of this chapter. If the same preset time is used for all line and background measurements, the accumulated counts may be used instead of intensities. If the same preset count is used for all measurements, the counting times may be used. In all cases, the calibration functions may be used graphically, or mathematical calibration factors may be derived.

14.3.2. Special Calibration Methods

14.3.2.1. *Single-Standard Method*

When the samples X have a very narrow range of analyte concentration C bracketing the analyte concentration of a standard S, the following simple proportionality holds reasonably well for the measured net intensities:

$$I_X/I_S = C_X/C_S \tag{14.11}$$

A refinement of this method is the *corrected single-standard method* (*S41*). Suppose that all components are to be determined in a certain type of alloy. Several chemically analyzed samples X' are obtained, one of which is designated as the standard S. A set of calibration curves—one for each analyte—is established as follows. Analyte-line net intensity is measured for each analyte from each analyzed sample $I_{X'}$ and from the standard I_S. Then for each analyte in each analyzed sample: (1) the intensity ratio $I_{X'}/I_S$ is calculated; (2) analyte concentration $C_{X'}$ is calculated by the single-standard method [Equation (14.11)]; and (3) the ratio of the x-ray and chemical concentrations $C_{X'}/C_{chem}$ is calculated. For each of the analyzed samples, the "net density" is calculated from the $C_{X'}$ values of all the analytes A and their densities ϱ: $\Sigma\,(C_{X'}\varrho)_A$. Finally, for each analyte A, a correction curve is plotted of $(C_{X'}/C_{chem})_A$ as a function of net density. The curve for each analyte has as many points as there are analyzed samples X′.

Thereafter, the procedure for analysis of each sample X is as follows. For each analyte, I_X and I_S are measured from the sample and standard, respectively; I_X/I_S is calculated; and C_X is derived from Equation (14.11). Net density is calculated for the sample, $\Sigma\,(C_X\varrho)_A$, and applied to the correction curve for each analyte A in turn to derive $(C_{X'}/C_{chem})_A$ for each analyte. The corrected analyte concentration is then $[C_X(C_{X'}/C_{chem})]_A$.

Single-standard methods are particularly suitable for analyses made with automatic spectrometers.

14.3.2.2. *Two-Standard Method* (*H33*)

When the analyte concentration in the sample X lies between that in two standards S1 and S2, and the net intensity *I vs.* concentration *C* function is substantially linear in the interval bracketed by the standards,

$$C_X = C_{S1} + \frac{I_X - I_{S1}}{I_{S2} - I_{S1}}\,(C_{S2} - C_{S1}) \tag{14.12}$$

The solid lines in Figure 14.5 show the intensity–concentration relationships for the two-standard method for the case of a linear calibration

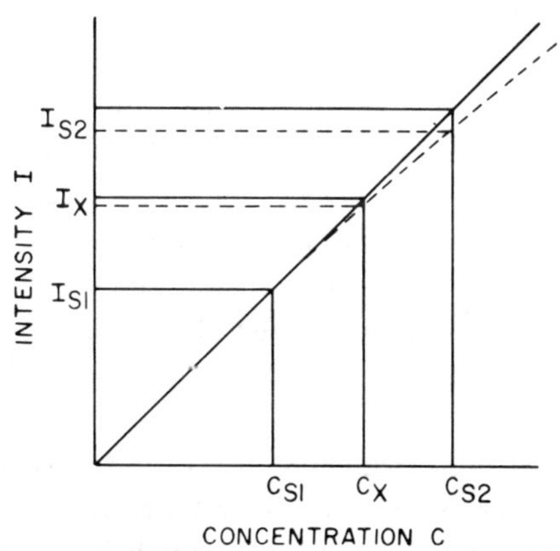

Fig. 14.5. Two-standard method.

function. The dashed lines show how nonlinearity introduces error. Another form of the equation for the two-standard method is the following (*H33*):

$$\frac{(I_{S1}/I_{S2})[(C_{S2}/C_{S1}) - 1]}{C_{S2} - C_{S1}} = \frac{(I_{S1}/I_{X})[(C_{X}/C_{S1}) - 1]}{C_{X} - C_{S1}} \tag{14.13}$$

14.3.2.3. *Binary-Ratio Method* (*B19*, *F4*)

The binary-ratio method is applicable to three classes of samples: (1) those in which only two elements A and B are present and both are determinable x-ray spectrometrically—for example, lead–tin alloys; (2) those in which only two such elements vary in an otherwise constant matrix—for example, indium gallium arsenide, (In,Ga)As; and (3) those in which only two elements having atomic number about 19 (potassium) or more are present in a very light matrix—for example, mixtures of chromium and silver oxides. For such specimens, the net intensities of the analyte lines of elements A and B are measured from each sample and standard. From the standard data, a calibration curve is established of the function $\log_{10}(I_A/I_B)$ *vs.* $\log_{10}(C_A/C_B)$, where I and C are net intensity and concentration, respectively.

The method has many advantages. It has greater sensitivity to change in concentration than the conventional I *vs.* C method because I_A increases as I_B decreases, and *vice versa.* Calibration curves are always linear, regardless of how distorted the individual curves (I_A *vs.* C_A and I_B *vs.* C_B) may be, that is, regardless of how severe the mutual absorption-enhancement effects are. The method is insensitive to reasonable variations in surface texture (porosity, chipping, cracking, grind marks), and the data points usually lie on the linear binary-ratio calibration curve regardless of how widely scattered they are on the individual curves. The method is insensitive to reasonable variations in specimen area and position and to surface contour of small irregular-shaped specimens and fabricated parts. Specimens need not be masked to constant area. In many cases, the standards need not necessarily have the same physical form as the samples. No additional measurements other than the two analyte lines and their backgrounds are required. The principal disadvantage of the method is its inapplicability near the concentration extremes where C_A/C_B and I_A/I_B approach zero or infinity.

The method has been evaluated for tungsten–rhenium, bismuth–tin, and niobium–tin alloys; Cr_2O_3–Ag_2O and Fe_3O_4–CuO powder mixtures; (Ga,In)As; and solutions containing molybdenum and tungsten. Figure 14.6 shows the effectiveness of the method as applied to a relatively un-

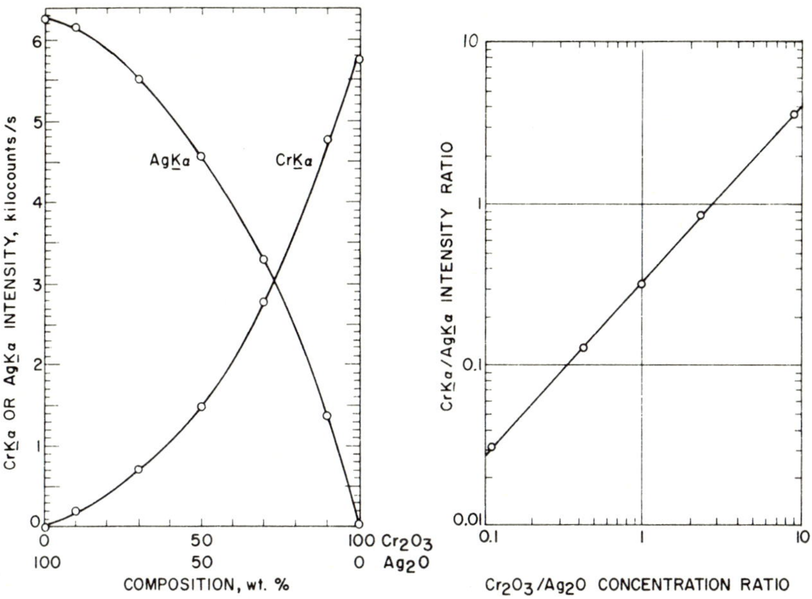

Fig. 14.6. Binary-ratio method. Intensity *vs.* concentration curves for the two individual elements (left) are compared with the binary-ratio curve (right). [E. P. Bertin, *Analytical Chemistry* **36**, 826 (1964); courtesy of the American Chemical Society.]

favorable system—mixtures of Cr_2O_3 and Ag_2O powders. It is evident that the individual *I vs. C* curves are very distorted, but the binary-ratio curve is linear.

For adjacent elements in the periodic table, or even for elements differing by ± 2 in atomic number, the spectral-line intensity ratio is substantially equal to the concentration ratio. Such systems correspond to curve *A* in Figure 12.1. Of course, the same line ($K\alpha$, $L\beta_1$, etc.) must be measured for both elements, and no absorption edge may lie between the measured lines. Rubidium–strontium concentration ratios in minerals have been determined in this way for estimation of geological age (*H27*), and the method has been applied to copper–zinc (brass), tungsten–rhenium, and tungsten–osmium alloys.

14.3.2.4. *Mutual Standards Method* (*C33*)

This method is applicable to substances in which: (1) the matrix is dominated by a single major constituent; (2) the other elements are known, have relatively low concentrations, and are determinable x-ray spectrometrically; and (3) it can be assumed that the sum of the concentrations of all the determined elements is 100%. The method has been applied to the determination of iron (88–100%) and seven other elements in meteoritic sulfide nodules. The nodules are "dissolved" in nitric acid, evaporated to dryness, and fired to expel sulfur trioxide and convert all eight elements to oxides. In the resulting specimens,

$$C_{\mathrm{Fe}} + \Sigma\, C_E = 100\% \tag{14.14}$$

where C is concentration as oxide (wt %), and E represents each of the seven minor and trace elements. Then,

$$C_{\mathrm{Fe}} = 100 - \Sigma\, C_E = \frac{100}{1 + \Sigma\,(C_E/C_{\mathrm{Fe}})} \tag{14.15}$$

$$= \frac{100}{1 + \dfrac{C_{\mathrm{V}}}{C_{\mathrm{Fe}}} + \dfrac{C_{\mathrm{Cr}}}{C_{\mathrm{Fe}}} + \dfrac{C_{\mathrm{Co}}}{C_{\mathrm{Fe}}} + \dfrac{C_{\mathrm{Ni}}}{C_{\mathrm{Fe}}} + \dfrac{C_{\mathrm{Cu}}}{C_{\mathrm{Fe}}} + \dfrac{C_{\mathrm{Zn}}}{C_{\mathrm{Fe}}} + \dfrac{C_{\mathrm{As}}}{C_{\mathrm{Fe}}}} \tag{14.16}$$

Seven working curves of line-intensity ratio I_E/I_{Fe} *vs.* C_E were established from measurements on the seven sets of binary mixtures. An analysis consisted of measurement of the line intensities I_E of all eight elements from the sample, calculation of the seven I_E/I_{Fe} ratios, and application of these ratios to their respective working curves to derive C_E/C_{Fe} concentration ratios. These ratios were substituted in Equation (14.16), which was then solved for C_{Fe}. Finally, the seven C_E values were calculated from their respective C_E/C_{Fe} ratios.

14.3.2.5. *Sets of Calibration Curves*

When analyte-line intensity depends on the concentration not only of the analyte A, but also of some matrix element B, a set of calibration curves of A-line intensity *vs.* A concentration may be established, each curve for a different B concentration (*D7*). An analysis for element A requires the previous determination of element B. The A-line intensity is then measured and applied to the particular I_{A} *vs.* C_{A} curve for the known B concentration. Those B concentrations not represented by a curve are dealt with by interpolation. Figure 14.7 shows a typical set of curves for determination of nickel in stainless steel, each curve for a different chromium concentration. In the analysis of petroleum products, sets of curves may be used, each curve for a different base-stock composition (*K19*).

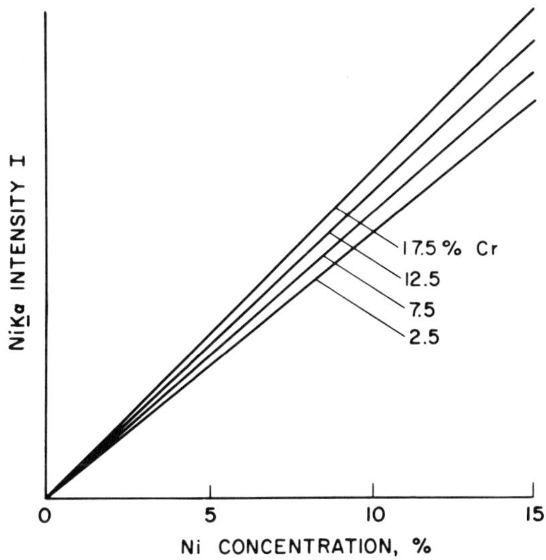

Fig. 14.7. Set of calibration curves for correction of absorption effect of chromium on determination of nickel in stainless steel.

14.4. INTERNAL STANDARDIZATION

14.4.1. Principles *(A10)*

If an element has a spectral line having excitation, absorption, and enhancement properties identical with those of the analyte line in a particular matrix, that element would be an ideal internal standard IS for that analyte A in that matrix, and

$$C_A/C_{IS} = I_A/I_{IS} \tag{14.17}$$

where C and I are concentration and intensity, respectively. In practice, absorption-enhancement effects may be compensated by the addition to all samples and standards of a constant concentration of an internal-standard element IS that fills these requirements as closely as possible. Fortuitously, a suitable internal-standard element may already be present in the specimens in a suitable constant concentration. Line intensities are measured from the analyte and internal standard from all samples and standards, and a calibration curve of I_A/I_{IS} intensity ratio *vs.* analyte concentration is established from the standard data. Table 14.2 gives some typical data demonstrating the effectiveness of the method.

TABLE 14.2

Comparison of Ratio Methods[a]

Iron concentration, mg/ml	Cu $K\alpha$ net intensity, counts/s	Ratio $I_{CuK\alpha}/I_{std}$: Internal standard (2 mg Zn/ml)	Ratio $I_{CuK\alpha}/I_{std}$: Background ratio (0.6 Å)	Ratio $I_{CuK\alpha}/I_{std}$: Coherently scattered target line (W $L\beta_1$)
0[b]	2117	1.045	3.261	3.541
5	1821	1.042	3.258	3.530
10	1628	1.040	3.247	3.522
15	1459	1.035	3.234	3.510
20	1308	1.032	3.221	3.503

[a] From Cullen (*C55*).
[b] Cu concentration in all solutions 2.0 mg/ml.

Alternatively, a calibration constant K may be derived from a standard having known concentrations of analyte C_A and internal standard C_{IS}:

$$I_A/I_{IS} = K(C_A/C_{IS}) \tag{14.18}$$

Thereafter, K can be used to calculate analyte concentration from the intensity ratio measured from the samples.

14.4.2. Selection of Internal-Standard Element

The basic requirements of the internal standard have already been given. In practice, the criterion is that the internal standard and analyte

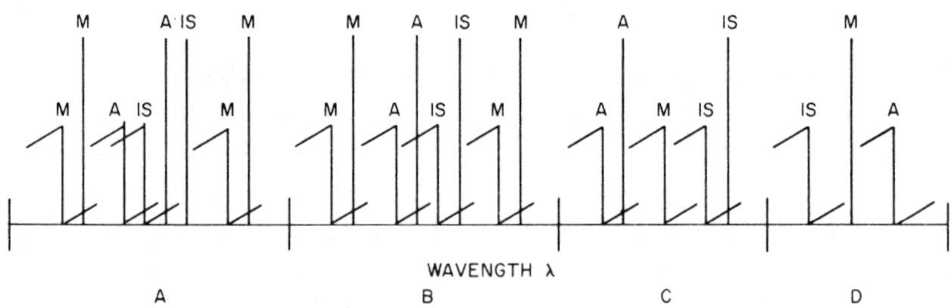

Fig. 14.8. Specific absorption-enhancement effects and the selection of an internal-standard element. The figure shows spectral lines and absorption edges of analyte A, matrix M, and prospective internal standard IS elements. Figure A is the ideal case, B is acceptable, C and D are unacceptable.

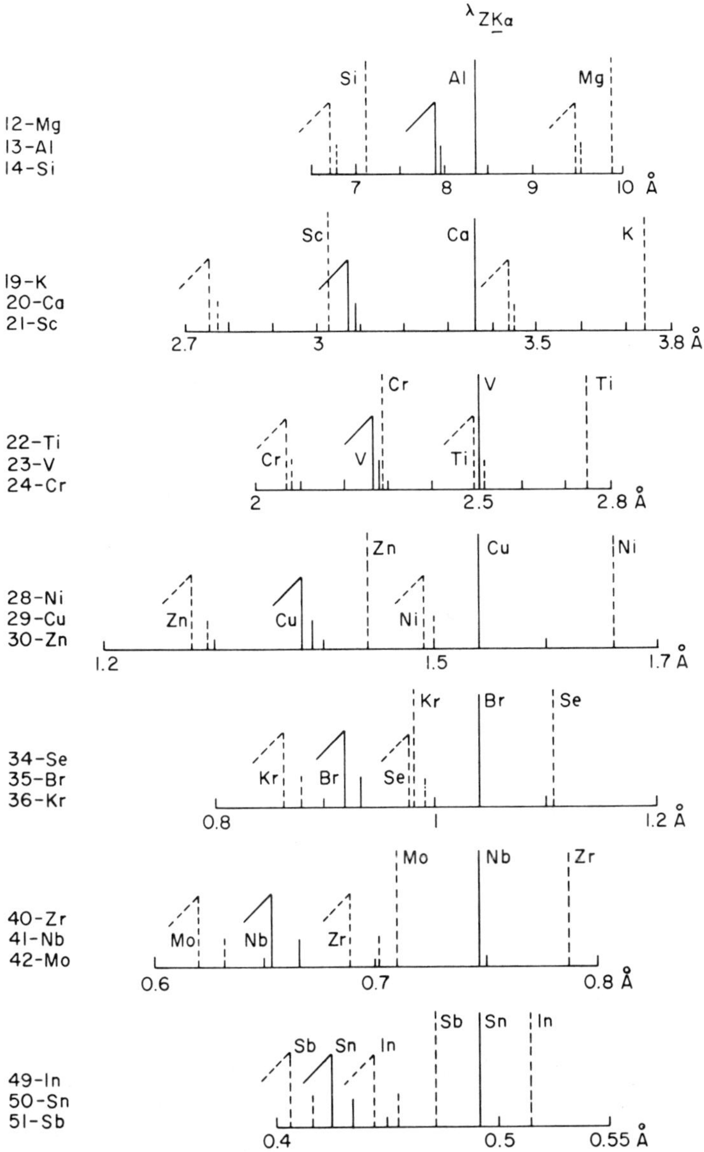

Fig. 14.9. Relationships among K spectral lines and absorption edges of adjacent elements in various regions of the periodic table. The long and short straight lines represent $K\alpha$ and $K\beta$ lines, respectively, the hooked lines K-absorption edges. The solid lines and edges represent the analyte.

lines and absorption edges should have wavelengths as close as possible. Figure 14.8 shows four conditions with respect to the relative wavelengths of the spectral lines and absorption edges of the analyte A, one or two matrix elements M, and the prospective internal standard IS. In every case, the positions of the analyte and internal standard could be interchanged without altering the discussion below. Figure 14.9 shows the wavelengths of the K-absorption edges and $K\alpha$ and $K\beta$ lines of triads of adjacent elements $Z + 1$, Z, and $Z - 1$ in several regions of the periodic table. In this figure, element Z (solid lines) may be regarded as the analyte, and elements $Z + 1$ and $Z - 1$ (dashed lines) as its prospective internal standards.

In Figure 14.8, case A represents the ideal condition. The analyte and internal-standard lines are close and on the same sides of both matrix absorption edges shown, and the analyte and internal-standard absorption edges are close and on the same sides of both matrix lines shown. The analyte and internal-standard lines are absorbed and enhanced to substantially the same extent by each of the two matrix elements. Incidentally, if the two matrix lines and edges are sufficiently remote from the analyte line and edge, internal standardization may not be required. The condition in Figure 14.8A prevails for the $K\alpha$ lines of analyte elements of atomic number down to 23 (vanadium) and internal standards $Z \pm 1$, as shown in Figure 14.9.

Case B is not as favorable, but is feasible, and many successful analyte–internal standard pairs are of this type. The same comments apply as for case A regarding the positions of the analyte and internal-standard lines and edges relative to the lines and edges of the matrix elements, and regarding the effect of each matrix element on the analyte and internal standard. However, here the analyte line falls between the analyte and internal-standard absorption edges, and the internal-standard edge falls between the analyte and internal-standard lines. Alternatively, the analyte and internal standard may be interchanged. One might expect case B to be unfavorable. However, difficulty is minimal because the internal-standard concentration is constant in all samples and standards and because the calibration function is based on the intensity ratio. Nevertheless, this condition should be avoided if case A is feasible. An example of case B is the use of Nb $K\beta$ as an internal standard for Mo $K\beta$ (*H6*). Figure 14.9 shows that the niobium K-absorption edge lies between these lines.

Case C is unacceptable: The matrix absorbs the analyte line strongly, the internal-standard line weakly. Interchanging of the analyte and internal standard would only reverse the effects and do no good. Similarly, case D is unacceptable. The matrix line is absorbed weakly by the internal standard, but strongly by the analyte, thereby enhancing the analyte line.

Again, interchanging of the analyte and internal standard would do no good.

The selection requirements are usually best filled by an element having atomic number one or two above or below that of the analyte. Then the analyte and internal-standard lines can be of the same series and have about the same intensity for the same concentration; also, the analyte and internal standard have about the same mass-absorption coefficients for the primary and secondary x-rays. As already mentioned, and as is evident in Figure 14.9, this "rule" applies for analytes down to atomic number 23 (vanadium). Among the lighter elements, element $Z - 1$ strongly absorbs the $K\alpha$ line of elements having atomic number 22 (titanium) or lower, so element $Z + 1$ is preferable. The $Z + 1$ elements enhance the Z K lines, but the effect is relatively small in the low-Z region. Also, among the lighter elements, case B (Figure 14.8) is unavoidable. In Figure 14.9, for analytes calcium and aluminum, an absorption edge lies between the analyte and internal-standard lines regardless of whether element $Z + 1$ or $Z - 1$ is chosen as the internal standard.

In selecting the internal standard, one must avoid the introduction of new absorption-enhancement effects or of spectral-line interference. For example, in a three-component system A–B–C, a fourth element D may be the ideal internal standard for A, but it may strongly absorb or enhance B, and its lines may interfere with the optimal line of C.

Rather surprisingly, there are many successful applications of an L internal-standard line for a K analyte line and *vice versa*; for example, Y $K\alpha$ (0.83 Å) serves as an internal-standard line for U $L\alpha_1$ (0.91 Å). However, such selections are to be avoided if possible. An internal standard having L lines at about the same wavelength as an analyte K line would necessarily have much higher atomic number than the analyte and therefore very different absorption coefficients for the primary and secondary x-rays. An analogous appraisal can be made for an internal standard having K lines at about the same wavelength as an analyte L line. Also, for the same analyte and internal-standard concentration, K lines are 5–10 times as intense as the L lines. An internal-standard line should be chosen from a series different from that of the analyte line only when a satisfactory internal-standard element cannot be found having a line in the same series.

14.4.3. Advantages and Limitations

The principal advantages of the internal-standard method are the following. It compensates absorption-enhancement and long-term instrument drift for all types of specimen. It partially compensates variations in density in powder and briquet specimens. It compensates density, evaporation, and bubble formation in liquids. Because the analyte and internal-standard lines are very close, their backgrounds are usually substantially the same and often need not be measured, in which case the peak-intensity ratio is used.

The principal disadvantages are the following. The method is not applicable to many types of sample, such as bulk solids, foils, and small fabricated parts. It is also not applicable to high analyte concentrations—above, say, 25%. Analysis time is greatly increased and the analysis is made much less convenient by the requirement for quantitative and thorough admixture of the internal standard and for measurement of a second line (and perhaps background) from each sample and standard. However, all measurements can be made simultaneously with a multichannel instrument. There is some loss of precision because of the additional measurement. The internal standard may be expensive; for example, the ideal internal standard for tantalum is hafnium; if large numbers of samples are to be analyzed, the cost of the highly tantalum-free hafnium would be substantial. Finally, although the method is fairly easily applied to one analyte in a multicomponent system, it is frequently impractical for more than one analyte in the same sample. However, multiple internal standardization is sometimes feasible: A single internal standard may serve for more than one analyte, or a mixture of two or more internal standards may be added to the specimens (*D26*).

14.4.4. Considerations

The internal standard must be intimately and homogeneously mixed with the specimens. This is most easily accomplished with solutions and fusion products. The internal standard may be admixed with powders in several ways. The specimens may be dissolved, treated with the internal standard, then precipitated. Specimen and internal-standard powders may be mixed and ground, with or without an inert abrasive (such as silicon carbide or aluminum oxide), to a very small particle size. The powders may be ground dry or repeatedly to dryness under ether or ethanol (*A7*). The internal standard in solution may be added to the specimen powder and the resulting slurry mixed, dried, and ground (*C42*). Even greater problems arise in introduction of the internal standard to specimens supported on ion-exchange resins or membranes.

The internal standard should be added at a concentration level equivalent to about the middle of the expected analyte concentration range. It is this requirement that limits the internal-standard method to analyte concentrations up to ~25% at most. The effectiveness of the internal standard in compensating absorption-enhancement effects increases as analyte concentration decreases, or as dilution increases. The method is most convenient and precise on multichannel instruments, where both lines, and their backgrounds, if necessary, can be measured simultaneously. On single-channel instruments, the method is more precise, the shorter the lapsed time between measurement of the analyte and internal-standard lines.

Interaction of the analyte and internal standard has already been mentioned (Section 14.4.2, Figure 14.8B). Actually, this condition probably always exists to a certain extent. For example, even if the analyte and internal-standard $K\alpha$ lines both lie on the same side of both their absorption edges, one or the other of their $K\beta$ lines is likely to lie between the edges and cause a slight interaction. In Figure 14.9, this condition is seen to be common. The probability of interaction is even greater with the more complex L spectra.

Hakkila *et al.* (*H6*) provide an example of how to deal with variations of the slope and intercept of the calibration curve caused by spectral interference of a matrix element line with the analyte or internal-standard line. They encountered such spectral interference from uranium L lines in the determination of zirconium and molybdenum in the presence of uranium, using niobium as the internal standard. For zirconium, the curve of Zr $K\alpha$/Nb $K\alpha$ intensity ratio R *vs.* zirconium concentration is linear. As uranium concentration changes, the slope changes, but the intercept on the R axis remains constant. Zirconium concentration is given by

$$C_{\mathrm{Zr}} = \frac{R - b}{a - c(d - 100)} \tag{14.19}$$

where C_{Zr} is zirconium concentration (mg/10 ml); a is the slope of the curve for solutions containing 100 mg U/10 ml; b is the constant intercept on the R axis; c is the change in slope per 1 mg U/10 ml; and d is uranium concentration (mg/10 ml). For molybdenum, the curve of Mo $K\beta$/Nb $K\beta$ intensity ratio R *vs.* molybdenum concentration is also linear. As uranium concentration changes, the slope remains constant, but the intercept on the R axis changes. Molybdenum concentration is given by

$$C_{\mathrm{Mo}} = \frac{R - b + c(100 - d)}{a} \tag{14.20}$$

where C_{Mo} is molybdenum concentration (mg/10 ml); a is the constant slope of the curve; b is the intercept on the R axis for solutions containing

100 mg U/10 ml; *c* is the change in intercept per 1 mg U/10 ml; and *d* is uranium concentration (mg/10 ml).

14.4.5. Special Internal-Standardization Methods

14.4.5.1. *Single-Standard Internal-Standard Method* (25)

A weighed portion of sample X is treated with a weighed portion of a substance containing internal-standard element IS, giving a treated sample X′ having internal-standard concentration $C_{IS,X'}$. A standard S is prepared having equal concentrations of analyte and internal-standard elements in a matrix similar to that of the sample. Intensities of the analyte line I_A and internal-standard line I_{IS} are measured from X′ and S. Then analyte concentration in the untreated sample is given by

$$C_{A,X} = C_{IS,X'} D \left(\frac{I_{IS}}{I_A}\right)_S \left(\frac{I_A}{I_{IS}}\right)_{X'} \tag{14.21}$$

where *D* is the dilution factor, weight of X′/weight of X.

14.4.5.2. *Variable Internal-Standard Method*

Strasheim and Wybenga (*S42*) compensated absorption-enhancement effects in the determination of iridium, platinum, and gold by a novel *variable internal-standard* method. Intensities of the Pt $L\alpha_1$ line and its background were measured from solutions containing platinum in concentration 400 μg/ml in pure 10% hydrochloric acid and in various concentrations of various other matrix constituents. Ratios were calculated of Pt $L\alpha_1$ intensity from 10% HCl to Pt $L\alpha_1$ intensity from each of the other matrixes M: $I_{Pt,HCl}/I_{Pt,M}$. Ratios were calculated of background intensity from 10% HCl to background intensity from each of the other matrixes: $I_{B,HCl}/I_{B,M}$. A curve was plotted of the function $I_{Pt,HCl}/I_{Pt,M}$ *vs.* $I_{B,HCl}/I_{B,M}$; this curve was linear. An analysis then consisted of measurements of Pt $L\alpha_1$ and background intensities from the samples and from a platinum standard in pure 10% hydrochloric acid. The ratio $I_{B,HCl}/I_{B,M}$ was calculated for each sample and applied to the curve to derive $I_{Pt,HCl}/I_{Pt,M}$, which was then used to correct the Pt $L\alpha_1$ intensity measured from the sample. The analytical platinum concentration was then obtained from the corrected Pt $L\alpha_1$ intensity for the sample and the Pt $L\alpha_1$ intensity measured from the standard.

The platinum already present in the samples was then used as a variable internal standard for determination of iridium and gold. Intensity ratios Ir $L\alpha_1$/Pt $L\alpha_1$ and Au $L\alpha_1$/Pt $L\alpha_1$ were measured, and iridium and gold concentrations determined relative to platinum. Ruthenium, rhodium, and palladium were determined the same way (*W12*). Palladium was determined the same way as platinum, then used as a variable internal standard for determination of ruthenium and rhodium.

14.4.6. Other Standardization Methods

14.4.6.1. *Internal Control-Standard Method*

An internal control standard is similar to an internal standard in some ways. It is added in constant concentration to all samples and standards, unless a suitable element is already present in the samples in appropriate concentration. The calibration function is the intensity ratio of the analyte and internal control-standard lines *vs.* analyte concentration. However, the control standard does not necessarily have the same excitation, absorption, and enhancement properties as the analyte in the specimen matrix, and it may serve for several analytes in the same specimen. The method has been applied to solutions where it compensates variations in volume, temperature, density, specimen plane, acid matrix, and, to a smaller degree, composition of the original matrix.

The method has been evaluated by Mitchell and O'Hear (*M25*) for the rapid, accurate, simultaneous determination of chromium, manganese, iron, cobalt, nickel, and copper in chromium and manganese alloys, Alnico, refractory alloys, ores, and slags. The sample concentration was 1 g/100 ml; analyte concentration was 0.1–99.9% of the original sample. The samples were put into solution by various acid and/or fusion treatments. All analyte lines and the internal control-standard line were measured simultaneously on an ARL X-Ray Quantometer. For each analyte in each original sample matrix, one set of standards suffices for all acid or fusion matrixes.

14.4.6.2. *Internal Intensity-Reference Standard Method*

In the determination of manganese in gasoline, Jones (*J14*) compensated variable carbon/hydrogen ratio by mounting an iron rod in the liquid-specimen cell (Figure 17.8) to serve as an internal intensity-reference standard. The calibration function was Mn $K\alpha$/Fe $K\alpha$ intensity ratio *vs.* manganese concentration. Gunn (*G17*) applied the same method in the determination of lead in gasoline by using a platinum disk in the cell and the Pb $L\alpha_1$/Pt $L\alpha_1$ intensity ratio. Ideal compensation is realized with the disk at such a depth that the intensity ratio is the same for the same lead concentration in liquids having different carbon/hydrogen ratios. Gunn gives a procedure for determination of optimal depth. By selection of a true internal-standard element to be the internal intensity-reference standard, a degree of true internal standardization is realized without the inconvenience of making a quantitative addition to all specimens.

14.4.6.3. *External-Standard Method*

An external standard, or external intensity-reference standard, is a specimen from which some intensity value is measured to be ratioed with the analyte-line intensity, principally to compensate long-term instrument drift. The standard may be one of the samples or standards retained for this purpose, or it may be any stable specimen not necessarily bearing any relation to the samples to be analyzed (*L18*)—such as a piece of copper alloy or nickel alloy. The measured intensity may be emitted or scattered from the standard. The standard may be measured once after each sample and calibration standard, or, more likely, from time to time during a long series of measurements. A particularly interesting example of the external-standard method is reported by Hirokawa and his colleagues, who used the emission lines from specimen-mask plates of zinc or lead as external-standard lines in the determination of chromium, nickel, and tungsten in steel (*H30*).

Another way to use an external standard is to set the excitation parameters to reproduce the intensity emitted or scattered from it. For example, in the determination of aluminum, silicon, and iron in clays and bauxites, day-to-day instrument drift was compensated by adjustment of the x-ray tube current and/or potential to reproduce intensity values established at the time of the original calibration from selected aluminum and iron plates (*S43*).

14.5. STANDARDIZATION WITH SCATTERED X-RAYS

Scattered x-rays, formerly regarded only as a nuisance (*J10*), provide the basis for several extremely convenient and effective means for dealing with absorption-enhancement effects and with certain other errors. These scatter methods are especially advantageous for specimens having low effective atomic number, for which scattered intensity is relatively high and can be measured with good statistical precision in relatively short counting time. Although scatter methods may reduce enhancement effects in some cases, in general, they are effective principally for absorption effects.

14.5.1. Background-Ratio Method (*A15, K2, K6*)

For a given minor or trace analyte A, the intensity ratio of the analyte line and scattered background I_A/I_B remains substantially constant as the mass-absorption coefficient of the specimen varies—that is, as the matrix varies. Andermann and Kemp (*A15*) were the first to present the theory of this phenomenon and demonstrate its great value in compensating absorption, instrumental, density, and particle-size effects. Of course, the background wavelength cannot be enhanced by the matrix, so the method is ineffective against enhancement.

The total scattered intensity at a specified wavelength λ may be represented by

$$I_{B,\lambda} \propto \frac{S_R}{2(\mu/\varrho)_\lambda} + \frac{S_C}{(\mu/\varrho)_\lambda + (\mu/\varrho)_{\lambda-\Delta\lambda}} \tag{14.22}$$

where S_R is the coherent (Rayleigh) scattering intensity per atom; S_C is the incoherent (Compton) scattering intensity per atom; $(\mu/\varrho)_\lambda$ is the mass-absorption coefficient of the specimen at wavelength λ; and $(\mu/\varrho)_{\lambda-\Delta\lambda}$ is the mass-absorption coefficient of the specimen at the incident wavelength $\lambda - \Delta\lambda$ that undergoes increase in wavelength $\Delta\lambda$ upon scattering incoherently [Equation (2.25)].

The emitted analyte-line intensity may be represented by

$$I_A \propto \frac{1}{(\mu/\varrho)_{\lambda_P} + (\mu/\varrho)_{\lambda_A}} \tag{14.23}$$

where $(\mu/\varrho)_{\lambda_P}$ and $(\mu/\varrho)_{\lambda_A}$ are the mass-absorption coefficients of the specimen for the primary and emitted analyte-line radiation, respectively.

All quantities in Equations (14.22) and (14.23) vary with atomic number Z. Thus, a change in matrix composition affects both the numerator and denominator of Equation (14.22), but only the denominator of Equation (14.23). The influence of the effective atomic number of the specimen on a specified analyte-line and scattered wavelength can be estimated from the following proportionalities:

$$S_P + S_C \propto Z^{1 \text{ to } 2} \tag{14.24a}$$

$$\mu/\varrho \propto Z^4 \tag{14.24b}$$

Therefore,

$$I_{B,\lambda} \propto Z^{-(3\text{ to }2)} \tag{14.24c}$$

$$I_{\mathrm{A}} \propto Z^{-4} \tag{14.24d}$$

Then,

$$\frac{I_{\mathrm{A}}}{I_{B,\lambda}} \propto Z^{-(1\text{ to }2)} \tag{14.24e}$$

A comparison of Equations (14.24e) and (14.24d) shows that the ratio is much less dependent on atomic number—and therefore on matrix composition—than analyte-line intensity.

Equation (14.24e) applies when the analyte-line and scattered wavelengths are about the same. Actually, scatter and absorption are also wavelength dependent, and it may be possible to select a scattered wavelength for which the ratio $I_{\mathrm{A}}/I_{B,\lambda}$ is substantially independent of matrix.

Analyte-line and scattered intensities are similarly affected by specimen position and x-ray tube current. If the analyte line is excited principally by the primary continuum, rather than by the target lines, analyte-line and scattered intensities are also similarly affected by x-ray tube potential. Thus, the ratio compensates variations in these parameters.

The background-ratio method is similar to the internal-standard method, except that the calibration function is $I_{\mathrm{A}}/I_{\mathrm{B}}$ *vs.* analyte concentra-

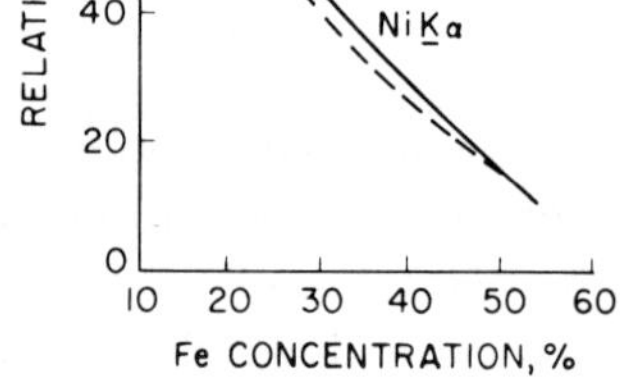

Fig. 14.10. Approximate constancy of relative intensities of Ni *Kα* and 0.6-Å scattered continuum from specimens having the same nickel concentration, but different iron concentrations. [G. Andermann and J. W. Kemp, *Analytical Chemistry* **30**, 1306 (1958); courtesy of the authors and the American Chemical Society.]

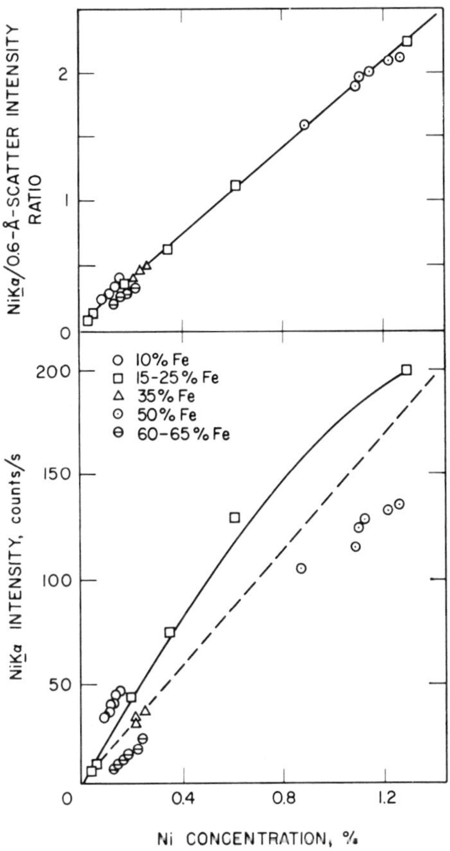

Fig. 14.11. Effectiveness of the background-ratio method in correcting for absorption-enhancement effects of iron on Ni $K\alpha$ intensity from iron–nickel minerals. [G. Andermann and J. W. Kemp, *Analytical Chemistry* **30**, 1306 (1958); courtesy of the authors and the American Chemical Society.]

tion. Here, I_A may be peak or net analyte-line intensity, and I_B may be the intensity of: (1) the scattered background at the analyte-line peak measured from a blank or in any of the ways described in Section 7.3.3; (2) the scattered background measured adjacent to the peak; or (3) a specified wavelength in the scattered continuum "hump." The denominator of the ratio may also be the intensity of a coherently or incoherently scattered target line (Section 14.5.3). A wavelength in the hump or a scattered target line is used when: (1) the background at or adjacent to the peak is too low to be measured with acceptable statistical precision in a reasonable counting time; or (2) such a wavelength gives better standardization than the background. The background-ratio method is sometimes described as "internal standardization with scattered x-rays," and, in a sense, this is a valid description, because the scattered intensity is dependent on the matrix [Equation (14.22)].

Figure 14.10 shows the relative intensities of emitted Ni $K\alpha$ and scattered 0.6-Å x-rays measured from minerals having the same nickel concentration, but 10–60% iron. Both individual intensities are strongly dependent on iron concentration, but the curves are both hyperbolic and more or less parallel, indicating that the Ni $K\alpha$/0.6 Å intensity ratio should be about constant with change in iron concentration. This prediction is strikingly confirmed in Figure 14.11. The data points on the Ni $K\alpha$ curve are widely scattered. Plotting of the same data as the intensity ratio of Ni $K\alpha$ and 0.6-Å scatter reduces the scatter in the points to insignificance.

Figure 14.12 strikingly shows the effectiveness of the method in compensating variations in x-ray tube potential and current, specimen plane, and particle size. Figures 14.10 and 14.11 are examples of an intensity ratio of an analyte line (Ni $K\alpha$, 1.66 Å) and a wavelength (0.6 Å) from the scattered continuum hump. Figure 14.12 is an example of an intensity ratio of an analyte line (Cu $K\alpha$, 1.54 Å) and its adjacent background (~1.4 Å). Table 14.2 gives some typical data for the scattered continuum method.

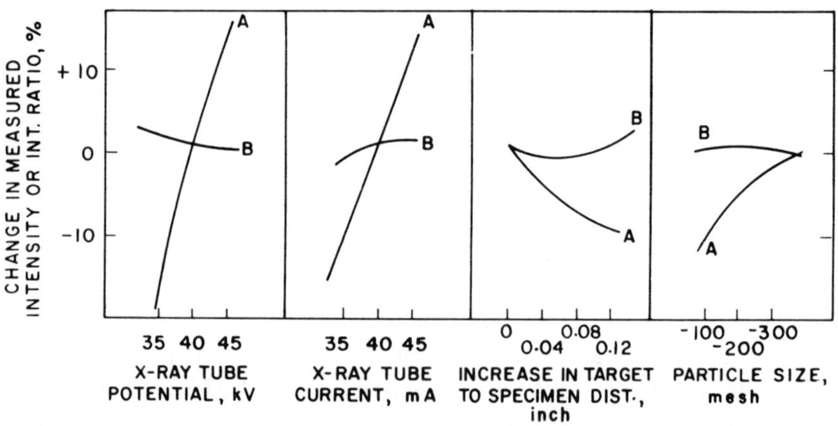

Fig. 14.12. Effectiveness of the background-ratio method in correcting for instrumental variables and the particle-size effect. Curves *A* show variation of Cu $K\alpha$ intensity. Curves *B* show variation of the intensity ratio of Cu $K\alpha$ and its adjacent background. [G. Andermann and J. W. Kemp, *Analytical Chemistry* **30**, 1306 (1958); courtesy of the authors and the American Chemical Society.]

Compared with the internal-standard method, the background-ratio method has the advantages that no addition need be made to the specimens, that it is applicable to specimens having practically any physical form, and that it is applicable to many analytes in the same specimen. The method is especially applicable to minor and trace analytes in low-Z matrixes, where the scattered intensity is high and can be measured precisely in short counting times. The method may not be as effective as internal standardization in compensating absorption effects in specific cases, and it is ineffective against enhancement. Also, all scattered wavelengths are not equally effective in the ratio, and principles for selection of the optimal wavelength are difficult to establish, so the choice is usually left to trial and error.

14.5.2. Graphic Method

Kalman and Heller (*K2*) developed a graphic background correction method for determination of minor and trace analytes in low-Z matrixes. For each analyte, a set of calibration curves is established of analyte-line intensity I_A *vs.* background intensity I_B, each curve for a different analyte concentration. All curves in the set are linear and have a more or less common origin. The curves were established from measurements on synthetic standards having matrixes consisting of various proportions of aluminum oxide (Al_2O_3), calcium carbonate ($CaCO_3$), sodium chloride (NaCl), and iron(III) oxide (Fe_2O_3) to give various scattered intensities. An analysis consists of measurement of I_A and I_B from the sample and location of their common point on the set of curves for the appropriate analyte. Analyte concentrations not represented by a curve are dealt with by interpolation.

14.5.3. Scattered Target-Line Ratio Method

This method is a special case of the background-ratio method. The calibration function is the intensity ratio of the analyte line and a scattered x-ray tube target line (*B50*, *C52*, *L26*, *S39*). Either the coherently (*C52*) or incoherently (*B50*) scattered line may be used. If only, say, the coherently scattered line is measured, Equation (14.22) reduces to

$$I_{B,\lambda} \propto S_R/[2(\mu/\varrho)_\lambda] \tag{14.25}$$

Coherent scatter S_R is proportional to Z^2, and μ/ϱ is proportional to Z^4, so that $I_{B,\lambda} \propto Z^{-2}$. Clearly, the scattered target-line method is most effective if the target line is a good internal-standard line for the analyte line. The limited number of targets available commercially and convenience considerations usually preclude the optimal choice. The next best alternative is the target having a line as close to the analyte line as possible. The target-line ratio method, like the background-ratio method, is ineffective against enhancement. Table 14.2 gives some typical data for this method.

14.5.4. Ratio of Coherent to Incoherent Scattered Intensities

For a given wavelength λ, both $(\mu/\varrho)_\lambda$ and the coherent-to-incoherent (Rayleigh-to-Compton) scatter ratio are related to atomic number, and therefore to each other. For light matrixes (*C15*),

$$I_{\mathrm{A}} = \frac{KC_{\mathrm{A}}}{(\mu/\varrho)_{\lambda_P} + (\mu/\varrho)_{\lambda_{\mathrm{A}}}} \tag{14.26}$$

where I_{A} and C_{A} are analyte-line intensity and concentration; $(\mu/\varrho)_{\lambda_P}$ and $(\mu/\varrho)_{\lambda_{\mathrm{A}}}$ are mass-absorption coefficients for the primary and analyte-line x-rays, respectively; and K is a constant. To a good approximation, this equation reduces to

$$I_{\mathrm{A}} \approx \frac{K'C_{\mathrm{A}}}{(\mu/\varrho)_{\lambda_{\mathrm{A}}}} \tag{14.27}$$

Ryland (*R19*) has shown that for light matrixes, μ/ϱ is a linear function of the ratio of the intensities of the coherently (Rayleigh) I_R and incoherently (Compton) I_C scattered target lines; that is,

$$(\mu/\varrho)_{\lambda_{\mathrm{A}}} \propto (I_R/I_C) = R \tag{14.28}$$

Then

$$I_{\mathrm{A}} \approx K''C_{\mathrm{A}}/R \tag{14.29}$$

Ryland has based a very successful absorption correction method on this equation. Analyte-line intensity is measured from the sample I_{X} and from a single standard I_{S} having known analyte concentration C_{S} in a matrix similar to that of the sample. Rayleigh and Compton scattered target-line intensities are measured from both the sample and standard and used to calculate R_{X} and R_{S} [Equation (14.28)]. Analyte concentration in the sample is then given by

$$C_{\mathrm{X}} = C_{\mathrm{S}}\left(\frac{I_{\mathrm{X}}}{I_{\mathrm{S}}}\right)\left(\frac{R_{\mathrm{X}}}{R_{\mathrm{S}}}\right) \tag{14.30}$$

Ryland applied the method to the determination of elements having atomic number 13–82 (aluminum to lead) in samples having average atomic number 5–10 in the form of briquets, glass disks, and water and organic solutions. Carman (*C21*) applied a similar technique to heavy-element matrixes diluted with polyethylene; he used calibration curves of $I_X R_X$ *vs.* analyte concentration.

Reynolds (*R2*) used a somewhat different approach. I_R and I_C are measured from a series of standards having known or readily measured or calculated absorption coefficients $(\mu/\varrho)_\lambda$ at some wavelength near the center of the wavelength range of the spectral lines of analytical interest. Reynolds chose 0.9 Å. The data were plotted in a graph of $(\mu/\varrho)_\lambda$ *vs.* R, and gave a linear curve. Analyses consisted of the measurement of analyte-line intensity from the sample I_X and from a single standard having known analyte concentration C_S, known $(\mu/\varrho)_\lambda$, and a matrix similar to that of the sample. I_R and I_C are measured from the sample, and R_X is calculated and applied to the curve to derive $(\mu/\varrho)_\lambda$. Analyte concentration in the sample is then given by

$$C_X = C_S\left(\frac{I_X}{I_S}\right)\left(\frac{(\mu/\varrho)_{\lambda,X}}{(\mu/\varrho)_{\lambda,S}}\right) \tag{14.31}$$

Later, Reynolds (*R2*) abandoned the ratio method and measured only the incoherently scattered target line I_C, but, except for use of I_C for R, the method is the same. The ratio method was abandoned because of difficulty in measurement of I_R due to diffraction effects from his highly crystalline powdered mineral samples. Compton scattered radiation, being incoherent, has no fixed phase relationship with the scattering centers, or between incident and emitted radiation, and is free from diffraction effects. Ryland's samples—solutions, glasses, organic solids—are free from diffraction effects, thereby accounting for her success with the ratio method.

Ryland used a chromium-target x-ray tube; Reynolds used molybdenum. Carman used a tungsten–chromium dual-target tube; the tungsten and chromium targets were used for high-Z and low-Z elements, respectively.

These techniques, in addition to providing effective absorption correction, may be used with a single standard and without calibration curves.

14.6. DILUTION METHODS

14.6.1. Principles

The term *dilution methods* is applied to a group of methods that: (1) correct for absorption-enhancement effects by leveling the absorption coefficients of the samples and standards to a common value determined by an added diluent; and/or (2) correct for inhomogeneity and particle-size effects by dissolution or fusion of the specimens.

The efficiency of the primary beam of effective wavelength λ_P for exciting analyte line λ_L is given by Equation (4.14):

$$\frac{I_L}{I_{0,\lambda_P}} = P_{\mathrm{A}} \frac{C_{\mathrm{A}}(\mu/\varrho)_{\mathrm{A},\lambda_P}}{\sum C_i[(\mu/\varrho)_{i,\lambda_P} + A(\mu/\varrho)_{i,\lambda_L}]}$$

where:

I_L is analyte-line intensity.

I_{0,λ_P} is incident primary intensity.

C_{A} is analyte concentration.

C_i is concentration of an individual matrix element *i*.

$(\mu/\varrho)_{\mathrm{A},\lambda_P}$ is mass-absorption coefficient of the analyte for the primary x-rays.

$(\mu/\varrho)_{i,\lambda_P}$ and $(\mu/\varrho)_{i,\lambda_L}$ are mass-absorption coefficients of an individual matrix element for the primary and analyte-line x-rays, respectively.

P_{A} and A are defined by Equations (4.7) and (4.8), respectively, and are constant for any given analysis and need not concern us here.

For the present discussion, Equation (4.14) may be expanded as follows.

$$\frac{I_L}{I_{0,\lambda_P}} = P_{\mathrm{A}} \frac{C_{\mathrm{A}}(\mu/\varrho)_{\mathrm{A},\lambda_P}}{C_{\mathrm{M}}[(\mu/\varrho)_{\mathrm{M},\lambda_P} + A(\mu/\varrho)_{\mathrm{M},\lambda_L}] + C_{\mathrm{D}}[(\mu/\varrho)_{\mathrm{D},\lambda_P} + A(\mu/\varrho)_{\mathrm{D},\lambda_L}] + \sum C_i[(\mu/\varrho)_{i,\lambda_P} + A(\mu/\varrho)_{i,\lambda_L}]} \tag{14.32}$$

where M refers to a specific matrix element, D to the added diluent, and *i* to the individual elements in the remainder of the matrix; M and D are discussed below.

If absorption-enhancement effects occur, variations in matrix composition among samples and standards may change the denominator in Equation (14.32) and thereby the value of $I_L/I_{0,\lambda_P}$ for the same analyte concentration C_{A}. Suppose that a certain matrix element M undergoes a change in concentration ΔC_{M} from specimen to specimen, causing a change in its contribution to the denominator of Equation (14.32), the first term in the denominator. If the change in this contribution is very small, the effect on the denominator—and therefore on the analyte-line intensity—may be negligible. This is the case if: (1) C_{M} is very small to begin with, so that even a *relatively* large ΔC_{M} and consequent *relatively* large change in its contribution may have little effect on the denominator; (2) ΔC_{M} is very small, so that the change in the term is small; or (3) $(\mu/\varrho)_{\mathrm{M},\lambda_P}$ and $(\mu/\varrho)_{\mathrm{M},\lambda_L}$ are not substantially different from the average absorption coefficient of the specimen.

However, the effect of ΔC_M on the analyte-line intensity is substantial when a relatively large ΔC_{M} occurs in a relatively large C_{M} of a matrix element having absorption coefficients very different—either higher or lower—from the average absorption coefficient of the specimen.

The denominator of Equation (14.32) may be made relatively insensitive to changes in concentration of the matrix element by judicious selection and addition of a diluent D in either or both of two ways: (1) A high concentration C_{D} may be added of a diluent having very low absorption coefficients $(\mu/\varrho)_{\mathrm{D},\lambda_P}$ and $(\mu/\varrho)_{\mathrm{D},\lambda_L}$ relative to the average absorption coefficient of the specimens (*G14*); and/or (2) a low concentration C_{D} may be added of a "diluent" having high absorption coefficients relative to the average absorption coefficients (*A3, C32, R11*). In either case, the diluent largely determines the absorption coefficient of the specimen for both primary and analyte-line radiation, and the variations in absorption coefficient caused by variations in concentrations of matrix elements are reduced, often to insignificance.

Dilution also minimizes enhancement, either by actual reduction of the concentration of the enhancing element, or by increasing the absorption of its enhancing spectral lines.

Frequently, when a low-absorption diluent (solvent or flux) is used to reduce inhomogeneity or particle-size effects, it is necessary to add a high-absorption "diluent" to level residual absorption-enhancement contributions of the original matrix (*R11*). The absorber is added in concentration such that reasonable variations in concentrations of elements in the specimen do not substantially affect the analyte-line intensity from the prepared specimen.

14.6.2. Discussion

The two basic forms of dilution mentioned above provide the basis for the several practical dilution methods. Low-absorption dilution is applied in solution, fusion, and glass-disk methods, and in the dry dilution of powders, with or without briqueting. In a sense, methods in which the specimen is supported on filter paper or ion-exchange resin or membrane are also dilution methods.

The technique of dilution to the extent that matrix absorption is about the same in all specimens is applicable only if, on such dilution, analyte-line intensity is still adequate. Cullen (*C55*) recommends the following procedure for evaluation of the technique as applied to solutions, although the principle is applicable to any specimen form. Two solutions are prepared having the same analyte concentration, one in pure solvent, the other in solution with the maximum expected concentrations of matrix elements. The solutions are diluted stepwise until analyte-line intensity is the same for both. It is then determined whether the analyte-line intensity is adequate for the required statistical precision in acceptable counting time. Table 14.3 gives data for such an evaluation for Cu $K\alpha$ in the presence of iron. Incidentally, the table also illustrates another feature of low-absorption dilution techniques: On dilution, analyte-line intensity decreases at a much lower rate than concentration, as is evident from comparison of the Cu $K\alpha$ intensities for 0 and 1:1 dilution, and for 20:1 and 100:1 dilution at 50 kV, 50 mA.

High-absorption dilution is applied in the high-absorption-fusion, absorption-buffer, and matrix-masking solution techniques. For addition to powders and for fusion techniques, lanthanum oxide (La_2O_3) is an ex-

TABLE 14.3

Effect of Dilution on Cu $K\alpha$ Intensity in the Presence of Iron[a]

Dilution[b]	Excitation kV, mA	Cu $K\alpha$ intensity, counts/s		Cu $K\alpha$ intensity ratio
		Cu[c]	(Cu + Fe)[d]	I_{Cu}/I_{Cu+Fe}
None	30, 20	8707	7194	1.210
1:1	30, 20	6562	6072	1.081
10:1	50, 20	4709	4568	1.033
20:1	50, 35	4866	4858	1.002
20:1	50, 50	7059	7029	1.004
100:1	50, 50	3549	3541	1.002

[a] From Cullen (*C55*).
[b] Volume of water:volume of solution (Cu or Cu + Fe).
[c] Cu concentration in undiluted solution 10 mg/ml.
[d] Cu and Fe concentrations in undiluted solution 10 mg/ml each.

TABLE 14.4

Effect of Matrix-Masking Agent on Cu $K\alpha$ Intensity in the Presence of Various Matrix Elements[a]

Element added and atomic no.	Cu $K\alpha$ net intensity, counts/s	Cu $K\alpha$ net intensity ratio I_{Cu}/I_{Cu+M}
None	6143 (I_{Cu})	1.000
19-K	6094	1.008
26-Fe	5999	1.024
28-Ni	6113	1.005
30-Zn	5964	1.030
47-Ag	6005	1.023
82-Pb	5895	1.042

[a] From Cullen (*C55*). Matrix masking agent $Ca(NO_3)_2$ in concentration 10 mg/ml. Cu concentration in all solutions 2 mg/ml; concentrations of matrix elements 2 mg/ml.

cellent high-absorption diluent. It has high absorption coefficients; it is stable and available in high purity; its *K* lines occur at relatively short wavelengths (~0.37 Å) and are relatively inefficiently excited by a 50-kV generator—especially a full-wave generator (V_{LaK} 39 kV); and its *L* lines occur at relatively long wavelengths (2–3 Å).

A *matrix-masking agent* in the form of a salt of a noninterfering, relatively light element may be added in constant concentration to solution specimens. Cullen (*C55*) demonstrated this technique by addition of calcium nitrate [$Ca(NO_3)_2$] in concentration 10 mg/ml to solutions of copper alone and with other elements. His data are given in Table 14.4. The method is often used to reduce absorption-enhancement effects in determination of plate metals in plating solutions, which have extremely variable matrixes.

Specific dilution techniques are discussed in Chapters 16 and 17.

In addition to correction of absorption-enhancement, inhomogeneity, and particle-size effects, dilution methods are useful when the same material must be analyzed in a variety of forms—such as bar, rod, wire, foil, drillings, shot, powder, and small fabricated parts. If all forms are fused or put into solution, a single set of standards serves for all. The principal disadvantages of the method are that specimen preparation is required and that low-concentration analytes may be diluted to the extent that analyte-line intensity is prohibitively low, especially with low-*Z* analytes. Other advantages and disadvantages of specific techniques are given in Chapters 16 and 17.

14.7. THIN-FILM METHODS

14.7.1. Principles

Absorption-enhancement effects substantially disappear in thin-film-type specimens because neither primary nor analyte-line x-rays are significantly absorbed in the extremely short path lengths in the thin layer. Each atom absorbs and emits substantially independently of the other atoms without significantly altering the incident beam. Thus, for very thin films of constant thickness, analyte-line intensity is directly proportional to analyte concentration, and for films of constant composition, intensity is proportional to thickness. These relationships provide the basis for a method free from absorption-enhancement effects for analysis of films containing more than one element, and a method for measurement of thickness of films of known composition. Of course, many samples are thin films to begin with, but for those that are not, reduction to a thin film is one way to deal with absorption-enhancement effects.

Gunn (*15H*) demonstrated the simple relationship between spectral-line intensity of an element and the number of atoms of that element in a thin film. The contribution dI of an incremental volume of element of constant area and thickness dt at depth t is

$$dI = k(\csc\phi)I_0 \exp - \{[(\mu/\varrho)_{\lambda_P} \csc\phi + (\mu/\varrho)_{\lambda_L} \csc\psi]\varrho t\}\, dt \qquad (14.33)$$

where:

- I_0 is incident primary intensity.
- ϕ is the angle between the central ray of the primary cone and the specimen.
- ψ is the angle between the central ray of the secondary cone and the specimen—the takeoff angle.
- λ_P and λ_L are primary and analyte-line wavelengths, respectively.
- μ/ϱ is the mass-absorption coefficient of the specimen (cm^2/g).
- ϱ is the density of the specimen (g/cm^3).
- t is film thickness (cm).
- k is a constant.

If t is very small, $(\mu/\varrho)_{\lambda_P}$ and $(\mu/\varrho)_{\lambda_L}$ both approximate 0, and

$$dI = k(\csc\phi)I_0\, dt \qquad (14.34)$$

or, incrementally,

$$\Delta I = k(\csc\phi)I_0\, \Delta t \qquad (14.35)$$

For a constant area, $\Delta N \propto \Delta t$, where N is the number of analyte atoms in the layer. Then,

$$\Delta I = k'(\csc \phi) I_0 \, \Delta N \tag{14.36}$$

Thus, for thin films, $\Delta I \propto \Delta N$.

Birks [p. 63 in (*6*)] demonstrated the absence of absorption-enhancement effects of diverse elements on the analyte-line intensity of an analyte in a thin film. The proof begins with Equation (15.29) derived in Section 15.4.2 for spectral-line intensity from element i in a thin film of thickness t:

$$I_{S,\lambda_L} = \frac{T_i \varrho_i I_0}{\bar{\mu}\varrho} \, (1 - \exp - \bar{\mu}\varrho t)$$

where:

I_{S,λ_L} is secondary analyte-line intensity.

T_i is an excitation constant relating incident primary intensity and emitted analyte-line intensity of element i.

ϱ_i is density of element i in the layer (g/cm³).

I_0, ϱ, and t have the same significance as in Equation (14.33).

$\bar{\mu}$ represents the term in square brackets in Equation (14.33), that is,

$$\bar{\mu} = (\mu/\varrho)_{\lambda_P} \csc \phi + (\mu/\varrho)_{\lambda_L} \csc \psi \tag{14.37}$$

In Section 15.4.2., Equation (15.29) provides the basis for the derivation of a method for mathematical correction of absorption-enhancement effects; here, it provides the basis for proof of the absence of these effects in thin films. In Section 15.4.2, it is shown that

$$T_i I_0 = I_{100i} \tag{14.38}$$

where I_{100i} is analyte-line intensity from pure i (100% i). It is also shown that the quantity $\bar{\mu}$ may be expressed in terms of influence coefficients a_{ij}, each representing the overall absorption-enhancement effect of a particular matrix element j in the film an element i. Then,

$$\bar{\mu}\varrho = \Sigma \, a_{ij} \varrho_j \tag{14.39}$$

where ϱ_j is the density of element j in the film. Substitution of Equations (14.38) and (14.39) in Equation (15.29) gives

$$I_i = \frac{I_{100i} \varrho_i}{\Sigma \, a_{ij} \varrho_j} (1 - \exp - t \, \Sigma \, a_{ij} \varrho_j) \tag{14.40}$$

where I_i is the I_{S,λ_L} of Equation (15.29). If t is very small, the exponential term may be approximated by use of the mathematical relationship,

$$\exp -y \approx 1 - y \qquad (y \text{ very small}) \tag{14.41}$$

giving

$$I_i = \frac{I_{100i}\varrho_i}{\sum a_{ij}\varrho_j}(1 - 1 + t \sum a_{ij}\varrho_j) \tag{14.42}$$

This equation reduces to

$$I_i = I_{100i}\varrho_i t \tag{14.43}$$

Equation (14.43) shows that the analyte-line intensity from a thin film is linearly related to thickness and is, in fact, simply the product of the intensity from massive pure element I_{100i} and area density $\varrho_i t$ (g/cm^2). The intensity is unaffected by other elements j in the film, since the a_{ij} values do not appear in the equation.

14.7.2. Infinite (Critical) Thickness

Figure 14.13 shows spectral-line intensity measured from a thin film as a function of thickness. The curve has three regions. For extremely thin films, attenuation of incident primary and emergent secondary spectral-line radiation is extremely small. The curve is linear, and intensity is simply proportional to thickness. For films of intermediate thickness, attenuation of both primary and secondary x-rays increases with depth. The longer-

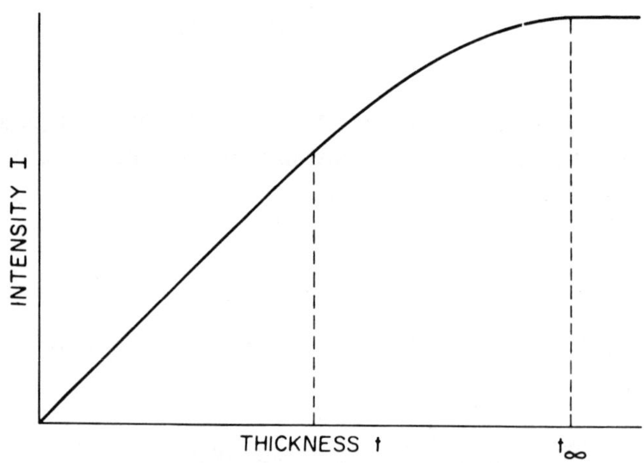

Fig. 14.13. Infinite (critical) thickness.

wavelength components of the primary beam are absorbed preferentially, so that as the beam penetrates progressively deeper, it becomes both weaker and of shorter effective wavelength. The emergent secondary spectral x-rays continue to increase in intensity with thickness, but at a decreasing rate. At infinite (critical) thickness, t_∞ or t_C, secondary x-rays are excited at depths from which they cannot emerge to the surface. Further increase in thickness results in no further increase in intensity.

Equation (14.33) may be used to calculate infinite thickness. From this equation, the intensity I_t from a film of thickness t is

$$I_t = kI_0 \csc\phi \int_0^t \exp[-(\bar{\mu}\varrho t)]\, dt \tag{14.44}$$

$$= \frac{kI_0 \csc\phi(1 - \exp -\bar{\mu}\varrho t)}{\bar{\mu}\varrho} \tag{14.45}$$

where the notation has the same significance as in Equation (14.33), except $\bar{\mu}$, which is defined in Equation (14.37). At infinite thickness, $t = \infty$, and

$$I_\infty = \frac{kI_0 \csc\phi}{\bar{\mu}\varrho} \tag{14.46}$$

and

$$\frac{I_t}{I_\infty} = 1 - \exp -\bar{\mu}\varrho t \tag{14.47}$$

or

$$\log_e\left(1 - \frac{I_t}{I_\infty}\right) = -\bar{\mu}\varrho t \tag{14.48}$$

At the infinite (critical) thickness in the x-ray sense (Figure 14.13), $I_t/I_\infty = 1$. To calculate the thickness at which I_t just becomes I_∞, a value of, say, 0.999 is arbitrarily chosen for I_t/I_∞, and Equation (14.47) or (14.48) solved for t.

In films consisting of more than one element, the absorption coefficient of the film is different for the spectral line of each element. Thus, infinite thickness is different for each spectral line.

14.7.3. Discussion

It has already been pointed out that some specimens already have the form of thin films. These include true evaporated films, platings, corrosion layers, and sublimation deposits, as well as microgram amounts of specimen evaporated on Mylar or absorbed in filter paper or ion-exchange membrane. Extremely small particles or flakes and extremely thin fibers also act as thin specimens. Birks refers to all specimens having less than infinite thickness as specimens of limited quantity. Alternatively, in order to reduce absorption-enhancement effects, bulk specimens may be reduced to thin films by vacuum evaporation, rolling and beating, etching, etc., or by grinding to extremely fine powder, suspension in an organic vehicle-binder, and casting to a thin film with a casting knife.

The intensities from extremely thin films are very low, so thin-film methods are applicable principally to major constituents. Moreover, both composition and thickness are variables. Thick films may begin to show absorption-enhancement effects. An internal standard may be added to such films by volumetric addition of a standard solution of an internal-standard element.

Thin-film standards can be prepared by vacuum evaporation, electrodeposition, or casting, or by evaporation of small volumes of dilute solutions in filter paper or on Mylar film. Standards on specific substrates can be prepared by deposition directly on the substrate, or the film may be deposited on thin Mylar film or Scotch tape, then backed by the required substrate. A set of graduated thin-film standards may be prepared by deposition of the thinnest film on a very thin x-ray-transparent substrate, such as Mylar or filter paper, then stacking various numbers of layers of these films.

A novel thin-film method requiring no standards and no quantitative addition or dilution has been developed for analysis of binary systems (*F6*). Filter-paper disks are impregnated with equal volumes of the solution of the binary sample. Spectral-line intensities are measured for both elements from increasing numbers of stacked papers. A curve is plotted of the intensity ratio as a function of thickness, that is, number of stacked papers, and extrapolated to zero thickness. The limiting ratio is directly equivalent to the molar concentration ratio. The method provides a means for correcting absorption-enhancement effects.

X-ray spectrometry has been applied to the determination of silver, chlorine, bromine, and iodine in photographic emulsions on plates, films, and papers (*B7*, *M30*). X-ray spectrometric determination of silver in developed films used for radiation dosimetry has extended their useful dosimetry range from ~100 R to ~2000 R.

If the thin-film specimen is unsupported, or if it is supported on an x-ray-transparent substrate, such as Mylar or filter paper, most of the primary beam passes through the specimen and is wasted. This lost radiation may be used to excite a secondary radiator having spectral lines on the short-wavelength side of the absorption edge of the analyte, and mounted under the specimen. The secondary radiation excites additional analyte-line radiation (*H35*).

X-ray spectrometric methods for determination of film and plating thickness are discussed in Chapter 18.

14.8. SPECIAL EXPERIMENTAL METHODS

This section describes a few examples of the many x-ray spectrometric methods that cannot be classified in any of the categories considered so far.

14.8.1. Emission-Absorption Methods

In emission-absorption methods (*D26, S1*), a portion of the sample itself is used as an absorption filter to obtain data for correction of absorption-enhancement effects on the analyte-line emission measured from the sample in the usual way. One such method is described briefly in Section 10.5, but probably the most sophisticated example of this method is that of Leroux and Mahmud (*L10*). This method is considered in detail in the following paragraphs.

Each analysis requires three briqueted specimen disks: X, the sample disk to be analyzed; S, a standard disk having known analyte concentration near the average of the expected analytical range; and R, a "radiator" disk containing analyte. The S and R disks may serve for all analyses of a given sample type. The X and S disks must be x-ray-transparent, that is, less than infinitely thick for the analyte line, so that analyte-line radiation emitted from the radiator disk can be measured after passing through an X or S disk. The weights of the X and S disks must be known. The R disk must have an analyte concentration high enough so that its analyte-line emission is measurable through the X and S disks.

Each determination requires ten x-ray intensity measurements—five analyte-line peak intensities and their backgrounds—to obtain the following five net intensities: I_R, net analyte-line intensity measured from the radiator disk exposed directly to the primary x-ray beam; I_X and I_S, net analyte-line intensities measured from the sample and standard disks, respectively, each exposed alone to the primary beam; and $I_{X,R}$ and $I_{S,R}$, net analyte-line intensities measured from the sample and standard disks, respectively, each backed by the radiator disk. In practice, I_R is measured behind an open-ring spacer having thickness about the same as that of pellets X and S.

The transmittance T of the sample and standard disks to analyte-line radiation is then calculated:

$$T_X = \frac{I_{X,R} - I_X}{I_R} \quad \text{and} \quad T_S = \frac{I_{S,R} - I_S}{I_R} \tag{14.49}$$

From the values of T_X and T_R, values of the function $\phi(T)$ are calculated for the sample and standard:

$$\phi(T_X) = \frac{-\log_e T_X}{1 - T_X} \quad \text{and} \quad \phi(T_S) = \frac{-\log_e T_S}{1 - T_S} \tag{14.50}$$

For frequently analyzed sample types, a curve of T as a function of $\phi(T)$ may be prepared. Analyte concentration is given by solution of the following equation for C_X.

$$\frac{C_X}{C_S} = \left(\frac{I_X}{I_S}\right)\left(\frac{w_S}{w_X}\right)\frac{\phi(T_X)}{\phi(T_S)} \tag{14.51}$$

where C_X and C_S are analyte concentration in the sample and standard disks, respectively, and w_X and w_S are their weights.

14.8.2. Indirect (Association) Methods

An extremely significant development in x-ray spectrometric analysis is the indirect determination—or determination by association—of the following classes of analytes: (1) elements relatively inconvenient to determine by x-ray spectrometry—atomic numbers 9–17 (fluorine through chlorine); (2) elements not determinable at all on standard commercial spectrometers—atomic numbers less than 9; and (3) organic compounds and other specific chemical compounds and substances. Such analyses are sometimes conveniently done by stoichiometric association with a readily determined element.

Examples of the first group (above) are the determination of trace chlorine in titanium by precipitation as silver chloride and measurement of the silver (*R17*) and the determination of silicon by formation of silico-molybdate-8-quinolinol complex (Si:Mo = 1:12) and measurement of the molybdenum (*S25*).

A striking example of the second group is the microdetermination of nitrogen in amounts 0.2-5 μg in biological materials by converting it to ammonia, absorbing the ammonia in filter paper impregnated with Nessler's reagent, washing out excess reagent, then determining the mercury in the mercury–iodine–amine complex thus formed (*M7*). A related example is the determination of carbon—or carbon-to-hydrogen ratio—in hydrocarbons by means of the intensity ratio of coherent to incoherent scattered target-line radiation (Section 20.2.2); of course, this is not an association method (*D24*, *D25*).

Three techniques have been evaluated for stoichiometric association of a determinable element with organic compounds (*P1*): bromination with bromine water, chelation with 5-chloro-7-iodo-8-quinolinol, and salt-formation with barium or lead. These techniques reverse the usual procedure of using an organic reagent, such as 8-quinolinol or dithizone, to specifically separate and precipitate an element prior to its determination by x-ray spectrometry.

Montmorillonite is determined in montmorillonite–kaolinite clay mixtures by prolonged agitation of the samples in strontium chloride solution, filtration, and measurement of the strontium in the recovered clay (*H28*). Only the montmorillonite undergoes ion exchange with the strontium to a significant extent. The method is calibrated by use of standards having known montmorillonite–kaolinite ratios. An alternative method would be to use strontium chloride solutions of known concentration, then determine the strontium depletion in the recovered filtrate.

The significance of these indirect methods cannot be overestimated. They extend the applicability of x-ray spectrometry to the entire periodic table, to organic chemistry, and to other specific materials—all with standard commercial instrumentation and accessories and with conventional preparation and presentation techniques.

14.8.3. Combinations of X-Ray Spectrometry and Other Methods

X-ray spectrometry has been used to complement many other analytical methods. Combined x-ray emission and absorption methods are discussed in Section 14.8.1. An example of the combination of x-ray emission and diffraction is provided by the measurement of the thickness of chromium and nickel double platings on brass (*G3*). The chromium (top) plating was determined by x-ray diffractometry. The underlying nickel plating was determined by measurement of Ni $K\alpha$ intensity and correction for absorption in the overlying chromium layer.

Thorium in concentration ~0.5% was determined in uranium–thorium ores by a combination of x-ray spectrometry and radiochemistry (*C10*). Total radioactivity was measured and expressed as equivalent uranium concentration. Then the U $L\alpha$/Th $L\alpha$ intensity ratio was measured and used to calculate the uranium-to-thorium weight ratio. Finally, the known relative radioactivity of uranium and thorium was used to calculate absolute uranium and thorium weight concentrations.

Alkylbenzene sulfonate and residual sodium sulfate were determined in water slurries by a combination of x-ray spectrometry and infrared spectrophotometry (*K22*). Total sulfur was determined by the x-ray method, the alkylbenzene sulfonate by the infrared method. Then the sulfur equivalent to the sulfonate was calculated and subtracted from total sulfur to give sodium sulfate concentration.

Sulfur in concentration down to 0.002% was determined in gasoline by a combined x-ray spectrometric standard-addition and density method (*J15*). Two 50-ml portions A and B were taken of samples having sulfur concentration $\leq 0.15\%$. Portion A was treated with 5 ml of sulfur-free iso-octane, B with 5 ml of a standard solution containing 0.13 g of sulfur. Net S $K\alpha$ intensities were measured from the treated samples, and the density ϱ of the untreated sample was measured with a hydrometer. Sulfur concentration in weight per cent is given by

$$C = \frac{(2)[0.13I_A/(I_B - I_A)]}{\varrho} \tag{14.52}$$

Jackwerth and Kloppenburg (*J1*, *J2*) describe an accessory for passing chromatographic paper strips continuously through the x-ray spectrometer specimen compartment. Small chromatographic spots are irradiated as a whole; large spots are passed under a slit as emitted analyte-line intensity is recorded. The area under the recorded curve is proportional to the amount of analyte. Organic compounds can be determined if combined with an equivalent of heavy element.

X-ray spectrometry has been used to complement gas chromatography for determination of halogens and sulfur in gas samples (*S9*). Chromatographic effluent—carrier gas and separated analytical constituents—is passed through a thin-walled capillary in the specimen compartment. With a 5-cm length of capillary, the detection limit was 0.5 μg of carbon disulfide per milliliter of carrier gas. X-ray spectrometry has also been used to complement paper electrophoresis (*W4*).

Chapter 15

Mathematical Correction of Absorption-Enhancement Effects

Having considered purely experimental methods of dealing with absorption-enhancement effects, we now consider primarily mathematical methods. Most of these methods are so complex that a comprehensive treatment cannot be given here. However, we can set forth the principles of some of the outstanding methods.

15.1. GEOMETRIC METHODS

Certainly the outstanding example of the geometric approach to the correction of absorption-enhancement effects is the method of Taylor *et al.* (*T3*) for ternary systems.

In a ternary system of elements A, B, and C having atomic weights M_A, M_B, and M_C in weight fractions W_A, W_B, and W_C and atom fractions A_A, A_B, and A_C, composition of any specimen is given by equations of the form

$$(W_A)_{ABC} = \frac{A_A M_A}{A_A M_A + A_B M_B + A_C M_C} \tag{15.1}$$

$$(A_A)_{ABC} = \frac{W_A/M_A}{(W_A/M_A) + (W_B/M_B) + (W_C/M_C)} \tag{15.2}$$

The corresponding relationships in the three binary systems A–B, A–C, and B–C have the forms

$$(W_{\mathrm{A}})_{\mathrm{AB}} = \frac{A_{\mathrm{A}}M_{\mathrm{A}}}{A_{\mathrm{A}}M_{\mathrm{A}} + A_{\mathrm{B}}M_{\mathrm{B}}} = \frac{A_{\mathrm{A}}M_{\mathrm{A}}}{A_{\mathrm{A}}M_{\mathrm{A}} + (1 - A_{\mathrm{A}})M_{\mathrm{B}}} \tag{15.3}$$

$$(A_{\mathrm{A}})_{\mathrm{AB}} = \frac{W_{\mathrm{A}}/M_{\mathrm{A}}}{(W_{\mathrm{A}}/M_{\mathrm{A}}) + (W_{\mathrm{B}}/M_{\mathrm{B}})} = \frac{W_{\mathrm{A}}/M_{\mathrm{A}}}{(W_{\mathrm{A}}/M_{\mathrm{A}}) + (1 - W_{\mathrm{A}})/M_{\mathrm{B}}} \tag{15.4}$$

All these equations are readily expressed in terms of per cent as well as fraction.

Suppose that the triangular composition diagram for the ternary system A–B–C is plotted in terms of atom concentration. Then Equation (15.3) and the analogous equations for the A–C and B–C binary systems can be used to mark off weight concentrations along the A–B, A–C, and B–C sides of the triangle. A skew grid may be constructed on the triangle by drawing lines joining equal or complementary concentrations on the three sides—for example, 90A–10C on A–C with 90A–10B on A–B and 90B–10C on B–C, and 50A–50C on A–C with 50A–50B on A–B and 50B–50C on B–C. The grid permits reading of weight concentrations directly on the atom composition diagram. Similarly, Equation (15.4) and its analogs can be used to construct a skew grid to permit reading of atom concentrations on a weight composition diagram.

Now, if the spectral-line intensities emitted by the three elements in the ternary system A–B–C are I_{A}, I_{B}, and I_{C}, the "count fraction" F for each intensity is given by an equation of the form

$$(F_{\mathrm{A}})_{\mathrm{ABC}} = I_{\mathrm{A}}/(I_{\mathrm{A}} + I_{\mathrm{B}} + I_{\mathrm{C}}) \tag{15.5}$$

In the corresponding binary systems, the count fraction equations have the form

$$(F_{\mathrm{A}})_{\mathrm{AB}} = I_{\mathrm{A}}/(I_{\mathrm{A}} + I_{\mathrm{B}}) \tag{15.6}$$

Like weight and atom fractions, count fractions can also be expressed in percent.

Taylor *et al.* (*T3*) have based a geometric method for correction of absorption-enhancement effects in ternary systems on the obvious similarity of the equations for weight, atom, and count fractions. Their procedure is as follows:

1. Three sets of standards are prepared, one for each of the three binary systems.
2. Three calibration curves are established from the spectral-line

intensity measurements on these standards: $(F_A)_{AB}$ *vs.* W_A in A–B, $(F_A)_{AC}$ *vs.* W_A in A–C, and $(F_B)_{BC}$ *vs.* W_B in B–C.

3. The x-ray data are used to calibrate a triangular coordinate graph having corners designated $100F_A$, $100F_B$, $100F_C$ and sides graduated in linear 0–100% F scales. Along the A–B and A–C sides, W_A (%) is plotted from the two F_A curves, and along the B–C side, W_B (%) is plotted from the F_B curve.

4. A weight–concentration skew grid is then drawn on the triangle in the manner described above.

The result is a ternary calibration graph. An analysis consists of measuring I_A, I_B, and I_C from the sample, calculating F_A, F_B, and F_C from Equation (15.5), and finding the composition corresponding to the three F values on the triangular diagram.

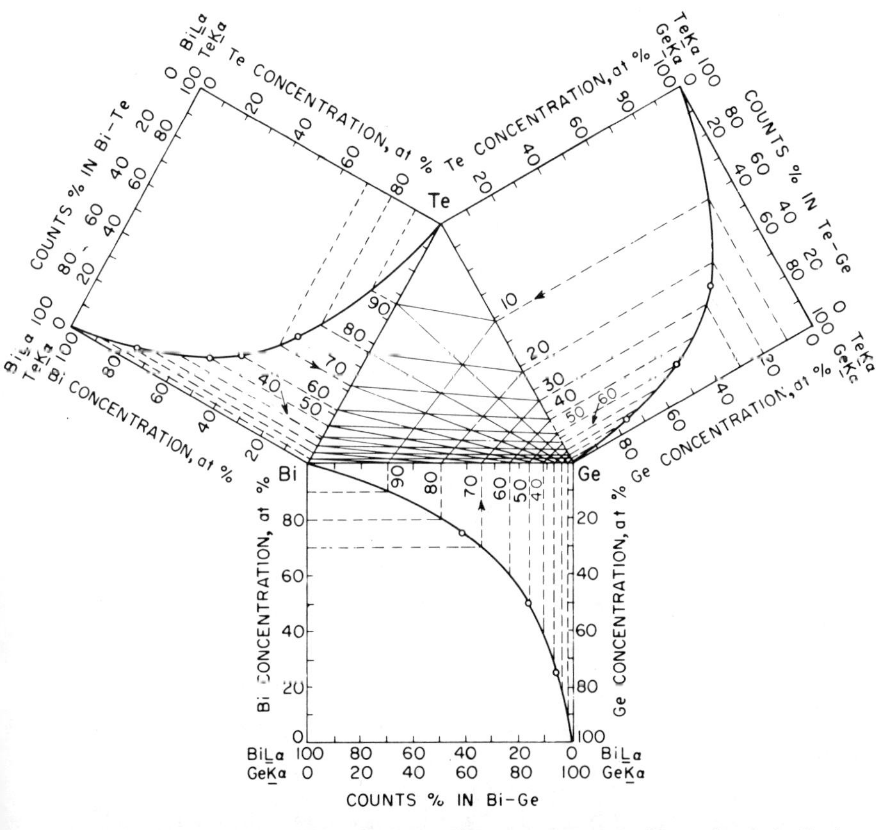

Fig. 15.1. Graphic method for analysis of ternary systems. [Data from A. Taylor, B. J. Kagle, and E. W. Beiter, *Analytical Chemistry* **33**, 1699 (1961).]

Figure 15.1 shows the application of the method to the germanium–tellurium–bismuth system. The sides of the triangle are graduated linearly in counts per cent, as shown by the equally spaced "tics." In practice, the triangle is a standard triangular coordinate diagram, but in the figure, the triangular grid is omitted for simplicity. X-ray intensity data measured from germanium–tellurium, germanium–bismuth, and tellurium–bismuth binary standards are used to construct calibration curves of counts per cent *vs.* atom per cent on the sides of the triangle. These curves are used to graduate the triangle in concentration units, as shown by the dashed lines and numbers along each side. Finally, a concentration skew grid is drawn on the triangle.

Another example of the geometric method is the application of nomographs (*M28*, *M29*).

15.2. ABSORPTION CORRECTION

Absorption-enhancement effects often consist simply of absorption effects, that is, the effects of the absorption coefficients of the matrix elements. In Table 15.1 and Figure 15.2B, the Sn $K\alpha$ intensity data from tin–lead alloys shows negative deviation from linearity (defined in Section 12.1) for this reason [pp. 136–8 in (*14*)]. At the primary wavelengths most efficient in exciting Sn $K\alpha$—those near the tin K-absorption edge (0.425 Å)—the mass-absorption coefficients of tin and lead are 40 and 25 cm^2/g, respec-

TABLE 15.1

X-Ray Intensity Data for Tin–Lead Alloys[a]

Sample	Sn concn., wt %	Sn $K\alpha$ net intensity, counts/s	Pb concn., wt %	Pb $L\alpha_1$ net intensity, counts/s
1	100	497	0	4
2	74.98	314	24.97	177
3	59.97	227	39.92	260
4	40.00	144	59.89	373
5	20.18	66	79.84	473
6	0	2	100	568

[a] From Jenkins and de Vries [p. 134 in (*14*)].

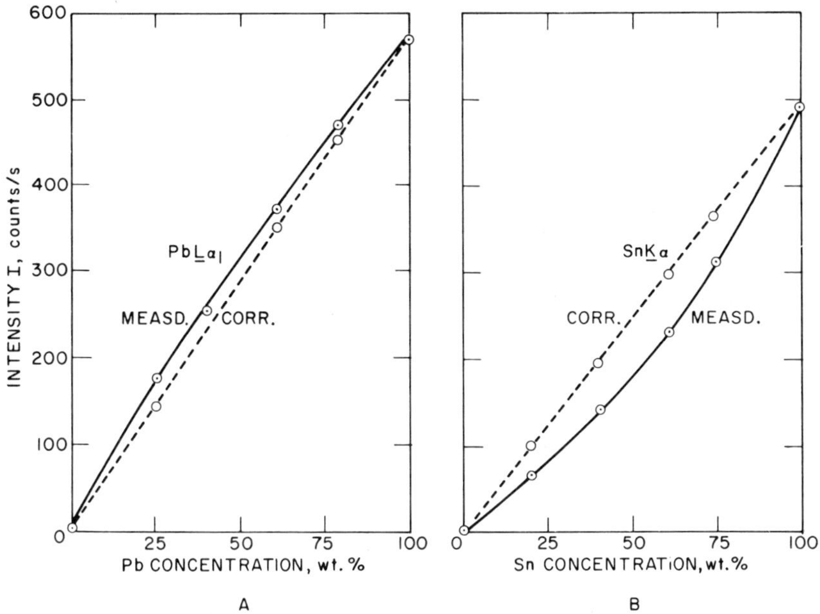

Fig. 15.2. Absorption-enhancement effects in the tin–lead system. [Data from R. Jenkins and J. L. de Vries, "Practical X-Ray Spectrometry," Springer-Verlag New York, Inc., pp. 133–8 (1967).]

tively; at Sn $K\alpha$, they are 12 and 52 cm^2/g. Then the total mass-absorption coefficients for both primary and Sn $K\alpha$ radiation are, for tin and lead, 52 and 77 cm^2/g, respectively.

If it is assumed that only absorption effects are present,

$$I_X = C_{Sn} I_{100Sn} \frac{(\mu/\varrho)_{100Sn}}{(\mu/\varrho)_X} \tag{15.7}$$

where C_{Sn} is tin concentration; I_X and I_{100Sn} are Sn $K\alpha$ net intensities from the sample X and pure tin 100Sn, respectively; $(\mu/\varrho)_X$ and $(\mu/\varrho)_{100Sn}$ are total mass-absorption coefficients of the sample and pure tin, respectively, for both primary and Sn $K\alpha$ radiation. Equation (15.7) is derived from Equation (4.13) by substitution of a monochromatic effective wavelength for the term $J(\lambda)$, which represents the total primary spectrum. Equation (15.7) may now be used to correct the Sn $K\alpha$ data in Table 15.1. Values of $(\mu/\varrho)_X$ for samples 1–6 are calculated by use of the μ/ϱ values for tin and lead given above. The Sn $K\alpha$ intensities from samples 2–5 are multiplied by the ratio $(\mu/\varrho)_X/(\mu/\varrho)_{100Sn}$, that is, $(\mu/\varrho)_X/(\mu/\varrho)_1$, where 1 refers to sample 1, which is pure tin. Then,

$$(I_X)_{cor} = (I_X)_{measd} \frac{(\mu/\varrho)_X}{(\mu/\varrho)_{100Sn}} \tag{15.8}$$

where $(I_X)_{cor}$ and $(I_X)_{measd}$ are, respectively, corrected and measured Sn $K\alpha$ net intensities from the sample. The $(I_X)_{cor}$ values are plotted in Figure 15.2B, giving the dashed linear curve. The Pb $L\alpha_1$ data in Table 15.1 and Figure 15.2A are discussed in Section 15.4.1.

Of course, μ/ϱ for a sample cannot be calculated before the sample composition is known. This problem may be circumvented in several ways, including the following:

1. If some accuracy may be sacrificed, an approximate analytical concentration may be derived from the uncorrected intensity data, and this value used to calculate $(\mu/\varrho)_X$ to a first approximation.
2. Analyte-line intensity may be measured from pure analyte placed behind a sample disk of known mass and thickness. Since both primary and analyte-line radiation must pass through the sample disk, total μ/ϱ for the matrix may be calculated.
3. If primary absorption can be neglected, the sample disk may be used to attenuate the analyte-line radiation from a sample of pure analyte.
4. This third technique can also be used to evaluate primary absorption if an effective wavelength can be assigned to the primary spectrum. A secondary target having a strong spectral line at this wavelength may be placed in the specimen compartment and its intensity measured through the sample disk.

Hower (*H43*) developed an absorption correction method for determination of trace heavy-element analytes in minerals. The method is based on the assumptions that: (1) of the three principal absorption-enhancement effects—primary-beam absorption, mutual enhancement, and secondary-beam absorption—only the first and last are significant for high-Z trace elements in minerals having light matrixes, and (2) these effects are determined by the mass-absorption coefficient of the mineral. Curves of mass-absorption coefficient as a function of wavelength were plotted for elements that are commonly major constituents in minerals. These curves showed that *relative* mass-absorption coefficient is approximately constant at all wavelengths. Curves of relative mass-absorption coefficient were plotted for various minerals relative to aluminum oxide. Matrix correction was then made for trace concentrations of nickel and heavier analytes by determination of the absorption of the sample mineral relative to aluminum oxide at a specified wavelength.

15.3. EMPIRICAL CORRECTION FACTORS

Mitchell (*M20*, *M21*), using empirical correction factors, successfully analyzed systems having up to seven components. The basic principle of the method is as follows. In the A–B–C–D–E–F–G system, line intensity of element A is measured, corrected for absorption-enhancement effects by C, D, E, F, and G, then applied to a calibration curve of A-line intensity *vs.* A concentration in a B matrix. In other words, after correction, A is determined as if it were in an A–B binary system. Mitchell applied the method to determination of all seven metallic elements in the system TiO_2–V_2O_5–Fe_2O_3–ZrO_2–Nb_2O_5–Ta_2O_5–WO_3. The procedure follows, but for simplicity, the discussion is in terms of the elements, rather than the oxides.

The calibration procedure is as follows. For each element, one of the other six is selected as its binary complement. The complement may be the element having highest concentration or most severe absorption-enhancement effect, or it may simply be an element for which binary standards are available. The complement is not necessarily the same for all seven analytes. Table 15.2 gives Mitchell's choice of binary complements. Standards are prepared for each of the seven binary systems in the composition range of interest, and the analyte-line intensities are measured from all standards. A calibration curve of analyte-line intensity *vs.* concentration is established for each of the seven systems. Then, for each analyte binary system, cor-

TABLE 15.2

Plan of Analysis of Seven-Component System[a]

Analyte	Binary complement for basic calibration curve	Interfering elements for which calculation of correction factors is required
Ti	Ta	V or Fe, Zr or Nb, W
V	Ti	Fe, Zr or Nb, Ta or W
Fe	Ta	Ti or V, Zr or Nb, W
Zr	Ta or W	Ti or V, Fe, Nb
Nb	Ta or W	Ti or V, Fe, Zr
Ta	Zr or Nb	Ti or V, Fe, W
W	Ta	Ti or V, Fe, Zr or Nb

[a] From Mitchell (*M21*).

rection factors are derived for the other five elements, except that similarity of certain elements may preclude the need for separate factors, as shown in column 3 of Table 15.2. The derivation of the correction factors is described below.

The analytical procedure is as follows. Line intensity is measured for each of the seven elements from the sample(s). Mitchell measured the intensities for all seven elements simultaneously on an Applied Research Laboratories model XIQ multichannel x-ray quantometer. The measured line intensity of each of the seven analytes is applied to its binary calibration curve to derive its approximate concentration. This approximate concentration is then corrected by the following procedure, outlined for niobium as an example.

For each interfering element for the niobium system (Table 15.2), the correction factor is derived by applying the approximate analyte and interferant concentrations to the appropriate table or curve in the manner discussed below. The total correction for niobium in tantalum F_{NbTa} is then calculated from

$$F_{\mathrm{NbTa}} = \Sigma F_{\mathrm{NbTa},i} = F_{\mathrm{NbTa,Ti}} + F_{\mathrm{NbTa,V}} + F_{\mathrm{NbTa,Fe}} + F_{\mathrm{NbTa,Zr}} \quad (15.9)$$

where F_{NbTa} is the correction factor to be applied to the Nb $K\alpha$ intensity measured from the sample, and $F_{\mathrm{NbTa},i}$ is the correction factor for the effect of element i on Nb $K\alpha$ intensity in a tantalum matrix. Because of similarity of the effect of titanium and vanadium on Nb $K\alpha$ intensity in a tantalum matrix (column 3 of Table 15.2),

$$F_{\mathrm{NbTa,V}} = F_{\mathrm{NbTa,Ti}} \quad (15.10)$$

The niobium-line intensity measured from the sample is then corrected:

$$I_{\mathrm{Nb,cor}} = I_{\mathrm{Nb,measd}}/F_{\mathrm{NbTa}} \quad (15.11)$$

This corrected intensity is applied to the niobium calibration curve to derive corrected niobium concentration. This same procedure is followed for each of the other six elements.

The correction factors are derived as follows. Figure 15.3A shows calibration curves of Nb $K\alpha$ intensity *vs.* Nb_2O_5 concentration in TiO_2 (or V_2O_5), Fe_2O_3, ZrO_2, and Ta_2O_5 (or WO_3) matrixes. These curves could be used for determination of Nb_2O_5 in the respective binary systems. For a Nb–Ti–Ta oxide ternary system, Nb $K\alpha$ intensities for all possible compositions lie on or between the Nb–Ti and Nb–Ta curves. Likewise, in

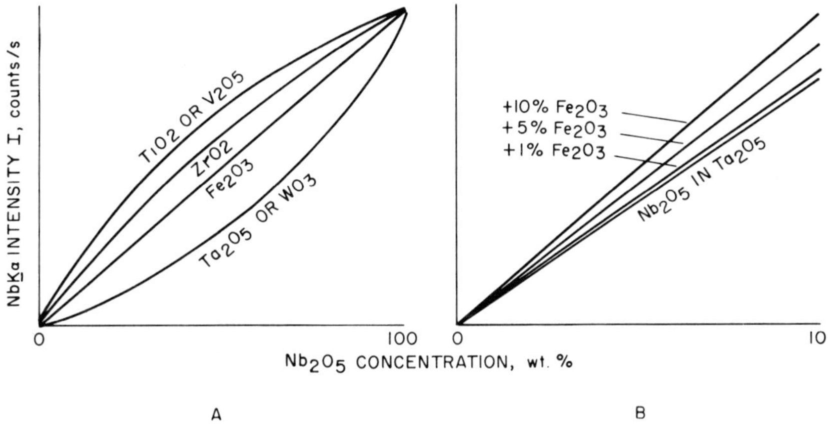

Fig. 15.3. Absorption-enhancement effects of several elements on Nb *Kα* intensity.

the Nb–Fe–Zr system, all possible Nb *Kα* intensities lie on or between the Nb–Fe and Nb–Zr curves. Of course, the entire "intensity area" may not be of analytical interest. Mitchell derived correction factors for multicomponent systems on the basis of such ternary relationships in two ways.

In general, for the multicomponent system A–B–C–D– . . ., let A be the analyte and B its designated binary complement. Then the correction factor for the effect of, say, C on A in a B matrix is

$$F_{\mathrm{AB,C}} = \frac{I_{\mathrm{A}} \text{ in a } B + C \text{ matrix}}{I_{\mathrm{A}} \text{ in a } B \text{ matrix}} \tag{15.12}$$

For absorption, $F < 1$; for enhancement, $F > 1$. Standards having appropriate A–B–C compositions are prepared, and the $F_{\mathrm{AB,C}}$ value is derived for each. A similar procedure is followed for A–B–D, A–B–E, etc.

The other method for derivation of correction factors is based on the fact that within a limited low-concentration region, curves of analyte-line intensity *vs.* concentration are substantially linear. Figure 15.3B shows a set of curves of Nb *Kα* net intensity as a function of Nb_2O_5 concentration in a Ta_2O_5 matrix having 0, 1, 5, and 10% Fe_2O_3 as the interferant (element C). These curves have the form

$$I = mC_{\mathrm{A}} + b \tag{15.13}$$

where m and b are slope and I-axis intercept, respectively. If I is net intensity, $b = 0$, as shown. Then, for a given C_{A}, correction factors are calculated for interferant concentrations 1, 2, 3, 4 (in Figure 15.3B, 0, 1, 5, 10% Fe_2O_3, respectively) as follows.

$$\begin{aligned} F_1 &= m_1/m_1 = I_1/I_1 = 1 \\ F_2 &= m_2/m_1 = I_2/I_1 \\ F_3 &= m_3/m_1 = I_3/I_1 \\ F_4 &= m_4/m_1 = I_4/I_1 \end{aligned} \tag{15.14}$$

Regardless of which method is used, for each analyte binary system, a set of curves is prepared for each interfering element in the multicomponent system. These curves relate F with interferant concentration, each curve for a different analyte concentration. The curves have the form shown in Figure 15.4. Alternatively, the data may be prepared in tabular form similar to Table 15.3.

In a later paper, Mitchell (*M22*) showed that data based on the relationships of absorption coefficient, atomic number, and wavelength permit prediction of spectral-line intensity for a particular element over the full concentration range in a matrix of any atomic number on a particular spectrometer. The spectral-line intensity of an element at each of several concentrations in each of several selected matrixes is measured and plotted. The form of the curves is shown in Figure 15.5B for 10% SnO_2 in matrixes of atomic number 10–92. The curve is essentially the reverse of the μ/ϱ *vs.* Z

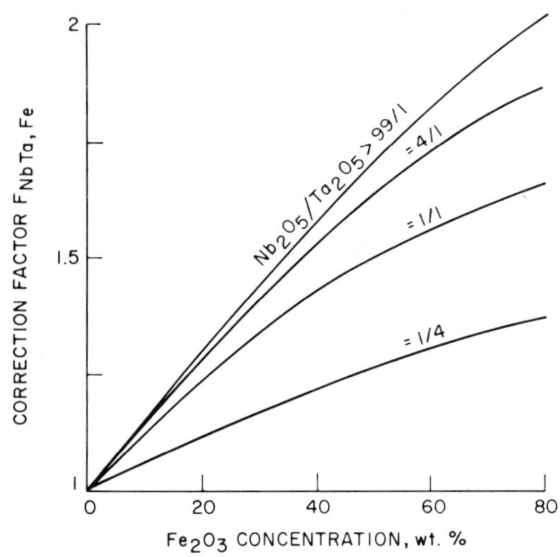

Fig. 15.4. Correction factors for iron on Nb $K\alpha$ intensity from a Ta_2O_5 matrix. [Data from B. J. Mitchell, *Analytical Chemistry* **30**, 1894 (1958).]

TABLE 15.3

Correction Factors for Effect of Fe_2O_3 on Nb $K\alpha$ Intensity from Nb_2O_5 in a Ta_2O_5 Matrix[a]

Fe_2O_3 concn., wt %	Correction factor F, for various Nb_2O_5/Ta_2O_5 weight ratios			
	>40/1	4/1	1/1	1/4
1	1.015	1.010	1.005	1.002
5	1.09	1.07	1.05	1.04
10	1.17	1.14	1.10	1.08
15	1.27	1.21	1.15	1.12
20	1.35	1.29	1.21	1.17
25	1.43	1.355	1.27	1.215
30	1.52	1.43	1.33	1.265
35	1.60	1.50	1.40	1.315
40	1.69	1.575	1.47	1.37
45	1.77	1.65	1.55	1.425
50	1.85	1.73	1.63	1.48

[a] From Mitchell (*M21*).

curve for Sn $K\alpha$ (Figure 15.5A). However, the peak of the I–Z curve occurs at Z 54, rather than Z 47, as in the (μ/ϱ)–Z curve, because of enhancement of Sn $K\alpha$ by the $K\beta$ lines of elements having $Z > 51$ and by the $K\alpha$ lines of elements having $Z > 54$. This enhancement effect is more complex in a matrix emitting strong L lines. The line of interest undergoes variations in absorption due to its position with respect to the three matrix L-absorption edges, and variations in enhancement by the matrix L lines. It is evident that Sn $K\alpha$ intensity is the same for certain pairs of elements having very different atomic numbers—for example, 26-Fe and 56-Ba, 30-Zn and 47-Ag, and 41-Nb and 73-Ta.

From such sets of curves, intensity may be calculated for other concentrations and for other matrixes. In this way, Mo $K\alpha$ data measured at several concentrations in eight matrixes permitted the establishment of the curves for 0–100% molybdenum in all matrixes from 20-Ca to 92-U. Representative curves from this set are shown in Figure 15.6. Table 15.4 shows the

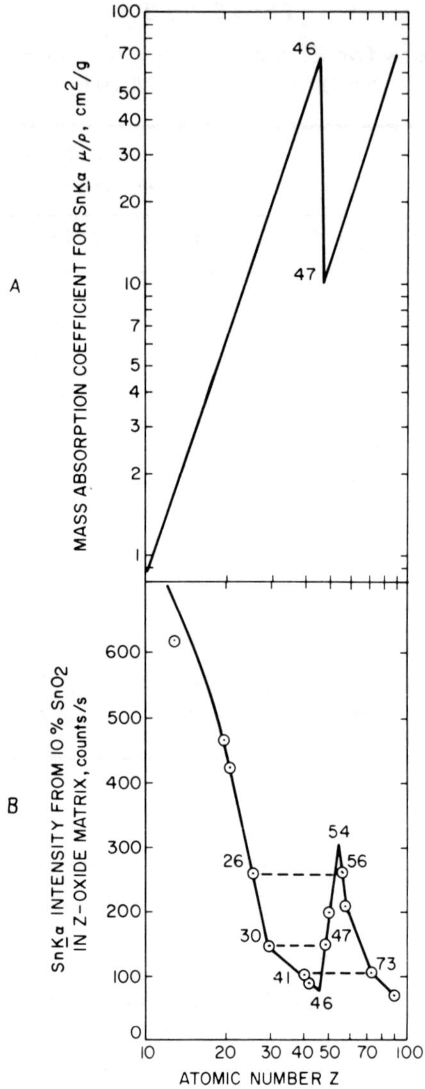

Fig. 15.5. Correlation of atomic number and (A) mass-absorption coefficient for Sn Kα and (B) Sn *Kα* intensity from 10% SnO_2. [B. J. Mitchell, *Analytical Chemistry* **33**, 917 (1961); courtesy of the author and the American Chemical Society.]

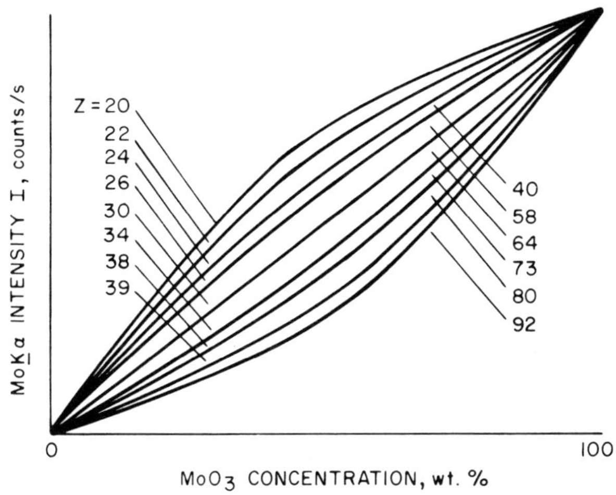

Fig. 15.6. Estimated Mo *Kα* intensity patterns for matrixes of calcium to uranium (atomic numbers 20–92). [B. J. Mitchell, *Analytical Chemistry* **33**, 917 (1961); courtesy of the author and the American Chemical Society.]

TABLE 15.4

Additive Nature of Absorption-Enhancement Effects on Nb *Kα* Intensity from Nb_2O_5 in a Ta_2O_5 Matrix[a]

Interferant and concentration, wt %	Individual effect of interferant: increase in Nb *Kα* intensity,[b] %	Measured combined effect,[b] %
1% TiO_2	2.5	
1% V_2O_5	2.5	
1% Fe_2O_3	1.5	
1% ZrO_2	1.5	
TOTAL	8.0	8.7
10% TiO_2	26	
10% V_2O_5	26	
10% Fe_2O_3	17	
10% ZrO_2	18	
TOTAL	87	86.8

[a] From Mitchell (*M21*).
[b] With respect to Nb *Kα* intensity from Nb_2O_5 in a Ta_2O_5 matrix in weight ratio >40/1.

additive nature of the correction factors for niobium in tantalum. The presence of titanium, vanadium, iron, and zirconium increases the Nb $K\alpha$ intensity relative to its intensity in a pure tantalum matrix: 1% each of TiO_2, V_2O_5, Fe_2O_3, and ZrO_2 gives an 8.7% enhancement; 10% of each gives an 86.8% enhancement. Mitchell proposes that these correction factors may be determined under standardized conditions readily attainable in any x-ray laboratory; then, in other laboratories, only a single basic calibration is required for each element in conjunction with sets of correction-factor tables or curves. In this respect, Mitchell's method is similar to that of Harvey (*H14*) in the field of optical emission spectrography.

15.4. INFLUENCE COEFFICIENTS

15.4.1. Introduction

Sherman (*S15–17*) attempted to derive general mathematical relationships between measured intensity and concentration in multicomponent systems. He first derived the intensity–concentration relationship for the element having highest atomic number. The *I–C* relationships for elements of successively lower atomic number were then corrected for enhancement by the spectra of elements of higher atomic number. However, the equations were successful only for limited concentration ranges in two- or three-component systems. Sherman's approach has been used also by Noakes (*N12*), Pluchery (*P11*), and Weyl (*W6*). Burnham *et al.* (*B60*) used graphic solutions to eliminate the need for solution of simultaneous equations.

The Pb $L\alpha_1$ data in Table 15.1 and Figure 15.2A provide a convenient and simple introduction to the influence-coefficient method for correction of absorption-enhancement effects [pp. 133–6 in (*14*)]. The Pb $L\alpha_1$ curve shows positive deviation from linearity (defined in Section 12.1), but for a reason different from that which explained the deviation in the Sn $K\alpha$ data (Section 15.2). The Pb $L\alpha_1$ line is not absorbed to very different degrees by tin and lead (μ/ϱ 134 and 128 cm^2/g, respectively). The Pb $L\alpha_1$ line, which is associated with the Pb LIII absorption edge (0.95 Å), is not strongly enhanced by Sn $K\alpha$ (0.49 Å). The deviation is due to the difference in absorption coefficients of tin and lead for the primary wavelengths most efficient in exciting Pb $L\alpha_1$—those near Pb $L\mathrm{III}_{\mathrm{ab}}$; for example, at 0.9 Å, the absorption coefficients of tin and lead are 64 and 145 cm^2/g, respectively; at 0.75 Å, they are 40 and 160. Thus, the excitation efficiency of the primary beam for Pb $L\alpha_1$ and the slope of the calibration curve increase with tin concentration. If it is assumed that the slope increases in proportion to tin concentration, the following equation obtains:

$$C_{\mathrm{Pb}} = \frac{I_{\mathrm{Pb}}}{m_{\mathrm{Pb}}}(1 + a_{\mathrm{PbSn}}C_{\mathrm{Sn}}) \tag{15.15}$$

where C is concentration; I is net intensity; m is the slope of the curve, or Pb $L\alpha_1$ intensity per 1% lead for pure lead; and a_{PbSn} is the absorption-enhancement influence factor for tin for the Pb $L\alpha_1$ line. From this,

$$(I_{\mathrm{Pb}})_{\mathrm{cor}} = (I_{\mathrm{Pb}})_{\mathrm{measd}}(1 + a_{\mathrm{PbSn}}C_{\mathrm{Sn}}) \tag{15.16}$$

The value of a_{PbSn} is determined by substitution of calibration data from standards into Equation (15.16). In the present example, a_{PbSn} was -0.0021. The $(I_{\mathrm{Pb}})_{\mathrm{cor}}$ values are plotted in Figure 15.2A, giving the dashed linear curve. In general terms,

$$C_i = \frac{I_i}{m_i}[1 + \Sigma\,(a_{ij}C_j)] \tag{15.17}$$

where C, I, and m have the significance given above; i is the analyte element; j is a matrix element; and a_{ij} is the absorption-enhancement effect of element j on the analyte i. The factor a is negative for calibration curves having negative deviation, $(\mu)\varrho)_j > (\mu/\varrho)_i$, positive for positive deviation, $(\mu/\varrho)_j < (\mu/\varrho)_i$. The influence factor corrects for both absorption and enhancement because the latter may be regarded as negative absorption.

For a ratio method,

$$C_{\mathrm{X}} = \left(\frac{I_{\mathrm{X}}}{I_{\mathrm{S}}}\right)_i [1 + \Sigma\,(a_{ij}C_j)]C_{\mathrm{S}} \tag{15.18}$$

where C and I are concentration and net intensity, respectively, and S and X refer to sample and standard, respectively.

Beattie and Brissey (*B10*) analyzed three- and four-component systems by establishing linear simultaneous equations involving empirical absorption parameters or influence coefficients a relating intensities and mass concentrations for pairs of elements. For a binary system A–B, these coefficients have the form

$$a_{\mathrm{AB}} = \frac{W_{\mathrm{A}}}{W_{\mathrm{B}}}(R_{\mathrm{A}} - 1) \tag{15.19}$$

$$a_{\mathrm{BA}} = \frac{W_{\mathrm{B}}}{W_{\mathrm{A}}}(R_{\mathrm{B}} - 1) \tag{15.20}$$

where a_{AB} is the influence coefficient of element B for element A; a_{BA} is the influence coefficient of element A for element B; W is weight fraction; and R is defined by

$$R_A = I_{AA}/I_{AS}; \qquad R_B = I_{BB}/I_{BS} \tag{15.21}$$

where I_{AA} and I_{AS} are A-line intensity from pure A and from the specimen, and I_{BB} and I_{BS} have similar significance for B-line intensity.

Many workers have applied, modified, refined, and extended Beattie's and Brissey's basic approach. The following derivation and treatment is based on Birks' development of the method [pp. 58–63 in (*6*)].

15.4.2. Derivation of Basic Equation

The derivation is based on the assumptions that: (1) the specimen is homogeneous, infinitely thick (in the sense of Section 14.7.2), and flat; (2) the primary x-radiation is monochromatic, or, alternatively, that a monochromatic beam has the same effect as the actual polychromatic primary spectrum for all specimen compositions of interest; and (3) enhancement by interelement secondary excitation within the specimen has the same effect as low matrix absorption—that is, enhancement is regarded as negative absorption.

In Figure 15.7, consider an incremental layer of thickness dt at depth t in a specimen of density ϱ. The incident angle of the primary beam O is ϕ, and the takeoff angle of the secondary beam S is ψ. Now consider a single wavelength λ_P having intensity I_{0,λ_P} in the incident primary beam. The intensity arriving at layer dt is

$$I_{t,\lambda_P} = I_{0,\lambda_P} \exp -[(\mu/\varrho)_{\lambda_P}\varrho t \csc \phi] \tag{15.22}$$

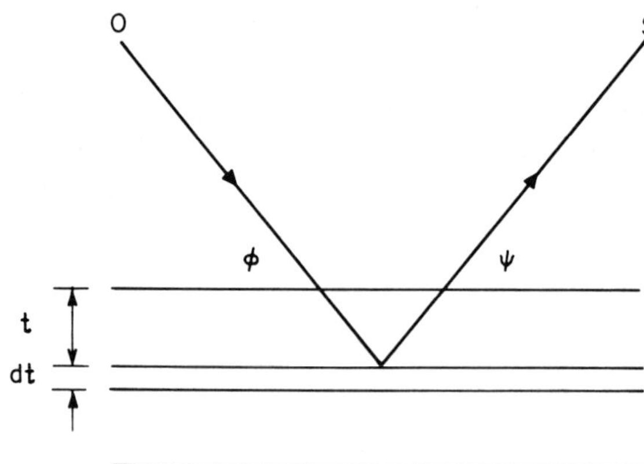

Fif. 15.7. Geometry of secondary excitation.

where $(\mu/\varrho)_{\lambda_P}$ is the mass-absorption coefficient of the specimen for λ_P. I_{t,λ_P} excites element i in layer dt to emit its characteristic spectrum, and the intensity of the measured line λ_L at layer dt is

$$dI_{t,\lambda_L} = T_i \varrho_i I_{t,\lambda_P}\, dt \tag{15.23}$$

where T_i is an excitation constant relating incident primary intensity and emitted spectral-line intensity, and ϱ_i is the density of element i in layer dt. The intensity of the spectral line emerging from the specimen is

$$dI_{S,\lambda_L} = dI_{t,\lambda_L} \exp -[(\mu/\varrho)_{\lambda_L} \varrho t \csc \psi] \tag{15.24}$$

where $(\mu/\varrho)_{\lambda_L}$ is the mass-absorption coefficient of the specimen for λ_L. Substitution of Equations (15.22) and (15.23) into (15.24) gives

$$dI_{S,\lambda_L} = T_i \varrho_i I_{0,\lambda_P} \{\exp -[(\mu/\varrho)_{\lambda_P} \varrho t \csc \phi]\}\, dt \exp -[(\mu/\varrho)_{\lambda_L} \varrho t \csc \psi] \tag{15.25}$$

$$= T_i \varrho_i I_{0,\lambda_P} \exp - \{\varrho t[(\mu/\varrho)_{\lambda_P} \csc \phi + (\mu/\varrho)_{\lambda_L} \csc \psi]\}\, dt \tag{15.26}$$

Since ϕ and ψ are constant for a given x-ray spectrometer, and $(\mu/\varrho)_{\lambda_P}$ and $(\mu/\varrho)_{\lambda_L}$ are constant for a given specimen, the term in square brackets in Equation (15.26) may be replaced with a single constant $\bar{\mu}$,

$$\bar{\mu} = (\mu/\varrho)_{\lambda_P} \csc \phi + (\mu/\varrho)_{\lambda_L} \csc \psi \tag{15.27}$$

The constant $\bar{\mu}$ includes both enhancement and absorption effects. Equation (15.26) then becomes

$$dI_{S,\lambda_L} = T_i \varrho_i I_{0,\lambda_P} (\exp -\bar{\mu} \varrho t)\, dt \tag{15.28}$$

Integration of Equation (15.28) from $t = 0$ to t gives

$$I_{S,\lambda_L} = \frac{T_i \varrho_i I_{0,\lambda_P}}{\bar{\mu} \varrho} (1 - \exp -\bar{\mu} \varrho t) \tag{15.29}$$

As t approaches infinity, Equation (15.29) becomes

$$I_{S,\lambda_L} = (T_i \varrho_i I_{0,\lambda_P})/\bar{\mu} \varrho \tag{15.30}$$

Equation (15.30) may now be rewritten by: (1) expressing $\bar{\mu}$ in terms of new coefficients a_{ij} representing the overall absorption-enhancement effect of element j on element i, so that

$$\bar{\mu} = \Sigma\, a_{ij} \tag{15.31}$$

(2) taking into account all wavelengths that can excite element i; and (3) noting that

$$\varrho = \sum \varrho_i \tag{15.32}$$

where ϱ_i is the density of element i in the specimen. Then, from Equation (15.30),

$$\frac{T_i \varrho_i I_0}{I_i} = \bar{\mu}\varrho = \sum a_{ij}\varrho_j = a_{i1}\varrho_1 + a_{i2}\varrho_2 + \cdots + a_{ii}\varrho_i + \cdots + a_{in}\varrho_n \tag{15.33}$$

where, for simplicity, I_0 is used for I_{0,λ_P}, and I_i is used for I_{S,λ_L}, the measured analyte-line intensity for element i. The term a_{ii} is always a true constant, and, in fact, has the value 1. This is readily shown by solving Equation (15.33) for the case of pure (100%) element i ($100i$): all the ϱ_j terms become 0, except that ϱ_i becomes ϱ, the specimen density; I_i becomes I_{100i}; and Equation (15.33) becomes

$$T_i \varrho I_0 / I_{100i} = a_{ii}\varrho \tag{15.34}$$

The ϱ's cancel, and from the definition of T_i above, $T_i I_0 = I_{100i}$. Equation (15.34) then reduces to $a_{ii} = 1$. Equation (15.33) may now be rewritten by substitution of these values for $T_i I_0$ and a_{ii} and gathering terms:

$$a_{i1}\varrho_1 + a_{i2}\varrho_2 + \cdots + [1 - (I_{100i}/I_i)]\varrho_i + \cdots + a_{in}\varrho_n = 0 \tag{15.35}$$

where I_i and I_{100i} represent analyte-line intensity from element i from the specimen and from pure i, respectively. If Equation (15.35) is divided through by ϱ, and it is noted that ϱ_i/ϱ is weight fraction W_i, one gets

$$a_{i1}W_1 + a_{i2}W_2 + \cdots + [1 - (I_{100i}/I_i)]W_i + \cdots + a_{in}W_n = 0 \tag{15.36}$$

A similar expression can be written for each element in the specimen, giving the set of simultaneous equations (15.38), in which, for simplicity,

$$I_{100i}/I_i = R_i \tag{15.37}$$

$$(1 - R_1)W_1 + a_{12}W_2 + a_{13}W_3 + \cdots + a_{1n}W_n = 0 \tag{15.38-1}$$

$$a_{21}W_1 + (1 - R_2)W_2 + a_{23}W_3 + \cdots + a_{2n}W_n = 0 \tag{15.38-2}$$

$$\cdots$$

$$a_{n1}W_1 + a_{n2}W_2 + a_{n3}W_3 + \cdots + (1 - R_n)W_n = 0 \tag{15.38-n}$$

$$W_1 + W_2 + W_3 + \cdots + W_n = 1 \tag{15.39}$$

Equation (15.39) is required because the n equations (15.38) are not mutually independent. In practice, if minor constituents are known to be present, but are of no interest or significance, Equation (15.39) may be set equal to the sum of the concentrations of the major and other significant constituents—for example, 0.98.

Because of the assumptions on which the derivation is based, Equations (15.38) are stated in terms of binary influence coefficients a_{ij}. Each of these coefficients represents the effect of only one element on another, disregarding all other elements present, and is therefore assumed to be constant. Unfortunately, the a_{ij} coefficients are not true constants, but are themselves functions of all other elements present. Sherman (*S17*) has analyzed this complex situation, but his equations can be solved only for specimens having only a few components. However, there are many systems for which acceptable accuracy is realized by considering the a_{ij} coefficients to be constant to a first-order or second-order approximation.

15.4.3. Evaluation of a_{ij} Coefficients to a First Approximation

To a first approximation, each a_{ij} coefficient may be assumed to depend only on elements i and j, and its value may be derived from the intensities of the spectral line of i from pure i and from a binary standard of i–j having known composition. The i/j concentration ratio in this standard may be 1/1 or, preferably, approximately the ratio of i and j in the samples to be analyzed. If the samples contain n elements, it is necessary to establish a set of n simultaneous equations of the form of Equations (15.38). These equations contain $n(n-1)$ coefficients of the form a_{ij} $(j \neq i)$. Evaluation of these coefficients requires the n pure elements and their $(n/2)(n-1)$ binary combinations.

To illustrate the procedure [pp. 60–61 in (*6*)], suppose alloys of Cr–Fe–Ni–Mo are to be analyzed. Let i (or j) = 1, 2, 3, 4, respectively, for these elements. To derive a_{CrMo}, the effect of Mo ($j = 4$) on Cr ($i = 1$), Equation (15.38-1) is used, but for a Cr–Mo binary system, all terms drop out except those containing W_{Cr} and W_{Mo}, leaving

$$(1 - R_{\text{Cr}})W_{\text{Cr}} + a_{\text{CrMo}}W_{\text{Mo}} = 0 \tag{15.40}$$

from which

$$a_{\text{CrMo}} = (R_{\text{Cr}} - 1)(W_{\text{Cr}}/W_{\text{Mo}}) \tag{15.41}$$

R_{Cr}, that is, I_{100Cr}/I_{Cr}, is obtained from the Cr $K\alpha$ intensities from pure chromium and the chromium-molybdenum binary alloy standard. W_{Cr} and W_{Mo} are the known concentrations in the standard. Similarly, a_{MoCr}, the effect of chromium on molybdenum, is derived from Equation (15.38-4) and Mo $K\alpha$ intensities from pure molybdenum and the chromium–molybdenum standard:

$$a_{MoCr}W_{Cr} + (1 - R_{Mo})W_{Mo} = 0 \tag{15.42}$$

$$a_{MoCr} = (R_{Mo} - 1)(W_{Mo}/W_{Cr}) \tag{15.43}$$

In general,

$$a_{ij} = (R_i - 1)W_i/W_j \tag{15.44}$$

Once all the a_{ij} coefficients are evaluated and substituted in Equations (15.38), these equations serve to calculate the composition of any quaternary alloy of these elements to a first approximation. The analysis of each sample consists of: (1) evaluation of R_i for each of the four elements by measurement of the i spectral-line intensity from pure i and from the sample; (2) substitution of the R_i values in Equations (15.38), which already contain the a_{ij} coefficients; and (3) solution of the simultaneous equations for the weight fractions W, usually by the method of determinants or with a computer.

The method of first approximations was used by Beattie and Brissey (*B10*) for analysis of chromium–iron–nickel–molybdenum alloys with relative errors of 0.5–6.5%, and by Hakkila and Waterbury (*H4*) for analysis of solutions of chromium–iron–nickel alloys. These analyses provide excellent practical examples of the application of the first-approximation method and are discussed in detail below.

The procedure of Hakkila and Waterbury (*H4*) is as follows:

1. Six solutions were prepared containing 200 mg of, respectively, chromium, iron, nickel, Cr–Fe (1:1), Cr–Ni (1:1), and Fe–Ni (1:1).

2. Cr $K\alpha$, Fe $K\alpha$, and Ni $K\alpha$ net intensities were measured from all six solutions; Cr $K\alpha$ background was obtained by averaging the intensities at 2θ for Cr $K\alpha$ from the pure iron and nickel solutions; Fe $K\alpha$ and Ni $K\alpha$ backgrounds were derived in a similar way.

3. Six net-intensity ratios were calculated: R_{CrFe}, R_{CrNi}, R_{FeCr}, R_{FeNi}, R_{NiCr}, and R_{NiFe}. These ratios had the form

$$R_{CrFe} = \frac{\text{Cr } K\alpha \text{ intensity from the (1:1) Cr–Fe solution}}{\text{Cr } K\alpha \text{ intensity from the pure Cr solution}} \tag{15.45}$$

4. Six interelement influence coefficients were then calculated: a_{CrFe}, a_{CrNi}, a_{FeCr}, a_{FeNi}, a_{NiCr}, and a_{NiFe}; for example, a_{CrFe} is the influence factor for the effect of iron on Cr $K\alpha$ intensity and is calculated from

$$\frac{1 - R_{\mathrm{CrFe}}}{R_{\mathrm{CrFe}}} = a_{\mathrm{CrFe}} \frac{1 - C_{\mathrm{Cr}}}{C_{\mathrm{Cr}}} \tag{15.46}$$

where C is concentration; since C_{Cr} in the (1:1) Cr–Fe solution is 0.5, the fractional term on the right side of the equation becomes 1, thereby permitting calculation of a_{CrFe}. Once the coefficients were evaluated, the analysis procedure was as follows.

5. The Cr $K\alpha$, Fe $K\alpha$, and Ni $K\alpha$ net intensities were measured from all sample solutions and from the pure chromium, iron, and nickel solutions.

6. Three intensity ratios were calculated for each sample: R_{Cr}, R_{Fe}, and R_{Ni}. These ratios had the form

$$R_{\mathrm{Cr}} = \frac{\text{Cr } K\alpha \text{ intensity from the sample}}{\text{Cr } K\alpha \text{ intensity from the pure Cr solution}} \tag{15.47}$$

7. Chromium concentration was then calculated from the equation

$$\frac{1 - R_{\mathrm{Cr}}}{R_{\mathrm{Cr}}} = \left(\frac{a_{\mathrm{CrFe}} R_{\mathrm{Fe}} + a_{\mathrm{CrNi}} R_{\mathrm{Ni}}}{R_{\mathrm{Fe}} + R_{\mathrm{Ni}}}\right)\left(\frac{1 - C_{\mathrm{Cr}}}{C_{\mathrm{Cr}}}\right) \tag{15.48}$$

Nickel concentration was calculated from a similar equation.

8. Using these calculated values for C_{Cr} and C_{Ni}, iron concentration was calculated from the equation

$$\frac{1 - R_{\mathrm{Fe}}}{R_{\mathrm{Fe}}} = \left(\frac{a_{\mathrm{FeCr}} C_{\mathrm{Cr}} + a_{\mathrm{FeNi}} C_{\mathrm{Ni}}}{C_{\mathrm{Cr}} + C_{\mathrm{Ni}}}\right)\left(\frac{1 - C_{\mathrm{Fe}}}{C_{\mathrm{Fe}}}\right) \tag{15.49}$$

Table 15.5 gives the data of Beattie and Brissey (*B10*) for analysis of chromium–iron–nickel–molybdenum alloys to a first approximation. Part *A* of the table gives the compositions of the 12 binary standards in weight per cent *W*. The influence coefficients were evaluated as follows. Cr $K\alpha$ net intensity was measured from the three chromium binaries and from pure chromium; Fe $K\alpha$ net intensity was measured from the iron binaries and from iron; etc. The values of R_i in part *B* of the table were calculated from these intensities. The influence coefficients in part *C* were obtained by insertion of the R_i values into a set of four simultaneous equations having the form of Equations (15.38). Then the simultaneous equations for analysis

TABLE 15.5

Data for Analysis of Cr–Fe–Ni–Mo Alloys to a First Approximation by the Influence-Coefficient Method[a]

Element i	Element j			
	Cr	Fe	Ni	Mo
A. Composition of binary standards, W				
Cr	—	48.24	48.07	74.28
Fe	50.83	—	51.53	65.33
Ni	48.19	46.65	—	53.70
Mo	23.53	34.44	46.27	—
B. Intensity ratios for binary standards, $R_i = I_{100i}/I_i$				
Cr	1.000	1.760	1.815	1.841
Fe	3.360	1.000	1.613	1.835
Ni	2.860	3.670	1.000	1.954
Mo	3.770	2.787	2.461	1.000
C. Influence coefficients a_{ij} from Equations (15.38)				
Cr	1.000	0.721	0.813	2.660
Fe	2.482	1.000	0.676	1.582
Ni	1.863	2.420	1.000	1.108
Mo	0.877	0.944	1.260	1.000
D. Intensity ratio for an analytical sample, $R_i = I_{100i}/I_i$				
	R_{Cr}	R_{Fe}	R_{Ni}	R_{Mo}
—	5.04	2.346	6.640	15.34

[a] From Beattie and Brissey (*B10*).

of chromium–iron–nickel–molybdenum alloys are

$$\begin{aligned} (1 - R_{Cr})W_{Cr} + 0.721W_{Fe} + 0.813W_{Ni} + 2.660W_{Mo} &= 0 \\ 2.482W_{Cr} + (1 - R_{Fe})W_{Fe} + 0.676W_{Ni} + 1.582W_{Mo} &= 0 \\ 1.863W_{Cr} + 2.420W_{Fe} + (1 - R_{Ni})W_{Ni} + 1.108W_{Mo} &= 0 \\ 0.877W_{Cr} + 0.944W_{Fe} + 1.260W_{Ni} + (1 - R_{Mo})W_{Mo} &= 0 \end{aligned} \tag{15.50}$$

An analysis of an alloy of this quaternary system consists of measurement of the Cr $K\alpha$, Fe $K\alpha$, Ni $K\alpha$, and Mo $K\alpha$ net intensities from the sample and from the four respective elements. The four intensity ratios R_i are then calculated; ratios for a typical alloy are given in Part *D* of Table 15.5. These ratios inserted in Equations (15.50) permit calculation of composition of the sample.

However, it is evident that because there are no constant terms in the simultaneous equations, only a ratio among the four unknowns can be obtained. Such a ratio can be obtained by dividing through the four equations by any of the four unknowns. In this way, one of the four equations —that corresponding to the divisor—can be omitted, leaving three equations to be solved for three unknown ratios. Thus, there are 16 independent ways to solve the four equations. These 16 solutions yield 16 composition ratios, which may differ somewhat because the equations are based on assumptions that are not wholly valid, and the influence coefficients are empirical and evaluated only approximately. Beattie and Brissey give a procedure for preselection of the solution likely to minimize the errors introduced by the assumptions and the coefficients.

15.4.4. Evaluation of a_{ij} Coefficients to a Second Approximation

To a second approximation, each a_{ij} coefficient may be evaluated from the intensities of the spectral line of *i* from pure *i*, two binary alloys, and a ternary alloy. The simplest case to which this approximation is applicable with high accuracy is that in which one major constituent has approximately the same concentration in all samples. To illustrate the procedure [pp. 61–62 in (*6*)], suppose that samples of a five-component alloy are to be analyzed in which the concentration of one major constituent is always ~65–75%. Let this element be designated *i* (or *j*) = 1. The coefficients for elements, say, 3 and 5 are then evaluated as follows:

1. A first approximation of $a_{31(1)}$ is made from the intensities of the spectral line of element 3 from pure 3 and from a 70:30 1–3 binary standard. From Equations (15.38-3) and (15.44),

$$a_{31(1)} = (R_3 - 1)W_3/W_1 \tag{15.51}$$

2. A first approximation of $a_{51(1)}$ is made from the intensities of the spectral line of element 5 from pure 5 and from a 70:30 1–5 binary standard. Again, from Equations (15.38-5) and (15.44),

$$a_{51(1)} = (R_5 - 1)W_5/W_1 \tag{15.52}$$

3. A second approximation of $a_{35(2)}$ is then made from $a_{31(1)}$ and from

the intensity of the spectral line of element 3 from a 70:15:15 1–3–5 ternary standard. From Equation (15.38-3),

$$a_{31(1)}W_1 + (R_3 - 1)W_3 + a_{35(2)}W_5 = 0 \qquad (15.53)$$

Similarly, from $a_{51(1)}$, the intensity of the spectral line of element 5 from the 1–3–5 standard, and Equation (15.38-5),

$$a_{51(1)}W_1 + a_{53(2)}W_3 + (R_5 - 1)W_5 = 0 \qquad (15.54)$$

All terms in Equations (15.53) and (15.54) are known except the two second approximations $a_{35(2)}$ and $a_{53(2)}$.

For a system of n elements, the following standards are required: the n pure elements, the $(n - 1)$ binaries of element 1 with each of the other elements, and the $[(n - 1)/2](n - 2)$ ternaries of element 1 with all pairs of the other elements.

15.4.5. Simultaneous Evaluation of All a_{ij} Coefficients *(C48)*

Equations (15.38) may be written in the somewhat more general form

$$R_i = C_i / \sum_{j=1}^{j=n} a_{ij}C_j \qquad (15.55)$$

where i is the analyte and j the interferant; C is concentration; and R_i is represented symbolically by

$$R_i = f(I_i) = f(C_1, C_2, \cdots, C_n) \qquad (15.56)$$

That is, R_i is a function of analyte-line intensity and of the mass concentrations of all elements present; it may be simply analyte-line intensity measured from the specimen I_X, or it may be that intensity expressed as a ratio to the analyte-line intensity from a reference standard I_S or from the pure element I_{100i}, that is,

$$R_i = I_X \quad \text{or} \quad I_X/I_S \quad \text{or} \quad I_i/I_{100i} \qquad (15.57)$$

Evaluation of the a_{ij} coefficients by first- and second-order approximations has been considered in the two preceding sections. However, it is best to determine all the coefficients simultaneously from multicomponent standards containing known concentrations of all elements of interest. This approach minimizes the shortcomings of the assumptions in the derivation of Equations (15.38) because secondary absorption effects are represented in the measured intensities.

The coefficients are evaluated by solution of sets of simultaneous equations. For example, in a three-component system, there are nine coefficients (if j may be i), and they are evaluated three at a time. Let 1, 2, and 3 be the three elements and A, B, and C the three required standards, which should cover the full range of composition expected in the samples. First, to evaluate coefficients a_{11}, a_{12}, a_{13}, the intensity of the spectral line of element 1 is measured from standards A, B, and C, and perhaps from a reference standard—which may be a specimen of pure element 1. The value of R_i is calculated for each standard from Equation (15.57) and used in the following set of simultaneous equations:

$$\begin{aligned} C_{1A}/R_{1A} &= a_{11}C_{1A} + a_{12}C_{2A} + a_{13}C_{3A} \\ C_{1B}/R_{1B} &= a_{11}C_{1B} + a_{12}C_{2B} + a_{13}C_{3B} \\ C_{1C}/R_{1C} &= a_{11}C_{1C} + a_{12}C_{2C} + a_{13}C_{3C} \end{aligned} \tag{15.58}$$

where C_{1A} indicates concentration of element 1 in standard A, etc.; R_{1A} indicates intensity function of the spectral line of element 1 from standard A [Equation (15.57)], etc.; and a is the influence coefficient. Since all the C's are known and all the R's are measured, the a's can be calculated.

Coefficients a_{21}, a_{22}, and a_{23} are evaluated in the same way by measurement of the line intensity of element 2 from standards A, B, and C, and from pure element 2, and application of the data to a set of equations similar to Equation (15.58); a_{31}, a_{32}, and a_{33} are evaluated from the line intensities of element 3 and a third set of equations. Somewhat better values of a_{ij} may be obtained by making measurements on more than the required number of standards, increasing the number of simultaneous equations in the three sets, and calculating a least-squares best-fit set of a_{ij} coefficients. Once derived, the a_{ij} coefficients are used in Equations (15.38) as before.

15.4.6. Method of Lucas-Tooth and Pyne

Lucas-Tooth and Pyne have applied the influence-coefficient method to multicomponent systems having up to 12 elements (*E8*, *L30*, *L31*). Their basic equation is derived from the same assumptions as Equations (15.38) and the following additional one: Over the concentration range of interest, the absorption of the spectral line of element A by element C is assumed to be proportional to the weight fraction of C; that is, if C is added to binary A–B in concentration W_C, A-line intensity will depart from its value in pure A–B, I_A, by $W_C I_A$.

The basic equation is

$$C_{iS} = k_i + I_{iS}[a_i + a_{ii}I_{iS} + \sum_{j=1}^{j=n} (a_{ij}I_{jS})] \qquad (j \neq i) \tag{15.59}$$

$$= k_i + I_{iS}(a_i + a_{ii}I_{iS} + a_{i1}I_{1S} + a_{i2}I_{2S} + \cdots + a_{in}I_{nS}) \qquad (j \neq i) \tag{15.60}$$

where C_{iS} is concentration of analyte i in specimen S; I_{iS} is analyte-line intensity from specimen S; I_{jS} is line intensity of matrix element j from specimen S; a_{ij} is the influence coefficient of element j on element i; and a_i and k_i are constants. The constant a_i may contain corrections for nonlinearity of the detector for element i.

If there are j elements (including i) in the specimen, there are $j + 2$ constants—k_i, a_i, a_{ii}, and $(j - 1)$ a_{ij}'s. Thus, in principle, evaluation of the constants requires a set of $j + 2$ simultaneous equations capable of rigorous solution, each for a different standard of accurately known composition. In practice, a larger number of equations—and standards—should be used and the constants evaluated by a least-squares fit. A different set of $j + 2$ constants, and therefore a different set of equations, is required for each analyte.

The constants can be evaluated by measurement of all intensities I_{iS} and I_{jS} from all $j + 2$ or more standards and solution of the $j + 2$ or more equations for the $j + 2$ constants. The evaluation may be simplified on the basis that, over reasonable concentration intervals, the line intensity of each element is approximately proportional to its concentration. A concept, *apparent concentration*, is then defined as follows:

$$C_{iS(ap)} \propto I_{iS}; \qquad C_{jS(ap)} \propto I_{jS} \tag{15.61}$$

Substitution of Equation (15.61) into (15.59) gives

$$C_{iS} = k_i + C_{iS(ap)}[a_i + a_{ii}C_{iS(ap)} + \sum_{j=1}^{j=n} (a_{ij}C_{jS(ap)})] \qquad (j \neq i) \tag{15.62}$$

The constants are evaluated by solution of $j + 2$ or more equations for $j + 2$ or more standards, by use of the *known* concentrations for $C_{iS(ap)}$ and $C_{jS(ap)}$. If $C_{(ap)}$ is in weight per cent, an additional equation may be added to the set:

$$C_{i\mathrm{S(ap)}} + \sum_{j=1}^{j=n} C_{j\mathrm{S(ap)}} = 100 \qquad (j \neq i) \tag{15.63}$$

Once the constants are known, determination of analyte i in sample X consists of measurement of intensity $I_{i\mathrm{X}}$ and all intensities $I_{j\mathrm{X}}$ from the sample X, then solution of a single equation of the form of Equation (15.59) for $C_{i\mathrm{X}}$:

$$C_{i\mathrm{X}} = k_i + I_{i\mathrm{X}}\left[a_i + a_{ii}I_{i\mathrm{X}} + \sum_{j=1}^{j=n} (a_{ij}I_{j\mathrm{X}})\right] \qquad (j \neq i) \tag{15.64}$$

15.5. FUNDAMENTAL-PARAMETERS METHOD

The fundamental-parameters method permits the calculation of analytical composition from the measured analyte-line intensity and the tabulated values of three fundamental parameters—primary spectral distribution, absorption coefficient, and fluorescent yield. No calibration standards are required. Several workers have developed fundamental-parameters methods, but the treatment in this section is taken largely from a paper by Criss and Birks (*C48*). This method, like the influence-coefficient method, assumes that the specimen is homogeneous and infinitely thick and has a plane surface. However, instead of assuming that the incident primary spectrum can be described by a single average effective wavelength, the method uses the spectral distribution of the primary beam. This distribution can be estimated theoretically, or it may be measured for a given x-ray tube target, potential, and type of power supply (full-wave or constant-potential). Matrix absorption and analyte-line excitation by matrix elements enter the equation explicitly for each specimen. Analyte concentration is calculated by iteration. The advantage of the fundamental-parameters method is that no intermediate standards or empirical coefficients are required for any matrix. The principal limitation is the present uncertainty in mass-absorption coefficient and fluorescent-yield data. However, this shortcoming is being overcome as more laboratories engage in measurement of these parameters accurately. The calculations are very complex and must be done on a computer.

The fundamental-parameters equation is derived from the geometry of Figure 15.7 and the basic excitation equations, such as Equations (4.6), (4.13), and (15.30). The equation used by Criss and Birks and given here is substantially similar to that derived by other workers, except that it uses a summation over wavelength instead of the more common integration:

$$[(I_{\lambda_L})_{\text{pri}} + (I_{\lambda_L})_{\text{matrix}}]_X = g_i C_i \sum_{\lambda_P} \left(\frac{D_{i,\lambda_P} (\mu/\varrho)_{i,\lambda_P} (I_{\lambda_P}\, \Delta\lambda_P)}{(\mu/\varrho)_{M,\lambda_P} \csc\phi + (\mu/\varrho)_{M,\lambda_L} \csc\psi} \right.$$

$$\times \left\{ 1 + \frac{1}{2(\mu/\varrho)_{i,\lambda_P}} \sum_{\lambda_J} [D_{j,\lambda_P} C_j K_j (\mu/\varrho)_{i,\lambda_J} (\mu/\varrho)_{j,\lambda_P}] \right.$$

$$\times \left[\frac{1}{(\mu/\varrho)_{M,\lambda_P} \csc\phi} \log_e \left(1 + \frac{(\mu/\varrho)_{M,\lambda_P} \csc\phi}{(\mu/\varrho)_{M,\lambda_J}} \right) \right.$$

$$\left.\left.\left. + \frac{1}{(\mu/\varrho)_{M,\lambda_L} \csc\psi} \log_e \left(1 + \frac{(\mu/\varrho)_{M,\lambda_L} \csc\psi}{(\mu/\varrho)_{M,\lambda_J}} \right) \right] \right\} \right) \tag{15.65}$$

where:

C_i is concentration of analyte element i.
C_j is concentration of matrix element j.
D_i has the value 1 if λ_P is short enough to excite λ_L; it has the value 0 for longer wavelengths.
D_j has the value 1 if λ_P is short enough to excite λ_J; it has the value 0 for longer wavelengths.
g_i is a function of the absolute intensity of λ_L from element i; g cancels if a relative intensity function is used (see below).
K_j is $[1 - (1/r)]\omega$ for the particular spectral line of element j that excites λ_L of element i. (See below for r and ω).
I_{λ_L} is intensity of analyte line of element i.
$(I_{\lambda_L})_{\text{pri}}$ is intensity of analyte line excited by the primary spectrum.
$(I_{\lambda_L})_{\text{matrix}}$ is intensity of analyte line excited by spectral lines of matrix elements j in the specimen.
I_{λ_P} is intensity of the primary beam in the wavelength interval $\Delta\lambda_P$.
i is the analyte element.
j is a matrix element.
M is the specimen matrix, including the analyte i.
r is absorption edge jump ratio.
$\Delta\lambda_P$ is a wavelength interval into which the primary spectrum is arbitrarily divided for summation purposes.
λ_J is the wavelength of a spectral line of element j capable of exciting λ_L.
λ_L is the wavelength of the analyte line of element i.
λ_P is a wavelength in the primary beam.
$(\mu/\varrho)_i$, $(\mu/\varrho)_j$, $(\mu/\varrho)_M$ are mass-absorption coefficients of analyte i, matrix element j, and matrix M, respectively.

$(\mu/\varrho)_{M,\lambda_J}$, $(\mu/\varrho)_{M,\lambda_L}$, $(\mu/\varrho)_{M,\lambda_P}$ are mass-absorption coefficients of matrix M for λ_J, λ_L, and λ_P, respectively; similar notation for $(\mu/\varrho)_{i,\lambda}$ and $(\mu/\varrho)_{j,\lambda}$ have analogous significance.

ϕ is the angle between the central ray of the primary cone and the specimen.

ψ is the angle between the central ray of the secondary cone and the specimen, the takeoff angle.

ω is fluorescent yield.

The primary spectrum is divided into wavelength intervals $\Delta\lambda_P$ each having intensity I_{λ_P}. The product of $g_i C_i$ and the first summation over λ_P represents $(I_{\lambda_L})_{\text{pri}}$, the analyte-line intensity excited only by the primary x-rays. The remainder of the equation represents $(I_{\lambda_L})_{\text{matrix}}$, the analyte-line intensity excited by spectral lines of matrix elements j; this interelement excitation contributes as much as 30% to the total measured analyte-line intensity.

The fundamental-parameters equation can be based on analyte-line intensity measured from the sample X, as in Equation (15.65). More often, the equation is based on the ratio R of this intensity with that measured from a reference standard S similar to X or from pure analyte 100i. In the remainder of this section, the symbol R and the term *intensity function* refer to any of these three alternatives, that is,

$$R = I_X \qquad \text{or} \qquad I_X/I_S \qquad \text{or} \qquad I_X/I_{100i} \tag{15.66}$$

where I is $(I_{\lambda_L})_{\text{pri}} + (I_{\lambda_L})_{\text{matrix}}$.

The denominator of the second and third forms of Equation (15.66) is represented by another equation identical with Equation (15.65), except that C_i, C_j, $(\mu/\varrho)_{M,\lambda_P}$, $(\mu/\varrho)_{M,\lambda_L}$, and $(\mu/\varrho)_{M,\lambda_J}$ refer to the standard or pure analyte. If the standard is pure analyte, $(I_{\lambda_L})_{\text{matrix}}$ in the denominator becomes 0. Analyte-line intensity ratios from the sample X and a multi-component standard S are readily converted to ratios relative to pure analyte as follows:

$$R = \frac{I_X}{I_{100i}} = \left(\frac{I_X}{I_S}\right)_{\text{measd}} \left(\frac{I_S}{I_{100i}}\right)_{\text{calcd}} \tag{15.67}$$

where I has the same significance as in Equation (15.66).

Equation (15.65) expresses intensity function in terms of specimen composition. However, it is intensity functions that are measured, whereas it is concentrations that are required. Unfortunately, algebraic operation cannot produce explicit equations for mass functions C_1, C_2, etc. Consequently, mass fractions are found by an iteration procedure that makes successively closer estimates of mass fractions until the intensity functions calculated from the estimated concentrations agree satisfactorily with the measured intensity functions. Obviously, the iteration procedure must be performed on a computer. Following is a simple iteration equation:

$$C_i' = \frac{(R_i)_{\text{measd}}}{R_i} C_i \qquad (15.68)$$

where C_i is the present estimate of concentration of element i; C_i' is the next estimate of i concentration in the iteration; $(R_i)_{\text{measd}}$ is the measured intensity function for the spectral line of i; and R_i is the intensity function calculated for C_i.

The iteration procedure is as follows: (1) Initial estimates are made of all analyte concentrations C_i by normalizing their measured intensity functions $(R_i)_{\text{measd}}$ to add to 1, or by some other method. (2) The corresponding intensity functions R_i are calculated. (3) The next estimate C_i' is made by application of the iteration equation to each element i. (4) The C_i' values are scaled to add to 1. (5) The corresponding expected intensity functions R_i' are calculated. (6) A decision is made whether further iteration is indicated. This decision is based on a comparison of values of R_i' and $(R_i)_{\text{measd}}$ or of C_i' and C_i, that is, the latest and the immediately preceding iterations. For example, it may be decided to stop the iteration when the value of C_i' for every element differs from C_i—the value at the preceding

TABLE 15.6

Composition of a Five-Component Alloy Calculated in Successive Iterations by the Fundamental-Parameters Method[a]

Iteration	Concentration, wt %				
	Cr	Fe	Co	Ni	Mo
1	26.81	18.77	1.41	39.68	13.33
2	22.90	19.60	1.47	46.07	9.96
3	22.11	18.59	1.44	47.55	10.31
4	22.20	18.38	1.43	47.58	10.41
5	22.23	18.39	1.43	47.53	10.42
6	22.23	18.40	1.43	47.53	10.41
7	22.23	18.40	1.43	47.53	10.41

[a] From Criss and Birks (*C48*).

iteration—by less than some arbitrary limit, say 0.1%. (7) If further iteration is indicated, the sequence is repeated starting with step 3 and using C_i' and R_i' as the "present" estimate and its calculated intensity function. Table 15.6 gives a series of seven successive iterations of an analysis of a five-component alloy.

15.6. MULTIPLE-REGRESSION METHOD

Multiple regression (*A13*, *M23*) is a statistical method used to study correlation among observed data and to form equations relating one dependent variable with several independent variables. In many x-ray analytical situations, regression methods can be applied to correlate analyte-line intensity I (the independent variable) with concentration C (the dependent variable). Mitchell and Hopper (*M23*) have developed this technique, and the following discussion is derived largely from their paper.

When plotted data obviously lie on a straight line, the dependent variable C may be predicted by use of the single independent variable I in a linear equation such as the following:

$$C = a_0 + a_1 I \tag{15.69}$$

When the plotted curve is nonlinear, other functional relationships must be assumed and the best fit determined, such as a three-term polynomial (parabolic or quadratic) curve:

$$C = a_0 + a_1 I + a_2 I^2 \tag{15.70}$$

In a multicomponent system, if the dependent variable C_1, the concentration of one of the elements, is a linear function of each of the independent variables $I_1, I_2, I_3, \cdots$, the line intensities of each of the component elements, a multiple linear equation may be estimated by the least-squares method:

$$C_1 = a_0 + a_1 I_1 + a_2 I_2 + a_3 I_3 + \cdots \tag{15.71}$$

However, if the I–C relationship is nonlinear, a higher-order polynomial may be required. In most cases, x-ray data fit an equation of the form

$$C_1 = a_0 + a_1 I_1 + a_2 I_2 + a_3 I_3 + a_4 I_1^2 + a_5 I_2^2 + a_6 I_3^2 + a_7 I_1 I_2 + a_8 I_1 I_3 + a_9 I_2 I_3 \tag{15.72}$$

The constants are evaluated from intensities measured from numbers of multicomponent standards having accurately known compositions. The calculations can be made on a desk calculator for two- and three-term equations and limited sets of I–C data. However, for more complex equations and large sets of data, a regression program using a digital computer is required. The result of the multiple regression procedure is an equation in which the constants a have numerical values giving the best fit to the measured intensity data and the known composition of the standards. Thereafter, the equation is used to provide analytical composition data from intensities measured from samples similar to the standards used to evaluate the constants.

Chapter 16

Specimen Preparation and Presentation—General; Solids, Powders, Briquets, Fusion Products

16.1. GENERAL CONSIDERATIONS

16.1.1. Introduction

X-ray secondary-emission spectrometry is probably the most versatile of all instrumental analytical chemical methods with respect to the variety of specimen forms to which it is readily applicable nondestructively. Many forms of specimen can be analyzed in the x-ray spectrometer without treatment and without any special mounting arrangement. Other specimens require treatment and/or mounting cells or supports.

In this chapter and the next, the term specimen *presentation* is used to indicate the physical form in which the specimen is placed in its compartment, that is, in which it is "presented" to the x-ray spectrometer. The term *preparation* indicates any treatment—other than simply cutting to appropriate size—to which the as-received sample is subjected prior to its presentation to the spectrometer. Specimen preparation may be physical, chemical, or both, and it may involve treatment of the specimen in its present form or conversion to another form. The terms *specimen*, *sample*, and *standard* are defined in Section 3.1, and the term *nondestructive*, as applied to x-ray spectrometric specimens, is discussed in Section 3.3.1.5.

The specimen presentation forms may be classified in seven categories as follows:

1. *Solids*, including (a) unfinished ingot, mineral, etc.; (b) finished bar, rod, sheet, foil, ceramic, glass, plastic, etc.; and (c) even rubber, wood, paper, and textile products. Bulk (large) specimens may also be included in this group.
2. *Small fabricated forms and parts*, including rod, wire, shot, screws, nuts, washers, lugs, gears, electron-tube and transistor components, and other parts of regular or irregular shape.
3. *Powders.*
4. *Briquets.*
5. *Fusion products* resulting from fusion of the sample material with fluxes, such as potassium pyrosulfate, or with glass-forming fluxes, such as lithium or sodium tetraborate.
6. *Liquids and solutions.*
7. *Supported specimens*, including those distributed on or in filter paper, Millipore filters, Fiberglas, Mylar, Scotch tape, ion-exchange membrane, etc., or on fine quartz or plastic fibers. Thin films and platings may also be included in this group.

In the preceding list, each form can be converted to a form listed below it, and some forms can also be converted to forms listed above them. For example, solutions may be converted to powders by precipitation, evaporation, or absorption in cellulose, and briquets and fusion products are converted to powder by grinding; some powders can be converted to solids by melting or sintering, or by reduction to metal followed by melting or compacting.

Usually, nondestructive x-ray spectrometric analyses of samples in their as-received forms are more rapid and convenient than analyses by any other method. And since the x-ray spectrometer can accommodate almost any specimen form, the analyst is well advised to strive to analyze samples in their original form whenever practical. In fact, one might wonder why it should ever be necessary to alter the specimen form at all—that is, why any preparation other than simple physical treatment of the specimen in its present form should ever be required.

Actually, for qualitative x-ray spectrometric analysis, usually the only requirement is that the specimen fit its compartment, and preparation other than, say, provision of a flat surface on a solid, may well be unnecessary. However, for quantitative analysis, elaborate preparation and conversion of form may be required to deal with one or more of a large number of conditions, which may be classified in the five groups listed below. When such elaborate preparation is required, much of the speed and convenience of the x-ray spectrometric method is lost.

1. *Specimen errors must be reduced* when: (a) samples are very inhomogeneous; (b) the surface layer is not representative of the bulk sample; (c) the average particle size or the particle-size distribution varies among the samples; (d) the amount of porosity or pore size varies among the samples; and/or (e) surface texture is unsuitable or varies among the samples.
2. *Additions must be made* to the samples when: (a) absorption-enhancement effects are severe and an internal standard must be added; (b)—or a matrix-masking agent must be added; (c)—or the sample must be diluted; (d) an internal control standard or intensity-reference standard must be added; or (e) a standard-addition or standard-dilution method is to be used. These addition methods are usually not feasible with specimens in the form of solids or fabricated parts.
3. *Chemical separation or other treatment is required* when: (a) the analyte(s) must be preconcentrated or separated; (b) interferants must be separated; or (c) the analyte(s) must be determined by an indirect or association method (Section 14.8.2).
4. *Standards must be matched.* Sometimes, it is necessary to change the as-received specimen form even though it is entirely suitable for analysis. This is the case when: (a) the sample shape is completely irregular and varies among the samples; (b) standards are not available or readily prepared in the as-received form; or (c) the same substance is routinely analyzed in a number of forms—such as bar, drillings, shot, parts, wire, powder—and conversion to a common form—such as solution or fusion product—permits all analyses to be made with a single set of standards.
5. The x-ray spectrometric method is to complement some other analytical method (Section 14.8.3).

Great ingenuity has been shown in the development of x-ray spectrometric preparation and presentation techniques. The number of such techniques reported in the literature is very large, and there are few new papers on x-ray spectrometry that do not report at least one more. Several comprehensive reviews of x-ray spectrometric preparation and presentation techniques have been published (*15F*, *B26*, *T7*).

The remainder of Section 16.1 gives some general considerations relative to all x-ray spectrometric specimen preparation and presentation techniques. Sections 16.2, 16.3, and 16.4 consider solids, powders and briquets, and fusion products, respectively. Chapter 17 deals with liquids, supported specimens, and micro and trace analysis. For each specimen type, there is a consideration of sources of samples, alternative methods of presentation, favorable and unfavorable features, treatment, precautions and considerations, and, perhaps, some detailed descriptions of certain selected techniques. However, the emphasis is on principles and techniques, rather than specific applications and materials.

16.1.2. Classification of Applications

Birks [p. 71 in (*6*)] classified x-ray spectrometric applications in 12 groups on the basis of similarities of physical form, nature and concentration range of analyte(s), wavelength region of analyte lines, general problems encountered, and applicable techniques. The following is an expansion of his list:

1. *Metals and alloys* (Section 16.2).

2. *Minerals and ores.* Carl and Campbell (*C20*) further classify minerals and ores in four groups as follows: (a) those in which one analyte varies in concentration in an otherwise substantially constant matrix; (b) those in which two elements vary in all proportions; (c) those in which two elements vary in an otherwise substantially constant matrix; and (d) complex mineral systems. Minerals and ores are characterized by a very wide range of composition, great inhomogeneity, and severe particle-size and absorption-enhancement effects. The analyte may be present in low concentration in a largely unknown and variable matrix. Severe spectral-line interference may result from the presence of many elements in complex minerals. Minerals and ores are usually analyzed as powders, briquets, or fusion products because of their great inhomogeneity. Standards are usually difficult to obtain.

3. *Ceramics and glasses* usually consist of minor and trace elements of relatively high atomic number in a low-Z matrix.

4. *Plastic, rubber, wood, textile, and paper products* usually consist of minor and trace elements of low and high atomic number in a carbon–hydrogen–oxygen matrix.

5. *Small fabricated forms and parts* may be analyzed as such or reduced to powder, briquet, fusion product, or solution.

6. *Powders* may be analyzed loose or briqueted, with or without diluents or binders (Section 16.3).

7. *Platings, coatings, and films* (Section 14.7, Chapter 18).

8. *Liquids and solutions* (Chapter 17).

9. *Gases and vapors* containing x-ray spectrometrically determinable elements may be taken up in solution or in a solid absorber. It may be feasible to determine gases not containing such an element by indirect or association analysis (Section 14.8.2). Even the group 0 ("inert") gases (helium, neon, argon, krypton, and xenon) may form stoichiometric association and clathrate compounds with certain substances. For helium and neon, an associated substance containing a determinable element is required. It might be possible to adsorb argon, krypton, and xenon in activated charcoal, which might then be placed in the spectrometer. Provision would be required to cool the charcoal to liquid-nitrogen temperature.

Hudgens and Pish (*H44*) analyzed mixtures of iodomethane (CH_3I, b.p. ~42°C) and bromoethane (CH_3CH_2Br, b.p. ~38°C) by direct measurements of I $K\alpha$ and Br $K\alpha$ from their vapors. The liquid samples were placed in a rubber balloon attached to the side tube of a liquid-specimen cell having a glass body and a 0.002-inch Mylar window. The measurements were made on the emission from the vapor in the cell. The I $K\alpha$/Br $K\alpha$ intensity ratios were calculated and applied to a calibration curve established from standard mixtures.

The $K\alpha$ lines of sulfur and halogens have been measured directly from gas-chromatograph effluents passing through thin-wall plastic tubes in the x-ray spectrometer specimen chamber (Section 14.8.3). Gases and vapors can also be measured by x-ray absorption methods (Section 20.1).

10. *Inhomogeneous materials—gross analysis.* The samples are rotated during measurement or homogenized by pulverization, fusion, or dissolution.

11. *Inhomogeneous materials—analysis of individual phases.* The individual phases may be analyzed after physical separation, or in place by use of selected-area techniques on the x-ray spectrometer (Chapter 19) or electron-probe microanalyzer (Chapter 21).

12. *Very low concentrations* (Chapter 17).

13. *Limited total amount* (Chapter 17).

Classes 12 and 13 together constitute *trace analysis* (*C5*, *C15*), which deals with trace concentrations in relatively large samples or with major concentrations in very small samples. Low-concentration analytes may be determined without preconcentration or after preconcentration by chemical (*D5*) or physical means. Preconcentration usually places the sample in the limited-total-amount class for which supported-specimen, selected-area, and curved-crystal techniques are usually applicable. Reduction of background is also very important for analysis of samples in these two classes.

14. *The lighter elements*, atomic number 9–22 (fluorine to titanium), having spectral lines in the long-wavelength spectral region (Section 16.1.3).

15. *The lightest elements*, atomic number 4–11 (beryllium to sodium), having spectral lines in the ultralong-wavelength spectral region (Section 9.4.4).

16. *Radioactive materials* (*M33*). The problem in this class, aside from personal safety, is to prevent radiation of radioactive origin from interfering with the x-rays to be measured. This is effected by a combination of the following measures: (a) encapsulation of the specimen, in Mylar, for example, to prevent contamination of the instrument and to absorb α and β radiation; (b) provision of lead shielding between the specimen compartment and detector to prevent direct γ-irradiation of the detector; and (c) pulse-height selection (*K17*) to eliminate pulses from γ-rays and x-rays of radioactive origin that have followed the x-ray optical path to the detector.

17. *Dynamic systems*. If special specimen-presentation systems are provided, x-ray spectrometry can be applied to the study of dynamic processes such as the following: (a) dissolution of a solid in a liquid; (b) precipitation of a solid from a liquid; (c) admixture of partially miscible liquids; (d) diffusion of solid in solid or liquid in liquid; (e) diffusion of an element to the surface; (f) diffusion of an element from the surface into the bulk; (g) sublimation or desorption; (h) deposition or adsorption; (i) preferential corrosion of a surface; (j) rate of electrodeposition (Section 18.3.8); and (k) continuous paper and gas chromatography (Section 14.8.3). For examples of (a) above, Cu $K\alpha$ intensity has been measured from a cell containing water standing undisturbed over copper(II) sulfate pentahydrate ($CuSO_5 \cdot 5H_2O$) crystals or from a window in a sidearm through which the supernatant solution was circulated from a vessel containing such crystals.

18. *Continuous on-stream analysis and process control* (Section 9.3.1).

16.1.3. Problems Specific to the Light Elements

In x-ray spectrometric analysis, the long-wavelength region extends from ~2.75 to 20 Å and contains the $K\alpha$ lines of elements 22 (titanium) to 9 (fluorine) and the $L\alpha_1$ lines of elements 56 (barium) to 25 (manganese).

The long-wavelength region is that in which a thin-window gas-flow proportional counter and helium or vacuum path may be regarded as essential, although $K\alpha$ lines can be measured in air path from relatively low concentrations of elements down to 19 (potassium). The ultralong-wavelength region is discussed in Section 9.4.4.

Standard commercial x-ray spectrometers are substantially useless for elements having atomic number less than 9 (fluorine), and they are subject to increasing difficulty and inconvenience for elements having atomic number 22 (titanium) to 9. In Section 13.1.3 and Table 13.2, it is shown that even for a very favorable analyte (copper), the x-ray spectrometer is extremely inefficient: For 10^{16} primary x-ray photons emitted by the x-ray tube target and capable of exciting the copper K spectrum, only 4×10^8 Cu $K\alpha$ photons are emitted by a monolayer of copper atoms having area 1 cm^2; and, of these, only ~250 are finally counted by the detector. For a "light element," the efficiency is much lower, and every component and process in the instrument contributes to this inefficiency. In fact, when one considers these individual contributions, one marvels that the light elements can be determined at all!

a. X-ray Tube. The x-ray tube target generates inefficiently the long wavelengths most effective in exciting spectra of the low-Z elements. Then the x-ray tube window has very low transmittance for these long wavelengths; for example, 0.040-inch beryllium transmits only 10, 1, 0.1, and 0.01% of the incident x-ray photons at 3.5, 4.5, 5, and 6 Å, respectively; these wavelengths correspond approximately to the K-absorption edges of potassium, chlorine, sulfur, and phosphorus, respectively; 0.010-inch beryllium transmits only 10 and 0.1% at 5 and 8 Å (sulfur and aluminum K edges), respectively, and even 0.001-inch beryllium transmits only 20% at 10 Å (magnesium K edge).

b. Specimen. As the atomic number of the analyte decreases, the wavelength of the analyte line increases, and: (1) infinite thickness decreases; (2) fluorescent yield decreases; (3) matrix absorption increases; (4) absorption in cell windows increases; and (5) displacement of wavelength with chemical state increases. Moreover, as the effective atomic number of the total specimen decreases, both coherent and incoherent scatter increase.

Because of the reduced infinite thickness, only a very thin surface layer contributes to the measured analyte-line radiation, and oxidation, corrosion, surface texture, and particle size have marked effect. Liebhafsky and his colleagues [p. 221 in (*16*)] cite a striking example of the effect of surface condition: a piece of transformer steel containing 3.3% silicon gave a Si $K\alpha$ intensity of 263 counts/s after polishing, 53 counts/s after oxidation, then 725 counts/s after a brief etch in nitric acid.

The fluorescent yield (Section 2.5.2, Figure 2.18) is only 15% for Ca $K\alpha$ (Z 20), only 0.1% for Na $K\alpha$ (Z 11). The absorption of the analyte line is high even by the analyte itself and by adjacent light elements. Heavier elements, even as minor constituents, cause high absorption. Although a commercial spectrometer may not resolve or permit accurate measurement of wavelength shifts due to chemical effects, the measured intensity may vary with these shifts, providing a source of error. The higher scatter increases the background and thereby the minimum detectable amount. Also, the intensity of the scattered target spectrum increases and with it the possibility of spectral interference. The increased incoherent scatter further complicates the spectrum.

Dilution methods (dry, fusion, solution) are difficult to apply to low-Z elements because analyte-line intensity, already relatively low, is reduced still further, both by dilution and by absorption of the analyte line by the diluent. Finally, internal standardization is less effective and more difficult to apply to low-Z elements.

c. Radiation Path. Absorption of the analyte line in the radiation path increases as the atomic number of the analyte decreases. Helium transmits only 80% at 10 Å (Mg $K\alpha$), and helium enclosures are frequently difficult to work with at the large 2θ angles required by some crystals for low-Z elements. Vacuum enclosures may have a limited 2θ range and force selection of a less favorable crystal.

d. Crystal. The crystals used at long wavelengths are frequently less "reflective" than those applicable at shorter wavelengths (*J5*). Higher background and energy interference may result from the emission of secondary spectra of the elements in the crystal. The temperature coefficient of expansion of crystals is highest at long wavelengths, and PET has the highest coefficient of any commonly used crystal (Figure 5.8).

e. Detector. Detector efficiency decreases at the long wavelengths (*J5*). The transmittance at 10 Å (Mg $K\alpha$) of 0.00025-inch Mylar and 0.001-inch beryllium are only 26.3% and 19.9%, respectively.

f. Spectrum. The increased possibility of spectral interference from scattered target lines and crystal emission has been mentioned already. The spectra of the low-Z elements are also highly prone to interference from higher-order lines of elements of higher atomic number.

16.1.4. Standards

It has already been mentioned that x-ray spectrometric analysis is a comparison method. Consequently, for a given analyte–matrix system, one, two, three, or more standards may be required, depending on the linearity of the calibration function, the analyte concentration range, and the severity of the matrix effects. It has also been mentioned that standards for x-ray spectrometry must be as similar as possible to the samples, not only in analyte concentration, but in matrix composition, physical form, surface texture, particle size, etc. Frequently, a rapid, convenient, nondestructive analysis of a metal, ceramic, or small part is precluded because of unavailability of standards in the required form.

In general, because of the substantially nondestructive nature of x-ray spectrometric analysis, once a set of standards has been obtained, the standards last indefinitely if handled and stored carefully. However, there are exceptions to this. Solution standards may deteriorate on long standing, and, in any case, the portions actually measured must usually be discarded. Repeated prolonged exposure to the intense primary beam and to the high ozone concentration in the specimen chamber may cause radiation damage and superficial oxidation, respectively. These processes are particularly serious for plated standards and for those having several phases that might deteriorate preferentially. Plastic standards may develop deep cracks and fissures. Briquets may powder, crumble, or break, but these effects are minimized by use of binders, backings, rings, or metal cups, and by spraying the edges and faces with clear Krylon plastic spray coating (Krylon, Inc.), or other plastic spray. Glass disks may devitrify or crack from stress.

In general, x-ray spectrometric standards are obtained in one of four ways. They may be purchased (*M14*), obtained by analysis of samples, prepared by concentration or dilution of a basic material of known composition, or synthesized.

Standards may be purchased from: (1) the National Bureau of Standards (*N10*) and from the corresponding agencies of other countries; (2) optical emission spectrographic supply firms; (3) manufacturers of the type of material to be analyzed; for example, standards for aluminum-base alloys are available from Aluminum Company of America; (4) professional technical societies and institutes concerned with the type of material to be analyzed; and (5) other analytical laboratories engaged in analysis of the same material.

Samples analyzed very carefully by independent methods may be used as standards. The "independent method" may be an x-ray spectrometric solution or fusion method using synthetic standards. The homogeneity of the prospective standards should be established; for example, if the samples are metal coupons, both sides should be finished, and x-ray spectrometric measurements of analyte-line intensity should be made from both sides with the coupon in various orientations, or with and without rotation. If the coupons are thus shown to be homogeneous, turnings or drillings are taken from one side for analysis. Alternatively, two or three consecutive adjacent slices may be cut from the sample and all sides finished and tested for uniformity in the same way (*F9*). Then one slice is retained, and the adjacent one(s) sacrificed for analysis. Standards for nondestructive analysis of small fabricated forms and parts are obtained in a similar way. Several parts from a lot are tested for uniformity of composition by x-ray spectrometric measurement of analyte-line intensities from each. If the parts all give the same intensities, several are sacrificed for analysis, the others retained as standards. Alternatively, standard parts may be fabricated from material of known composition.

If a single powder or liquid of the type to be analyzed and of accurately known composition is available, several standards may be prepared from it by quantitative addition of analyte(s) or of analyte-free matrix material or solvent to form more and less concentrated standards, respectively. Analyte(s) may be added to the powder by: (1) quantitative admixture of powder having accurately known analyte concentration and substantially the same particle size as the base powder; or (2) volumetric addition of a standard analyte solution, followed by homogenizing, drying, and grinding (*D28*).

Synthesis of standards is most convenient for powders, briquets, fusion products, and solutions. However, small melts of alloys—especially low-melting alloys—or small firings of ceramics can be made. For metals, small electric vacuum or inert-atmosphere furnaces, or small arc-furnaces having argon atmosphere may be used to prevent oxidation (*F5*). In some cases, metal standards can be prepared from synthetic solutions containing all major, minor, and trace elements in proper relative concentrations by evaporation to dryness, firing and reducing to metal sponge, briqueting, and sintering. This technique has been applied to preparation of nickel alloys (*L12*).

Powder standards may be prepared by admixture of individual powder components or from solution by precipitation, evaporation, or absorption in cellulose and firing. Briquet standards may be prepared from the powders. Briquets pressed in metal cups are more rugged and easier to store, and unless low-Z elements are to be determined, the surface should be lightly sprayed with clear Krylon.

Fusion products and solutions are most readily synthesized because all inhomogeneities and particle-size effects disappear completely. The fusion-product standards may be in the form of glass disks, or they may be ground and briqueted. Solution standards are usually the easiest to synthesize. Metallic elements are added to organic liquids as organometallic compounds, such as napthenates, or as salts of organic acids. Silicon, phosphorus, and sulfur, and fluorine, chlorine, bromine, and iodine are added as their organic compounds (*D26*, *S10*, *Y1*).

Standards for analysis of films and platings are prepared in several ways. An analyte solution may be sprayed from an atomizer on a blank substrate and area concentration calculated from gain in weight and composition of the sprayed chemical species. The analyte may be vacuum-evaporated on a blank substrate, and area concentration derived from interference colors or fringes, optical transmission, or gain in weight. Alternatively, several substrate blanks may be exposed at geometrically equivalent positions, the uniformity of the deposits verified by x-ray spectrometric measurement, and one or more of the blanks sacrificed for chemical analysis. Finally, a weighed portion of analyte may be evaporated completely and area concentration calculated from the geometry, that is, from the solid angle intercepted by the substrate.

A measured volume of a slurry having a known weight/volume analyte concentration can be applied to a large, optically flat glass plate and spread into a thin film by means of a casting knife ("doctor blade") (*B22*, *F7*). The area of the film is measured with a planimeter or by tracing the outline of the film on graph paper and counting squares. The area concentration is then calculated. Blanks can be removed from the film by scoring, flooding with water, and "coaxing" squares of the film to float off. Films having various area concentrations are made by varying either the analyte concentration in the slurry or the film thickness, or by stacking various numbers of layers of film. A film may be placed on a substrate and moistened with solvent to affix it. The substrate may be plane, cylindrical, or of other form. Films may also be vacuum-evaporated on Scotch tape and affixed to substrates of various materials and forms.

Plated standards are prepared by electroplating a relatively large area of substrate having its back, edges, and a 1-cm border on its front surface masked with insulating tape. These precautions avoid useless plating of the back and nonuniform plating of the edge (edge effect). Calculations based on the electrolysis laws and the exposed area permit plating to approximately the required thickness. Several blanks are cut from the plated area. The uniformity of the plating is verified by x-ray spectrometric measurement of the intensity of a plate-metal line. Then one or more blanks are sacrificed for chemical analysis.

Standards for analysis of samples supported on filter paper, Millipore filters, Fiberglas, Mylar, ion-exchange membrane, etc. are prepared by volumetric addition of standard analyte solutions to blanks of the support material. Addition of a wetting agent to the solution ensures more uniform distribution, especially on Mylar.

16.1.5. Effective Layer Thickness

In general, x-ray spectrometric, analysis involves a relatively thin surface layer, and this layer is thinner the longer the analyte wavelength and the higher the absorption coefficient of the specimen for that wavelength. However, short-wavelength analyte lines in light matrixes, such as water or lithium tetraborate glass, may have much greater effective layers.

Togel (*T7*) has shown that the depth of the effective specimen layer, that is, the layer in which, say, 99.9% of the measured analyte-line intensity originates, $t_{99.9}$, is given by

$$t_{99.9} = \frac{6.91}{[(\mu/\varrho)_P/\sin\phi] + [(\mu/\varrho)_A/\sin\psi]} \text{ cm} \tag{16.1}$$

where $(\mu/\varrho)_P$ and $(\mu/\varrho)_A$ are the mass-absorption coefficients of the specimen for the primary x-rays and analyte line, respectively; and ϕ and ψ are the angles made with the specimen plane by the central rays of the primary and secondary beams, respectively. If $\phi = \psi = 45°$,

$$t_{99.9} = \frac{6.91 \sin 45°}{(\mu/\varrho)_P + (\mu/\varrho)_A} = \frac{4.9}{(\mu/\varrho)_P + (\mu/\varrho)_A} \text{ cm} \tag{16.2}$$

Three cases may be defined on the basis of the relationship of λ_P, the wavelength of a monochromatic primary beam or the effective wavelength of a polychromatic beam; λ_A, the wavelength of the analyte line; and λ_M, the absorption edge of the principal matrix constituent:

Case 1. $\lambda_P \approx \lambda_A$, so that both are either $>\lambda_M$ or $<\lambda_M$, and $(\mu/\varrho)_P \approx (\mu/\varrho)_A$; then,

$$t_{99.9} \approx \frac{4.9}{2(\mu/\varrho)_A} \text{ cm} \tag{16.3}$$

Case 2. $\lambda_P < \lambda_M < \lambda_A$; then, if δ is the absorption-edge jump ratio of λ_M,

$$(\mu/\varrho)_P \approx \delta(\mu/\varrho)_A \tag{16.4}$$

and

$$t_{99.9} \approx \frac{4.9}{\delta(\mu/\varrho)_A} \quad \text{cm} \tag{16.5}$$

Case 3. $\lambda_P \ll \lambda_A$, but both are either $>\lambda_M$ or $<\lambda_M$, and $(\mu/\varrho)_P \ll (\mu/\varrho)_A$; then,

$$t_{99.9} \approx \frac{4.9}{(\mu/\varrho)_A} \quad \text{cm} \tag{16.6}$$

Case 3 gives the greatest effective thickness. Figure 16.1 shows effective

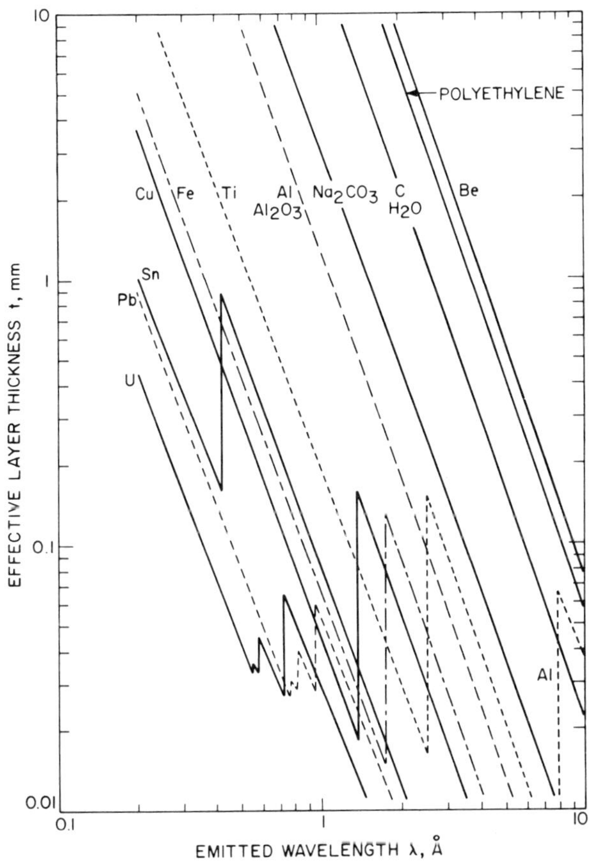

Fig. 16.1. Effective layer thickness (layer that emits 99.9% of the measured analyte line intensity) as a function of wavelength for various matrixes. [K. Togel, in "Zerstörungsfreie Materialprüfung," R. Oldenbourg, Munich, Germany (1961); courtesy of the author and publisher.]

specimen layer thickness, calculated from Equation (16.6), as a function of analyte-line wavelength λ_A for several matrix materials. Figure 16.2 shows effective specimen mass, that is, mass of material in the effective specimen layer, as a function of effective thickness for various densities. The calculations are based on an effective specimen area 2.8×2.8 cm (7.84 cm^2). The data in Figures 16.1 and 16.2 must be used carefully. For example, the first figure shows that the effective layer for Cu $K\alpha$ (1.54 Å) in water is ~1 cm. However, the effective layer in a water *solution* of heavy solutes would be much less.

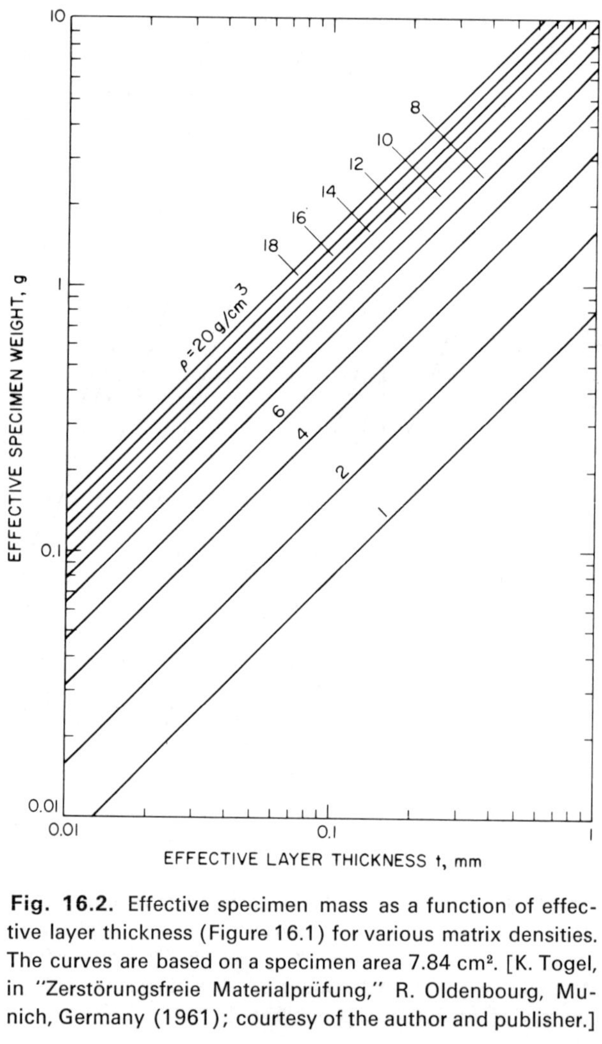

Fig. 16.2. Effective specimen mass as a function of effective layer thickness (Figure 16.1) for various matrix densities. The curves are based on a specimen area 7.84 cm^2. [K. Togel, in "Zerstörungsfreie Materialprüfung," R. Oldenbourg, Munich, Germany (1961); courtesy of the author and publisher.]

16.1.6. The Specimen-Preparation Laboratory

Many x-ray spectrometric laboratories deal with only a limited number of sample types. For example, a laboratory in a brass foundry may be required only to cut and finish coupons of copper-base alloys, a laboratory in a cement plant only to grind and briquet cement powders. Such laboratories need only limited specimen-preparation equipment. However, as the diversity of specimen types increases, so does the diversity of equipment required. A laboratory that must deal with the full range of x-ray spectrometric specimens—say, the customer application laboratory of a manufacturer of x-ray spectrometers—requires a very well equipped specimen-preparation laboratory. Following is a list of most of the equipment required by such a laboratory, in addition to the basic chemicals and apparatus of the analytical chemistry laboratory:

- Water wheel.
- Diamond wheel.
- Belt grinder.
- Paper-strip grinder.
- Disk grinder.
- Rotating polishing wheel, with various abrasive papers and cloths, and abrasive powders of various materials and grits.
- Rotating polishing wheel, vibratory type.
- Lapping wheel.
- Spiral-grooved, lead-covered polishing disk, for use with diamond dust.
- Small end mill.
- Small lathe.
- Small drill press.
- Agate, tungsten carbide, silicon carbide, and boron carbide mortars and pestles (in order of increasing hardness).
- Mixer/Mill (Spex Industries, Inc., Metuchen, N. J.) or equivalent, with plastic, tungsten carbide, and alundum vials and balls.
- Wig-L-Bug mixer.
- Freezer/Mill (Spex).
- Mechanical grinder, preferably of the type that grinds to uniform particle size.
- Wire-cloth sieves.
- Briqueting press for pressures up to 100,000 lb/inch^2, with high-temperature and vacuum dies of various sizes.
- Drying oven.

Muffle furnace.
Crucible furnace.
Vacuum and inert-atmosphere furnace.
Casting mold.
Roller ball mill.
Film-casting knives ("doctor blades").
Hot plates.
Electroplating apparatus for preparation of plated standards.
Bell-jar vacuum system for evaporation of films.
Ion-exchange columns.
Binocular microscope, 1–100×.
Balance.
Filter flasks and small Buchner funnels (1-inch, $\frac{1}{2}$-inch diameters).

In addition, the preparation laboratory often uses techniques and accessories of related disciplines, including microchemistry, optical emission spectrography, radiochemistry, electron microscopy, and metallography. Finally, the laboratory requires a supply of specimen cells and mounting devices of various sizes and for all types of specimen.

16.2. SOLIDS

16.2.1. Scope, Advantages, Limitations

The term *solid* is used here to indicate bulk solid, as distinguished from turnings, drillings, chips, small pellets, shot, granules, and powders. The solid may be of metal, rock, mineral, ore, slag, ceramic, glass, plastic, etc.; or it may be of any material that can be formed into a "solid" specimen, such as rubber, wood, textile, and paper. Briquets are not included. The solid may have the form of unfinished ingot, rock, etc.; finished plate, bar, rod, tube, sheet, or even wire; or fabricated parts. It is stated in Section 16.1.1 that solids may be presented to the instrument as such, or they may be reduced to powders, briquets, solutions, or supported specimens. This section is limited to presentation as solid.

If the material can be analyzed in solid form, the analyses are usually very rapid and convenient. Small fabricated parts may require no preparation. Finished materials may require only cutting to appropriate size. Unfinished bulk materials may require only the cutting of a coupon and finishing of one of its surfaces.

The principal disadvantage of the solid form is that any analytical method involving an addition is likely not to be feasible—standard addition and dilution, low- and high-absorption dilution, internal standardization, or internal intensity-reference standardization. The last two methods may be feasible if a suitable element is already present in all specimens in appropriate constant concentration. Chemical concentration or separation is also not feasible. Surface texture and composition are sometimes difficult to reproduce. Standards may not be available and are usually difficult to synthesize.

16.2.2. Presentation

For purely qualitative analysis, it is usually sufficient that the sample fit in the specimen chamber or in a large-specimen accessory. Little attention need be given surface treatment other than, perhaps, to provide a roughly flat surface. Irregular-shaped pieces may be pressed into a Lucite block in a metallographic press for convenience in mounting in the instrument. However, for quantitative and even semiquantitative analysis, it is essential that for all samples and standards in a given analysis, the effective specimen area (Section 4.5) be identical in all respects—area, shape, dimensions, orientation, inclination (ϕ, ψ), position in the plane defined by the mask, position above or below this plane, and surface contour or topography.

16.2.2.1. *Flat Specimens*

Most solid x-ray spectrometric specimens are objects already having a flat surface, or rectangular or circular coupons cut from bulk solids and finished on one side. Some specimens of this form are shown in *A–G* in Figure 16.3. For such specimens that fit in the specimen chamber, all the conditions stated above are met by the simple means of placement behind the mask provided in the chamber. Even if the specimen does not fit the chamber, and it is not permissible to cut it, the conditions are met by use of a large-area specimen accessory. For specimens of small area—say 1–50 mm^2—or for small areas of interest on larger specimens, a collimated specimen mask may be used (Section 4.5). For even smaller areas, selected-area techniques may be used, in which a pinhole or slit aperture is placed in the primary or secondary beam (Chapter 19).

16.2.2.2. *Fabricated Forms and Parts*

For fabricated forms and parts that are not flat and cannot provide flat coupons, other measures than masking must be taken to ensure reproducibility of specimen area, position, and topography.

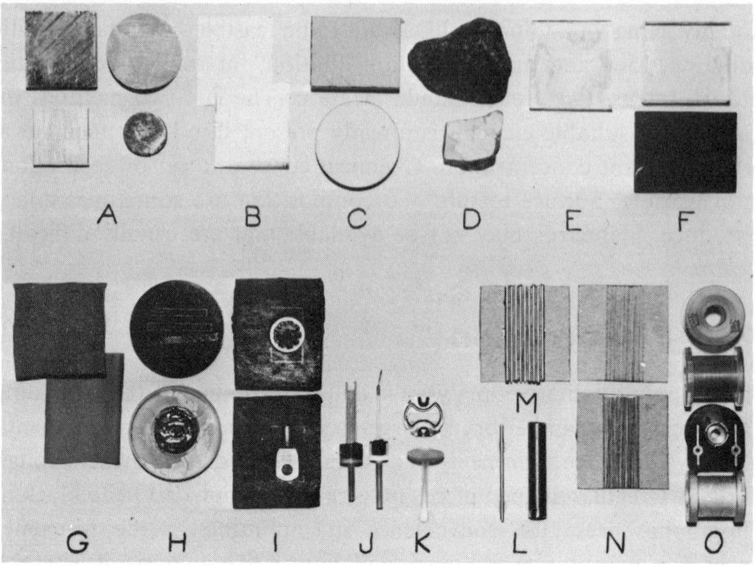

Fig. 16.3. Typical solid-specimen forms. [E. P. Bertin and R. J. Longobucco, *Norelco Reporter* **9**, 31 (1962); courtesy of Philips Electronic Instruments, Inc.] *A*. Metal coupons and stock. *B*. Metal sheet and foil. *C*. Ceramic stock. *D*. Rock, mineral, ore, etc. *E*. Glass stock. *F*. Plastic stock. *G*. Rubber stock. *H*. Metallographic sections. *I*. Small parts held in wax-impression molds. *J*. Small parts held in wax-filled plastic cups (rotable). *K*. Small parts held on plastic pedestals (rotable). *L*. Small rod. *M*. Small rod or stiff wire cemented on card. *N*. Wire wound on cards. *O*. Wire wound on small spools (rotable).

If the part does have a flat portion, it may be presented behind a mask, or with a selected-area aperture. A single very small part can be analyzed by use of a pinhole aperture. However, higher analyte-line intensity—and therefore greater statistical precision—are realized by cementing a number of parts in a row on a small card and using a slit aperture. In this way, the writer has analyzed lead–tin solders in small (~2 mm square) solder-foil preforms and on small irregular-shaped solder-coated nickel and phosphor-bronze components for transistors. Because of the inevitable variation of effective area of such specimens, both Sn $K\alpha$ and Pb $L\alpha_1$ lines are measured and their ratio taken. Strips of solder foil or of solder-coated nickel or phosphor-bronze are used as standards.

Wax-impression molds (*Z11*, *Z13*) (*I* in Figure 16.3) are made by cutting or casting rectangular blocks of sealing wax, warming them on a hot plate to the softening point, and pressing a part of the type to be analyzed into the soft wax. After the wax has cooled and hardened, the part is removed, leaving its impression. The mold is mounted in the spectrometer drawer with the impression centered in the mask window. The mold permits a series of sample and standard parts to be placed reproducibly. The method has been used for determination of thickness of plating or solder coating on small plated and "tinned" parts.

Components for small electron tubes and semiconductor devices and other small parts may be analyzed in a rotating-specimen drawer of the type shown in Figure 16.4 (*B21*). The drawer has two rotors. Small filamentary or cylindrical parts are implanted in plastic cups (*J* in Figure 16.3) filled with stiff wax. The cups are placed in the upper rotor in Figure 16.4 and rotated about their own axes during measurement. Small, more or less flat parts are affixed to plastic pedestals (*K* in Figure 16.3). The pedestals are placed in the lower rotor and rotated in their own planes.

Some parts may be flattened in a press or vise and treated as ordinary disk specimens. Some parts may be melted to small buttons in a small vacuum electric furnace (*F5*). Of course, there is risk of oxidation, or sublimation of volatile constituents.

Stiff wire and rod of small diameter (*L* in Figure 16.3) may be presented by the wax-impression mold or rotating-spindle techniques described above. Alternatively, V-grooves may be machined on the under side of the specimen mask. If greater intensity is required, a number of rods may be cemented side-by-side on a stiff card (*M* in Figure 16.3). Individual wires or rods

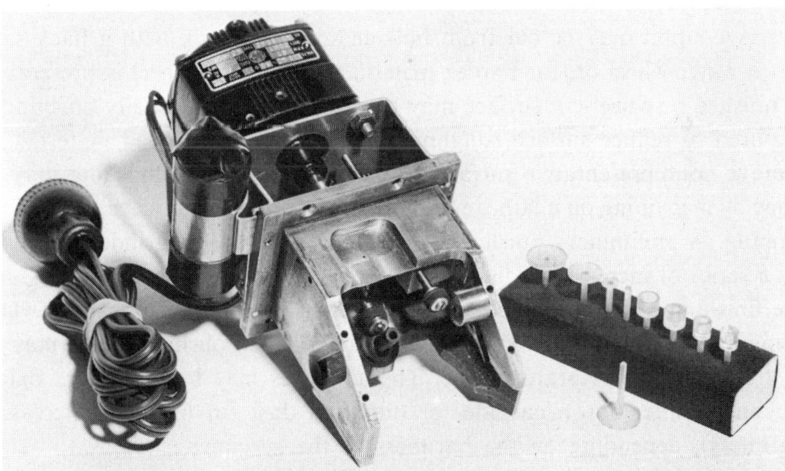

Fig. 16.4. Rotating-specimen drawer and specimen-mounting spindles. Plane specimens are affixed to pedestals (shown), inserted in the center rotor, and rotated in their own plane. Cylindrical or filamentary specimens are supported in cup-type spindles (shown) filled with stiff wax, inserted in the side rotor, and rotated about their axes.

may also be analyzed with a selected-area slit aperture (Chapter 19), especially if a curved analyzer crystal is used.

A specified number of turns of fine wire may be wound on a stiff card (*N* in Figure 16.3), open frame, or small spool (*O* in Figure 16.3) (*B23*). Provision may be made to rotate the spools during measurement. If the wire is always received on identical spools, these may be placed in the instrument specimen chamber or on a large-area accessory, depending on the spool size. In this way, the need for taking samples is avoided. The effective areas and the surfaces presented by all these wire specimens may vary somewhat among samples and standards, and it is preferable, if not essential, to use some kind of a ratio method in their analysis. For example, if plate thickness is to be determined, the intensity ratio of plate and substrate lines should be used; or if molybdenum–tungsten alloy wires are to be analyzed, the intensity ratio of a line of each constituent should be used. Another way to deal with wire is to gather it into a wad and briquet it.

Metallographic potting and sectioning techniques may be applicable in preparation of small parts. For example, a cross section of an electron tube permits identification of the material of each component by use of selected-area techniques (*H* in Figure 16.3).

16.2.3. Preparation (*M15*)

A coupon may be cut from bulk material or stock with a hack saw, but a water wheel or, for harder materials, a diamond wheel is preferable. A finished or wheel-cut surface may be satisfactory, but usually finishing is required to reduce surface roughness, remove work-damaged material, or remove nonrepresentative surface layers (see below). The finishing may be done by machining on a lathe, end mill, or planer, or by abrasive grinding or lapping. A preliminary finish may be imparted on a belt grinder, followed by a series of increasingly finer wet papers on a strip grinder. If necessary, the finishing may then be concluded on a disk grinder or polishing wheel using appropriate abrasive papers or slurries. The polishing wheel may be of the manual or vibratory type. The abrasives may be aluminum oxide, silicon carbide, boron carbide, or diamond dust (in order of increasing hardness), depending on the hardness of the specimen.

A final finish of $\sim$100 μm is usually adequate for measurement of analyte lines of shorter wavelength, 20–50 μm for longer wavelengths down to Al $K\alpha$ or Mg $K\alpha$. Table 16.1 gives approximate abrasive particle sizes for various nominal "grit" or "mesh" designations. In selection of the final abrasive, it is preferable to use a finer abrasive than the required finish. For

TABLE 16.1

Approximate Particle Size of Abrasives[a]

Nominal grit or mesh	Approximate particle size of abrasive, μm	
	Grit	Mesh
500	20	32
400	25	40
300	35	50
200	60	75
100	150	150
80	200	—
50	350	—
40	450	—
30	650	—
20	1000	—

[a] Buehler, Ltd., Evanston, Ill., "The Metal Analyst," catalog of metallurgical and mineralogical preparation equipment (1962).

example, if a 100-μm finish is required, a 200-grit (60-μm) or 200-mesh (75-μm) abrasive should be used.

If the specimen is a multiphase system and smearing of soft constituents occurs, the smear may be removed by final light polishes with 6-μm, then 0.25-μm diamond dust with kerosene lubricant on a cloth polishing wheel (*K9, M15*). Alternatively, the smear may be removed by a light chemical etch specific for the smeared constituent(s). In general, chemical etches and electropolishes are avoided because of possible preferential action on certain constituents.

16.2.4. Precautions and Considerations

When the specimen mask is used to define the effective specimen area, the mask window must be filled by all samples and standards, and the finished surface must lie flat against the under side of the mask. No burrs or other protrusions are permissible, because these might tilt the specimen slightly, thereby changing ϕ and ψ.

With thin sheet metals and foils, care must be taken to ensure a plane surface free of warping, wrinkles, and creases. Particular care is necessary to avoid warping on heating under prolonged irradiation. Sheet and foil

—as well as thin rubber, paper, or textile materials—must be backed by a rigid support, possibly even cemented to it.

For low-Z specimens, care must be taken that the critical thickness is exceeded or that all samples and standards have the same thickness. For very low-Z specimens—such as plastics—care must be taken that spectral emission from the specimen retaining clip or other structures under the specimen do not emerge through the specimen and enter the optical path.

Coarse-grained, inhomogeneous, and rough-finished specimens should be rotated during measurement, and no harm is done by rotating any flat specimen. If rotation is provided, care must be taken to ensure that the irradiated surfaces of all samples and standards lie in the same plane, and that the rotation mechanism does not impart a wobble to the plane.

If rotation is not provided, the specimen should always be loaded in its chamber with linear parallel grind marks oriented the same way, preferably parallel to the plane defined by the directions of the primary and secondary beams. This precaution is more necessary the longer is the wavelength of the analyte line to be measured. Figure 16.5 shows the effect of orientation and roughness of grind marks for lines of short (Sn $K\alpha$) and long (Si $K\alpha$) wavelength. If the grind marks are circular or random, orientation may not matter, but rotation is still beneficial.

Although the specimen surface need not be given a high polish, it must be plane and smooth, and the roughness should be reduced to a certain maximum value. The secondary analyte-line radiation leaves the surface at $\sim 90°$ to the incident primary radiation, so there is possibility of absorption or shielding of either primary or secondary x-rays by surface roughness. The longer the wavelength of the analyte line and/or the higher the absorption coefficient of the specimen for this wavelength, the finer the surface finish must be. In general, the depth of surface roughness or grind marks should be the same for all samples and standards in a given analysis, and less than the critical thickness of the specimen for the analyte line of longest wavelength. Figure 16.6 shows the effect of surface roughness and absorption coefficient on analyte-line intensity for several wavelengths. The figure shows that a 30–50-μm finish is likely to be adequate for the K lines of elements down to atomic number 13 (aluminum).

The surface of a solid may not be representative of the bulk composition, and surface composition is sometimes difficult to reproduce from specimen to specimen. Smearing of soft constituents may occur during polishing, causing high analytical results for the smeared constituent(s) and low results for the covered constituents. The effect is more serious the higher the absorption coefficient of smeared constituent, and the longer the wave-

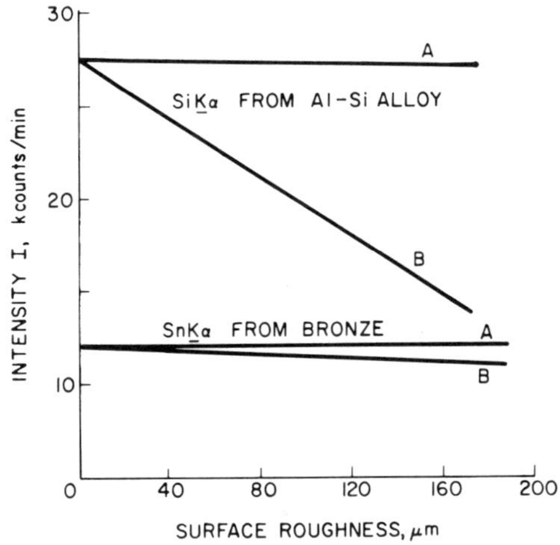

Fig. 16.5. Effect of roughness and direction of parallel grind marks on intensity of analyte lines of long (Si $K\alpha$) and short (Sn $K\alpha$) wavelengths measured from plane solid surfaces. Grind marks for curves *A* were parallel, for curves *B* perpendicular to the plane defined by the central rays of the primary and secondary x-ray beams. [K. Togel, in "Zerstörungsfreie Materialprüfung," R. Oldenbourg, Munich, Germany (1961); courtesy of the author and publisher.]

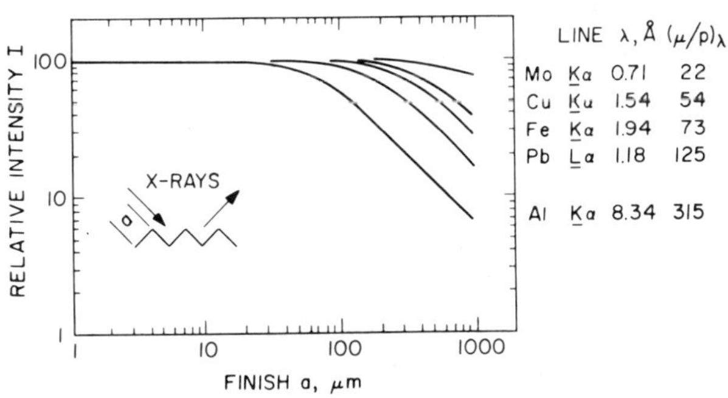

Fig. 16.6. Effect of roughness of parallel grind marks on intensity for analyte lines of various wavelengths measured from plane solid surfaces. The grind marks were perpendicular to the plane defined by the central rays of the primary and secondary x-ray beams as shown. [R. O. Müller, "Spektrochemische Analysen mit Roentgenfluoreszenz," R. Oldenbourg, Munich, Germany (1961); courtesy of the author and publisher.]

length of the spectral line of the covered constituent. Smearing is minimized by machining, rather than polishing the surface. Methods for removing the smear are given in Section 16.2.3. Removal of inclusions may occur during the surface treatment, causing low analytical results for the removed constituent(s). Pickup of surface contamination or abrasive may also occur, and all traces of abrasive, lubricant, and cutting fluid must be removed. Oxidation and other corrosion must be removed. This oxidation and corrosion may be preferential, so that their removal may result in a nonrepresentative surface. Pores and other cavities usually decrease analyte-line intensity in proportion to the surface area they represent. However, cavities may increase analyte-line intensity if the walls of the cavities happen to be enriched in the analyte. Cast and rolled surfaces are very poor for x-ray spectrometric analysis. They are very inhomogeneous; they contain oxide layers, impurities, and inclusions; and their texture may be very coarse-grained and porous. When metal specimens are heated, preferential diffusion of one or more constituents to the surface and/or preferential oxidation may occur. If the metal is heated strongly, sublimation of volatile constituents may occur. Finally, if small buttons of alloys are prepared as standards, their surfaces cool and solidify first and may not be representative of the bulk. Consequently, it is essential to remove sufficient surface material to ensure that the surface is representative of the material to be analyzed.

16.3. POWDERS AND BRIQUETS

16.3.1. Scope, Advantages, Limitations

Powder samples are derived from: (1) powders; (2) bulk solids—especially minerals, ores, slags, etc.—pulverized because they cannot be analyzed in solid form due to gross inhomogeneity, the need to add an internal standard, or some other reason; (3) metal filings, chips, drillings, and turnings; (4) metals converted to oxide (*A19*, *D3*); (5) metals put into solution, then precipitated with a suitable reagent (*F3*); (6) precipitates and residues from solutions (*K7*); (7) ground-up fusion products; and (8) ashed organic matter.

Powders may be presented to the instrument as: (1) loose powder packed in cells (Section 16.3.4.1); (2) briquets (Section 16.3.4.2); or (3) thin layers (Section 16.3.4.3). These three specimen forms are discussed together in this section because they have in common the feature that the powder particles retain their individual identities. Fusion products are usually prepared from powders, but because the particles disappear on fusion, these melts constitute a basically different type and are considered separately in Section 16.4.

The principal disadvantage of the solid specimen form—the inapplicability of addition methods—does not apply to powders. Standard addition and dilution, low- and high-absorption dilution, internal standardization, and internal intensity-reference standardization are all readily applied to powders. The applicability of these methods permits great flexibility in dealing with absorption-enhancement effects. Standards are prepared relatively easily (Section 16.3.2). Powder standards loose-packed in open cells can be stored, but must be handled carefully, and may require occasional repacking. Powders packed in Mylar-covered cells are more nearly permanent, but the Mylar may require occasional replacement. Briquets pressed in metal cups are substantially permanent, and the absence of a Mylar window is beneficial in the long-wavelength region. Wherever applicable, powder methods are usually rapid and convenient.

The principal disadvantages of powder specimens are the following: Trace impurities may be introduced by the grinding and briqueting operations, especially when the specimen powder is abrasive. The surface texture of loose-packed powders may be difficult to reproduce, but the difficulty is surmounted by use of back-loading cells (see below) and substantially eliminated by briqueting. Some powders are hygroscopic or react with oxygen or carbon dioxide in the air; such powders are best handled in Mylar-covered cells. Other powders have low cohesion and flow when the specimen plane is inclined; such powders are placed in Mylar-covered cells, mixed with binder and packed in open cells, or briqueted with binder. Powders may outgas in vacuum spectrometers, sometimes ruining the specimen surface and spattering into the chamber; this difficulty is dealt with by use of helium path, Mylar-covered cells, previous outgassing, or vacuum briqueting.

However, the most serious difficulties with powders arise from particle-size effects. If the powder is not homogeneous, particles of different size, shape, density, or composition tend to segregate. The effect is discussed further in Section 16.3.5.1.

16.3.2. Powder and Briquet Standards

General methods for preparation of standards for various specimen forms are given in Section 16.1.4. However, a more detailed discussion of preparation of loose powder and briquet standards is in order here. It is essential that in any analysis, all powder or briquet samples and standards have the same average particle size and particle-size distribution. This condition is achieved in one of three ways: The standards and samples may be prepared in precisely the same way, or they may be ground to the same particle size or to a particle size so small that differences are no longer significant—say, <30–50 μm. Powder and briquet standards are prepared in any of the following ways.

a. Dry Synthesis. The standards may be wholly synthesized from individual component powders (*M31*, *S43*). When this is done, both the standards and samples should be ground to the same particle size or to a size at which particle-size effects disappear.

b. From a Single Standard. A single standard powder of the type to be analyzed may be varied quantitatively in concentration. Standards having higher analyte concentrations than the base powder are prepared by quantitative addition of constituents of analytical interest. Standards having lower analyte concentrations than the base powder are prepared by quantitative addition of matrix constituents to dilute the analyte(s).

c. From Bulk Solids. A bulk solid of known composition may be ground to powder. Samples analyzed by independent methods may be used. This method is particularly advantageous when the samples to be analyzed are also obtained by grinding solids.

d. From Solutions. Powder standards may be derived from solutions by precipitation, evaporation, or absorption in cellulose. These methods are useful when the samples are obtained the same way. The standard solutions may be prepared synthetically by dissolution of individual constituents in proper relative concentrations, or they may be prepared by dissolution of solids of known composition, including analyzed samples. If only one such standard material is available, the addition technique described above may be applied.

The powder may be recovered by precipitation or by evaporation to dryness. In either case, the recovered solid is dried and fired to convert it to oxides. In the evaporation technique, the solids come out of solution in order of increasing solubility. A much more homogeneous residue is obtained by evaporating the solution to near saturation, then absorbing it in cellulose (filter-paper or chromatographic paper pulp). The cellulose is then dried, heated to expel solvent and acids, then fired to burn off the cellulose and convert the salts to oxides (*C11*).

e. From Fusion Products. If the powder samples are derived from fusion products, the standards may be prepared the same way (*M31*). The fusion product may be prepared by fusion of standard materials, analyzed samples, or synthetic mixtures of individual constituents in proper proportion. If only a single standard material is available, the addition technique may be applied prior to fusion.

Methods for mixing, grinding, and making additions to powders are given in the next section.

16.3.3. Preparation of Powders

In the preparation of powder and briquet specimens, powders must be subjected to one or more of the following treatments: drying, calcining, addition, mixing, grinding, casting, dispersion to thin layers, and briqueting.

a. Drying. Prior to any treatment, powders must be dried at a temperature to expel residual solvent from precipitation, etc. Inorganic powders are usually dried at ~120°C for several hours or overnight to expel residual, absorbed, and adsorbed moisture.

b. Calcining. Powders may be heated at temperatures of several hundred degrees to expel combined water, convert to oxide, and break down crystal structure.

c. Addition. The admixture of powders in the preparation of synthetic standards has already been mentioned. It may also be necessary to add one or another type of material to a powder sample for some required effect. These additives include internal standards, low-absorption diluents, high-absorption diluents, abrasives, and binders.

An internal standard may be added to compensate absorption-enhancement effects. With powders and briquets, internal standards may perform the additional function of compensating packing density effects and may even preclude the need to briquet. If the internal standard is added as powder, the particle size of the standard and sample materials must be the same, or they must be ground together until they are. The internal standard is best added as a solution that wets the sample powder. In one of the earliest applications of internal standardization to x-ray spectrometry, the sample, internal standard, aluminum binder, and silicon carbide abrasive were mixed, repeatedly mulled to dryness under diethyl ether, then briqueted (*A7*).

A low-absorption diluent (Section 14.6.1,2), such as lithium carbonate

(Li_2CO_3), boric acid (H_3BO_3), carbon (*D3*), aluminum (*A6*), starch, powdered sugar, and filter-paper or chromatographic-paper pulp, may be added to reduce absorption-enhancement effects. A high-absorption buffer (Section 14.6.1,2), such as lanthanum oxide (La_2O_3) (*R11*) or tungstic acid (H_2WO_4) (*C32*), may be added to powders having a high-Z analyte in a low-Z matrix to make the calibration function more nearly linear. Powdered inert abrasives, such as aluminum oxide and silicon carbide, may be added to facilitate grinding (*A7*). This is done most frequently when the powder is to be ground in a mortar. A binder may be required to permit powders having low cohesion to be packed into cells, especially for instruments having inclined specimen planes, and to permit briqueting. The amount of binder used is as small as possible, usually 5–20% of the total specimen weight. Among the binders used are the following: lithium carbonate (Li_2CO_3) (*G14*), boric acid (H_3BO_3), aluminum, stearic acid ($C_{17}H_{35}COOH$), starch (*M31*), nitrocellulose, ethyl cellulose, Lucite (*D3*), carnauba wax, and filter-paper or chromatographic-paper pulp (*R12*). Some additives may serve more than one purpose, as is evident from the duplication in the lists of low-absorption diluents and binders.

Low-absorption diluents, abrasives, and binders are always added in powder form. However, individual constituents of synthetic standards, internal standards, and high-absorption diluents may be added in powder or solution form. If the powders to be mixed all have small particle size (say, −325 mesh or smaller), or substantially the same particle size, shape, and density, they may simply be mixed in dry form. If they are coarse, or differ in particle size or shape, they must be ground separately before admixture, or, preferably, mixed and ground together. If they differ substantially in density, a standard solution of the heavier constituent(s) may be added volumetrically to a weighed portion of the lighter constituent in powder form. Alternatively, standard solutions of the minor and trace constituents may be added to the major constituent(s) in powder form. In either case, the wet powder is homogenized and dried, then ground and mixed.

d. Mixing. After the additions have been made, the powder may be mixed or homogenized by spatulation or shaking in a bottle, but best by a mechanical mixer, such as the Spex Mixer/Mill using a plastic vial and ball. Small volumes of powder may be mixed in a dental mixer, such as the Wig-L-Bug, again using a plastic vial and ball.

e. Grinding. Prior to being packed into cells or briqueted, the powder should be ground to −325 or −400 mesh. Coarser powders may be acceptable if only short-wavelength analyte lines are to be measured. If any of the additives mentioned above are to be used, they should be added prior to the grinding operation so that the base powder and additives can be ground to substantially the same particle size.

Grinding may be done manually with an agate, silicon carbide, or boron carbide mortar and pestle. The powder may be ground dry, or it may be ground under ethanol or diethyl ether to dryness several times. However, grinding is preferably done mechanically in a Spex Mixer/Mill using a plastic, alundum, or tungsten carbide vial and ball, depending on the hardness of the material. The Spex Freezer/Mill, which grinds under liquid nitrogen, may be used for soft and resilient materials. In some cases, a Pitchford or equivalent mill may be used, that grinds to a substantially uniform particle size. For extremely small particle size, 1 μm or less, the powder is slurried in an organic vehicle and rolled in a ball mill overnight or longer.

Casting, briqueting, and fusion are discussed in subsequent sections.

16.3.4. Presentation

16.3.4.1. *Loose Powders*

For upright and inclined specimen planes, loose powders are usually simply placed in shallow, metal or plastic, circular or rectangular dishlike cells. Most manufacturers of x-ray spectrometers supply powder cells for use with their instruments. Thin-walled aluminum cups for briqueting (Somar Laboratories, New York), Caplugs (Protective Closures Co., Buffalo), and various stainless steel, glass, and plastic "planchets" used in radiochemistry all make excellent disposable powder cells.

The cell may be slightly overfilled, then leveled by drawing the *edge* of a stainless steel spatula across its rim, or it may be made heaping full, then packed and surfaced with the *flat* side of a wide spatula. If care is taken always to load the same measured weight or volume of powder and to avoid any loss, reasonably uniform packing density among the specimens is achieved. If specimen rotation is desired, circular cells are used and placed in rimmed pedestals in a specimen rotator (*B21*). The pedestals are similar to those shown in Figure 16.4, but larger in diameter to closely fit the powder cell.

For inverted specimen planes, the filled circular cells just described may be covered with 0.00025-inch Mylar secured with an O-ring or other retaining ring. More conveniently, a liquid-specimen cup is used. The liquid-specimen cells supplied with inverted-geometry instruments consist essentially of open cylinders covered on the bottom with 0.00025-inch Mylar secured with a suitable ring or collar. In fact, a short length of plastic tubing fitted with a Mylar bottom may be used. Enough powder is poured into the cell to exceed infinite thickness, and the powder is settled by tapping the cell sharply on the table. Alternatively, disposable plastic liquid-specimen *cups* (Spex Industries, Metuchen, N. J.) may be used. The powder is placed in the cup, the Mylar window applied and secured, and the cup inverted (Mylar down) and tapped to settle the powder. Most modern instruments having inverted geometry provide for specimen rotation.

A smoother, more uniform surface is obtained by backloading powder cells (*B21*, *L18*). For this purpose, the cells have the form of cylindrical rings or rectangular frames open on both sides. They are loaded by laying them on small glass plates and filling them with powder. The powder is packed firmly in the frames with a stainless steel spatula, and all powder is removed from the rims. Measured weights or volumes of powder may be used to realize uniform packing density. Lucite back covers $\frac{1}{8}$ inch thick and of the same outside diameter or length and width as the cells are now placed on the filled cells. The glass–cell–cover assemblies are then inverted with the glass plates up. The glass plates are removed, exposing the useful surface, and the cells with their backing plates are placed in the specimen drawer with the smooth powder surface up. Sometimes, when the glass plate is removed, particles of nonmetallic powders are attracted from the surface by electrostatic charge. This effect is avoided by use of a metal plate or by sliding the glass plate off sideways, rather than lifting it off. If the backloaded cell is to be rotated, the cylindrical ring is used and loaded as before, but instead of a Lucite plate, a rimmed Lucite pedestal is placed on the cell (*B21*). The glass–cell–pedestal assembly in then inverted, the glass plate removed, and the pedestal shaft inserted in the specimen rotator.

For powders having low cohesion, the tendency to flow in an inclined specimen plane is avoided by use of binders and/or Mylar-covered cells similar to the liquid-specimen cell shown in Figure 17.3. These cells consist of rectangular Lucite dishes having cemented-on 0.00025-inch Mylar windows and filling ports and plugs. Thicker Mylar may be used if only short-wavelength lines are to be measured. When the cell is being filled, a small funnel is fitted to the filling port and a protective Lucite plate is held on the window by a small C-clamp to prevent bulging. The powder is poured through the funnel and settled by tapping the cell on the table top. Again, measured weights or volumes of powder may be used. In general, the plugs are required only for storing standard cells. These covered cells are also good for general-purpose use—upright, inclined, or inverted geometry, and vacuum path. They are much easier to handle, and standards are stored more conveniently. The Mylar cover does substantially attenuate the *K* lines of fluorine, sodium, magnesium, and aluminum. Instructions for preparation of the cells are given in Section 17.1.4.2.

For special applications, loose powders may be presented in other ways. Powders in which only two elements vary in concentration may simply be placed in small cellophane envelopes (*D1*). The variable nature of the specimen is compensated by use of the intensity ratio of the analyte lines of the two constituents. The technique has been used for the 8-quinolinolates of two coprecipitated elements (*D2*). Powders that are unstable in air may be placed in small polyethylene envelopes or in a small folded sheet of polyethylene and sealed in by thermal "welding" of the open edges (*T7*). Again, unless only qualitative analysis is required, some kind of ratio method must be used. A cell designed to maintain x-ray diffractometer specimens in special atmospheres may also have application in x-ray spectrometry (*M26*). The cell is sealed and has a tubulation to attach to a small cylinder or other source of the required gas.

16.3.4.2. *Briquets*

The prepared powder may be formed into a self-supporting pellet or disk without loss of identity of the particles by admixture with wax or by suspension in a plastic monomer (*B55*) in a cylindrical cup, followed by polymerization. However, the most common and most convenient way is by briqueting at high pressure in a press. The briqueting technique is also applicable to metal granules, chips, shot, drillings, turnings (*C51*), and wire. Usually, a binder is admixed with the powder, but cohesive powders, including metals, may be briqueted without binder.

The powder, with or without binder, is placed in a cylindrical die in a hydraulic press capable of developing up to 100,000 lb/inch². The die usually consists of a heavy outer retaining shell surrounding a cylinder or piston chamber in which slides a piston. When the powder is loaded in the die, the piston is withdrawn, and a loading cylinder of the same outside diameter is inserted in its place. The powder is introduced through the loading cylinder to avoid getting powder on the inside wall of the piston chamber, where it would soon score the chamber and piston. The loading cylinder is then withdrawn and the piston replaced. Dies of several diameters should be available. It is preferable that provision be made to heat and evacuate the die for certain applications. For any given series of samples and standards, uniform packing density is achieved by compacting the same weight of powder at the same pressure for each specimen.

Briquets may be formed in several ways as follows:

1. *Without backing or other support*. The powder is simply compacted as is.

2. *With a backing of pure binder*. A layer of pure binder (*R11*) or cellulose (*A14*) is placed in the die and compacted lightly with a manually inserted piston or at full pressure in the press. Then the powder is loaded and the two-layer briquet compacted at full pressure. Alternatively, a supply of binder or cellulose briquets may be made in advance. One of these disks is placed in the die prior to loading each powder charge.

3. *With a backing and jacket of pure binder*. A special die arrangement permits the analytical powder to be ringed as well as backed with pure binder or cellulose (*B8*).

4. *With a metal or plastic jacket*. A thin-walled cylindrical metal (*M21*) or Bakelite (*B1*) ring having outside diameter just less than the inside diameter of the piston chamber is placed in the die prior to loading of the powder.

5. *With a metal jacket and binder backing*. This technique is a combination of 2 and 4 above.

6. *In a metal cup*. The technique is the same as 4 above, except that a thin-walled cylindrical metal cup (Somar Laboratories, New York) is used (*M8*). The die may be operated so as to slightly crimp over the top rim of the cup.

If a glass disk is placed in the die prior to loading of the powder, the briquet surface that was in contact with the glass is exceptionally smooth (*V8*).

All the foregoing techniques are for conversion of a powder to a briquet. The solution–briquet technique (*R12*), in effect, permits substantially direct conversion of a solution to a briquet. The prepared solution is evaporated to remove some solvent, then taken up in 500 mg of chromatographic-paper pulp. The pulp is then homogenized, dried, ground in a boron carbide mortar, and pressed into a double-layer disk against a backing of previously pelletized chromatographic paper.

Briquet surfaces and edges may be sprayed with clear Krylon to prevent flaking and crumbling (*A14*). However, some attenuation of wavelengths greater than 6 or 7 Å results from spraying the measured surface.

Some considerations relevant to briqueting are given in Section 16.3.5.3.

16.3.4.3. *Thin Layers*

The effects of particle size and inhomogeneity and absorption-enhancement effects disappear if the powder is ground to extremely fine particle size and distributed in a thin layer. Under these conditions, the powder assumes, to a certain extent, the nature of a thin film. Another advantage of the thin-layer presentation is realized when only a small piece of material is available. The measured analyte-line intensity is greater if the small piece is ground to fine powder and spread out, as shown in Figure 16.7. Powders may be spread into thin layers dry or as slurries.

In the dry technique, the powder is lightly dusted on the adhesive surface of a piece of Scotch tape and all excess powder gently blown off (*A2*). A better technique is as follows (*S2*, *S3*): A 0.00025-inch Mylar film is tightly stretched, in the manner of a drum head, across an empty plastic liquid-specimen cup or a short length of plastic tubing 2–3 cm in diameter. The Mylar is secured with an O-ring or other suitable retainer. A measured weight or volume of fine powder is placed on the Mylar and distributed as uniformly as possible. A second Mylar film is now stretched over the cup and secured, enclosing the powder between the two tightly stretched films. Standards and samples are prepared in the same way.

In the slurry technique, a small amount of fine powder is slurried with a solution of 2–5% (wt/vol) nitrocellulose in amyl acetate and spread over a scrupulously clean microscope slide (*D3*). The film may be presented to the spectrometer thus supported on glass, or it may be scored around the

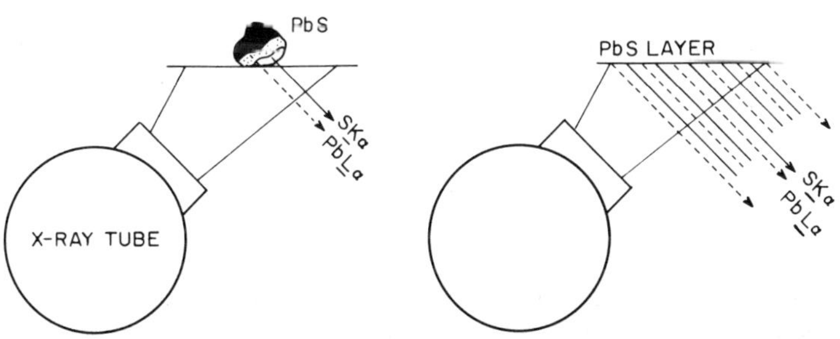

Fig. 16.7. Effect of specimen presentation on S $K\alpha$ (5.37 Å) and Pb $L\alpha_1$ (1.18 Å) line intensity measured from lead sulfide. For the particulate specimen, only the unshaded region contributes to the measured S $K\alpha$ intensity, the unshaded and dotted regions to the Pb $L\alpha_1$ intensity. If the specimen is dispersed as a film, the entire mass contributes to the measured intensity of both lines. [R. Jenkins and J. L. de Vries, "Practical X-Ray Spectrometry," Springer-Verlag N. Y. (1967); courtesy of the authors and publisher.]

edge and floated off on water, then scooped up and supported some other way.

A more versatile slurry technique is that of Finnegan (*F7*). A weighed portion of powder is mixed with a weighed portion of ethylcellulose or nitrocellulose binder and added to a measured volume of amyl or butyl acetate. The slurry is best mixed by rolling on a roller mill, and the powder may be ground further to very small particle size if the slurry is rolled with grinding balls. A measured volume of the slurry is placed on an optically flat glass plate and spread in a uniform thin film with a casting knife (doctor blade). Blanks are removed from the film by scoring, flooding with water, and coaxing squares of the film to float off. Preparation of standards for analysis of such films is described in Section 16.1.4.

16.3.5. Precautions and Considerations

16.3.5.1. *Particle-Size Effects*

With regard to surface roughness, the particle-size effect in powders and briquets is analogous to the surface roughness effect in solids (Section 16.2.4, Figure 16.6). Just as the surface roughness of solids should be less than the critical thickness of the solid for the longest analyte line to be measured, so should the roughness—that is, the particle size—of powders and briquets. In general, particle-size effects substantially disappear, even for Mg $K\alpha$, at about -400 mesh. Coarser particles may be satisfactory for short wavelengths, and finer particles are required for Na $K\alpha$ and F $K\alpha$. If the particle size is such that it still affects the analyte-line intensity, then all samples and standards in a given analysis must have the same average particle size and particle-size distribution, and measures should be taken to avoid segregation (*A2*, *H2*).

In powders having nonuniform particle size, segregation may occur by size, and, in nonhomogeneous powders, segregation may occur by density. In either case, sifting is likely to accentuate the separation.

In filling cells for instruments having inverted geometry, the powder is usually poured onto the cell window. The small particles tend to flow through gaps between the large particles and collect on the window, forming a layer behind which lie the large particles. In filling cells for instruments having upright geometry, the powder is placed in the cell and packed by exertion of a small manual pressure. The larger particles are pressed toward the bottom, leaving a layer of small particles on top, which is the analytical surface. Thus, in either case, the specimen surface is likely to consist of an abnormally high proportion of small particles. However, in filling back-loaded cells (Section 16.3.4.1), these effects are not as marked. When the powder is poured into the cell, the first effect occurs, forming a layer of small particles on the bottom. Then, when the powder is packed and leveled, the second effect occurs, tending to push large particles into the layer of small particles on the bottom, which is inverted and becomes the specimen surface. These effects are substantially eliminated by briqueting or by grinding to −325 mesh or smaller.

Consider a heterogeneous powder consisting of large A-rich particles and small B-rich particles. If the effective specimen depth contributing to the measured A-line and B-line intensities is less than the average size of the large A-rich particles, B-line intensity will be abnormally high, A-line intensity abnormally low. This condition is minimized by further grinding to reduce the A-rich particle size.

In a completely homogeneous powder, the measured analyte-line intensity decreases as the path length of the measured wavelength in the specimen approaches the particle size. This phenomenon is due to mutual shielding of particles and is partially dealt with by dilution.

For a given briqueting pressure, the smaller the particle size, the higher is the analyte-line intensity. For a given particle size, the higher the briqueting pressure, the higher is the analyte-line intensity. These relationships are

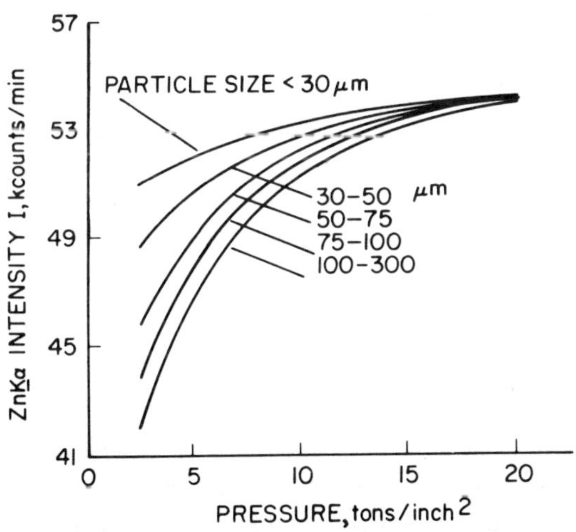

Fig. 16.8. Effect of briqueting pressure on Zn $K\alpha$ intensity measured from briquets of zinc powder having various particle sizes. [K. Togel, in "Zerstörungsfreie Materialprüfung," R. Oldenbourg, Munich, Germany (1961); courtesy of the author and publisher.]

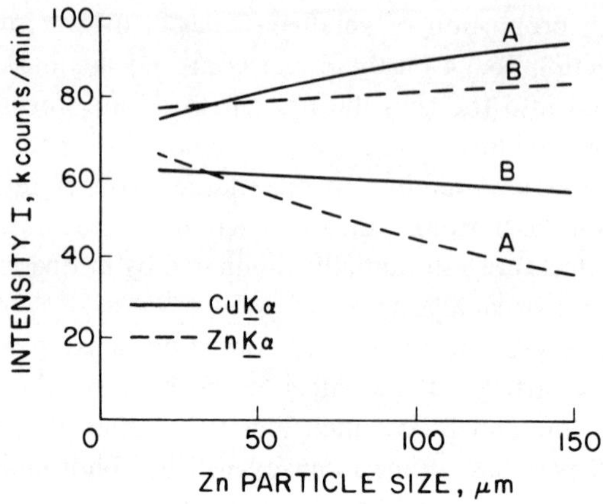

Fig. 16.9. Effect of particle size and surface facing on analyte-line intensity measured from briqueted 1 : 1 mixtures of copper and zinc powders. For all briquets, copper particle size was constant at a few micrometers, and briqueting pressure was 4 tons/inch². Briquets were measured as-pressed (curves A) and after removal of the surface to a depth of 0.1 mm (curves B). [K. Togel, in "Zerstörungs-freie Materialprüfung," R. Oldenbourg, Munich, Germany (1961); courtesy of the author and publisher.]

shown in Figure 16.8. The use of a binder or diluent accentuates the decreased intensity with increased particle size, and reduces the effect of pressure in overcoming this decrease.

On briqueting, ductile particles tend to flow around less ductile ones. This effect is most marked with metals, and, for this reason, briquets of metal filings, granules, chips, drillings, turnings, or wire should be surfaced to a depth of 0.1–0.5 mm. Figure 16.9 illustrates this effect.

16.3.5.2. *Additives*

Binders have the following advantages: (1) Stable briquets are formed even with powders having low cohesion; (2) better homogeneity is realized in packed and briqueted powders of nonuniform particle size or density; (3) higher packing densities and smoother surfaces are obtained; and (4) absorption-enhancement effects are reduced by the dilution. However, binders also have disadvantages. A few per cent of binder may not significantly affect the absorption of a medium-Z matrix, but it may reduce the

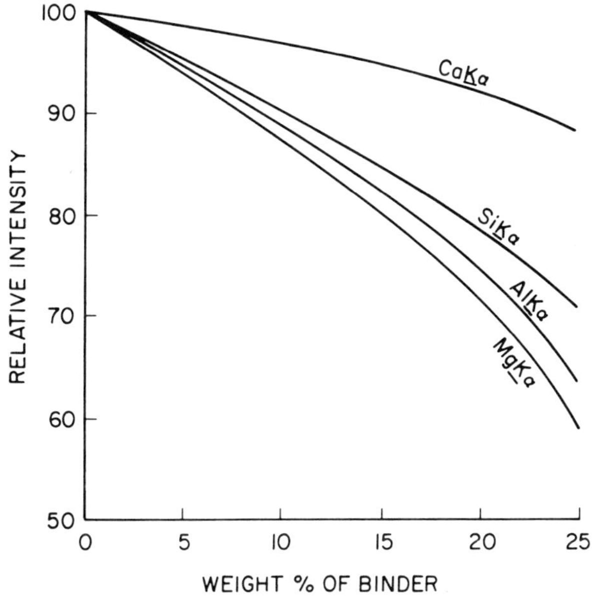

Fig. 16.10. Effect of concentration of binder or diluent on measured intensity for analyte lines of various wavelengths. The specimen material was a slag containing 6.5% MgO, 16.8% Al_2O_3; 39.7% SiO_2, and 29.5% CaO. The binder was starch. [K. Togel, in "Zerstörungsfreie Materialprüfung," R. Oldenbourg, Munich, Germany (1961); courtesy of the author and publisher.]

effective atomic number enough to increase primary-beam scatter and thereby the background. With low-Z analytes, binders reduce analyte-line intensity, both by dilution and by absorption. Figure 16.10 shows the effect of binder concentration on analyte-line intensity in the long-wavelength region. Other adverse effects of binders are cited above. In general, it is advisable to mix the minimum effective amount of binder with the specimen, then briquet the mixed powder against a backing of pure binder.

16.3.5.3. *Briqueting*

Figure 16.8 shows that the smaller the particle size, the lower is the briqueting pressure required to realize a specified analyte-line intensity. However, even for very small particle size, pressures of 20 tons/inch^2 are required to attain maximum Zn $K\alpha$ intensity. Even higher pressures may be required in the long-wavelength region, and lower pressures may be satisfactory for short wavelengths.

The briqueting die should be evacuated before and after pressing so that occluded air does not effuse and spoil the surface when pressure is released or later when the briquet is used in a vacuum spectrometer. Maximum pressure should be maintained for a minute or so to overcome any elasticity of the powder. Powders should be briqueted in a thin-walled metal cup, or at least in a metal or plastic cylindrical jacket with a backing of cellulose or pure binder.

16.3.5.4. *Other Considerations*

Variations in packing density may cause variations in analyte-line intensity unless an internal standard is used, or unless the same weight of each specimen in the analysis is briqueted at the same pressure. Mixing, grinding, and briqueting may introduce trace impurities, especially when the specimen has an abrasive nature. Powders and briquets may outgas or even mildly explode in vacuum spectrometers unless previously outgassed in a vacuum chamber or vacuum oven, or unless briqueted in a vacuum die. Rotation of powder cells and briquets is to be recommended. If rotation is provided, care must be taken to ensure that the irradiated surfaces of all samples and standards lie in the same plane and that the rotation mechanism does not impart a wobble to the specimen.

Campbell and Thatcher (*C14*) have compared briqueting, fusion, and solution techniques for determination of calcium in wolframite concentrates.

16.4. FUSION PRODUCTS

16.4.1. Scope, Advantages, Limitations

The material to be analyzed may be fused with a flux for one of three reasons: (1) to render an insoluble or difficultly soluble material readily soluble; (2) to homogenize a material with respect to composition and density and to eliminate its particulate structure so that it may be ground to a homogeneous powder that may be used as such or as a briquet; or (3) to provide a homogeneous solid or glass pellet suitable for direct use as a specimen. The first application is the concern of Chapter 17; the other two are the subject of this section.

Fusion methods have many advantages regardless of whether the "melt" is used directly or powdered or briqueted. All inhomogeneity of composition, density, and particle size is eliminated. Standard addition and dilution methods and internal standardization are readily feasible. Absorption-enhancement effects are reduced or even eliminated by the leveling effect of the flux. Standards are easily prepared—synthetically, if necessary. If the melt is ground and briqueted, it need be ground only to −200 mesh or so because of the homogeneity. If the solidified melt is used as the specimen, additional advantages are realized. The surface texture is smooth and uniform, although some polishing may be required, and standards are very conveniently stored.

The most serious disadvantages are that analyte-line intensities from low-*Z* elements are reduced by dilution and absorption, and that trace and minor constituents are highly diluted. The fusion requires considerable time and effort and, if glass disks are to be prepared, skill. Glass-disk standards may break in storage due to stress or devitrification. The pieces can be remelted and recast if a crucible is used from which the glass pours cleanly without adherence.

16.4.2. Materials

The principal fluxes used in fusions for x-ray spectrometric specimen preparation are potassium pyrosulfate ($K_2S_2O_7$) and the glass-formers, sodium tetraborate ($Na_2B_4O_7$) and lithium tetraborate ($Li_2B_7O_4$). In place of lithium tetraborate, a mixture of lithium carbonate (Li_2CO_3) and boron oxide (B_2O_3) in 1:2 molar ratio can be used. A supply of each flux that is to be used should be fused in a large crucible, then ground to a powder and stored in a tightly stoppered bottle. The most commonly used high-absorption diluent (absorption buffer) is lanthanum oxide (La_2O_3).

If a lighter matrix than sodium tetraborate is required, but also a lower fusion temperature than lithium tetraborate (m.p. 930°C, compared with 741°C for $Na_2B_4O_7$), the eutectic formed by fusion of lithium tetraborate and lithium carbonate in weight ratio 6:1 may be used. This eutectic actually has a lower melting point than sodium tetraborate. Unfortunately, the fused eutectic is hygroscopic, and so are the specimen disks prepared from it.

The borate glasses wet quartz, nickel, and platinum crucibles, and they are unsuitable. Graphite is satisfactory. Palau (80Au–20Pd) and platinum–gold alloy (97Pt–3Au) are not wet by the glasses, which pour cleanly from crucibles of these alloys. Potassium pyrosulfate fusions may be done in Vycor or even Pyrex.

16.4.3. Specific Fusion Procedures

In the *potassium pyrosulfate fusion* (*C50*), the charge typically consists of 200 mg of sample and 10 g of potassium pyrosulfate. If graphite is present,

200 mg of potassium persulfate ($K_2S_2O_8$) is added. For silicate minerals, a few grams of potassium fluoride (KF) or sodium fluoride (NaF) is added to volatilize silicon as silicon tetrafluoride (*C53*). The charge is heated in a Pyrex or Vycor crucible at $<700°C$ for $\sim$3 min. The melt is allowed to cool, ground to -200 mesh, and briqueted at $>12{,}000$ lb/inch^2. There is probably no reason why disks cannot be cast from the melt, but this is not done.

In the first reported application of the *borate-glass fusion method* (*A16*), the charge consisted of 300 mg of sample, 1.5 g of lithium carbonate, and 1.5 g of boron oxide (or about 3 g of lithium tetraborate). The charge was heated in a graphite crucible at 1150°C for $\sim$20 min, then poured into a brass ring 5/8 inch in diameter on a copper-surfaced hot plate. Immediately on solidifying, the pellet was imbedded in asbestos for $\sim$5 min to anneal. It was then removed from the asbestos and allowed to cool. The flat surface was polished, if necessary, on a belt grinder.

In the *Claisse sodium tetraborate fusion* (*C34*, *M4*), the charge typically consists of 100 mg of sample and 10 g of sodium tetraborate. For sulfide ores, 1.5–2 g of barium peroxide (BaO_2) is added; alternatively, the sulfides are fused with about ten times their weight of potassium pyrosulfate, and then the sodium tetraborate is added. Metals may be converted to sulfides by heating with sulfur, then treated as just described (*S6*). If a high-absorption diluent is required, 100 mg of lanthanum oxide may be added. Claisse used platinum crucibles, but graphite, Palau, or 97Pt–3Au alloy are all preferable (Section 16.4.2). The charge is fused over a Meker burner or in a muffle furnace at 800–1000°C for 5–15 min. By means of platinum-tipped tongs, the crucible is tipped and agitated occasionally to ensure homogeneity and to aid the expulsion of gas bubbles. The melt is poured on a polished aluminum plate heated at 450°C on a hot plate. If the melt is very fluid, an aluminum ring of $\sim 1\frac{1}{4}$ inch inside diameter and at room temperature is placed on the aluminum plate and the melt poured into it. The pellet is kept on the hot plate for $\sim$2 min, then pushed with a Transite rod onto a Transite plate previously heated to 450°C. The pellet is allowed to cool with the Transite to room temperature. The ring is removed. The bottom surface of the glass pellet may be satisfactory as is; if not, it may be polished.

Strain-free glass disks are prepared as follows (*D15*). The charge consists of 3 g of sample and 9 g of fused lithium tetraborate, or, if a high-absorption diluent is required, 1 g of sample, 1 g of lanthanum oxide, and 8 g of lithium tetraborate. The charge is fused in a 97Pt–3Au alloy crucible at 1250°C for $\sim$10 min. The crucible is kept covered, but is stirred occasionally with a rod of the same alloy. The crucible and charge are allowed to cool to room temperature. The crucible is then inverted, and the glass button drops out. The button is ground to powder, which is returned to the crucible, remelted at 1250°C for 5 min, then allowed to cool. The button is recovered as before, replaced in the crucible upside down, and again remelted at 1250°C for 5 min. The button, still in the crucible, is annealed at 650°C for 30 min, then allowed to cool with the furnace. The flat side of the button is polished to a final 600-grit surface with silicon carbide.

In the *minimum-flux technique* (*A14*), 1 g of sample and 1 g of lithium tetraborate are mixed and fused in a graphite crucible at 1400°C for 1.5 min, then poured on a copper block and allowed to cool. This glass is unsuitable for use as a disk. It is ground and briqueted without binder, but with a cellulose backing. The low dilution ratio makes this technique favorable for low-Z analytes and minor and trace constituents.

In the *high-concentration, glass-disk method*, Longobucco (*L23*) was able to achieve high sample-to-flux ratio and still use the melt in the form of glass disks. She applied the method to determination of magnesium, aluminum, silicon, potassium, and calcium in clay minerals. A charge consisting of 5 g of sample, 5.5 g of sodium tetraborate, and 2 g of lithium carbonate is thoroughly mixed, placed in a platinum crucible, and heated slowly over a Meker burner until fusion begins. The temperature is then gradually increased until the crucible and charge are at a bright red heat. The crucible is heated at this temperature with constant swirling until fusion is complete, then placed in a muffle furnace at 1350°C for 3 min to eliminate gas bubbles. The melt is then poured on a polished aluminum plate maintained at ~260°C on a hot plate and allowed to cool slowly. Finally, the flat side of the disk is polished on a wet abrasive paper.

Great care is required in preparation of the disks. If the aluminum plate is too hot, the disk will wet and adhere to it; if it is not hot enough, the bottom of the disk may be rippled. If the hot disk comes in contact with a cold (room temperature) object, it will crack. If it cools too quickly, it will be so strained that it may "explode" like a Prince Rupert drop when handled. Nevertheless, standard disks prepared in this way have withstood years of use and storage. Longobucco later applied the same method to preparation of lithium tetraborate glass disks.

In the *heavy-absorber, fusion-briquet method* (*R11*), 125 mg of sample, 125 mg of lanthanum oxide, and 1.000 g of lithium tetraborate are mixed in a boron carbide mortar, transferred to a graphite crucible, and fused at 1100°C for ~10 min. The crucible is then removed from the furnace, and the melt is allowed to cool in the crucible. This glass is also unsuitable for use as a disk. The pellet is removed and weighed. Enough boric acid (H_3BO_3) is added to bring the total weight to 1.300 g to level the final sample weight and compensate ignition weight loss. The bead and boric acid are placed in a tungsten carbide Mixer/Mill grinding vial, and the bead is broken up. The vial is then placed on the Mixer/Mill and ground for ~10 min to about −300 mesh. The ground powder is briqueted at 50,000 lb/inch^2 against a backing of a previously briqueted boric acid disk. Here again, the low dilution factor makes this technique favorable for low-Z and trace analytes. The lanthanum oxide reduces absorption-enhancement effects.

In the *chemical-separation, borax-disk method* (*L32*), the glass-disk method is applied to trace and micro analysis. Net intensities of 1–5 counts/s per μg is obtained for the $K\alpha$ lines of elements 20–42 (calcium to molybdenum). In order to avoid the need for internal standardization, the analytes are separated by chemical means prior to fusion. Each separated analyte, as salt or oxide, is placed in a separate 8-ml platinum crucible, and 2 g of $Na_2B_4O_7$ is added and fused. An aluminum plate $\frac{1}{4}$ inch thick having several plane, poished, circular depressions 1.5 inch in diameter and 1/16 inch deep is heated on a hot plate to 225°C. The melt for each analyte is poured into a depression and immediately flattened with a hot $2 \times 2 \times \frac{1}{4}$ inch aluminum plate having a wood handle. Thus, all disks have uniform 1/16-inch thickness. When a disk has been cast in each depression, the aluminum plate is removed from the hot plate and allowed to cool with the disks in place. Standard disks are prepared the same way from known weights of analyte salts or oxides.

Chapter 17

Specimen Preparation and Presentation—Liquids; Supported Specimens

17.1. LIQUID SPECIMENS

17.1.1. Introduction

In Chapter 16, it was pointed out that usually x-ray spectrometric analyses of samples in their as-received forms are more rapid and convenient than analyses by any other method. Thus, the analyst strives to analyze samples in their original forms whenever this is practical. However, it is often necessary to change the original form—especially of solid samples and parts—when: (1) specimen errors arising from inhomogeneity, particle size, etc. must be eliminated; (2) a method involving addition is to be used; (3) chemical separation or other treatment is required; (4) the samples have form different from that of the available standards, or the same substance is received in many forms, and it is desired to use a single set of standards for all; or (5) the x-ray spectrometric method is to complement some other analytical method. In such cases, all specimens are usually reduced to powder, fusion product, or solution. If the decision is made to put the specimens into solution, many alternative possibilities for subsequent preparation and treatment, presentation, and analysis become feasible that would not be so for other forms. Many general papers dealing with x-ray spectrometric solution techniques have appeared (*15F*, *B17*, *B20*, *C5*, *C15*, *C17*, *C55*, *G13*, *L4*, *T7*).

Of course, many samples are received in liquid form, including the following: (1) liquids and solutions, aqueous and nonaqueous, including plating solutions and hydrocarbons; (2) dissolved solids, powders, and parts; (3) dissolved fusion products; (4) leach (*L37*) and wash liquids; (5) brines; (6) eluates from ion-exchange columns; (7) liquid–liquid and liquid–solid extractions; (8) biological fluids; (9) emulsions and gels; (10) sols (*G12*); (11) slurries (*H45*); (12) greases, waxes, etc.; (13) plastic monomers; (14) liquid ion-exchange resins; and (15) melts.

A specimen in liquid or solution form may be readily subjected to various treatments, including the following: (1) dilution; (2) evaporation (concentration); (3) absorption in cellulose or other material and firing to powder (*C11*); (4) evaporation to dryness; (5) precipitation of analyte or matrix constituents, selectively or collectively; (6) liquid–liquid extraction (*B45*, *B46*, *K4*, *W11*); (7) ion exchange with granular resin by batch or column technique, or with ion-exchange liquid, membrane, or paper (*C18*, *V1*, *V2*); (8) paper chromatography (*J1*, *J2*); (9) gas chromatography (*S9*); (10) paper electrophoresis (*W4*); (11) electrodeposition on solid (*L24*) or mercury cathode; (12) distillation, including steam distillation; (13) dry or wet ashing (*D6*, *L38*, *R16*, *S20*); (14) gas formation—for example, dissolved sulfur may be converted to hydrogen sulfide and absorbed in paper impregnated with lead acetate, and arsenic may be converted to arsine and absorbed in mercury(II) bromide paper (*J3*); and (15) quantitative addition.

Quantitative additives include: (1) analyte (standard-addition method); (2) solvent (standard-dilution method); (3) a metal salt or a matrix constituent of no analytical interest (matrix-masking agent); (4) low- or high-absorption diluent; (5) internal standard; (6) internal control or intensity-reference standard; or (7) a relatively heavy element to combine stoichiometrically with a low-Z element, organic compound, or nondeterminable substance to make its determination feasible or more convenient (indirect or association method).

Solutions may be converted to powders by precipitation, evaporation, or absorption in cellulose; and the powders may be converted to briquets or fusion products. Solutions may also be presented as supported specimens of various types (Section 17.2). Solutions are perhaps more readily applicable than other forms to x-ray analytical methods other than x-ray secondary-emission spectrometry, including polychromatic and monochromatic absorptiometry, absorption-edge spectrometry, and methods based on x-rays scattered from the specimen. Finally, when x-ray spectrometry is to supplement another wholly independent analytical method, the samples must usually be in solution form.

The remainder of Section 17.1 deals with the presentation of specimens in liquid form, Section 17.2 with supported specimens of various types.

17.1.2. Advantages

The principal advantages of specimen presentation in liquid form are the following:

1. Solution techniques are applicable to almost any type of sample. They are directly applicable to liquids, solutions, slurries, and sols; and solids, fabricated parts, powders, and fusion products may be put into solution.

2. Solution techniques are particularly advantageous when the same material is received in various forms, such as bar, foil, drillings, powder, wire, and parts. A single specimen procedure and set of standards can deal with all sample forms.

3. For samples already in liquid, sol, or slurry form, often little or no preparation is required; what preparation is required is usually simple and convenient. For other sample forms, preparation is also usually simple and convenient.

4. Samples and standards are perfectly homogeneous.

5. The analytical results are representative of the bulk sample, not just the surface; the effects of nonrepresentative surface composition, surface texture, and particle size are eliminated.

6. Standards are usually prepared easily by dissolution of the individual constituents in proper concentrations, dissolution of standard materials, or quantitative variation of the composition of a single dissolved standard or of a standard solution.

7. Blanks for measurement of background and for compensation of contamination are usually prepared easily.

8. Standard addition and dilution techniques are applicable which preclude the need for standards; such analyses are particularly useful for one-of-a-kind analyses.

9. The generally low effective atomic number of liquid matrixes results in relatively high scattered background. This higher background is usually disadvantageous, but may be beneficial in facilitating the use of methods for compensation of absorption-enhancement effects and analytical methods based on scattered x-rays.

10. Chemical separations, if required, are usually made conveniently (*D5*).

11. Concentration or dilution to appropriate analyte concentrations is usually feasible.

12. The absorption-enhancement effects of the total solute are reduced or eliminated by the leveling effect of the solvent.

13. The matrix (solvent) usually consists of elements which have low atomic number and which contribute little absorption and no enhancement of the analyte lines.

14. When residual absorption-enhancement effects from the original sample persist, a wide choice of techniques is available to deal with them (*C17*, *C55*, *G18*). Some of these techniques are applicable only to solutions; others are more convenient with solutions. These techniques are as follows: (a) separation of analyte or interferant; (b) use of sets of calibration curves; (c) dilution with low or high absorber; (d) internal standardization; (e) internal control or intensity-reference standardization; (f) background-ratio method; (g) coherently or incoherently scattered target-line ratio methods; (h) Reynolds–Ryland scatter method; (i) standard addition and dilution methods; (j) special experimental methods, such as those described in Section 18.2.2; and (k) mathematical methods (*D25*, *K3*).

15. The low absorption by the matrix leads to high penetration by the primary beam and low absorption of the emitted secondary radiation, and, consequently, high absolute sensitivity. Unfortunately, this advantage may be canceled by the dilution of the analyte and by the increased background and consequent decreased line-to-background ratio.

16. On dilution of a solution, analyte-line intensity decreases at a much slower rate than analyte concentration.

17. The concentrations of heavy elements in their compounds with low-Z elements can be determined in solution. For example, the Fe $K\alpha$ intensity from FeO, Fe_2O_3, and Fe_3O_4 is about the same for the solids due to low contrast or slope in the calibration function; if equal weights of these substances are put into equal volumes of solutions, the Fe $K\alpha$ intensities have much greater spread. Differences as small as 0.1 in the uranium:oxygen ratio of various samples of uranium(IV) oxide (UO_2) can be determined in nitric acid solution (*H42*).

18. Solution and slurry techniques are particularly applicable to dynamic systems and to on-line process control.

19. Dissolution of solid and powder samples opens up many alternative possibilities for subsequent preparation, presentation, and analysis (Section 17.1.1).

20. Even when the samples are to be analyzed in another form, a solution method may be used to provide standards. Samples of the required form are selected as prospective standards and analyzed by a solution method using synthetic solution standards.

17.1.3. Disadvantages

The principal disadvantages of specimen presentation in liquid form are the following:

1. Except for samples already in the form of liquids, solutions, sols, and slurries, liquid-specimen methods are destructive.

2. For solid and powder samples, a solution technique is usually less rapid and convenient than a technique that uses the sample in its original form.

3. Some elements are very difficult to put into and/or keep in solution. Fusion may be required.

4. The portions of liquids standards actually used for the analysis are usually discarded. When large numbers of samples are to be analyzed and the analysis requires many hours or all day, standards must be measured at intervals throughout the analysis. In such cases, sometimes a cell of standard can be measured only a few times because of deterioration on repeated prolonged exposure to the primary beam. Thus, relatively large volumes of standards must be prepared and stored, or else standards must be prepared frequently. Moreover, liquid standards may not be stable in long storage. Conversely, solid, powder, briquet, and glass-disk standards are substantially permanent and require little storage space. It is sometimes feasible to analyze liquid samples against solid—especially plastic—standards (*C1*).

5. Minor and trace constituents and micro samples may be more difficult to analyze because of the high dilution.

6. The intensities from analytes of low atomic number are seriously reduced by dilution and by absorption in the liquid matrix and cell windows.

7. Selected-area analyses are not feasible.

8. Because of the low effective atomic number of the solvents and cell windows, the scattered primary background is relatively high. This increases

the minimum detectable concentration, and the coherently and incoherently scattered target lines increase the possibility of spectral-line interference.

9. Evaporation of solvent may occur during irradiation, especially with volatile solvents and in vacuum spectrometers, with consequent change in concentration and, possibly, in specimen plane.

10. The specimen compartment, drawer, mask, and, worse, the beryllium window of the x-ray tube may be subjected to leakage and corrosive vapors. With inverted specimen planes, perforation or rupture of the cell window may allow specimen liquid in quantity to flood these components.

11. Liquids may outgas vigorously, spatter, or rupture their cell windows when used in vacuum spectrometers.

12. Liquids may heat under prolonged irradiation, causing expansion and, in closed cells, distension of the cell window.

13. Because of the outgassing, heating, and expansion, the specimen plane may vary during the irradiation.

14. Some liquid cells are difficult to fill properly or are leaky and messy.

15. Bubbles often form in the liquid during irradiation, adhere to the cell window, and reduce the measured analyte-line intensity. The effect is reduced, but by no means eliminated, by use of inverted specimen plane.

16. Some solutions undergo x-ray-induced radiolysis, with consequent precipitation. In inverted geometry, the precipitate settles on the window, which is the specimen surface. In upright geometry, the precipitate settles to the bottom of the cell, away from the measured surface. Thus, assuming that the precipitate is analyte-rich, the measured analyte-line intensity is increased in inverted geometry, decreased in upright geometry.

Many of these unfavorable features may be eliminated or minimized by suitable precautions (Section 17.1.5) or treatment (Section 17.1.1), or by use of alternative specimen-presentation techniques (Section 17.2).

17.1.4. Liquid-Specimen Cells

17.1.4.1. *Forms and Materials*

The early Philips x-ray spectrometers had flat specimen drawers and a horizontal, upright specimen plane. Several liquid-specimen drawers having a liquid chamber as an integral part of the drawer were made for these instruments. Some of these drawers were water-cooled (*D13*). However, present-day instruments all use a liquid-specimen cell having a body consisting of a cup, dish, or open cylinder and a thin window attached semipermanently or with some form of retaining ring or cap.

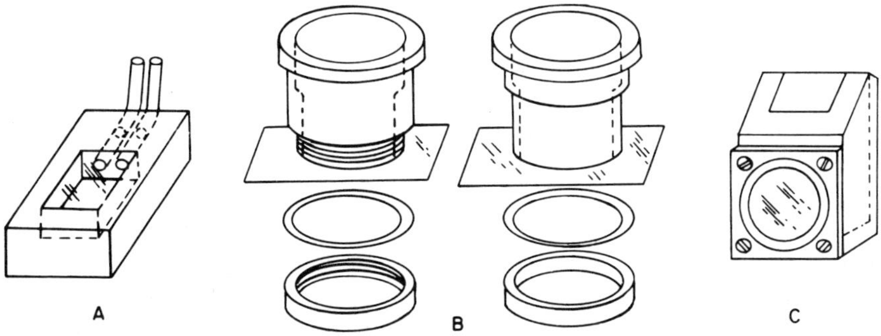

Fig. 17.1. Liquid-specimen cells for some commercial x-ray spectrometers. A. Early Philips spectrometer; the specimen plane is horizontal; the window is cemented on semipermanently. B. Philips Inverted Three-Position and Universal-Vacuum spectrometers; the rings are gaskets; the retaining collars may screw (left) or push (right) on; the drawings are schematic only. C. Early Siemens spectrometer; the window is held in place by the mask, which, in turn, is secured by the four screws.

Each x-ray spectrometer manufacturer provides cells for its own instruments (Figure 17.1), and disposable cells that fit all commercial instruments are available from Spex Industries (Figure 17.2A). However, great ingenuity has always been shown in the design and construction of cells in the laboratory. Cells for upright, horizontal or inclined specimen planes must have the form of cups or dishes. Cells for inverted geometry may have the same form, or they may be simple open cylinders.

Cell bodies have been made of aluminum (*D26*), stainless steel or gold-plated steel (*D20*), glass (*K4*, *P9*), and plastic (Lucite, polyethylene, polypropylene, nylon, and Teflon). Cylindrical cells have been made from tubing of these materials cut to appropriate length (Figure 17.2C). Cup

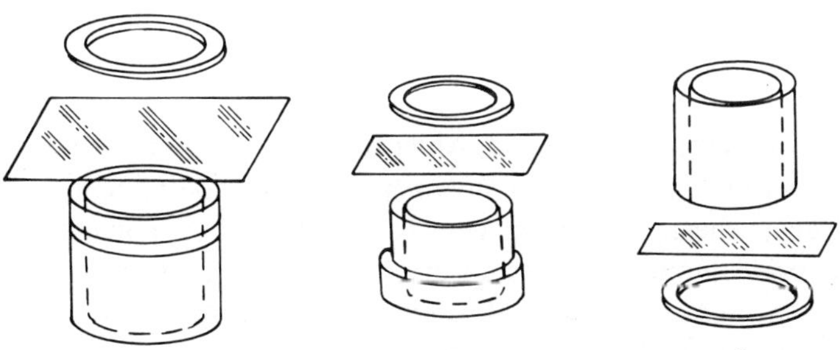

Fig. 17.2. Disposable liquid-specimen cells made from (A) Spex cups, (B) Caplugs, and (C) cylinders cut from plastic tubing. The rings slip over the outside of the cylindrical bodies.

cells have been made of polyethylene vial caps of various sizes and of polyethylene or polystyrene vials of various diameters cut off at appropriate depths. Caplugs (Protective Closures Co., Buffalo, N. Y.) provide excellent disposable cups of various sizes (Figure 17.2B). Only white Caplugs should be used because the pigments in colored plugs contain traces of heavy elements (*B44*).

Rigid cell windows have-been made of thin mica, glass microscope-slide covers, polystyrene, methyl methacrylate, and beryllium sheet (*D26*). Such windows are usually permanently or semipermanently cemented to the cell body (*K4*) or sealed to it with a gasket. However, usually windows are made by stretching thin plastic film over the rim of the cell body. Formerly, cellophane (*B43*) and even Scotch tape were used. However, today, the most commonly used films are 0.00025–0.005-inch Mylar and 0.001–0.005-inch polyethylene or polypropylene. Saran must be avoided because its chlorine content results in relatively high absorption and, of course, emission of chlorine *K* lines. The longer the wavelengths to be measured, the thinner the window must be; 0.00025-inch Mylar has only 40% transmission for Al $K\alpha$, only 20% for Mg $K\alpha$. However, when only short-wavelength lines are to be measured, thicker films may be used. There is no advantage to thicker films for windows replaced after each specimen, but thick films have longer life for semipermanently attached windows.

Mylar is the most widely used film for cell windows, and, in general, it is excellent for this purpose. However, when compared with polypropylene, Mylar has some disadvantages. Mylar having thickness 0.00025 inch (6 μm) is most commonly used. Mylar having thickness 0.00012 inch (3 μm) is available and would be advantageous for the $K\alpha$ lines of elements 9–14 (fluorine to silicon). However, the thickness of these thin Mylar films is not as uniform as it is for thicker films (*A12*), and pinholes are likely to be present in the 0.00012-inch film. Mylar has a higher content of high-*Z* trace impurities than polypropylene; the ash contents of the two films are ~0.2% and <0.005%, respectively. Moreover, the trace impurities in Mylar are not evenly distributed, while those in polypropylene are. Mylar is polyethyleneterephthalate, a polymer consisting of carbon, hydrogen, and oxygen, whereas polypropylene is a hydrocarbon polymer, and has effective atomic number ~1/2 and mass absorption coefficient ~1/5 those of Mylar. A greater thickness of polypropylene is required to prevent sag under the weight of the specimen, but, due to its lower absorption, 0.0005-inch (12-μm) polypropylene has less than half the attenuation of 0.00025-inch (6-μm) Mylar. For maximum strength, only fully oriented polypropylene film should be used. Both these plastics, as well as most others, have the property of adsorbing metal ions and, under intense x-radiation, actually incorporating them into their polymer structures. This applies to film used for windows and to the heavier stock used for cell bodies.

Thin film windows are usually stretched over the rims of cylinder or cup cells in the manner of drum heads and secured with plastic curtain rings, O-rings, or rings of nylon, Teflon, or metal. Windows have also been affixed to cells with grease, and semipermanently cemented on with Pliobond rubber cement (Goodyear Tire and Rubber Co.). Some liquid-specimen cells require gaskets; these are usually made of polyethylene, nylon, or Teflon. In practice, the cell window is usually replaced after use with each specimen, and some cells are so inexpensive (Spex cells, Caplugs) that the entire cell may be discarded after one use.

The following sections describe some liquid-specimen cells worthy of special note.

17.1.4.2. *General-Purpose Cell*

A completely clean, leak-free cell especially useful for instruments having upright, inclined specimen planes is shown in Figure 17.3 (*B21*, *B24*). The cell body *A* is $\frac{3}{8}$ inch deep and may be of Lucite, polyethylene, or Teflon. The depression *B* is $\frac{1}{4}$ inch deep and holds 2–3 ml. The Mylar cover

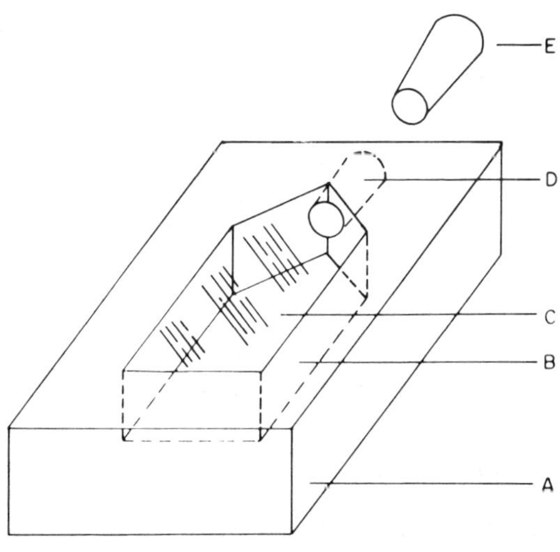

Fig. 17.3. Leak-free liquid-specimen cell for upright or inverted, horizontal or inclined specimen planes (Bertin). *A*. Cell body. *B*. Depression. *C*. Mylar cover. *D*. Tapered filling port. *E*. Tapered plug.

C may be 0.00025–0.005 inch thick, depending on the wavelength to be measured, and is cemented to the cell with Pliobond rubber cement. When the cell is in place in the specimen drawer, the triangular space at the neck of the cell is upward and covered by the specimen mask. This space serves to conceal the air volume at the tapered filling port *D* and provides an expansion chamber for the liquid. The plug *E* is required only for volatile or corrosive liquids. The cells cannot be made of any readily available component and must be made in a machine shop.

Five to ten cells may be covered at a time with Mylar 0.00025–0.005 inch thick as follows. A strip of Mylar, somewhat wider than the cell length and long enough to accommodate five to ten cells side by side, is stretched extremely tightly on a smooth plane surface and secured at both ends. Each cell is coated on its top face with a thin layer of Pliobond and laid face down on the Mylar. A heavy weight—2 kg or so—is placed on each cell and left undisturbed for an hour or so. The Mylar is then cut between the cells, and any wrinkles in the window are stretched out. The cells are then replaced face down on the flat surface, weighted again, and left undisturbed overnight. Thicker windows are applied the same way, but less care is required to avoid wrinkling. Windows 0.010 inch thick never do sag or wrinkle, and windows 0.00025 inch thick have been used up to 100 times at 50 kV, 50 mA before they finally sagged or perforated. When this happens, the window is peeled off, the cell resurfaced on a 400-grit wet paper, and a new window applied.

Filling pipets are made from 5-ml pipets by drawing down the tips and blowing spherical chambers in the upper part of the enlarged chambers. One of these pipets is shown in Figure 20.3. After each use, the cell is emptied, rinsed with one or two portions of the next specimen, then filled with the next specimen.

The principal disadvantage of the cells is that some specimens tend to form adherent deposits on the window. However, often, complexing agents can be added to the specimen to prevent this, or a suitable rinse can be used to remove the deposits between fillings.

17.1.4.3. *Cell for Use in Vacuum*

General Electric supplies a liquid-specimen cell capable of presenting a reproducible surface in the vacuum x-ray spectrometer. The cell, shown in Figure 17.4, consists of a cylindrical aluminum housing *A*, aluminum or copper window grid *B*, specimen chamber *C*, lower chamber *D*, and plug *E*. The assembled cell also has two 0.00025-inch Mylar windows *F*, and two

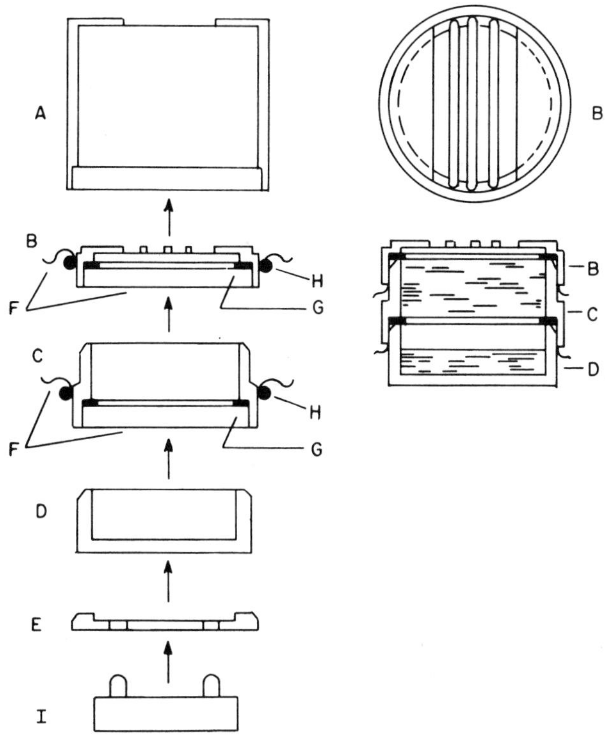

Fig. 17.4. Liquid-specimen cell for use in vacuum spectrometers. [Courtesy of General Electric Co., X-Ray Dept.] *A*. Aluminum housing. *B*. Window-retaining grid (aluminum or copper). *C*. Specimen chamber. *D*. Lower chamber. *E*. Screw plug. *F*. 0.00025-inch Mylar film. *G*. Gasket. *H*. O-ring used temporarily during filling and assembly of the cell. *I*. Key for screw plug *E*.

flat gaskets *G*. During the assembly and filling of the cell, temporary use is made of two retaining rings *H* and a two-pronged key *I* for inserting and tightening the plug *E*, which has a threaded edge, into the bottom of the housing, which has an inside thread. The cells are assembled and filled as follows.

Flat gaskets *G* are placed in the grid *B* and specimen cell *C*, which are then fitted with tightly stretched Mylar covers *F*, temporarily secured with rings *H*, as shown. The lower chamber *D* provides a sort of expansion chamber or cushion for the specimen. It may be empty for specimens having low vapor pressure, or it may be filled ~1/3 full of specimen liquid or ethanol for specimens having higher vapor pressure. The specimen chamber *C* is pressed over the lower chamber *D* until the rim of *D* deforms the Mylar film on the bottom of *C* and seats against its gasket. The specimen chamber is filled with specimen ($\sim$10 ml) until a convex meniscus protrudes just slightly above the rim. The grid is now pressed over the specimen chamber *C* until the rim of *C* deforms the Mylar film on the bottom of the grid and seats against its gasket. Any wrinkles in either Mylar film are smoothed by tugging at the excess protruding around the edge. The retaining rings *H* are now removed, the excess Mylar is trimmed away, and any spilled liquid is blotted. The assembly, appearing as shown at the right in Figure 17.4, is inserted in the housing *A* and secured with the plug *E*, which is tightened with the key *I*.

The filling procedure is obviously complex, but, by application of the gaskets and Mylar covers to a number of specimen chambers and grids at a time, much time is saved. It is claimed that four cells can be filled and assembled in 10 min.

In the vacuum spectrometer, the top Mylar window bulges through the openings in the grid, but no air space is formed, because the air or vapor in the lower chamber expands against the Mylar on the bottom of the specimen cell. Apparently, the bulging is uniform from cell to cell, because excellent precision is realized with the cells.

17.1.4.4. *Uncovered Cell*

Mylar cell covers trap bubbles, scatter the primary beam, and absorb the long-wavelength spectral lines of elements of low atomic number. Elimination of the cover should improve precision, line-to-background ratio, and detection limit, or permit reduced counting times for a given precision, with consequent reduction of thermal or radiolytic effects. Beard and Proctor (*B9*) describe an uncovered cell for use with the Picker spectrometer, in which an end-window x-ray tube and the spectrogoniometer constitute an integral unit, which may be mounted in any position. Ordinarily, the instrument is mounted with the x-ray tube axis vertical; the specimen plane is then inclined to the horizontal at 25°. Beard and Proctor tipped the instrument so that the specimen plane is horizontal.

The cell, shown in Figure 17.5, consists of a cylindrical Plexiglas body *A* having a polyethylene film *B* stretched across its bottom and secured with a retaining ring *C*, and is mounted on a Micarta specimen mask *D*. The cell is filled nearly to the brim, and a frame *E* fitted with two stretched parallel nylon threads *F* is placed over the mask temporarily. The mask has grooves to receive the threads, which then lie in the plane of the rim of

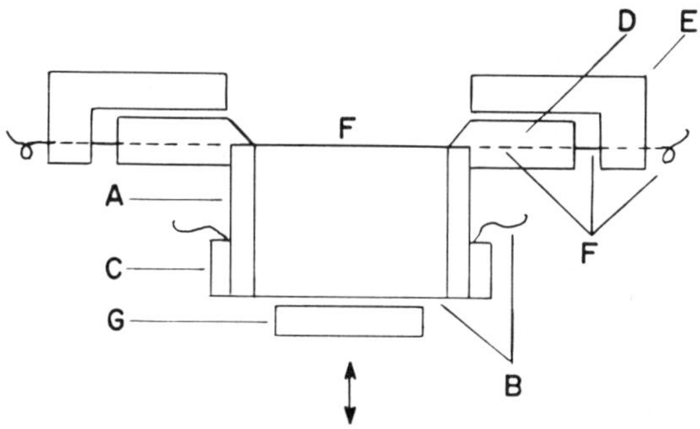

Fig. 17.5. Uncovered liquid-specimen cell (Beard and Proctor). *A*. Cylindrical Plexiglas cell body. *B*. Polyethylene film. *C*. Cylindrical Plexiglas retaining ring. *D*. Micarta specimen mask. *E*. Removable frame. *F*. Nylon threads stretched on *E* and, when *E* is in place on the mask *D*, lying in grooves in *D*. *G*. Movable piston for adjustment of liquid level by deformation of *B*.

the cell. The movable disk *G* presses against the polyethylene cell bottom *B* and is adjusted to bring the liquid surface into contact with the threads. Then *E* is removed, and the cell is ready for x-ray measurements.

Comparison of results for solutions in the uncovered cells and in the same cells with 0.00025-inch Mylar covers showed net intensity gains for Al $K\alpha$, Si $K\alpha$, and S $K\alpha$ by factors of 6, 3.3, and 2, respectively.

17.1.4.5. *Frozen-Specimen Cell*

Freezing eliminates many of the disadvantages of liquid specimens, including bubble formation, expansion, radiolysis, inconvenience with inclined specimen planes and vacuum spectrometers, and scatter from and absorption in Mylar cell covers. Birks [p. 79 in (*6*)] was probably the first to suggest that certain liquids could be analyzed in the x-ray spectrometer in frozen form. He suggested an arrangement in which nitrogen gas is circulated through a coil immersed in liquid nitrogen, then through a channel in a metal block having a shallow depression to hold the specimen. He also suggested that the frozen specimen might be more homogeneous if the specimen liquid were sprayed, rather than pipeted, onto the cold block, and that such frozen specimens would be particularly advantageous for low-*Z* analytes. For example, at room temperature, liquid water has a vapor pressure of ~17.5 mm Hg and absorbs ~58% of Mg $K\alpha$ intensity in a 20-cm path. However, ice at $-20°C$ has a vapor pressure of <1 mm Hg and absorbs only ~4% of the Mg $K\alpha$.

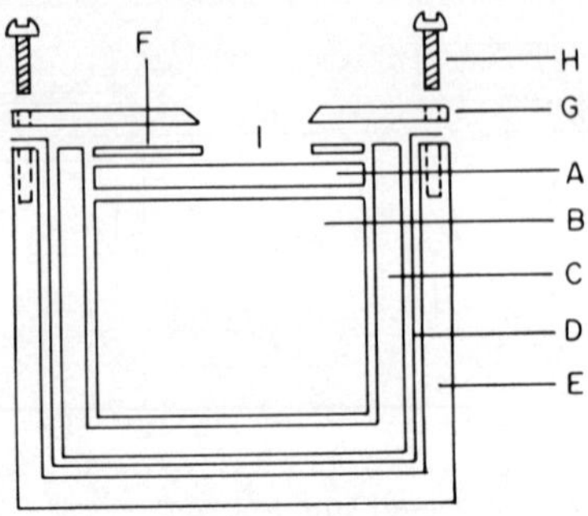

Fig. 17.6. Frozen liquid-specimen cell (Chan). *A*. Frozen specimen (see text). *B*. Precooled aluminum cylindrical block. *C*. Plexiglas inner thermal-insulation cup. *D*. Aluminum thermal-reflection cup. *E*. Plexiglas outer thermal-insulation cup. *F*. Cork disk. *G*. Metal specimen-mask disk. *H*. Retaining screws. *I*. Rectangular specimen window in *F* and *G*.

Figure 17.6 shows a frozen-specimen cell built by Chan (*C30*). An aluminum cylinder *B*, previously cooled in liquid nitrogen or a freezing mixture, is placed in a thermal insulating cup consisting of a Plexiglas inner cup *C*, aluminum reflecting cup *D*, and Plexiglas outer cup *E*. The frozen specimen *A* (see below) is placed on *B*, followed by an insulating cork disk *F* and metal mask *G*. The assembly is secured with the screws *H*. *F* and *G* have rectangular windows *I*.

Three types of specimen are described: (1) A metal disk 33 mm in diameter and 3 mm thick has a rectangular depression $20 \times 12 \times 2$ mm. Liquids requiring no Mylar cover (low vapor pressure, noncorrosive) are placed in the depression. The disk is then placed between cold metal blocks to freeze the specimen. (2) A shallow metal planchet 33 mm in diameter and 0.5 mm deep holds a thick filter-paper disk impregnated with ~0.5 ml of solution. A 0.00025-inch Mylar film may be placed on the planchet and secured with a retaining ring. The specimen is frozen as before. (3) A shallow metal planchet 33 mm in diameter and 3 mm deep is filled with aqueous solution thickened with gelatin or with organic solution thickened with alkali metal stearate, covered with Mylar, and frozen.

All specimen preparation is done in a dry-box fitted with rubber gloves and provided with dewar vessels containing the supply of aluminum cylinders and freezing blocks. A continuous stream of thoroughly dried nitrogen is passed through the box. The box communicates directly with the spectrometer specimen chamber through a flexible plastic tunnel. The helium and air inlets to the spectrometer are provided with drying trains.

17.1.4.6. *Cell for Slurries*

X-ray spectrometric analyses of powders and briquets are limited to thin surface layers, and may be influenced by inhomogeneity and particle size. In such cases, especially if the powder is difficult to dissolve or fuse, it may be advantageous to analyze it as a slurry. Hudgens and Pish (*H45*) describe a slurry cell consisting of a sealed body having 5–6-ml capacity, 0.003-inch Mylar cover, externally powered stirrer mounted asymmetrically for maximum turbulence, and a bleeder valve to allow escape of gases produced by radiolysis and heating. The sample powder and internal standard are placed in the disassembled cell, covered with liquid vehicle, and stirred with a toothpick to remove occluded air. The cell is then assembled and filled with vehicle through the bleeder hole by use of a hypodermic syringe. The cell was evaluated for determination of the thorium and uranium in mixtures of their oxides using strontium sulfate ($SrSO_4$) as internal standard.

17.1.4.7. *High-Temperature Liquid-Specimen Cell*

Figure 17.7 shows a high-temperature liquid-specimen cell built by Campbell and Leon (*C13*). A ceramic or metal crucible *A* rests in a depres-

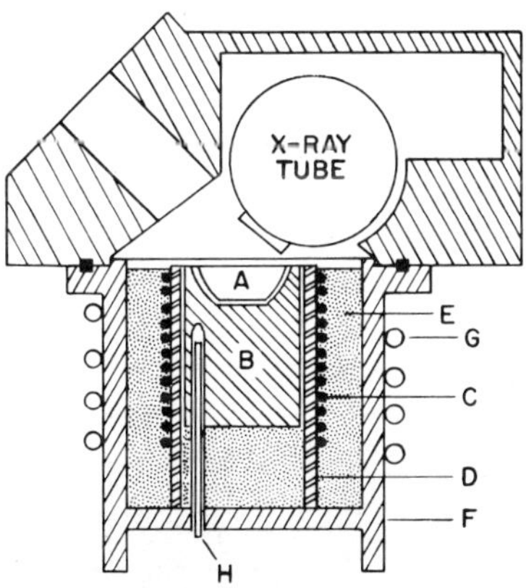

Fig. 17.7. High-temperature liquid specimen cell (Campbell and Leon). *A*. Ceramic or metal crucible. *B*. Inconel cylinder with depression for *A*. *C*. Nichrome heater coil. *D*. Refractory cylinder, support for *C*. *E*. Thermal insulation. *F*. Brass shell. *G*. Water cooling coils. *H*. Thermocouple tube.

sion in an Inconel block *B* surrounded by a Nichrome heating element *C* wound on a refractory ceramic cylinder *D* in a brass housing *F* provided with water-cooling coils *G* and packed with high-temperature insulation *E*. A thermocouple well *H* and an inert gas inlet (not shown) are provided. With this arrangement, specimen temperatures up to 900°C are attainable, but Campbell and Leon suggest that with a platinum or alumina crucible, platinum block, platinum–rhodium alloy heater coil, and alumina support, temperatures up to 1400°C should be attainable. The x-ray tube window was $\frac{1}{2}$ inch away from the specimen surface and was protected by a thin sheet of mica. No adverse effect on the x-ray tube was observed after 50 runs of up to 5 h at up to 800°C. The cell was used to make dynamic studies of the selective oxidation rates of traces of arsenic and antimony in molten lead.

17.1.4.8. *Other Liquid-Specimen Cells*

For determination of traces of manganese in gasoline, Jones (*J14*) used the cell shown in Figure 17.8. The screw in the cell cover has an adjustable iron cap or sleeve, which serves as an internal intensity-reference standard. The analysis consists of measurement of Mn $K\alpha$ intensity from the analyte and Fe $K\alpha$ intensity from the cap. The Mn $K\alpha$/Fe $K\alpha$ intensity ratio is applied to a calibration curve prepared from standards measured in the same cell. Later, Gunn (*G17*) described a similar cell having an internal intensity-reference standard in the form of a platinum disk. The cell was used for determination of lead in hydrocarbon oils from the Pb $L\alpha_1$/Pt $L\alpha_1$ intensity ratio. The position of the intensity-reference

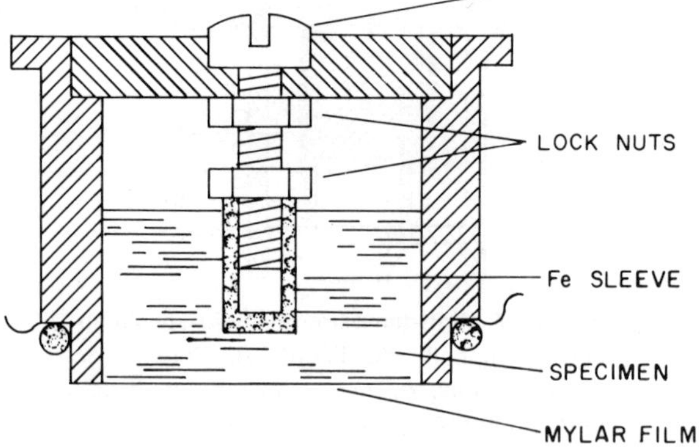

Fig. 17.8. Liquid-specimen cell with built-in iron intensity-reference standard (Jones).

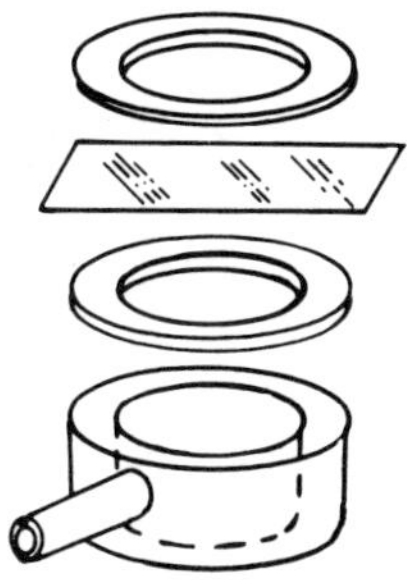

Fig. 17.9. All-glass liquid-specimen cell for upright, inclined specimen plane. The two rings are gaskets. The window may be thin borosilicate glass or Mylar film. The assembled cell is tightly clamped to the under side of the specimen mask and is filled through the side tube with a hypodermic syringe. Alternatively, a thin glass window may be cemented to the cell rim.

standard is adjusted to give reference intensity about the same as the analyte-line intensity from the samples.

An all-glass cell for upright, inclined specimen planes is shown in Figure 17.9 (*K4*, *P9*). An all-polyethylene cell for use in the same geometry with hydrofluoric acid solutions is shown in Figure 17.10 (*W1*). A cell through which liquid can be circulated continuously is shown in Figure 17.11 (*C6*).

Cells for liquid micro specimens (*T7*) are made by pressing concave depressions into polyethylene disks warmed on a hot plate. The drop of specimen is placed in the depression, and a 0.001-inch polyethylene film is hotwelded to the rim to enclose the drop. The welding may be done with a small pencil soldering iron or with a small heated cylinder.

Liquid specimens may be cooled by provision of a water-cooled jacket for the cell (*L3*), by water-cooling the specimen drawer or chamber (*D27*), or by directing on the cell a stream of gas cooled by passage through a coil immersed in liquid nitrogen (*M27*).

Hydrocarbon liquids may be thickened to self-supporting semisolids by addition of heavy grease or silicone, wax, or lithium or sodium stearate. Inorganic solutions may be thickened by addition of gelatin. The thickened liquids may be placed in shallow cylindrical cups and covered with Mylar. The gelatin preparations can be dried to form small disks (*R14*). Waxes can be melted and poured into plastic cups (Spex cups or Caplugs), which are then covered with Mylar and inverted to permit the wax to set with a smooth, flat surface.

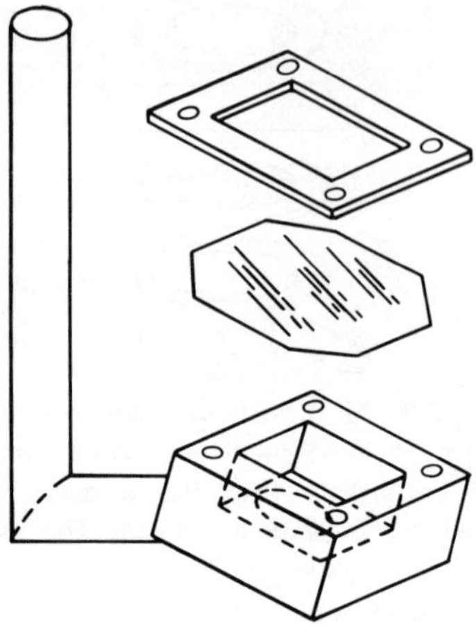

Fig. 17.10. All-polyethylene liquid-specimen cell for use in specimen drawers having upright, inclined specimen planes and for use with hydrofluoric acid solutions (Waehner). The assembled cell is affixed to the under side of the specimen mask by screws in the four holes shown in the gasket and cell body. The horizontal portion of the reservoir tube passes through a hole in the specimen drawer so that the upright tube is outside the drawer.

17.1.5. Precautions and Considerations

17.1.5.1. *Interaction of the Primary Beam with Liquid Specimens*

In general, only liquid specimens interact with the primary beam in such a way as to significantly affect the measured analyte-line intensity. The primary beam causes heating and radiolysis with consequent expansion, bubble formation, precipitation, and displacement of the specimen plane. These interactions may be reduced by freezing the specimen and by reduction of excitation power if the intensity is not reduced below values that give acceptable statistical precision. In absorption-edge spectrometry, another means of reducing incident flux is to derive the monochromatic bracketing wavelengths from secondary targets. Other ways of dealing with interaction effects are given by Cullen (*C55*), and the following discussion is based largely on his paper.

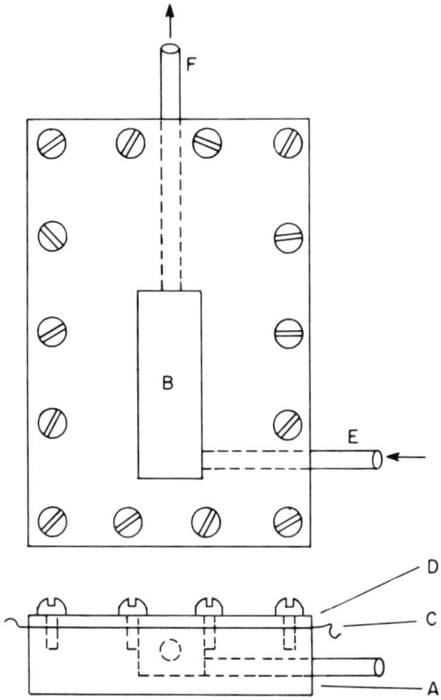

Fig. 17.11. Continuous-flow liquid-specimen cell (Campbell). *A*. Plexiglas body. *B*. Depression. *C*. Mylar window. *D*. Aluminum cover plate. *E*. Specimen inlet. *F*. Outlet.

a. Specimen Heating. Cullen reports that water in a liquid cell was warmed only 4°C in 10 min by a Machlett OEG-50 tungsten-target tube operating at 50 kV, 45 mA in air path. The corresponding expansion is only ~0.02%, and the effect on analytical results would be less than the statistical counting error. However, specimen heating increases with concentration and absorption coefficient of solute, but may be reduced by use of: (1) newer, cooler-operating x-ray tubes; (2) water-cooled jacket on the x-ray tube head; (3) water-cooling of the specimen drawer and/or compartment; (4) water-cooled jacket for the specimen cell; (5) thermal insulation of the specimen by use of plastic cells and specimen masks to reduce heat conduction from the drawer or compartment; and (6) continuous helium flow.

b. Bubble Formation. Bubbles usually form in liquid specimens during irradiation. The bubbles originate from three causes—expulsion of dissolved gases, volatilization of the solvent, and/or x-ray-induced radiolysis of solvent or solute. Measures to reduce the invariably adverse effects of bubbles regardless of origin are of two basic types—preventive and remedial.

The preventive measures include: (1) reduction of specimen heating by the means listed above; (2) use of previously boiled or otherwise outgassed water and acids for specimen dilution and preparation; (3) avoidance of gas-forming reagents insofar as is feasible, such as peroxides, ammonium hydroxide (NH_4OH), and ammonium persulfate [$(NH_4)_2S_2O_8$]; (4) use of gas inhibitors—for example, perchlorates reduce gas formation in iron(III) chloride ($FeCl_3$) solutions; and (5) outgassing of specimen solutions by heating or exposure to vacuum in a vacuum desiccator prior to loading them in cells; the filled cells may also be exposed to vacuum, but the bubbles must then be dislodged by sharply jarring the cells.

The remedial measures include: (1) use of inverted optics; (2) rotation of cells in inverted geometry to possibly dislodge some of the bubbles from the cell window; (3) in upright geometry, use of cells having a space covered by the specimen mask and into which bubbles may rise (Figure 17.3); (4) use of antiwetting agent on cell windows to reduce bubble formation and adherence to the window; and (5) preirradiation for a period of ~60 s after the cell is placed in the primary beam, to permit the bubbles to reach equilibrium before the measurements are taken; also, (6) when two or more analytes and, perhaps, their backgrounds are to be measured from each specimen, they should always be measured in the same sequence from each specimen; and (7) the measuring time should be as short as possible consistent with the accumulation of enough counts for the required statistical precision.

It is often assumed that the adverse affects of bubble formation disappear with inverted geometry. This is not necessarily true. Very small bubbles formed on the cell window may be very adherent and may not dislodge and rise to the liquid surface until they have grown relatively large. However, bubbles are always less troublesome with inverted geometry.

c. Precipitation. Avoidance of high concentrations of radiolytically labile constituents and use of complexing agents reduce precipitation. For example, cobalt, nickel, indium, and gold plating solutions are more stable in the presence of potassium citrate than with potassium cyanide. When it does occur, precipitation may equilibrate after ~60 s, and a preirradiation period like that recommended above may be beneficial. When there is any possibility of even minimal adherence of precipitate on the cell window, cells must be used in which the window is replaced for each specimen, or the cell must be rinsed with a reagent that removes any precipitate from the cell window.

d. Specimen Plane. Variation in specimen plane occurs with inverted geometry in open or closed cells from sagging of the specimen window under the weight of the specimen, and in closed cells from thermal expansion or outgassing of the specimen. Clearly, specimen-plane variation is minimized by use of open cells and by masking of the specimen window to support part of the specimen.

Variation in specimen plane occurs with upright geometry in closed cells from expansion and outgassing. Use of cells having open side-arms, filling ports, or breather holes is indicated. There is no problem of sagging under specimen weight with upright geometry. However, if the volume of liquid in a closed cell is slightly smaller than the volume of the cell, the window may be drawn inward slightly.

17.1.5.2. *Composition*

Concentrations of all analytes should be as high as possible to give high intensities. Solutions must be stable and not prone to decomposition during irradiation or, for standards, in storage. Complexing agents should be used to improve the stability of solutions. Volatile and corrosive solvents are avoided if possible.

The nature and concentration of the acid used in preparation of the

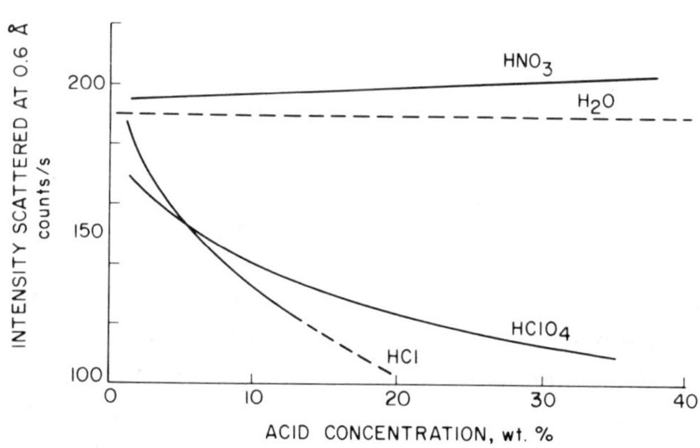

Fig. 17.12. Effect of acid concentration on intensity of scattered 0.6-Å primary x-rays for various acids and concentrations. The intensity scattered from water is shown for comparison. [K. Togel, in "Zerstörungsfreie Materialprüfung," R. Oldenbourg, Munich, Germany (1961); courtesy of the author and publisher.]

solutions must be considered from two aspects. The higher the mass-absorption coefficient and concentration, the lower is the measured analyte-line intensity, especially for long-wavelength lines; in general, the decreasing order of preference in this respect is nitric, phosphoric, sulfuric, perchloric, and hydrochloric. The higher the effective atomic number and concentration of the acid, the lower is the scattered primary x-ray intensity, as shown in Figure 17.12.

In preparation of standards for analysis of hydrocarbon liquids, the carbon:hydrogen ratio of the solvent must be substantially the same as that in the samples. For any kind of solution, the matrix must be substantially the same in all samples and standards. However, substantially greater latitude is permissible if an internal standard is added.

Emulsified liquid and suspended solid must be removed by centrifugation, coagulation, or other means. If the emulsified liquid is volatile, such as residual organic liquid in an aqueous solution, it may be distilled off.

17.1.5.3. *Miscellany*

With inverted specimen planes, there is no problem with air bubbles entrapped during filling and covering of the cells, because the bubbles rise to the top surface of the liquid, where they do no harm. With upright, horizontal specimen planes, care must be taken to avoid entrapping any air. This is done by filling the cell until the meniscus rises just above the rim, then carefully applying the Mylar window. With upright, inclined specimen planes, an air bubble is permissible if the specimen surface is masked in such a way that the bubble is hidden.

Care is taken when liquid specimens are used in vacuum spectrometers to avoid rupture of the thin cell windows. Special cells may be used (Section (17.1.4.3) or nylon grids may be used on the specimen mask.

Infinite thickness is much greater for liquids than for solids and varies greatly with the wavelength of the measured line, and with the nature and concentration of the analyte and matrix. Consequently, it is desirable that the depth of the liquid cells be infinite for the shortest wavelength to be measured. Otherwise, the specimen depth must be constant for all specimens. Lambert used an immiscible heavy organic liquid, such as carbon tetrachloride, to raise the specimen level in cells for upright horizontal geometry.

For inverted geometry, it may be prudent to provide a protective cover for the x-ray tube window in case of rupture of a cell window. Campbell states that Mylar covers are too short-lived because of embrittlement by the extremely high intensities at the x-ray tube window; he recommends a thin, removable beryllium cover.

17.2. SUPPORTED SPECIMENS

17.2.1. General

The term *supported specimen* as used here indicates a specimen in which a relatively small amount of analytical material—less than enough to provide an infinitely thick layer—is supported on some type of substrate. Usually, the analytical material is derived from a solution or some other material by a preconcentration technique, and supported specimens are the most commonly used forms in trace and micro analysis. Davis *et al.* have reviewed these preconcentration techniques (*D5*).

Commonly used support materials include the following: (1) filter paper, (2) black filter paper (for better visibility of small white precipitates), (3) chromatographic paper, (4) Fiberglas or other glass-fiber filters, (5) Millipore filters, (6) thin Mylar and polyethylene film, (7) ultrathin collodion or Formvar film, (8) microscope-slide cover glasses, (9) ion-exchange resin granules, (10) ion-exchange resin membrane, (11) ion-exchange resin-impregnated filter paper, (12) abrasive paper and cloth, and (13) glass, quartz, and plastic capillaries and fibers. Filter-paper, glass-fiber, and Millipore filters are available in disks 2.1 cm in diameter, and disks of any diameter can be cut from the stock. Confined spot-test papers consisting of small filter-paper squares having impregnated wax rings ~1 cm in inside diameter are available in several thicknesses. Bioassay disks of filter paper ~0.020 inch thick and ½ or ¼ inch in diameter are available. Certain chromatographic papers have the advantage that they absorb a greater volume of liquid per unit weight and do not warp when dry. Micro Buchner funnels and suction flasks are useful in collecting specimens on small filter disks.

The nature of supported specimens leads to several problems peculiar to this specimen form. Such specimens are not as uniform or reproducible as other specimen forms. Consequently, it is usually necessary to provide internal standardization and/or, when feasible, to rotate the specimen in its own plane during measurement. When the specimen is supported on an x-ray-transparent support, such as filter paper or Mylar film, the side containing the material, or the side to which it was pipeted or on which it was collected, should face the primary x-ray beam. Supported specimens tend to be loose and fluffy, and loss of material can occur. Specimens supported on thin Mylar or filter disks may be sprayed lightly with clear Krylon or covered with or sandwiched between 0.00025- or 0.00012-inch Mylar films. Such sandwiched specimens may be stretched across the end of a plastic cylinder or the rim of a plastic cup and secured with an O-ring.

Because most supported specimens are highly transparent to the primary x-ray beam, enhancement techniques (Section 4.4.3) are applicable.

17.2.2. Specimens Derived from Solutions

Small volumes of solutions may be pipeted on filter-paper disks, glass-fiber filters, or Millipore filters, and evaporated to dryness (*L34*, *M3*, *M6*). The dried disks constitute the specimens. Natelson and his co-workers (*N2*, *N3*) have developed highly refined filter-paper techniques for analysis of biological fluids, and the techniques are readily applicable to other solution samples: (1) Filter paper is treated with hydrochloric acid to remove trace impurities, and is imprinted with wax rings by use of cork borers dipped in melted wax or ballpoint pens filled with a hexanone solution of wax (*N6*, *N9*). Samples are micropipeted inside the rings. Similar confined-spot papers are available commercially (*J11*). (2) It is well known that when drops of solution are placed on filter paper, each drop tends to dissolve the residue from the preceding drops and diffuse outward, forming concentric rings of residues of various solubilities around a relatively residue-free center. More-uniform deposits are obtained by pipeting the sample solutions onto the filter paper in an apparatus combining a ring oven and aspirator (*N4*, *N8*). (3) The analyses are automated by use of rolls of filter-paper ribbon imprinted with wax rings and driven mechanically through the spectrometer sample compartment by a mechanical accessory (*N5*). (4) Even greater speed is realized by incorporation of a modified 35-mm photographic slide changer into the sample compartment (*N7*). Individual filter-paper squares are placed in the slide changer and indexed automatically or manually through the sample compartment. In both automated techniques, sample concentration is derived from peak heights on ratemeter recorder charts.

Possibly, very small volumes of solutions could be absorbed in a textile fiber, or evaporated on a quartz fiber, or taken up in a plastic, quartz, or glass capillary and used as a line source in a curved-crystal spectrometer. In a similar technique, metallic analytes are electrodeposited on the tip or along the length of a platinum or other metal wire or carbonized quartz fiber and used in a curved-crystal spectrometer. The solution may be subjected to paper-chromatographic techniques to separate the analytes (Section 14.8.3). The paper chromatogram constitutes the specimen.

Small volumes of solutions may be pipeted or sprayed into small dishes or planchets and evaporated to dryness (*M5*). The dishes or planchets constitute the specimens. Birks (*B34*) evaporated 1-ml portions of solutions drop by drop into small depressions in electrically heated aluminum-foil strips. Small volumes of solutions may be sprayed or distributed on microscope-slide cover glasses or thin Mylar (*G15*, *M2*) or even ultrathin collodion or Formvar films (*M13*). The plastic films have very low background compared with filter paper. Desicote or other wetting agents effect more uniform distribution on metal, glass, and plastic. Oils may be spread in thin films on metal substrates (*S10*).

The analytes may be precipitated, with or without gathering agent, and collected on a small filter disk in a micro Buchner funnel. The dried disks constitute the specimens. Alternatively, the powders may be transferred to small plastic envelopes or distributed on Scotch tape, or used in the micro borax-glass disk fusion technique (Section 16.4.3). The technique of homogeneous precipitation may be used: Chan (*C28*) allowed homogeneously precipitated selenium to settle for 1–2 days onto filter-paper disks secured to the bottoms of Pyrex reaction tubes 7 inches high. The disks were dried and mounted between protective 0.00025-inch Mylar films.

Organic reagents such as 8-quinolinol, cupferron (ammonium nitrosophenylhydroxylamine), dithizone (diphenylthiocarbazone), and tannin may be used as general, group, or specific precipitants (*D2*, *H31*, *H32*, *R18*, *T9*). Individual analytes may be precipitated with specific reagents. Several organic or inorganic reagents may be added to the same solution, each under appropriate conditions, to precipitate various analytes. The precipitates may be collected separately or all together. Complexing agents, such as EDTA (ethylenediaminetetraacetic acid) may be used to prevent precipitation of selected analyte or matrix constituents. Indirect or association analysis may be effected by stoichiometric precipitation of a low-Z element or organic compound with an easily determined heavier element.

Extremely dilute solutions may be reduced in volume and used as a liquid, or the analyte(s) may be recovered on a support (*P7*). Probably the best technique is to take up the analyte(s) by prolonged agitation with a piece of ion-exchange membrane (*G11*). Suspended solids in liquids and gases are collected on filter disks (*H36*).

17.2.3. Specimens Derived from Solids

Surface deposits can be analyzed without removal from the substrate if the substrate does not contain elements present in the deposit. Otherwise, the deposit can be removed in any of several ways. It may be gently abraded off with a disk of fine abrasive paper, wiped off with a filter-paper disk moistened with a suitable reagent, or stripped off with Scotch tape. Another stripping technique is to coat the surface with a solution of collodion in amyl acetate, allow the plastic to dry, then strip it off with the adherent deposit, perhaps by use of Scotch tape. Some coatings may be removed by scoring them with a pattern of crossed lines ~5 mm apart, then immersing the specimen in a reagent that dissolves the substrate; squares of the coating float off. Finally, a drop of reagent can be spread on the surface to dissolve the deposit, then absorbed in a small filter-paper disk or even taken up in a micropipet and delivered to some other substrate.

Segregations can be analyzed without removal from the specimen surface by selected-area techniques. Segregations may be dislodged under a microscope by prying with a stainless steel or platinum needle, or by drilling with a small standard drill or with a microdrill. A standard number 80 drill is 0.0134 inch (0.34 mm) in diameter. Microdrills 0.005–0.001 inch in diameter are available. Alternatively, selective etching techniques may be used to etch the matrix away and leave the segregations either loose or in relief. The protruding segregations may be dislodged with an ultrasonic vibrator. The loose segregations may be picked up on a glass, quartz, or plastic fiber coated with vaseline and used in a curved-crystal spectrometer. Or a "micro vacuum cleaner" may be made by placing a hypodermic needle in the end of a rubber tube and applying gentle suction. A small piece of filter paper or glass-fiber paper is held at the end of the needle, and the particles are drawn into the paper. Finally, the extraction-replica technique may be used (*B48*). The segregations are etched into relief as before, but the surface is then coated with collodion or Formvar to imbed the protruding particles. The plastic is then backed with Scotch tape and stripped off with the particles.

17.2.4. Ion-Exchange Techniques

The analytes may be transferred from a solution to an ion-exchange resin. The resin may be of the cationic- or anionic-exchange form, and the analytes to be absorbed must be in the same form. The resin may have the physical form of granules, liquid, membrane, or impregnated paper. The analytes may be taken up by one of three processes (*V1*, *V2*):

1. In the *batch technique*, the solution is shaken with a portion of ion-exchange granules or liquid, or with a square or circle of membrane. The granules are separated and washed by filtration, then placed in a powder cell wet or dry, or briqueted (*C38*). The liquid resin is separated and washed by decantation and used in a liquid cell (*V1*). The membrane is scooped out, rinsed, dried, and mounted as is (*C5*, *G11*, *Z8*).

2. In the *filtration technique*, the solution is filtered several times through a resin-impregnated filter paper on a Buchner funnel in a suction flask until quantitative recovery of the analytes is realized (*C18*).

3. In the *column technique* (*V1, V2*), the solution is passed through a column of ion-exchange granules packed in a glass apparatus resembling a short buret. The analytes are retained on the column and removed collectively, or individually in order of increasing retentivity by elution with a solution of complexing agent. The eluate constitutes a liquid specimen.

In the batch and filter techniques, it is wise to establish the agitation time or number of filtrations required to effect complete recovery prior to undertaking the actual analysis. This is done by plotting analyte-line intensity *vs.* time or number of filtrations. The point is noted beyond which no further increase in intensity occurs. Equilibrium is established more rapidly with liquid resins than with granules, but sensitivity is lower with liquids. Equilibrium with granules is expedited by presoaking. Care must be taken not to exceed the capacity of the resin, which is usually stated in milliequivalents per gram. For use with very dilute analyte solutions, it may be necessary to purify the ion-exchange resin by repeated washing or elution with pure eluant. Campbell and Thatcher (*C5, C15*) compare mechanical and ultrasonic agitation of ion-exchange membranes, describe a specimen holder for the membranes, and give mass-absorption coefficients of ion-exchange membranes and papers.

Internal standards may be added to ion-exchange resins in three ways: An internal standard may be added to the sample solution prior to treatment with the resin (*C38*); this technique is not recommended, because the internal standard may not coabsorb reproducibly under conditions optimal for the analyte. A solution of internal standard in exchangeable form may be added volumetrically to the dry or moist analyte-loaded resin in the specimen cell. Probably the best way is to quantitatively add to the dry analyte-loaded resin in the cell a solution of internal standard in *non*exchangeable form. The solution fills the voids between the granules. The specimen cell is then covered with Mylar and measured in the wet form.

Ion-exchange techniques have been applied to trace and micro analysis. Grubb and Zemany (*G11*) determined 1 μg of cobalt in 1 liter of water by collecting it on 1 cm^2 of ion-exchange membrane stirred in the water for $\sim$48 h. Zemany *et al.* (*Z8*) determined microgram quantities of potassium by a similar technique. Campbell *et al.* (*C18*) determined 11 trace elements in tungsten by dissolution of the samples in tartrate medium, passage of the solutions through an ion-exchange column to retain the trace analytes, elution of the analytes, and "filtration" of the eluate through disks of ion-exchange paper. Seven passes through the disks gave >99% retention of all trace analytes. Luke (*L33*) determined analytes in quantities as small as 0.01 μg by isolating them chemically, taking them up in 1 ml of acid or water, then collecting them in individual disks of ion-exchange membrane $\frac{1}{8}$ inch in diameter. The disks were measured in a curved-crystal spectrometer.

17.2.5. Ashing Techniques

Metallic analytes in organic and biological materials in concentration range 0.01–10 ppm may be preconcentrated by ashing. One of four essentially distinct processes may be used.

a. Dry Ashing or Combustion. This process is applicable only to combustible samples, especially hydrocarbons. The hydrocarbon is actually ignited and burns to a carbonaceous mass, which may be collected for analysis or ignited further on a burner or in a muffle furnace to a clean ash. This method is the most prone of the four to loss of volatile analytes. However, the soot may be collected on a cold surface and analyzed by itself or combined with the carbonaceous residue.

b. Wet Ashing (*D6*, *L38*). The sample is heated with a mixture of a strong dehydrating acid and a strong oxidizing acid, usually concentrated sulfuric and nitric acids, respectively. In some applications, concentrated perchloric acid or fuming nitric acid may perform both functions. It may be necessary to heat the sample repeatedly with successive portions of acid until, finally, a clean ash remains after all the acid has been fumed off. This method results in no loss of analytes if carefully conducted, but is very time-consuming.

c. Dehydration–Ignition. The sample is reduced to a carbonaceous residue by heating with a dehydrating acid alone—usually sulfuric. The carbon residue is then ignited on a burner or in a muffle furnace to burn off the carbon and form a clean ash. This method also results in no loss of analytes and is much more rapid than wet ashing.

d. Wet Ashing and Combustion (*R16*, *S20*). This process, like dry ashing, is also applicable only to combustible samples. It has been shown that when hydrocarbons containing trace metallic impurities are heated with benzenesulfonic acid ($C_6H_5SO_3H$) or a homologous compound —such as xylenesulfonic acid [$(CH_3)_2C_6H_3SO_3H$]—the metal–organic complexes are destroyed, leaving the metallic elements in a nonvolatile condition. It is then feasible to burn the sample without loss of analytes. The procedure consists of heating the sample with benzenesulfonic acid, then burning it to a carbonaceous residue, as in dry ashing. The residue may be collected for analysis or ignited on a burner or in a muffle furnace to a clean ash.

Prior to ashing, the sample may be treated with additives. If the ash is to be collected and analyzed as is—rather than after dissolution— a gathering agent may be added to impart some bulk and facilitate the collection. Magnesium or aluminum, added as naphthenate, salt of an organic acid, or other organic compound, are usually used because they provide a light matrix. Because of the generally nonreproducible nature of the applicable specimen-presentation forms (see below), an internal standard or internal intensity-reference standard is usually added, again a suitable metal as an organic compound.

The ash may be presented in several ways. The dry ash may be transferred to a Mylar film (*R16*) or to a paper, glass-fiber, or Millipore filter disk, or it may be washed with water or a hydrocarbon liquid onto a filter disk in a Buchner funnel in a suction flask (*S20*). The Mylar or filter disk may be used as is, or sprayed with Krylon, or sandwiched between Mylar films. Alternatively, the ash may be dissolved. The solution may be absorbed in a filter disk and dried, or evaporated on an aluminum plate or microscope-slide cover glass (*D6*), or evaporated to dryness and used as a powder. The analytes may also be taken up in ion-exchange resin or membrane, or precipitated individually or collectively and filtered (*L38*).

Chapter 18

Measurement of Thickness of Films and Platings

18.1. PRINCIPLES AND BASIC METHODS

Actually, the x-ray spectrometer directly provides only two types of information about films and platings—composition (Section 14.7) and area density. Thickness is derived from these quantities indirectly and only after assumptions are made about the uniformity of the film thickness and about the actual densities of the film and the constituent elements in the film, as distinguished from bulk densities. Section 14.7 discusses x-ray spectrometric determination of film composition, particularly as a means of dealing with absorption-enhancement effects. This chapter discusses determination of film thickness (*15J*, *B25*, *L15*).

The following notation is used throughout this chapter:

A is the measured analytical secondary x-ray beam.

O is the incident primary x-ray beam.

P refers to the plating or film.

S refers to the substrate or base metal.

I_0 is the intensity of the incident primary beam.

$I_{P,t}$ is the intensity of the plate-element line from a plating of thickness t.

$I_{P,\infty}$ is the intensity of the plate-element line from infinitely thick plate metal.

$I_{P,S}$ is the intensity of the plate-element line from plate element in the substrate after emerging through the overlying plating.

$I_{P,S,0}$ is the intensity of the plate-element line from plate element in the substrate before emerging through the overlying plating.

I_S is the intensity of the substrate-element line from unplated substrate.

$I_{S,P}$ is the intensity of the substrate-element line after emerging through the overlying plate.

$I_{K\alpha}$, $I_{K\beta}$, $I_{L\alpha}$, $I_{L\beta}$ are the intensities of the indicated specific lines.

t is thickness of the film or plate.

$\bar{\mu}$ is defined in Equation (14.37).

The film may be presented to the instrument in any of three ways: (1) after removal from the substrate without regard to preservation of film integrity; (2) after removal from the substrate intact; and (3) nondestructively, that is, in place on the substrate. Unsupported films and films on thin x-ray-transparent substrates belong by their nature in the second group. The first technique is not considered here. The work of Rhodin (*R5*) is an outstanding example of the second technique. He studied films prepared in three ways: (1) Chromium, iron, nickel, and stainless steel films up to 300 Å thick were vacuum-evaporated on 0.00025-inch Mylar. (2) Oxide films were prepared by evaporation of chromium, iron, nickel, and stainless steel on Formvar and floating the Formvar-backed films on a solution of potassium dichromate in nitric acid for various times to form oxide films of various thicknesses. Unoxidized metal was removed in a bromine–methanol solution in an inert atmosphere, and the Formvar was removed in ethylene chloride. The oxide films were scooped up on Mylar film. (3) Oxide films were removed from bulk metals by chemical stripping in bromine–methanol solution, then scooped up on Mylar. Rhodin obtained sensitivities for chromium, iron, and nickel of 8.5, 16.5, and 27 counts/s per $\mu g/cm^2$, respectively.

The remainder of this section is devoted to consideration of x-ray spectrometric determination of thickness of films and platings in place on the substrate. Such methods may be classified in five categories as follows:

1. The *substrate-line attenuation method* (*B12*, *L8*, *L9*, *Z6*), shown in Figure 18.1A, consists of measurement of the attenuation of a substrate

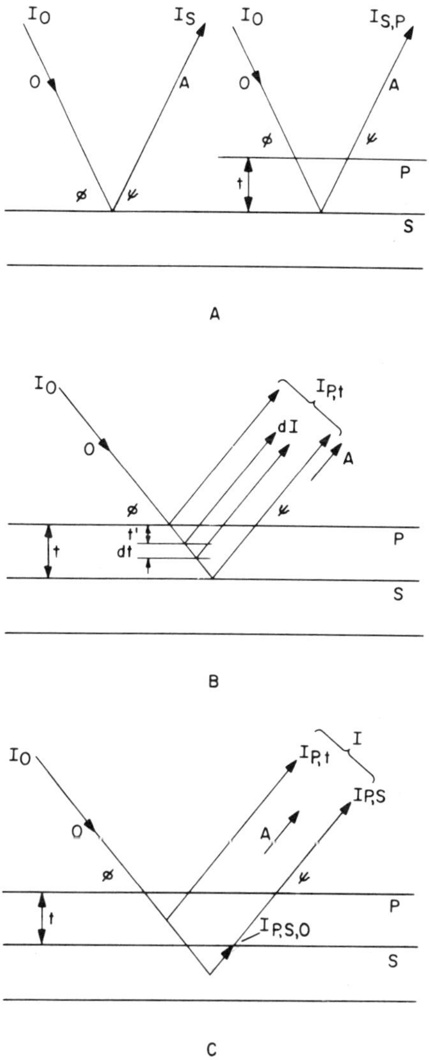

Fig. 18.1. Basic methods for x-ray spectrometric determination of film and plating thickness. A. Attenuation of the line of a substrate element absent in the plate. B. Emission of the line of a plate element absent in the substrate. C. Emission of the line of a plate element also present in the substrate.

line on passing through the overlying plating. From Equation (14.33),

$$\log_e(I_S/I_{S,P}) = \bar{\mu}\varrho t \tag{18.1}$$

where ϱ is film density (g/cm³). A plot of $I_{S,P}$ *vs.* thickness may be prepared from standards having known plate thickness.

2. The *plate-line emission method* (*K16*), for systems in which the plate element is absent from the substrate, is shown in Figure 18.1B, and consists of measurement of the intensity of the plate line. From Equation (14.33),

$$\log_e[1 - (I_{P,t}/I_{P,\infty})] = -\bar{\mu}\varrho t \tag{18.2}$$

A plot of $I_{P,t}$ *vs.* thickness may be prepared from standards.

3. The *plate-line emission method* (*K1*), for systems in which the plate element is present in the substrate, is shown in Figure 18.1C. If I_T is the total measured plate-element line intensity,

$$I_T = I_{P,S} + I_{P,t} \tag{18.3}$$

in which

$$I_{P,S} = I_{P,S,0} \exp -\bar{\mu}\varrho t \tag{18.4}$$

$$I_{P,t} = I_{P,\infty} - I_{P,\infty} \exp -\bar{\mu}\varrho t \tag{18.5}$$

Then,

$$I_T = I_{P,S,0} \exp(-\bar{\mu}\varrho t) + I_{P,\infty} - I_{P,\infty} \exp -\bar{\mu}\varrho t \tag{18.6}$$

$$I_T - I_{P,\infty} = (I_{P,S,0} - I_{P,\infty}) \exp -\bar{\mu}\varrho t \tag{18.7}$$

$$\log_e \frac{I_T - I_{P,\infty}}{I_{P,S,0} - I_{P,\infty}} = -\bar{\mu}\varrho t \tag{18.8}$$

4. Many ratio methods have been used for determination of plate thickness. Calibration curves or mathematical relationships are established from standards. Any of the following quantities may be plotted as a function of thickness:

$I_{P,t}/I_B$, where B refers to the background at or adjacent to the peak or in the continuum hump (*G7*, *K6*).

$(I_{P,\infty} - I_{P,t})/I_{P,\infty}$ (*Z11*, *Z13*).

$I_{P,\infty}/(I_{P,\infty} - I_{P,t})$ (*S14*).

$1 - (I_{P,t}/I_{P,\infty})$ (*H34*).

$(I_{K\alpha}/I_{K\beta})_P$ (for low-Z platings) (*Z11*, *Z13*).

$(I_{L\alpha}/I_{L\beta})_P$ (for high-Z platings) (*Z11*, *Z13*).

$I_{P,t}/I_{S,P}$ (*B23*).

$I_{S,P}/I_S$ (*Z11, Z13*).
$(I_{K\alpha}/I_{K\beta})_S$ (for low-Z substrates) (*Z11, Z13*).
$(I_{L\alpha}/I_{L\beta})_S$ (for high-Z substrates) (*Z11, Z13*).

The intensities may be peak or net. For a single plating on a substrate, the functions $I_{P,t}$ or $I_{S,P}/I_S$, or even $I_{P,t}/I_{S,P}$, are of comparable effectiveness. Multiple platings are considered below. The $K\alpha/K\beta$ or $L\alpha/L\beta$ intensity-ratio methods, whether from plate or substrate, usually give calibration curves having relatively shallow slopes.

5. The *transmission-absorption method* may be used if the substrate is transparent to x-rays. The specimen is placed between the analyzer crystal and the detector, and a monochromatic beam diffracted by the crystal is passed through the specimen. Transmitted intensity is given by

$$I = I_0 \exp[-(\mu/\varrho)\varrho t] \tag{18.9}$$

where μ/ϱ is the mass-absorption coefficient (cm^2/g) of the film for the monochromatic wavelength; ϱ is the density (g/cm^3) of the film; and I_0 is the intensity transmitted by uncoated substrate. The monochromatic wavelength may be derived from the direct primary beam, from primary x-rays scattered from a low-Z scatter target in the specimen compartment, or from the secondary-emission spectrum of an appropriate element in the compartment.

18.2. MULTIPLE-LAYER AND ALLOY PLATINGS

18.2.1. Multiple-Layer Platings

Birks *et al.* (*B41*) determined the thickness of both platings in nickel/copper-plated steel. Nickel thickness was obtained from Ni $K\alpha$ intensity. Copper thickness was then obtained from a set of curves of Fe $K\alpha$ intensity *vs.* copper thickness, each curve for a different overlying nickel thickness. Keesaer (*K6*) determined the thickness of chromium and nickel in "triplate" (chromium/nickel/copper-plated steel). Chromium thickness was obtained from Cr $K\alpha$ intensity, nickel thickness from Ni $K\alpha$ intensity corrected for absorption in the overlying chromium plate. Gerold (*G3*) combined x-ray diffraction and spectrometry to determine chromium and nickel in chromium/nickel-plated brass. Chromium was determined by diffractometry, nickel by spectrometry of the Ni $K\alpha$ line. Ebel and Klugel (*E1, E2*) calculated film thickness from equations based on variation of emitted spectral-line intensity as a function of takeoff angle from the film and from infinitely thick film element. Calibration with films of known thickness is not required. Under certain conditions, the method is useful for multiple layers, and was applied to layers of ~5570 Å of tin on ~940 Å of copper.

18.2.2. Alloy Platings—Composition and Thickness

Niobium and tin have been determined in cryogenic Nb_3Sn coatings on metal and ceramic substrates (*B19*) by application of the binary-ratio method (Section 14.3.2.3). Most of the coatings were infinitely thick for both Nb $K\alpha$ and Sn $K\alpha$; those that were not had constant thickness. For such films, a log-log plot of Nb $K\alpha$/Sn $K\alpha$ intensity ratio *vs.* Nb/Sn concentration ratio was established from standards having coatings of known composition. If the thickness was less than infinite for the *K* lines and variable, the *L* lines were used, so that infinite thickness was smaller.

Many methods have been reported for simultaneous determination of both composition and thickness of films. Most of these methods were developed for analysis of Ni–Fe (Permalloy) films.

Schindler *et al.* (*S8*) report solution and nondestructive methods for determination of composition and thickness of nickel–iron films on gold in the ranges 0–40% iron, 10^2–10^5 Å. Five stock solutions were prepared having concentration 10 mg Ni + Fe per ml and Ni:Fe weight ratios 10:0, 9:1, 8:2, 7:3, and 6:4. Aliquots of each of these five solutions were diluted to give five solutions having concentrations 1, 2, 3, 4, and 5 mg Ni + Fe per ml for each Ni:Fe ratio. Ni $K\alpha$ and Fe $K\alpha$ intensities were measured from all 25 solutions, and the data were plotted on a graph of Fe $K\alpha$ intensity *vs.* Ni $K\alpha$ intensity. A grid consisting of two sets of lines was drawn on the plot. One set represented the same iron concentration for different total weights of Ni–Fe. The other set represented the same total weight of Ni–Fe for different iron concentrations.

Next, Ni $K\alpha$ and Fe $K\alpha$ intensities were measured on 150 Ni–Fe films having various compositions and thicknesses on gold disks. The films were then dissolved quantitatively, and Ni $K\alpha$ and Fe $K\alpha$ intensities measured on the solutions. Composition (Ni:Fe) of each film was derived from the grid. Thickness was derived from the film weight, area, and composition, and from the densities of nickel and iron. The film weight was obtained from the weights of the disks before and after dissolution of the film. A plot of Fe $K\alpha$ *vs.* Ni $K\alpha$ intensity was made of the data from the *films.* A grid consisting of two sets of lines was drawn on the plot. One set represented the same iron concentration for different thicknesses. The other set represented the same thickness for different iron concentrations. Finally, the grid on the plot of *solution* data was redrawn in terms of thickness rather than total weight of Ni–Fe, so that thickness could be derived directly either nondestructively or by the solution technique. The form of the calibration grid is shown in Figure 18.2.

Verderber (*V6*) developed a method for determination of composition and thickness of Ni–Fe films on magnetic tapes in the ranges ~81Ni–19Fe and 1000–4000 Å. Three stock solutions were prepared having concentration 3 mg Ni + Fe per ml and Ni:Fe weight ratios 85:15, 80:20, and 75:25. Aliquots of each of these three solutions were diluted to give seven solutions of different concentration for each Ni:Fe ratio, the most dilute having 0.4 mg Ni + Fe per ml. Ni $K\alpha$ and Fe $K\alpha$ intensities were measured from all 21 solutions, and two calibration curves were established: Ni $K\alpha$ intensity *vs.* mg Ni/ml and Fe $K\alpha$ intensity *vs.* mg Fe/ml.

Next, 21 evaporated Ni–Fe films were selected, all having nominal 81Ni–19Fe composition, but varying in thickness from 1000 to 4000 Å. About 100 cm² of each film was dissolved quantitatively, and Ni $K\alpha$ and Fe $K\alpha$ intensities were measured from each solution. Nickel and iron con-

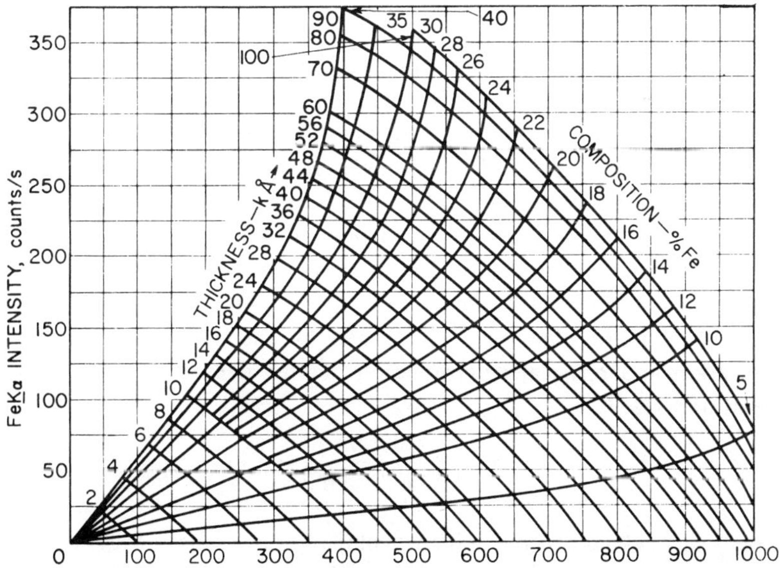

Fig. 18.2. Nondestructive determination of composition and thickness of nickel–iron films.

centrations were derived from the calibration curves and used to calculate film composition.

On an unused portion of each film, thickness was measured by a physical method, and Ni *Kα* and Fe *Kα* intensities were measured. Two new calibration curves were now established from the data from the films: Ni *Kα* intensity *vs.* (thickness × %Ni) and Fe *Kα* intensity *vs.* (thickness × %Fe). Thereafter, these curves were used for nondestructive analysis of films.

Weyl (*W6*) and Pluchery (*P11*) developed absolute mathematical methods for determining thickness and composition of thin alloy films without calibration standards other than the pure constituent elements. Their equations correct for spectral intensity distribution in the primary beam, fluorescent yields of the constituent elements, secondary excitation of low-atomic-number elements by high-atomic-number elements, and attenuation of the spectral emission on its way to the detector. Pluchery used filters to remove target lines from the primary beam, and corrected for scattered continuous background. Weyl and Pluchery evaluated their equations for simple Ni–Fe films.

Hirokawa *et al.* (*H34*) prepared a series of nickel, iron, and 80Ni–20Fe films having thickness up to ~3 μm. Ni *Kα* and Fe *Kα* intensities were measured from all the standards ($I_{P,t}$) and from infinitely thick nickel and iron ($I_{P,\infty}$). Four curves were plotted on a graph of the function $\log[1 - (I_{P,t}/I_{P,\infty})]$ *vs.* thickness: Ni *Kα* intensity from nickel and from Ni–Fe, and Fe *Kα* intensity from iron and from nickel–iron. All the curves were linear. From these curves, two pairs of simultaneous equations were derived for determination of composition and thickness near the composition 80Ni–20Fe. The two pairs of equations were for thickness >0.1 μm and <0.1 μm, respectively.

Silver and Chow (*S22*) compared optical interferometric, stylus, and x-ray spectrometric methods for determination of thickness of Ni–Fe films having composition 81–92% nickel and thickness 400–3500 Å. Thicknesses of several standard films were established by chemical determination of composition and physical determination of area density by measurement of mass and area:

$$t_S = \frac{100(m/a)}{(W\varrho)_{Ni} + (W\varrho)_{Fe}} \qquad (18.10)$$

where t_S is thickness of the standard film (Å); m is mass (μg); a is area (cm²); W is weight fraction; and ϱ is bulk density (g/cm³). The quantity (m/a) is area density (μg/cm²). An x-ray spectrometric thickness measurement consisted of measurement of the Ni *Kα* and Fe *Kα* intensities from the samples and from one of the standards. Then.

$$t_X = t_S \frac{(kI_{Ni} + I_{Fe})_X}{(kI_{Ni} + I_{Fe})_S} \tag{18.11}$$

where t is thickness (Å); I is net intensity (counts/s); X and S refer to the sample and standard, respectively; and k is calculated from the data measured from the standard:

$$k = \frac{I_{Fe}/I_{Ni}}{C_{Ni}/C_{Fe}} \tag{18.12}$$

where C is concentration (wt %). Note than in Equations (18.10) and (18.11), the subscript S refers to a standard, rather than to the substrate; the subscript italic S elsewhere in this chapter refers to the substrate.

Chan (*C29*) analyzed films of semiconductive II–VI compounds, that is, compounds of zinc, cadmium, and/or mercury with sulfur, selenium, and/or tellurium. He also compared the x-ray spectrometric results with those from analyses by substrate weight gain, neutron activation, and optical emission spectrography with laser-probe excitation.

18.3. SPECIAL TECHNIQUES

18.3.1. Special Specimen Forms

Zimmerman (*Z11, Z13*) applied three techniques to determination of plate thickness on a wide variety of small, irregular-shaped, plated parts, such as washers, solder lugs, and connectors: selected-area apertures (see below), masks, and impression molds in sealing wax to permit reproducible positioning of irregular-shaped parts in the spectrometer specimen drawer (*I* in Figure 16.3).

Bertin and Longobucco (*B23*) developed two methods for determining plate thickness on wires having diameter 0.0004–0.010 inch. One method is applicable to large numbers of specimens consisting of infinitely thick layers of wire (≥3–5 layers) wound on identical spools that fit the spectrometer specimen compartment. With the instrument set at a plate-metal line, each spool is placed in turn in the specimen drawer and the ratemeter allowed to trace plate-metal intensity for 15 s or more. A set of spools wound with standard wires is treated the same way. As many as four specimens can be analyzed each minute. Figure 18.3 shows a typical ratemeter chart.

For more accurate analysis, a few meters of wire is wound on a small

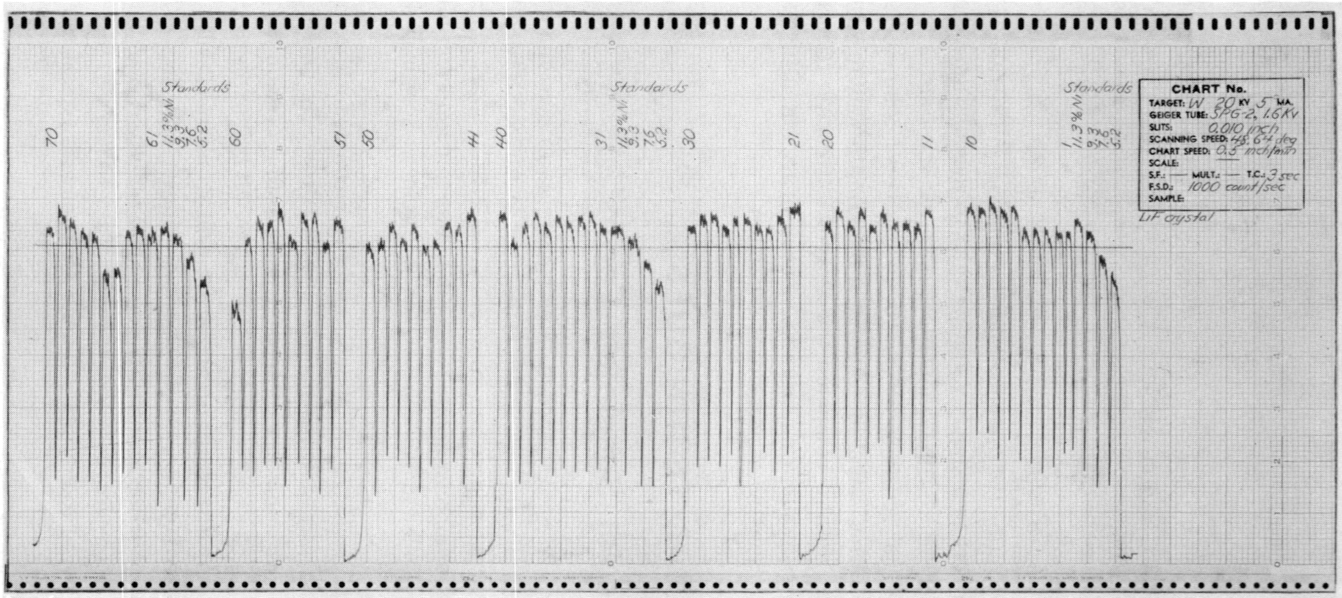

Fig. 18.3. Determination of plate metal on plated wire by the ratemeter method. [E. P. Bertin and R. J. Longobucco, *Analytical Chemistry* **34**, 804 (1962); courtesy of the American Chemical Society.]

open frame, card, or spool. The intensities of a plate and substrate line are measured and their ratio calculated. A calibration curve of plate/substrate intensity ratio *vs.* plate thickness is established from wires of known plate thickness on the same type of support. The ratio method corrects for variations in number, spacing, and orientation of the turns of wire in the specimen. A set of standards is required for each combination of plate metal and substrate, wire diameter, and specimen mount (card, spool diameter, etc.).

Selected-area methods have also been used to evaluate uniformity of plating thickness around the rims of small disks and around small cylindrical rods (Section 19.5).

18.3.2. Selected-Area Analysis

By use of selected-area techniques (Chapter 19), analysis of small selected areas and point-by-point measurement of uniformity of composition and/or thickness of films become feasible. With the spectrometer set at the spectral line of an element in the film, the specimen may be translated continuously with a motor drive while intensity is recorded as a function of distance in a line across the specimen. Alternatively, the specimen may be moved manually in small increments and intensity measured at each point with the scaler. Successive scans can be made at different places and/or at the same place for different elements. Using such techniques and a 0.5-mm secondary-beam pinhole aperture, Sloan (*S24*) evaluated point-by-point variation of composition and thickness of Nichrome (nickel–chromium) films ~1500 Å thick. Ni $K\alpha$ and Cr $K\alpha$ intensities were measured at 1-mm intervals over the film. The individual intensities indicated film thickness; their ratio indicated composition. Zimmerman (*Z11*, *Z12*) measured variation in plate thickness over plated surfaces. He also investigated wear of gold-plated nickel by scanning a wear track and recording Ni $K\alpha$ intensity along or across the track. Bertin and Longobucco (*B22*) analyzed and mapped the distribution of sublimates on bulbs, stems, micas, grids, and plates of electron tubes.

18.3.3. Selective Excitation

Beeghley (*B12*) measured tin-plate thickness by operation of the x-ray tube below the *K*-excitation potential of tin (28 kV) so as to excite only Fe $K\alpha$ (excitation potential 7 kV) from the base metal. The selective-excitation method has since been applied by others (*B41*, *L15*).

18.3.4. Enhancement

If the specimen is an unsupported film or a film on a highly x-ray-transparent substrate, most of the primary radiation passes through the specimen and is lost. Hirokawa *et al.* (*H35*) placed under such specimens a secondary radiator having a strong spectral line of wavelength just shorter than that of the absorption edge of the element to be excited in the film. The primary x-rays transmitted by the film excite the radiator spectrum, which in turn excites additional emission from the film. Enhancements up to 50% were obtained for chromium and nickel films. Even a simple low-*Z* scatter target to direct the transmitted primary x-rays back through the film is very beneficial.

Alternatively, for x-ray-opaque film-substrate systems, the radiator may be placed in the primary beam (*G16*). However, this technique reduces the total primary intensity incident on the specimen and therefore reduces the measured intensity by ~50%. The only advantage is improved line-to-background ratio.

18.3.5. Excitation by Radioactive Sources

Cook *et al.* (*C40*) used radioactive isotopes to excite, and a nondispersive spectrometer to measure, plate-metal emission. Excitation of base-metal emission by radioactive isotopes is described by the same workers, by Cameron and Rhodes (*C3*), and by Zemany (*Z4*).

Cameron and Rhodes (*C3*) report the use of tritium sources for excitation of Fe *K* radiation in galvanized (zinc-coated) steel and in copper-plated steel, Ti *K* radiation in platinum-clad titanium, and Cu *K* radiation in silver-plated copper wire. Their work was done on a nondispersive instrument.

Zemany's work (*Z4*) is particularly noteworthy because it combines the techniques of excitation by radioactive sources, selective excitation, and nondispersive detection. He used Fe-55, a *K*-capture isotope emitting only Mn *K* x-rays, which can excite only *K* spectra of elements *below* atomic number 25 (manganese), and *L* spectra of elements *below* 61 (promethium). If the measurements are made in air, absorption eliminates the *K* spectra of elements below 19 (potassium) and *L* spectra of elements below 47 (silver). If the method is to be applicable to one element plated on another, it is necessary that the plate be among the elements potassium through chromium, or silver through neodymium, but that the base metal not be among these elements. Zemany applied the method to titanium-plated Kovar (Fe–Co–Ni), using a gas-flow proportional counter as detector. He suggests that other *K*-capture isotopes might extend the utility of the technique.

18.3.6. Nondispersive Operation

Use of specially built nondispersive spectrometers has been mentioned above in connection with radioactive sources. Standard commercial x-ray spectrometers can be operated in nondispersive mode by setting the goniometer at $0° \ 2\theta$, removing the crystal, and using a pulse-height selector. Elimination of the crystal may increase the measured intensity by as much as two orders. Further increase in sensitivity may be realized by moving the detector closer to the specimen compartment. Achey and Serfass (*A1*) used balanced (Ross) filters (*K14*, *R13*) to isolate the spectral line to be measured. The disadvantages are that the x-rays from the specimen may contain secondary emission from both plate and substrate, and/or primary x-rays diffracted from crystalline plate or substrate. Selective excitation may eliminate the emission from either plate or substrate, depending on which has the higher excitation potential. Balanced filters may isolate the line to be measured from either emitted or diffracted radiation.

18.3.7. Decoration

Lundquist (*L35*) evaluated porosity of nickel electroplatings on uranium as follows. The nickel-plated uranium was anodized in a solution containing potassium ferrocyanide [$K_4Fe(CN)_6$] under conditions that minimize attack of nickel, but precipitate uranyl ferrocyanide [$(UO_2)_2Fe(CN)_6$] wherever uranium is exposed. Fe $K\alpha$ intensity is then measured. This technique should be valuable for investigating porosity and discontinuity of other kinds of films and platings.

18.3.8. Dynamic Studies

A special electroplating cell designed by Burbank (*B56*) for dynamic x ray diffractometric studies of the plating process is also applicable in principle to dynamic x-ray spectrometric studies. A rotating-disk cathode turns half-immersed in the cell containing the electrolyte and anode. The unimmersed surface constitutes the diffractometer specimen surface. Only crystalline material diffracts x-rays, so the thin layer of plating solution on the inspected surface is not detrimental. However, during the early stage of the plating, when the plate is still very thin, the intensity of the plate-metal line from the solution would exceed that from the plating. Provision would be required to remove the plating solution continuously. The cell could be modified so that a continuous-belt cathode would pass in turn through the plating cell, a rinse, and the spectrometer specimen chamber.

18.4. CONSIDERATIONS

X-ray spectrometric methods for determination of composition and thickness of films and platings are prone to several sources of error. The methods are impractical when there are more than two or three layers to be determined, and are more difficult when the plate element is also present in the substrate. Intensities from thin films are very low, and those from thick films are affected by surface texture—rough, matte, polished, porous, etc. Thickness and composition may not be uniform. Films and platings may be single-crystalline or highly oriented and thus produce diffraction effects.

In the measurement of plate thickness, serious errors arise from the presence of very thin platings ("strikes" or "flashes") and/or interfacial alloying between plate and substrate. The alloying occurs on annealing or other heat treatment. The effects of interfacial strikes and alloying are discussed below in terms of nickel-plated molybdenum wire as an example (*B23*). However, the principles apply as well to any plated specimen.

Figure 18.4A shows the simple case of a plate on a base wire with no strike or alloy formation at the interface. If such a wire is placed in the

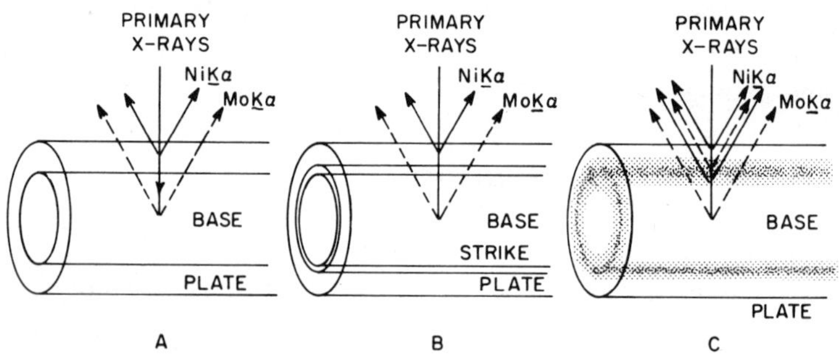

Fig. 18.4. Effect of interfacial strike and alloying on determination of plate thickness. [E. P. Bertin and R. J. Longobucco, *Analytical Chemistry* **34**, 804 (1962); courtesy of the American Chemical Society.] A. Absence of strike and alloying. B. Presence of strike. C. Presence of alloying.

x-ray spectrometer, the primary x-radiation in turn excites the plate metal emission, undergoes some attenuation as it penetrates the plate, and excites the base-metal emission. The base-metal emission also undergoes some attenuation as it emerges through the plate, and possibly enhances the plate emission. Plate or substrate-line intensity or their ratio may be used as the calibration function.

Suppose now that there is a strike between the plate and base wire, as shown in Figure 18.4B. The intensity of the plate emission remains unchanged, unless the strike consists of an element capable of enhancing the plate emission, or unless the plate emission is enhanced by the base emission, and the strike attenuates the base emission. Thus, calibration functions involving only the plate-element line intensity are usually not substantially affected by strikes. However, both the base-metal emission and the primary radiation which excites it must pass through the strike. If the strike consists of an element of low atomic number (e.g., chromium), and the base-metal emission is of relatively short wavelength (e.g., Mo $K\alpha$, 0.7 Å), the base emission undergoes little reduction in intensity. Conversely, if the strike consists of an element of high atomic number (e.g., silver or gold), and/or the base-metal emission is of relatively long wavelength (e.g., Ni $K\alpha$, 1.66 Å), the base emission may undergo substantial reduction in intensity. This reduction increases with increasing wavelength of the base-metal emission, and with increasing absorption coefficient and/or thickness of the strike. Of course, reduction in base-metal intensity increases the intensity ratio.

Because of these effects, calibration functions involving substrate lines should be applied very cautiously to platings having strikes. In some cases, the effect of the strike can be compensated by use of standards having similar strikes or by determination of the strike thickness and correcting the substrate line intensity.

Now suppose that alloying has occurred between the plate and base wire, as shown in Figure 18.4C. Consider what happens, using nickel-plated molybdenum as an example. That part of the Ni $K\alpha$ emission originating in the unalloyed part of the nickel layer is excited in the same manner as in unalloyed wire. That part of the Ni $K\alpha$ emission originating in the alloyed layer and the primary radiation which excites it must pass through a certain amount of molybdenum. In doing so, they both undergo some absorption, with resultant reduction in Ni $K\alpha$ intensity. This phenomenon becomes more marked the deeper the nickel has diffused into the molybdenum. That part of the Mo $K\alpha$ emission originating in the unalloyed "core" of the base wire is excited in the same manner as before, that is, by primary radiation that has passed through all the nickel. Similarly, the

Mo $K\alpha$ must emerge through all the nickel. Finally, that part of the Mo $K\alpha$ radiation originating in the alloyed layer and the primary radiation which excites it pass through only a part of the nickel layer. Consequently, they undergo less absorption than in unalloyed wire, with resultant increase in Mo $K\alpha$ intensity.

For a given per cent nickel, the Ni $K\alpha$ intensity is lower and the Mo $K\alpha$ intensity higher for alloyed wire than for unalloyed wire. In the intensity-ratio method, the Ni $K\alpha$/Mo $K\alpha$ intensity ratio is smaller. Thus, for nickel-plated molybdenum, all three analytical methods give low results. If the substrate were a light element having relatively low absorption for Ni $K\alpha$, a similar consideration would show that for a given per cent nickel, both Ni $K\alpha$ and substrate-line intensity are higher. The change in the ratio would depend on the relative increases in the individual intensities.

Chapter 19

Selected-Area Analysis

19.1. PRINCIPLE AND SCOPE

19.1.1. Principle

If the conventional aluminum-specimen-mask technique is applied to very small specimens or to small areas on large specimens, analyte line-to-background ratio may decrease prohibitively. As analyte-line intensity decreases with specimen area, primary scatter from the increasing mask area increases, and primary scatter from air in the specimen compartment assumes greater significance. Scatter from the mask is reduced by use of a high-Z metal, but then the spectrum of this element is strongly excited, increasing the probability of spectral-line interference. Small specimens can be supported on thin Mylar or fine fibers, but air scatter remains, unless helium or vacuum is used. A better approach is to confine the primary beam to the area of interest, or to exclude all secondary radiation from the optical path except that originating in the area of interest.

If a small aperture is placed in the primary beam, only that incremental specimen area lying in the projection of the aperture receives primary radiation. Alternatively, if the aperture is placed in the secondary beam, the entire specimen is irradiated, but only secondary radiation originating in the projection of the aperture enters the optical path. Such apertures provide the basis for selected-area analysis. This chapter is devoted to analysis of small specimens and areas by x-ray secondary-emission spectrometry. Emphasis is placed on techniques that are applicable to commercial instruments, particularly with relatively simple accessories and minimal modification. The subject has been reviewed by Adler [pp. 101–8 in (*1*)] and (*A4*), Duncumb (*D19*), and Loomis and Vincent (*15N*).

An x-ray spectrometer arranged for selected-area analysis is often termed an x-ray "macroprobe" or "milliprobe" (compare with electron "microprobe") because: (1) excitation of the "probed" (selected) area is by x-rays rather than electrons, and (2) the probed area has dimensions of the order of millimeters or mils rather than micrometers.

Four other x-ray methods for evaluating point-to-point composition are discussed in other sections: the Von Hamos image spectrograph (Section 5.5.2.4), x-ray contact microradiography (Section 20.1.5), x-ray absorptiometry (Section 20.1.3), and the electron-probe microanalyzer (Chapter 21).

19.1.2. Applications

Selected-area techniques are applicable to two classes of specimens: very small specimens, which may be regarded as "preselected" areas, and small selected areas on extensive specimens.

The first group includes the following: (1) particles, flakes, very small chips, etc.; (2) fibers, filaments, fine wires, etc.; (3) submilligram weights of powders on fibers or in fine plastic capillary tubes (these are the types of specimen used in x-ray powder diffraction analysis, and selected-area techniques may serve as a preliminary to a powder analysis); (4) metallic trace analytes electroplated on fine wire or on carbonized quartz fibers (*L24*); (5) flecks of purified trace analyte after microchemical separation from the matrix; (6) very small disks of filter paper or ion-exchange membrane supporting microgram amounts of analyte (*L33*); (7) small spots of analyte residue formed by evaporating droplets of solution on thin Mylar film; and (8) segregations, grains, and inclusions extracted from materials by micro drilling, extraction replication (*B48*), or preferential dissolution or etching away of the matrix.

The second group includes samples submitted for the following types of work:

1. Mapping of composition distribution of analyte(s) and evaluation of homogeneity of composition (*C31*).

2. Analysis of individual grains of various phases in heterogeneous alloys, minerals, ceramics, etc. (*A5*, *A8*).

3. Analysis of inclusions, segregations, etc. in the same materials (*J12, J13*).

4. Analysis of particles filtered from gases or liquids (*L13*).

5. Evaluation of composition, thickness, and uniformity of thin films and platings (*B22, R1, S24, Z11-14*).

6. Measurements of extent of diffusion in diffusion couples (*R7*).

7. Identification of layers and measurement of extent of diffusion in sections of metal-to-metal and metal-to-ceramic interfaces (*B15, B18*).

8. Investigation of sections of plated, coated, or corroded surfaces (*Z11-14*).

9. Analysis of parts of electronic devices or other small parts (*G5*).

10. Determination and distribution of natural and foreign elements in biological materials (*H10, H11, L22*).

11. Determination of minor and trace constituents in bulk specimens. Sometimes, greater sensitivity is realized by analyzing a small area of specimen with a curved crystal than by analyzing a large area with a flat crystal. For example, in the determination of manganese and iron in aluminum alloys, Mn $K\alpha$ and Fe $K\alpha$ intensity, resolution, and line-to-background ratio are better from a 1-mg sample (containing <10 μg of manganese and iron) in a curved-crystal spectrometer than from a 10-g sample in a flat-crystal spectrometer (*B35*).

Selected-area apertures may serve in place of constant-area masks for small and irregular-shaped specimens. For example, the writer inspected samplings from all incoming lots of about 20 different parts for an electron device to ascertain that they were of Invar (64Fe 36Ni). The parts ranged in size from a bent wire 0.5 mm in diameter and 5 mm long to a tube 2 cm in diameter and 20 cm long. A secondary-beam aperture was selected having a projection lying entirely on the smallest parts—the wire after flattening, and a small hexagonal nut. Ni $K\alpha$ intensity was observed on the panel ratemeter for each sample and from pieces of Invar. A sample could be verified in less than 2 min, most of the time being spent in aligning it with the pinhole.

Specimen areas somewhat greater than $\sim$1 mm^2 and substantially smaller than $\sim$1 cm^2 represent a case intermediate between conventional and selected-area techniques and may be measured in a *collimated specimen drawer* (*C31, G5*). The primary beam is conducted to the specimen and the secondary beam to the optical path in a continuous L-shaped tunnel in a block on the mask plate. This arrangement reduces primary scatter from air and mask.

19.1.3. Advantages and Limitations

The technique of selected-area analysis on the x-ray secondary-emission spectrometer has not assumed the degree of application that its capabilities and advantages warrant. The principal reason is the ready availability of many models of electron-probe microanalyzers. However, the two techniques should be complementary rather than competitive, and, for certain applications, the x-ray probe is actually preferable.

The advantages of the x-ray probe over the electron probe include the following:

1. There are applications where the extremely high resolution of the electron probe is actually detrimental. If the average composition of an area, say, 0.5 mm in diameter or of a narrow layer ~1 cm long is required, a single x-ray measurement for each analyte is sufficient. However, if the area is inhomogeneous on a 1–100-μm scale, the average of several electron-probe measurements for each analyte is required, or the electron beam must be defocused to cover the larger area.

2. Specimen translations up to 2 cm are usually possible in x-ray probes; some electron probes permit translations of only ~0.5 cm or less. Moreover, it is practical to map concentration profiles over distances of 1–2 cm with the x-ray probe.

3. The equipment is much simpler and less expensive. Electron-optical and vacuum equipment are not required. Many techniques are applicable to standard commercial x-ray spectrometers with simple accessories.

4. Specimen preparation, manipulation, and measurement are more rapid and convenient. Electrically nonconductive specimens require no provision of conductive film or paint.

5. Since primary x-rays penetrate deeper into the specimen than electrons, surface condition and takeoff angle are not as critical, and analyses are more representative. The relatively high takeoff angle (30–60°) of commercial x-ray spectrometers is favorable.

6. Secondary excitation gives lower background than primary.

7. Biological and volatile specimens are usually not damaged because the specimen is not in vacuum and is not bombarded by electrons.

8. The instrument can be arranged to accommodate larger specimens than the electron probe, and so can deal with objects that may not be damaged or marred (*B2*, *H9*, *Y2*). One instrument has been designed with the specimen plane vertical at the back of the instrument to permit paintings, art objects, and antiquities to be studied (*Y2*).

9. The instrument is applicable to routine analysis of large numbers

of samples and has been applied to incoming inspection—an example of which is cited above—and even to production control of plate thickness on very small critical parts (*G5, Z11–14*).

The disadvantages of the x-ray probe include the following:

1. Secondary excitation is two to three orders less intense than primary. As a result, second- and higher-order lines are useful only for relatively high concentrations, and, when L lines must be used, analyte sensitivity decreases sharply. Sensitivity for the low-Z elements (11–18) is very poor.
2. Spacial resolution is poorer by two to three orders (10–100 μm compared with $\sim$1 μm).
3. The electron probe is useful for elements down to beryllium.
4. The electron probe can be used for all elements without change of accessories. Two simultaneously operable spectrometers are provided, each of which may be arranged for a different spectral region. Because of the high primary excitation efficiency, M, as well as K and L lines, are useful.
5. The electron probe may be operated in many modes based on phenomena other than x-ray emission, the two principal ones being scattered primary electrons and specimen current. Each of these phenomena gives a particular kind of information.
6. In the electron probe, analyte-line intensity and scattered electron and specimen current distribution along a line or over an area can be displayed on a cathode-ray oscilloscope.

Compared with conventional secondary-emission spectrometry, selected-area techniques give improved spectral resolution, reduced background, and, when a curved crystal is used, increased intensity and sensitivity.

19.2. SELECTED-AREA ANALYSIS ON STANDARD COMMERCIAL X-RAY SPECTROMETERS

19.2.1. Specimen Irradiation

The primary radiation intensity at a point i on the specimen is given by

$$I_i = (J_{\text{pri}} \sin^3\phi_i)/L^2 \tag{19.1}$$

where I_i is primary intensity at point i; J_{pri} is x-ray flux emitted by the x-ray tube per unit solid angle; ϕ_i is the angle between the specimen plane

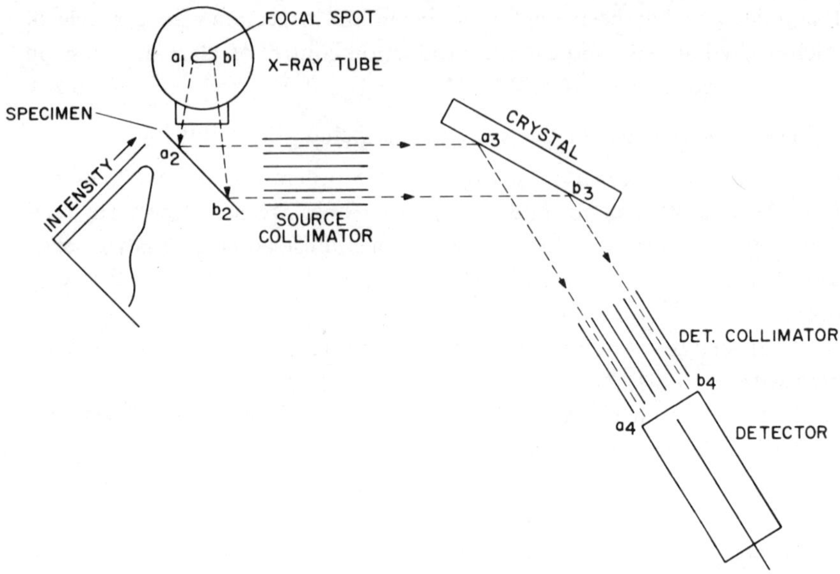

Fig. 19.1. Effect of specimen position and goniometer angle (2θ) on measured analyte-line intensity from small specimens. [R. Jenkins and J. L. de Vries, "Practical X-Ray Spectrometry," Springer-Verlag New York, Inc., p. 153 (1967); courtesy of the authors and publisher.]

and the line joining i and the center of the x-ray tube focal spot; and L is the perpendicular distance from the specimen surface to the center of the focal spot.

For a given x-ray optical system, measured analyte-line intensity from a very small specimen varies with position in the specimen plane, position above or below the specimen plane, and goniometer angle 2θ. Primary intensity distribution over the specimen plane is nonuniform because the x-ray tube focal spot distance is different to different parts of the plane. This is shown in Figure 19.1, where primary intensity at the specimen plane decreases from a_2 to b_2 because the focal spot distance a_1a_2 is smaller then b_1b_2. Even if the specimen plane were uniformly irradiated, the measured intensity from a small specimen would vary with position. Different parts of the analyzer crystal diffract radiation from different parts of the specimen plane, and the crystal surface may not be uniformly "reflective." Finally, secondary x-rays originating at a_2 follow path $a_2a_3a_4$ to the detector, while rays originating at b_2 follow path $b_2b_3b_4$. Movement of a small specimen from a_2 to b_2 has the effect of decreasing the goniometer setting(2θ) at which maximum measured intensity occurs. This effect is minimized by use of a detector with a large useful solid angle, but may assume values as high as 0.5° 2θ. All these factors apply as well to variations in specimen position above and below the specimen plane.

Because primary intensity varies over the specimen plane, small specimens and areas must be placed so as to receive maximum primary intensity. Reproducibility of this position is particularly important when a series of specimens or of standards and samples are to be compared quantitatively. The use of glass microscope slides and unexposed x-ray film to delineate the area of primary irradiation is described in Section 4.2.5.4. However, other techniques are required to establish intensity distribution within this area. Figure 19.2A shows a typical primary-beam distribution pattern on the specimen plane in the form of "isointensity" contour lines. The smallest (central) contour line defines the area of greatest primary intensity and therefore the optimal position for a small specimen or selected area. Successive contour lines represent progressively lower intensities. The pattern is obtained by mounting a piece of graph paper (10 × 10 divisions per inch) in the specimen window and scaling Ni $K\alpha$ intensity from a very small square of nickel foil placed successively in each square. The intensities are written in their relative positions, and contour lines are drawn through positions having about the same intensity.

Figure 19.2B shows the effect of position on intensity measured from

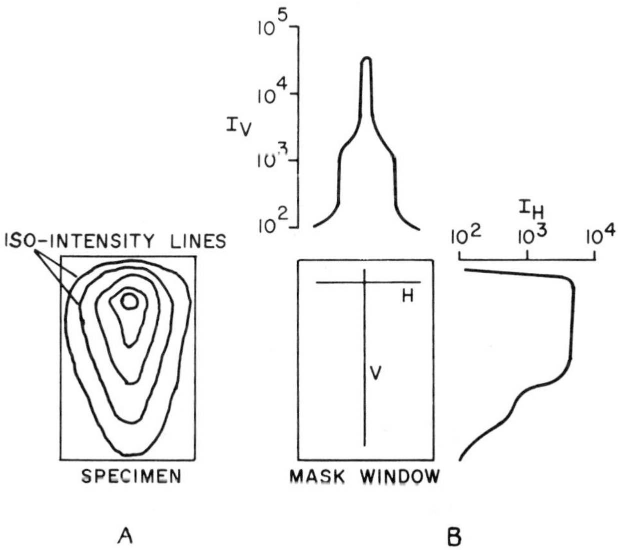

Fig. 19.2. (A) Variation of primary intensity over the specimen plane and (B) effect of specimen position on measured analyte-line intensity from filamentary specimens.

linear specimens by use of curved crystals. A nickel wire 0.0025 inch in diameter is oriented parallel to the top edge of the specimen window and translated downward, then oriented parallel to a side edge and translated laterally. Ni $K\alpha$ intensity is recorded during the translations. Figure 19.2B shows the Ni $K\alpha$ intensity profiles and the optimal specimen positions for horizontally H and vertically V oriented filamentary specimens or linear areas. The strong intensity maximum for vertically oriented specimens does not necessarily lie at the center of the window, but may be placed there by slight rotation of the x-ray tube about its axis. The data for Figure 19.2 were obtained on an instrument having the geometry of the bottom row of Figure 9.2; the long edge of the window (Figure 19.2) is the inclined edge. No aperture was used in making any of the measurements. When selected-area work is to be done, intensity profiles of these types should be plotted each time the x-ray tube is changed.

19.2.2. Selected-Area Apertures

19.2.2.1. *General*

Possibly the simplest selected-area technique is to place a metal foil having an aperture on the specimen with the aperture over the area of interest (*A5*), but this is simply the masking technique. This section is devoted to more versatile arrangements in which provision is made to translate the specimen to place the area of interest under a fixed aperture.

The aperture can be mounted on the specimen compartment or, preferably, the specimen drawer, which can be removed to permit convenient

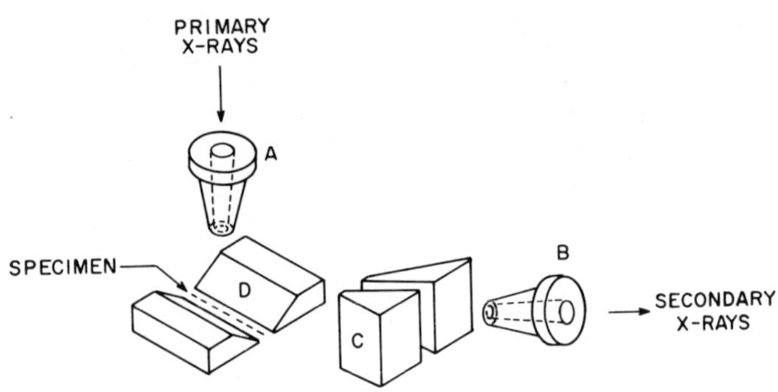

Fig. 19.3. Forms and positions of selected-area apertures. *A*. Primary pinhole. *B*. Secondary pinhole. *C*. Secondary slit. *D*. Primary-secondary slit or edge.

alignment of the selected area with the aperture. The aperture can be mounted in the primary (*A8*, *L24*, *L25*, *T6*) or secondary (*H17*, *M17*) beam, or both (*B16*, *P6*, *R7*). With a primary aperture, only the selected area receives primary radiation. With the secondary aperture, only secondary x-rays originating in the selected area enter the optical path. The aperture can have the form of a pinhole, slit, or tunnel. Figure 19.3 shows some typical apertures.

19.2.2.2. *Pinholes*

Pinhole apertures have been made from drilled metal disks (*A5*); heavy-wall glass capillary, thermometer, and barometer tubing (*H17*); hypodermic needles (*H17*); and Leroy lettering pen points (*S24*, *W9*, *W10*). Square-bore "pinholes" (*C31*) have been used for evaluation of homogeneity to permit complete point-by-point coverage of the specimen surface with minimum overlap from adjacent incremental areas. The optimal pinhole aperture has a conically tapered bore, as shown in *A* and *B* of Figure 19.3 (*15N*, *A5*, *A8*, *L24*, *L25*). In the primary beam, the tapering allows the selected area on the specimen to receive primary radiation from the entire

Fig. 19.4. Specimen drawer fitted with secondary-beam pinhole. [Courtesy of General Electric Co., X-Ray Dept.]

x-ray tube focal spot. It is possible to irradiate the area with a demagnified pinhole image of the x-ray tube focal spot, giving optimal use of primary energy. In the secondary beam, the tapering allows the specimen to emit analyte-line radiation into the spectrometer through a relatively large solid angle.

The end of a secondary pinhole toward the specimen should be inclined as shown in *B* in Figure 19.3 to shield the pinhole itself from the primary beam (*H17*). Otherwise, the characteristic radiation of the aperture will be excited, and some of it will pass into the spectrometer. For maximum resolution, a secondary pinhole must approach the specimen plane as closely as possible without blocking the primary beam; a primary aperture must approach as closely as possible without blocking the secondary beam.

Figure 19.4 shows a secondary pinhole accesory available commercially for the General Electric x-ray spectrometer (*H17*). A brass block on the mask plate has a bore to receive pinholes having diameters 1, 0.1, and

Fig. 19.5. Philips Macro-Probe Attachment in place on the Philips Universal Vacuum spectrometer. Specimen-translation micrometers, specimen illuminator, and viewing microscope are visible on the principal unit on the spectrometer specimen platform. The crystal-bending mechanism and curved crystal are mounted in the spectrometer chamber. [Courtesy of Philips Electronic Instruments, Inc.]

0.05 mm. Under the mask is a specimen stage that can be translated in two mutually perpendicular directions by the micrometer drives shown. A simple slit to replace the detector collimator and a specimen motor-drive are also supplied. The motor-drive couples to a micrometer drive by means of a flexible shaft. A collimated specimen drawer is also available for the GE instrument (*C31*, *G5*). A somewhat more sophisticated secondary pinhole accessory is available commercially for the Picker spectrometer. The device provides for specimen alignment in place in its compartment while it is observed through a 32× microscope. These accessories are used with flat or semifocusing curved crystals. A full-focusing primary pinhole accessory is available commercially for the Philips Universal Vacuum spectrometer. Features of the device include 0.5- and 0.05-mm pinholes, 70× microscope for observation and alignment of the specimen in its compartment, and continuous variation of curved-crystal radius with 2θ. Lithium fluoride and mica crystals are supplied. The accessory is shown in place on the spectrometer in Figure 19.5.

19.2.2.3. *Slits*

Slits are preferable to pinholes for some specimens such as sections of long linear inclusions; diffusion couples; metal-to-metal and metal-to-ceramic interfaces; plated, coated, or corroded surfaces; and other layered structures where composition varies in the direction across, but not along, the layers (*B15*, *B16*, *B18*). Figure 19.6 shows a section of a typical ceramic-to-metal system. The ceramic has been metallized, plated, and brazed to the metal. The metallizer may diffuse into the ceramic, the braze into the metal, and the metallizer, plating, and braze into each other. Inclusions or segregations occur in the ceramic and metal. Selected-area techniques can be applied to identify the metallizer, plating, braze, and inclusions, and to evaluate the extent of diffusion. The figure shows the projections of pinhole and slit apertures on the specimen.

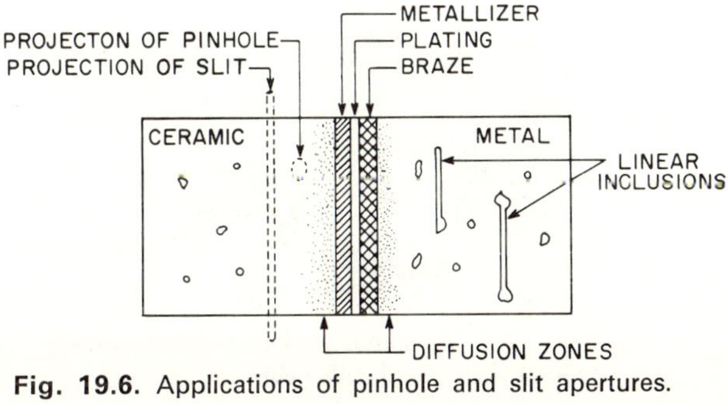

Fig. 19.6. Applications of pinhole and slit apertures.

The roughly circular inclusions are identified by use of the pinhole. However, if a pinhole is used to identify the linear inclusions or interfacial layers, or to evaluate the extent of diffusion, only that incremental specimen layer under the pinhole contributes to the measured analyte-line intensity. A slit is advantageous for these applications because it passes secondary radiation from the entire length of the incremental layer. Increased intensity, and therefore sensitivity and statistical precision, are realized without loss of resolution by replacing a pinhole of given diameter with a slit of the same width. Increased resolution is realized without loss of intensity by replacing the pinhole with a slit of smaller width. Moreover, a more representative sampling of a layer is obtained with a slit than with a pinhole. For example, the writer, using the binary-ratio technique (Section 14.3.2.3) determined the lead/tin ratio for a wide variety of very small solder-coated parts and solder-foil preforms used in making transistors. It would have been feasible to use a pinhole and make measurements on individual parts. However, higher intensities, improved precision, and more-representative sampling were realized by using a slit and making measurements on a row of parts mounted on two-sided masking tape.

Figure 19.7 shows a slit accessory consisting of the drawer shown in

Fig. 19.7. Specimen drawer in Figure 19.4 fitted with secondary-beam slit aperture.

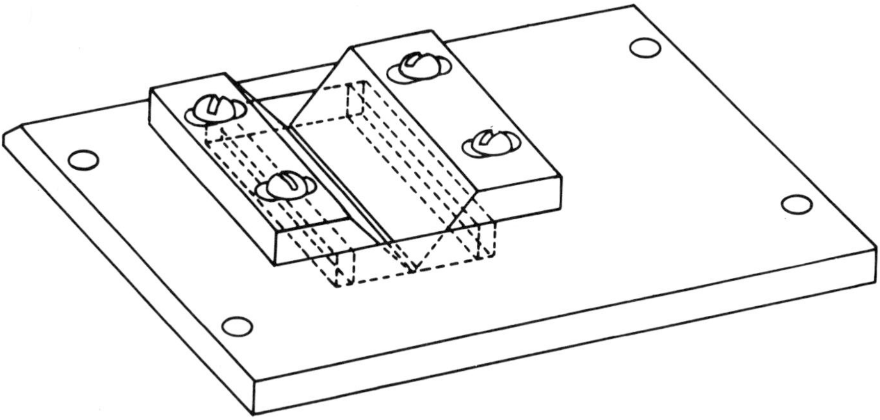

Fig. 19.8. Specimen mask plate fitted with slit aperture lying in both primary and secondary beams.

Figure 19.4 in which the pinhole mounting block has been replaced with a secondary-beam slit (*B15*). Figure 19.8 shows the mask plate of the same drawer fitted with a slit lying in both the primary and secondary beams (*B16*).

A slit of the type shown in *C* of Figure 19.3 and in Figure 19.7 has many limitations that do not apply to pinholes (*B15*):

1. Its applicability is limited to the types of specimen listed above.

2. It is extremely wasteful of secondary intensity. The jaws may not be tapered in a way analogous to the tapering of a pinhole because the jaws cannot approach the specimen surface without blocking the primary beam. Consequently, tapered jaws would accept secondary radiation from a relatively wide specimen area, and resolution would be lost.

3. Effective slit width (see below) and resolution vary from top to bottom of the slit because distance from slit to specimen plane varies.

4. Analyte-line intensity measured from a small specimen decreases as the specimen is moved from opposite the top to opposite the bottom of the slit, because distance between x-ray tube focal spot and sample plane increases.

5. It follows from the preceding two considerations that if several specimens, or samples and standards, are to be compared quantitatively, they must have the same height (in the direction parallel to the projection of the slit) and the same position with respect to the slit.

6. The lateral position of the slit is very critical if full advantage is to be taken of the intense central "hot spot" (*V* in Figure 19.2).

7. Finally, the linear specimen area must be precisely aligned with the

slit. Nevertheless, a simple slit accessory used with a semifocusing curved crystal on a commercial spectrometer permits much sophisticated work to be done that would require an elaborate full-focusing curved-crystal accessory with a pinhole.

The two-jaw primary-secondary slit (*B16*) shown in *D* of Figure 19.3 and in Figure 19.8 gives analyte-line intensity about an order higher than the slit discussed above. The beveled jaws permit the selected area to receive full primary irradiation and to emit secondary radiation into the optical system through a wide solid angle. Resolution is excellent because the slit jaws lie nearly in contact with the specimen surface. However, this arrangement constitutes little more than a mask. Background is high because of primary scatter and secondary emission from the jaws. With narrow slit widths, one of the jaws must be withdrawn to permit observation of the specimen for alignment; this is no problem with an individual specimen. However, when a number of specimens are to be compared quantitatively, the slit must be opened and reset for each, and the analytical results are impaired by slight nonreproducibility of slit width.

Slit width is set with shims. Mylar film and sheet are available in a range of thicknesses from 0.003 mm (0.00012 inch) up and make excellent shims that do not wear the slit jaws as much as metal shims. When setting slits to very narrow width—0.125 mm (0.005 inch) or less—the Mylar must be free of wrinkles and dust.

Rizzo (*R7*) studied diffusion of electroplated metal into its substrate with a similar slit having only one jaw. The sectioned specimen was mounted with the plate–substrate interface parallel to the edge, than translated incrementally as analyte-line intensity was scaled or recorded.

Proctor (*P14*) has shown that a single pinhole in a thin mask acts as a slit when mounted perpendicular to the central ray of the secondary beam between specimen and crystal; this is because the crystal diffracts rays from a row of points on the specimen surface, the row being coplanar with the crystal's axis of rotation. This arrangement is extremely simple and convenient, but gives lower intensity than a true slit because the pinhole passes only one ray from each point on the specimen to the crystal, whereas the slit passes a V-shaped shaft of rays from each point. Two such pinholes in register a short distance apart act as a pinhole.

19.2.2.4. *Resolution*

The resolution attainable with a pinhole or slit increases as the pinhole diameter or slit width decreases. The *effective* aperture width (pinhole diameter or slit width) is measured by translating a very sharp two-element interface past the aperture while recording intensity of a spectral line of either element. The effective width is defined as the distance between the two points on the specimen where response is just distinguishable above background and where it attains its maximum value. A test specimen is made by bolting highly polished flat pieces of copper and nickel together, sectioning the couple, and polishing the section. Alternatively, a copper or nickel wire can be imbedded in a lead disk and polished flat. Thermally bonded couples are unsuitable because diffusion may blur the sharpness of the interface. The effective width is always greater than the true width, but the closer the aperture approaches the specimen, the more nearly equal are the two widths.

The projection of a pinhole on an inclined specimen plane (Figure 19.3) is an ellipse having its major axis lying in the direction of the inclination. Thus, concentration profiles should be scanned in the other direction, where effective pinhole diameter is narrower and resolution better.

Slits of the type shown in *C* of Figure 19.3 and in Figure 19.7 cannot approach the specimen plane as close as a pinhole without blocking the primary beam. Therefore, effective slit width is always substantially greater than the true width. Fortunately, this condition can be compensated by use of narrower slit widths. The distance from the slit jaws to the specimen plane increases from bottom to top of the slit, and so does effective slit width.

19.2.3. Dispersion

An incremental area on a specimen emits secondary x-radiation in all directions. A flat crystal diffracts only that narrow V-shaped shaft of these x-rays that is incident on the crystal at the Bragg angle θ. With an extensive specimen, the Soller collimator permits diffraction of such a V-shaped shaft of x-rays from each incremental area, and high intensity reaches the detector. But when the specimen area is very small, only one such shaft is effective, and all other secondary radiation is wasted. However, with a curved crystal of appropriate radius, all divergent secondary x-rays intercepted by the crystal strike the tangent to its surface at the point of intersection at the Bragg angle and are diffracted. The diffracted rays converge to a "focus" at the focusing circle, where the detector slit is placed.

Selected-area analysis is feasible with flat crystals. The collimators are removed to avoid possible blocking of the small effective secondary beam by a Soller foil. A simple slit may be placed before the detector window to reduce background and improve resolution. However, flat crystals are satisfactory only for pinholes 0.05–1 mm in diameter, and the smaller diameters are useful only for relatively high concentrations of elements having intense lines. In general, sensitivity falls off sharply when *L* lines or crystals having "reflectivity" substantially lower than lithium fluoride must be used. For linear specimens or areas, slits may permit measurements to be made where pinholes would be useless. Slit widths as narrow as 0.003 mm can be used in favorable cases.

Measured intensity from pinholes and slits can be increased as much as an order by use of semifocusing curved-crystal optics, that is, by simply replacing the flat crystal with a fixed-radius curved crystal. The collimators are removed, but for a different reason—because the parallel foils would intercept the divergent rays. Again, a slit before the detector window may be beneficial. One crystal having radius optimal for the middle of the 2θ range may suffice if some line broadening at high and low 2θ can be tolerated. For high resolution over the full 2θ range, two or three crystals having different radii are required.

Selected-area analysis on a commercial spectrometer is best accomplished by fitting it with a full-focusing curved-crystal accessory. Such accessories have been designed for General Electric (*15N*, *L24*, *L25*), Philips (*D21*, *L13*, *M17*, *P6*) and Siemens (*J12*, *J13*) instruments, and an accessory for the Philips Universal vacuum x-ray spectrometer is available commercially (*H26*). For selected-area analysis, a full-focusing x-ray optical system must "view" the same point or line on the specimen at all values of 2θ, that is, either (1) the instrument must have a constant takeoff angle, or (2) the specimen point or line must lie on the focusing circle. The first condition is met by instruments having the ARL design or provision for continuous variation of crystal radius with 2θ; these instruments are shown in Figures 5.20 and 5.18, respectively. The second condition is met by instruments having the NRL design shown in Figure 5.19 with the x-ray tube and specimen in the alternate positions indicated.

Figure 19.9 (*15N*) shows an excellent accessory having the modified NRL design for use on the basic General Electric spectrogoniometer. A 75-kV x-ray tube irradiates the selected area on the specimen through a primary-beam pinhole having a tapered bore 0.02–1.6 mm in minimum diameter. Secondary radiation emerges from the specimen chamber through a rectangular exit port that intercepts radiation that would miss the crystal, but permits maximum coverage of the crystal. The crystals (LiF and EDDT) are curved to a 14-inch radius, and, for optimal resolution, should also be ground to a 7-inch radius. The diffracted rays converge on an adjustable

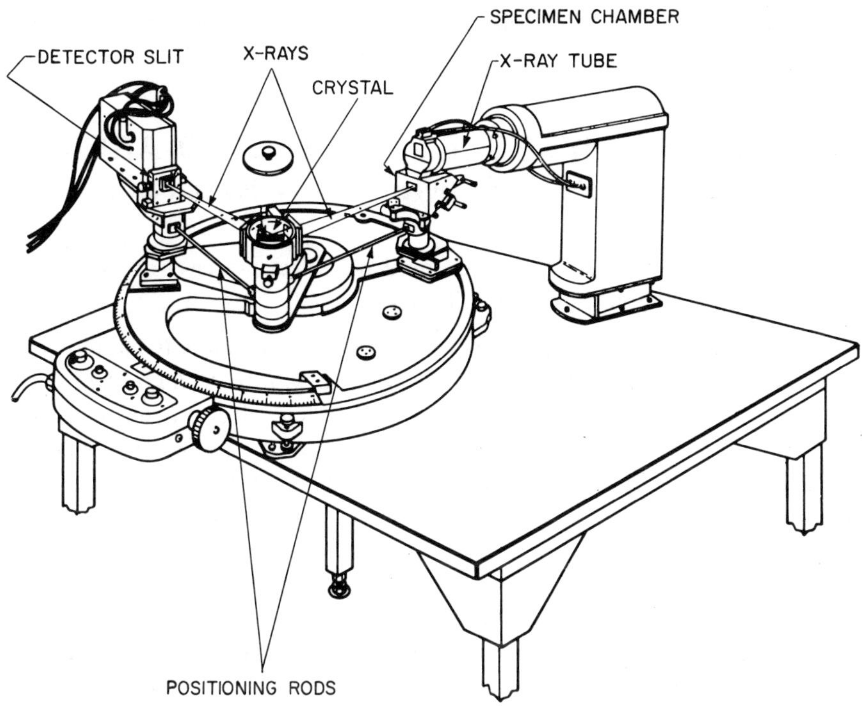

Fig. 19.9. Full-focusing curved-crystal accessory for use with General Electric x-ray spectrogoniometer. [T. C. Loomis and S. M. Vincent, in "Handbook of X-Rays," E. F. Kaelble, ed.; McGraw-Hill Book Co., p. **37**-12 (1967); courtesy of the authors and publisher.]

detector slit behind which is the conventional detector. As 2θ is varied, the entire specimen chamber rotates about a vertical axis concentric with the primary pinhole, and a system of rods and couplings maintains the selected area, rectangular exit port, crystal, and detector in the proper geometric relationship. The specimen-stage assembly, including the translation mechanism and aperture, can be removed for loading the specimen and for aligning it under a microscope. Bellowed plastic tunnels can be placed between the specimen and crystal chamber and the crystal chamber and detector to permit use of helium path.

In general, curved-crystal transmission instruments are avoided for selected-area work because of the loss of intensity due to absorption in the crystal.

Incidentally, when the specimen is homogeneous and selected-area technique is used simply to take advantage of curved-crystal optics to realize increased sensitivity, the instrument need not necessarily meet the two conditions cited above.

19.3. INSTRUMENTS FOR SELECTED-AREA X-RAY SPECTROMETRIC ANALYSIS

The instruments that have been used for selected-area x-ray spectrometric analysis may be classified as follows:

1. Electron-probe x-ray primary-emission spectrometers (Chapter 21).

2. Microfocus x-ray tubes (*C43*, *C44*, *D8*, *L20*, *L21*, *L22*, *O1*, *R15*, *Z1*, *Z2*). An electron-optical system forms a fine-focus electron image of its filament on a thin metal foil serving as both target and x-ray window. Thin specimens are placed just outside this window, and the emitted characteristic emission is measured in a curved-crystal or nondispersive spectrometer.

3. Laboratory-built, full-focusing, curved-crystal spectrometers (*A5*, *A8*, *A11*, *B2*, *B6*, *B35*, *N1*, *S18*, *T5*, *Y2*).

4. Standard commercial x-ray spectrometers fitted with apertures and full-focusing curved-crystal accessories (*15N*, *D21*, *J12*, *J13*, *L24*, *L25*, *M17*, *P6*). The instrument shown in Figure 19.9 (*15N*) is of this type.

5. Standard commercial x-ray spectrometers fitted with apertures and semifocusing curved-crystal accessories (*B15*, *B16*, *B18*).

6. Standard commercial x-ray spectrometers fitted with apertures and flat-crystal accessories (*B21*, *C31*, *H17*, *L36*, *M17*, *R7*, *S24*, *T6*, *W9*, *W10*, *Z11-14*).

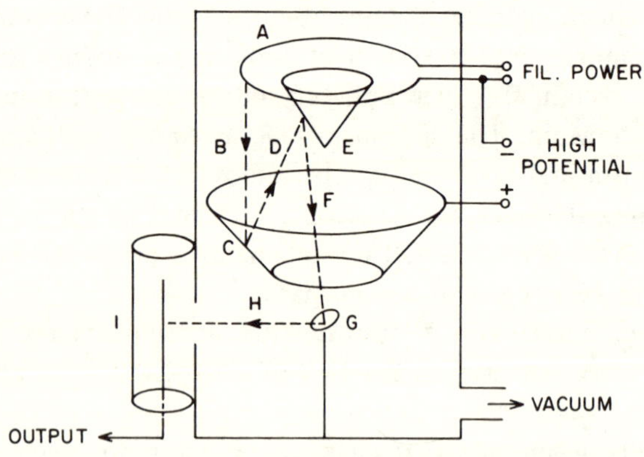

Fig. 19.10. Nondispersive high-intensity x-ray spectrometer.

7. Nondispersive spectrometers having x-ray sources (*D8*, *H10*, *H11*). An example of this category is the instrument shown in Figure 19.10 (*H10*, *H11*). The circular filament *A* emits electrons *B*, which bombard the inside surface of the bottomless saucer-shaped primary target *C* and excite intense primary x-radiation *D*. Because of the shape of *C*, the primary x-rays are directed to the interchangeable conic secondary target *E*, where they excite its characteristic x-rays *F*. Again because of the shape of the emitter, these x-rays converge on the specimen *G*. The secondary target is selected to have a strong spectral line on the short-wavelength side of the analyte absorption edge. In this way, efficient selective excitation of the analyte spectrum *H* is realized. The x-rays from the specimen are detected by a proportional counter *I*, the output of which is subjected to pulse-height analysis. The instrument was designed primarily for biological applications, where the interest is principally in light elements, up to atomic number ~30. Consequently, the primary target is designed for currents up to 500 mA at relatively low potentials, up to ~15 kV.

8. Nondispersive spectrometers having radioisotope sources.

9. Instruments based on special principles. Two examples follow (*D8*). In the method of *selective excitation*, a single wavelength on the short side of the analyte absorption edge is selected from the primary beam, and diffracted and focused by a doubly curved aluminum-crystal monochromator on a selected area ~1 mm² on the specimen. The analyte-line emission is measured in a curved-crystal spectrometer in the usual way. Unless a target line can be used for the selected wavelength, the technique is very wasteful of primary intensity, and gives low intensity. In the method of *differential excitation*, the primary beam is dispersed into a spectrum by a cylindrically curved aluminum crystal. A mask plate having a small aperture can be placed at any position—and therefore any wavelength—around the focusing circle. The specimen is placed behind the mask with the aperture over the area of interest. The detector window is situated to receive x-rays emitted from the selected area. The mask-specimen-detector assembly is placed to allow the selected area to be irradiated in turn by two wavelengths λ_S and λ_L on, respectively, the short- and long-wavelength sides of the analyte absorption edge. The x-rays emitted and/or scattered by the specimen pass directly into the detector. The detector receives analyte-line emission at λ_S, but not at λ_L. The difference in the intensities at the two wavelengths gives analyte-line intensity—*without pulse-height analysis*. The technique has been applied to areas of ~0.5 mm². The disadvantage cited above for the selective excitation technique applies here as well, but is partially compensated by the nondispersive arrangement.

19.4. SPECIMEN TECHNIQUES

19.4.1. Preparation

Very small specimens require no preparation and are usually supported on Scotch tape, thin Mylar, or fine fibers. If a curved crystal is used, an aperture, although preferable, may not be required, because the specimen itself constitutes a point or line source. Linear (filamentary) specimens must be oriented in the plane defined by the directions of the central rays of the primary and secondary beam cones, even when a flat crystal is used.

Extensive specimens are supported in the specimen compartment, and an aperture must be used. The projection of the aperture on the specimen defines the selected area and constitutes the source of measured secondary radiation. Such specimens usually require some kind of preparation, usually the metallographic techniques of potting, sectioning, and polishing. Oblique sectioning increases layer width in layered specimens; however, if the cut is too oblique, x-rays from lower layers may penetrate the thin edges of overlying layers and give erroneous layer widths and diffusion depths. A dull matte finish (~600 grit) is preferable to a high metallographic polish because, with the latter, it is difficult to see the light beam when aligning the specimen with the aperture (see below). Specimens should be etched only to eliminate surface smear or contamination, and preferential etching must be avoided. If layered specimens are to be measured with a slit, all layers must have the same length, or else be masked, and the layers must be oriented in the same way as filamentary specimens (see above).

19.4.2. Alignment of Selected Area and Aperture

In principle, it is preferable to align the specimen with the aperture in place in the spectrometer. However, the alignment is much more convenient and nearly as accurate if the aperture is mounted on a specimen drawer that can be removed and placed on a microscope stage. Final adjustment can then be made with the drawer in the instrument. If the major constituent of the selected area is known, the goniometer can be set at its line and the specimen micrometers adjusted slightly for maximum response. If the composition is unknown, a 2θ scan can be made and alignment refined later.

Five alignment techniques are commonly used:

1. Sighting by eye. If the aperture is large and the specimen area can

be illuminated, the specimen can be aligned while being observed through the aperture with the unaided eye.

2. Light-beam technique. A light beam is directed through the aperture onto the specimen as the latter is viewed through a microscope. The specimen is translated to bring the selected area into the light beam. This technique is most convenient with drawer-type accessories of the types shown in Figures 19.4 and 19.7, which can be removed and placed on a microscope stage. However, a microscope and light source can also be provided on accessories in which the aperture is mounted on the specimen compartment. With very small pinholes and slits, the light beam is extremely difficult to see.

3. Fine, stiff wires and thin, stiff shims may be passed through, respectively, pinholes and slits and advanced to contact with the specimen, which is then translated to bring the selected area into alignment with the wire or shim.

4. Fluorescent technique. A small grain of fluorescent material is placed directly over the selected area. With the specimen in place in the x-ray spectrometer, x-rays are turned on and the specimen translated to give maximum fluorescence. The particle is then removed. This technique is applicable only to primary apertures.

5. A microscope having cross-hairs in the eyepiece can be adjusted so that the cross-hairs mark the projection of a pinhole or so that one of the hairs lies parallel to the projection of a slit. Thereafter, it is simply necessary to bring the selected area into alignment with the cross-hairs.

19.5. ANALYTICAL TECHNIQUES

The small specimen or area can be placed in the projection of the pinhole or slit and a qualitative analysis made by recording a 2θ scan on the ratemeter-recorder, or a quantitative analysis made by scaling the analyte-line intensities. The 2θ scan charts are similar to those shown in Figure 10.1. The procedure is repeated at other points of interest.

The concentration distribution profile of a selected analyte along a line across the specimen can be mapped by setting the spectrometer at the analyte line, moving the specimen manually in small increments, and scaling the intensity at each point. For layered specimens, a slit is preferable; otherwise, a pinhole must be used. The data are plotted manually. The same distribution profile can be mapped by translating the specimen with a motor-drive while recording analyte-line intensity on the recorder. The two techniques are compared below. Successive scans can be made for different analytes, and, with pinholes, scans can be made along different lines for the same analyte.

The motor-drive should be slow enough to allow the recorder to follow small irregularities in concentration. In the accessory shown in Figures 19.4 and 19.7, two turns of the specimen-translation micrometers displace the specimen 1 mm. A motor speed of 20 revolutions per hour moves the specimen at a rate 1 cm/h. The writer uses a motor having speeds 1, 2, 4, 12, 30, and 60 revolutions per hour.

The writer has evaluated the uniformity of molybdenum metallizing and nickel plating on the edges of small ceramic disks as follows. The disks were mounted on suitable spindles in the *side* rotor of the rotating specimen drawer shown in Figure 16.4 so that the plane of the disks coincided with the plane defined by the central rays of the primary and secondary x-ray beams. A secondary pinhole was mounted on the drawer in line with the edge of the disk. The intensity of the Mo $K\alpha$ or Ni $K\alpha$ line was recorded as the disk slowly rotated. In a similar manner, Lublin (*L28*) used a primary slit aperture to determine uniformity of cladding thickness around the circumference of small rotating rods.

Campbell *et al.* (*C4*) describe an accessory for photography of a color mosaic picture of the analyte-line intensity distribution over the specimen surface. By means of X and Y stepping motors, the specimen stage is translated in mutually perpendicular directions in such a way that, in effect, a primary-beam aperture scans a raster on the specimen surface. The X motor advances the specimen for a predetermined number of 0.3- or 1-mm steps; then the Y motor moves one step. The X motor reverses and scans the second line in the opposite direction; then the Y motor moves another step. The X motor scans the third line in the forward direction, etc. until the raster is scanned. Simultaneously, a Polaroid color-film holder is driven by two other X and Y stepping motors in synchronism with the specimen drive motors.

At each point on the specimen scan, the corresponding point on the color film is illuminated by a beam of light passing through a mechanical shutter and a 10-sector color disk, each sector of which is a different color. The disk is rotated by a repeater motor driven by the ratemeter output. Thus, at each point, the position of the disk—and the color of the light beam—are determined by the ratemeter, and thereby the analyte-line intensity. In this way, a color mosaic photograph of the analyte distribution is built up, the colors representing a 10-step gradation of the intensity range. By alteration of the ratio of the specimen and film scanning linkages, magnification or reduction of the photograph may be realized. The authors include a striking color photograph in their paper.

Secondary-emission selected-area techniques are best suited to qualitative and semiquantitative analysis, and to mapping of concentration profiles. Standards are usually not required for such work. If the identity of the selected area is suspected or at least narrowed to a few possibilities, it may be identified by comparison with specimens of the likely substances. The presence or absence of some distinguishing element or combination of elements, or the concentration level of some key element, may also permit identification. Sometimes, the intensities of several elements in the sample can be measured, ratioed with the highest intensity, and compared with similar ratios from specimens of likely substances.

Quantitative analysis is also feasible, and high accuracy is attainable when an aperture and curved crystal are used to realize higher intensity from a trace element in a homogeneous sample. Also, quantitative analysis may be simplified for very small isolated quantities and thin films, for which matrix effects may be disregarded. However, when selected-area techniques are applied to quantitative analysis of heterogeneous samples, difficulties arise that are not encountered in conventional x-ray spectrometric analysis. The selected area under the aperture is not likely to be uniform in composition. The aperture may cover matrix or layers adjacent to the spot or layer of interest. A diffusion zone has a range of compositions, and a layer may contain grains of different phases. Intensity may be low, especially for minor constituents, with consequent low statistical precision. Long counting times may be risky because prolonged exposure to the primary beam may cause the specimen to heat and shift slightly. Standards must be homogeneous on a microscopic scale. Nevertheless, accurate quantitative analyses can often the made when suitable calibration standards are available.

19.6. PERFORMANCE

The sensitivity attainable by selected-area techniques varies widely with the particular system. However, Table 19.1 gives some idea of the performance for several typical arrangements. The data in part *C* of the table for a full-focusing instrument probably represent maximum attainable performance from an accessory for a commercial spectrometer. This accessory is capable of producing measurable intensities from areas approaching those involved in electron-probe microanalysis.

The effectiveness of a selected-area analysis in excluding surrounding matrix is shown in Figure 19.11 (*L25*) for a nickel wire 0.04 mm (0.0015

TABLE 19.1

Performance of X-Ray Secondary-Emission Spectrometers for Small Selected Areas

Z-line[a]	Target, kV, mA	Path	Detector	Aperture	Crystal[b]	Intensity, counts/s	
						Line	Background
A. Cylindrical secondary pinhole aperture, flat crystal [Heinrich (*H17*)]							
14-Si $K\alpha$	W, 50, 50	He	Flow	0.8-mm	EDDT	18	0
16-S $K\alpha$	W, 50, 50	He	Flow	0.8-mm	NaCl	360	2
22-Ti $K\alpha$	W, 50, 50	He	Flow	0.8-mm	LiF	3,000	10
22-Ti $K\alpha$	W, 50, 50	He	Flow	0.1-mm	LiF	40	0
26-Fe $K\alpha$	W, 50, 50	Air	Scint.	0.8-mm	LiF	6,250	7
26-Fe $K\alpha$	W, 50, 50	Air	Scint.	0.1-mm	LiF	125	0
29-Cu $K\alpha$	W, 50, 50	Air	Scint.	0.8-mm	LiF	11,000	12
29-Cu $K\alpha$	W, 50, 50	Air	Scint.	0.1-mm	LiF	160	0
40-Zr $K\alpha$	W, 50, 50	Air	Scint.	0.8-mm	LiF	4,700	12
40-Zr $K\alpha$	W, 50, 50	Air	Scint.	0.1-mm	LiF	80	0
47-Ag $K\alpha$	W, 50, 50	Air	Scint.	0.8-mm	LiF	2,300	15
47-Ag $K\alpha$	W, 50, 50	Air	Scint.	0.1-mm	LiF	43	0
50-Sn $K\alpha$	W, 50, 50	Air	Scint.	0.8-mm	LiF	1,500	12
50-Sn $K\alpha$	W, 50, 50	Air	Scint.	0.1-mm	LiF	25	0

TABLE 19.1 *(continued)*

Z-line[a]	Target, kV, mA	Path	Detector	Aperture	Crystal[b]	Intensity, counts/s	
						Line	Background
73-Ta $L\alpha_1$	W, 50, 50	He	Flow	0.8-mm	LiF	1,350	3
73-Ta $L\alpha_1$	W, 50, 50	He	Flow	0.1-mm	LiF	21	0
82-Pb $L\alpha_1$	W, 50, 50	Air	Scint.	0.8-mm	LiF	1,150	3
82-Pb $L\alpha_1$	W, 50, 50	Air	Scint.	0.1-mm	LiF	15	1
B. Tapered-bore secondary pinhole aperture, semifocusing curved crystal [Loomis (*L24*)]							
12-Mg $K\alpha$	Cr, 50, 40	He	Flow	0.030-inch	ADP, 8 inch	253	—
13-Al $K\alpha$	Cr, 50, 40	He	Flow	0.030-inch	EDDT, 8 inch	1,418	—
14-Si $K\alpha$	Cr, 50, 40	He	Flow	0.030-inch	EDDT, 8 inch	2,386	—
16-S $K\alpha$[a]	Cr, 50, 40	He	Flow	0.030-inch	NaCl, 7 inch	9,785	—
17-Cl $K\alpha$[a]	Cr, 50, 40	He	Flow	0.030-inch	NaCl, 7 inch	14,098	—
19-K $K\alpha$[a]	Cr, 50, 40	He	Flow	0.030-inch	LiF, 8 inch	7,748	—
20-Ca $K\alpha$[a]	Cr, 50, 40	He	Flow	0.030-inch	LiF, 8 inch	29,290	—
22-Ti $K\alpha$	Cr, 50, 40	He	Flow	0.030-inch	LiF, 8 inch	124,500	—
24-Cr $K\alpha$	Cr, 50, 40	He	Flow	0.030-inch	LiF, 8 inch	6,827	—
28-Ni $K\alpha$	Cr, 50, 40	He	Flow	0.030-inch	LiF, 14 inch	14,610	—
30-Zn $K\alpha$	Cr, 50, 40	Air	Scint.	0.030-inch	LiF, 14 inch	14,390	—
39-Y $K\alpha$	Cr, 50, 40	Air	Scint.	0.030-inch	LiF, 24 inch	32,450	—
42-Mo $K\alpha$	Cr, 50, 40	Air	Scint.	0.030-inch	LiF, 24 inch	13,420	—

TABLE 19.1 *(continued)*

Z-line[a]	Target, kV, mA	Path	Detector	Aperture	Crystal[b]	Intensity, counts/s	
						Line	Background
C. Tapered-bore primary pinhole aperture, full-focusing curved crystal [Loomis (*L24*)]							
12-Mg $K\alpha$	Cr, 50, 40	Vacuum	Flow	0.030-inch	ADP, 8 inch/4 inch	320	—
13-Al $K\alpha$	Cr, 50, 40	Vacuum	Flow	0.030-inch	EDDT, 8 inch/4 inch	1,560	—
14-Si $K\alpha$	Cr, 50, 40	Vacuum	Flow	0.030-inch	EDDT, 8 inch/4 inch	2,040	—
15-P $K\alpha$[a]	Cr, 50, 40	Vacuum	Flow	0.030-inch	EDDT, 8 inch/4 inch	640	—
16-S $K\alpha$[a]	Cr, 50, 40	Vacuum	Flow	0.030-inch	NaCl, 8 inch/4 inch	9,600	—
17-Cl $K\alpha$[a]	Cr, 50, 40	Vacuum	Flow	0.030-inch	NaCl, 8 inch/4 inch	10,960	—
19-K $K\alpha$[a]	Cr, 50, 40	Vacuum	Flow	0.030-inch	LiF, 8 inch/4 inch	28,600	—
20-Ca $K\alpha$[a]	Cr, 50, 40	Vacuum	Flow	0.030-inch	LiF, 8 inch/4 inch	154,000[c]	—
22-Ti $K\alpha$	Cr, 50, 40	Vacuum	Flow	0.030-inch	LiF, 8 inch/4 inch	303,600[c]	—
24-Cr $K\alpha$	W, 50, 40	Vacuum	Flow	0.030-inch	LiF, 8 inch/4 inch	152,000[c]	—
26-Fe $K\alpha$	W, 50, 40	Vacuum	Flow	0.030-inch	LiF, 8 inch/4 inch	216,400[c]	—
28-Ni $K\alpha$	W, 50, 40	Vacuum	Flow	0.030-inch	LiF, 8 inch/4 inch	208,800[c]	—
47-Ag $L\alpha_1$	Cr, 50, 40	Vacuum	Flow	0.030-inch	LiF, 8 inch/4 inch	4,640	—
D. Secondary slit aperture, semifocusing curved crystal [Bertin (*B15*)]							
24-Cr $K\alpha$	W, 50, 45	Air	Flow	0.00012-inch	LiF, 14 inch	71	0
28-Ni $K\alpha$	W, 50, 45	Air	Kr prop.	0.00012-inch	LiF, 14 inch	186	1
30-Zn $K\alpha$	W, 50, 45	Air	Kr prop.	0.00012-inch	LiF, 14 inch	223	1

TABLE 19.1 *(continued)*

Z-line[a]	Target, kV, mA	Path	Detector	Aperture	Crystal[b]	Intensity, counts/s	
						Line	Background
32-Ge $K\alpha$	W, 50, 45	Air	Kr prop.	0.00012-inch	LiF, 14 inch	213	1
73-Ta $L\alpha_1$	W, 50, 45	Air	Kr prop.	0.00012-inch	LiF, 14 inch	26	1
78-Pt $L\alpha_1$	W, 50, 45	Air	Kr prop.	0.00012-inch	LiF, 14 inch	33	1
82-Pb $L\alpha_1$	W, 50, 45	Air	Kr prop.	0.00012-inch	LiF, 14 inch	35	1
E. Comparison of secondary slit and pinhole apertures, flat crystal [Bertin (*B18*)]							
29-Cu $K\alpha$	W, 50, 45	Air	Kr prop.	0.1-mm pinhole	LiF	898	11
29-Cu $K\alpha$	W, 50, 45	Air	Kr prop.	0.05-mm pinhole	LiF	183	3
29-Cu $K\alpha$	W, 50, 45	Air	Kr prop.	0.025-mm pinhole	LiF	30	3
29-Cu $K\alpha$	W, 50, 45	Air	Kr prop.	0.025-mm slit	LiF	3,152	40
82-Pb $L\alpha_1$	W, 50, 45	Air	Kr prop.	0.1-mm pinhole	LiF	86	1
82-Pb $L\alpha_1$	W, 50, 45	Air	Kr prop.	0.05-mm pinhole	LiF	19	1
82-Pb $L\alpha_1$	W, 50, 45	Air	Kr prop.	0.025-mm pinhole	LiF	5	1
82-Pb $L\alpha_1$	W, 50, 45	Air	Kr prop.	0.025-mm slit	LiF	406	6

[a] All secondary targets were pure element except the following: P—ADP crystal (27% P). S—Rubber (40% S). Cl—Polyvinyl chloride (35% Cl). K—Fused $K_2S_2O_7$ (30% K); Ca—Calcite (40% Ca).

[b] Numbers are crystal radii; for example, 8 inch indicates that the crystal is curved to 8-inch radius; 8 inch/4 inch indicates that the crystal is curved to 8-inch radius, ground to 4-inch radius.

[c] Loomis (*L24*) reports the data in terms of counts/s per mA; the data here are all reported in terms of 40 mA to allow direct comparison with the other parts of the table.

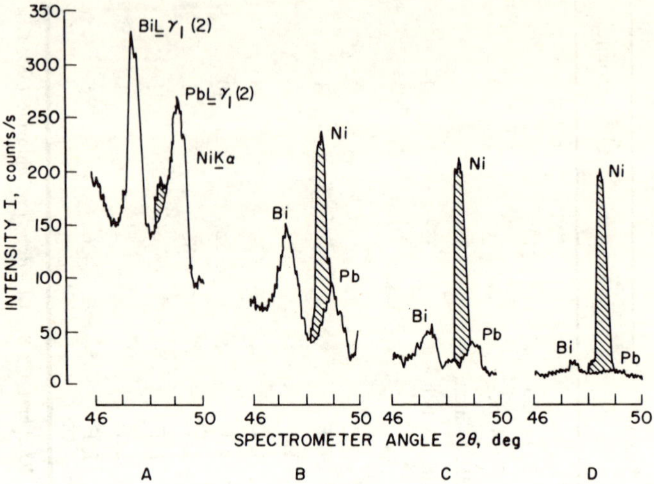

Fig. 19.11. Partial spectra from a 0.04-mm nickel wire imbedded in Wood's metal (Bi–Pb–Cd–Sn). [T. C. Loomis and K. H. Storks, *Bell Laboratories Record* **45**, 2(1967); courtesy of the authors and the Bell Telephone Laboratories.] A. Conventional flat-crystal x-ray spectrometer. B. Full-focusing curved-crystal x-ray probe (Figure 19.9) with 1.6-mm pinhole. C. Same as B with 0.75-mm pinhole. D. Same as B with 0.35-mm pinhole.

inch) in diameter imbedded in Wood's metal (Bi–Pb–Cd–Sn). Ratemeter–recorder 2θ scans are shown in the Ni $K\alpha$ spectral region for conventional flat-crystal operation and for selected-area operation with a curved-crystal and three pinhole apertures having different diameters. Note the progressive decrease in intensity of matrix lines and background without loss of Ni $K\alpha$ intensity.

Typical performance of a secondary slit aperture (Figure 19.7) with a semifocusing curved crystal is shown in Figures 19.12 and 19.13 (*B15*). The slit probe was applied to an investigation of a disk-shaped system involving both metal-to-metal and metal-to-ceramic bonds. Figure 19.12 shows the various concentric layers in the disk as established by metallography and x-ray-probe spectrometry, and a photomicrograph of the region of specific interest in the x-ray study. Particular curiosity was aroused by the layer indicated by the question mark on the photomicrograph; only a braze was expected between the Kovar and the molybdenum. The disk was sectioned on a chord just intersecting the molybdenum center area, rather than on the diameter as shown in Figure 19.12; this section results in somewhat wider layers in the zone of interest and facilitates the x-ray

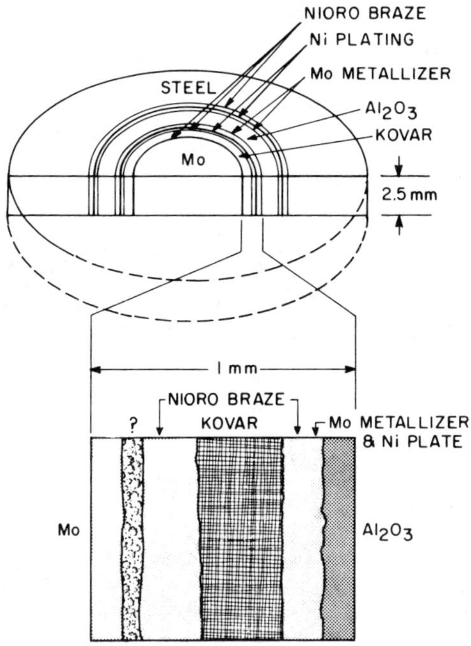

Fig. 19.12. Diagram and photomicrograph of circular metal-ceramic bonded system, which was sectioned as shown and subjected to selected-area analysis over the 1-mm width indicated (Figure 19.13). On the diagram, thicknesses of metallizer, plating, and braze layers are exaggerated.

study. The disk was ~2.5 mm thick, covering substantially less than the full useful height of the slit. The specimen was scanned with a 0.006-mm (0.00025-inch) slit for Fe $K\alpha$, Co $K\alpha$, Ni $K\alpha$, Mo $K\alpha$, and Au $L\alpha_1$ both manually and with the motor drive. The results of both series of scans are shown in Figure 19.13, where the following features are evident: (1) The ceramic was metallized with molybdenum, which diffused a considerable distance into the ceramic; (2) the molybdenum was nickel-plated to improve bonding; (3) the brazing metal was a gold–nickel alloy, probably Nioro (82Au–18Ni); (4) separation of gold and nickel in the brazing alloy has occurred; and (5) the well-defined phase indicated by the question mark on the photomicrograph in Figure 19.12 is seen to contain cobalt and iron, which appear to have been dissolved from the Kovar by the Nioro during brazing and deposited near the molybdenum.

Figure 19.13 permits comparison of the manual and motor-drive

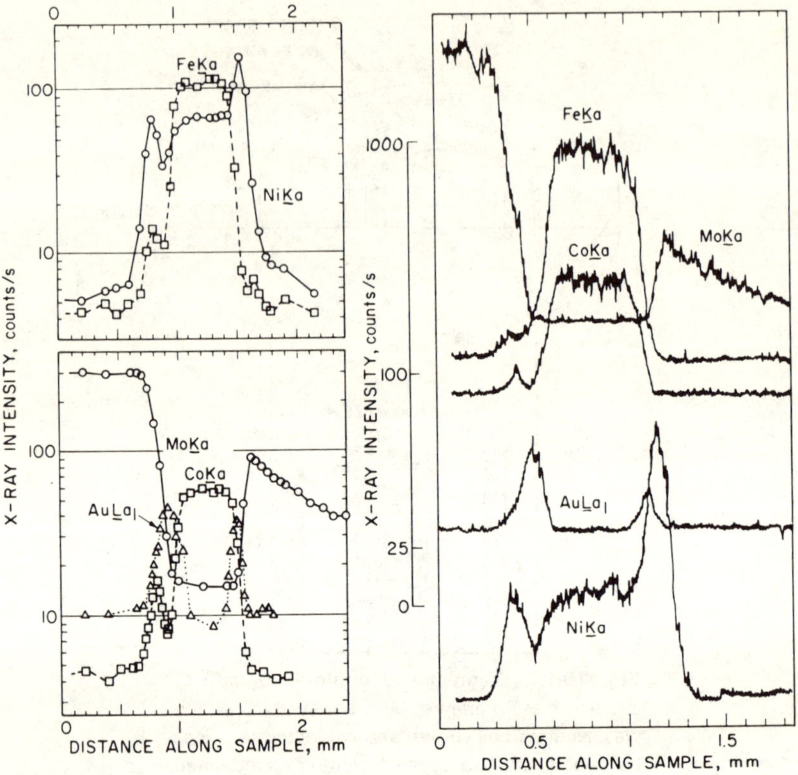

Fig. 19.13. Intensity profiles across the indicated portion of the system shown in Figure 19.12 made with manual incremental sample translation and the scaler–timer (left) and with motor-driven sample translation and the ratemeter–recorder (right). The selected-area accessory was that shown in Figure 19.7 set at 0.006-mm (0.00025-inch) slit width. To avoid confusion, the curves at left were traced from the ratemeter charts and displaced vertically with respect to one another. The low-intensity portion of each curve corresponds to 0 on the intensity scale. The scale is linear up to ~25 counts/s, logarithmic above.

methods. It is evident that the information available from the two methods is substantially the same. However, whereas the manual study required nearly half a work day, each of the motor-drive scans required ~18 min, the entire study ~1.5 h.

More recently, the writer has identified eight distinct layers involving six elements in a distance of ~0.25 mm using a 0.003-mm (0.00012-inch) secondary slit (Figure 19.7) and a specimen translation speed of 0.5 mm/h. Comparable performance is to be expected with pinholes and full-focusing curved-crystal accessories.

Chapter 20

Other Analytical Methods Based on Emission, Absorption, and Scatter of X-Rays; Other Spectrometric Methods Involving X-Rays

The x-ray secondary-emission spectrometer is an extremely versatile instrument and need not necessarily be operated in the mode for which it was designed. Use of the instrument for nondispersive and selected-area analysis has already been described, and this chapter is devoted principally to still other modes of operation. Brief consideration is also given to some other spectrometric methods involving x-rays, but for which special instrumentation is required.

20.1. X-RAY ABSORPTION METHODS

Whereas x-ray emission spectrometry is based on measurement of the characteristic x-rays *emitted* by the specimen, x-ray absorption methods are based on measurement of the x-rays *transmitted* by the specimen. We have already considered emission-absorption methods. In this section, we consider methods involving only absorption (*L14*).

20.1.1. Polychromatic X-Ray Absorptiometry

20.1.1.1. *Principles and Instrumentation*

In polychromatic x-ray absorptiometry (*15M*), the primary x-ray beam is directed through samples and standards of equal thickness and similar composition. Transmitted intensity, which varies inversely with analyte concentration, is measured. Calibration curves or mathematical relationships are established from the standard data and permit analysis of the samples. Sensitivity is proportional to the difference in mass-absorption coefficient of the analyte across its absorption edge and to the difference in mass-absorption coefficients of analyte and matrix on the short side of the edge. From Equation (2.11),

$$I = I_0 \exp - \{[C_A(\mu/\varrho)_A + (1 - C_A)(\mu/\varrho)_M]t\} \qquad (20.1)$$

where I_0 and I are incident and transmitted intensity, respectively; μ/ϱ is mass-absorption coefficient (cm^2/g); C is mass concentration (g/cm^3); and t is specimen thickness (cm); A and M refer to analyte and matrix, respectively; and $(1 - C_A)$ is matrix concentration.

The instrument arrangements are shown in Figures 9.2 and 9.3 for goniometers having horizontal and vertical planes, respectively. The x-ray tube is rotated on its axis so the window is in position *W*2, where it faces directly into the source collimator. The absorption specimen is placed in position 7 or 8. Selected-area apertures *PA* and *SA* are omitted, but a tunnel of appropriate area and shape may be used to confine the x-ray beam to the specimen area. The goniometer is set at 0° 2θ so the x-ray tube window, specimen, collimator, and detector are in line, and the crystal is removed so it won't partially block the x-ray beam.

Polychromatic absorptiometry is most conveniently practiced on a diffractometer, which already has the x-ray beam path described above. Alternatively, a spectrogoniometer may be arranged for polychromatic absorptiometry in one of three ways: (1) It may be fitted with diffractometer accessories; (2) the collimators and detector may be elevated on spacers so they are in line with the x-ray tube window after rotation to position *W*2; (3) the spectrogoniometer may be arranged for conventional x-ray secondary-emission spectrometry with the x-ray tube window in position $W1$, but with the crystal removed and the goniometer set at 0° 2θ. A very low-Z scatter target is placed in the specimen compartment in position 2, the same position as an emission specimen. The absorption specimen is placed in position 5 or 6. The scatter target scatters primary x-rays into the x-ray optical path, but at greatly reduced intensity.

In general, the x-ray tube is operated at a potential such that the peak of the continuum lies on the short-wavelength side of the analyte absorption edge, where analyte absorption is maximum and transmitted intensity most sensitive to analyte concentration. Transmitted intensity may be measured without pulse-height selection. However, some improvement is realized by using the pulse-height selector with the window set to pass a relatively wide band of wavelengths on the short side of the analyte absorption edge. It may be feasible to determine two or three analytes having widely separated absorption edges by using a relatively narrow window set on the short side of each edge in turn.

Polychromatic absorption measurements are also made on specially built x-ray photometers and on small, compact instruments consisting of a radioisotope source, detector, and electronic system.

20.1.1.2. *Advantages and Limitations*

The method is simple and convenient. When some accuracy can be sacrificed, transmitted intensity can be read on the panel ratemeter or on the recorder; otherwise, the scaler is used. The high primary-beam intensity gives relatively high transmitted intensity, so good statistical precision is attained, or else, some precision may be sacrificed so that specimens may be thicker or have higher effective atomic number. If very long path length is feasible, gas and vapor specimens can be analyzed.

However, the method has serious limitations:

1. It is substantially nonspecific and limited to a single analyte having relatively high absorption with respect to the matrix, or at best to a very few such analytes having widely separated absorption edges.

2. The method is useless for qualitative analysis and for quantitative analysis of samples having substantial concentrations of unknown elements or variable matrix composition.

3. Applicability to low-Z elements is very limited. Absorption edges of these elements lie at long wavelengths, where primary intensity is low and matrix absorption high. Consequently, the transmitted beam is mostly shorter-wavelength radiation largely unaffected by the analyte absorption edge.

4. The high primary-beam intensity, although advantageous for statistical precision, causes serious difficulty with liquid specimens.

5. Because of unequal absorption of long and short wavelengths, the effective wavelength of the polychromatic beam changes continuously with distance through the specimen and with analyte concentration. Therefore, calculations of corrections for matrix absorption and other effects are complex and inaccurate, and such corrections are best derived empirically.

20.1.2. Monochromatic X-Ray Absorptiometry

20.1.2.1. *Principles and Instrumentation*

The principle of monochromatic x-ray absorptiometry is the same as that of polychromatic absorptiometry except that monochromatic radiation is used having wavelength on the short side of the analyte absorption edge (*15G*, *L11*). The monochromatic radiation is usually obtained in one of three ways: (1) The primary beam can be directed through the specimen as in polychromatic absorptiometry, but a filter is placed in the beam to pass the principal target line and attenuate other lines and continuum; (2) the primary beam can be directed through the specimen as before, but the goniometer is set to diffract and detect only the selected wavelength; (3) the primary beam can be directed on a secondary target having a strong spectral line at the appropriate wavelength.

The instrument arrangements are shown in Figures 9.2 and 9.3 for goniometers having horizontal and vertical planes, respectively. If the monochromatic wavelength is to be derived from the primary beam, the instrument arrangement is the same as for polychromatic absorptiometry, with the x-ray tube window in position *W*2 and the absorption specimen in position 7 or 8. For filtered primary beams, the crystal is removed and the goniometer set at 0° 2θ; this technique avoids the severe intensity loss incurred by crystal monochromatization, but is generally limited to the target lines of commercially available tubes. For unfiltered beams, the crystal is replaced and the goniometer set to diffract the selected wavelength, which may be a target line or a wavelength in the continuum. If the monochromatic wavelength is to be derived from a secondary target, the instrument is arranged for conventional x-ray spectrometry, with the x-ray tube window in position *W1*. The secondary target is placed in the specimen compartment in position 2, the same position as an emission specimen. The absorption specimen is placed in position 5 or 6. The goniometer may be set at 0° 2θ with the crystal removed or at 2θ for a specific secondary target line with the crystal in place. The former technique (0° 2θ)

gives higher intensity, but all secondary target lines are measured, and so all but the weakest ones must have wavelengths shorter than the analyte absorption edge. The latter technique gives strictly monochromatic radiation, but at greatly reduced intensity. In any of these arrangements, a tunnel should be provided to confine the incident x-rays.

Monochromatic absorption measurements are also made on special x-ray photometers and on instruments having radioactive sources (*G9*, *H46*).

20.1.2.2. *Advantages and Limitations*

Monochromatic absorptiometry eliminates or minimizes most of the limitations of polychromatic absorptiometry:

1. It is more specific, and several analytes can be determined if their absorption edges are relatively widely spaced and there are no interfering edges of matrix elements.
2. When the incident wavelength is derived from a secondary target, the method is better than polychromatic absorptiometry for low-Z elements. The secondary emission line is likely to be more intense than the primary beam in the long-wavelength region. Nevertheless, high matrix absorption for long wavelengths still limits the utility of the method for low-Z elements.
3. When the incident wavelength is derived from a secondary target, the specimen is not subjected to the direct primary beam, and the reduced incident intensity minimizes heating and radiolysis effects in liquid specimens.
4. The "effective wavelength" does not change with distance through the specimen or with analyte concentration, so calculations of matrix absorption coefficient and other corrections are simpler and analytical results more accurate.

The principal disadvantage, compared with polychromatic absorptiometry, is the much lower incident intensity at medium and short wavelengths. In polychromatic absorptiometry, a wide band of the primary continuum is effective, whereas in monochromatic absorptiometry, only a single wavelength is effective—and even this may be derived by secondary excitation.

20.1.3. Applications of X-Ray Absorptiometry

A typical x-ray absorptiometric analysis consists of the measurement of incident I_0 and transmitted I intensities [Equation (20.1)] from samples and standards. The I_0 may be measured through an empty specimen cell or support, or from a specimen of analyte-free matrix or solvent. Analytical data may be derived from a curve of I/I_0 *vs.* C_A.

X-ray absorptiometry is well suited to determination of a relatively high-Z analyte in solutions having relatively constant low-Z matrix compositions; for example, phosphorus, sulfur, chlorine, lead tetraethyl, ethylene bromide, or heavy-metal additives in gasoline and other hydrocarbon liquids; plate metals in plating solutions; and heavy elements in mineral processing effluents.

Porosity of solid and compacted materials can be evaluated (*C36*):

$$P = \frac{v_{\text{ap}} - v_{\text{tr}}}{v_{\text{ap}}} = 1 - \frac{v_{\text{tr}}}{v_{\text{ap}}} = 1 - \frac{(\varrho t)_{\text{tr}}}{(\varrho t)_{\text{ap}}} \tag{20.2}$$

Since $I = I_0 \exp[-(\mu/\varrho)\varrho t]$ [Equation (2.9)],

$$2.303 \log_{10}(I_0/I) = (\mu/\varrho)\varrho t \tag{20.3}$$

Therefore

$$P = 1 - \frac{2.303 \log_{10}(I_0/I)}{(\mu/\varrho)(\varrho t)_{\text{ap}}} \tag{20.4}$$

where P is porosity; v is volume (cm^3); I_0 and I are incident and transmitted intensities, respectively; μ/ϱ is mass-absorption coefficient (cm^2/g); ϱ is density (g/cm^3); t is thickness (cm); ϱt is mass thickness (g/cm^2); and subscripts "ap" and "tr" indicate apparent and true values, respectively. Thus, if the mass-absorption coefficient of the material is known, porosity is derived from the measured values of I_0, I, and $(\varrho t)_{\text{ap}}$. Even if mass-absorption coefficient is not known, porosity is readily compared for specimens having the same composition and thickness. The technique is especially applicable to paper and wood products, asbestos products, ceramics and glasses, plastics, filled materials, etc.

Absorptiometry is also convenient for: (1) measurement of thickness of relatively high-Z coatings on x-ray-transparent substrates, such as paper and wood products, plastics, and ceramics (*S30*); (2) determination of relatively high-Z impurities and fillers in similar products and in textiles; (3) evaluation of purity of organic chemicals; (4) identification of plastics and ceramics by comparison of transmittance of specimens of equal thickness; and (5) analysis of gases. Finally, because of the high primary-beam intensities in polychromatic absorptiometry, useful measurements may be obtained with very small beams, and selected-area and point-by-point analysis is feasible. The uniformity of cadmium-, barium-, and lead-salt impregnation of carbon used for making electric motor brushes has been evaluated in this way.

20.1.4. X-Ray Absorption-Edge Spectrometry (Differential X-Ray Absorptiometry)

20.1.4.1. *Principles and Instrumentation*

Referring to Figure 2.2, suppose that monochromatic x-ray beams of wavelength 0.40 Å (Cs $K\alpha$) and 0.45 Å (Te $K\alpha$) are passed in turn through a thin foil of ship-nail brass (64Cu–25Zn–8.5Pb–2.5Sn). Only tin, whose K-absorption edge (0.42 Å) is bracketed by these wavelengths, undergoes a marked change in absorption coefficient, and therefore in transmittance. Similarly, if beams of wavelength 1.34 Å (Ga $K\alpha$) and 1.54 Å (Cu $K\alpha$) were passed through the foil, only copper (K_{ab} 1.38 Å) would undergo a marked change in transmittance. X-ray absorption-edge spectrometry (or differential x-ray absorptiometry) is based on this phenomenon (*15G, D10*). One or more wavelengths on each side of the analyte absorption edge are passed in turn through each sample and through an empty cell or blank. Analyte concentration is calculated from these measured intensities, the absorption coefficient of the analyte at the bracketing wavelengths, and the specimen thickness. In general, standards are not required.

An expression having the form of Equation (20.1) can be written for each of the two wavelengths λ_S and λ_L bracketing the analyte absorption edge on the short and long sides, respectively, as shown in Figure 20.1. If the expression for λ_S is divided by that for λ_L, and if we let

$$\Delta(\mu/\varrho) = (\mu/\varrho)_S - (\mu/\varrho)_L \tag{20.5}$$

the following equation results:

$$I_S/I_L = (I_S/I_L)_0 \exp\{-[C_A\, \Delta(\mu/\varrho)_A + (1 - C_A)\, \Delta(\mu/\varrho)_M]t\} \tag{20.6}$$

where the notation is as defined in Equations (20.1) and (20.5). The exponential term may be separated as follows:

$$I_S/I_L = (I_S/I_L)_0 \exp[-(1 - C_A)\, \Delta(\mu/\varrho)_M t] \exp[-C_A\, \Delta(\mu/\varrho)_A t] \tag{20.7}$$

The matrix undergoes little change in absorption coefficient between the bracketing wavelengths, that is, $\Delta(\mu/\varrho)_M \approx 0$, so the first exponential term becomes exp($\sim$0), or $\sim$1. Also, for a given series of measurements, the

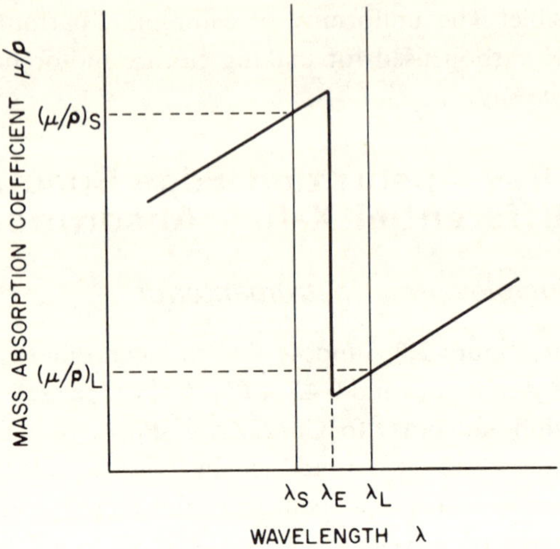

Fig. 20.1. Principle of absorption-edge spectrometry.

incident intensities at the bracketing wavelengths, and therefore $(I_S/I_L)_0$, are constant. Then,

$$I_S/I_L = K \exp[-C_A \Delta(\mu/\varrho)_A t] \tag{20.8}$$

It is evident that analytical sensitivity—the slope of the curve of Equation (20.8)—increases with path length t and with $\Delta(\mu/\varrho)_A$, which, in turn, increases with the analyte absorption-edge jump difference (Figure 2.4) and with proximity of the bracketing lines to the analyte edge. The bracketing wavelengths should also be as close together—and therefore as close to the edge—as possible to minimize difference in matrix absorption and to avoid bracketing an absorption edge or spectral line of a matrix element. However, the two wavelengths should be far enough from the edge, especially on the short-wavelength side, to avoid absorption-edge fine structure (see below).

The instrument arrangements are identical with those for monochromatic absorptiometry with the crystal in place. The bracketing wavelengths may be obtained from the primary beam or from secondary targets in the specimen compartment. In either case, the goniometer is set at 2θ for each of the wavelengths in turn. The secondary wavelengths may be derived from separate targets or from the same target. Often, an element can be found that has $K\alpha$ and $K\beta$ or $L\alpha_1$ and $L\beta_1$ lines bracketing the analyte absorption edge. Otherwise, an alloy target or a briquet of mixed element, oxide, or other compound powders may be used.

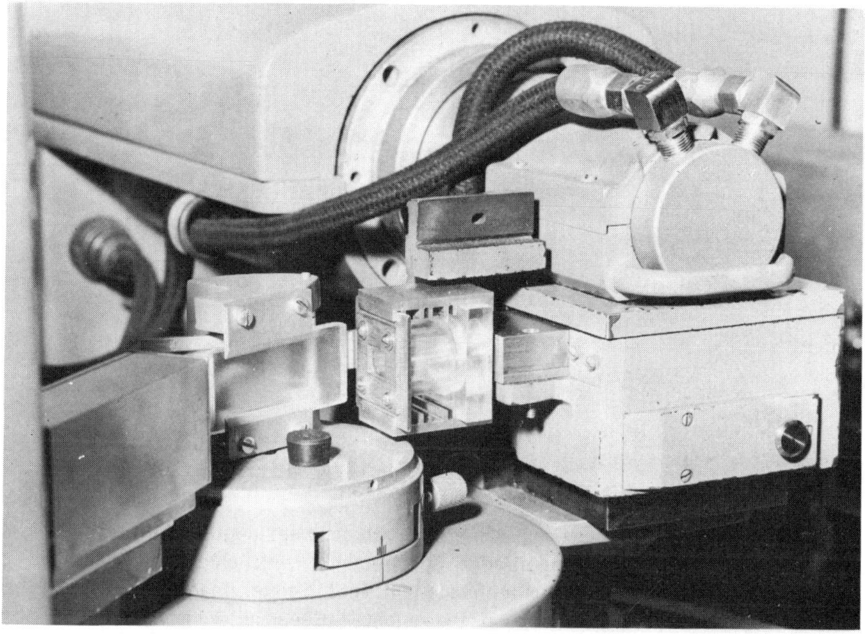

Fig. 20.2. Accessory for x-ray absorptiometry and absorption-edge spectrometry on the secondary-beam exit port of an x-ray spectrometer. [E. P. Bertin, R. J. Longobucco, and R. J. Carver, *Analytical Chemistry* **36**, 641 (1964); courtesy of the American Chemical Society.]

Figure 20.2 shows an accessory for absorption-edge spectrometry—or for polychromatic or monochromatic absorptiometry—on the secondary-beam exit port of the specimen compartment of a commercial x-ray spectrometer (*B28*). Figure 20.3 shows a series of liquid-specimen absorption cells having path length 0.5–5 cm (*B28*). The construction of these cells is shown in Figure 20.4. All the cells are made from Lucite blocks *A* $1\frac{1}{2}$ inch square and 1 cm thick: 0.5-cm cells are made by machining cylindrical holes $\frac{3}{4}$ inch in diameter through the blocks and concentric circular recesses *C* $1\frac{1}{8}$ inch in diameter and 0.25 cm deep on both sides of the blocks; 1-cm cells are made the same way without the recesses; cells having longer path length are made by machining through the blocks cylindrical holes 1 inch in diameter in which are cemented cylindrical Lucite inserts *B* having 1-inch outside diameter, $\frac{3}{4}$-inch inside diameter, and 2-, 3-, or 5-cm length. Circular Mylar windows *D* 0.005 inch thick are cemented on both sides of all cells with Pliobond rubber cement. All the cells have tapered filling ports *E* and plugs *F*. The cells may also be made of polyethylene or Teflon and the

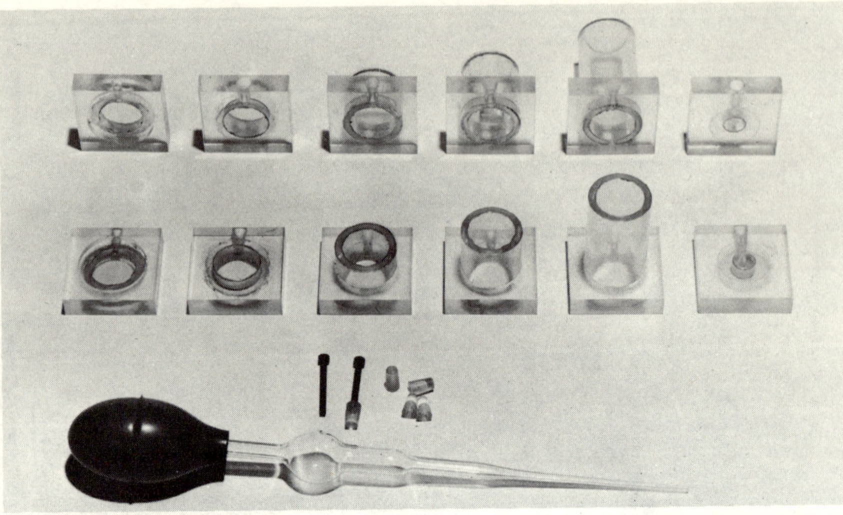

Fig. 20.3. Liquid-specimen cells having various path lengths for x-ray absorptiometry and absorption-edge spectrometry, with plugs and filling pipet. [E. P. Bertin, R. J. Longobucco, and R. J. Carver, *Analytical Chemistry* **36**, 641 (1964); courtesy of the American Chemical Society.]

windows of polystyrene or even thin glass or quartz. Figure 20.5 shows an all-polystyrene cell having windows $\frac{1}{32}$ inch thick (*H5*). Supports for other specimen forms are described in (*B28*).

Multichannel x-ray spectrometers may be used to determine more than one element simultaneously. For example, lead and bromine can be deter-

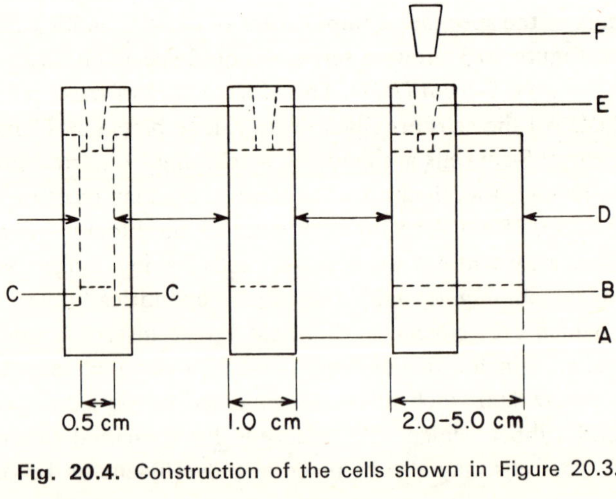

Fig. 20.4. Construction of the cells shown in Figure 20.3.

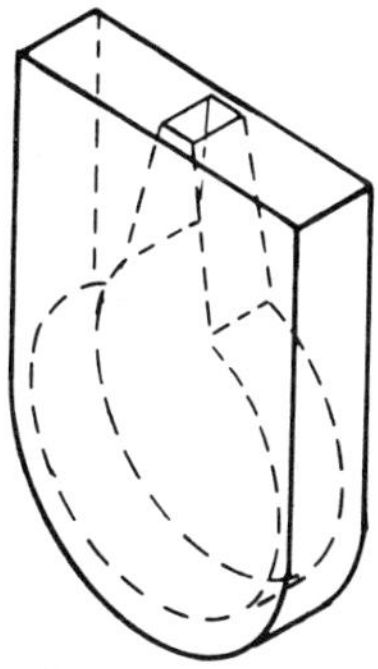

Fig. 20.5. All-polystyrene liquid-specimen cell for x-ray absorptiometry and absorption-edge spectrometry. [E. A. Hakkila and G. R. Waterbury, *Developments in Applied Spectroscopy* **2**, 297 (1963); courtesy of the authors and Plenum Press.]

mined in gasoline by placing a sample cell in each of three channels set for Sr $K\alpha$ (0.88 Å), Rb $K\alpha$ (0.93 Å), and Br $K\alpha$ (1.04 Å), respectively. The first two wavelengths bracket the bromine K edge (0.92 Å), the last two the lead LIII edge (0.95 Å). The lines are derived from a briquet of compounds of the three elements.

For maximum accuracy, the x-ray tube power should be highly filtered constant potential, and the operating potential should be well above the potential corresponding to the analyte absorption edge plus the ripple. High-order diffraction from the crystal at the 2θ angles of the bracketing wavelengths must be avoided by: (1) use of secondary targets; (2) operation of the x-ray tube at a potential too low to generate the interfering wavelengths; (3) use of a crystal not having the interfering order; or (4) use of pulse-height analysis. The elements in the crystal, detector window, and scintillation crystal or detector gas must have no absorption edge between the bracketing wavelengths. Detector escape peaks must be avoided or excluded unless they arise from the wavelength being measured, in which case they must be counted.

20.1.4.2. *Advantages and Limitations*

In view of the many advantages of absorption-edge spectrometry and its ready applicability to commercial x-ray spectrometers, it is difficult to understand why the method is not more widely used:

1. The preparation of standards and calibration curves for x-ray

fluorescence spectrometric analysis of a multicomponent system is warranted if there are many samples to be analyzed from time to time, but usually not for one or a few samples on a single occasion. Absorption-edge spectrometry eliminates the need for preparation and storage of standards and for their measurement each time an analysis is made. Consequently, the method is particularly well suited to one-of-a-kind analyses.

2. The method is applicable to liquids, solids, briquets, powders, foils, films, and even gases.

3. Liquid volumes as small as 0.2 ml and very small solid samples may be sufficient.

4. Matrix effects are minimized because only the analyte undergoes a marked change in absorption coefficient at the bracketing wavelengths.

5. The method is highly specific; as many analytes can be determined in each sample by absorption-edge spectrometry as by secondary-emission spectrometry.

6. Because the calculations involve ratios of the intensities transmitted by the empty and sample-filled cells, the method has many of the advantages of ratio methods in general.

7. The disproportionate intensity-concentration relationship for an element having short-wavelength spectral lines in a low-Z matrix (Section 12.5) does not occur in absorption-edge spectrometry.

The technique of deriving the bracketing wavelengths from secondary targets brings additional advantages:

8. Most primary-beam absorption-edge procedures are carried out on diffractometers. However, determination of chemical elements is usually the concern of x-ray spectrometric laboratories, many of which are not equipped for diffractometry, or at least cannot afford the time required to rearrange their instruments from time to time for diffractometric operation. The secondary target method is readily applied on x-ray spectrometers with little modification.

9. The total x-ray intensity incident on the sample is only a fraction of that in the primary-beam method; this reduces heating and radiolysis of liquid samples, destruction of thin-film samples, and secondary emission and scatter from the sample.

10. If the bracketing lines are derived from different secondary targets, they are not coexistent in the incident radiation and may be very close without exceeding the resolution capability of the instrument.

11. In the primary-beam method, it may be necessary to operate the x-ray tube at a potential below that at which harmonics of the analyte

absorption edge appear (unless pulse-height selection or a suitable crystal is used). There is no such restriction on x-ray tube potential in the secondary-target method.

12. Finally, a relatively high x-ray intensity is available at all wavelengths required for bracketing the analyte absorption edge; this is not necessarily true for the primary-beam method, particularly when the tube is operated at reduced potentials to avoid generating harmonics of the absorption edge, as mentioned above.

The principal limitations of the method are the following:

1. Compared with x-ray fluorescence spectrometry, the method is not as readily applicable to elements of low atomic number (fluorine through titanium), because of high absorption of the long wavelengths in the matrix and cell windows.

2. It is not readily made as sensitive to low concentrations as x-ray emission.

3. Sensitivity is reduced by the presence in the matrix of elements having high atomic number, and the reduction is more severe the higher the atomic number and concentration of the interfering element.

4. The principal source of specific interference is the relatively infrequent occurrence of absorption edges of matrix elements between the bracketing wavelengths.

5. The transmitted intensities may be very low and require long counting times for acceptable statistical precision.

6. Many of the routine procedures of wide general applicability require more or less elaborate preliminary experimental evaluation of mathematical constants whenever the procedure is applied to a new analyte for the first time. However, the increasing availability of tables of accurate absorption coefficients is eliminating the need for much of this work.

7. Finally, the calculations are complex—but no more so than elaborate matrix corrections in x-ray emission spectrometry.

20.1.4.3. *Applications*

Many absorption-edge spectrometric procedures have been reported, and some of these are routine methods of wide general applicability. A few of these are considered as examples (*B3*, *B28*, *D20*, *K15*, *S40*).

Dunn's procedure (*D20*) is for analysis of solutions and uses a diffractometer, deriving the bracketing wavelengths from the primary beam. The analytical equation is

$$C_A = \frac{2303}{\Delta(\mu/\varrho)_E t} [m_S \log_{10} I_S - m_L \log_{10} I_L - a] \tag{20.9}$$

where C_A is analyte concentration (mg/ml); $\Delta(\mu/\varrho)_E$ is analyte absorption-edge jump difference (cm^2/g); t is cell length (cm); I is transmitted intensity; S and L refer to the bracketing wavelengths on the short and long sides of the analyte absorption edge, respectively; and m_S, m_L, and a are experimentally evaluated constants. The method gives good results, but the preliminary work required to evaluate the constants may be time-consuming.

The writer (*B28*) developed a simple routine procedure for solutions, thin solids, foils, briquets, powders, and films. The measurements are made on a standard commercial spectrometer with only minor modification and simple accessories, some of which are shown in Figures 20.2, 20.3, and 20.4. The bracketing wavelengths are derived from secondary targets.

Putting Equation (20.1) in common logarithmic form gives

$$2.303 \log_{10}(I_0/I) = C_A(\mu/\varrho)_A t + (1 - C_A)(\mu/\varrho)_M t \tag{20.10}$$

where I_0 and I are incident and transmitted intensities, respectively; C_A and $1 - C_A$ are analyte and matrix concentrations, respectively (g/cm^3); μ/ϱ is mass-absorption coefficient (cm^2/g); t is sample thickness (cm); and A and M refer to analyte and matrix, respectively. If now an equation of this form is written for each of the bracketing wavelengths λ_S and λ_L, the equation for λ_L is subtracted from that for λ_S, it is assumed that $(\mu/\varrho)_M$ is the same at λ_S and λ_L, and the resulting expression is solved for C_A, one gets

$$C_A = \frac{2.303}{[(\mu/\varrho)_S - (\mu/\varrho)_L]_A} \left[\log_{10}\left(\frac{I_0}{I}\right)_S - \log_{10}\left(\frac{I_0}{I}\right)_L\right] \tag{20.11}$$

where S and L refer to λ_S and λ_L, respectively.

In practice, matrix absorption is somewhat different at λ_S and λ_L, and if Equation (20.11) is to give high accuracy, a correction must be made for this difference. For this purpose, a correction method devised by Barieau (*B3*) was adopted. In a matrix having components 1, 2, . . ., n, the ratios of mass-absorption coefficients $(\mu/\varrho)_S/(\mu/\varrho)_L$ at λ_S and λ_L are related as follows.

$$\left[\frac{(\mu/\varrho)_S}{(\mu/\varrho)_L}\right]_1 = \left[\frac{(\mu/\varrho)_S}{(\mu/\varrho)_L}\right]_2 = \cdots = \left[\frac{(\mu/\varrho)_S}{(\mu/\varrho)_L}\right]_n = \left(\frac{\lambda_S}{\lambda_L}\right)^a = k \tag{20.12}$$

where k and a are experimentally evaluated constants which are different for each λ_S–λ_L line pair. Equation (20.12) assumes that no matrix element has an absorption edge between λ_S and λ_L. Incorporation of the correction constant into Equation (20.11) and conversion of C_A to mg/cm³ gives

$$C_{\mathrm{A}} = \frac{2303}{[(\mu/\varrho)_S - k(\mu/\varrho)_L]_{\mathrm{A}} t} \left[\log_{10}\left(\frac{I_0}{I}\right)_S - k \log_{10}\left(\frac{I_0}{I}\right)_L\right] \tag{20.13}$$

where C_A is in mg/cm³ (or mg/ml). For foils and films, the equation may be put in the form

$$C_{\mathrm{A}} t = \frac{2303}{[(\mu/\varrho)_S - k(\mu/\varrho)_L]_{\mathrm{A}}} \left[\log_{10}\left(\frac{I_0}{I}\right)_S - k \log_{10}\left(\frac{I_0}{I}\right)_L\right] \tag{20.14}$$

where $C_A t$ is area concentration in mg/cm². The first term on the right in Equations (20.13) and (20.14) is constant for each combination of analyte, path length t, and λ_S–λ_L pair.

The correction factor k is evaluated from the intensities I_C transmitted at λ_S and λ_L by an analyte-free correction standard. The author used an aluminum plate for all analyses, but greater accuracy would result if the correction standard were similar to the matrix in composition. Since the first term on the right in Equations (20.13) and (20.14) is finite, if $C_A = 0$, it follows that

$$\log_{10}(I_0/I_C)_S - k \log_{10}(I_0/I_C)_L = 0 \tag{20.15}$$

from which

$$k = \frac{\log_{10}(I_0/I_C)_S}{\log_{10}(I_0/I_C)_L} \tag{20.16}$$

An analysis consists of the measurement of the intensities transmitted by the sample-filled cell I, empty cell I_0, and correction standard I_C at λ_S and λ_L. The I_0 and I_C data are used in Equation (20.16) to calculate k; the I_0 and I data and k are used in Equations (20.13) and (20.14) to calculate the analytical result. The absorption coefficients are obtained from the literature directly or by interpolation.

Hakkila *et al.* (*H7*) developed a three-wavelength method for determination of plutonium. Intensities transmitted by the absorption cell filled in turn with water and a standard solution of known plutonium concentration are measured at three wavelengths: Nb $K\beta$ (λ_1) and Mo $K\alpha$ (λ_2), which bracket the Pu LIII absorption edge, and Nb $K\alpha$ (λ_3). Then the intensities transmitted by the same cell filled with the sample(s) are measured at the two bracketing wavelengths (λ_1 and λ_2). The measurements on the water and standard are used to calculate k_1/k_2:

$$\frac{k_1}{k_2} = \left[\frac{\log_e(I_0/I)_3}{\log_e(I_0/I)_2}\right]^{\log_e(\lambda_2/\lambda_1)/\log_e(\lambda_3/\lambda_2)}. \tag{20.17}$$

Then,

$$C = k_1 \log_e(I_0/I)_1 - k_2 \log_e(I_0/I)_2 \tag{20.18}$$

I_0 is the intensity transmitted by the water-filled cell, and I is the intensity transmitted by the cell filled with standard [Equation (20.17)] or sample [Equation (20.18)].

Knapp *et al.* (*K15*) developed an absorption-edge spectrometric method for standard commercial x-ray spectrometers. The bracketing wavelengths are derived from secondary targets, usually the $K\alpha$–$K\beta$ lines of a single element. The transmitted intensity is measured at these two wavelengths from a series of standards in a certain solvent, and a calibration curve is established having the equation

$$\log_e(I_L/I_S) = m_\mathrm{M}C_\mathrm{A} + b_\mathrm{M} \tag{20.19}$$

where I is transmitted intensity; C_A is analyte concentration; m_M is the slope of the curve; b_M is the intercept; L and S refer to the bracketing wavelengths; and M refers to the matrix, that is, the solvent. Calibration curves for other solvents M′ are calculated simply from the absorption coefficients of the pure solvents M and M′ at λ_S and λ_L.

Cullen (*C54*) describes three novel *internal differential absorption-edge* spectrometric methods in which the wavelengths bracketing the analyte absorption edge are either generated within or scattered by the specimen itself: (1) An element having bracketing spectral lines may already be present in the sample; for example, nickel can be determined in nickel–copper alloys from the intensities of the Cu $K\alpha$ and Cu $K\beta$ lines. (2) The element having the bracketing lines may be added to the sample; for example, selenium can be determined in solutions by adding bromine and measuring Br $K\alpha$ and Br $K\beta$ intensities. (3) The two bracketing wavelengths can be derived from the continuous spectrum originating in the x-ray tube and scattered by the sample; for example, silver (K_ab 0.49 Å) can be determined from the intensities scattered at 0.46 Å and 0.63 Å. Scattered characteristic target lines may be used if they have suitable wavelengths. In all three methods, calibration curves of the function (I_L/I_S) *vs.* C_A are established from measurements on standards. These methods, like that of Knapp *et al.* above, lack one of the advantages of most absorption-edge methods, in that they do require standards.

20.1.5. X-Ray Contact Microradiography

If a uniformly thin piece of metal, ceramic, mineral, plastic, wood, paper or other material is placed in contact with the emulsion of a fine-grain photographic plate and irradiated with x-rays, the emulsion records the distribution pattern of the x-rays transmitted by the specimen. Such contact microradiographs may be magnified up to 200× and reveal microscopic voids, cracks, inclusions, segregations, inhomogeneities, and other irregularities in the specimen, or they may reveal the distribution of selected elements over the specimen. An x-ray spectrometer constitutes a convenient instrument for contact microradiography (*15B, T11*).

In polychromatic contact microradiography, the primary x-ray beam is used to expose the microradiograph. A small microradiography camera may be placed in the specimen compartment in the position usually occupied by the emission specimen—position 2 in Figures 9.2 and 9.3. The inclination of the camera to the primary beam may prove advantageous, in that the target-to-film distance varies over the specimen, giving a range of exposure conditions. Alternatively, a specimen drawer can be made in which the camera is perpendicular to the x-ray beam—position 4. Provision may be made to incline the camera for stereomicroradiography.

In monochromatic contact microradiography (*S35*), the specimen is photographed with one or more single wavelengths. In Figure 2.2, suppose that 0.4-Å (Cs $K\alpha$) x-rays are passed through a thin foil of Cu–Zn–Pb–Sn alloy in contact with a photographic plate; then suppose a new plate is provided and 0.45-Å (Te $K\alpha$) x-rays are passed through the foil. Only tin, whose K absorption edge lies between these wavelengths, would appear different on the two photographs; if the alloy were inhomogeneous, tin-rich areas would appear lighter (relatively unexposed) in the 0.4-Å photograph, where tin absorption is high, darker in the 0.45-Å photograph. The distribution of the other elements can be revealed by photographing the alloy with other appropriate pairs of wavelengths, as indicated in Figure 2.2.

For monochromatic microradiography, the primary x-ray beam is used to excite various secondary targets placed in the specimen drawer (position 2 in Figures 9.2 and 9.3) in the same manner as specimens for secondary-emission spectrometry. These targets are chosen to have strong spectral lines at the wavelengths required for the microradiograph. A small microradiography camera can be mounted directly outside the secondary-beam exit window of the specimen compartment, or in position 5. In instruments having specimen drawers, a special drawer can be made in which the secondary target is inclined in the opposite direction, away from the crystal, toward the front of the drawer. A window may be cut in the front of the drawer and the camera mounted at this window in position 1. Commercial direct or stereo cameras may be used in either case.

For strictly monochromatic microradiography, the camera may be mounted in position 6 or 8, beyond the crystal, which is set to diffract the required wavelength from a secondary target in the specimen drawer or from the primary beam. Of course, the diffracted intensities are relatively low, so that very long exposure times are required. Also, any pattern of nonuniform diffracted intensity over the crystal surface is superposed on the microradiograph.

20.1.6. X-Ray Absorption-Edge Fine Structure

Throughout this book, absorption edges have been represented as simple abrupt discontinuities in absorption coefficient at a photon energy just sufficient to expel electrons from a specific inner level in the atom. Fine structure (*15Q*) consists of deviations from this simple step function and includes deviations both in the abrupt rise and in the region on the high-energy side of the rise. In general, fine structure is confined to within ~200 eV of the edge. Figure 20.6 is a composite absorption-edge fine-

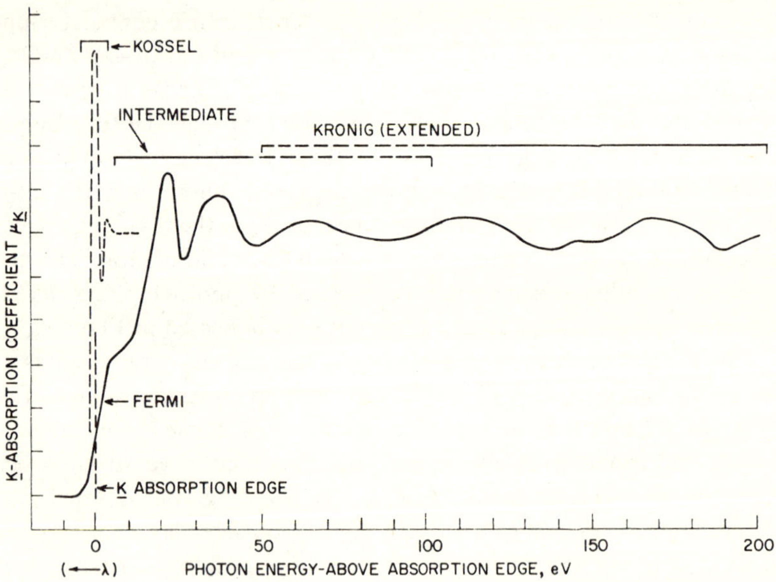

Fig. 20.6. Composite absorption-edge fine-structure curve showing four principle types of fine structure.

structure curve showing the four principal recognized types. All these types do not occur simultaneously.

1. *Fermi-level* fine structure is the first rise in absorption at the K edge of a metal, and is attributed to elevation of a K electron to the Fermi level in the conduction band.

2. *Kossel* fine structure occurs within ~3 eV of the high-energy (short-wavelength) side of the edge, and is attributable to electron transitions to bound atomic states in accordance with the selection rules. This type of fine structure has been observed only in monatomic gases.

3. *Intermediate* fine structure occurs in the region 5–100 eV on the high-energy side of the edge, and is attributable to short-range order, valence state, and neighboring atoms.

4. *Konig* (or extended) fine structure consists of low-amplitude fluctuations in absorption coefficient in the region 50 to ~200 eV on the high-energy side of the edge, and is related to lattice parameter and temperature.

In the abscissa of Figure 20.6, energy increases to the right, wavelength to the left. This sense is opposite to that used throughout this book, but is conventional for fine-structure work. The term *K-absorption coefficient* μ_K is defined below.

Fine-structure measurements may be made on diffractometers, flat-crystal spectrometers, or full-focusing curved-crystal spectrometers; quartz or silicon crystals are used because of their narrow half-width and consequent high resolution. The most precise measurements are made with two-crystal spectrometers.

Incident I_0 and transmitted I intensities are measured at small increments of photon energy (or wavelength) from 10–20 eV below the K edge of the element of interest to 100–200 eV above; $\log_e(I_0/I)$ is calculated for each increment. The K-absorption coefficient μ_K is derived from $\log(I_0/I)$ by (1) correcting it for all absorption other than that responsible for expelling K electrons from the element of interest, and then (2) normalizing the corrected values relative to unity at an energy beyond the fine-structure region. Matrix absorption is substantially constant in the region of the absorption edge (if no matrix element has an edge in that region), and no K-excitation occurs below the edge. Consequently, $\log(I_0/I)$ is relatively low and constant below the edge. This constant value—$\log(I_0/I)_{\text{low}}$—is used to correct for absorption from all sources other than K excitation of the element of interest. Somewhere beyond 50 to ~200 eV above the edge, $\log(I_0/I)$ becomes substantially constant at some relatively high value. It is this value—$\log(I_0/I)_{\text{high}}$—to which the corrected values are normalized. At each incremental photon energy i, K-absorption coefficient is calculated from the following equation:

$$(\mu_K)_i = \frac{\log_e(I_0/I)_i - \log_e(I_0/I)_{\text{low}}}{\log_e(I_0/I)_{\text{high}}} \tag{20.20}$$

The applications of absorption-edge fine structure are based principally on its sensitivity to chemical state. The method is used to establish the nature of chemical bonds; the electronic band structure in solids; identity or class of compounds, especially when the absorbing element is present in more than one oxidation state; oxidation state(s) of the absorbing element; and the nature of catalysts on x-ray-transparent substrates.

20.1.7. Specimen Preparation for X-Ray Absorption

Specimen preparation for x-ray absorption is somewhat more critical than for emission because the x-ray beam passes through the specimen. For quantitative analysis by absorptiometry and absorption-edge spectrometry, the specimen must be thin enough to transmit x-radiation sufficiently intense to be measured with acceptable precision in a reasonable counting time. For microradiography, it must be thin enough to permit photography of the image in a reasonable exposure time. The thickness must be uniform, and opposite surfaces must be plane and parallel. All specimens in a given analysis must have the same thickness. For absorption-edge fine structure, the requirement for uniformity of thickness may be relaxed somewhat.

Films on x-ray-transparent substrates, foils, and uniformly thin sheets of metal, ceramic, plastic, rubber, paper, etc. may usually be used without preparation. Metals, ceramics, minerals, etc. may be reduced to uniformly thin plates by combinations of sawing, rolling, abrading, etching, and electropolishing. Care must be taken not to etch certain elements preferentially. Powders should have small, uniform particle size to minimize packing voids. High-Z powders may be diluted with carbon, lithium carbonate, starch, etc. In a given analysis, the same accurately weighed amount of all specimens may be packed into open, windowless frames or windowed cells having the same volume, or pressed into briquets of the same diameter and thickness. Weighed amounts of powder may be slurried and cast into films of uniform thickness with a casting blade. These films may be stripped and mounted in open frames, or they may be cast on x-ray-transparent substrates. For fine-structure studies, powders may be dusted on adhesive cellulose tape. Liquids are held in cells having windows rigid enough not to bow on filling, yet transparent to x-rays, or small constant volumes of liquid may be absorbed in filter-paper disks. The lower the analyte concentration, the longer the cell length must be. Gases, having very low absorption coefficients, require very long cell lengths—up to 100 cm—and it may be necessary to calculate the mass of gas from temperature and pressure and the gas laws.

20.2. X-RAY SCATTER METHODS

The occurrence of x-ray scatter in spectrometers was formerly regarded only as a nuisance (*J10*), and, in fact, scatter is the principal source of background, limits analytical sensitivity, and may cause spectral-line interference. However, more recently, scattered x-rays have proved to be of great value in x-ray spectrometry. Their use in correction of absorption-enhancement and other errors has already been discussed. This section describes two analytical methods applicable to commercial x-ray spectrometers and based on difference in efficiency of coherent and incoherent scatter by elements of high and low atomic number.

20.2.1. Coherent Scatter

Ziegler and his co-workers (*M9, Z9*) have developed a method for determination of heavy elements in light matrixes based on coherent (Rayleigh) scatter. The technique may be regarded as x-ray nephelometry. A high-energy (80–150-kV) polychromatic x-ray beam irradiates the specimen solution, but, rather than an analyte spectral line, coherently scattered radiation is measured at an appropriate angle. Maximum differentiation of coherent and incoherent (Compton) scatter is realized by filtering the long-wavelength components from the primary beam, measuring the scatter at a large angle ($\sim$140° 2θ), and using a discriminator to pass only detector output pulses at or near the maximum primary energy. Since coherent and incoherent scatter are proportional, respectively, to Z^3 and Z, the scattered intensity is highly dependent on concentration of high-Z elements.

The method was used to determine uranium in concentrations 1–10 g/l in solutions having high concentrations of aluminum and stainless steel. Scattered intensity was measured from the sample(s) and from two standards having known uranium concentrations bracketing that of the samples. The analytical equation is

$$C_X = C_1 + (C_2 - C_1)\frac{I_X - I_1}{I_2 - I_1} \tag{20.21}$$

where C is concentration; I is scattered intensity at the selected angle; subscripts 1 and 2 refer to the standards; and X refers to the sample. The work apparently was done on a specially constructed instrument using a full-wave x-ray radiography unit as the x-ray source. However, the technique should be feasible on a 75- or 100-kV x-ray spectrometer, particularly one having a constant-potential generator.

20.2.2. Coherent/Incoherent Scatter Ratio

Dwiggins (*D24*, *D25*) has made an extremely significant contribution with his method of determining carbon in hydrocarbon liquids using an unmodified conventional x-ray spectrometer. The intensity ratio R of the coherently and incoherently scattered W $L\alpha_1$ line from the x-ray tube is measured from the sample(s) and from a series of pure hydrocarbon liquid compounds having various known carbon/hydrogen ratios. A calibration curve of R *vs.* %C is established from the known liquids; analytical concentrations are derived from this curve.

C—H bond type has no effect. If a few per cent of nitrogen and/or oxygen is present, the method gives %(C + N + O), and independent determinations of nitrogen and oxygen are required to obtain carbon concentration. In the presence of sulfur in concentrations up to ~10%, if the C/H ratio is constant, R is proportional to sulfur concentration; if sulfur concentration is constant, R is proportional to C/H ratio. Dwiggins describes ways to correct for sulfur. An uncovered water-cooled liquid-specimen drawer was used in the original work (*D27*), but later, inverted geometry and cells covered with 0.00025-inch Mylar were used. Toussaint and Vos (*T10*) have extended the method to solid hydrocarbons.

20.3. X-RAY PHOTOELECTRON AND AUGER-ELECTRON SPECTROMETRY

20.3.1. X-Ray Photoelectron Spectrometry

In secondary excitation of K x-ray spectra, the energy of each incident primary photon is expended as follows:

$$E_0 = E_K + \phi + E_{\text{pe}} \tag{20.22}$$

where E_0 is energy of the incident photon; E_K is the energy required to expel a K electron (K-excitation potential); ϕ is the energy required to remove an electron from the specimen surface (work function); and E_{pe} is the kinetic energy of the free photoelectron. All energies are in electron volts. Since ϕ is only a few electron volts,

$$E_{pe} \sim E_0 - E_K \tag{20.23}$$

Similar expressions apply to photoelectron emission from *L*I, *L*II, *L*III, and the five *M* shells. Thus, for monoenergetic primary radiation, photoelectron energy should be characteristic of the secondary emitter. However, only photoelectrons emitted from atoms on the surface have the energy indicated by Equation (20.23). Those emitted from atoms beneath the surface undergo stepwise deceleration and emerge with less than their initial energy, and only those emitted in an extremely thin surface layer can emerge from the surface at all.

Thus, a monochromatically excited x-ray photoelectron energy spectrum should show a series of peaks, one for each atom energy level (*K*, *L*I, *L*II, etc.). The peaks should have vertical edges on the high-energy side, as given by Equation (20.23), and fall off exponentially on the low-energy side. The lower the excitation potential of the shell, the smaller is the fraction of the primary energy expended in ionizing it, and the greater is the balance remaining to impart kinetic energy to the photoelectron.

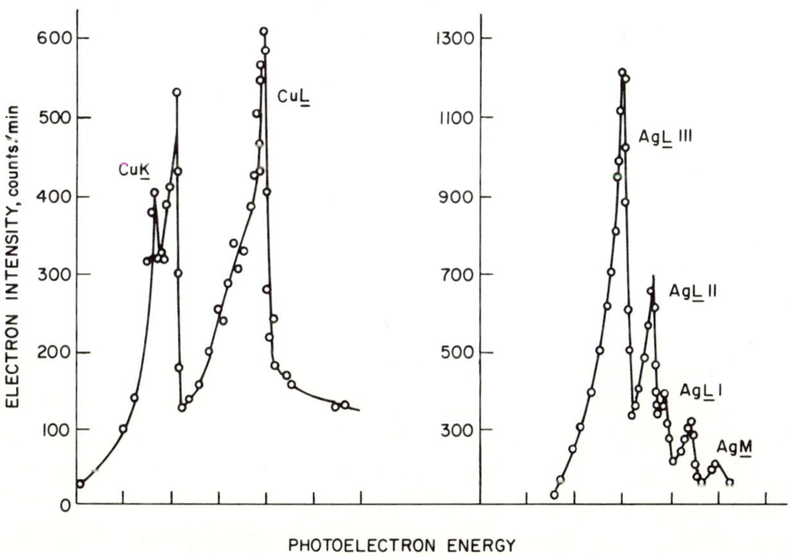

Fig. 20.7. X-ray photoelectron energy spectra of copper and silver excited by Mo $K\alpha$ x-rays. [R. G. Steinhardt, Jr., and E. J. Serfass, *Analytical Chemistry* **23**, 1585 (1951); courtesy of the authors and the American Chemical Society.]

Figure 20.7 (*S36*) shows the photoelectron energy spectra of copper and silver excited by Mo $K\alpha$ x-rays. The spectra have essentially the predicted form, but may be complicated by several phenomena. Absorption and scatter of photoelectrons in the specimen may broaden the peaks. The incident x-rays are usually only filtered, rather than truly monochromatic, to avoid the prohibitive loss of intensity resulting from crystal monochromatization. Although the photoelectrons originate in inner atomic orbitals and are therefore relatively independent of chemical state, they are extremely sensitive to surface conditions, which may broaden and displace the peaks and reduce their intensity. Finally, Auger electrons are also produced, and their spectral distribution is superposed on that of the photoelectrons. For a given intensity and spectral distribution of primary x-rays, photoelectron intensity is approximately proportional to Z^5.

Little application has been made of x-ray photoelectron spectrometry (*F1*, *S36–38*), but perhaps interest in the method may be renewed by recent developments in Auger-electron spectrometry (see below). Many of the hoped-for applications of Auger spectrometry would be feasible for photoelectron spectrometry as well.

20.3.2. Auger-Electron Spectrometry

Energies of Auger electrons (Section 2.5), like those of photoelectrons, are characteristic of the emitting element, but Auger-electron spectra are less intense and less sharply defined than photoelectron spectra.

Assuming monochromatic excitation, a photoelectron originating in, say, the *L*III shell of an atom can have only one *initial* energy—the difference between incident photon energy and *L*III excitation energy. An Auger electron originating in the same shell may have any of a large number of energies. A vacancy in, say, the *K* shell can be filled by an electron transition from the *L*II, *L*III, *M*II–*M*V, or *N*II–*N*VII shells. Each such transition produces a photon of different energy, each of which can undergo internal conversion in the *L*III shell, producing an Auger electron. Moreover, the *L*III Auger electron may be ejected by internal conversion of photons originating from electron transitions to the *L*I or *L*II shells. In each case, the initial energy of the Auger electron is different. In practice, the most common Auger energies correspond to the most probable electron transitions—those giving rise to the strongest x-ray spectral lines. Auger electrons, having somewhat less energy than photoelectrons, undergo more severe absorption in emerging from the emitter, and probably only the top several atom layers contribute to the observed Auger spectrum. For ex-

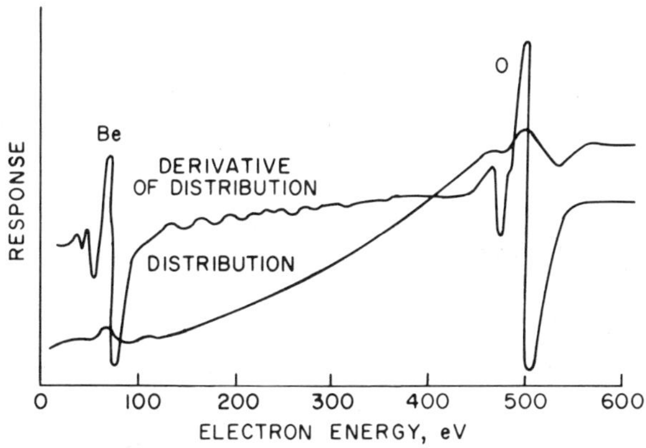

Fig. 20.8. Secondary-electron energy spectra of beryllium, showing Auger-electron spectra of beryllium and oxygen. [L. A. Harris, *Industrial Research* **10**(2), 52 (1968); courtesy of the author and publisher.]

ample, Figure 20.8 shows the Auger-electron spectrum of beryllium metal; it is evident that the oxygen peak from the surface oxide layer is more prominent than the beryllium peak.

The observed Auger-electron energy spectrum consists of relatively weak broad peaks slightly below the corresponding transition energies of the emitting atoms. The peaks are superposed on photoelectron spectra when excited by x-rays, on secondary and scattered electron spectra when excited by electrons. Auger peaks, like photoelectron peaks, taper off toward low energies. The Auger-electron spectrum of beryllium is shown in Figure 20.8. Fortunately, a way has been found to accentuate this poorly defined Auger spectrum (see below). The Auger spectra of low-Z elements are simple and have well separated and easily identified peaks. The observed Auger spectra of high-Z elements are much more complex, and overlap. The high-Z elements also have simple Auger spectra, but they lie outside the energy range convenient for observation.

Auger-electron spectrometry promises to be particularly advantageous for elements of low atomic number. It has been shown (Section 2.5) that the utility of x-ray emission spectrometry decreases with Z because of decreasing fluorescent yield. The reason for the low fluorescent yield is internal conversion of characteristic x-ray photons, with consequent production of Auger electrons. Thus, as Z decreases, x-ray spectral line intensity decreases and Auger electron intensity increases.

Auger-electron spectrometry (*H12, H13*) is a very sensitive method for chemical analysis of extremely thin surface layers. The entire observed Auger-electron spectrum probably arises in the top 5–10 atom layers, and as little as 1/50 atom layer of cesium has been detected and identified on a silicon surface. The method is being applied, or at least has promise of application, in such areas as the following: (1) analysis of extremely thin films; (2) oxidation and corrosion; (3) adsorption and desorption; (4) material transfer by friction, evaporation, etc.; (5) diffusion to the surface; (6) evaluation of superclean surfaces; (7) analysis of residues from wash solutions; (8) determination of dopants and identification of contaminants on semiconductor and other electronically active surfaces; and (9) analysis of catalytic surfaces. The method is also useful in studies of oxidation states and electron-energy levels. Finally, bulk analyses have been made for certain elements with previously unattainable sensitivity when the surface and bulk compositions are the same. Carbon in concentration <40 μg/g (ppm) in iron and sulfur in concentration <10 μg/g in nickel have been determined, and the method is also applicable to oxygen.

20.3.3. Photo- and Auger-Electron Spectrometer

The instrument is essentially the same for both photoelectron and Auger-electron spectrometry, except for the excitation. A typical arrangement is shown in Figure 20.9 and consists of four parts—exciter, specimen, electron-energy analyzer, and measurement system. The drawing shows both x-ray and electron excitation for convenience in explaining the function of the instrument. The use of both in the same instrument is unlikely.

Photoelectrons are produced only during secondary excitation of x-ray spectra, so, for photoelectron spectrometry, the exciter is an x-ray tube. The beam is filtered to provide substantially monochromatic target-line radiation. Of course, some Auger electrons are produced along with the photoelectrons. Auger electrons are produced during both primary and secondary excitation of x-ray spectra. However, primary excitation is more efficient by two or three orders, and the greater the number of orbital electron vacancies created, the greater is the probability of Auger-electron production. Thus, for Auger-electron spectrometry, the exciter is an electron gun. Many scattered and secondary electrons are produced along with the Auger electrons, and instruments having electron-gun excitation are often referred to as secondary-electron spectrometers.

The specimen is usually a solid having a plane, relatively smooth surface. A rotable wheel holding many specimens is provided to permit chang-

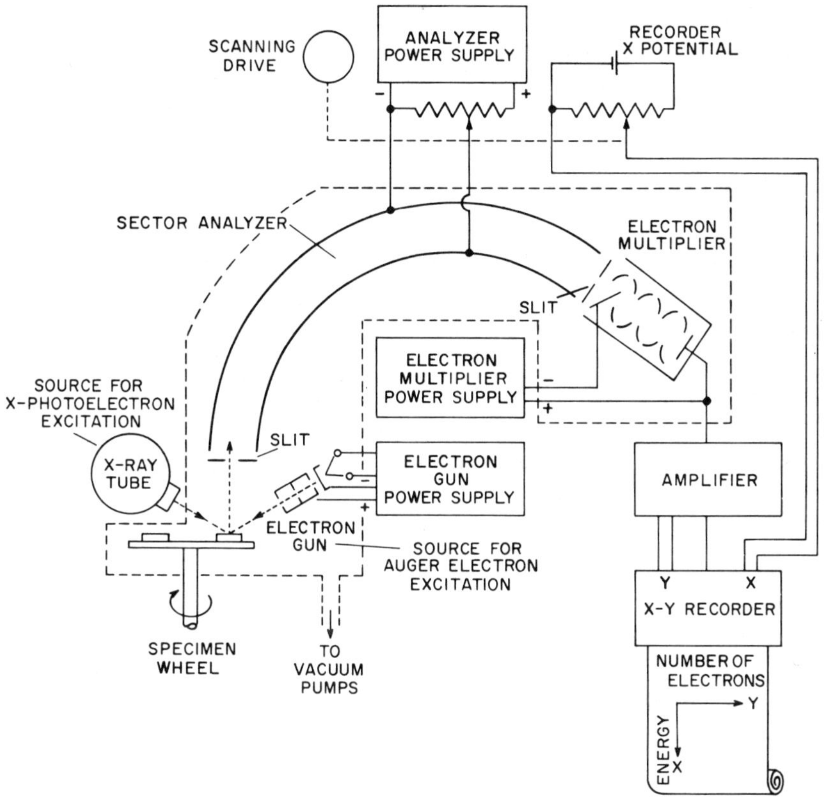

Fig. 20.9. Instrument for x-ray photoelectron spectrometry and Auger-electron spectrometry. The instrument is shown with both x-ray and electron excitation for illustration only.

ing specimens without breaking the vacuum. Provision may be made for heating the specimen.

The photo, Auger, secondary, and scattered electrons ejected from the specimen by the x-ray tube or electron gun enter the electron analyzer through an entrance slit. The analyzer shown in the figure is a 127° electrostatic electron analyzer in which electrons are sorted according to energy by passing them between two charged, parallel, curved plates. With a given potential across the plates, only electrons having a certain velocity pass through the analyzer; all others strike the wall of the positive plate. The potential applied to the plates is varied continuously over a specified range by a motor-driven potentiometer, thereby permitting electrons having a range of energies to pass in sequence. The energy scan requires ~5 min.

The electrons that pass through the analyzer enter the electron multiplier, where each produces an output pulse of 10^6 or more electrons. The electron multiplier functions the same way as that in a multiplier phototube. The multiplier output is amplified, electronically differentiated, and passed to the Y input of the X–Y recorder. The X input comes from a potentiometer coupled to the electron-energy motor drive as shown.

The entire electron path, including the electron gun, is enclosed in a vacuum housing. Although some work is done with oil-diffusion pumps and rubber or other greased gaskets, Auger electrons are so extremely sensitive to surface contamination that metal gaskets, ion pumps, and pressures down to 10^{-8} mm-Hg (torr) or less are preferable.

The recorder may plot a simple electron-energy distribution curve—electron current *vs.* energy. Such curves are satisfactory for photoelectron spectrometry because secondary excitation produces only photo and relatively few Auger electrons. However, in Auger-electron spectrometry, the Auger peaks are small, poorly defined, and superposed on a relatively high background of scattered and secondary electrons. The Auger peaks are accentuated by displaying the derivative of the energy distribution curve with respect to energy—in effect, a plot of the instantaneous slope of the energy distribution curve. The electronic differentiation mentioned above produces this plot, and the great improvement is evident in Figure 20.8. Alternatively, the energy spectrum may be scanned manually in small increments and the amplified electron multiplier output pulses counted by a scaler–timer. The photoelectron spectra in Figure 20.7 were plotted from data obtained in this way.

An Auger-electron spectrometer is available commercially from Varian.

20.4. X-RAY-EXCITED OPTICAL-FLUORESCENCE SPECTROMETRY

X-rays excite optical fluorescence in many solids and gases. The phenomenon is especially prominent among trivalent rare-earth ions. X-rays excite most rare-earth impurities in Y_2O_3, CeO_2, La_2O_3, Gd_2O_3, and Lu_2O_3 to emit intense, highly characteristic optical line fluorescence in the 3–10-kÅ region. The excitation mechanism is not fully understood, but it is believed that x-ray-excited photoelectrons preferentially transfer their energy to the impurity atoms by inelastic scattering. Minimum detectable concentrations in any of these five matrixes are of the order 1 ng/g (ppb) for neodymium, terbium, and dysprosium; 10 ng/g for samarium, europium, and thulium;

and 100 ng/g for praseodymium, gadolinium, holmium, erbium, and ytterbium. Yttrium, lanthanum, and lutetium emit no such spectra; the cerium spectrum is farther in the infrared region; and the gadolinium spectrum is relatively simple. In rare-earth oxide matrixes other than these five, the impurity spectra are severely quenched by the host oxide. This difficulty is overcome by 1000-fold dilution with Y_2O_3, but the sensitivity is decreased 1000-fold by the dilution.

The instrument is shown in Figure 20.10 (*B57*, *L17*). X-rays from an x-ray spectrometer excitation tube operating at 50 kV, 50 mA irradiate the specimen packed in a shallow planchet. The optical fluorescence is reflected by a series of mirrors through a rotating chopper into an optical grating monochromator. The monochromator disperses and scans the optical spectrum, permitting the multiplier phototube to measure the intensity at each wavelength in turn. The detector output is amplified and plotted on the recorder as a graph of intensity *vs.* wavelength. The chopper interrupts the fluorescent beam at a rate ~13 times per second, so that the detector output can be amplified by an ac, rather than dc, amplifier.

X-ray excitation has the advantage that the rare-earth ions are excited to higher energy states than would be the case with ultraviolet excitation. Thus, all of the allowed optical transitions can be observed. Another advan-

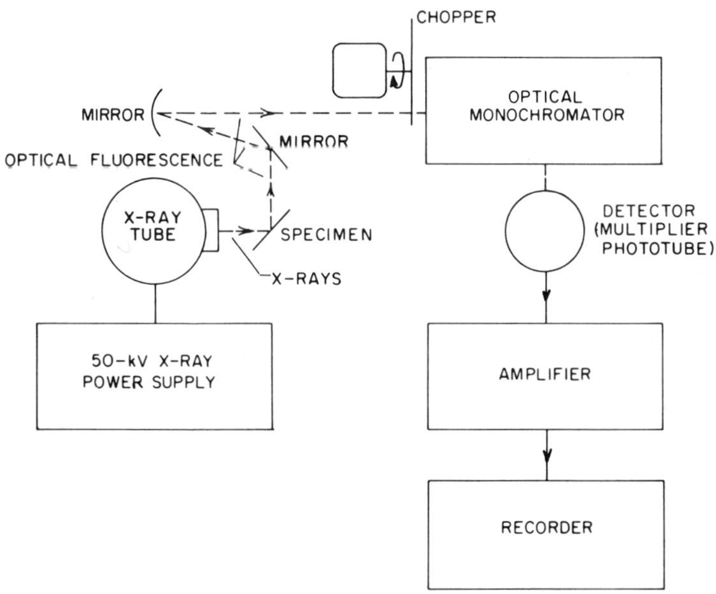

Fig. 20.10. X-ray-excited optical fluorescence spectrometer.

tage is that there is no interference of excitation and fluorescence radiation, because the x-rays are not transmitted by the optical system. At present, the method is applied principally to analysis of rare-earth oxides and to investigation of energy levels in solids.

Chapter 21

The Electron-Probe Microanalyzer

The electron-probe microanalyzer (sometimes termed electron microprobe analyzer) is basically an electron-probe x-ray primary-emission spectrometer. X-ray spectra of the elements in a specimen area ~1 μm in diameter are excited by a fine-focus electron beam. The x-radiation is then dispersed, detected, and read out in a more or less conventional x-ray spectrometer system. However, the interaction of the electron beam with the specimen results in several other phenomena that provide the basis for other types of analytical information. Thus, the more general term electron-probe microanalyzer is more appropriate. For good introductions to the method, see Chapters 8–11 in (*1*), Chapter 7 in (*6*), and (*5*) and (*15C*).

21.1. THE INSTRUMENT

A highly simplified diagram of an electron-probe microanalyzer is shown in Figure 21.1. It consists of an evacuated electron-optical column very similar to that of an electron microscope, one or more attached x-ray spectrometers, and an electronics console.

At the top of the column is an electron gun. Electrons emitted by an incandescent tungsten hairpin filament are formed into a thin beam and accelerated to high velocity by an anode-aperture arrangement operated at 5–60 kV dc. Two electromagnetic electron lenses demagnify and focus the beam to a final cross-over ~1 μm in diameter at the specimen plane. The lenses consist of cylindrical soft-iron housings having cross-sectional form similar to those shown. The lower portions of the cylinders are formed into pole pieces having apertures to pass the electron beam. The cylinders house electromagnetic coils through which the lens current passes.

A light microscope, usually $\sim$400×, is provided to view the specimen. Provision is made to translate the specimen in two mutually perpendicular (X and Y) directions either manually, or by motor drive at rates ranging typically from 1 to 100 μm/min. Such translation is required to bring a specified area into position for analysis and for making line-scan profiles (see below). Provision is also made to elevate and lower the specimen on the column (Z) axis and to rotate the specimen in its own plane. The specimen must be electrically conductive; otherwise, electrons will accumulate on it until it is charged sufficiently negative to repel the electron beam. Nonconductive specimens are made conductive by application of an evaporated film of carbon and by painting the area not to be investigated with a dag of carbon, aluminum, or silver.

The electron beam can be made to scan the specimen surface by means of two horizontal, mutually perpendicular deflection coils connected to saw tooth sweep (time-base) generators. The X-sweep coils deflect the electron beam forth and back in the X direction, say, 200 times a second, while the Y-sweep coils deflect it in the Y direction once each second. Actually, two sets of deflection coils like the one shown are required, one set above the other. If only one set were used as shown, the electron beam would be deflected away from the second aperture. The lower coils deflect the beam back, that is, in the direction opposite that imparted to it by the upper coils, so that it passes through the second aperture, but at a small angle to the column axis. Alternatively, a single set of coils like the one shown is sufficient if it is placed below the second lens aperture. By appropriate use of the deflection system, the specimen area covered by the electron beam can be made to assume three forms—point, line, or rectangular area:

1. With neither X nor Y deflection, the intersection of the x-ray beam with the specimen surface is a fixed point.

2. With only X deflection, the electron beam scans a line across the specimen. By varying the X-size control shown, which varies the amplitude of the X-sweep current, the length of the scanned line may be varied from zero to perhaps 500 μm. The line scan rate can be varied; typical fast and slow limits are 100/s to once every 6 s.

3. With both X and Y deflection, the electron beam scans a rectangular (usually square) line raster similar to that on a television picture tube.

The number of lines in the raster may be varied, typically from 100 to 500 lines per frame. By varying the amplitude of the X and Y sweep current with the X and Y size controls, the size of the raster may be varied from a point to an area perhaps 500 μm square. The frame scan rate can be varied; typical fast and slow limits are 1–150 s/frame.

The X and Y sweep generators also provide deflection for the cathode-ray oscilloscope. Thus, the electron beam in the cathode-ray tube scans a line or line raster on its fluorescent display screen in synchronism with the electron beam on the specimen surface. The cathode-ray display is typically 5–10 cm square.

The electronics console houses the following components: (1) filament and 60-kV dc supply for the electron gun; these supplies must be extremely well regulated ($\pm$0.01% or less), not so much for constancy of x-ray emission, but for constancy of electron spot diameter and position; (2) electron lens current supplies; (3) X and Y sweep generators; (4) backscattered electron current amplifier; (5) specimen current amplifier; (6) a complete electronic readout system for each x-ray spectrometer, including amplifier, pulse-height selector, ratemeter, scaler, and timer; (7) detector tube power supply; (8) multichannel recorder, having a channel for each of the x-ray ratemeters, the backscatter current, and the specimen current; and (9) cathode-ray oscilloscope; the display tube phosphor must have high persistence because of the relatively slow scanning rates, and highly actinic fluorescence, to permit photography of the displays in reasonably short exposure times.

21.2. INTERACTION OF THE ELECTRON BEAM AND SPECIMEN

At the point of intersection of the electron beam and specimen, eight phenomena occur that have been used to provide analytical information.

a. X-Rays. The incident electron beam excites a continuous x-ray spectrum and the characteristic x-ray spectra of the elements in the bombarded incremental area. In fact, the vacuum column and specimen constitute an x-ray tube having a very small focal spot.

b. Backscattered Electrons. Some of the incident electrons are backscattered. The fraction so scattered varies with the effective atomic number of the incremental area; about half the electrons are scattered by the heaviest elements, very few by the lightest. Backscatter is also strongly influenced by surface topography.

c. Secondary Electrons. The incident electrons eject some true secondary electrons from the specimen surface; the number is influenced by composition, topography, and surface potential.

d. Absorbed Electrons. Some of the incident electrons are collected by the specimen and constitute what is variously termed the absorbed-electron, specimen, target, or beam current, analogous to the x-ray tube current. The fraction so collected varies inversely with atomic number; nearly all are collected by beryllium, ~50% by uranium. The backscattered and absorbed electrons are essentially complementary, and the observed current is the beam current minus the backscattered electrons.

e. Transmitted Electrons. If the specimen is extremely thin, electrons may pass through it in numbers inversely proportional to its thickness and atomic number.

f. Electromotive Force. If the specimen is a semiconductor, the incident electrons may create a charge pattern over the surface. This charge pattern is influenced by composition, dopant, oxide and diffusion layers, and semiconductor type (*n*, *p*, or *i*), etc. If the specimen is a semiconductor device, it may be observed in operation, in which case the observed potential distribution is affected by applied potentials, electric connections, etc.

g. Kossel Patterns. The electron beam excites a spherically divergent point source of x-rays on the top surface of the specimen, and a cone of x-rays is transmitted through the specimen if it is thin enough. In the absence of diffraction, a photographic plate placed below the specimen would be uniformly exposed by the transmitted x-rays. However, for thin single crystals, a pattern of dark and light circles, parabolas, and hyperbolas results from complex diffraction effects. These *Kossel lines* convey information about crystallographic perfection, orientation, lattice parameter, stress, etc.

h. Cathodoluminescence. The incident electron beam may excite luminescence in the visible and ultraviolet regions, particularly with mineral specimens.

Appropriate means may be provided to measure any of these phenomena.

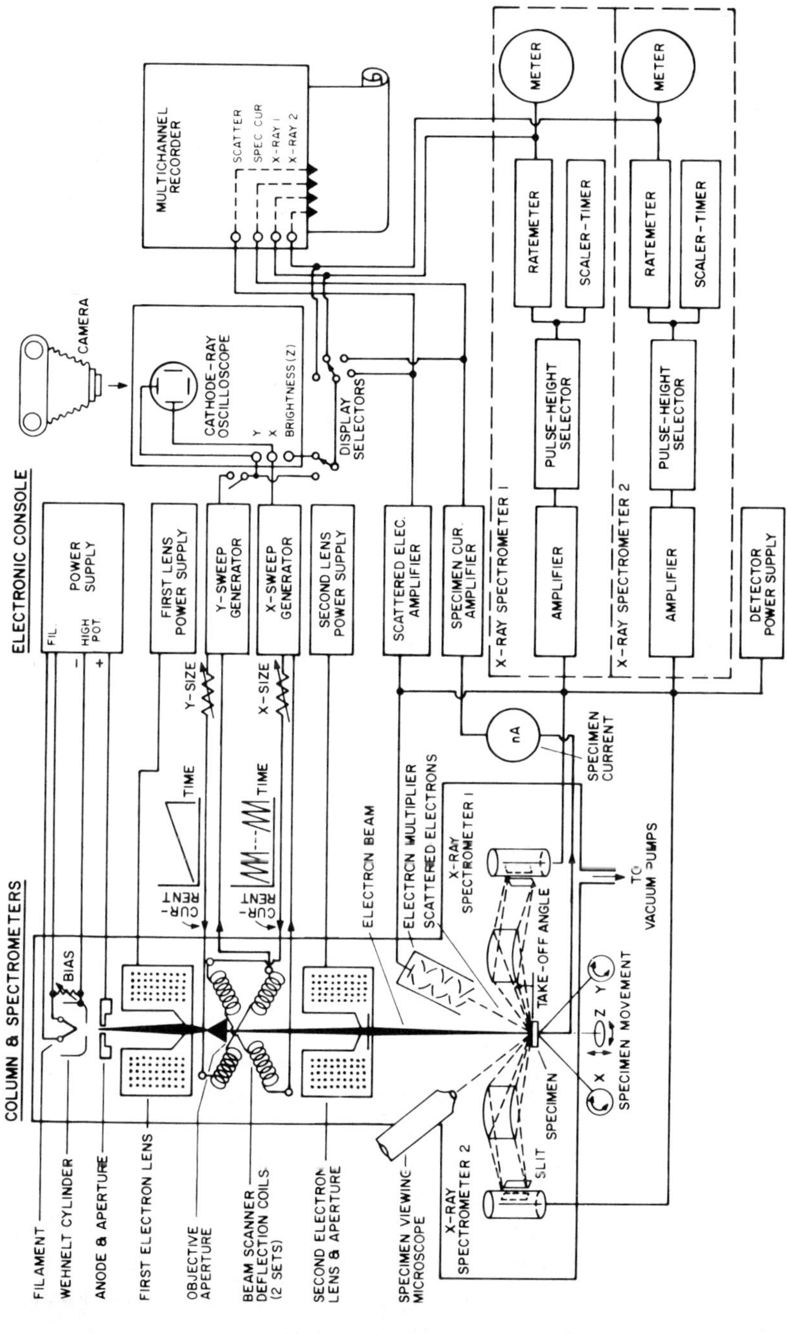

Fig. 21.1. Electron-probe microanalyzer.

Characteristic x-ray spectra are detected and measured by one or (more often) two x-ray spectrometers arranged as shown in Figure 21.1, each complete with electronic detection and readout systems. The point source of x-rays is ideal for use with curved crystals, and Soller collimators are not required. Each spectrometer has its detector, usually a sealed or flow-proportional counter. Each spectrometer may be set manually or motor-driven to permit making 2θ scans. The entire path from specimen to detector can be evacuated, and, with appropriate crystals and detectors, K spectra of elements down to beryllium can be measured. The combination of primary excitation, curved-crystal optics, and vacuum path results in high excitation and detection efficiency. As a result, L lines of elements of intermediate atomic number and M lines of elements of high atomic number are used more frequently than in x-ray secondary-emission spectrometry. Alternatively, one or both spectrometers may be operated with air path, in which case a vacuum-tight beryllium x-ray window is provided between the column and the spectrometer. The crystal spectrometers can also be operated in nondispersive mode, or a separate nondispersive channel may be provided.

The backscattered electrons may be detected by an electron multiplier placed at a relatively large angle with respect to the specimen surface as shown. The multiplier output is proportional to the atomic number of the incremental area under the beam. However, a much more useful arrangement is the following. Two small solid-state electron detectors are mounted near the specimen surface, facing each other on opposite sides of the electron beam, and inclined at the same angle to the specimen surface. This is the arrangement for stereo viewing. The sum of the signals from the two detectors gives the same information as the single electron multiplier—signal current approximately proportional to atomic number. The difference signal from the two detectors can be used to provide stereoscopic displays of surface topography (see below).

Secondary electrons are detected by an electron multiplier at a smaller angle with respect to the surface. The absorbed electrons, of course, simply constitute the beam or specimen current. Transmitted electrons are collected in a Faraday cup placed below the specimen. Electromotive force is detected by potential sensors placed above and near the specimen surface.

Kossel patterns are recorded on a photographic plate placed below the specimen. In practice, a Kossel camera assembly is inserted in the specimen compartment. Cathodoluminescence is detected by conveying the luminescence from the specimen surface to a multiplier phototube either through a Lucite light pipe or by means of a lens–mirror system.

21.3. MEASUREMENTS AND DISPLAYS

Electron-probe x-ray microanalyzers may be fitted to detect, measure, and display any or all of the phenomena listed above. However, the following discussion is limited to the three most commonly applied phenomena—characteristic x-rays, backscattered electrons, and specimen current. The instrument is capable of great versatility in measurement and display of these quantities. X-ray intensities can be measured with the scaler–timers of the spectrometers. X-ray intensity from both ratemeters and scatter and specimen current can all be recorded simultaneously on the multichannel recorder or one at a time on the cathode-ray oscilloscope, as selected by the display selector switch shown. Of course, additional oscilloscopes can be provided. Specimen current may be read on the beam current panel nanoammeter.

The measurements and displays can be made in many different ways as follows.

a. Mode I. No X or Y Deflection. The electron beam intersects the specimen plane at a fixed point. An electron-probe primary-emission x-ray spectrometer operating in this mode is exactly analogous to a secondary-emission spectrometer fitted with a selected-area aperture (Chapter 19) and provides the same types of information in the same forms. However, the selected are a is ~1 μm in diameter for the electron probe, 0.01–1 mm with the "x-ray milliprobe."

1. The specimen can be placed so the electron beam falls on a point of interest and a qualitative analysis made by recording a 2θ scan on one of the spectrometers, or a quantitative analysis made by scaling the analyte-line intensities. The 2θ-scan charts are similar to those shown in Figure 10.1.

2. The concentration distribution profile of a selected analyte along a line across the specimen can be mapped by setting one of the spectrometers at the analyte spectral line, moving the specimen manually in small increments, and scaling the intensity at each point. Profiles of two analytes can be mapped simultaneously, one on each of the two spectrometers. Subsequent scans can be made along the same line for one or two different analytes, or along different lines for the same analyte(s). The data are plotted manually, giving a plot similar to that shown in Figure 19.13 (left) except for the distances.

3. The same distribution profiles can be mapped by translating the specimen with the motor drive while recording analyte-line intensity on the recorder. Profiles of two analytes, one on each spectrometer, can be re-

corded simultaneously on the multichannel recorder. Except for distances, the profiles are similar to those shown in Figure 19.13 (right).

4. When the motor drive is used, the backscatter or specimen current profiles can be recorded simultaneously with one or two analyte-line intensities. These profiles reveal major changes in atomic number or surface topography, which may be correlated with the concentration profiles.

b. Mode II. X Deflection Only (One-Dimensional Analysis). The electron beam scans a line across the specimen surface. The length of the line can be varied from 0 to ~500 μm by use of the *X*-size control.

1. The concentration distribution profile of a selected analyte along a line can be mapped by setting one of the spectrometers at the analyte line, allowing the electron beam to line-scan the specimen, and displaying analyte-line intensity on the oscilloscope. The *X*-sweep coil deflects the electron beam in the cathode-ray tube across the display screen in synchronism with the electron beam across the specimen surface. The spectrometer ratemeter output then provides the *Y* deflection for the cathode-ray tube, that is, deflects the cathode-ray beam vertically in proportion to instantaneous analyte-line intensity. The result is a bright-line cathode-ray trace analogous to the ratemeter recorder chart in technique I-3 above. Such line scans are shown in Figures 21.3 and 21.5.

2. The backscatter and specimen current profiles along the line can be displayed in the same way.

c. Mode III. X and Y Deflection (Two-Dimensional Analysis). The electron beam scans a line raster on the specimen surface. The dimensions of the raster can be varied from nearly a point to ~500 μm square by use of the *X*- and *Y*-size controls.

1. The concentration distribution of a selected analyte over a specimen area can be displayed on the oscilloscope. The *X*- and *Y*-sweep coils deflect the electron beam in the cathode-ray tube in a line raster in synchronism with the electron beam on the specimen surface. The spectrometer ratemeter output then modulates the *intensity* of the cathode-ray beam in proportion to instantaneous analyte-line intensity, so the brightness of each incremental area on the display is proportional to analyte-line intensity—and concentration—at the corresponding incremental area on the specimen.

2. Backscatter and specimen current can be displayed in the same way. The higher the effective atomic number of the incremental specimen area under the electron beam, the brighter is the corresponding point in the backscatter pattern, the darker is the point in the specimen-current pattern.

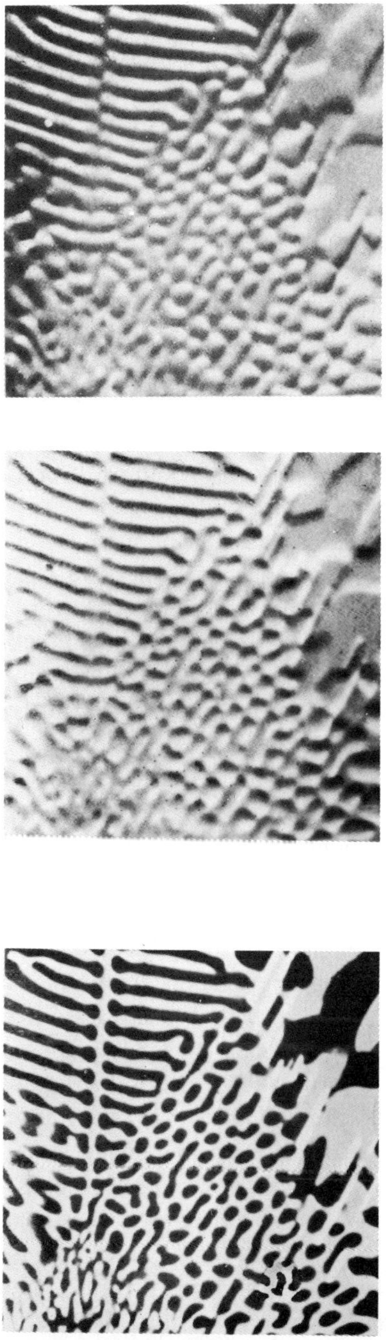

COMPOSITION

(+) TOPOGRAPHY (−)

Fig. 21.2. Backscattered electron images of 67Al–33Cu alloy. [Courtesy of JEOLCO (USA), Inc.]

When the electron-probe microanalyzer is operated with the backscattered electrons displayed in this way, the instrument is, in effect, a scanning electron microscope.

3. If two backscatter detectors are provided as described above, and if their difference signal is used to modulate the brightness of the cathode-ray display, the backscatter display has the appearance of a stereoscopic map of specimen surface topography. In some cases, the relief is very striking. The sum signal of the two detectors gives the same display as a single detector. Figure 21.2 shows backscattered electron displays in composition and topography modes.

The magnification of the area displays can be varied by changing the dimensions of the scanned area on the specimen while leaving the dimensions of the display substantially the same. A typical magnification range is 300–10,000×.

In some instruments, mixed signals from both x-ray spectrometers can be displayed simultaneously on the same oscilloscope. Likewise, mixed signals from an x-ray spectrometer and backscatter or specimen current can be so displayed. Provision may be made to reverse the positive–negative polarity of the area displays to aid in their visual conception and interpretation. This feature is particularly beneficial in topographic backscattered electron displays.

On most instruments it is possible to: (1) select on an x-ray, electron, or specimen current raster display a line of particular interest; (2) mark it electronically with a bright line on the display; (3) superpose on the display a line scan of an x-ray intensity profile of a selected analyte or of a scattered or absorbed electron profile. Such superpositions facilitate correlation of concentration and topography. X-ray intensity profiles superposed on topographic displays are particularly beneficial in studies of interfacial diffusion zones, as shown in Figure 21.3.

4. A particularly interesting type of oscilloscope raster display is provided by combination of the line and raster modes. Each line in the raster is deflected to represent the analyte x-ray line-intensity profile as already explained—that is, to represent the analyte concentration profile along the corresponding line across the specimen. Figure 21.4C shows such an "oblique-projection, deflection-modulation" raster image of the Au $M\alpha$ intensity from a section of a structure consisting of alternate gold and copper layers. The gold layers have thicknesses 0.2, 0.3, 0.6, 0.8, 1, 2, and 4 μm, respectively. The display conveys a sort of three-dimensional representation of gold concentration. An individual Au $M\alpha$ line scan and an absorbed electron image of the sandwich are shown in Figures 21.4B and 21.4A, respectively, for comparison.

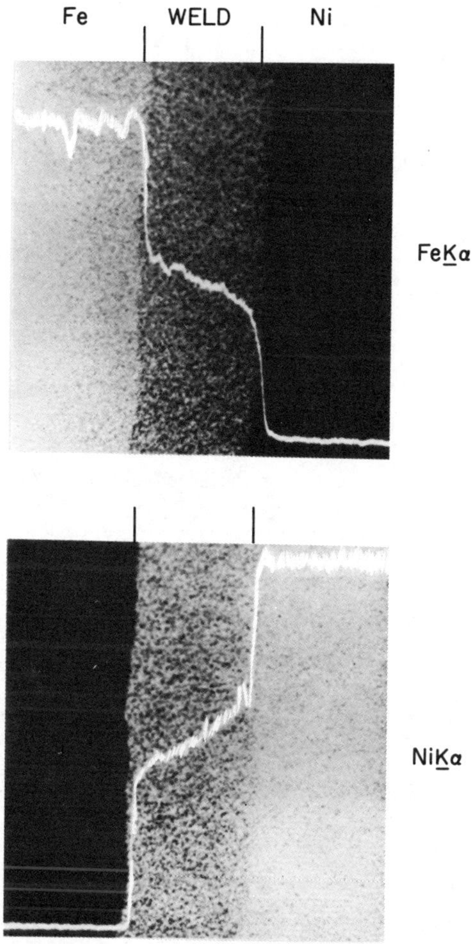

Fig. 21.3. Fe $K\alpha$ and Ni $K\alpha$ raster images with superposed line scans of an iron-to-nickel weld. [Courtesy of JEOLCO (USA), Inc.]

The light-microscope image and any of the oscilloscope displays can be photographed. Usually, a Polaroid camera is used. Figures 21.2 to 21.4 have been cited already. Figure 21.5 shows a photomicrograph and most of the types of oscilloscope display for a heterogeneous mineral.

For all displays, the region of interest is selected approximately by use of the light microscope and specimen translators. Then the spot, line, or area is selected accurately and marked electronically on the oscilloscope raster displays. After the electron beam has bombarded the selected region

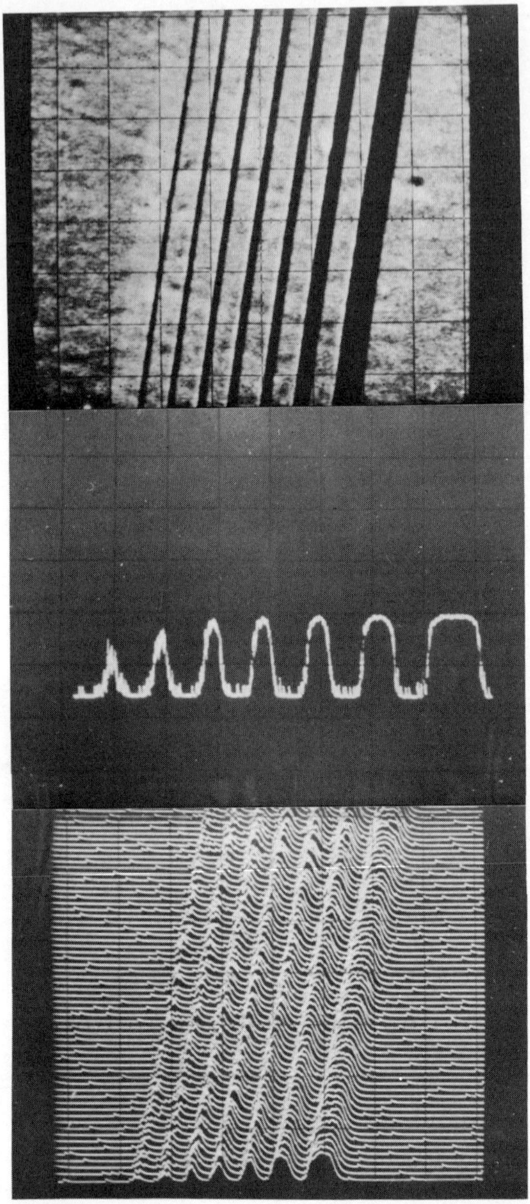

A

B

5 μm

C

Fig. 21.4. Electron-probe microanalyzer displays of a section of a layered gold–copper specimen in which the gold layers have thicknesses 0.2, 0.3, 0.6, 0.8, 1, 2, and 4 μm, respectively. [Courtesy of Materials Analysis Company.] A. Absorbed electron image. B. Au $M\alpha$ x-ray line scan. C. Au $M\alpha$ oblique-projection, deflection-modulation raster image.

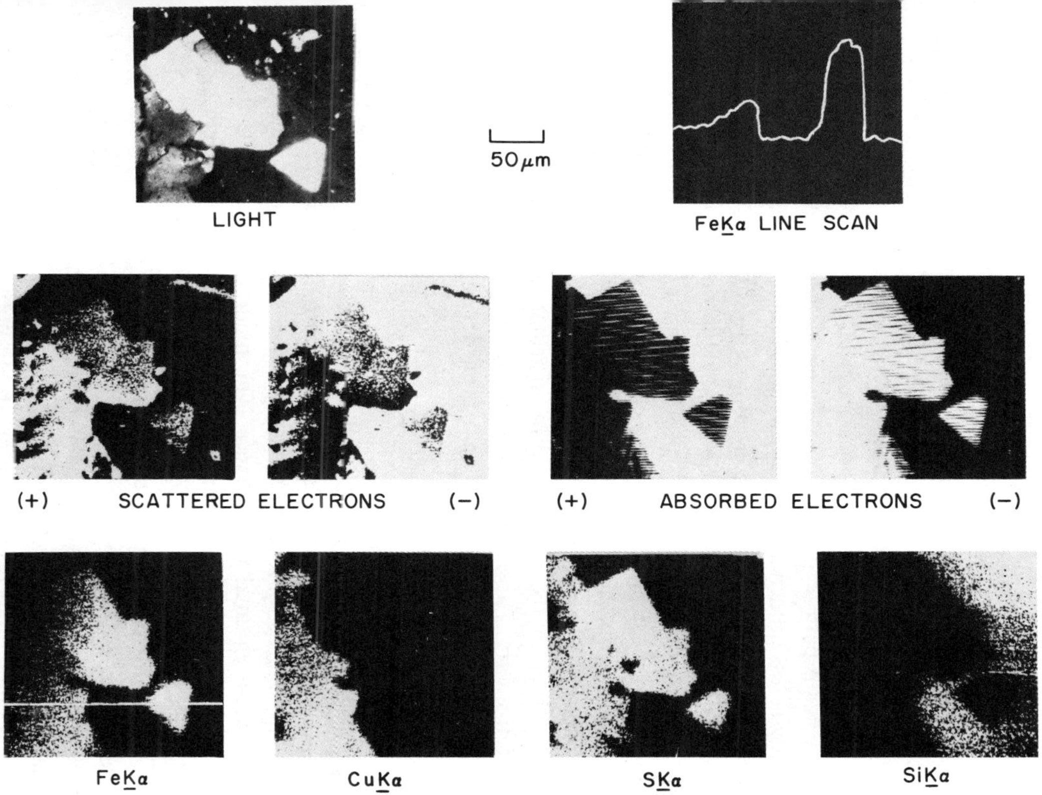

Fig. 21.5. Electron-probe microanalyzer displays. The specimen was a heterogeneous mineral consisting mostly of pyrite (FeS_2) and chalcopyrite [$(Cu,Fe)S_2$] with some siliceous material. The Fe $K\alpha$ line scan was made at the position indicated by the bright-line marker on the Fe $K\alpha$ raster display. [Courtesy of JEOLCO (USA), Inc.]

for a short time, a visible dark spot, line, or square appears on the specimen viewed through the microscope. This darkening results from carbonization, by the electron beam, of adsorbed organic vapors from the vacuum pumps. These deposits interfere with the x-ray measurements only at very long wavelengths and may be beneficial in that they locate precisely the spot, line, or area investigated. The deposits appear on photomicrographs of the specimen and can be correlated with the analytical data and displays.

The angle between the specimen plane and the central ray of the x-ray cone accepted by the spectrometer is the *takeoff angle*. In general, relatively large ($\geq 15°$) takeoff angles are advantageous for two reasons: the intensity–concentration relationship is more linear, especially for low-Z elements, because x-ray beam path length in the specimen is shorter. Also, the measured intensity is less influenced by shadowing effects of surface topography.

The depth of the surface layer contributing to the measured x-rays in secondary-emission spectrometry is limited by the depth from which the characteristic x-rays can emerge. In primary-emission spectrometry, the limit is imposed by the depth to which the electrons penetrate, which is only a few micrometers, compared with $\sim 10^{-2}$ cm for secondary excitation. As a result, absorption-enhancement effects are *less* serious and surface texture effects *more* serious in the electron-probe instrument than in secondary-emission spectrometry. However, absorption-enhancement effects do occur and elaborate correction procedures have been devised for them.

21.4. APPLICATIONS

Table 21.1 gives performance data for a modern electron-probe microanalyzer functioning as an x-ray primary-emission spectrometer. The data probably represent optimal performance. The method is applicable to all chemical elements down to beryllium (Z 4). Background is relatively high with electron excitation, but line-to-background ratios are excellent. An increase of the beam spot diameter from 1 μm to, say, 3–5 μm increases the beam current from ~ 1 μA to ~ 10 μA and increases the net analyte-line intensity proportionally. However, background also increases proportionally, and sensitivity remains substantially unimproved.

If beam diameter and penetration depth are both ~ 1 μm, the effective specimen volume is of the order 1 μm^3 or 10^{-12} cm^3. For a density of 10 g/cm^3, the effective specimen mass is of the order 10^{-11} g, and, at the detection limit, say 0.01%, the contributing analytical mass is only 10^{-15} g.

The electron-probe x-ray microanalyzer fills the great need for in-place

TABLE 21.1
Performance of a Modern Electron-Probe Microanalyzer[a]

Atomic number	Element	Specimen	Crystal[b]	Line intensity,[c] I_L, $\frac{\text{counts/s}}{\mu\text{A}}$	Line to background ratio	Limit of detectability,[d] %
4	Be	Be	PbLig	6.0×10^2	50	0.055
5	B	B	PbSt	1.5×10^4	50	0.051
6	C	C	PbSt	4.0×10^4	70	0.027
8	O	SiO_2	PbSt	3.0×10^4	20	0.031
9	F	LiF	KHP	1.1×10^4	280	0.017
11	Na	NaCl	KHP	3.2×10^4	550	0.004
12	Mg	Mg	KHP	4.0×10^5	1500	0.0016
13	Al	Al	KHP	8.0×10^5	1300	0.0014
14	Si	Si	KHP	1.2×10^6	800	0.0014
22	Ti	Ti	PET	1.8×10^6	700	0.0013
26	Fe	Fe	LiF	2.5×10^5	460	0.0042
29	Cu	Cu	LiF	1.5×10^5	230	0.0066

[a] Courtesy of JEOLCO (USA), Inc.
[b] The symbols indicate, respectively, lead lignocerate, lead stearate, potassium hydrogen phthalate, pentaerythritol, and lithium fluoride.
[c] All $K\alpha$ lines; accelerating potential 25 kV.
[d] Defined as $[2(2I_BC)^{1/2}]/(I_L - I_B)\sqrt{T}$; where I_L and I_B are analyte-line and background intensities, respectively; C is analyte concentration (wt %); and T is counting time (s).

nondestructive determination of composition and distribution on a microscopic scale in analytical chemistry, metallurgy, ceramics, mineralogy, geology, and biology and medicine. Among the many classes of applications are the following: (1) analysis of crystallites or grains of individual phases in heterogeneous alloys, ceramics, and minerals; (2) analysis of microscopic precipitates, inclusions, and segregations in the same materials; (3) analysis of precipitates at grain boundaries; (4) analysis of individual layers in sections of coated, plated, or corroded surfaces or of bonded surfaces; (5) concentration gradients at metal-to-metal or metal-to-ceramic interfaces, diffusion couples, heterogeneous grain boundaries, etc.; (6) evaluation of micro-scale homogeneity; (7) mapping distribution of elements over the surfaces of heterogeneous alloys, ceramics, minerals, etc.; (8) analysis of very small particles filtered from gases and liquids; (9) determination and distribution of natural and foreign elements in biological materials; and (10) analysis of specific areas on small semiconductor electron devices and other very small parts.

APPENDIX 1

Wavelengths of the Principal X-Ray Spectral Lines of the Chemical Elements—*K* Series (*51; 49*)

Wavelength, Å

Line	α	α_1	α_2	β_1	β_3	β_2
Electron transition $K \leftarrow$		*L*III	*L*II	*M*III	*M*II	*N*II,*N*III
Approximate relative intensity	150	100	50	20		5
3-Li	240	—	—	—	—	—
4-Be	113	—	—	—	—	—
5-B	67	—	—	—	—	—
6-C	44	—	—	—	—	—
7-N	31.603	—	—	—	—	—
8-O	23.707	—	—	—	—	—
9-F	18.307	—	—	—	—	—
10-Ne	14.615	—	—	14.460	—	—
11-Na	11.909	—	—	11.617	—	—
12-Mg	9.889	—	—	9.558	—	—
13-Al	8.339	8.338	8.341	7.981	—	—
14-Si	7.126	7.125	7.127	6.769	—	—
15-P	6.155	6.154	6.157	5.804	—	—
16-S	5.373	5.372	5.375	5.032	—	—
17-Cl	4.729	4.728	4.731	4.403	—	—
18-Ar	4.192	4.191	4.194	3.886	—	—
19-K	3.744	3.742	3.745	3.454	—	—
20-Ca	3.360	3.359	3.362	3.090	—	—
21-Sc	3.032	3.031	3.034	2.780	—	—
22-Ti	2.750	2.749	2.753	2.514		—
23-V	2.505	2.503	2.507	2.285		—
24-Cr	2.291	2.290	2.294	2.085		—
25-Mn	2.103	2.102	2.105	1.910	—	—
26-Fe	1.937	1.936	1.940	1.757	—	—
27-Co	1.791	1.789	1.793	1.621	—	—
28-Ni	1.659	1.658	1.661	1.500	—	1.489
29-Cu	1.542	1.540	1.544	1.392	1.393	1.381
30-Zn	1.437	1.435	1.439	1.296	—	1.284
31-Ga	1.341	1.340	1.344	1.207	1.208	1.196
32-Ge	1.256	1.255	1.258	1.129	1.129	1.117
33-As	1.177	1.175	1.179	1.057	1.058	1.045

APPENDIX 1 (*continued*) Wavelength, Å

Line	α	α_1	α_2	β_1	β_3	β_2
Electron transition $K \leftarrow$		LIII	LII	MIII	MII	NII,NIII
Approximate relative intensity	150	100	50	20		5
34-Se	1.106	1.105	1.109	0.992	0.993	0.980
35-Br	1.041	1.040	1.044	0.933	0.933	0.921
36-Kr	0.981	0.980	0.984	0.879	0.879	0.866
37-Rb	0.927	0.926	0.930	0.829	0.830	0.817
38-Sr	0.877	0.875	0.880	0.783	0.784	0.771
39-Y	0.831	0.829	0.833	0.740	0.741	0.728
40-Zr	0.788	0.786	0.791	0.701	0.702	0.690
41-Nb	0.748	0.747	0.751	0.665	0.666	0.654
42-Mo	0.710	0.709	0.713	0.632	0.633	0.621
43-Tc	0.674	0.673	0.676	0.601	0.602	0.590
44-Ru	0.644	0.643	0.647	0.572	0.573	0.562
45-Rh	0.614	0.613	0.617	0.546	0.546	0.535
46-Pd	0.587	0.585	0.590	0.521	0.521	0.510
47-Ag	0.561	0.559	0.564	0.497	0.498	0.487
48-Cd	0.536	0.535	0.539	0.475	0.476	0.465
49-In	0.514	0.512	0.517	0.455	0.455	0.445
50-Sn	0.492	0.491	0.495	0.435	0.436	0.426
51-Sb	0.472	0.470	0.475	0.417	0.418	0.408
52-Te	0.453	0.451	0.456	0.400	0.401	0.391
53-I	0.435	0.433	0.438	0.384	0.385	0.376
54-Xe	0.418	0.416	0.421	0.369	—	0.360
55-Cs	0.402	0.401	0.405	0.355	0.355	0.346
56-Ba	0.387	0.385	0.390	0.341	0.342	0.333
57-La	0.373	0.371	0.376	0.328	0.329	0.320
58-Ce	0.359	0.357	0.362	0.316	0.317	0.309
59-Pr	0.346	0.344	0.349	0.305	0.305	0.297
60-Nd	0.334	0.332	0.337	0.294	0.294	0.287
61-Pm	0.322	0.321	0.325	0.283	0.284	0.276
62-Sm	0.311	0.309	0.314	0.274	0.274	0.267
63-Eu	0.301	0.299	0.304	0.264	0.265	0.258
64-Gd	0.291	0.289	0.294	0.255	0.256	0.249
65-Tb	0.281	0.279	0.284	0.246	0.246	0.239
66-Dy	0.272	0.270	0.275	0.237	0.238	0.231
67-Ho	0.263	0.261	0.266	0.230	0.231	0.224
68-Er	0.255	0.253	0.258	0.222	0.223	0.217
69-Tm	0.246	0.244	0.250	0.215	0.216	0.210

APPENDIX 1 (*continued*) Wavelength, Å

Line	α	α_1	α_2	β_1	β_3	β_2
Electron transition $K \leftarrow$		LIII	LII	MIII	MII	NII,NIII
Approximate relative intensity	150	100	50	20		5
70-Yb	0.238	0.236	0.241	0.208	0.209	0.203
71-Lu	0.231	0.229	0.234	0.202	0.203	0.197
72-Hf	0.224	0.222	0.227	0.195	0.196	0.190
73-Ta	0.217	0.215	0.220	0.190	0.191	0.185
74-W	0.211	0.209	0.213	0.184	0.185	0.179
75-Re	0.204	0.202	0.207	0.179	0.179	0.174
76-Os	0.198	0.196	0.201	0.173	0.174	0.169
77-Ir	0.193	0.191	0.196	0.168	0.169	0.164
78-Pt	0.187	0.185	0.190	0.163	0.164	0.159
79-Au	0.182	0.180	0.185	0.159	0.160	0.155
80-Hg	0.177	0.175	0.180	0.154	0.155	0.150
81-Tl	0.172	0.170	0.175	0.150	0.151	0.146
82-Pb	0.167	0.165	0.170	0.146	0.147	0.142
83-Bi	0.162	0.161	0.165	0.142	0.143	0.138
84-Po	0.158	0.156	0.161	0.138	0.139	0.134
85-At	0.154	0.152	0.157	0.134	0.135	0.131
86-Rn	0.150	0.148	0.153	0.131	0.132	0.127
87-Fr	0.146	0.144	0.149	0.127	0.128	0.124
88-Ra	0.142	0.140	0.145	0.124	0.125	0.120
89-Ac	0.138	0.136	0.141	0.121	0.121	0.117
90-Th	0.135	0.133	0.138	0.117	0.118	0.114
91-Pa	0.131	0.129	0.134	0.114	0.115	0.111
92-U	0.128	0.126	0.131	0.111	0.112	0.108
93-Np	0.125	0.123	0.128	0.109	—	0.105
94-Pu	0.122	0.120	0.125	0.106	—	0.103
95-Am	0.119	0.117	0.122	0.103	—	0.100
96-Cm	0.116	0.114	0.119	0.101	—	0.098
97-Bk	0.113	0.111	0.116	0.098	—	0.095
98-Cf	0.110	0.108	0.113	0.097	—	0.093

APPENDIX 2

Wavelengths of the Principal X-Ray Spectral Lines of the Chemical Elements—*L* Series (*51; 37, 49*)

Wavelength, Å

Line	α_1	α_2	β_1	β_2	β_3	β_4	β_5	γ_1	γ_2	γ_3	γ_4	γ_6	l	η
Electron transition														
*L*III ←	*M*V	*M*IV		*N*V			*O*IV,*O*V						*M*I	
*L*II ←			*M*IV					*N*IV				*O*IV		*M*I
*L*I ←					*M*III	*M*II			*N*II	*N*III	*O*III			
Approximate relative intensity	100	10	80	60	30	20	30	40	10	10	10	20	30	10
16-S	—	—	—	—	—	—	—	—	—	—	—	—	83.400	
17-Cl	—	—	—	—	—	—	—	—	—	—	—	—	67.840	67.250
18-Ar	—	—	—	—	—	—	—	—	—	—	—	—	56.212	56.813
19-K	—	—	—	—	—	—	—	—	—	—	—	—	47.835	47.325
20-Ca	36.393		36.022	—	—	—	—	—	—	—	—	—	41.042	40.542
21-Sc	31.393		31.072	—	—	—	—	—	—	—	—	—	35.671	35.200
22-Ti	27.445		27.074	—	—	—	—	—	—	—	—	—	31.423	30.942
23-V	24.309		23.898	—	21.890		—	—	—	—	—	—	27.826	27.375
24-Cr	21.713		21.323	—	19.429		—	—	—	—	—	—	24.840	24.339
25-Mn	19.489		19.158	—	17.575		—	—	—	—	—	—	22.315	21.864

26-Fe	17.602		17.290	—	15.710		—	—	—	—	—	—	20.201	19.730
27-Co	16.000		15.698	—	14.240		—	—	—	—	—	—	18.358	17.860
28-Ni	14.595		14.308	—	13.146		—	—	—	—	—	—	16.693	16.304
29-Cu	13.357		13.079	—	12.115		—	—	—	—	—	—	15.297	14.940
30-Zn	12.282		12.009	—	11.185		—	—	—	—	—	—	14.081	13.719
31-Ga	11.313		11.045	—	10.365		—	—	—	—	—	—	12.976	12.620
32-Ge	10.456		10.194	—	9.580	9.640	—	—	—	—	—	—	11.944	11.608
33-As	9.671		9.414	—	8.930		—	—	—	—	—	—	11.069	10.732
34-Se	8.990		8.735	—	8.321		—	—	—	—	—	—	10.293	9.959
35-Br	8.375		8.126	—	7.767		—	—	—	—	—	—	9.583	9.253
36-Kr	7.817		7.576	—	7.264	7.304	—	—	—	—	—	—	8.946	8.626
37-Rb	7.318	7.325	7.075	—	6.788	6.821	—	—	6.045		—	—	8.363	8.042
38-Sr	6.863	6.870	6.623	—	6.367	6.403	—	—	5.644		—	—	7.836	7.517
39-Y	6.449	6.456	6.211	—	5.983	6.018	—	—	5.283		—	—	7.356	7.040
40-Zr	6.070	6.077	5.836	5.586	5.632	5.668	—	5.384	4.953		—	—	6.918	6.606
41-Nb	5.725	5.732	5.492	5.238	5.310	5.346	—	5.036	4.654		—	—	6.517	6.210
42-Mo	5.406	5.414	5.176	4.923	5.013	5.048	—	4.726	4.380		—	—	6.150	5.847
43-Tc	5.114	5.123	4.887	4.636	4.737	4.773	—	4.440	4.138		—	—	5.819	5.518
44-Ru	4.846	4.854	4.620	4.372	4.487	4.523	—	4.182	3.897		—	—	5.503	5.204
45-Rh	4.597	4.605	4.374	4.130	4.253	4.289	—	3.944	3.685		—	—	5.217	4.922
46-Pd	4.368	4.376	4.146	3.909	4.034	4.071	—	3.725	3.489		—	—	4.952	4.660
47-Ag	4.154	4.162	3.935	3.703	3.834	3.870	—	3.523	3.307		—	—	4.707	4.418
48-Cd	3.956	3.965	3.739	3.514	3.644	3.681	—	3.336	3.137		—	—	4.480	4.193
49-In	3.752	3.781	3.555	3.339	3.470	3.507	—	3.162	2.980		2.926	—	4.269	3.983
50-Sn	3.600	3.609	3.385	3.175	3.306	3.344	—	3.001	2.835		2.778	—	4.071	3.789
51-Sb	3.439	3.448	3.226	3.023	3.152	3.190	—	2.852	2.695		2.639	—	3.888	3.607
52-Te	3.290	3.299	3.077	2.882	3.009	3.046	—	2.712	2.567		2.511	—	3.716	3.438
53-I	3.148	3.157	2.937	2.751	2.874	2.912	—	2.582	2.447		2.391	—	3.557	3.280

APPENDIX 2 (*continued*)

Wavelength, Å

Line	α_1	α_2	β_1	β_2	β_3	β_4	β_5	γ_1	γ_2	γ_3	γ_4	γ_6	l	η
Electron transition														
*L*III ←	*M*V	*M*IV		*N*V			*O*IV,*O*V						*M*I	
*L*II ←			*M*IV					*N*IV				*O*IV		*M*I
*L*I ←					*M*III	*M*II			*N*II	*N*III	*O*III			
Approximate relative intensity	100	10	80	60	30	20	30	40	10	10	10	20	30	10
54-Xe	3.015	3.025	2.803	2.626	2.745	2.784	—	2.462	2.338	2.331	—	—	3.421	3.143
55-Cs	2.892	2.902	2.683	2.511	2.628	2.666	—	2.348	2.237	2.233	2.174	—	3.267	2.994
56-Ba	2.776	2.785	2.567	2.404	2.516	2.555	—	2.242	2.138	2.134	2.075	—	3.135	2.862
57-La	2.665	2.674	2.458	2.303	2.410	2.449	—	2.141	2.046	2.041	1.983	—	3.006	2.740
58-Ce	2.561	2.570	2.356	2.208	2.311	2.349	—	2.048	1.960	1.955	1.899	—	2.892	2.620
59-Pr	2.463	2.473	2.259	2.119	2.216	2.255	—	1.961	1.879	1.874	1.819	—	2.784	2.512
60-Nd	2.370	2.382	2.166	2.035	2.126	2.166	—	1.878	1.801	1.797	1.745	1.855	2.675	2.409
61-Pm	2.283	2.292	2.081	1.956	2.042	2.081	—	1.799	1.729	1.724	—	—	2.591	2.322
62-Sm	2.199	2.210	1.998	1.882	1.962	2.000	1.779	1.726	1.659	1.655	1.606	—	2.482	2.218
63-Eu	2.120	2.131	1.920	1.812	1.887	1.926	—	1.657	1.597	1.591	1.544	—	2.395	2.131
64-Gd	2.046	2.057	1.847	1.746	1.815	1.853	—	1.592	1.534	1.529	1.485	—	2.312	2.049
65-Tb	1.976	1.986	1.777	1.682	1.747	1.785	1.577	1.530	1.477	1.471	1.427	—	2.234	1.973
66-Dy	1.909	1.920	1.710	1.623	1.681	1.720	—	1.473	1.423	1.417	1.374	—	2.158	1.898
67-Ho	1.845	1.856	1.647	1.567	1.619	1.658	—	1.417	1.371	1.364	1.323	—	2.086	1.826
68-Er	1.785	1.796	1.587	1.514	1.561	1.601	—	1.364	1.321	1.315	1.276	—	2.019	1.757
69-Tm	1.726	1.738	1.530	1.463	1.505	1.544	—	1.316	1.274	1.268	1.229	—	1.955	1.695
70-Yb	1.672	1.682	1.476	1.416	1.452	1.491	1.387	1.268	1.228	1.222	1.185	1.243	1.894	1.635

71-Lu	1.619	1.630	1.424	1.370	1.402	1.441	1.342	1.222	1.185	1.179	1.143	1.198	1.836	1.478
72-Hf	1.569	1.580	1.374	1.327	1.353	1.392	1.298	1.179	1.144	1.138	1.103	1.155	1.782	1.523
73-Ta	1.522	1.533	1.327	1.285	1.307	1.346	1.256	1.138	1.105	1.099	1.065	1.114	1.728	1.471
74-W	1.476	1.487	1.282	1.245	1.263	1.302	1.215	1.098	1.068	1.062	1.028	1.074	1.678	1.421
75-Re	1.433	1.444	1.238	1.206	1.220	1.260	1.177	1.061	1.032	1.026	0.993	1.037	1.630	1.374
76-Os	1.391	1.402	1.197	1.169	1.179	1.218	1.140	1.025	0.998	0.992	0.959	1.001	1.585	1.328
77-Ir	1.352	1.363	1.158	1.135	1.141	1.179	1.106	0.991	0.966	0.959	0.928	0.967	1.541	1.285
78-Pt	1.313	1.325	1.120	1.102	1.104	1.142	1.072	0.958	0.934	0.928	0.897	0.934	1.499	1.243
79-Au	1.277	1.288	1.083	1.070	1.068	1.106	1.040	0.927	0.905	0.898	0.867	0.903	1.460	1.202
80-Hg	1.242	1.253	1.049	1.040	1.034	1.072	1.010	0.897	0.876	0.869	0.839	0.873	1.422	1.164
81-Tl	1.207	1.218	1.015	1.010	1.001	1.039	0.981	0.868	0.848	0.842	0.812	0.845	1.385	1.127
82-Pb	1.175	1.186	0.982	0.983	0.969	1.007	0.953	0.840	0.822	0.815	0.786	0.817	1.350	1.092
83-Bi	1.144	1.155	0.952	0.955	0.939	0.977	0.926	0.814	0.796	0.790	0.761	0.791	1.317	1.058
84-Po	1.114	1.125	0.922	0.929	0.909	0.948	0.900	0.788	0.772	0.765	0.736	0.765	1.283	1.024
85-At	1.085	1.097	0.894	0.905	0.881	0.919	0.875	0.763	0.747	0.740	0.713	0.741	1.256	0.997
86-Rn	1.057	1.069	0.866	0.881	0.854	0.892	0.852	0.739	0.725	0.718	0.692	0.717	1.223	0.968
87-Fr	1.030	1.042	0.840	0.858	0.828	0.867	0.829	0.716	0.703	0.696	0.670	0.695	1.199	0.938
88-Ra	1.005	1.017	0.814	0.836	0.803	0.841	0.807	0.694	0.682	0.675	0.649	0.673	1.167	0.908
89-Ac	0.980	0.992	0.789	0.814	0.778	0.816	0.786	0.671	0.662	0.655	0.630	0.653	1.144	0.882
90-Th	0.956	0.968	0.766	0.794	0.755	0.793	0.765	0.653	0.642	0.635	0.611	0.632	1.115	0.855
91-Pa	0.933	0.945	0.742	0.774	0.732	0.770	0.746	0.634	0.624	0.617	0.594	0.613	1.091	0.830
92-U	0.911	0.923	0.720	0.755	0.710	0.748	0.726	0.615	0.605	0.598	0.577	0.595	1.067	0.806
93-Np	0.889	0.901	0.698	0.736	0.689	0.727	0.708	0.597	0.587	0.581	0.558	0.577	1.043	0.781
94-Pu	0.869	0.880	0.678	0.720	0.669	0.707	0.691	0.579	0.571	0.564	0.542	0.560	1.023	0.759
95-Am	0.849	0.860	0.658	0.701	0.649	0.686	0.674	0.562	0.554	—	—	0.543	—	—
96-Cm	0.829	0.841	0.639	0.685	—	—	—	0.546	—	—	—	—	—	—
97-Bk	0.810	0.822	0.621	0.669	—	—	—	0.530	—	—	—	—	—	—
98-Cf	0.792	0.804	0.603	0.653	—	—	—	0.515	—	—	—	—	—	—

APPENDIX 3

Wavelengths of the Principal X-Ray Spectral Lines of the Chemical Elements—*M* Series (*35; 37*)

Wavelength, Å

Line	α_1	α_2	β	γ	ζ_1	ζ_2	—
Electron transition							
M V ←	*N* VII	*N* VI			*N* III		
M IV ←			*N* VI			*N* II	
M III ←				*N* V			
M II ←							*N* IV
Approximate relative intensity	100	100	80	5	1	1	5
35-Br	—	—	—	—	192.6	191.1	—
36-Kr	—	—	—	—	—	—	—
37-Rb	—	—	—	—	128.7	127.7	—
38-Sr	—	—	—	—	108.7	108.0	—
39-Y	—	—	—	—	93.400		—
40-Zr	—	—	—	38.390	82.100		37.000
41-Nb	—	—	—	34.900	72.190		33.100
42-Mo	—	—	—	32.700	64.380		31.400
43-Tc	—	—	—	30.100	59.500		28.800
44-Ru	—	—	—	26.900	52.340		25.500
45-Rh	—	—	—	25.010	47.670		24.450
46-Pd	—	—	—	23.300	43.600		22.100
47-Ag	—	—	—	21.820	39.770		20.660
48-Cd	—	—	—	20.470	36.800		19.400
49-In	—	—	—	19.210	33.200		18.240
50-Sn	—	—	—	17.940	31.240		16.930
51-Sb	—	—	—	16.920	28.880		15.980
52-Te	—	—	—	15.930	26.720		15.020
53-I	—	—	—	15.010	24.650		14.150
54-Xe	—	—	—	14.180	23.020		13.310
55-Cs	—	—	—	13.420	21.690		12.580
56-Ba	—	—	—	12.750	20.640		11.890
57-La	14.880		14.510	12.080	19.440		11.280
58-Ce	14.040		13.750	11.530	18.350		10.690
59-Pr	13.343		13.060	10.998	17.380		10.180
60-Nd	12.680		12.440	10.505	16.460		9.700
61-Pm	—	—	—	10.050	15.680		9.260
62-Sm	11.470		11.270	9.600	14.910		8.840

APPENDIX 3 (*continued*) Wavelength, Å

<table>
<tr><th>Line</th><th>α_1</th><th>α_2</th><th>β</th><th>γ</th><th>ζ_1</th><th>ζ_2</th><th>—</th></tr>
<tr><td>Electron transition</td><td></td><td></td><td></td><td></td><td></td><td></td><td></td></tr>
<tr><td>MV ←</td><td>NVII</td><td>NVI</td><td></td><td></td><td>NIII</td><td></td><td></td></tr>
<tr><td>MIV ←</td><td></td><td></td><td>NVI</td><td></td><td></td><td>NII</td><td></td></tr>
<tr><td>MIII ←</td><td></td><td></td><td></td><td>NV</td><td></td><td></td><td></td></tr>
<tr><td>MII ←</td><td></td><td></td><td></td><td></td><td></td><td></td><td>NIV</td></tr>
<tr><td>Approximate relative intensity</td><td>100</td><td>100</td><td>80</td><td>5</td><td>1</td><td>1</td><td>5</td></tr>
<tr><td>63-Eu</td><td colspan="2">10.960</td><td>10.750</td><td>9.211</td><td colspan="2">14.220</td><td>8.450</td></tr>
<tr><td>64-Gd</td><td colspan="2">10.460</td><td>10.254</td><td>8.844</td><td colspan="2">13.570</td><td>8.120</td></tr>
<tr><td>65-Tb</td><td colspan="2">10.000</td><td>9.792</td><td>8.486</td><td colspan="2">12.980</td><td>7.740</td></tr>
<tr><td>66-Dy</td><td colspan="2">9.590</td><td>9.357</td><td>8.144</td><td colspan="2">12.430</td><td>7.460</td></tr>
<tr><td>67-Ho</td><td colspan="2">9.200</td><td>8.965</td><td>7.865</td><td colspan="2">11.860</td><td>7.160</td></tr>
<tr><td>68-Er</td><td colspan="2">8.820</td><td>8.592</td><td>7.546</td><td colspan="2">11.370</td><td>6.860</td></tr>
<tr><td>69-Tm</td><td colspan="2">8.480</td><td>8.249</td><td>7.318</td><td colspan="2">10.920</td><td>6.540</td></tr>
<tr><td>70-Yb</td><td colspan="2">8.149</td><td>7.909</td><td>7.024</td><td colspan="2">10.480</td><td>6.270</td></tr>
<tr><td>71-Lu</td><td colspan="2">7.840</td><td>7.601</td><td>6.768</td><td colspan="2">10.060</td><td>6.020</td></tr>
<tr><td>72-Hf</td><td colspan="2">7.539</td><td>7.303</td><td>6.544</td><td colspan="2">9.686</td><td>5.770</td></tr>
<tr><td>73-Ta</td><td colspan="2">7.252</td><td>7.023</td><td>6.312</td><td>9.316</td><td>9.330</td><td>5.570</td></tr>
<tr><td>74-W</td><td>6.983</td><td>6.992</td><td>6.757</td><td>6.092</td><td>8.962</td><td>8.993</td><td>5.357</td></tr>
<tr><td>75-Re</td><td colspan="2">6.729</td><td>6.504</td><td>5.885</td><td>8.629</td><td>8.664</td><td>5.150</td></tr>
<tr><td>76-Os</td><td colspan="2">6.490</td><td>6.267</td><td>5.682</td><td>8.310</td><td>8.359</td><td>4.955</td></tr>
<tr><td>77-Ir</td><td>6.262</td><td>6.275</td><td>6.038</td><td>5.500</td><td>8.021</td><td>8.065</td><td>4.780</td></tr>
<tr><td>78-Pt</td><td>6.047</td><td>6.058</td><td>5.828</td><td>5.319</td><td>7.738</td><td>7.790</td><td>4.601</td></tr>
<tr><td>79-Au</td><td>5.840</td><td>5.854</td><td>5.624</td><td>5.145</td><td>7.466</td><td>7.523</td><td>4.432</td></tr>
<tr><td>80-Hg</td><td>5.648</td><td>5.677</td><td>5.432</td><td>4.984</td><td>7.232</td><td>7.250</td><td>4.266</td></tr>
<tr><td>81-Tl</td><td>5.460</td><td>5.472</td><td>5.249</td><td>4.823</td><td>6.974</td><td>7.032</td><td>4.116</td></tr>
<tr><td>82-Pb</td><td>5.286</td><td>5.299</td><td>5.076</td><td>4.674</td><td>6.740</td><td>6.802</td><td>3.968</td></tr>
<tr><td>83-Bi</td><td>5.118</td><td>5.130</td><td>4.909</td><td>4.532</td><td>6.521</td><td>6.585</td><td>3.834</td></tr>
<tr><td>84-Po</td><td>4.955</td><td>4.958</td><td>4.736</td><td>4.361</td><td>6.290</td><td>6.349</td><td>3.680</td></tr>
<tr><td>85-At</td><td>4.802</td><td>4.802</td><td>4.581</td><td>4.234</td><td>6.096</td><td>6.156</td><td>3.559</td></tr>
<tr><td>86-Rn</td><td>4.655</td><td>4.657</td><td>4.436</td><td>4.124</td><td>5.911</td><td>5.971</td><td>3.448</td></tr>
<tr><td>87-Fr</td><td>4.515</td><td>4.521</td><td>4.303</td><td>4.008</td><td>5.737</td><td>5.801</td><td>3.322</td></tr>
<tr><td>88-Ra</td><td>4.383</td><td>4.392</td><td>4.178</td><td>3.892</td><td>5.579</td><td>5.642</td><td>3.220</td></tr>
<tr><td>89-Ac</td><td>4.256</td><td>4.270</td><td>4.060</td><td>3.798</td><td>5.389</td><td>5.489</td><td>3.118</td></tr>
<tr><td>90-Th</td><td>4.138</td><td>4.151</td><td>3.941</td><td>3.679</td><td>5.245</td><td>5.340</td><td>3.011</td></tr>
<tr><td>91-Pa</td><td>4.022</td><td>4.035</td><td>3.827</td><td>3.577</td><td>5.092</td><td>5.193</td><td>2.910</td></tr>
<tr><td>92-U</td><td>3.910</td><td>3.924</td><td>3.716</td><td>3.479</td><td>4.946</td><td>5.050</td><td>2.817</td></tr>
</table>

APPENDIX 4

X-Ray Excitation Potentials of the Chemical Elements (*37*, *39*)

Excitation potential, kV

Element	*K*	*L*I	*L*II	*L*III	*M*I	*M*II	*M*III	*M*IV	*M*V
1-H	0.014	—	—	—	—	—	—	—	—
2-He	0.025	—	—	—	—	—	—	—	—
3-Li	0.055	—	—	—	—	—	—	—	—
4-Be	0.116	—	—	—	—	—	—	—	—
5-B	0.192	—	—	—	—	—	—	—	—
6-C	0.283	—	—	—	—	—	—	—	—
7-N	0.399	—	—	—	—	—	—	—	—
8-O	0.531	—	—	—	—	—	—	—	—
9-F	0.687	—	—	—	—	—	—	—	—
10-Ne	0.874	0.048	0.022	0.022	—	—	—	—	—
11-Na	1.080	0.055	0.034	0.034	—	—	—	—	—
12-Mg	1.303	0.063	0.050	0.049	—	—	—	—	—
13-Al	1.559	0.087	0.073	0.072	—	—	—	—	—
14-Si	1.838	0.118	0.099	0.098	—	—	—	—	—
15-P	2.142	0.153	0.129	0.128	—	—	—	—	—
16-S	2.470	0.193	0.164	0.163	—	—	—	—	—
17-Cl	2.819	0.238	0.203	0.202	0.020	—	—	—	—
18-Ar	3.203	0.287	0.247	0.245	0.026	—	—	—	—
19-K	3.607	0.341	0.297	0.294	0.033	—	—	—	—

20-Ca	4.038	0.399	0.352	0.349	0.040	—	—	—	—
21-Sc	4.496	0.462	0.411	0.406	0.046	—	—	—	—
22-Ti	4.964	0.530	0.460	0.454	0.054	—	—	—	—
23-V	5.463	0.604	0.519	0.512	0.061	—	—	—	—
24-Cr	5.988	0.679	0.583	0.574	0.072	—	—	—	—
25-Mn	6.537	0.762	0.650	0.639	0.082	—	—	—	—
26-Fe	7.111	0.849	0.721	0.708	0.093	—	—	—	—
27-Co	7.709	0.929	0.794	0.779	0.104	—	—	—	—
28-Ni	8.331	1.015	0.871	0.853	0.120	—	—	—	—
29-Cu	8.980	1.100	0.953	0.933	0.135	0.090	—	0.015	—
30-Zn	9.660	1.200	1.045	1.022	0.151	0.106	—	0.022	—
31-Ga	10.368	1.300	1.134	1.117	0.169	0.125	0.115	0.030	—
32-Ge	11.103	1.420	1.248	1.217	0.190	0.137	0.132	0.041	—
33-As	11.863	1.529	1.359	1.323	0.211	0.156	0.150	0.052	—
34-Se	12.652	1.652	1.473	1.434	0.234	0.177	0.170	0.066	—
35-Br	13.475	1.794	1.599	1.552	0.265	0.198	0.191	0.082	—
36-Kr	14.323	1.931	1.727	1.675	0.294	0.225	0.217	0.095	—
37-Rb	15.201	2.067	1.866	1.806	0.328	0.250	0.240	0.114	0.112
38-Sr	16.106	2.221	2.008	1.941	0.358	0.280	0.270	0.136	0.134
39-Y	17.037	2.369	2.154	2.079	0.394	0.312	0.300	0.159	0.156
40-Zr	17.998	2.547	2.305	2.220	0.435	0.348	0.335	0.187	0.184
41-Nb	18.987	2.706	2.467	2.374	0.468	0.379	0.362	0.207	0.204
42-Mo	20.002	2.884	2.627	2.523	0.507	0.412	0.394	0.232	0.228
43-Tc	21.054	3.054	2.795	2.677	0.551	0.449	0.429	0.260	0.257
44-Ru	22.118	3.236	2.966	2.837	0.591	0.486	0.467	0.290	0.288
45-Rh	23.224	3.419	3.145	3.002	0.637	0.531	0.506	0.321	0.315
46-Pd	24.347	3.617	3.329	3.172	0.684	0.573	0.546	0.354	0.349
47-Ag	25.517	3.810	3.528	3.352	0.734	0.619	0.588	0.389	0.383

APPENDIX 4 *(continued)*

Excitation potential, kV

Element	*K*	*L*I	*L*II	*L*III	*M*I	*M*II	*M*III	*M*IV	*M*V
48-Cd	26.712	4.019	3.727	3.538	0.781	0.666	0.632	0.423	0.420
49-In	27.928	4.237	3.939	3.729	0.839	0.716	0.678	0.464	0.456
50-Sn	29.190	4.464	4.157	3.928	0.894	0.772	0.720	0.506	0.497
51-Sb	30.486	4.697	4.381	4.132	0.952	0.822	0.774	0.546	0.536
52-Te	31.809	4.938	4.613	4.341	1.010	0.873	0.822	0.586	0.575
53-I	33.164	5.190	4.856	4.559	1.071	0.929	0.873	0.630	0.618
54-Xe	34.579	5.452	5.104	4.782	1.147	0.989	0.926	0.677	0.662
55-Cs	35.959	5.720	5.358	5.011	1.199	1.048	0.981	0.722	0.704
56-Ba	37.410	5.995	5.623	5.247	1.266	1.111	1.036	0.770	0.750
57-La	38.931	6.283	5.894	5.489	1.330	1.173	1.092	0.823	0.801
58-Ce	40.449	6.561	6.165	5.729	1.401	1.240	1.152	0.870	0.851
59-Pr	41.998	6.846	6.443	5.968	1.476	1.305	1.210	0.923	0.898
60-Nd	43.571	7.144	6.727	6.215	1.544	1.372	1.266	0.969	0.946
61-Pm	45.207	7.448	7.018	6.466	1.642	1.439	1.327	1.019	0.994
62-Sm	46.846	7.754	7.281	6.721	1.689	1.512	1.388	1.073	1.048
63-Eu	48.515	8.069	7.624	6.983	1.767	1.584	1.450	1.129	1.101
64-Gd	50.229	8.393	7.940	7.252	1.849	1.653	1.511	1.185	1.153
65-Tb	51.998	8.724	8.258	7.519	1.937	1.737	1.583	1.245	1.211
66-Dy	53.789	9.083	8.621	7.850	2.019	1.805	1.642	1.304	1.266
67-Ho	55.615	9.411	8.920	8.074	2.104	1.886	1.715	1.365	1.327
68-Er	57.483	9.776	9.263	8.364	2.184	1.973	1.783	1.430	1.385
69-Tm	59.335	10.144	9.628	8.652	2.291	2.071	1.861	1.498	1.451
70-Yb	61.303	10.486	9.977	8.943	2.387	2.165	1.948	1.566	1.518
71-Lu	63.304	10.867	10.345	9.241	2.488	2.262	2.025	1.637	1.586
72-Hf	65.313	11.264	10.734	9.556	2.601	2.366	2.109	1.718	1.664

73-Ta	67.400	11.676	11.130	9.876	2.698	2.459	2.184	1.783	1.725
74-W	69.508	12.090	11.535	10.198	2.812	2.566	2.273	1.864	1.803
75-Re	71.662	12.522	11.955	10.531	2.926	2.676	2.361	1.946	1.879
76-Os	73.860	12.965	12.383	10.869	3.047	2.792	2.453	2.033	1.963
77-Ir	76.097	13.413	12.819	11.211	3.171	2.908	2.551	2.119	2.040
78-Pt	78.379	13.873	13.268	11.559	3.296	3.036	2.649	2.204	2.129
79-Au	80.713	14.353	13.733	11.919	3.379	3.149	2.744	2.307	2.220
80-Hg	83.106	14.841	14.212	12.285	3.566	3.287	2.848	2.392	2.291
81-Tl	85.517	15.346	14.697	12.657	3.702	3.418	2.957	2.483	2.389
82-Pb	88.001	15.870	15.207	13.044	3.853	3.558	3.072	2.586	2.484
83-Bi	90.521	16.393	15.716	13.424	4.003	3.709	3.186	2.694	2.586
84-Po	93.112	16.935	16.244	13.817	4.147	3.863	3.312	2.798	2.681
85-At	95.740	17.490	16.784	14.215	4.350	4.008	3.428	2.905	2.780
86-Rn	98.418	18.058	17.337	14.618	4.524	4.156	3.536	3.014	2.882
87-Fr	101.147	18.638	17.904	15.028	4.678	4.324	3.654	3.125	2.986
88-Ra	103.927	19.233	18.481	15.442	4.811	4.477	3.779	3.237	3.093
89-Ac	106.759	19.842	19.078	15.865	5.019	4.637	3.892	3.352	3.202
90-Th	109.630	20.460	19.588	16.296	5.176	4.810	4.030	3.474	3.313
91-Pa	112.581	21.102	20.311	16.731	5.355	4.993	4.164	3.597	3.416
92-U	115.591	21.753	20.943	17.163	5.532	5.177	4.293	3.712	3.533
93-Np	118.619	22.417	21.596	17.614	—	—	—	—	—
94-Pu	121.720	23.097	22.262	18.066	—	—	—	—	—
95-Am	124.876	23.793	22.944	18.525	—	—	—	—	—
96-Cm	128.088	24.503	23.640	18.990	—	—	—	—	—
97-Bk	131.357	25.230	24 352	19.461	—	—	—	—	—
98-Cf	134.683	25.971	25 080	19.938	—	—	—	—	—
99-Es	138.067	26.729	25.824	20.422	—	—	—	—	—
100-Fm	141.510	27.503	26.584	20.912	—	—	—	—	—

APPENDIX 5

Wavelengths of the X-Ray Absorption Edges of the Chemical Elements *(37; 16)*

Wavelength, Å

Element	*K*	*L*I	*L*II	*L*III	*M*I	*M*II	*M*III	*M*IV	*M*V
3-Li	226.950	—	—	—	—	—	—	—	—
4-Be	107.200	—	—	—	—	—	—	—	—
5-B	65.604	—	—	—	—	—	—	—	—
6-C	43.887	—	—	—	—	—	—	—	—
7-N	31.220	—	—	—	—	—	—	—	—
8-O	23.233	—	—	—	—	—	—	—	—
9-F	17.897	—	—	—	—	—	—	—	—
10-Ne	14.170	—	—	—	—	—	—	—	—
11-Na	11.475	—	—	—	—	—	—	—	—
12-Mg	9.512	197.300	—	220.534	—	—	—	—	—
13-Al	7.951	142.500	163.022	163.942	—	—	—	—	—
14-Si	6.745	105.000	122.028	122.761	—	—	—	—	—
15-P	5.787	81.000	95.070	95.801	—	—	—	—	—
16-S	5.018	64.100	75.314	75.868	—	—	—	—	—
17-Cl	4.397	52.100	60.739	61.222	591.900	—	—	—	—
18-Ar	3.871	43.200	50.025	50.446	476.587	—	—	—	—
19-K	3.436	36.400	41.754	42.179	373.456	—	—	—	—
20-Ca	3.070	30.700	35.797	36.173	309.146	—	—	—	—
21-Sc	2.757	26.800	30.943	31.296	264.440	—	—	—	—
22-Ti	2.497	23.400	26.936	27.290	228.816	—	—	—	—

23-V	2.269	19.803	23.877	24.229	201.484	—	—	—	—
24-Cr	2.070	17.840	21.294	21.637	171.626	—	—	—	—
25-Mn	1.896	16.138	19.086	19.417	150.739	—	—	—	—
26-Fe	1.743	14.650	17.188	17.504	132.991	—	—	—	—
27-Co	1.608	13.333	15.545	15.843	118.175	—	—	—	—
28-Ni	1.488	12.201	14.104	14.387	103.062	—	—	—	—
29-Cu	1.380	11.172	12.841	13.113	91.685	137.152	—	793.498	—
30-Zn	1.283	10.262	11.725	11.987	81.898	116.000	—	546.224	—
31-Ga	1.196	9.416	10.728	10.930	73.146	98.670	107.500	405.000	—
32-Ge	1.116	8.692	9.840	10.089	65.116	89.857	93.488	299.437	—
33-As	1.045	8.067	9.056	9.298	58.581	79.123	82.205	238.357	—
34-Se	0.980	7.456	8.347	8.584	52.835	69.999	72.578	187.662	—
35-Br	0.920	6.920	7.721	7.952	46.679	62.341	64.705	150.218	—
36-Kr	0.866	6.444	7.158	7.377	42.071	54.882	57.020	129.756	—
37-Rb	0.816	5.995	6.642	6.861	37.696	49.562	51.527	108.466	109.986
38-Sr	0.770	5.591	6.173	6.387	34.553	44.124	45.866	90.985	92.251
39-Y	0.728	5.226	5.752	5.961	31.453	39.733	41.308	77.850	78.973
40-Zr	0.689	4.889	5.378	5.565	28.474	35.564	36.972	66.136	67.184
41-Nb	0.653	4.595	5.031	5.230	26.482	32.708	34.155	59.815	60.608
42-Mo	0.620	4.323	4.716	4.913	24.413	30.083	31.401	53.277	54.201
43-Tc	0.589	4.068	4.431	4.623	22.500	27.578	28.853	47.507	48.140
44-Ru	0.560	3.837	4.169	4.358	20.945	25.461	26.510	42.660	43.038
45-Rh	0.534	3.623	3.928	4.113	19.453	23.341	24.492	38.561	39.286
46-Pd	0.509	3.425	3.706	3.889	18.109	21.603	22.699	34.940	35.494
47-Ag	0.486	3.243	3.501	3.680	16.877	20.011	21.061	31.793	32.299
48-Cd	0.464	3.073	3.312	3.488	15.873	18.602	19.614	29.269	29.507

APPENDIX 5 *(continued)*

Wavelength, Å

Element	*K*	*L*I	*L*II	*L*III	*M*I	*M*II	*M*III	*M*IV	*M*V
49-In	0.444	2.916	3.137	3.311	14.764	17.314	18.285	26.717	27.165
50-Sn	0.425	2.648	2.975	3.146	13.856	16.050	17.200	24.500	24.900
51-Sb	0.407	2.634	2.824	2.994	13.020	15.072	16.014	22.699	23.114
52-Te	0.390	2.508	2.685	2.853	12.274	14.185	15.079	21.124	21.528
53-I	0.374	2.390	2.555	2.721	11.575	13.344	14.193	19.670	20.050
54-Xe	0.358	2.277	2.434	2.597	10.800	12.532	13.381	18.300	18.716
55-Cs	0.345	2.175	2.321	2.482	10.338	11.824	12.638	17.158	17.607
56-Ba	0.331	2.078	2.214	2.374	9.787	11.156	11.957	16.087	16.513
57-La	0.318	1.988	2.115	2.273	9.316	10.560	11.343	15.060	15.470
58-Ce	0.306	1.902	2.022	2.178	8.844	9.997	10.758	14.242	14.560
59-Pr	0.295	1.822	1.934	2.089	8.394	9.498	10.242	13.429	13.792
60-Nd	0.284	1.747	1.852	2.007	8.026	9.033	9.787	12.789	13.092
61-Pm	0.274	1.675	1.775	1.928	7.550	8.610	9.342	12.160	12.461
62-Sm	0.265	1.608	1.703	1.855	7.338	8.199	8.926	11.547	11.828
63-Eu	0.256	1.545	1.634	1.785	7.013	7.827	8.548	10.975	11.257
64-Gd	0.247	1.485	1.569	1.719	6.703	7.496	8.200	10.457	10.747
65-Tb	0.238	1.428	1.507	1.656	6.398	7.135	7.829	9.958	10.233
66-Dy	0.230	1.375	1.449	1.597	6.139	6.865	7.547	9.506	9.791
67-Ho	0.223	1.323	1.393	1.540	5.892	6.571	7.225	9.079	9.342
68-Er	0.216	1.274	1.341	1.487	5.675	6.283	6.950	8.664	8.950
69-Tm	0.209	1.227	1.291	1.436	5.410	5.985	6.661	8.275	8.541
70-Yb	0.202	1.183	1.243	1.387	5.192	5.724	6.361	7.912	8.164
71-Lu	0.196	1.140	1.198	1.341	4.981	5.480	6.119	7.573	7.813

72-Hf	0.190	1.100	1.155	1.297	4.765	5.238	5.876	7.216	7.450
73-Ta	0.184	1.062	1.114	1.255	4.594	5.041	5.675	6.951	7.183
74-W	0.178	1.025	1.074	1.215	4.407	4.831	5.453	6.649	6.875
75-Re	0.173	0.990	1.037	1.177	4.237	4.632	5.250	6.371	6.597
76-Os	0.168	0.956	1.001	1.140	4.067	4.440	5.052	6.097	6.313
77-Ir	0.163	0.924	0.966	1.105	3.909	4.262	4.858	5.848	6.076
78-Pt	0.158	0.893	0.933	1.071	3.761	4.082	4.679	5.622	5.820
79-Au	0.153	0.863	0.902	1.039	3.669	3.936	4.518	5.374	5.584
80-Hg	0.149	0.835	0.872	1.008	3.476	3.771	4.352	5.182	5.409
81-Tl	0.145	0.807	0.843	0.979	3.348	3.626	4.192	4.992	5.187
82-Pb	0.141	0.781	0.815	0.950	3.217	3.484	4.035	4.793	4.989
83-Bi	0.137	0.756	0.788	0.923	3.097	3.342	3.890	4.601	4.793
84-Po	0.133	0.731	0.762	0.897	2.989	3.209	3.742	4.430	4.624
85-At	0.129	0.708	0.738	0.872	2.850	3.092	3.616	4.267	4.458
86-Rn	0.125	0.686	0.715	0.848	2.740	2.982	3.506	4.112	4.301
87-Fr	0.122	0.665	0.592	0.825	2.650	2.866	3.392	3.966	4.151
88-Ra	0.119	0.644	0.571	0.803	2.576	2.769	3.280	3.829	4.008
89-Ac	0.116	0.625	0.650	0.782	2.470	2.673	3.185	3.698	3.871
90-Th	0.113	0.606	0.629	0.761	2.394	2.577	3.076	3.568	3.742
91-Pa	0.110	0.587	0.610	0.741	2.314	2.483	2.977	3.446	3.628
92-U	0.108	0.570	0.592	0.722	2.240	2.394	2.887	3.339	3.509

APPENDIX 6

X-Ray Mass-Absorption Coefficients of the Chemical Elements (*16*)

Mass-absorption coefficient μ/ϱ, cm²/g

At. No.	Element	0.1	0.15	0.2	0.25	0.3	0.4	0.5	0.6	0.7	0.8	0.9	1.0
		Wavelength, Å											
1	H	0.29	0.32	0.34	0.36	0.37	0.38	0.40	0.42	0.43	0.44	0.44	0.45
2	He	0.114	0.124	0.132	0.140	0.146	0.159	0.173	0.186	0.203	0.222	0.241	0.255
3	Li	0.124	0.132	0.143	0.153	0.163	0.180	0.198	0.223	0.254	0.302	0.358	0.428
4	Be	0.131	0.142	0.153	0.162	0.171	0.185	0.210	0.240	0.292	0.362	0.445	0.57
5	B	0.138	0.152	0.164	0.173	0.182	0.198	0.222	0.277	0.355	0.470	0.61	0.76
6	C	0.142	0.155	0.170	0.186	0.204	0.240	0.305	0.41	0.55	0.75	1.05	1.40
7	N	0.144	0.159	0.175	0.195	0.216	0.288	0.395	0.62	0.89	1.25	1.73	2.20
8	O	0.145	0.162	0.181	0.206	0.236	0.345	0.508	0.87	1.25	1.80	2.40	3.20
9	F	0.147	0.165	0.192	0.228	0.270	0.417	0.675	1.20	1.85	2.60	3.40	4.40
10	Ne	0.149	0.169	0.208	0.256	0.319	0.508	0.865	1.55	2.50	3.50	4.75	6.50
11	Na	0.150	0.175	0.228	0.287	0.380	0.630	1.18	2.05	3.25	4.75	6.70	8.80
12	Mg	0.152	0.190	0.251	0.330	0.445	0.78	1.47	2.70	4.30	6.50	9.10	11.7
13	Al	0.155	0.205	0.277	0.380	0.525	0.97	1.82	3.70	5.75	8.80	11.8	15.2
14	Si	0.159	0.215	0.310	0.442	0.615	1.22	2.25	4.65	7.20	11.1	14.2	18.2
15	P	0.165	0.228	0.346	0.510	0.725	1.48	2.80	5.45	8.30	13.3	16.5	21.7
16	S	0.170	0.241	0.392	0.592	0.855	1.78	3.45	6.40	9.6	15.5	19.3	26.0
17	Cl	0.176	0.260	0.433	0.667	1.02	2.13	4.25	7.45	11.3	18.5	22.6	29.7
18	A	0.183	0.278	0.490	0.76	1.21	2.62	5.08	8.60	13.2	21.1	26.4	34.4
19	K	0.191	0.301	0.542	0.86	1.39	3.02	5.75	10.2	15.3	24	30.7	40
20	Ca	0.200	0.327	0.601	0.98	1.63	3.45	6.50	11.5	18.0	26.5	36	48
21	Sc	0.210	0.358	0.667	1.12	1.85	3.95	7.4	13.1	20.2	30	41.0	55
22	Ti	0.221	0.395	0.740	1.26	2.10	4.50	8.4	14.8	22.7	34	47	62.5
23	V	0.231	0.431	0.830	1.42	2.37	5.15	9.6	16.8	25.5	38	54	71.5
24	Cr	0.241	0.480	0.925	1.60	2.68	5.88	10.8	18.9	28.7	42.5	60.3	80.5
25	Mn	0.253	0.528	1.03	1.79	3.05	6.65	12.3	21.4	32.2	47.5	66.6	90
26	Fe	0.265	0.58	1.16	2.02	3.45	7.6	14.1	23.8	36.0	53.2	74	100
27	Co	0.285	0.64	1.28	2.25	3.80	8.35	15.5	26.6	39.8	58.2	80.5	110
28	Ni	0.303	0.71	1.42	2.51	4.15	9.20	17.0	29.6	43.6	64	87.5	121
29	Cu	0.328	0.785	1.57	2.79	4.55	10.1	18.6	32.4	48.5	70	95.5	130
30	Zn	0.350	0.818	1.76	3.07	5.00	11.1	20.6	35.3	53	76.5	106	141
31	Ga	0.378	0.94	1.91	3.36	5.50	12.2	22.3	38.7	59	82	112	152
32	Ge	0.404	1.02	2.06	3.67	6.15	13.3	24.4	42.1	63.5	88	120	163
33	As	0.427	1.12	2.22	4.01	6.45	14.6	26.3	45.1	68.5	94	129	175
34	Se	0.472	1.21	2.41	4.33	7.00	15.8	28.5	48.4	74	101	138	33.2
35	Br	0.502	1.31	2.60	4.70	7.60	17.0	30.8	52.0	79.5	108	149	35.2
36	Kr	0.535	1.41	2.81	5.05	8.30	18.3	33.3	55.5	86	116	26.5	37.5
37	Rb	0.572	1.50	3.03	5.45	9.05	19.7	35.9	59.5	93	125	28.0	40.2
38	Sr	0.61	1.61	3.28	5.82	9.80	21.3	38.7	63	100	22	29.8	43
39	Y	0.65	1.72	3.52	6.20	10.3	22.6	41.3	67.5	108	23.2	32	46
40	Zr	0.69	1.83	3.73	6.60	11.0	24.0	43.5	72	16.8	24.5	33.8	49
41	Nb	0.74	1.93	4.00	7.02	11.7	25.4	46.0	76	17.8	26	36.0	51.8
42	Mo	0.79	2.04	4.29	7.50	12.4	26.9	49.0	81	19.0	27.7	38.4	54.8
43	Tc	0.84	2.15	4.55	7.95	13.1	28.6	50.5	14.2	20.3	29.5	41.0	58
44	Ru	0.89	2.27	4.70	8.4	13.8	30.5	53	15.0	21.6	31.4	43.7	61.2
45	Rh	0.94	2.40	5.01	8.9	14.6	32.3	55.5	15.8	23.0	33.5	46.5	65
46	Pd	1.00	2.53	5.25	9.4	15.5	34.3	58.2	16.6	24.5	35.5	49.5	69
47	Ag	1.05	2.67	5.50	9.9	16.3	36.5	10.0	17.5	26.0	38.0	52.8	72.7
48	Cd	1.10	2.80	5.74	10.3	17.2	38.5	10.7	18.5	27.7	40.2	56.2	76.2
49	In	1.15	2.94	5.98	10.8	17.8	40.6	11.4	19.5	29.5	42.7	60.0	80
50	Sn	1.20	3.07	6.22	11.3	18.7	43.1	12.0	20.5	31	45.5	63.5	84

K

APPENDIX 6 (*continued*)

Mass-absorption coefficient μ/ϱ, cm²/g

At. No.	Element	1.5	2.0	2.5	3	4	5	6	7	8	9	10
		Wavelength, Å										
1	H	0.49	0.52	0.62	0.75	1.25	2.12	3.28	4.85	7.1	10.0	13.7
2	He	0.355	0.715	1.04	1.48	3.55	6.9	11.6	18.1	26.6	37.7	51
3	Li	1.02	2.18	3.98	6.6	15.2	28.8	48.8	76	113	157	213
4	Be	1.55	3.38	6.2	10.3	22.7	43.7	74	118	174	245	333
5	B	2.31	5.05	9.6	15.8	36.0	69	116	187	285	405	560
6	C	4.2	9.7	14.0	32	74	145	250	390	570	810	1100
7	N	6.9	16.0	30.5	52	123	235	400	620	910	1290	1800
8	O	10.5	24.0	45.5	78	180	350	580	920	1350	1900	2600
9	F	14.2	32	62	102	240	450	760	1190	1750	2450	3300
10	Ne	19.8	43	90	145	325	630	1040	1580	2350	3200	4300
11	Na	25.5	53.5	132	250	520	920	1450	2200	3100	4100	5300
12	Mg	32.5	69	185	280	610	1100	1800	2700	3900	5200	340
13	Al	41.5	87	235	360	780	1400	2250	3300	280	390	520
14	Si	52	112	290	440	950	1700	2800	270	400	550	740
15	P	66	140	330	490	1050	1900	245	380	550	770	1070
16	S	78.5	170	375	590	1250	2200	320	500	730	1040	1400
17	Cl	95	202	430	680	1400	250	410	620	900	1250	1700
18	Ar	112	240	485	780	170	300	490	730	1050	1450	1900
19	K	130	280	540	860	205	370	600	910	1300	1800	2350
20	Ca	150	328	625	950	255	460	730	1100	1550	2100	2800
21	Sc	172	375	720	145	310	550	880	1320	1850	2520	3300
22	Ti	195	425	113	175	360	630	1000	1450	2050	2750	3550
23	V	222	485	125	195	400	710	1120	1620	2250	3000	3700
24	Cr	245	540	136	210	440	760	1200	1720	2350	3200	4100
25	Mn	277	73	148	232	480	820	1260	1850	2550	3400	4350
26	Fe	312	80	162	255	510	880	1400	2050	2800	3700	4700
27	Co	345	87	177	270	550	960	1500	2200	3000	4000	5200
28	Ni	44	96	193	295	600	1050	1570	2350	3300	4300	5600
29	Cu	48.5	101	212	320	660	1150	1750	2550	3500	4600	6000
30	Zn	53.5	116	227	350	700	1200	1870	2700	3750	4900	6400
31	Ga	59	126	248	385	760	1300	2020	2950	4050	5400	5500
32	Ge	63	138	267	410	820	1400	2150	3100	4300	4700	4000
33	As	69	150	287	440	870	1500	2320	3330	4600	5050	2000
34	Se	74	165	311	470	940	1600	2500	3600	3900	1650	2150
35	Br	80	180	335	510	1020	1720	2700	2180	1250	1750	2300
36	Kr	87	192	359	550	1100	1860	2900	2280	1350	1870	2450
37	Rb	94	207	386	590	1170	2000	3100	950	1440	2000	2620
38	Sr	102	223	412	640	1260	2150	2600	1020	1540	2150	2800
39	Y	109	238	442	690	1380	2320	800	1110	1650	2350	3000
40	Zr	118	255	478	740	1500	1840	850	1200	1770	2450	3200
41	Nb	127	271	507	800	1600	1900	910	1300	1900	2600	3400
42	Mo	136	290	544	860	1700	600	980	1400	2050	2800	3650
43	Tc	147	310	583	920	1850	650	1030	1510	2200	3000	3900
44	Ru	158	342	625	960	1530	700	1090	1650	2350	3200	4150
45	Rh	170	355	670	1050	1050	750	1160	1770	2500	3400	4450
46	Pd	183	380	720	1140	440	800	1230	1900	2700	3650	4750
47	Ag	196	405	770	1250	480	850	1350	2050	2900	3900	5100
48	Cd	210	432	820	1340	510	900	1450	2150	3100	4100	5400
49	In	222	460	880	1120	535	940	1520	2250	3200	4300	5700
50	Sn	236	495	940	775	560	990	1600	2360	3350	4500	5900

K (edge, right margin at rows 12–13); *LI*, *LII*, *LIII* (edges, right margin at rows 31–34); *LI LII LIII* (edges, below table)

APPENDIX 6 (*continued*)

Mass-absorption coefficient μ/ϱ, cm²/g

At. No.	Element	0.1	0.15	0.2	0.25	0.3	0.4	0.5	0.6	0.7	0.8	0.9	1.0
		Wavelength, Å											
51	Sb	1.25	3.22	6.48	11.8	19.5	45.5	12.8	21.6	33.1	48.5	67	88
52	Te	1.30	3.37	6.76	12.3	20.3	7.4	13.6	22.7	35.2	51.5	70.6	93
53	I	1.36	3.52	7.05	12.9	21.2	7.8	14.3	24	37.5	54.5	74.8	97
54	Xe	1.42	3.68	7.33	13.4	21.9	8.2	15.2	25.2	39.8	58	78.5	101
55	Cs	1.48	3.85	7.65	13.9	22.7	8.6	16.2	26.5	42.0	61.3	82.5	106
56	Ba	1.53	4.03	7.98	14.5	23.1	9.05	17.2	27.9	44.2	64	87	111
57	La	1.60	4.22	8.27	15.0	24.5	9.5	18.1	29.3	46.7	67.2	91	116
58	Ce	1.66	4.37	8.55	15.6	25.2	10.0	19.1	30.9	49.2	70.5	95	122
59	Pr	1.72	4.53	8.88	16.1	4.7	10.5	20.1	32.5	51.6	73.8	100	127
60	Nd	1.80	4.69	9.20	16.7	4.95	11.0	21.1	34.2	54.0	77.2	104	132
61	Pm	1.86	4.86	9.57	17.3	5.20	11.5	22.1	36	56.9	80.7	108	139
62	Sm	1.93	5.03	9.95	17.7	5.45	12.0	23.2	37.7	59.5	84	114	144
63	Eu	2.02	5.20	10.3	18.3	5.71	12.5	24.2	39.8	62.5	87.8	119	150
64	Gd	2.09	5.37	10.5	3.95	6.00	13.2	25.3	41.8	65.5	91.5	125	157
65	Tb	2.18	5.52	10.8	4.15	6.28	13.7	26.6	44.0	69.5	95.5	130	163
66	Dy	2.26	5.70	11.2	4.36	6.57	14.4	27.2	46.2	72	99.5	135	170
67	Ho	2.33	5.87	11.6	4.55	6.90	15.1	28.2	48.6	75	103	141	177
68	Er	2.42	6.03	11.9	4.75	7.2	15.8	30.2	50.8	79	107	147	185
69	Tm	2.50	6.23	12.3	5.02	7.6	16.5	31.4	53.2	82.8	112	152	193
70	Yb	2.58	6.41	12.7	5.27	7.95	17.3	32.7	55.8	87	115	158	201
71	Lu	2.66	6.61	3.07	5.50	8.4	18.2	34.2	58.4	90.5	120	165	210
72	Hf	2.75	6.80	3.22	5.80	8.8	19.0	35.5	61	94.2	124	172	219
73	Ta	2.82	7.02	3.36	6.05	9.2	19.9	37.2	64	99	129	178	229
74	W	2.90	7.24	3.50	6.27	9.7	20.8	38.7	66.5	103	134	185	239
75	Re	2.96	7.45	3.63	6.60	10.2	21.8	40.5	69	106	139	192	213
76	Os	3.03	7.65	3.78	6.9	10.6	22.8	42.2	72	110	144	200	222
77	Ir	3.10	7.86	3.92	7.3	11.1	23.8	44.0	74.5	113	150	208	157
78	Pt	3.17	8.06	4.08	7.6	11.6	25.0	46.0	77.6	117	155	170	163
79	Au	3.23	8.28	4.22	8.0	12.2	26.2	47.5	80.5	122	160	176	170
80	Hg	3.30	2.17	4.38	8.3	12.7	27.4	49.5	83.7	126	165	135	178
81	Tl	3.36	2.24	4.53	8.7	13.3	28.6	51.5	86.6	130	170	140	72
82	Pb	3.41	2.32	4.67	9.2	13.9	30.0	53.5	89.2	135	147	145	75
83	Bi	3.45	2.37	4.81	9.6	14.5	31.3	55.5	92	140	109	150	77
84	Po	3.52	2.43	4.88	10.0	15.2	32.8	57.4	94.5	144	112	58.2	79.5
85	At	3.56	2.49	5.02	10.5	15.7	34.0	59.3	97	149	116	60	82
86	Rn	3.61	2.55	5.14	10.9	16.3	35.4	61	100	135	120	62	84
87	Fr	3.66	2.60	5.21	11.2	17.0	36.7	63	102	88	123	64	87
88	Ra	3.70	2.66	5.28	11.7	17.7	38.3	65	105	91	127	66.5	90
89	Ac	3.75	2.71	5.38	12.1	18.6	39.8	67	107	94	131	68	92
90	Th	3.81	2.76	5.45	12.4	19.5	41.2	69	110	97	52	70	95
91	Pa	3.86	2.80	5.53	12.8	20.1	42.6	71	101	100	53.3	72	98
92	U	3.91	2.85	5.61	13.1	20.9	44.0	83.1	79	103	55	74.4	101
93	Np	3.95	2.90	5.68	13.4	21.7	45.2	85.3	82	106	56.8	76.6	104
94	Pu	4.00	2.95	5.75	13.7	22.5	46.7	87.5	85	51	58.7	78.8	107
95	Am	4.05	3.00	5.81	14.1	23.3	47	89.5	87	52	60.5	81	110
96	Cm	2.50	3.04	5.89	14.5	24.2	49.8	91.5	90	53.5	62.3	83.6	113
97	Bk	2.56	3.08	5.97	14.8	25.0	51.2	84.4	92	55	64.4	86.2	117
98	Cf	2.62	3.12	6.04	15.2	25.8	52.8	65	94	57	66.4	89	121
99	Es	2.68	3.16	6.11	15.6	26.6	54.0	67	97	59	68.5	91.6	125
100	Fm	2.74	3.20	6.18	15.9	27.5	55.5	69	46	60	70.5	94.5	128

K   *L*I   *L*II   *L*III

APPENDIX 6 (*continued*)

Mass-absorption coefficient μ/ϱ, cm²/g

At. No.	Element	1.5	2.0	2.5	3	4	5	6	7	8	9	10	
		Wavelength, Å											
51	Sb	250	525	1000	805	590	1050	1660	2480	3500	4700	6200	
52	Te	265	557	1070	295	620	1100	1730	2600	3650	4900	6400	
53	I	280	590	880	310	650	1160	1810	2700	3700	5100	6700	
54	Xe	296	622	710	340	710	1230	1930	2850	4000	5300	7000	
55	Cs	312	656	215	360	740	1300	2050	3000	4200	5600	7200	
56	Ba	330	690	227	390	790	1370	2150	3150	4400	5800	5400	*M*I
57	La	347	650	240	415	840	1450	2250	3300	4600	6100	5600	
58	Ce	365	685	252	440	880	1570	2360	3400	4800	4750	5800	
59	Pr	385	540	267	465	920	1600	2470	3650	5000	4950	6000	
60	Nd	405	179	280	485	980	1670	2580	3750	3900	5100	4500	*M*II
61	Pm	425	187	296	515	1040	1750	2700	3900	3500	5300	3300	*M*III
62	Sm	446	196	313	535	1080	1830	2810	4100	3600	3900	3400	
63	Eu	470	206	330	555	1120	1900	2910	3300	3700	2900	3500	
64	Gd	440	215	348	575	1170	1980	3020	3370	2700	3000	2800	*M*IV
65	Tb	465	225	368	610	1230	2090	3200	3440	2800	3100	2170	*M*V
66	Dy	365	236	388	635	1300	2200	3400	3530	2200	3200	2270	
67	Ho	390	247	409	665	1350	2280	2500	2800	2260	2500	2380	
68	Er	136	258	432	690	1410	2350	2570	2300	2330	1890	2490	
69	Tm	142	272	456	720	1480	2500	2650	2390	2400	1970	2500	
70	Yb	148	285	482	750	1540	2600	2740	2480	1950	2050	2600	
71	Lu	156	298	510	780	1600	1980	2060	2570	1580	2130	2700	
72	Hf	162	313	540	825	1660	2050	2100	2650	1660	2210	2800	
73	Ta	168	327	570	855	1720	2210	1520	1900	1740	2290	2900	
74	W	176	342	600	890	1800	2280	1660	1350	1820	2370	3020	
75	Re	184	358	636	925	1900	1750	1700	1400	1900	2450	3130	
76	Os	193	376	672	970	1980	1300	1740	1460	1980	2560	3250	
77	Ir	200	392	710	1010	1450	1340	1400	1530	2060	2670	3370	
78	Pt	208	410	750	1070	1500	1380	1160	1590	2150	2780	3490	
79	Au	217	430	795	1130	1550	1430	1200	1650	2250	2900	3600	
80	Hg	225	445	825	1190	1150	1480	1250	1700	2300	2950	3700	
81	Tl	233	460	865	1250	1200	1120	1300	1790	2370	3110	3900	
82	Pb	242	478	900	1310	1250	1150	1360	1880	2440	3270	4100	
83	Bi	252	495	940	1380	920	910	1410	1970	2560	3430	4300	
84	Po	262	507	980	1050	950	960	1460	2060	2680	3590	4500	
85	At	272	532	1020	1080	980	1000	1510	2150	2900	3750	4700	
86	Rn	283	550	1070	1120	1020	1040	1550	2240	3010	3900	4900	
87	Fr	294	572	1110	840	710	1080	1590	2330	3120	4050	5100	
88	Ra	305	592	1160	860	750	1120	1630	2420	3230	4200	5500	
89	Ac	316	615	890	890	620	1160	1660	2510	3450	4350	4300	*N*I
90	Th	328	638	930	920	730	1200	1700	2600	3460	4500	4470	
91	Pa	342	660	725	720	760	1250	1790	2700	3570	4650	4650	
92	U	354	683	755	750	790	1300	1890	2790	3690	3600	4820	
93	Np	367	709	785	780	820	1350	1980	2890	3800	3700	3500	*N*II
94	Pu	382	733	820	810	850	1400	2080	2980	2900	3850	3650	
95	Am	395	760	850	610	880	1450	2170	3070	3000	4000	3800	
96	Cm	412	530	880	640	910	1500	2250	3170	3080	3100	3950	
97	Bk	427	420	570	500	950	1560	2330	3270	3180	3200	4100	
98	Cf	443	435	595	515	980	1620	2420	2500	3300	3330	3400	*N*III
99	Es	461	452	620	530	1020	1690	2520	2580	3400	3450	3550	
100	Fm	480	470	640	550	1050	1730	2600	2600	2600	3570	3700	

*M*I *M*II *M*III *M*IV *M*V *N*I *N*II *N*III

APPENDIX 7

Periodic Table of the Chemical Elements

Showing atomic number Z and wavelengths of selected emission lines

Lines	IA	IIA	IIIB	IVB	VB	VIB	VIIB	VIII	VIII	VIII	IB	IIB	IIIA	IVA	VA	VIA	VIIA	0	Lines
	H 1																	He 2	
$K\alpha$	Li 3 230.0	Be 4 113.0											B 5 67.00	C 6 44.00	N 7 31.60	O 8 23.71	F 9 18.31	Ne 10 14.61	$K\alpha$
$K\alpha$ $K\beta_1$	Na 11 11.91 11.62	Mg 12 9.889 9.570											Al 13 8.339 7.982	Si 14 7.126 6.778	P 15 6.155 5.804	S 16 5.373 5.032	Cl 17 4.728 4.403	Ar 18 4.192 3.886	$K\alpha$ $K\beta_1$
$K\alpha$ $K\beta_1$ $L\alpha$	K 19 3.742 3.454	Ca 20 3.360 3.090 36.39	Sc 21 3.032 2.780 31.39	Ti 22 2.750 2.514 27.45	V 23 2.505 2.284 24.31	Cr 24 2.291 2.085 21.71	Mn 25 2.103 1.910 19.49	Fe 26 1.937 1.756 17.60	Co 27 1.790 1.621 16.00	Ni 28 1.659 1.500 14.60	Cu 29 1.542 1.392 13.36	Zn 30 1.436 1.295 12.28	Ga 31 1.341 1.208 11.31	Ge 32 1.255 1.129 10.46	As 33 1.177 1.057 9.671	Se 34 1.106 0.992 8.990	Br 35 1.041 0.933 8.375	Kr 36 0.982 0.878 7.817	$K\alpha$ $K\beta_1$ $L\alpha$
$K\alpha$ $K\beta_1$ $L\alpha_1$	Rb 37 0.927 0.829 7.318	Sr 38 0.876 0.783 6.862	Y 39 0.830 0.741 6.448	Zr 40 0.787 0.702 6.070	Nb 41 0.748 0.666 5.724	Mo 42 0.711 0.632 5.406	Tc 43 0.676 0.601 5.114	Ru 44 0.644 0.572 4.846	Rh 45 0.615 0.546 4.597	Pd 46 0.587 0.520 4.368	Ag 47 0.561 0.497 4.154	Cd 48 0.536 0.475 3.956	In 49 0.514 0.454 3.772	Sn 50 0.492 0.435 3.600	Sb 51 0.472 0.417 3.439	Te 52 0.453 0.400 3.289	I 53 0.435 0.384 3.148	Xe 54 0.417 0.368 3.015	$K\alpha$ $K\beta_1$ $L\alpha_1$
$K\alpha$ $L\alpha_1$ $M\alpha$ N	Cs 55 0.402 2.892	Ba 56 0.386 2.775	La 57 0.372 2.665 14.88	Hf 72 0.224 1.570 7.539	Ta 73 0.217 1.522 7.252 58.20	W 74 0.211 1.476 6.983 55.80	Re 75 0.204 1.433 6.728	Os 76 0.198 1.391 6.490 51.90	Ir 77 0.193 1.351 6.262 50.20	Pt 78 0.187 1.313 6.046 48.10	Au 79 0.182 1.276 5.840 46.80	Hg 80 0.177 1.241 5.666 45.20	Tl 81 0.172 1.207 5.461 46.50	Pb 82 0.167 1.175 5.285 42.30	Bi 83 0.162 1.144 5.118 13.36	Po 84 0.158 1.114	At 85 0.154 1.085	Rn 86 0.150 1.057	$K\alpha$ $L\alpha_1$ $M\alpha$ N
$K\alpha$ $L\alpha_1$ $M\alpha$ N	Fr 87 0.146 1.030	Ra 88 0.142 1.005	Ac 89 0.138 0.980																

Lines															Lines
$K\alpha$ $L\alpha_1$ $M\alpha$ N	Ce 58 0.359 2.561 14.06	Pr 59 0.346 2.463 13.34	Nd 60 0.340 2.370 12.68	Pm 61 0.322 2.283	Sm 62 0.310 2.199 11.43	Eu 63 0.300 2.120 10.95	Gd 64 0.290 2.046 10.41	Tb 65 0.280 1.975 9.937 86.00	Dy 66 0.271 1.909 9.543 83.00	Ho 67 0.262 1.845 9.161	Er 68 0.254 1.784 8.801 72.70	Tm 69 0.246 1.726 8.460	Yb 70 0.238 1.672 8.146 65.10	Lu 71 0.231 1.619 7.840 63.00	$K\alpha$ $L\alpha_1$ $M\alpha$ N
$K\alpha$ $L\alpha_1$ $M\alpha$ N	Th 90 0.134 0.956 4.138 9.442	Pa 91 0.131 0.933 4.022	U 92 0.128 0.910 3.910 8.808	Np 93 0.125 0.889	Pu 94 0.122 0.869	Am 95 0.119 0.849	Cm 96 0.116 0.829	Bk 97 0.113 0.810	Cf 98 0.110 0.792	Es 99	Fm 100	Md 101	No 102		$K\alpha$ $L\alpha_1$ $M\alpha$ N

[E. W. White, G. V. Gibbs, G. G. Johnson, Jr., and G. R. Zechman, Jr., "X-Ray Emission Line Wavelength and Two-Theta Tables," *ASTM Data Series* **DS-37**, American Society for Testing and Materials (1965); courtesy of the authors and ASTM.]

Bibliography

BOOKS

(*1*) Adler, I., "X-Ray Emission Spectrography in Geology," Elsevier Publishing Co., N. Y., 258 pp. (1966).
(*2*) American Society for Testing and Materials, "Symposium on Fluorescent X-Ray Spectrographic Analysis," *Spec. Tech. Publ.* **157**, 68 pp. (1954); Xerox copies available from University Microfilms, Inc.
(*3*) American Society for Testing and Materials, "Symposium on X-Ray and Electron-Probe Analysis," *Spec. Tech. Publ.* **349**, 209 pp. (1964).
(*4*) Bertin, E. P., "Principles and Practice of X-Ray Spectrometric Analysis," Plenum Press, N. Y., 679 pp. (1970).
(*5*) Birks, L. S., "Electron Probe Microanalysis," Interscience Publishers, N. Y., 253 pp. (1963).
(*6*) Birks, L. S., "X-Ray Spectrochemical Analysis," Interscience Publishers, N. Y., 137 pp. (1959).
(*7*) Blokhin, M. A., (translated from Russian by F. L. Curzon), "Methods of X-Ray Spectroscopic Research," Pergamon Press, N. Y., 448 pp. (1965).
(*8*) Clark, G. L., "Applied X-Rays," McGraw-Hill Book Co., N. Y., ed. 4, 843 pp. (1955).
(*9*) Compton, A. H., and S. K. Allison, "X-Rays in Theory and Experiment," D. Van Nostrand Co., N. Y., 828 pp. (1935).
(*10*) Dewey, R. D., R. S. Mapes, and T. W. Reynolds, "Handbook of X-Ray and Microprobe Data"; Progress in Nuclear Energy; Series 9, Analytical Chemistry; vol. 9; Pergamon Press, N. Y., 353 pp. (1969).
(*11*) Elion, H. A., "Instrument and Chemical Analysis Aspects of Electron Microanalysis and Macroanalysis," Progress in Nuclear Energy; Series 9, Analytical Chemistry; vol. 5; Pergamon Press, N. Y., 256 pp. (1966).
(*12*) Glasser, O., "Dr. W. C. Rontgen," Charles C. Thomas, Publisher, Springfield, Ill., ed. 2, 169 pp. (1958).
(*13*) von Hevesy, G., "Chemical Analysis by X-Rays and Its Applications," McGraw-Hill Book Co., N. Y., 333 pp. (1932); Xerox copies available from University Microfilms, Inc.
(*14*) Jenkins, R., and J. L. de Vries, "Practical X-Ray Spectrometry," Philips Technical Library, Springer-Verlag New York, N. Y., 182 pp. (1967).

(*15*) Kaelble, E. F., ed., "Handbook of X-Rays," McGraw-Hill Book Co., N. Y. (1967); chapters pertinent to this book:

A) Adler, I., [X-Ray] Optics, ch. 31, 20 pp.
B) Bertin, E. P., and R. L. Samber, Contact Microradiography, ch. 45, 28 pp.
C) Birks, L. S., Electron-Probe Microanalysis, ch. 39, 12 pp.
D) Buhrke, V. E., Detection and Measurement of X-Rays, ch. 3, 46 pp.
E) Cave, W. T., Effects of Chemical State of Specimen, ch. 34, 3 pp.
F) Croke, J. F., and W. R. Kiley, Specimen-Preparation Techniques, ch. 33, 22 pp.
G) Dodd, C. G., Absorptiometry with Monochromatic X-Rays, ch. 42, 14 pp.
H) Gunn, E. L., Quantitative Techniques, ch. 36, 48 pp.
I) Holliday, J. E., Soft X-Ray Emission Spectroscopy in the 10–150-Å Region, ch. 38, 41 pp.
J) Johns, W. D., Measurement of Film Thickness, ch. 44, 12 pp.
K) Kaelble, E. F., Introduction to X-Ray Emission Spectroscopy, ch. 30, 5 pp.
L) Kemp, J. W., High-Speed Automatic Spectrometers and Continuous Analyzers, ch. 40, 8 pp.
M) Lambert, M. C., Absorptiometry with Polychromatic X-Rays, ch. 41, 24 pp.
N) Loomis, T. C., and S. M. Vincent, Trace and Microanalysis, ch. 37, 16 pp.
O) Salmon, M. L., Factors which Determine Sensitivity, ch. 32, 8 pp.
P) Salmon, M. L., Qualitative and Semiquantitative Techniques, ch. 35, 18 pp.
Q) Van Nordstrand, R. A., Fine Structure in Absorption-Edge Spectra, ch. 43, 18 pp.

(*16*) Liebhafsky, H. A., H. G. Pfeiffer, E. H. Winslow, and P. D. Zemany, "X-Ray Absorption and Emission in Analytical Chemistry," John Wiley and Sons, N. Y., 357 pp. (1960).

(*17*) Müller, R. O., "Spektrochemische Analysen mit Roentgen Fluoreszenz," R. Oldenbourg, Munich, 315 pp. (1967); in German.

(*18*) Parrish, W., ed., "X-Ray Analysis Papers," Centrex Publishing Co., Eindhoven, Netherlands, 310 pp. (1965).

(*19*) Polonio, J. Bermudez, "Teoria y Practica de la Espectroscopia de Rayos X," Editorial Alhambra, S. A., Madrid, 267 pp. (1967); in Spanish.

(*20*) Siegbahn, M., (translated from German by G. A. Lindsay), "The Spectroscopy of X-Rays," Oxford University Press, London, 287 pp. (1925); Xerox copies available from University Microfilms, Inc.

(*21*) Sproull, W. T., "X-Rays in Practice," McGraw-Hill Book Co., N. Y., 615 pp. (1946).

(*22*) "Advances in X-Ray Analysis" (various editors), Proceedings of the Annual Conference on Applications of X-Ray Analysis, Denver; sponsored by Denver Research Institute, University of Denver; Plenum Press, N. Y., vols. 1–12 (1958–69).

(*23*) "Developments in Applied Spectroscopy" (various editors), Proceedings of the Annual Mid-America Spectroscopy Symposium, Chicago; Plenum Press, N. Y., vols. 1–8 (1962–69).

GENERAL REVIEWS

(*24*) Adler, I., and H. S. Rose, Jr., X-Ray Emission Spectrography; in "Trace Analysis, Physical Methods," C. H. Morrison, ed.; Interscience Publishers, N. Y., ch. 8, pp. 271–324 (1965).

(*25*) Campbell, W. J., X-Ray Emission and Absorption; in "Standard Methods of Chemical Analysis," F. J. Welcher, ed.; D. Van Nostrand Co., N. Y., ed. 6, vol. 3, part A, ch. 9, pp. 169–85 (1966).

(*26*) Hall, T., X-Ray Spectroscopy; in "Trace Analysis," J. H. Yoe and H. J. Koch, eds.; John Wiley and Sons, N. Y., ch. 17, pp. 458–92 (1957).

(*27*) Sandström, A. E., Experimental Methods of X-Ray Spectroscopy, Ordinary Wavelengths; in "Handbuch der Physik," S. Flügge, ed.; Springer-Verlag, Berlin, vol. 30, pp. 78–245 (1957).

(*28*) Shalgosky, I., Fluorescent X-Ray Spectrography; in "Methods in Geochemistry," A. A. Smales and L. R. Wager, eds.; Interscience Publishers, N. Y., pp. 111–47 (1960).

(*29*) X-Ray Absorption and Emission, Annual (1949–52) and Biennial (1954–) Reviews, *Anal. Chem.*:

H. A. Liebhafsky, E. H. Winslow, and H. G. Pfeiffer, **21**, 17–24 (1949); **22**, 15–16 (1950); **23**, 14–16 (1951); **24**, 16–20 (1952); **26**, 26–31 (1954); **28**, 583–91 (1956); **30**, 580–9 (1958); **32**, 240R–248R (1960); **34**, 282R–294R (1962).

W. J. Campbell, J. D. Brown, and J. W. Thatcher, **36**, 312R–328R (1964); **38**, 416R–439R (1966); **40**, 364R–375R (1968).

See also (*2*), (*3*), and (*15*) above and (*L4*) and (*P4*) below.

BIBLIOGRAPHIES

(*30*) American Society for Testing and Materials, "Index to the Literature on X-Ray Spectrographic Analysis. Part 1. 1913–1957," *Spec. Tech. Publ.* **292**, 42 pp. (1961).

(*31*) N. V. Philips Gloeilampenfabrieken, Scientific Instruments Dept., "Review of Literature, X-Ray Spectrometry," ed. 2, 55 pp. (1965).

(*32*) Stoddart, H. A., and W. A. Dowden, "X-Ray Fluorescence Analysis: Part 1. Bibliography. Part 2. Indexes," *At. Energy Res. Estab. Gt. Brit., Rep.* **AERE-Bib 136**, 203 pp. (1966).

(*33*) Yakowitz, H., and J. R. Cuthill, "Annotated Bibliography on Soft X-Ray Spectroscopy," *Nat. Bur. Stand.* (*U.S.*) *Monogr.* **52**, 109 pp. (1962).

TABLES OF WAVELENGTHS, 2θ ANGLES, AND MASS-ABSORPTION COEFFICIENTS

(*34*) Amsbury, W. P., W. W. Lee, J. H. Rowan, and G. E. Walden, "X-Ray Fluorescence Tables," *U. S. At. Energy Comm. Rep.* **Y-1470** (1964); in 9 parts: A. Program Description, 47 pp.; B. Topaz, 116 pp.; C. LiF, 135 pp.; D. EDDT, 171 pp.; E. Gypsum, 187 pp.; F. Mica (2*d* 19.896 Å), 191 pp.; G. Mica (2*d* 19.96 Å), 192 pp.; H. KHP, 196 pp.; I. Ba Stearate, 206 pp.

(*35*) Bearden, J. A., "X-Ray Wavelengths," *U. S. At. Energy Comm. Rep.* **NYO-10586**, 533 pp. (1964).

(*36*) Cauchois, Y., and H. Hulubei, "Tables of Constants and Numerical Data. I. Selected Constants—Wavelengths of X-Ray Emissions and of X-Ray Absorption Discontinuities," Hermann & Cie., Paris, 200 pp. (1947).

(*37*) Dewey, R. D., R. S. Mapes, and T. W. Reynolds, "Computed X-Ray Wavelengths with Wavelength Tables," Reynolds Metals Co., Metallurgical Research Div., Richmond, Va., 198 pp. (1966).

(*38*) Dewey, R. D., R. S. Mapes, and T. W. Reynolds, "A Study of X-Ray Mass Absorption Coefficients with Tables of Coefficients," Reynolds Metals Co., Metallurgical Research Div., Richmond Va., 51 pp. (1967).

(*39*) Fine, S., and C. F. Hendee, Table of X-Ray *K* and *L* Emission and Critical Absorption Energies for All the Elements, *Nucleonics* **13**(3), 36–7 (1955); *Norelco Reptr.* **3**, 113–15 (1956).

(*40*) General Electric Co., X-Ray Dept., "X-Ray Wavelengths for Spectrometer," Cat. no. A4961DA, ed. 5 (1969): ADP, EDDT, Ge, Gypsum, KHP, LiF(200), LiF(220), NaCl, PET, Quartz (2*d* 2.75, 6.68), Si, Topaz; Pb myristate, Pb stearate.

(*41*) Heinrich, K. F. J., X-Ray Absorption Uncertainty [Including Table of Mass Absorption Coefficients]; in "The Electron Microprobe," E. D. McKinley, K. F. J. Heinrich, and D. B. Wittry, eds.; John Wiley and Sons, N. Y., pp. 296–377 (1966).

(*42*) Heinrich, K. F. J., and M. A. M. Giles, "X-Ray Wavelength Conversion Tables and Graphs for Qualitative Electron-Probe Microanalysis," *Nat. Bur. Stand.* (*U. S.*) *Tech. Note* **406**, 58 pp. (1967).

(*43*) Henke, B. L., R. L. Elgin, R. E. Lent, and R. B. Ledingham, "X-Ray Absorption in the 2 to 200 Å Region," *U. S. Air Force*, *Office Sci. Res. Rep.* **AFOSR 67-1254**, 35 pp. (1967).

(*44*) International Union of Crystallography, Absorption; in "International Tables for X-Ray Crystallography," Kynoch Press, Birmingham, Eng.; vol. 3, ch. 3.2, p. 157–200 (1962).

(*45*) N. V. Philips Gloeilampenfabrieken, Application Lab., "Table of X-Ray Mass Absorption Coefficients," *Norelco Reptr.* **9**(3) (1962).

(*46*) Powers, M. C., "X-Ray Fluorescent Spectrometer Conversion Tables," Philips Electronic Instruments, Inc., Mt. Vernon, N. Y. (1957): ADP, EDDT, LiF, NaCl, Topaz.

(*47*) Sagel, K., "Tabellen zur Roentgen Emissions und Absorptions Analyse," Springer-Verlag, Berlin, 135 pp. (1959).

(*48*) Stainer, H. M., "X-Ray Mass Absorption Coefficients—A Literature Survey," *U. S. Bur. Mines Inform. Circ.* **8166**, 124 pp. (1963).

(*49*) White, E. W., G. V. Gibbs, G. G. Johnson, Jr., and G. R. Zechman Jr., "X-Ray Emission Line Wavelength and Two-Theta Tables," *Amer. Soc. Test. Mater. Data Ser.* **37** (1965): ADP, $CaCO_3$, EDDT, Gypsum, KHP, Lead stearate decanoate, LiF, Mica, NaCl, PET, Quartz (2*d* 6.68).

(*50*) White, E. W., G. V. Gibbs, G. G. Johnson, Jr., and G. A. Zechman, Jr., "X-Ray Wavelengths and Crystal Interchange Settings for Wavelength-Geared Curved-Crystal Spectrometers," Mineral Industries Expt. Sta., Pa. State Univ., University Park, Pa., *Spec. Publ.* **3-64**, ed. 2 (1965).

(*51*) Zingaro, P. W., X-Ray Spectrographic Data [Emission Lines and Absorption Edges], *Norelco Reptr.*: I. *K*-Series, **1**, 67–8 (1953); II. *L*-Series, **1**, 70, 78–9 (1953); III. X-Ray Emission Lines of the Elements Arranged in Order of Their Wavelengths, **2**, 92–5 (1955).

PAPERS AND REPORTS

(*A1*) Achey, F. A., and E. J. Serfass, A New Approach to Coating Thickness by Fluorescent X-Ray Absorption, *J. Electrochem. Soc.* **105**, 204–5 (1958).

(*A2*) Addink, N. W. H., Note on Analysis of Small Quantities of Material by X-Ray Fluorescence, *Rev. Universelle Mines* **15**, 530–2 (1959).

(*A3*) Addink, N. W. H., H. Kraay, and A. W. Witmer, Putting to Advantage the Absorbing Qualities of Diluting Agents for Obtaining 45-Degree Calibration Curves in X-Ray Fluorescence Analysis, *Colloq. Spectrosc. Int., 9th, Lyons, 1961* **3**, 368–84 (1962).

(*A4*) Adler, I., Intermediate X-Ray Probe; in "Encyclopedia of X-Rays and Gamma Rays," G. L. Clark, ed.; Reinhold Publishing Corp., N. Y., pp. 834–40 (1963).

(*A5*) Adler, I., and J. M. Axelrod, The Curved-Crystal X-Ray Spectrometer—A Mineralogical Tool, *Amer. Mineral.* **41**, 524–7 (1956).

(*A6*) Adler, I., and J. M. Axelrod, Aluminum Powder as a Binder in Sample Preparation for X-Ray Spectrometry, *Anal. Chem.* **26**, 931–2 (1954).

(*A7*) Adler, I., and J. M. Axelrod, Determination of Thorium by Fluorescent X-Ray Spectrometry, *Anal. Chem.* **27**, 1002–3 (1955).

(*A8*) Adler, I., and J. M. Axelrod, The Reflecting Curved-Crystal X-Ray Spectrograph—A Device for the Analysis of Small Mineral Samples, *Econ. Geol.* **52**, 694–701 (1957).

(*A9*) Adler, I., and J. M. Axelrod, A Multi-Wavelength Fluorescence Spectrometer, *J. Opt. Soc. Amer.* **43**, 769–72 (1953).

(*A10*) Adler, I., and J. M. Axelrod, Internal Standards in Fluorescent X-Ray Spectroscopy, *Spectrochim. Acta,* **7**, 91–9 (1955).

(*A11*) Adler, I., J. M. Axelrod, and J. R. Branco, Further Application of the Intermediate X-Ray Probe, *Advan. X-Ray Anal.* **2**, 167–73 (1959).

(*A12*) Alley, B. J., and J. H. Higgins, Empirical Corrections for Variable Absorption of Soft X-Rays by Mylar, *Norelco Reptr.* **10**, 77–80 (1963).

(*A13*) Alley, B. J., and R. H. Meyers, Corrections for Matrix Effects in X-Ray Fluorescence Analysis Using Multiple Regression Methods, *Anal. Chem.* **37**, 1685–90 (1965).

(*A14*) Andermann, G., and J. D. Allen, X-Ray Emission Analysis of Finished Cements, *Anal. Chem.* **33**, 1695–9 (1961).

(*A15*) Andermann, G., and J. W. Kemp, Scattered X-Rays as Internal Standards in X-Ray Emission Spectroscopy, *Anal. Chem.* **30**, 1306–9 (1958).

(*A16*) Applied Research Laboratories, Inc., X-Ray Analysis of Metallic Elements in Nonmetallic Material, *ARL Spectrogr. News Lett.* **7**(3), 1–3 (1954).

(*A17*) Arndt, U. W., W. A. Coates, and A. R. Crathorn, Gas-Flow X-Ray Diffraction Counter, *Proc. Phys. Soc.* (*London*) **67B**, 357–9 (1954).

(*A18*) Ashby, W. D., and E. M. Proctor, Measurement of X–Ray Source Stability, *Picker Analyzer* **4**(1), 1–8 (1965).

(*A19*) Ashley, R. W., and R. W. Jones, Application of X-Ray Fluorescence Methods to the Analysis of Zircaloy, *Anal. Chem.* **31**, 1632–5 (1959).

(*A20*) Auger, P., Compound Photoelectric Effect. *Ann. Phys.* **6**, 183–253 (1926).

(*A21*) Auger, P., Secondary β-Rays Produced in O Gas by X-Rays, *C. R. Acad. Sci.* (*Paris*) **180**, 65–8 (1925).

(*A22*) Auger, P., Compound Photoelectric Effect, *J. Phys. Radium* **6**, 205–8 (1925).

(*A23*) Axelrod, J. M., and I. Adler, X-Ray Spectrographic Determination of Cesium and Rubidium, *Anal. Chem.* **29**, 1280–1 (1957).

(*B1*) Baird, A. K., A Pressed-Specimen Die for the Norelco Vacuum-Path X-Ray Spectrograph, *Norelco Reptr.* **8**, 108–9 (1961).

(*B2*) Banks, M., and E. T. Hall, X-Ray Fluorescent Analysis in Archaeology—The "Milliprobe," *Archaeometry* **6**, 31–6 (1963).

(*B3*) Barieau, R. E., X-Ray Absorption-Edge Spectrometry as an Analytical Tool, *Anal. Chem.* **29**, 348–52 (1957).

(*B4*) Barkla, C. G., Spectra of the Fluorescent Röntgen Radiation, *Phil. Mag.* [6] **22**, 396–412 (1911).

(*B5*) Barkla, C. G., and C. A. Sadler, Absorption of Roentgen Rays, *Phil. Mag.* [6] **17**, 739–60 (1909).

(*B6*) Barstad, G. E. B., and I. N. Refsdal, Sensitive Quantitative Recording X-Ray Spectrometer, *Rev. Sci. Instrum.* **29**, 343–8 (1958).

(*B7*) Baumgartner, W. V., X-Ray Spectrometry Extends Film-Badge Dosimetry, *Nucleonics* **18**(8), 76, 78–9 (1960).

(*B8*) Bean, L., Method of Producing Sturdy Specimens of Pressed Powders for Use in X-Ray Spectrochemical Analysis, *Appl. Spectrosc.* **20**, 191–3 (1966).

(*B9*) Beard, D. W., and E. M. Proctor, A Method of Liquid Analysis Providing Increased Sensitivity for Light Elements, *Advan. X-Ray Anal.* **10**, 506–19 (1967).

(*B10*) Beattie, H. J., and R. M. Brissey, Calibration Method for X-Ray Fluorescence Spectrometry, *Anal. Chem.* **26**, 980–3 (1954).

(*B11*) Bedo, D. E., and D. H. Tomboulian, Photon Counting Spectrometer for Attenuation Measurements in the Soft X-Ray Region, *Rev. Sci. Instrum.* **32**, 184–8 (1961).

(*B12*) Beeghley, H. F., X-Ray Method for Determining Tin Coating Thickness, *J. Electrochem. Soc.* **97**, 152–7 (1950).

(*B13*) Behr, F. A,. and P. W. Zingaro, Notes on the Selection of Analyzing Crystals in X-Ray Spectrography, *Norelco Reptr.* **1**, 26–8 (1953).

(*B14*) Bernstein, F., Relationship Between X-Ray Tube Target Materials and X-Ray Emission Intensities, *Develop. Appl. Spectrosc.* **4**, 45–55 (1965); *ISA* (*Instrum. Soc. Amer.*) *Trans.* **6**, 99–102 (1967).

(*B15*) Bertin, E. P., Evaluation and Application of an Improved Slit Probe for the X-Ray Secondary-Emission Spectrometer, *Advan. X-Ray Anal.* **8**, 231–47 (1965).

(*B16*) Bertin, E. P., Comparison of Four Slit Apertures for Selected-Area Analysis with the X-Ray Secondary-Emission Spectrometer, *Advan. X-Ray Anal.* **10**, 462–73 (1967).

(*B17*) Bertin, E. P., Recent Advances in Quantitative X-Ray Spectrometric Analysis by Solution Techniques, *Advan. X-Ray Anal.* **11**, 1–22 (1968)

(*B18*) Bertin, E. P., X-Ray Probe with Slit Aperture in the Secondary Beam, *Anal. Chem.* **36**, 441–3 (1964).

(*B19*) Bertin, E. P., Intensity-Ratio Technique for X-Ray Spectrometric Analysis of Binary Samples, *Anal. Chem.* **36**, 826–32 (1964).

(*B20*) Bertin, E. P., Solution Techniques in X-Ray Spectrometric Analysis, *Norelco Reptr.* **12**, 15–26 (1965).

(*B21*) Bertin, E. P., and R. J. Longobucco, Some Special Sample Mounting Devices for the X-Ray Fluorescence Spectrometer, *Advan. X-Ray Anal.* **5**, 447–56 (1962).

(*B22*) Bertin, E. P., and R. J. Longobucco, X-Ray Spectrometric Determination of Composition and Distribution of Sublimates in Receiving-Type Electron Tubes, *Advan. X-Ray Anal.* **7**, 566–83 (1964).

(*B23*) Bertin, E. P., and R. J. Longobucco, X-Ray Spectrometric Determination of Plate Metals on Plated Wires, *Anal. Chem.* **34**, 804–11 (1962).

(*B24*) Bertin, E. P., and R. J. Longobucco, X-Ray Spectrometric Analysis—Nickel, Copper, Silver, and Gold in Plating Baths, *Metal. Finish.* **60**(3), 54–8 (1962).
(*B25*) Bertin, E. P., and R. J. Longobucco, X-Ray Methods for Determination of Plate Thickness, *Metal. Finish.* **60**(8), 42–4 (1962).
(*B26*) Bertin, E. P., and R. J. Longobucco, Sample Preparation Methods for X-Ray Fluorescence Emission Spectrometry, *Norelco Reptr.* **9**, 31–43 (1962).
(*B27*) Bertin, E. P., and R. J. Longobucco, Spectral-Line Interference in X-Ray Fluorescence Spectrometry, *Norelco Reptr.* **9**, 64–72 (1962).
(*B28*) Bertin, E. P., R. J. Longobucco, and R. J. Carver, Simplified Routine Method for X-Ray Absorption-Edge Spectrometric Analysis, *Anal. Chem.* **36**, 641–55 (1964).
(*B29*) Bessen, I. I., Principles and Applications of Proportional Counters for X-Ray Spectrographs, *Advan. X-Ray Anal.* **1**, 455–68 (1958).
(*B30*) Birks, L. S., Apparatus for Vacuum X-Ray Fluorescence Analysis of Light Elements, *Rev. Sci. Instrum.* **22**, 891–1004 (1951).
(*B31*) Birks, L. S., Fluorescent X-Ray Excitation Efficiencies, *Spectrochim. Acta* **17**, 148–54 (1961).
(*B32*) Birks, L. S., and A. P. Batt, Use of a Multichannel Analyzer for Electron-Probe Microanalysis, *Anal. Chem.* **35**, 778–82 (1963).
(*B33*) Birks, L. S., and E. J. Brooks, Hafnium–Zirconium and Tantalum–Columbium Systems—Quantitative Analysis by X-Ray Fluorescence, *Anal. Chem.* **22**, 1017–20 (1950).
(*B34*) Birks, L. S., and E. J. Brooks, Analysis of Uranium Solutions by X-Ray Fluorescence, *Anal. Chem.* **23**, 707–9 (1951).
(*B35*) Birks, L. S., and E. J. Brooks, Applications of Curved-Crystal X-Ray Spectrometers, *Anal. Chem.* **27**, 437–40 (1955).
(*B36*) Birks, L. S., and E. J. Brooks, Miniature Fluorescence X-Ray Spectrograph, *Anal. Chem.* **27**, 1147–9 (1955).
(*B37*) Birks, L. S., and D. M. Brown, Precision in X-Ray Spectrochemical Analysis—Fixed Time *vs.* Fixed Count, *Anal. Chem.* **34**, 240–1 (1962).
(*B38*) Birks, L. S., and D. L. Harris, Unusual Matrix Effects in Fluorescent X-Ray Spectrometry, *Anal. Chem.* **34**, 943–5 (1962).
(*B39*) Birks, L. S., and R. T. Seal, X-Ray Properties of Plastically Deformed LiF, *J. Appl. Phys.* **28**, 541–3 (1957).
(*B40*) Birks, L. S., and J. M. Siomkajlo, Long-Spacing Metal-Organic Crystals for X-Ray Spectrometry, *Spectrochim. Acta* **18**, 363–6 (1962).
(*B41*) Birks, L. S., E. J. Brooks, and H. Friedman, Fluorescent X-Ray Spectroscopy, *Anal. Chem.* **25**, 692–7 (1953).
(*B42*) Birks, L. S., E. J. Brooks, and G. W. Gourlay, Compact Curved-Crystal X-Ray Spectrometer, *Rev. Sci. Instrum.* **29**, 425–6 (1958).
(*B43*) Birks, L. S., E. J. Brooks, H. Friedman, and R. M. Roe, X-Ray Fluorescence Analysis of Ethyl Fluid in Aviation Gasoline, *Anal. Chem.* **22**, 1258–61 (1950).
(*B44*) Blank, G. R., and H. A. Heller, X-Ray Spectrometric Determination of Copper, Tin, and Uranium in Bronze Heat-Treating Material, *Advan. X-Ray Anal.* **4**, 457–73 (1961).
(*B45*) Blank, G. R., and H. A. Heller, X-Ray Spectrometric Determination of Thorium in Refined Uranium Materials, *Norelco Reptr.* **8**, 112–15 (1961); *Develop. Appl. Spectrosc.* **1**, 3–15 (1962).
(*B46*) Blank, G. R., and H. A. Heller, X-Ray Spectrometric Determination of Zirconium in Refined Uranium Materials, *Norelco Reptr.* **9**, 23–7 (1962).

(*B47*) Blodgett, K. B., Films Built by Depositing Successive Unimolecular Layers on a Solid Surface, *J. Amer. Chem. Soc.* **57**, 1007–22 (1935).

(*B48*) Booker, G. R., and J. Norbury, Extraction Replica Method for Large Precipitates and Nonmetallic Inclusions in Steel, *Brit. J. Appl. Phys.* **8**, 109–13 (1957).

(*B49*) Bowman, H. R., E. K. Hyde, S. G. Thompson, and R. C. Jared, Application of High-Resolution Semiconductor Detectors in X-Ray Emission Spectrography, *Science* **151**, 562–8 (1966).

(*B50*) Boyd, B. R., and H. T. Dryer, Analysis of Nonmetallics by X-Ray Fluorescence Techniques, *Develop. Appl. Spectrosc.* **2**, 335–49 (1963).

(*B51*) Bragg, W. L., Diffraction of Short Electromagnetic Waves by a Crystal, *Proc. Cambridge Phil. Soc.* **17**, 43–57 (1912).

(*B52*) Bragg, W. H., and W. L. Bragg, Reflection of X-Rays by Crystals, *Proc. Phys. Soc.* (*London*) **A88**, 428–38 (1913).

(*B53*) Brentano, J. C. M., and I. Ladany, Scintillation Counting of Low-Energy Photons and Its Use in X-Ray Diffraction, *Rev. Sci. Instrum.* **25**, 1028–9 (1954).

(*B54*) Brown, G. A., Applications of an X-Ray Pulse Spectroscope; in "X-Ray and Electron Microscopy News," *Siemens Rev.* **34** (Special Issue), 69–73 (1967).

(*B55*) Brown, O. E., H. L. Austin, and D. M. Schaefer, Preparation of Powdered Materials in Plastic Mounts for X-Ray Analysis, *Norelco Reptr.* **7**, 140 (1960).

(*B56*) Burbank, J., Identification and Characterization of Electrochemical Reaction Products by X-Ray Diffraction, *Naval Res. Lab.* (*U. S.*) *Rep.* **6626**, 15 pp. (1967).

(*B57*) Burke, W. E., and D. L. Wood, Rare Earth Analysis by X-Ray-Excited Optical Fluorescence, *Advan. X-Ray Anal.* **11**, 204–13 (1968).

(*B58*) Burkhalter, P. G., and W. J. Campbell, Comparison of Detectors for Isotopic X-Ray Analyzers; in "Proceedings of the Second Symposium on Low-Energy X and Gamma Sources and Applications," P. S. Baker and M. Gerrard, eds.; *Oak Ridge Nat. Lab. Rep.* **ORNL-IIC-10**, vol. 1, pp. 393–423 (1967).

(*B59*) Burley, R. A., Diffraction Effects in X-Ray Fluorescence Analysis of Semiconductor-Grade Silicon and Other Monocrystalline Materials, *Appl. Spectrosc.* **21**, 416–19 (1967).

(*B60*) Burnham, H. D., J. Hower, and L. C. Jones, Generalized X-Ray Emission Spectrographic Calibration Applicable to Varying Compositions and Sample Forms, *Anal. Chem.* **29**, 1827–34 (1957).

(*C1*) Caley, J. L., Use of X-Ray Emission Spectrography for Petroleum Product Quality and Process Control, *Advan. X-Ray Anal.* **6**, 396–402 (1963).

(*C2*) Callendar, M. V., and D. F. White, Aspects of the Emission of X-Rays from Television Receivers, *J. Brit. Inst. Radio Eng.* **21**, 389–400 (1961).

(*C3*) Cameron, J. F., and J. R. Rhodes, Measurement of Tinplate Thickness Using Fluorescent X-Rays Excited by a Radioactive Source, *Brit. J. Appl. Phys.* **11**, 49–52 (1960).

(*C4*) Campbell, J. T., F. W. J. Garton, and J. D. Wilson, An X-Ray Milliprobe Analyzer with Photographic Recording of the Spatial Distribution of Elements, *Talanta* **15**, 1205–16 (1968).

(*C5*) Campbell, W. J., Fluorescent X-Ray Spectrographic Analysis of Trace Elements, Including Thin Films, *Amer. Soc. Test. Mater. Spec. Tech. Publ.* **349**, 48–69 (1964).

(*C6*) Campbell, W. J., Apparatus for Continuous Fluorescent X-Ray Spectrographic Analysis of Solutions, *Appl. Spectrosc.* **14**, 26–7 (1960).

(*C7*) Campbell, W. J., Fluorescent X-Ray Spectrographic Analysis—Studies of Low-Energy *K*, *L*, and *M* Spectral Lines, *U. S. Bur. Mines Rep. Invest.* **5538**, 20 pp. (1959).

(*C8*) Campbell, W. J., Energy-Dispersion X-Ray Analysis Using Radioactive Sources; in "X-Ray and Electron Methods of Analysis," H. Van Olphen and W. Parrish, eds.; Plenum Press, N. Y., pp. 36–54 (1968).

(*C9*) Campbell, W. J., and H. F. Carl, Quantitative Analysis of Niobium and Tantalum in Ores by Fluorescent X-Ray Spectroscopy, *Anal. Chem.* **26**, 800–5 (1954).

(*C10*) Campbell, W. J., and H. F. Carl, Combined Radiometric and Fluorescent X-Ray Spectrographic Method of Analyzing for Uranium and Thorium, *Anal. Chem.* **27**, 1884–6 (1955).

(*C11*) Campbell, W. J., and H. F. Carl, Fluorescent X-Ray Spectrographic Determination of Tantalum in Commercial Niobium Oxides, *Anal. Chem.* **28**, 960–2 (1956).

(*C12*) Campbell, W. J., and H. F. Carl, Quantitative Analysis by Fluorescent X-Ray Spectrography—Determination of Germanium in Coal and Coal Ash, *Anal. Chem.* **29**, 1009–17 (1957).

(*C13*) Campbell, W. J., and M. Leon, Fluorescent X-Ray Spectrograph for Dynamic Selective Oxidation Rate Studies—Design and Principles, *U. S. Bur. Mines Rep. Invest.* **5739**, 21 pp. (1961).

(*C14*) Campbell, W. J., and J. W. Thatcher, Determination of Calcium in Wolframite Concentrates by Fluorescent X-Ray Spectrography, *U. S. Bur. Mines Rep. Invest.* **5416**, 18 pp. (1958); *Advan. X-Ray Anal.* **2**, 313–32 (1959).

(*C15*) Campbell, W. J., and J. W. Thatcher, Fluorescent X-Ray Spectrography—Determination of Trace Elements, *U. S. Bur. Mines Rep. Invest.* **5966**, 29 pp. (1962); *Develop. Appl. Spectrosc.* **1**, 31–62 (1962).

(*C16*) Campbell, W. J., M. Leon, and J. W. Thatcher, Flat-Crystal X-Ray Optics, *Advan. X-Ray Anal.* **1**, 193–206 (1958).

(*C17*) Campbell, W. J., M. Leon, and J. W. Thatcher, Solution Techniques in Fluorescent X-Ray Spectrography, *U. S. Bur. Mines Rep. Invest.* **5497**, 24 pp. (1959).

(*C18*) Campbell, W. J., E. F. Spano, and T. E. Green, Micro and Trace Analysis by a Combination of Ion-Exchange Resin-Loaded Papers and X-Ray Spectrography, *Anal. Chem.* **38**, 987–96 (1966).

(*C19*) Canon, M., Polycrystalline X-Ray Analyzing Crystal, *Advan. X-Ray Anal.* **8**, 285–300 (1965).

(*C20*) Carl, H. F., and W. J. Campbell, Fluorescent X-Ray Spectrographic Analysis of Minerals, *Amer. Soc. Test. Mater. Spec. Tech. Publ.* **157**, 63–8 (1954).

(*C21*) Carman, C. J., X-Ray Fluorescent Determination of Major Constituents in Multielement Matrices by Use of Coherent to Incoherent Scattering Ratios, *Develop. Appl. Spectrosc.* **5**, 45–58 (1966).

(*C22*) Carson, J. S., Timesaving with a Parallel Analyzing Channel Attachment, *Norelco Reptr.* **10**, 84, 120 (1963).

(*C23*) Carter, G. F., X-Ray Fluorescence Analysis of Roman Coins, *Anal. Chem.* **36**, 1264–6 (1964).

(*C24*) Caticha-Ellis, S., A. Ramos, and L. Saravia, Use of Primary Filters in X-Ray Spectrography—A New Method for Trace Analysis, *Advan. X-Ray Anal.* **11**, 105–13 (1968).

(*C25*) Cauchois, Y., Spectrograph Giving High Intensity from Transmission of Non-Collimated X-Rays through Curved Mica Sheets, *C. R. Acad. Sci.* (*Paris*) **194**, 362–5 (1932).
(*C26*) Cauchois, Y., Universal X-Ray Spectrograph for Use up to 20 Å, *J. Phys. Radium*, **6**, 89–96 (1945).
(*C27*) Champion, K. P., and R. N. Whittem, Utilization of Increased Sensitivity of X-Ray Fluorescence Spectrometry Due to Polarization of the Background Radiation, *Nature* **199**, 1082 (1963).
(*C28*) Chan, F. L., Detection, Confirmation, and Determination of Trace Amounts of Selenium by X-Ray Methods, *Advan. X-Ray Anal.* **7**, 542–54 (1964).
(*C29*) Chan, F. L., Study of X-Ray Fluorescence Method with Vacuum and Air-Path Spectrographs for the Determination of Film Thickness of II–VI Compounds, *Advan. X-Ray Anal.* **11**, 191–203 (1968).
(*C30*) Chan, F. L., Apparatus for Analysis of Liquid Samples by the X-Ray Fluorescence Method with a Vacuum Spectrograph, *Develop. Appl. Spectrosc.* **5**, 59–75 (1966).
(*C31*) Chan, F. L., and D. A. Brooks, Study of Homogeneity of Solid Solutions of Cadmium Sulfide and Cadmium Selenide by X-Ray Fluorescence, *Advan. X-Ray Anal.* **8**, 420–30 (1965).
(*C32*) Chodos, A. A., and C. G. Engel, Fluorescent X-Ray Spectrographic Analysis of Amphibolite Rocks, *Amer. Mineral.* **46**, 120–33 (1961).
(*C33*) Chodos, A. A., and W. Nichiporuk, Application of the Mutual Standards Concept to X-Ray Fluorescence Spectroscopy—Analysis of Meteoritic Sulfide Nodules for Eight Elements, *Advan. X-Ray Anal.* **2**, 247–55 (1959).
(*C34*) Claisse, F., Accurate X-Ray Fluorescence Analysis without Internal Standard, *Prov. Que., Can., Dept. Mines, Prelim. Rep.* **327**, 24 pp. (1956); *Norelco Reptr.* **4**, 3–7, 17, 19 (1957).
(*C35*) Clark, G. L., and W. Duane, Theory of the Tertiary Radiation Produced by Impacts of Photoelectrons, *Proc. Nat. Acad. Sci.* **10**, 41–7, 191–6 (1924).
(*C36*) Clark, G. L., and C. H. Liu, Quantitative Determination of Porosity by X-Ray Absorption, *Anal. Chem.* **29**, 1539–41 (1957).
(*C37*) Cockroft, A. L., and S. C. Curran, Elimination of the End Effects in Counters, *Rev. Sci. Instrum.* **22**, 37–42 (1951).
(*C38*) Collin, R. L., X-Ray Fluorescence Analysis Using Ion Exchange Resin for Sample Support—Determination of Strontium in 0.1 *M* Calcium Acetate Solutions, *Anal. Chem.* **33**, 605–7 (1961).
(*C39*) Compton, A. H., Quantum Theory of the Scattering of X-Rays by Light Elements, *Phys. Rev.* **21**, 483–502 (1923).
(*C40*) Cook, G. B., C. E. Mellish, and J. A. Payne, Measurement of Thin Metal Layers—Fluorescent X-Ray Production by Radioisotope Sources, *Anal. Chem.* **32**, 590–3 (1960).
(*C41*) Coolidge, W. D., A Powerful Roentgen-Ray Tube with a Pure Electron Discharge, *Phys. Rev.* [2] **2**, 409–30 (1913).
(*C42*) Cope, J. H., Quantitative X-Ray Spectrographic Analysis for Uranium with the Use of an Internal Standard, *Norelco Reptr.* **3**, 41–3 (1956).
(*C43*) Cosslett, V. E., and W. C. Nixon, "X-Ray Microscopy," Cambridge Univ. Press, Cambridge, Eng., pp. 206–11 (1960).
(*C44*) Cosslett, V. E., P. Duncumb, J. V. P. Long, and W. C. Nixon, Microanalysis by X-Ray Absorption, Fluorescence, Emission, and Diffraction Using Ultra-Fine X-Ray Sources, *Advan. X-Ray Anal.* **1**, 329–37 (1958).
(*C45*) Coster, D., and G. von Hevesy, Missing Element of Atomic Number 72, *Nature* **111**, 79 (1923).

(*C46*) Coster, D., and G. von Hevesy, The New Element Hafnium, *Nature* **111**, 182 (1923).

(*C47*) Cowgill, U. M., X-Ray Emission Spectroscopy in the Chemical Analysis of Lake Sediments—Determining 41 Elements, *Develop. Appl. Spectrosc.* **5**, 3–23 (1966).

(*C48*) Criss, J. W., and L. S. Birks, Calculation Methods for Fluorescent X-Ray Spectrometry—Empirical Coefficients *vs.* Fundamental Parameters, *Anal. Chem.* **40**, 1080–6 (1968).

(*C49*) Croke, J. F., W. J. Pfozer, and M. J. Solazzi, New Automatic X-Ray Spectrometer in Industrial Control, *Norelco Reptr.* **11**, 129–31 (1964).

(*C50*) Cullen, T. J., Potassium Pyrosulfate Fusion Technique—Determination of Copper in Mattes and Slags by X-Ray Spectroscopy, *Anal. Chem.* **32**, 516–17 (1960).

(*C51*) Cullen, T. J., Briqueted Copper Alloy Drillings as a Sample for X-Ray Spectroscopy, *Anal. Chem.* **33**, 1342–4 (1961).

(*C52*) Cullen, T. J., Coherent Scattered Radiation Internal Standardization in X-Ray Spectrometric Analysis of Solutions, *Anal. Chem.* **34**, 812–14 (1962).

(*C53*) Cullen, T. J., Addition of Sodium Fluoride to Potassium Pyrosulfate Fusions for X-Ray Spectrometric Analysis of Siliceous Samples, *Anal. Chem.* **34**, 862 (1962).

(*C54*) Cullen, T. J., Internal Differential X-Ray Absorption-Edge Spectrometry, *Anal. Chem.* **37**, 711–13 (1965).

(*C55*) Cullen, T. J., X-Ray Spectrometric Analysis of Solutions—Comparison of Techniques, *Pittsburgh Conf. Anal. Chem. Appl. Spectrosc.* (1963), unpublished; reviewed in detail in Ref. (*B17*).

(*C56*) Curran, S. C., and E. R. Rae, Some Properties of Counters with Beaded Wires, *J. Sci. Instrum.* **24**, 233–8 (1947).

(*D1*) Daugherty, K. E., and R. J. Robinson, X-Ray Fluorescence Spectrometric Analysis of the Iron(II), Cobalt(II), Nickel(II), and Copper(II) Chelates of 8-Quinolinol, *Anal. Chem.* **36**, 1869–70 (1964).

(*D2*) Daugherty, K. E., R. J. Robinson, and J. I. Mueller, X-Ray Fluorescence Spectrometric Analysis of the Copper(II) and Mercury(II) Complexes of 6-Chloro-2-Methoxy-9-Thioacridine, *Anal. Chem.* **36**, 1098 1100 (1964).

(*D3*) Davis, C. M., and G. R. Clark, X-Ray Spectrographic Analysis of Nickel-Containing Alloys with Varied Sample Forms, *Advan. X-Ray Anal.* **1**, 351–66 (1958).

(*D4*) Davis, C. M., and M. M. Yanak, Performance of an Unattended Automated X-Ray Spectrograph, *Advan. X-Ray Anal.* **7**, 644–52 (1964).

(*D5*) Davis, C. M., K. E. Burke, and M. M. Yanak, X-Ray Spectrographic Analysis of Traces in Metals by Preconcentration Techniques, *Advan. X-Ray Anal.* **11**, 56–62 (1968).

(*D6*) Davis, E. N., and B. C. Hoeck, X-Ray Spectrographic Method for the Determination of Vanadium and Nickel in Residue Fuels and Charging Stocks, *Anal. Chem.* **27**, 1880–4 (1955).

(*D7*) Davis, E. N., and R. A Van Nordstrand, Determination of Barium, Calcium, and Zinc in Lubricating Oils, *Anal. Chem.* **26**, 973–7 (1954).

(*D8*) Despujols, J., H. Roulet, and G. Senemaud, X-Ray Fluorescence Analysis with a Focused Primary Beam; in "X-Ray Optics and X-Ray Microanalysis," H. H. Pattee, V. E. Cosslett, and A. Engstrom, eds.; Academic Press, N. Y., pp. 445–9 (1963).

(*D9*) Dinklage, J., and R. Frericks, X-Ray Diffraction and Diffusion in Metal-Film-Layered Structures, *J. Appl. Phys.* **34**, 2633–5 (1963).

(*D10*) Dodd, C. G., and D. J. Kaup, Optimum Conditions for Chemical Analysis by X-Ray Absorption-Edge Spectrometry, *Anal. Chem.* **36**, 2325–9 (1964).

(*D11*) Dolby, R. M., Methods for Analyzing Unresolved Proportional Counter Curves of X-Ray Line Spectra, *Proc. Phys. Soc.* (*London*) **73**, 81–96 (1959).

(*D12*) Dothie, H. J., "Filterscan"—Novel Technique in X-Ray Fluorescence Analysis, *Nature* **196**, 984–5 (1962).

(*D13*) Doughman, W. R., R. C. Hirt, and R. T. Langdon, Water-Cooled Liquid Sample Cell for the X-Ray Emission Spectrograph, *Norelco Reptr.* **4**, 36, 40 (1957).

(*D14*) Dowling, P. H., C. F. Hendee, T. R. Kohler, and W. Parrish, Counters for X-Ray Analysis, *Philips Tech. Rev.* **18**, 262–75 (1956–57); *Norelco Reptr.* **4**, 23–33 (1957).

(*D15*) Drummond, C. H., Preparation of Glass Blanks by Fusion with Lithium Tetraborate for X-Ray Spectrographic Analysis, *Appl. Spectrosc.* **20**, 252 (1966).

(*D16*) Duane, W., and F. L. Hunt, X-Ray Wave-Lengths, *Phys. Rev.* **6**, 166–71 (1915).

(*D17*) DuMond, J. W. M., A High Resolving-Power Curved-Crystal Focusing Spectrometer for Short-Wavelength X-Rays and Gamma Rays, *Rev. Sci. Instrum.* **18**, 626–38 (1947).

(*D18*) DuMond, J. W. M., and H. A. Kirkpatrick, Multiple-Crystal X-Ray Spectrograph, *Rev. Sci. Instrum.* **1**, 88–105 (1930).

(*D19*) Duncumb, P., Absorption and Emission Methods of Microanalysis, *Rev. Universelle Mines* **17**, 240–46 (1961).

(*D20*) Dunn, H. W., X-Ray Absorption-Edge Analysis, *Anal. Chem.* **34**, 116–21 (1962).

(*D21*) Dunne, J. A., A New Focusing Vacuum X-Ray Macroprobe Analyzer, *Advan. X-Ray Anal.* **8**, 223–30 (1965).

(*D22*) Dunne, J. A., Components for X-Ray Fluorescence Spectroscopy in the 5–65-Å Wavelength Region, *Norelco Reptr.* **11**, 109–12 (1964).

(*D23*) Dunne, J. A., and W. R. Muller, Demountable X-Ray Tube for Light-Element Fluorescence Analysis, *Develop. Appl. Spectrosc.* **4**, 33–44 (1965).

(*D24*) Dwiggins, C. W., Jr., Quantitative Determination of Low Atomic Number Elements Using Intensity Ratio of Coherent to Incoherent Scattering of X-Rays—Determination of Hydrogen and Carbon, *Anal. Chem.* **33**, 67–70 (1961).

(*D25*) Dwiggins, C. W., Jr., Automated Determination of Elements in Organic Samples Using X-Ray Emission Spectrometry, *Anal. Chem.* **36**, 1577–81 (1964).

(*D26*) Dwiggins, C. W., Jr., and H. N. Dunning, Quantitative Determination of Traces of Vanadium, Iron, and Nickel in Oils by X-Ray Spectrography, *Anal. Chem.* **32**, 1137–41 (1960).

(*D27*) Dwiggins, C. W., Jr., J. R. Lindley, and B. H. Eccleston, Cooled Sample Holder for the X-Ray Spectrograph, *Anal. Chem.* **31**, 1928 (1959).

(*D28*) Dyroff, G. V., and P. Skiba, Determination of Trace Amounts of Iron, Nickel, and Vanadium on Catalysts by Fluorescent X-Ray Spectrography, *Anal. Chem.* **26**, 1774–8 (1954).

(*E1*) Ebel, H., Thickness Determination of Planar Thin Films by X-Ray Fluorescence Analysis, *Z. Metallk.* **56**, 802–4 (1965).

(*E2*) Ebel, H., and J. Klugel, Determination of Thickness of Thin Planar Multilayers by X-Ray Fluorescence, *Z. Naturforsch.* **21**, 2108–9 (1966).

(*E3*) Ebert, F., and A. Wagner, Explanation of "Accidental Interferences" as Systematic Accompanying Phenomena in X-Ray Fluorescent Spectroscopic Measurements and the Possibility of Avoiding the Interferences, *Z. Metallk.* **48**, 646–9 (1957).

(*E4*) Eddy, C. E., and T. H. Laby, Analysis by X-Ray Spectroscopy, *Proc. Roy. Soc.* (*London*) **124A**, 249–68 (1929).

(*E5*) Eddy, C. E., and T. H. Laby, Quantitative Analysis by X-Ray Spectroscopy, *Proc. Roy. Soc.* (*London*) **127A**, 20–42 (1930).

(*E6*) Ehlert, R. C., Diffraction of X-Rays by Multilayer Stearate Soap Films, *Advan. X-Ray Anal.* **8**, 325–31 (1965).

(*E7*) Ehlert, R. C., and R. A. Mattson, Multilayer Soap Film Structures, *Advan. X-Ray Anal.* **10**, 389–97 (1967).

(*E8*) Eick, J. D., H. J. Caul, D. L. Smith, and S. D. Rasberry, Analysis of Gold and Platinum Group Alloys by X-Ray Emission and Correction for Interelement Effects, *Appl. Spectrosc.* **21**, 324–8 (1967).

(*E9*) Elian, H. A., and R. E. Ogilvie, Reflecting Variable Bent-Crystal Spectrometer, *Rev. Sci. Instrum.* **33**, 753–5 (1962).

(*F1*) Fadley, C. S., S. B. M. Hagstrom, J. M. Hollander, M. P. Klein, and D. A. Shirley, Chemical Bonding Information from Photoelectron Spectroscopy, *Science* **157**, 1571–3 (1967).

(*F2*) Faessler, A., and M. Goehring, X-Ray Spectrum and State of Combination from the *K*-Fluorescence Radiation of Sulfides, *Naturwiss.* **39**, 169–77 (1952).

(*F3*) Fagel, J. E., Jr., E. W. Balis, and L. B. Bronk, Organometallic Precipitates and Briqueting for X-Ray Spectrography, *Anal. Chem.* **29**, 1287–9 (1957).

(*F4*) Fagel, J. E., Jr., H. A. Liebhafsky, and P. D. Zemany, Determination of Tungsten or Molybdenum in X-Ray Emission Spectrography, *Anal. Chem.* **30**, 1918–20 (1958).

(*F5*) Fahlbusch, W. A., Preparation of Metallic Buttons and Analysis by X-Ray Spectrography, *Appl. Spectrosc.* **17**, 72–3 (1963).

(*F6*) Felten, E. J., I. Fankuchen, and J. Steigman, A Possible Solution to the Matrix Problem in X-Ray Fluorescence Spectrometry, *Anal. Chem.* **31**, 1771–6 (1959).

(*F7*) Finnegan, J. J., Thin-Film X-Ray Spectroscopy, *Advan. X-Ray Anal.* **5**, 500–11 (1962).

(*F8*) Fischer, D. W., and W. L. Baun, Experimental Dispersing Devices and Detection Systems for Soft X-Rays, *Advan. X-Ray Anal.* **7**, 489–96 (1964).

(*F9*) Flikemma, D. S., and R. V. Schablaske, Direct Assay for Molybdenum and Ruthenium in Uranium Alloys by X-Ray Emission Spectrometry, *Advan. X-Ray Anal.* **1**, 387–98 (1958).

(*F10*) Fonda, G. R., and G. B. Collins, The Cathode-Ray Tube in X-Ray Spectroscopy and Quantitative Analysis, *J. Amer. Chem. Soc.* **53**, 113–25 (1931).

(*F11*) Fox, J. G. M., X-Ray Spectrometric Determination of Light Elements Using Direct Electron Excitation, *J. Inst. Metals* **91**, 239–41 (1963).

(*F12*) Francis, J. E., P. R. Bell, and J. C. Gundlach, Single-Channel Analyzer, *Rev. Sci. Instrum.* **22**, 133–7 (1951).

(*F13*) Franks, A., Optical Techniques Applied to the Study of X-Ray Wavelengths; in "X-Ray Optics and X-Ray Microanalysis," H. H. Pattee, V. E. Cosslett, and A. Engstrom, eds.; Academic Press, N. Y., pp. 199–213 (1963).

(*F14*) Franks, A., and R. F. Braybrook, Analysis of the Lighter Elements by Total

Reflection of Their Characterisic X-Ray Wavelengths, *Brit. J. Appl. Phys.* **10**, 190–1 (1959).

(*F15*) Frerichs, R., Cadmium Sulfide X-Ray Detector, *J. Appl. Phys.* **21**, 312–7 (1950).

(*F16*) Friedman, H., Geiger Counter Tubes, *Proc. I.R.E.* (*Inst. Radio Eng.*) **37**, 791–808 (1949).

(*F17*) Friedman, H., and L. S. Birks, Geiger-Counter Spectrometer for X-Ray Fluorescence Analysis, *Rev. Sci. Instrum.* **19**, 323–30 (1948).

(*F18*) Friedman, H., L. S. Birks, and E. J. Brooks, Basic Theory and Fundamentals of Fluorescent X-Ray Spectrographic Analysis, *Amer. Soc. Test. Mater. Spec. Tech. Publ.* **157**, 3–26 (1954).

(*F19*) Friedrich, W., P. Knipping, and M. Von Laue, Interference Phenomena with Roentgen Rays, *Ber. Bayer. Akad. Wiss. Munchen* **1912**, 303–22; *Ann. Phys.* **41**, 971–88 (1913).

(*G1*) Gaylor, D. W., Precision of Fixed-Time *vs.* Fixed-Count Measurements, *Anal. Chem.* **34**, 1670–1 (1962).

(*G2*) Geiger, H., and W. Muller, Mode of Action and Construction of a Counter Tube, *Phys. Z.* **29**, 839–41 (1928).

(*G3*) Gerold, V., X-Ray Thickness Gaging of Galvanized Layers with a Counter-Tube Diffractometer, *Z. Angew. Phys.* **4**, 247–54 (1952).

(*G4*) Gilfrich, J. V., and L. S. Birks, Spectral Distribution of X-Ray Tubes for Quantitative X-Ray Fluorescence Analysis, *Anal. Chem.* **40**, 1077–80 (1968).

(*G5*) Glade, G. H., Production Control of Gold and Rhodium Plating Thickness on Very Small Samples by X-Ray Spectroscopy, *Advan. X-Ray Anal.* **11**, 185–90 (1968).

(*G6*) Glocker, R., and H. Schrieber, Quantitative Roentgen Spectrum Analysis by Means of Cold Excitation of the Spectrum, *Ann. Phys.* **85**, 1089–1102 (1928).

(*G7*) Goodwin, P. S., and C. L. Winchester, Continuous Measurement of Plating Thickness, *Plating* **46**, 41–4 (1959).

(*G8*) Graeber, E. J., Direct Identification of X-Ray Spectra, *Appl. Spectrosc.* **16**, 24 (1962).

(*G9*) Griffin, L. H., Iron-55 X-Ray Absorption Analysis of Organically Bound Chlorine Using Conventional Proportional Counting Facilities, *Anal. Chem.* **34**, 606–9 (1962).

(*G10*) Grodski, J. J., Photoelectric Detectors for 10–400-Å Photons, *Norelco Reptr.* **14**, 107–11 (1967).

(*G11*) Grubb, W. T., and P. D. Zemany, X-Ray Emission Spectrography with Ion-Exchange Membranes, *Nature* **176**, 221 (1955).

(*G12*) Gunn, E. L., X-Ray Fluorescence of Suspended Particles in a Liquid Hydrocarbon, *Advan. X-Ray Anal.* **11**, 164–76 (1968).

(*G13*) Gunn, E. L., X-Ray Spectrometric Analysis of Liquids and Solutions, *Amer. Soc. Test. Mater. Spec. Tech. Publ.* **349**, 70–85 (1964).

(*G14*) Gunn, E. L., Fluorescent X-Ray Spectral Analysis of Powdered Solids by Matrix Dilution, *Anal. Chem.* **29**, 184–9 (1957).

(*G15*) Gunn, E. L., X-Ray Fluorescence Intensity of Elements Evaporated from Solution onto Thin Films, *Anal. Chem.* **33**, 921–7 (1961).

(*G16*) Gunn, E. L., Problems of Direct Determination of Trace Nickel in Oil by X-Ray Emission Spectrography, *Anal. Chem.* **36**, 2086–90 (1964).

(*G17*) Gunn, E. L., Determination of Lead in Gasoline by X-Ray Fluorescence Using an Internal Intensity Reference, *Appl. Spectrosc.* **19**, 99–102 (1965).

(*G18*) Gunn, E. L., Practical Methods of Solving Absorption and Enhancement Problems in X-Ray Emission Spectrography, *Develop. Appl. Spectrosc.* **3**, 69–96 (1964).

(*H1*) Hadding, A., Mineral Analysis by X-Ray Spectroscopic Methods, *Z. Anorg. Chem.* **122**, 195–200 (1922).

(*H2*) Haftka, F. J., X-Ray Fluorescence Analysis of Powders, *Rev. Universelle Mines* **15**, 549–54 (1959).

(*H3*) Hagg, G., and N. Karlsson, Aluminum Monochromator with Double Curvature for High-Intensity X-Ray Powder Photographs, *Acta Cryst.* **6**, 728–30 (1952).

(*H4*) Hakkila, E. A., and G. R. Waterbury, X-Ray Fluorescence Spectrometric Determination of Chromium, Iron, and Nickel in Ternary Alloys, *Anal. Chem.* **37**, 1773–5 (1965).

(*H5*) Hakkila, E. A., and G. R. Waterbury, Applications of X-Ray Absorption-Edge Analysis, *Develop. Appl. Spectrosc.* **2**, 297–307 (1963).

(*H6*) Hakkila, E. A., R. G. Hurley, and G. R. Waterbury, X-Ray Fluorescence Spectrometric Determination of Zirconium and Molybdenum in the Presence of Uranium, *Anal. Chem.* **36**, 2094–7 (1964).

(*H7*) Hakkila, E. A., R. G. Hurley, and G. R. Waterbury, A Three-Wavelength X-Ray Absorption-Edge Method for Determination of Plutonium in Nitrate Media, *Anal. Chem.* **38**, 425–8 (1966).

(*H8*) Hale, C. C., and W. H. King, Jr., Direct Nickel Determination in Petroleum Oils by X-Ray at the 0.1-ppm Level, *Anal. Chem.* **33**, 74–7 (1961).

(*H9*) Hall, E. T., M. S. Banks, and J. M. Stern, Uses of X-Ray Fluorescent Analysis in Archaeology, *Archaeometry* **7**, 84–9 (1964).

(*H10*) Hall, T., Nondispersive X-Ray Fluorescence Unit for Analysis of Biological Tissue Sections, *Advan. X-Ray Anal.* **1**, 297–313 (1958).

(*H11*) Hall, T., X-Ray Fluorescence Analysis in Biology, *Science* **134**, 449–55 (1961).

(*H12*) Harris, L. A., "Secondary Electron Spectroscopy," *Ind. Res.* **10**(2), 52–6 (1968).

(*H13*) Harris, L. A., Analysis of Materials by Electron-Excited Auger Electrons, *J. Appl. Phys.* **39**, 1419–27 (1968).

(*H14*) Harvey, C. E., "A Method of Semiquantitative Spectrographic Analysis," Applied Research Labs., Glendale, Calif., 285 pp. (1947).

(*H15*) Hasler, M. F., and J. W. Kemp, Use of Multichannel Recording in X-Ray Fluorescent Analysis, *Amer. Soc. Test. Mater. Spec. Tech. Publ.* **157**, 34–42 (1954).

(*H16*) Heinrich, K. F. J., Pulse-Height Selection in X-Ray Fluorescence, *Advan. X-Ray Anal.* **4**, 370–81 (1961).

(*H17*) Heinrich, K. F. J., X-Ray Probe with Collimation of the Secondary Beam, *Advan. X-Ray Anal.* **5**, 516–26 (1962).

(*H18*) Hendee, C. H., and S. Fine, Moseley's Law Applied to Proportional Counter Resolution of Adjacent Elements, *Phys. Rev.* **95**, 281–3 (1954).

(*H19*) Hendee, C. F., S. Fine, and W. B. Brown, Gas-Flow Proportional Counter for Soft X-Ray Detection, *Rev. Sci. Instrum.* **27**, 531–5 (1956).

(*H20*) Henke, B. L., Microanalysis with Ultrasoft X-Radiations, *Advan. X-Ray Anal.* **5**, 285–305 (1962).

(*H21*) Henke, B. L., X-Ray Fluorescence Analysis for Sodium, Fluorine, Oxygen, Nitrogen, Carbon, and Boron, *Advan. X-Ray Anal.* **7**, 460–88 (1964).

(*H22*) Henke, B. L., Some Notes on Ultrasoft X-Ray Fluorescence Analysis—10 to 100 Å Region, *Advan. X-Ray Anal.* **8**, 269–84 (1965).

(*H23*) Henke, B. L., Production, Detection, and Application of Ultrasoft X-Rays; in "X-Ray Optics and X-Ray Microanalysis," H. H. Pattee, V. E. Cosslett, and A. Engstrom, eds.; Academic Press, N. Y., pp. 157–72 (1963).

(*H24*) Henke, B. L., Spectroscopy in the 15–150 Å Ultrasoft X-Ray Region; in "X-Ray Optics and Microanalysis," R. Castaing, P. Deschamps, and J. Philibert, eds.; Hermann, Paris, pp. 440–53 (1966).

(*H25*) Herglotz, H., Roentgen Tube with Transparent Anode for Secondary Excitation of Roentgen Spectra, *Osterr. Ing.-Arch.* **6**, 135–40 (1952).

(*H26*) Hermes, O. D., and P. C. Ragland, Quantitative Chemical Analysis of Minerals in Thin Section with the X-Ray Macroprobe, *Amer. Mineral.* **52**, 493–508 (1967).

(*H27*) Herzog, L. F., Age Determination by X-Ray Fluorescence Rubidium–Strontium Measurement in Lepidolite, *Science* **132**, 293–5 (1960).

(*H28*) Hinckley, D. N., and T. F. Bates, X-Ray Fluorescence Method for Quantitative Determination of Small Amounts of Montmorillonite in Kaolin Clays, *Amer. Mineral.* **45**, 239–41 (1960).

(*H29*) Hirokawa, K., Determination of Impurities in Some Nonferrous Metals by Fluorescent X-Ray Spectroscopy, *Sci. Rep. Res. Inst. Tohuku Univ., Ser. A* **13**, 263–70 (1961).

(*H30*) Hirokawa, K., Determination of Alloyed Elements in Special Steels by X-Ray Fluorescent Spectroscopy. I. Nickel and Chromium in Nickel–Chromium Steel and Tungsten in Tungsten Steel, *Sci. Rep. Res. Inst. Tohuku Univ., Ser. A* **14**, 278–87 (1962).

(*H31*) Hirokawa, K., and H. Goto, Determination of Small Amounts of Copper and Iron with X-Ray Fluorescent Spectroscopy—Combination with Organometallic Precipitation, *Bull. Chem. Soc. Jap.* **35**, 961–4 (1962).

(*H32*) Hirokawa, K., and H. Goto, Determination of Uranium in Thorium by X-Ray Fluorescent Spectroscopy—Organometallic Precipitation Method, *Bull. Chem. Soc. Jap.* **35**, 964–6 (1962).

(*H33*) Hirokawa, K., T. Shimanuki, and H. Goto, Determination of Manganese in Ferromanganese and Chromium in Ferrochromium by X-Ray Fluorescent Spectroscopy, *Sci. Rep. Res. Inst. Tohuku Univ., Ser. A* **15**, 124–32 (1963).

(*H34*) Hirokawa, K., T. Shimanuki, and H. Goto, Simultaneous Determination of Alloy Film Thickness and Its Composition with X-Ray Fluorescent Spectroscopy—Determination of the Composition and Thickness of Nickel–Iron Alloy Films on Copper, *Z. Anal. Chem.* **190**, 309–15 (1962).

(*H35*) Hirokawa, K., M. Suzuki, and H. Goto, Effect of Backing Metals in the X-Ray Fluorescence Spectral Analysis of Metal Films and Its Application—Absorption-Edge Effect, *Z. Anal. Chem.* **199**, 89–94 (1964).

(*H36*) Hirt, R. C., W. R. Doughman, and J. B. Gisclard, Application of X-Ray Emission Spectrography to Air-Borne Dusts in Industrial Hygiene Studies, *Anal. Chem.* **28**, 1649–51 (1956).

(*H37*) Hofstadter, R., and J. A. McIntyre, Gamma-Ray Spectroscopy with Crystals of NaI(Tl), *Nucleonics* **7**(3), 32–7 (1950).

(*H38*) Holliday, J. E., Interpretation of Changes in Shape of *K* Emission Bands of Light Elements with Chemical Combination, *Develop. Appl. Spectrosc.* **5**, 77–105 (1966).

(*H39*) Holliday, J. E., Soft X-Ray Spectroscopy in the 13 to 44 Å Region, *J. Appl. Phys.* **33**, 3259–65 (1962).

(*H40*) Holliday, J. E., Soft X-Ray Spectrometer Using a Flow Proportional Counter, *Rev. Sci. Instrum.* **31**, 891–5 (1960).

(*H41*) Hoskins, J. H., Universal Vacuum X-Ray Spectrograph, *Norelco Reptr.* **7**, 111–13, 124 (1960).

(*H42*) Houk, W. W., and L. Silverman, Determination of Oxygen to Uranium Ratio in Uranium Dioxides by Fluorescent X-Ray Analysis, *Pittsburgh Conf. Anal. Chem. Appl. Spectrosc., 10th* (1959); Abstr.: *Spectrochim. Acta.* **15**, 308 (1959).

(*H43*) Hower, J., Matrix Corrections in the X-Ray Spectrographic Trace-Element Analysis of Rocks and Minerals, *Amer. Mineral.* **44**, 19–32 (1959).

(*H44*) Hudgens, C. R., and G. Pish, Gas Analysis by X-Ray Fluorescence Emission Spectrometry, *Anal. Chem.* **37**, 414 (1965).

(*H45*) Hudgens, C. R., and G. Pish, X-Ray Fluorescence Emission Analysis of Slurries, *Develop. Appl. Spectrosc.* **5**, 25–9 (1966).

(*H46*) Hughes, H. K., and J. W. Wilczewski, *K*-Capture Spectroscopy—Iron-55 X-Ray Absorption Determination of Sulfur in Hydrocarbons, *Anal. Chem.* **26**, 1889–93 (1954).

(*J1*) Jackwerth, E., and H. G. Kloppenburg, Quantitative Evaluation of Paper Chromatograms of Inorganic Ions by Means of X-Ray Fluorescence Analysis, *Naturwiss.* **47**, 444 (1960).

(*J2*) Jackwerth, E., and H. G. Kloppenburg, Quantitative Evaluation of Paper Chromatograms in Trace Analysis by X-Ray Fluorescence Spectroscopy, *Z. Anal. Chem.* **179**, 186–96 (1961).

(*J3*) Jackwerth, E., and H. G. Kloppenburg, X-Ray Fluorescence Spectroscopic Determination of Traces of Arsenic by Using a Modified Gutzeit Method, *Z. Anal. Chem.* **186**, 428–36 (1962).

(*J4*) Jacob, L., R. Noble, and H. Yee, Photomultiplier Soft X-Ray Spectrometer, *J. Sci. Instrum.* **37**, 460–3 (1960).

(*J5*) Jecht, U., and I. Petersohn, Counters and Analyzer Crystals for the X-Ray Fluorescence Analysis of Elements of Low Atomic Number, *Siemens-Z.* **41**, 59–63 (1967).

(*J6*) Jenkins, R., Fractionation Effects in the Gas Supply of Flow Proportional Counters, *J. Sci. Instrum.* **41**, 696–7 (1964).

(*J7*) Jenkins, R., Modification of a Philips PW-1540 X-Ray Spectrometer for Nondispersive Analysis, *J. Sci. Instrum.* **42**, 480–2 (1965).

(*J8*) Johann, H. H., Intense X-Ray Spectra Obtained with Curved Crystals, *Z. Phys.* **69**, 185–206 (1931).

(*J9*) Johansson, T., New Focusing X-Ray Spectrometer, *Z. Phys.* **82**, 507–28 (1933).

(*J10*) Johnson, C. M., and P. R. Stout, Fluorescent X-Ray Spectrometry—Interferences from Compton Scattering from Matrices of Low Atomic Number, *Anal. Chem.* **30**, 1921–3 (1958).

(*J11*) Johnson, J. L., and B. E. Nagel, Microanalysis on Controlled Spot Test Paper by X-Ray Fluorescence, *Mikrochim. Acta.* **3**, 525–31 (1963).

(*J12*) Johnson, M. P., P. R. Beeley, and J. Nutting, Application of X-Ray Fluorescence Probe Analysis to the Study of Segregation in Steels, *Advan. X-Ray Anal.* **8**, 259–68 (1965).

(*J13*) Johnson, M. P., P. R. Beeley, and J. Nutting, X-Ray Fluorescence Technique for the Study of Segregation in Steel, *J. Iron Steel Inst.* **201**, 32–5 (1967); in "X-Ray and Electron Microscopy News," *Siemens Rev.* **34** (Special Issue) 46–50 (1967).

(*J14*) Jones, R. A., Determination of Manganese in Gasoline by X-Ray Emission Spectrography, *Anal. Chem.* **31**, 1341–4 (1959).

(*J15*) Jones, R. A., Determination of Sulfur in Gasoline by X-Ray Emission Spectrography, *Anal. Chem.* **33**, 71–4 (1961).

(*K1*) Kaelble, E. F., and G. J. McEwan, X-Ray Spectrographic Measurement of Wax Film Thickness, *11th Ann. Symp. Spectrosc.*, Chicago (1960), unpublished; reviewed in detail in (*15J*).

(*K2*) Kalman, Z. H., and L. Heller, Theoretical Study of X-Ray Fluorescent Determination of Traces of Heavy Elements in a Light Matrix, *Anal. Chem.* **34**, 946–51 (1962).

(*K3*) Kang, C. C., E. W. Keel, and E. Solomon, Determination of Traces of Vanadium, Iron, and Nickel in Petroleum Oils by X-Ray Emission Spectrometry, *Anal. Chem.* **32**, 221–5 (1960).

(*K4*) Karttunen, J. O., Separation and Fluorescent X-Ray Spectrometric Determination of Zirconium, Molybdenum, Ruthenium, Rhodium, and Palladium in Solution in Uranium-Base Fissium Alloys, *Anal. Chem.* **35**, 1044–9 (1963).

(*K5*) Karttunen, J. O., H. B. Evans, D. J. Henderson, P. J. Markovich, and R. L. Niemann, Portable Fluorescent X-Ray Instrument Utilizing Radioisotope Sources, *Anal. Chem.* **36**, 1277–82 (1964).

(*K6*) Keesaer, W. C., Establishment of *Q* Rating Factor for Decorative Chromium Plate by X-Ray Fluorescence, *Advan. X-Ray Anal.* **3**, 77–94 (1959).

(*K7*) Kehl, W. L., and R. G. Russell, Fluorescent X-Ray Spectrographic Determination of Uranium in Waters and Brines, *Anal. Chem.* **28**, 1350–1 (1956).

(*K8*) Kemp, J. W., M. F. Hasler, J. L. Jones, and L. Zeitz, Multichannel Instruments for Fluorescent X-Ray Spectroscopy, *Spectrochim. Acta* **7**, 141–8 (1955).

(*K9*) Kilday, B. A., and R. E. Michaelis, Determination of Lead in Leaded Steels by X-Ray Spectroscopy, *Appl. Spectrosc.* **16**, 136–8 (1962).

(*K10*) Kiley, W. R., Function and Application of Counters and the Pulse-Height Analyzer, *Norelco Reptr.* **7**, 143–9 (1960).

(*K11*) Kiley, W. R., and R. W. Deichert, Design Considerations for On-Stream X-Ray Analysis, *Advan. X-Ray Anal.* **6**, 429–35 (1963).

(*K12*) Kiley, W. R., and J. A. Dunne, X-Ray Detectors, *Amer. Soc. Test. Mater. Spec. Tech. Publ.* **349**, 24–40 (1964).

(*K13*) Kimoto, S., M. Sato, H. Kamada, and T. Ui, The Primary X-Ray Analyzer, *Advan. X-Ray Anal.* **9**, 508–14 (1966).

(*K14*) Kirkpatrick, P., Theory and Use of Ross Filters, *Rev. Sci. Instrum.* **10**, 186–91 (1939); **15**, 223–9 (1944).

(*K15*) Knapp, K. T., R. H. Lindahl, and A. J. Mabis, An X-Ray Absorption Method for Elemental Analysis, *Advan. X-Ray Anal.* **7**, 318–24 (1964).

(*K16*) Koh, P. K., and B. Caugherty, Metallurgical Applications of X-Ray Fluorescent Analysis, *J. Appl. Phys.* **23**, 427–33 (1952).

(*K17*) Kohler, T. R., and W. Parrish, X-Ray Diffractometry of Radioactive Samples, *Rev. Sci. Instrum.* **26**, 374–9 (1955).

(*K18*) Kohler, T. R., and W. Parrish, Conversion of Quantum Counting Rate to Roentgens, *Rev. Sci. Instrum.* **27**, 705–6 (1956).

(*K19*) Kokotailo, G. T., and G. F. Damon, Determination of Bromine in Liquids by X-Ray Fluorescent Spectroscopy, *Anal. Chem.* **25**, 1185–7 (1953).

(*K20*) Kramers, H. A., Theory of X-Ray Absorption and of the Continuous X-Ray Spectrum, *Phil. Mag.* **46**, 836–71 (1923).

(*K21*) Kulenkampff, H., Continuous X-Ray Spectrum, *Ann. Phys.* **69**, 548–96 (1923).

(*K22*) Kullbom, S. D., W. K. Pollard, and H. F. Smith, Combined Infrared and X-Ray Spectrometric Method for Determining Sulfonate and Sulfate Concentration of Detergent Range Alkylbenzene Sulfonate Solutions, *Anal. Chem.* **37**, 1031–4 (1965).

(*K23*) Kustner, H., Intense Monochromatic X-Rays Obtained from Ordinary Tubes without Spectroscopic Apparatus, *Z. Phys.* **70**, 324–47 (1931).

(*K24*) Kustner, H., Monochromatic X-Rays, *Z. Phys.* **77**, 52–9 (1932).

(*L1*) Ladell, J., and W. Parrish, Determination of Contamination of X-Ray Tubes, *Philips Res. Rep.* **14**, 401–20 (1959).

(*L2*) Ladell, J., and N. Spielberg, Laue Spectrometer for Multichannel X-Ray Spectrochemical Analysis, *Rev. Sci. Instrum.* **31**, 23–9 (1960).

(*L3*) Lamb, F. W., L. M. Niebylski, and E. W. Kieffer, Determination of Tetraethyllead in Gasoline by X-Ray Fluorescence, *Anal. Chem.* **27**, 129–32 (1955).

(*L4*) Lambert, M. C., Some Practical Aspects of X-Ray Spectrography, *Norelco Reptr.* **6**, 37–51 (1959).

(*L5*) Long, A. R., Design and Performance of X-Ray Proportional Counters, *J. Sci. Instrum.* **33**, 96–102 (1956).

(*L6*) Langheinrich, A. P., and J. W. Forster, Application of Radioisotope Nondispersive X-Ray Spectrometry to Analysis of Molybdenum, *Advan. X-Ray Anal.* **11**, 275–86 (1968).

(*L7*) Lee, F. S., and W. J. Campbell, Variation of X-Ray Spectral Line Position with Ambient Temperature Change—A Source of Error in X-Ray Spectrography, *Advan. X-Ray Anal.* **8**, 431–42 (1965).

(*L8*) Legrand, C., Measurement of the Thickness of Thin Layers by X-Ray Fluorescence, *J. Chim. Phys.* **53**, 587–92 (1956).

(*L9*) Legrand, C., and J. J. Trillat, Investigation of Surface Films by X-Ray Fluorescence, *Rev. Met.* **53**, 645–8 (1956).

(*L10*) Leroux, J., and M. Mahmud, X-Ray Quantitative Analysis by an Emission-Transmission Method, *Anal. Chem.* **38**, 76–82 (1966).

(*L11*) Leroux, J., P. A. Maffett, and J. L. Monkman, Microdetermination of Heavy Elements Such as Mercury and Iodine in Solution by X-Ray Absorption, *Anal. Chem.* **29**, 1089–92 (1957).

(*L12*) Lewis, C. L., W. L. Ott, and N. M. Sine, "The Analysis of Nickel," Pergamon Press, N. Y., ch. 4, pp. 85–102 (1966).

(*L13*) Lewis, R., S. Moll, R. E. Ogilvie, and S. Shur, A Method for Identification of Stratospheric Particles, *Advan. Metals Res. Corp., Res. Rep.* **AMR-1005**, 65 pp. (1960).

(*L14*) Liebhafsky, H. A., Analytical Methods Based on X-Ray Absorption, *Anal. Chem.* **25**, 689–92 (1953).

(*L15*) Liebhafsky, H. A., and P. D. Zemany, Film Thickness by X-Ray Emission Spectrography, *Anal. Chem.* **28**, 455–9 (1956).

(*L16*) Liebson, S. H., and H. Friedman, Self-Quenching Halogen-Filled Counters, *Rev. Sci. Instrum.* **19**, 303–6 (1948).

(*L17*) Linares, R. C., J. B. Schroeder, and L. A. Hurlbut, Applications of X-Ray-Excited Optical Fluorescence to Analytical Chemistry, *Spectrochim. Acta* **21**, 1915–20 (1965).

(*L18*) Lincoln, A. J., and E. N. Davis, Determination of Platinum in Alumina-Base Reforming Catalyst by X-Ray Spectroscopy, *Anal. Chem.* **31**, 1317–20 (1959).

(*L19*) Long, A., Detectors for Low-Energy X-Radiation, *J. Brit. Inst. Radio Eng.* **19**, 273–87 (1959).

(*L20*) Long, J. V. P., Applications of and Some Sources of Error in X-Ray Microchemical Analysis; in "X-Ray Microscopy and Microradiography," V. E. Cosslett, A. Engstrom, and H. H. Pattee, eds.; Academic Press, N. Y., pp. 628–35 (1957).

(*L21*) Long, J. V. P., and V. E. Cosslett, Some Methods of Microchemical Analysis; in "X-Ray Microscopy and Microradiography," V. E. Cosslett, A. Engstrom, and H. H. Pattee, eds.; Academic Press, N. Y., pp. 435–42 (1957).

(*L22*) Long, J. V. P., and H. O. E. Rockert, X-Ray Fluorescence Microanalysis and the Determination of Potassium in Nerve Cells; in "X-Ray Optics and X-Ray Microanalysis," H. H. Pattee, V. E. Cosslett, and A. Engstrom, eds.; Academic Press, N. Y., pp. 513–21 (1963).

(*L23*) Longobucco, R. J., X-Ray Spectrometric Determination of Major and Minor Constituents in Ceramic Materials, *Anal. Chem.* **34**, 1263–7 (1962).

(*L24*) Loomis, T. C., X-Ray Spectroscopy as an Analytical Tool, *Ann. N. Y. Acad. Sci.* **137**, 284–96 (1966).

(*L25*) Loomis, T. C., and K. H. Storks, X-Ray Microanalysis in Chemistry, *Bell. Labs. Rec.* **45**, 2–7 (1967).

(*L26*) Lopp, V. R., and C. G. Claypool, Direct Determination of Vanadium and Nickel in Crude Oils by X-Ray Fluorescence, *Advan. X-Ray Anal.* **3**,131–7 (1960).

(*L27*) Lublin, P., Novel Approach to Discrimination in X-Ray Spectrographic Analysis, *Advan. X-Ray Anal.* **2**, 229–38 (1958).

(*L28*) Lublin, P., Determination of Cladding Thickness of Nuclear Fuel Elements by X-Rays, *Norelco Reptr.* **6**, 57–9 (1959).

(*L29*) Lucas-Tooth, H. J., Description of the Solartron Vacuum Automatic X-Ray Spectrometer and Program Selection by Punched Card, *Rev. Universelle Mines* **17**, 180–3 (1961).

(*L30*) Lucas-Tooth, H. J., and B. J. Price, A Mathematical Method for Investigation of Interelement Effects in X-Ray Fluorescent Analysis, *Metallurgia* **64**, 149–52 (1961).

(*L31*) Lucas-Tooth, H. J., and E. C. Pyne, Accurate Determination of Major Constituents by X-Ray Fluorescent Analysis in the Presence of Large Interelement Effects, *Advan. X-Ray Anal.* **7**, 523–41 (1964).

(*L32*) Luke, C. L., Trace Analysis of Metals by Borax Disk X-Ray Spectrometry, *Anal. Chem.* **35**, 1551–2 (1963).

(*L33*) Luke, C. L., Ultratrace Analysis of Metals with a Curved-Crystal X-Ray Milliprobe, *Anal. Chem.* **36**, 318–22 (1964).

(*L34*) Lund, P. K., and J. C. Mathies, X-Ray Spectroscopy in Biology and Medicine. I. Total Iron (Hemoglobin) Content in Whole Human Blood, *Norelco Reptr.* **7**, 127–30 (1960).

(*L35*) Lundquist, L. R., Porosity Test for Thin Electrodeposits on Uranium by X-Ray Fluorescence, *Plating* **52**, 1316 (1965).

(*L36*) Lynch, J. T., X-Ray Techniques as a Tool to Analyze Intermetallic Coatings, *Plating* **51**, 1173–7 (1964).

(*L37*) Lytle, F. W., J. I. Botsford, and H. A. Heller, X-Ray Emission Spectrographic Analysis of Bastnaesite Rare Earths, *Advan. X-Ray Anal.* **1**, 367–86 (1958).

(*L38*) Lytle, F. W., W. B. Dye, and H. J. Seim, Determination of Trace Elements in Plant Materials by Fluorescent X-Ray Analysis, *Advan. X-Ray Anal.* **5**, 433–46 (1962).

(*M1*) Mack, M., and N. Spielberg, Statistical Factors in X-Ray Intensity Measurements, *Spectrochim. Acta* **12**, 169–78 (1958).

(*M2*) MacKenzie, A. R., T. Hall, and W. F. Whitmore, Zinc Content of Expressed Human Prostatic Fluid, *Nature* **193**, 72–3 (1962).

(*M3*) MacNevin, W. M., and E. A. Hakkila, Fluorescent X-Ray Spectroscopic Estimations of Palladium, Platinum, Rhodium, and Iridium, *Anal. Chem.* **29**, 1019–22 (1957).

(*M4*) Maneval, D. R., and H. R. Lovell, Determination of Lanthanum, Cerium, Praseodymium, and Neodymium as Major Components by X-Ray Emission Spectroscopy, *Anal. Chem.* **32**, 1289–92 (1960).

(*M5*) Mathies, J. C., and P. K. Lund, X-Ray Spectroscopy in Biology and Medicine. II. Calcium Content of Human Blood Serum, *Norelco Reptr.* **7**, 130–3 (1960).

(*M6*) Mathies, J. C., and P. K. Lund, X-Ray Spectroscopy in Biology and Medicine. III. Bromide (Total Bromine) in Human Blood, Serum, Urine, and Tissues, *Norelco Reptr.* **7**, 134–5, 139 (1960).

(*M7*) Mathies, J. C., P. K. Lund, and W. Eide, X-Ray Spectroscopy in Biology and Medicine. IV. A Simple, Indirect, Sensitive Procedure for Determination of Nitrogen (Ammonia) at the Microgram and Submicrogram Level, *Anal. Biochem.* **3**, 408–14 (1962); *Norelco Reptr.* **9**, 93–5 (1962).

(*M8*) Matocha, C. K., New Briqueting Technique, *Appl. Spectrosc.* **20**, 252–3 (1966).

(*M9*) McCue, J. C., L. L. Bird, C. A. Ziegler, and J. J. O'Connor, X-Ray Rayleigh Scattering Method for Determination of Uranium in Solution, *Anal. Chem.* **33**, 41–3 (1961).

(*M10*) McCue, R. A., W. M. Mueller, and P. J. Dunton, Feasibility of the X-Ray Spectrograph as a Continuous Analytical Instrument for Process Control, *Advan. X-Ray Anal.* **1**, 399–407 (1958).

(*M11*) Mecke, P., Quantitative X-Ray Fluorescence Analysis of Small Amounts of Mixtures by a Filter Difference Method, *Z. Anal. Chem.* **193**, 241–7 (1963).

(*M12*) Mellish, C. E., X-Ray Fluorescence Spectroscopy—Applications of a Simple Apparatus Using Radioactive Sources, *Research* (*London*) **12**, 212–17 (1959).

(*M13*) Merrit, J., and E. J. Agazzi, Casting Formvar Films on Water, *Rev. Sci. Instrum.* **38**, 127 (1967).

(*M14*) Michaelis, R. E., "Report on Available Standard Samples and Related Materials for Spectrochemical Analysis," *Amer. Soc. Test. Mater. Spec. Tech. Publ.* **58-E**, 150 pp. (1963).

(*M15*) Michaelis, R. E., and B. A. Kilday, Surface Preparation of Solid Metallic Samples for X-Ray Spectrochemical Analysis, *Advan. X-Ray Anal.* **5**, 405–11 (1962).

(*M16*) Miller, D. C., Norelco Portable Spectrometer (Portospec), *Advan. X-Ray Anal.* **3**, 57–67 (1960).

(*M17*) Miller, D. C., Norelco Pinhole Attachment, *Advan. X-Ray Anal.* **4**, 513–20 (1961).

(*M18*) Miller, D. C., Some Considerations in the Use of Pulse-Height Analysis with X-Rays, *Norelco Reptr.* **4**, 37–40 (1957).

(*M19*) Miller, D. C., and P. W. Zingaro, Universal Vacuum Spectrograph and Comparative Data on the Intensities Observed in Air, Helium, and Vacuum Paths, *Advan. X-Ray Anal.* **3**, 49–56 (1960).

(*M20*) Mitchell, B. J., X-Ray Spectrographic Determination of Tantalum, Niobium, Iron, and Titanium Oxide Mixtures Using Simple Arithmetic Corrections for Interelement Effects, *Anal. Chem.* **30**, 1894–1900 (1958).

(*M21*) Mitchell, B. J., X-Ray Spectrographic Determination of Zirconium, Tungsten, Vanadium, Iron, Titanium, Tantalum, and Niobium Oxides—Application of the Correction Factor Method, *Anal. Chem.* **32**, 1652–6 (1960).

(*M22*) Mitchell, B. J., Prediction of X-Ray Fluorescent Intensities and Interelement Effects, *Anal. Chem.* **33**, 917–21 (1961).

(*M23*) Mitchell, B. J., and F. N. Hopper, Digital Computer Calculation and Correction of Matrix Effects in X-Ray Spectroscopy, *Appl. Spectrosc.* **20**, 172–80 (1966).

(*M24*) Mitchell, B. J., and J. E. Kellam, Unusual Matrix Effects in X-Ray Spectroscopy—A Study of the Range and Reversal of Absorption-Enhancement, *Appl. Spectrosc.* **22**, 742–8 (1968).

(*M25*) Mitchell, B. J., and H. J. O'Hear, General X-Ray Spectrographic Solution Method for Iron-, Chromium-, and/or Manganese-Bearing Materials, *Anal. Chem.* **34**, 1620–5 (1962).

(*M26*) Mitzner, B. M., Cell for X-Ray Examination of Compounds Requiring a Special Atmosphere, *Appl. Spectrosc.* **11**, 45–6 (1957).

(*M27*) Moak, W. D., Cooling Liquid Samples for X-Ray Fluorescence Analysis, *Anal. Chem.* **29**, 1906 (1957).

(*M28*) Momoki, K., Nomographic Method of X-Ray Fluorescence Routine Analysis for Daily Variations of Analytical Line Intensities with a Standard X-Ray Spectrograph, *Bunseki Kagaku* **10**, 40–7 (1961); *Chem. Abstr.* **58**, 6165*c* (1963).

(*M29*) Momoki, K., Nonlinear Calibration Curves in Nomographic Fluorescent X-Ray Spectrometry—Comparison of Various Calibration Curves, *Bunseki Kagaku* **10**, 330–6 (1961); *Chem. Abstr.* **55**, 25583*f* (1961).

(*M30*) Moore, J. E., G. P. Happ, and D. W. Stewart, Automatic Direct-Reading X-Ray Spectrometry—Application to Determinations of Silver, *Anal. Chem.* **33**, 61–4 (1961).

(*M31*) Mortimore, D. M., P. A. Romans, and J. L. Tews, X-Ray Spectroscopic Determination of Columbium and Tantalum in Rare Earth Ores, *Norelco Reptr.* **1**, 107–10 (1954).

(*M32*) Moseley, H. G. J., High-Frequency Spectra of the Elments, *Phil. Mag.* **26**, 1024–34 (1912); **27**, 703–14 (1913).

(*M33*) Mueller, J. I., V. G. Scotti, and J. J. Little, Fluorescent X-Ray Analysis of Highly Radioactive Samples, *Advan. X-Ray Anal.* **2**, 157–66 (1959).

(*M34*) Muller, E. A. W., Attenuation of X-Rays in the Wavelength Range 0.01–10 mμ. I. General Principles, *Arch. Tech. Messen* **9114-17**, 37–8 (1956); II. Mass Absorption Coefficients μ/ϱ and Position of Absorption Edges; *ibid.* **9114-18**, 107–10 (1956).

(*M35*) Mulvey, T., and A. J. Campbell, Proportional Counters in X-Ray Spectrochemical Analysis, *Brit. J. Appl. Phys.* **9**, 406–10 (1958).

(*M36*) Munch, R. H., Process Stream Analysis by X-Ray Spectroscopy, *Develop. Appl. Spectrosc.* **3**, 45–57 (1964).

(*N1*) Narbutt, K. I., X-Ray Fluorescence Microanalyzer, *Izv. Akad. Nauk SSSR, Ser. Fiz.* **26**, 423–8 (1962); *Chem. Abstr.* **57**, 7877*e* (1962).

(*N2*) Natelson, S., Microanalysis in the Modern Analytical Laboratory of Clinical Chemistry, *Anal. Chem.* **31**(3), 17A–30A (1959).

(*N3*) Natelson, S., Present Status of X-Ray Spectrometry in the Clinical Laboratory, *Trans. N. Y. Acad. Sci.* [2] **26**, 3–26 (1963).

(*N4*) Natelson, S., and B. Sheid, [X-Ray Spectroscopy in the Clinical Laboratory. V.] X-Ray Spectrometric Determination of Strontium in Human Serum and Bone, *Anal. Chem.* **33**, 396–401 (1961).

(*N5*) Natelson, S., and B. Sheid, X-Ray Spectroscopy in the Clinical Laboratory. II. Chlorine and Sulfur—Automatic Analysis of Ultramicro Samples, *Clin. Chem.* **6**, 299–313 (1960).

(*N6*) Natelson, S., and B. Sheid, X-Ray Spectroscopy in the Clinical Laboratory.

III. Sulfur Distribution in the Electrophoretic Protein Fractions of Human Serum, *Clin. Chem.* **6**, 314–26 (1960).

(*N7*) Natelson, S., and B. Sheid, X-Ray Spectroscopy in the Clinical Laboratory, IV. Phosphorus—Total Blood Iron as a Measure of Hemoglobin Content, *Clin. Chem.* **7**, 115–29 (1961).

(*N8*) Natelson, S., and B. Sheid, X-Ray Spectroscopy in the Clinical Laboratory. VI. Acid-Labile Protein-Bound Iodine, *Clin. Chem.* **8**, 17–34 (1962).

(*N9*) Natelson, S., M. R. Richelson, B. Sheid, and S. L. Bender, X-Ray Spectroscopy in the Clinical Laboratory. I. Calcium and Potassium, *Clin. Chem.* **5**, 519–31 (1959).

(*N10*) National Bureau of Standards, "Standard Materials Issued by the National Bureau of Standards," *Nat. Bur. Stand.* (*U. S.*) *Circ.* **552**, ed. 3, 27 pp. (1959).

(*N11*) Nelson, J. T., and R. T. Ellickson, Scintillation Counter for Detection of X-Rays, *J. Opt. Soc. Amer.* **45**, 984–6 (1955).

(*N12*) Noakes, G. E., An Absolute Method of X-Ray Fluorescence Analysis Applied to Stainless Steels, *Amer. Soc. Test. Mater. Spec. Tech. Publ.* **157**, 57–62 (1954).

(*O1*) Ong, S. P., Projection Microscopy and Microanalysis, *Advan. X-Ray Anal.* **5**, 324–34 (1962).

(*P1*) Papariello, G. J., H. Letterman, and W. J. Mader, X-Ray Fluorescent Determination of Organic Substances through Inorganic Association, *Anal. Chem.* **34**, 1251–3 (1962).

(*P2*) Parrish, W., Escape Peaks in X-Ray Diffractometry, *Advan. X-Ray Anal.* **8**, 118–33 (1965).

(*P3*) Parrish, W., X-Ray Intensity Measurements with Counter Tubes, *Philips Tech. Rev.* **17**, 206–21 (1956).

(*P4*) Parrish, W., X-Ray Spectrochemical Analysis, *Philips Tech. Rev.* **17**, 269–86 (1956); *Norelco Reptr.* **3**, 24–36 (1956).

(*P5*) Parrish, W., and T. R. Kohler, Use of Counter Tubes in X-Ray Analysis, *Rev. Sci. Instrum.* **27**, 795–808 (1956).

(*P6*) Passell, T. O., J. B. Swedlund, and S. L. Taimuty, Development of X-Ray Fluorescence Analysis Techniques for Small Samples—A Prototype Atmospheric Electron Probe, *Stanford Res. Inst. Final Rep.* **AFCRL-62-933**, 69 pp. (1962).

(*P7*) Pfeiffer, H. G., and P. D. Zemany, Trace Analysis by X-Ray Emission Spectrography, *Nature* **174**, 397 (1954).

(*P8*) Pike, E. R., Introduction to Soft X-Ray Spectroscopy, *Amer. J. Phys.* **28**, 235–42 (1960).

(*P9*) Pish, G., and A. A. Huffmann, Quantitative Determination of Thorium and Uranium in Solutions by Fluorescent X-Ray Spectrometry, *Anal. Chem.* **27**, 1875–8 (1955).

(*P10*) Plesch, R., Statistics and Probability in Radiation Measuring Technique; in "X-Ray and Electron Microscopy News," *Siemens Rev.* **34** (Special Issue), 73–9 (1967).

(*P11*) Pluchery, M., Absolute Method for X-Ray Fluorescence Analysis Applicable to the Films Obtained by Vacuum Evaporation of Iron–Nickel Alloys, *Spectrochim. Acta* **19**, 533–40 (1963).

(*P12*) Priestly, E. F., New Method for Increasing X-Ray Reflection Power of Lithium Fluoride Crystals, *Brit. J. Appl. Phys.* **10**, 141–2 (1959).

(*P13*) Priestly, E. F., X-Ray Spectrographs for Rapid X-Ray Fluorescence Analysis with Photographic Recording, *J. Sci. Instrum.* **35**, 409–13 (1958).

(*P14*) Proctor, E. M., Picker X-Ray Corp., Cleveland, Ohio, unpublished work.

(*R1*) Raag, V., E. P. Bertin, and R. J. Longobucco, Distribution of Cathode Sublimation Deposits in a Receiving Tube as Determined by X-Ray Spectrometric Scanning, *Advan. Electron Tube Techniques* **2**, 249–59 (1963).

(*R2*) Reynolds, R. C., Matrix Corrections in Trace Element Analysis by X-Ray Fluorescence—Estimation of Mass Absorption Coefficient by Compton Scattering, *Amer. Mineral.* **48**, 1133–43 (1963).

(*R3*) Rhodes, J. R., Radioisotope X-Ray Spectrometry, a Review, *Analyst* **91**, 683–99 (1966).

(*R4*) Rhodes, J. R., and T. Furuta, Applications of a Portable Radioisotope X-Ray Fluorescence Spectrometer to Analysis of Minerals and Alloys, *Advan. X-Ray Anal.* **11**, 249–74 (1968).

(*R5*) Rhodin, T. N., Chemical Analysis of Thin Films by X-Ray Emission Spectrography, *Anal. Chem.* **27**, 1857–61 (1955).

(*R6*) Riggs, F. B., Simple Aid to Pulse-Height Selection with Scanning X-Ray Spectrometers, *Rev. Sci. Instrum.* **34**, 312 (1963).

(*R7*) Rizzo, F. E., Diffusion of Electroplated Couples, Ph. D. Thesis, Univ. Cincinnati, Cincinnati, Ohio, 168 pp. (1964).

(*R8*) Roentgen, W. C., *Ann. Phys. Chem.* **64**, 1–37 (1898).

(*R9*) Rogers, T. H., High-Intensity Radiation from Beryllium-Window X-Ray Tubes, *Radiology* **48**, 594–603 (1947).

(*R10*) Rose, H. J., Jr. and F. Cuttitta, X-Ray Fluorescence Spectroscopy in the Analysis of Ores, Minerals, and Waters, *Advan. X-Ray Anal.* **11**, 23–39 (1968).

(*R11*) Rose, H. J., Jr., I. Adler, and F. J. Flanagan, X-Ray Fluorescence Analysis of Light Elements in Rocks and Minerals, *Appl. Spectrosc.* **17**, 81–5 (1963).

(*R12*) Rose, H. J., Jr., F. Cuttitta, and R. R. Larson, Use of X-Ray Fluorescence in Determination of Selected Major Constituents in Silicates, *U. S. Geol. Surv. Prof. Pap.* **525-B**, 155–9 (1965).

(*R13*) Ross, P. A., New Method of Spectroscopy for Faint X-Radiations, *J. Opt. Soc. Amer.* **16**, 433–7 (1928).

(*R14*) Rothe, G., and A. Koster-Pflugmacher, The Gelatin Method, a New Sample Preparation Technique for X-Ray Fluorescence Spectroscopy, *Z. Anal. Chem.* **201**, 241–5 (1964).

(*R15*) Roulet, H., and J. Despujols, New Method of Analysis by X-Ray Fluorescence Applicable to Small Samples, *C. R. Acad. Sci.* (*Paris*) **253**, 641–3 (1961).

(*R16*) Rowe, W. A., and K. P. Yates, X-Ray Fluorescence Method for Trace Metals in Refinery Fluid Catalytic Cracking Feedstocks, *Anal. Chem.* **35**, 368–70 (1963).

(*R17*) Rudolph, J. S., and R. J. Nadalin, Determination of Microgram Quantities of Chloride in High-Purity Titanium by X-Ray Spectrochemical Analysis, *Anal. Chem.* **36**, 1815–17 (1964).

(*R18*) Rudolph, J. S., O. H. Kriege, and R. J. Nadalin, Applications of Chemical Precipitation Methods for Improving Sensitivity in X-Ray Fluorescent Analysis, *Develop. Appl. Spectrosc.* **4**, 57–64 (1965).

(*R19*) Ryland, A. L., A General Approach to the X-Ray Spectroscopic Analysis of Samples of Low Atomic Number, *147th Natl. Meeting*, *Am. Chem. Soc.*, Phila. (1964); unpublished work.

(*S1*) Salmon, M. L., Recent Developments in Fluorescent X-Ray Spectrographic-Absorptiometric Analysis of Mineral Systems, *Advan. X-Ray Anal.* **2**, 305–12 (1959).

(*S2*) Salmon, M. L., Effects of Operating Variables in the Application of Multi-

Element Calibration Systems for Fluorescent X-Ray Spectrographic Analyses of Mineral Samples, *Advan. X-Ray Anal.* **4**, 433–56 (1961).

(*S3*) Salmon, M. L., Simple Multielement-Calibration System for Analysis of Minor and Major Elements in Minerals by Fluorescent X-Ray Spectrography, *Advan. X-Ray Anal.* **5**, 389–404 (1962).

(*S4*) Salmon, M. L., Practical Applications of Filters in X-Ray Spectrography, *Advan. X-Ray Anal.* **6**, 301–12 (1963).

(*S5*) Salmon, M. L., Improved Trace Analysis with the Use of Synchronized Electronic Discrimination in an X-Ray Scanning Procedure, *Advan. X-Ray Anal.* **7**, 604–14 (1964).

(*S6*) Sarian, S., and H. W. Weart, X-Ray Fluorescence Analysis of Polyphase Metals—Use of Borax Glass Matrix, *Anal. Chem.* **35**, 115 (1963).

(*S7*) Sayce, L. A., and A. Franks, N. P. L. Gratings for X-Ray Spectrometry, *Proc. Roy. Soc.* (*London*) **282A**, 353–7 (1964).

(*S8*) Schindler, M. J., A. Month, and J. Antalec, An X-Ray Fluorescence Technique for Analysis of Iron–Nickel Films over an Extremely Wide Range of Composition and Thickness, *Pittsburgh Conf. Anal. Chem. Appl. Spectrosc.* (1961), unpublished; reviewed in detail in Ref. (*B17*).

(*S9*) Schnell, E., Roentgen Fluorescence Analysis—Use of Secondary Radiation in the Detection of Halogen and Sulfur in Gas Samples, *Monatsschr. Chem.* **96**, 1302–7 (1965).

(*S10*) Schreiber, T. P., A. C. Ottolini, and J. L. Johnson, X-Ray Emission Analysis of Thin Films Produced by Lubricating Oil Additives, *Appl. Spectrosc.* **17**, 17–19 (1963).

(*S11*) Schumacher, B. W., and J. J. Grodiszewski, Research on a Versatile X-Ray Detector, *U. S. Air Force Mater. Lab. Rep.* **AFML-TR-64-414**, 51 pp. (1964).

(*S12*) Seaman, W., H. C. Lawrence, and H. C. Craig, Solid-Sample Techniques in *K*-Capture Spectroscopy—Determination of Chlorine and Bromine, *Anal. Chem.* **29**, 1631–2 (1957).

(*S13*) Seibel, G., Nondispersive X-Ray Spectrometry—Applications to Elementary Chemical Analysis, *Int. J. Appl. Radiat. Isotop.* **15**, 25–41 (1964).

(*S14*) Sellers, W. W., Jr. and K. G. Carroll, Gaging of Thin Nickel Coatings by X-Ray Fluorescence, *Proc. Amer. Electroplaters Soc.* **43**, 97–100 (1956).

(*S15*) Sherman, J., Theoretical Derivation of the Composition of Mixable Specimens from Fluorescent X-Ray Intensities, *Advan. X-Ray Anal.* **1**, 231–50 (1958).

(*S16*) Sherman, J., Correlation between Fluorescent X-Ray Intensity and Chemical Composition, *Amer. Soc. Test. Mater. Spec. Tech. Publ.* **157**, 27–33 (1954).

(*S17*) Sherman, J., Theoretical Derivation of Fluorescent X-Ray Intensities from Mixtures, *Spectrochim. Acta* **7**, 283–306 (1955).

(*S18*) Shiraiwa, T., and N. Fujino, Micro Fluorescent X-Ray Analyzer, *Advan. X-Ray Anal.* **11**, 95–104 (1968).

(*S19*) Shirley, D. A., Applications of Germanium Gamma-Ray Detectors, *Nucleonics* **23**(3), 62–6 (1965).

(*S20*) Shott, J. E., Jr., T. J. Garland, and R. O. Clark, Determination of Traces of Nickel and Vanadium in Petroleum Distillates—An X-Ray Emission Spectrographic Method Based on a New Rapid Ashing Procedure, *Anal. Chem.* **33**, 506–10 (1961).

(*S21*) Siemes, H., X-Ray Fluorescence Spectrometric Determination of Light Elements in Low Concentrations in Water Solutions, *Z. Anal. Chem.* **199**, 321–34 (1964).

(*S22*) Silver, M. D., and E. T.-K. Chow, Thickness Measurement of Thin Permalloy Films—Comparison of X-Ray Emission Spectroscopy, Interferometry, and Stylus Methods, *J. Vac. Sci. Technol.* **2**, 203–7 (1965).

(*S23*) Slack. C. M., Leonard-Ray Tube with Glass Window, *J. Opt. Soc. Amer. and Rev. Sci. Instrum.* **18**, 123–6 (1929).

(*S24*) Sloan, R. D., X-Ray Spectrographic Analysis of Thin Films by the Milliprobe Technique, *Advan. X-Ray Anal.* **5**, 512–15 (1962).

(*S25*) Smith, G. S., A Useful Technique in X-Ray Fluorescence Spectrography, *Chem. Ind.* (*London*) **1963**(22), 907–9.

(*S26*) Soller, W., A New Precision X-Ray Spectrometer, *Phys. Rev.* **24**, 158–67 (1924).

(*S27*) Spielberg, N., Tube Target and Inherent Filtration as Factors in the Fluorescence Excitation of X-Rays, *J. Appl. Phys.* **33**, 2033–5 (1962).

(*S28*) Spielberg, N., Intensities of Radiation from X-Ray Tubes and the Excitation of Fluorescence X-Rays, *Philips Res. Rep.* **14**, 215–36 (1959).

(*S29*) Spielberg, N., Orientation of Topaz Crystals for X-Ray Spectrochemical Analysis, *Rev. Sci. Instrum.* **36**, 1377–8 (1965).

(*S30*) Spielberg, N., and G. Abowitz, Calibration Techniques for X-Ray Fluorescence Analysis of Thin Nickel–Chromium Films, *Anal. Chem.* **38**, 200–3 (1966).

(*S31*) Spielberg, N., and M. Brandenstein, Instrumental Factors and Figure of Merit in the Detection of Low Concentrations by X-Ray Spectrochemical Analysis, *Appl. Spectrosc.* **17**, 6–9 (1963).

(*S32*) Spielberg, N., and J. Ladell, Crystallographic Aspects of Extra Reflections in X-Ray Spectrochemical Analysis, *J. Appl. Phys.* **31**, 1659–64 (1960).

(*S33*) Spielberg, N., and J. Ladell, Scanning Single-Crystal Multichannel X-Ray Spectrometer, *Rev. Sci. Instrum.* **34**, 1208–12 (1963).

(*S34*) Spielberg, N., W. Parrish, and K. Lowitzsch, Geometry of the Non-Focusing X-Ray Fluorescence Spectrograph, *Spectrochim. Acta* **13**, 564–83 (1959).

(*S35*) Splettstosser, H. R., and H. E. Seeman, Application of Fluorescence X-Rays to Metallurgical Microradiography, *J. Appl. Phys.* **23**, 1217–22 (1952).

(*S36*) Steinhardt, R. G., Jr. and E. J. Serfass, X-Ray Photoelectron Spectrometer for Chemical Analysis, *Anal. Chem.* **23**, 1585–90 (1951).

(*S37*) Steinhardt, R. G., Jr. and E. J. Serfass, Surface Analysis with the X-Ray Photoelectron Spectrometer, *Anal. Chem.* **25**, 697–700 (1953).

(*S38*) Steinhardt, R. G., Jr., F. A. D. Granados, and G. L. Post, X-Ray Photoelectron Spectrometer with Electrostatic Deflection, *Anal. Chem.* **27**, 1046–50 (1955).

(*S39*) Stever, K. R., J. L. Johnson, and H. H. Heady, X-Ray Fluorescence Analysis of Tungsten-Molybdenum Metals and Electrolytes, *Advan. X-Ray Anal.* **4**, 474–87 (1961).

(*S40*) Stewart, J. H., Jr., Determination of Thorium by Monochromatic X-Ray Absorption, *Anal. Chem.* **32**, 1090–2 (1960).

(*S41*) Stoner, G. A., Rapid Automatic Analysis of Magnesium Alloys, *Anal. Chem.* **34**, 123–6 (1962).

(*S42*) Strasheim, A., and F. T. Wybenga, Determination of Certain Noble Metals in Solution by Means of X-Ray Fluorescence Spectroscopy. I. Determination of Platinum, Gold, and Iridium, *Appl. Spectrosc.* **18**, 16–20 (1964).

(*S43*) Sun, S. C., Fluorescent X-Ray Spectrometric Estimation of Aluminum, Silicon, and Iron in Flotation Products of Clays and Bauxites, *Anal. Chem.* **31**, 1322–4 (1959).

(*S44*) Suoninen, E. J., Simultaneous Use of an Analyzer Crystal for Reflection and Transmission in X-Ray Fluorescence Spectrometry, *Suomen Kemistilehti* **34B**, 45–7 (1961); *Chem. Abstr.* **55**, 24223*e* (1961).

(*T1*) Tabikh, A. A., Use of Sample Rotators for Fluorescence Analysis of Light Elements in Cement Raw-Mix, *Anal. Chem.* **34**, 303–4 (1962).

(*T2*) Tanemura, T., Nondispersive X-Ray Spectroanalysis with Filter and Proportional Counter, *Rev. Sci. Instrum.* **32**, 364–6 (1961).

(*T3*) Taylor, A., B. J. Kagle, and E. W. Beiter, Geometrical Representation and X-Ray Fluorescence Analysis of Ternary Alloys, *Anal. Chem.* **33**, 1699–1706 (1961).

(*T4*) Taylor, J., and W. Parrish, Absorption and Counting Efficiency Data for X-Ray Detectors, *Rev. Sci. Instrum.* **26**, 367–73 (1955); **27**, 108 (1956).

(*T5*) Thatcher, J. W., and W. J. Campbell, Fluorescent X-Ray Spectrographic Probe—Design and Applications, *U. S. Bur. Mines Rep. Invest.* **5500**, 23 pp. (1959).

(*T6*) Togel, K., X-Ray Fluorescence Analysis of Small Quantities and Areas, *Siemens-Z.* **36**, 497–501 (1962).

(*T7*) Togel, K., Preparation Technique for X-Ray Spectrometry; in "Zerstörungsfreie Materialprüfung" ("Nondestructive Material Testing"), E. A. W. Muller, ed.; R. Oldenbourg, Munich, Ger., sec. U152 (1961).

(*T8*) Tomaino, M., and A. De Pietro, [Philips] Inverted-Sample Three-Position Spectrograph, *Norelco Reptr.* **3**, 57, 90, 94 (1956).

(*T9*) Tomkins, M. L., G. A. Borun, and W. A. Fahlbusch, Quantitative Determination of Tantalum, Tungsten, Niobium, and Zirconium in High-Temperature Alloys by X-Ray Fluorescent Solution Method, *Anal. Chem.* **34**, 1260–3 (1962).

(*T10*) Toussaint, C. J., and G. Vos, Quantitative Determination of Carbon in Solid Hydrocarbons Using the Intensity Ratio of Incoherent to Coherent Scattering of X-Rays, *Appl. Spectrosc.* **18**, 171–4 (1964).

(*T11*) Trillat, J.-J., Metallurgical Aspects of Microradiography, *Met. Rev.* **1**, 3–30 (1956).

(*V1*) Van Niekerk, J. N., and J. F. De Wet, Trace Analysis by X-Ray Fluorescence Using Ion Exchange Resins, *Nature* **186**, 380–1 (1960).

(*V2*) Van Niekerk, J. N., J. F. De Wet, and F. T.Wybenga, Trace Analysis by Combination of Ion-Exchange and X-Ray Fluorescence—Determination of Uranium in Sulfate Effluents, *Anal. Chem.* **33**, 213–15 (1961).

(*V3*) Van Rennes, A. B., Pulse-Amplitude Analysis in Nuclear Research. I. Voltage Discriminators, *Nucleonics* **10**(7), 20–7 (1952).

(*V4*) Van Rennes, A. B., Pulse-Amplitude Analysis in Nuclear Research. II. Single-Channel Differential Analyzers, *Nucleonics* **10**(8), 22–8 (1952).

(*V5*) Van Rennes, A. B., Pulse-Amplitude Analysis in Nuclear Research. III, IV. Multichannel Analyzers, *Nucleonics* **10**(9), 32–8; (10) 50–6 (1952).

(*V6*) Verderber, R. R., X-Ray Fluorescence Analysis of the Composition of Ni–Fe Thin Films, *Norelco Reptr.* **10**, 30–34 (1963).

(*V7*) Victoreen, J. A., Calculation of X-Ray Mass-Absorption Coefficients, *J. Appl. Phys.* **20**, 1141–7 (1949).

(*V8*) Volborth, A., Total Instrumental Analysis of Rocks, *Nevada Bur. Mines Rep.* **6A**, 1–72 (1963).

(*V9*) Von Hamos, L., X-Ray Image Method of Chemical Analysis, *Amer. Mineral.* **23**, 215–26 (1938).

(*V10*) Von Hamos, L., Determination of Very Small Quantities of Substances by the X-Ray Microanalyzer, *Arkiv. Mat. Astron. Fys.* **31**, 1–11 (1945).

(*V11*) Von Hamos, L., X-Ray Micro-Analyzer, *J. Sci. Instrum.* **15**, 87–94 (1938).

(*W1*) Waehner, K. A., Analyzing Refractory Metal Alloys—Fansteel 82 Metal, *Fansteel Metallurgy* (Dec. 1961).

(*W2*) Wagner, J. C., and F. R. Bryan, X-Ray Spectrographic Analysis of Automotive Combustion Deposits without the Use of Calibration Curves, *Advan. X-Ray Anal.* **9**, 528–32 (1966).

(*W3*) Weber, K., and J. Marchal, Device for Coupling a Pulse-Height Discriminator to a Scanning X-Ray Spectrometer, *J. Sci. Instrum.* **41**, 15–22 (1964).

(*W4*) Wenger, P. E., I. Kapetanidis, and W. Van Janstein, X-Ray Fluorescence as a Supplement to Paper Electrophoresis, *Pharm. Acta Helv.* **37**, 472–89 (1962).

(*W5*) West, H. I., Jr., W. E. Meyerhof, and R. Hofstadter, Detection of X-Rays by Means of NaI(Tl) Scintillation Counters, *Phys. Rev.* **81**, 141–2 (1951).

(*W6*) Weyl, R., Nondestructive Measurement of Composition and Layer Thickness of Thin Films by X-Ray Fluorescence, *Z. Angew. Phys.* **13**, 283–8 (1961).

(*W7*) White, E. W., H. A. McKinstry, and T. F. Bates, Crystal Chemical Studies by X-Ray Fluorescence, *Advan. X-Ray Anal.* **2**, 239–45 (1959).

(*W8*) White, J. E., X-Ray Diffraction by Elastically Deformed Crystals, *J. Appl. Phys.* **21**, 855–9 (1950).

(*W9*) Wittig, W. J., Uses of an X-Ray Fluorescence Semi-Microprobe Attachment in Metallurgy, *Advan. X-Ray Anal.* **8**, 248–58 (1965).

(*W10*) Wittig, W. J., Development and Use of a Semimicro X-Ray Fluorescence Attachment in Metallurgy, *Develop. Appl. Spectrosc.* **3**, 36–44 (1964).

(*W11*) Wolfe, R. A., and R. G. Fowler, Application of Spectrographic Methods to Chemical Concentration of Trace Elements in Iron and Steel Analysis, *J. Opt. Soc. Amer.* **35**, 86–91 (1945).

(*W12*) Wybenga, F. T., and A. Strasheim, Determination of Certain Noble Metals in Solution by Means of X-Ray Fluorescence Spectroscopy. II. Determination of Palladium, Rhodium, and Ruthenium. *Appl. Spectrosc.* **20**, 247–50 (1966).

(*W13*) Wytzes, S. A., Theoretical Considerations on the Collimators of an X-Ray Spectrograph, *Philips Res. Rep.* **16**, 201–24 (1961).

(*W14*) Wytzes, S. A., Automatic X-Ray Spectrometer, *Philips Tech. Rev.* **27**, 300–10 (1966).

(*Y1*) Yao, T. C., and F. W. Porsche, Determination of Sulfur and Chlorine in Petroleum Liquids by X-Ray Fluorescence, *Anal. Chem.* **31**, 2010–12 (1959).

(*Y2*) Young, W. J., Applications of the Electron Microbeam Probe and Micro X-Rays in Nondestructive Analysis; in "Recent Advances in Conservation," G. Thomson, ed.; Butterworths, London, pp. 33–8 (1961).

(*Z1*) Zeitz, L., Potentialities of an X-Ray Microprobe, *Rev. Sci. Instrum.* **32**, 1423–4 (1961).

(*Z2*) Zeitz, L., and A. V. Baez, Microchemical Analysis by Emission Spectrographic and Absorption Methods; in "X-Ray Microscopy and Microradiography," V. E. Cosslett, A. Engstrom, and H. H. Pattee, eds.; Academic Press, N. Y., pp. 417–34 (1957).

(*Z3*) Zemany, P. D., Effects Due to Chemical State of the Samples in X-Ray Emission and Absorption, *Anal. Chem.* **32**, 595–7 (1960).

(*Z4*) Zemany, P. D., Use of Fe^{55} for Measuring Titanium Coating Thickness, *Rev. Sci. Instrum.* **30**, 292–3 (1959).

(*Z5*) Zemany, P. D., Line Interference Corrections for X-Ray Spectrographic Determination of Vanadium, Chromium, and Manganese in Low-Alloy Steels, *Spectrochim. Acta* **16**, 736–41 (1960).

(*Z6*) Zemany, P. D., and H. A. Liebhafsky, Plating Thickness by the Attenuation of Characteristic X-Rays, *J. Electrochem. Soc.* **103**, 157–9 (1956).

(*Z7*) Zemany. P. D., H. G. Pfeiffer, and H. A. Liebhafsky, Precision in X-Ray Emission Spectrography—Background Present, *Anal. Chem.* **31**, 1776–8 (1959).

(*Z8*) Zemany. P. D., W. W. Welbon, and G. L. Gaines, Determination of Microgram Quantities of Potassium by X-Ray Emission Spectrography of Ion-Exchange Membranes, *Anal. Chem.* **30**, 299–300 (1958).

(*Z9*) Ziegler, C. A., L. L. Bird, and D. J. Chleck, X-Ray Rayleigh Scattering Method for Analysis of Heavy Elements in Low-*Z* Media, *Anal. Chem.* **31**, 1794–8 (1959).

(*Z10*) Zingaro, P. W., Helium-Chamber X-Ray Spectrograph, *Norelco Reptr.* **1**, 45–6 (1953–54).

(*Z11*) Zimmerman, R. H., Industrial Applications of X-Ray Methods for Measuring Plating Thickness, *Advan. X-Ray Anal.* **4**, 335–50 (1961).

(*Z12*) Zimmerman, R. H., X-Rays Check Plating Thickness, *Iron Age* **186** (15), 84–7 (1960).

(*Z13*) Zimmerman, R. H., Measuring Plating Thickness—Industrial Applications of X-Ray Methods, *Metal Finish.* **59**(5), 67–73 (1961).

(*Z14*) Zimmerman, R. H., Industrial Applications of X-Ray Methods for Measuring Plating Thickness, *Qual. Assurance* **2**(3), 27–32 (1963).

Index